Encyclopedia of CONSUMPTION and WASTE

Encyclopedia of CONSUMPTION and WASTE

The Social Science of Garbage

Carl A. Zimring
Roosevelt University
General Editor

William L. Rathje
University of Arizona
Consulting Editor

Margaret E. Heggan Public Library
606 Delsea Drive
Sewell, NJ 08080

Los Angeles | London | New Delhi
Singapore | Washington DC

Los Angeles | London | New Delhi
Singapore | Washington DC

FOR INFORMATION:

SAGE Publications, Inc.
2455 Teller Road
Thousand Oaks, California 91320
E-mail: order@sagepub.com

SAGE Publications India Pvt. Ltd.
B 1/I 1 Mohan Cooperative Industrial Area
Mathura Road, New Delhi 110 044
India

SAGE Publications Ltd.
1 Oliver's Yard
55 City Road
London EC1Y 1SP
United Kingdom

SAGE Publications Asia-Pacific Pte. Ltd.
33 Pekin Street #02-01
Far East Square
Singapore 048763

Vice President and Publisher: Rolf A. Janke
Senior Editor: Jim Brace-Thompson
Project Editor: Tracy Buyan
Cover Designer: Candice Harman
Editorial Assistant: Michele Thompson
Reference Systems Manager: Leticia Gutierrez
Reference Systems Coordinator: Laura Notton

Golson Media
President and Editor: J. Geoffrey Golson
Author Manager: Joseph K. Golson
Senior Layout Editor: Mary Jo Scibetta
Layout Editors: Mary Sudul, Lois Rainwater, Oona Patrick
Copy Editor: Carl Atwood
Proofreaders: Barbara Paris, Mary LeRouge
Indexer: J S Editorial

Copyright © 2012 by SAGE Publications, Inc.

All rights reserved. No part of this book may be reproduced or utilized in any form or by any means, electronic or mechanical, including photocopying, recording, or by any information storage and retrieval system, without permission in writing from the publisher.

Printed in the United States of America

Library of Congress Cataloging-in-Publication Data

Encyclopedia of consumption and waste : the social science of garbage / Carl A. Zimring, general editor ; William L. Rathje, consulting editor.
v. cm.
Includes bibliographical references and index.
ISBN 978-1-4129-8819-3 (cloth)
1. Refuse and refuse disposal–Social aspects--Encyclopedias. 2. Consumption (Economics)–Social aspects–Encyclopedias. I. Zimring, Carl A., 1969- II. Rathje, William L.
HD4482.E53 2012
363.72'803–dc23

2011034826

MIX
Paper from responsible sources
FSC® C014174

12 13 14 15 16 10 9 8 7 6 5 4 3 2 1

Contents

List of Articles

Reader's Guide

Archaeology of Garbage
Archaeological Techniques, Modern Day
Archaeology of Garbage
Dating of Garbage Deposition
Dump Digging
Garbology
Trash as History/Memory

Consumption and Waste, Industrial/Commercial
Acid Rain
Aluminum
Celluloid
Coal Ash
Computers and Printers, Business Waste
Construction and Demolition Waste
Copper
Emissions
Farms
Fusion
Garbage Project
Hanford Nuclear Reservation
High-Level Waste Disposal
Hospitals
Incinerator Waste
Incinerators
Incinerators in Japan
Industrial Revolution
Industrial Waste
Iron
Malls
Medical Waste
Midnight Dumping
Mineral Waste
Mining Law
Noise
Noise Control Act of 1972
Nuclear Reactors
Ocean Disposal
Pesticides
Power Plants
Producer Responsibility
Radioactive Waste Disposal
Restaurants
Rubber
Sanitation Engineering

Scrubbers
Solid Waste Data Analysis
Stadiums
Sugar Shortage, 1975
Supermarkets
Sustainable Waste Management
Thallium
Uranium
Waste Disposal Authority

Consumption and Waste, Personal

Adhesives
Aerosol Spray
Air Filters
Alcohol Consumption Surveys
Audio Equipment
Automobiles
Baby Products
Beverages
Books
Candy
Car Washing
Carbon Dioxide
Certified Products (Fair Trade or Organic)
Children
Cleaning Products
Composting
Computers and Printers, Business Waste
Computers and Printers, Personal Waste
Consumption Patterns
Cosmetics
Dairy Products
Disposable Diapers
Disposable Plates and Plastic Implements
Dumpster Diving
Engine Oil
Environmental Tobacco Smoke
Fast Food Packaging
Fish
Floor and Wall Coverings
Food Consumption
Food Waste Behavior
Fuel
Funerals
Funerals/Corpses
Furniture
Garden Tools and Appliances
Gasoline
Gluttony
Hoarding and Hoarders
Home Appliances
Home Shopping
Household Consumption Patterns
Household Hazardous Waste
Human Waste
Internet Purchasing
Junk Mail
Lighting
Linen and Bedding
Magazines and Newspapers
Marketing, Consumer Behavior, and Garbage
Meat
Microorganisms
Mobile Phones
NIMBY (Not in My Backyard)
Open Burning
Packaging and Product Containers
Paint
Paper Products
Personal Products
Pets
Post-Consumer Waste
Pre-Consumer Waste
Recyclable Products
Recycling Behaviors
Residential Urban Refuse
Seasonal Products
Septic System
Sewage
Shopping
Shopping Bags
Slow Food
Sports
Street Scavenging and Trash Picking
Styrofoam
Swimming Pools and Spas
Television and DVD Equipment
Tires
Tools
Toys
Ventilation
Water, Consumption
Water, Treatment
Wood
Yardwaste

Geography, Culture, and Waste

Global Cities: Consumption, Waste Collection, and Disposal

History of Consumption and Waste

Issues and Solutions

People

Sociology of Waste

U.S. States: Consumption, Waste Collection, and Disposal

Waste, Municipal/Local

About the Editors

General Editor Carl A. Zimring is assistant professor of social science at Roosevelt University's Evelyn T. Stone College of Professional Studies, where he co-founded the school's sustainability studies program in 2010. He is an environmental historian who has published on several topics relating to waste and urban environments, including the history of municipal smoke control efforts in the early 20th century and the unintended consequences of shredding junked automobiles. His book, *Cash for Your Trash: Scrap Recycling in America,* investigates changing ideas about material reuse from colonial times to the end of the 20th century. Zimring earned his B.A. in history from the University of California at Santa Cruz, his M.A. in social sciences from the University of Chicago, and his Ph.D. in history from Carnegie Mellon University. He is an Environmental Protection Agency Science to Achieve Results Fellow and a scholar-in-residence at the Smithsonian Institute Libraries. He received the 2010 American Society for Environmental History Samuel P. Hays Research Fellowship and serves on the board of directors of the Chicago Recycling Coalition.

Consulting Editor William L. Rathje is the founder and director of the Garbage Project, which conducts archaeological studies of modern refuse. Rathje received his B.A. from the University of Arizona in 1967 and his Ph.D., which focused on the archaeology of the ancient Maya, from Harvard University in 1971; he is currently professor emeritus at the University of Arizona and a consulting professor at Stanford University.

Since 1973, the Garbage Project has studied fresh refuse to document household-level food waste, diet and nutrition, recycling, and discard of hazardous wastes; in addition, since 1987, the project has excavated 21 landfills across North America to record the quantities of various types of buried refuse and what happens to these materials over time. *Garbology*, the term coined to describe Rathje's research, is now in the *Oxford English Dictionary.* In 1991, Rathje won the prestigious AAAS/Westinghouse Award for Public Understanding of Science and Technology, and in 1992 he received the AAA Solon T. Kimball Award for Public and Applied Anthropology.

List of Contributors

Maaheen Ahmed
Jacobs University
Akanni Ibukun Akinyemi
Obafemi Awolowo University
Melike Aktas Yamanoglu
Ankara University
Jeanne E. Arnold
University of California, Los Angeles
Hasret Balcioglu
Cyprus International University
Tryfon Bampilis
University of Leiden
Lori Barcliff Baptista
University of Chicago
Sarah Besky
University of Wisconsin, Madison
Malgorzata Olimpia Bielenia
Gdansk University of Technology
Mostaem Billah
Memorial University
Thomas Birch
University of Aberdeen
Rachel Black
Boston University
Bennis Blue
Virginia State University
Sarah Boslaugh
Washington University
Robin Branson
University of Sydney
Deborah Breen
Boston University
Barrett Brenton
St. John's University
Robert Brinkmann
University of South Florida
Keri Vacanti Brondo
University of Memphis
Marita Bullock
Macquarie University
Jakob Calice
Leeds Metropolitan University
Stacey Lynn Camp
University of Idaho

Benedetta Cappellini
Royal Holloway University
Christine Caruso
Graduate Center, City University of New York
Laura Chambers
Jacksonville University
Karlen Chase
D'Youville College
Marie-France Chevron
University of Vienna
David Chicoine
Louisiana State University
Jill M. Church
D'Youville College
Creighton Connolly
Memorial University of Newfoundland
John Cook
University of California, Riverside
Michelle Coyne
York University
Christopher Cusack
Keene State College
Aimee Dars Ellis
Ithaca College
Gareth Davey
Hong Kong Shue Yan University
Carmen De Michele
Ludwig-Maximilians University Munich
Kim De Wolff
University of California, San Diego
Nancy G. DeBono
Texas A&M University
Will Delavan
Lebanon Valley College
Jean-Francois Denault
Independant Scholar
Paul Dobraszczyk
University of Reading
Tracy Duvall
Universitas Indonesia
Leslie Elrod
University of Cincinnati
Braden Engel
University of Greenwich
Julie H. Ernstein
Northwestern State University
Rebecca Estrada
University of Manchester
Blaise Farina
Rensselaer Polytechnic Institute
David Fazzino
University of Alaska
Michael P. Ferber
The King's University College
Rosalind Fisher
University of West Florida
Erin Elizabeth Fitz-Henry
University of Melbourne
Jonathan R. Fletcher
Knox College
Stephanie Foote
University of Illinois
Leonardo Freire de Mello
State University of Campinas
Stephanie Joy Friede
Columbia University
Divine Fuh
Universität Basel
Jamie Furniss
Oxford University
Anne Galvin
St. John's University
Alessandro Gandini
Milan University
Stephen Gasteyer
Michigan State University
Shamira M. Gelbman
Illinois State University
Zsuzsa Gille
University of Illinois
Curt Gilstrap
Drury University
Andrew Glover
University of Technology, Sydney
Nisse Goldberg
Jacksonville University
Jennifer Ellen Good
Brock University
Joan Gross
Oregon State University
Velma Grover
York University
Emilie Guitard
Paris Ouest La Défense University
Angela Gumm
Iowa State University

Britt Halvorson
Colby College
Andrew Hao
University of California, Berkeley
Cyrille Harpet Sr.
EHESP
Joseph Haynes
UK Government Economic Service
Gisela Heffes
Rice University
Shana Heinricy
University of New Mexico
Jason A. Helfer
Knox College
Laura Rosel Herrera
Independent Scholar
Max Holleran
New York University
Maartje Hoogsteyns
Open University Netherlands
Emily Huddart Kennedy
University of Alberta
Alison Hulme
Goldsmiths College
Cherry Lei Hunsaker-Clark
Independent Scholar
Linda Hylkema
Santa Clara University
Tomoaki D. Imamichi
City University of New York
Richard Iveson
Goldsmiths College
Sarah Edith James
University College London
Helena Mateus Jerónimo
Technical University of Lisbon
Dolly Jørgensen
Norwegian University of Science & Technology
Finn Arne Jørgensen
Umeå University
Hye Yoon Jung
Ewha Womans University
Arn Keeling
Memorial University of Newfoundland
Helen Kopnina
University of Amsterdam
Daniela Korbas-Magal
Ben Gurion University of the Negev
Dana Kornberg
University of Michigan
Robert Gerald Krausz
Lincoln University
Naomi Krogman
University of Alberta
Bill Kte'pi
Independent Scholar
Tim Kubal
California State University, Fresno
Petra Kuppinger
Monmouth College
Anastasia Kurilova
Peoples' Friendship University of Russia
Frank LaFone
West Virginia University
Jordan K. Lanfair
Knox College
Patrick Laviolette
Tallinn University
Mary Lawhon
Clark University
Mauricio Leandro
Graduate Center, City University of New York
Josh Lepawsky
Memorial University of Newfoundland
Christopher A Lepczyk
University of Hawaii at Manoa
Max Liboiron
New York University
Peter Lindenmann
Universitat Basel
Ingmar Lippert
University of Augsburg
Andriko Lozowy
University of Alberta
Scott Lukas
Lake Tahoe Community College
Samantha MacBride
New York University
Barbara MacLennan
West Virginia University
Graeme MacRae
Massey University
Ignatius Ani Madu
University of Nigeria
Nuno Luis Madureira
University of Lisbon

Paolo Magaudda
University of Padova
Christian D. Mahone
Knox College
Tom Mallard
Independent Scholar
Rafael D'Almeida Martins
State University of Campinas
Scott James Massen
University of Guelph
Neil Maycroft
University of Lincoln
Susan Mazur-Stommen
Independent Scholar
Mylinda McDaniel
West Virginia University
Allison Reilly McGrath
Western Washington University
Chris McNabb
Memorial University of Newfoundland
Fran Mentch
Cleveland State University
Skye K. Moody
Independent Scholar
Paul S. Moore
Ryerson University
Alex Nading
University of Wisconsin, Madison
Robin Nagle
New York University
Hazel Nash
Cardiff University
Jamal A. Nelson
Knox College
Lolita Nikolova
International Institute of Anthropology
Thaddeus Chidi Nzeadibe
University of Nigeria
Lionel Obadia
Université Lyon 2
Michael J. O'Brien
University of Missouri
Kevin Kjell Olsen
Montclair State University
Angela Orlando
University of California, Los Angeles
Laura Orlando
Boston University School of Public Health
Maggie Ornstein
Graduate Center, City University of New York
Elizabeth Parsons
Keele University
Michael Pearce
University of Queensland
Melissa Fletcher Pirkey
University of Notre Dame
Matthew Piscitelli
University of Illinois at Chicago
Aneta Podkalicka
Swinburne University of Technology
Gordon C. Pollard
State University of New York, Plattsburgh
Alexandre Pólvora
University Paris 1 Panthéon-Sorbonne
Jerry Ratcliffe
Temple University
Josh Reno
Goldsmiths College
Héctor R. Reyes
Harold Washington College
Michael V. Rienti Jr.
State University of New York, University at Buffalo
Wesley W. Roberts
University of Pittsburgh
Abhijit Roy
University of Scranton
Mousumi Roy
Penn State University, Worthington Scranton
Stephen Curtis Sambrook
Centre for Business History in Scotland
Jen Schneider
Colorado School of Mines
Stephen T. Schroth
Knox College
Tomas Moe Skjølsvold
Norwegian University of Science and Technology
Alexia Smith
University of Connecticut
Monica L. Smith
University of California, Los Angeles
Suzanne M. Spencer-Wood
Oakland University

Laura J. Stiller
Monongalia County Solid Waste Authority
Ray Stokes
University of Glasgow
Donovan Storey
University of Queensland
Sarah Surak
Virginia Tech
Christopher Sweet
Illinois Wesleyan University
Leah Tang
Monash University
Caroline Tauxe
Le Moyne College
Erin Brooke Taylor
University of Sydney
Gretchen Thompson
North Carolina State University
Dimitry Tikhaze
Russian Peoples' Friendship University
Jude Todd
University of California, Santa Cruz
Rebecca Tolley-Stokes
East Tennessee State University
Davide Torsello
University of Bergamo, Italy
Aloisia de Trafford
Independent Scholar
Kevin Trumpeter
University of South Carolina
Eric Venbrux
Radboud University Nijmegen
Sintana Eugenia Vergara
University of California, Berkeley
Lucia Vodanovic
Goldsmiths College
Travis Wagner
University of Southern Maine
Beth Ellen Walton
Walton Enterprises
Scott Webel
University of Texas, Austin
Heike Weber
Technische Universität Berlin
Daniel Arthur Weissman
Harvard University
Jon Welsh
AAG Archaeology
Ann White
Michigan State University
Richard Wilk
Indiana University
John R. Wood
Rose State College
Melike Aktas Yamanoglu
Ankara University
Hiroko Yoshida
University of Wisconsin, Madison
Amy Zhang
Yale University
Gerald Zhang-Schmidt
Independent Scholar
Carl A. Zimring
Roosevelt University

Introduction

So much of our stuff lacks worth or merit. That, at least, is what we assume and establish with our routines. Every day, we put unwanted material in toilets and garbage bins, regularly flushing it away or taking it out in bags to be transported far away from our homes by others. The names we give this material—*waste*, *garbage*, *refuse*, *trash*, *rubbish*—have pejorative definitions. Worthless. Rejected and useless matter of any kind. Unimportant.

This material is certainly rejected by someone, but it is far from unimportant. What we classify and dispose of as wastes provides rich insight into our behavior, social structures, and treatment of our environment. In the 1966 book *Purity and Danger*, anthropologist Mary Douglas stated that dirt is matter that is out of place and that a common thread in all human societies is the development of taboos to regulate waste and establish order from chaos. Over the half-century since that book was published, social scientists have challenged and advanced theories of waste and value, using observed behavior and the materials humans leave behind as evidence. Archaeologists have long studied artifacts of refuse from the distant past as a portal into ancient civilizations lacking written testaments, but examining what we throw away today tells a story in real time and becomes an important and useful tool for academic study. Our trash is a testament; what we throw away says much about our values, our habits, and our lives.

Sometimes what we throw away is uncomfortably revealing, as Bob Dylan found out in 1971. The singer had moved to New York's Greenwich Village when he found fanatic A. J. Weberman rifling through his garbage can in search of clues about Dylan's life. (What Weberman found were bills, receipts, correspondence, coffee grounds and diapers—evidence of a young family and a mind in need of stimulation.) Weberman called his investigations into Dylan's discards "garbagology," claiming his research would provide insight into the songwriter's art.

Weberman's research method and findings may have strained credulity, but his attention to the value of garbage did not. In the early 1970s, substantial developments in the social sciences advanced our understandings of waste in the modern world. Historians deepened their investigations of public health and pollution; in particular, Joel Tarr analyzed the political and technological systems that societies developed to find appropriate sinks for sewage, garbage, and industrial wastes. Young archaeologist William Rathje began an excavation of a landfill in Tuscon, Arizona, in 1973 in order to analyze the waste stream of that community. His research became known as the Garbage Project and spawned concerted examination of the record modern societies leave in the trash.

Close examination of waste practices reveals rich complexities. While dictionary definitions of garbage describe it as "filth" and "worthless," scholars are careful to note that perceptions of waste and the value of material are neither static nor universally shared. Discarded objects may become antiques, embarking on a journey from valued new object to disvalued old object to newly valued vintage object. Beverage companies take old cans and bottles removed from households to make new cans and bottles. Material from demolished buildings is used to construct new ones. Even human excrement, widely feared and flushed away to prevent disease, is collected to fertilize fields. (If filth is one definition of waste, another definition is "the inefficient squandering of resources.") One man's trash may be another man's treasure; how and why both parties classify the material is a subject of ample study.

We do not know what Bob Dylan thinks of this research field, but he certainly protested Weberman snooping through what the singer felt was his private property. Whatever questions Weberman's research method and findings raised, they did not, at least in a legal sense, invade Dylan's privacy. Waste as property was the subject of the 1988 legal decision *California v. Greenwood*, in which the U.S. Supreme Court ruled that a person should have no reasonable expectation of privacy concerning any material that person knowingly threw away. Given the billions of dollars spent by municipalities each year to landfill, incinerate, recycle, or otherwise handle waste, the question of who owns these discards is not trivial. The industries and public programs devoted to waste around the world range from sophisticated technology capturing methane from landfills to open pits where workers risk their health harvesting precious metals by burning old computers shipped from thousands of miles away. In a globalized economy, waste is a truly global commodity—and a global burden. At the same time, the burdens of waste management often are local, with municipal governments responsible for making sure the streets are free of waste. When they fail to perform this task, as the Italian city of Naples infamously did in 2008, the results include physical dangers to the community and widespread ridicule of local political leaders. Effective waste management is an expectation of modern society.

Today, academic investigations into garbage range widely in method, geographic scope, and chronology. Excavations at landfills join close examination of municipal waste management systems, policy history, industrial research, marketing, design, and psychology. All of these approaches allow us to better understand the complexities of our consumption and waste, complexities regularly on display in our minds, our homes, and our communities.

Mine is no exception. I live in a village located just west of Chicago that prides itself on its progressive values. The residents enjoy curbside recycling services, tree-lined streets, and several parks. Seasonal farmers' markets allow residents to purchase locally grown organic produce. While many suburbanites have to drive to work in the city, we can commute via two Chicago Transit Authority train lines that run through the village, reducing carbon emissions we might otherwise produce with automobiles.

In 2010, the elevated line nearest where I live painted several panels on its viaducts with images promoting the international "350.org" campaign. Established by environmental scholar Bill McKibben in 2007, the campaign attempts to get carbon consumption under 350 parts per million carbon dioxide in order to combat global climate change before it produces devastating effects on the atmosphere, oceans, disease vectors, food sources, coastlines and myriad other entities vital to life on Earth. A successful campaign requires substantial reforms to the consumption of energy, packaging, food, and materials in the industrialized world, and the village's embrace of the 350 campaign demonstrates a hopeful awareness of the challenges ahead.

One walks less than a block from the westernmost train station with a 350.org panel before encountering an upscale boutique with a sign in its window promising "inner peace through impulse purchasing." A joke, to be sure, but one that resonates as an uncomfortable truth. We seek fulfillment through the goods and services we acquire, consume, and dispose of, often blurring the line between needs and wants. Throughout the village, plastic bags carrying goods manufactured all over the world are regularly bought and sold. (But not, ultimately, disposed of locally. Garbage and recyclables leave the village on trucks hauling them to waste management facilities in poorer communities many miles away.) The proximity of the impulse-purchasing sign to images promoting the 350 campaign indicates the complexity of our challenges in the early 21st century: We consume to fulfill our needs and wants, yet our consumption has effects that may be terribly consequential to the land, air, water, other species, other people, and ourselves.

The tensions in my village are ones found across the planet. Consumption and its concomitant waste are defining aspects of our societies. What we consume, why we consume it, and what we do with the remnants of that consumption reveal how we organize our landscapes, our economies, our social structures, and our values. The wastes we leave behind, in the form of landfills, atmospheric pollution, estate inventories, and the ruins of civilization, are sources for social scientists to interpret.

Even attempts to erase our wastes are revealing. After the terrorist Osama bin Laden was found and killed in a residential neighborhood in Abbottabad, Pakistan, after years in hiding, neighbors remarked that the most unusual aspect of his compound was that its residents never put trash out for collection. Instead, bin Laden had all waste incinerated on site so as not to leave clues to his whereabouts. The absence of a waste stream aroused suspicion, just as the presence of particular items tell us about the habits of the consumers who generate a waste stream. Our trash is part of us, whether or not we choose to acknowledge it.

In the *Encyclopedia of Consumption and Waste*, you will read the perspectives of anthropologists, archaeologists, historians, philosophers, policy analysts, and sociologists, just to name a few. The interdisciplinary lens of the volume reveals the complexity of our relationship to the world of goods, services, and wastes. This is evident whether you read every entry alphabetically, or follow the listings of related entries from one to another. Contributing editor William Rathje initially planned the encyclopedia, apparent in the array of entries on garbage archaeology and his appendix "Garbology 101." I then became general editor, and readers may find evidence of my background as an environmental historian present in the organization of individual entries and the encyclopedia as a whole. Our goal was to bring together scholars working on waste from many perspectives, so that we all may better understand the dynamics of consumption and waste that affect our households, cities such as Shanghai, nations such as Brazil, the garbage patch growing in the Pacific, and the ecosystems around the world that we pollute every day. Entries on each of these topics await you. Whether you find this encyclopedia in a library or satiated an impulse to purchase it, we hope it encourages conversation about the patterns and consequences of our consumption.

Carl A. Zimring
General Editor

Chronology

600 B.C.E.: Construction is completed on the Cloaca Maxima sewage collection system in Rome.

1346 C.E.: The second pandemic of the bubonic plague begins. It is later concluded by scientists that the plague was caused by infected rats accidentally carried onboard ships traveling throughout Europe and the Middle East.

1349: King Edward III of England complains to the mayor of London in a letter, saying, "The streets and lanes through which people had to pass were foul with human faeces and the air of the city poisoned to the great danger of men passing, especially in this time of infectious disease."

1374: The first above-ground sewer in the city of Paris is constructed.

1555: In his book *De Re Metallica*, German doctor Georgius Agricola states that the environmental degradation brought on by mining excavation is only a temporary phenomenon.

1649: In an effort to protect navigating boats, the Japanese government enacts laws designed to prevent the dumping of trash into rivers and canals.

1657: Residents of the New York City borough of Manhattan are prohibited from throwing "any rubbish, filth, oyster shells, dead animals or anything like it" into the streets.

1690: The Rittenhouse Mill, the first paper mill established in the United States, begins operations near the city of Philadelphia.

1702: An outbreak of yellow fever in New York City results in the deaths of one-tenth of the city's population.

1750–73: The availability of imported goods into the U.S. market for personal consumption increases by 120 percent.

1790–1840: The percentage of Americans living in urban areas increases from 5.1 to 10.8 percent.

1800–50: The population of London nearly triples, leading to increased concerns over the city's sanitation system.

1827: Using electrolysis, chemists Hans Orsted and Frederich Wohler become the first people to isolate pure aluminum.

1834: In the West Virginia city of Charleston, local officials pass a law that sets penalties for the shooting of vultures; the law is passed due to the vultures' tendency to eat the city's garbage.

1848: The city of London establishes a centralized governing body called the Metropolitan Commission of Sewers, with reformer Edwin Chadwick selected to head the new organization.

1849: Walter Hunt invents the safety pin, significantly increasing the ease of the diaper-changing process.

1853: Condensed milk, an infant feeding alternative for mothers who either cannot afford a wet nurse or cannot nurse themselves, becomes available to U.S. consumers.

1860: The Argentine city of Buenos Aires establishes its first *quema* (burn) waste incineration system.

1866: Health conditions in New York City worsen to the point where officials compare to medieval London.

1872: The U.S. Congress passes the Mining Law, placing governance over 270 million acres of public domain lands that are potentially suitable for mining.

1874: The first waste incineration system is constructed in England.

1874: The world's first curbside recycling program is introduced in the city of Baltimore.

1895: City waste and human disease expert George Waring is appointed as New York City's street cleaning commissioner.

1900: The Japanese government passes the Waste Cleaning Act.

1900: The Japanese government passes the National Waste Disposal Law, requiring local municipalities to manage their own waste.

1904: In the metalworks area of the city of Chicago, Illinois, the first large-scale aluminum recycling program begins operation.

1904: King C. Gillette patents his safety razor kit.

1905: *Waste Trade Journal* publishes its first issue.

1913: The National Association of Waste Material Dealers is founded.

1916: The first Japanese draft furnace incinerator is constructed.

1918: The U.S. War Food Administration reports that, during World War I, American households threw away nearly 30 percent of food that was purchased for consumption.

1919: The U.S. Bureau of Mines is established as an agency under the U.S. Department of the Interior.

1921: The world's first textbook dealing with urban waste management, *Collection and Disposal of Municipal Refuse*, is published by civil engineers Rudolph Hering and Samuel Greeley.

1939: The chemical compound DDT is discovered and its use as an insecticide begins.

1941: Researchers release reports showing that over 20 percent of food served in mess halls to American soldiers is discarded.

1946–84: Dumped refrigerators with poorly made mechanical latches lead to the deaths of over 400 young children.

1947: Construction of the Fresh Kills Landfill, which would become the largest municipal waste dump in the world, is completed.

1948: The U.S. Congress passes the Federal Water Pollution Control Act, establishing the authority for the federal government to regulate water quality.

1950–70: The U.S. population rises by 30 percent, but total waste output for the country increases by 60 percent.

1950–2000: The percentage of people in sub-Saharan Africa living in urban areas increases from 15 to 42 percent.

1951: The number of "automobile graveyards" in the United States rises to 25,000.

1955: President Dwight Eisenhower signs the Air Pollution Control Act into law, providing federal funding for research into air pollution.

1958–72: The amount of beef the average American consumes increases from 80.5 to 115.9 pounds annually.

1958–98: The amount of concentration of carbon dioxide in the Earth's atmosphere increases from 316 to 369 ppmv.

1959: Procter & Gamble introduces the world's first disposable diaper.

1959: The first fully automated Japanese waste incineration machine is installed in the Sumiyoshi plant in the city of Osaka.

1962: American biologist Rachel Carson publishes *Silent Spring*, a heavily researched book on how the pesticide DDT is causing widespread environmental degradation.

1963: The U.S. Congress passes the Clean Air Act, the first piece of U.S. legislation to federally require the control of air pollution.

1964: U.S. President Lyndon B. Johnson signs the Highway Beautification Act into law.

1965: The Solid Waste Disposal Act is passed by the U.S. Congress.

1970: The United States' first Earth Day is celebrated nationwide.

1970: The U.S. Congress passes the Resource Recovery Act.

1971: The famous "Crying Indian" public service announcement begins to air on U.S. television stations as part of an effort to reduce pollution.

1972: The U.S. Congress passes the Clean Water Act, not only requiring the end of dangerous pollution practices, but also aiming to restore America's water quality to passable levels.

1972: The U.S. Congress passes the Marine Protection, Research, and Sanctuaries Act.

1972: The London Convention, an international environmental meeting with delegates from dozens of countries, is held.

1972: A team of researchers led by Mathis Wackernagel concludes that the Earth's ecosystems are at 85 percent of total sustainability, meaning that humans have consumed resources to the point where only 15 percent of sustainability remains. In 2008, that percentage would increase to 125 percent, meaning that humans have consumed resources beyond the Earth's total capacity.

1973: The Garbage Project, an in-depth analysis of consumers' garbage in the Arizona area, is launched at the University of Arizona by professor William Rathje.

1973: In an effort to curb the rising cost of beef, U.S. President Richard Nixon announces that a price ceiling will be placed on all beef products.

1973: The International Convention for the Prevention of Pollution From Ships is held.

1974: U.S. President Gerald Ford signs the Safe Drinking Water Act into law.

1976: U.S. President Gerald Ford signs the Toxic Substances Control Act into law.

1976: The Resource Conservation and Recovery Act is passed by the U.S. Congress.

1978: Residents of the New York neighborhood of Love Canal experience toxic waste seeping into their houses' basements, leading to the discovery of what some have called the worst toxic waste disaster in U.S. history.

1978: An explosion occurs at the Consol #9 coal mine in West Virginia, resulting in the deaths of more than 78 people.

1978: Deng Xiaoping rises to the top of leadership in the government of China. Xiaoping's reforms would significantly alter China's position toward economic modernization, forever changing the country's consumer culture.

1979: British social scientist Michael Thompson introduces rubbish theory, a philosophy that attempts to address how value is placed on material objects.

1980: The U.S. Congress passes the Infant Formula Act, forcing manufacturers of baby formula to meet certain nutritional requirements in their products.

1980: The Comprehensive Environmental Response, Compensation, and Liability Act, also known as the Superfund, is passed by Congress.

1980: As large, imposing landfills threaten to overrun the available living space of the small island nation of Japan, its government begins a shift toward incineration of garbage, investing funds into research. Over time, 99.9 percent of combustible waste produced in of Osaka will be destroyed using incineration methods.

1981: As pollution concerns stemming from industrial expansion continue to mount in Japan, representatives of one of the country's most prosperous industrial centers, Osaka, begin the Osaka Phoenix Project.

1984: Cairo, Egypt, founds the Cairo Cleansing and Beautification Authorities organization.

1984: An insecticide manufacturing plant in Bhopal, India, accidentally releases tons of toxic gas, resulting in the deaths of 15,000 to 20,000 residents who live in the neighborhoods surrounding the plant.

1987: The Garbage Project begins its excavation of landfills for the purpose of studying garbage. Over time, nearly 30,000 pounds of garbage are stored and analyzed.

1987: Japan's first waste incinerator is constructed in the port city of Tsuruga. Over time, the number of incinerators in the island nation will increase to over 1,500.

1987: The *Mobro 4000* barge makes several unsuccessful attempts to unload 3,168 tons of New York trash along Atlantic coast communities in the United States, Mexico, and Belize, before returning to Islip, NewYork.

1988: The Supreme Court rules in the case of *California v. Greenwood* that "the Fourth Amendment's protection against unwarranted search and seizure does not extend to one's trash."

1988–98: The number of municipal composting programs in the United States increases from 700 to 3,800.

1989: Curbside recycling is mandated in New York City.

1991: Turkey establishes a Ministry of Environment.

1991–2001: In a 10-year period, the production of waste in the densely populated Indian city of Mumbai increases by 50 percent.

1999: A gasoline pipeline in the state of Washington ruptures and breaks, spilling 236,000 gallons of gasoline into Hanna Creek and resulting in the deaths of three people.

1999: The European Union establishes the Landfill Directive with the purpose of reducing "negative effects on the environment, in particular the pollution of surface water, groundwater, soil and air, and on the global environment, including the greenhouse effect, as well as any resulting risk to human health, from the landfilling of waste, during the whole life-cycle of the landfill."

1999–2008: The percentage of Chileans covered by Chile's sewage treatment system rises from 20 percent to 84 percent.

2000: An open dump landslide in the Philippines results in the deaths of nearly 200 people.

2001: Reports show that Nigeria consumes nearly one-third of all energy used by all sub-Saharan African nations.

2005: A survey released by a CBS Poll shows that 23 percent of Americans consume fast food every day.

2005: A study released by Global Issues shows that the wealthiest 20 percent of the world accounted for 76.6 percent of total private consumption, while the world's middle 60 percent consumed 21.9 percent, and the poorest 20 percent consumed only 1.5 percent. Furthermore, the study adds, the richest 20 percent consume 45 percent of all meat and fish, 58 percent of total energy, 84 percent of all paper, and own 87 percent of the world's vehicle fleet.

2005–06: During a two-year period, over 2.5 million cases of illegal dumping of garbage are reported in the United Kingdom.

2006: Researchers report that American consumers annually purchase a total of more than 57 billion liters of carbonated soft drinks, 31 billion liters of bottled water, 24 billion liters of beer, and 21 billion liters of milk.

2006: The 16th Nationwide Survey of Municipal Solid Waste Management in the United States is released, showing that Wyoming ranks 49th in terms of overall municipal solid waste production.

2007: San Francisco becomes the first city in the United States to place an outright ban on plastic grocery bags.

2007: The documentary film *Trashed* is released.

2008: A survey is released from the National Survey on Drug Use and Health (NSDUH) concluding that about 51 percent of Americans age 12 or older consume alcohol. Other data from the survey shows that 23 percent of people from the same group are binge drinkers, while 7 percent are heavy drinkers. Also, Asian Americans have the lowest alcohol consumption rate at 37 percent, and whites have the highest at 56 percent.

2008: The retailers Walmart, Target, and Toys "R" Us recall approximately 25 million toys that were made in China and found to contain dangerous levels of lead or other heavy metals.

2008: The Population Reference Bureau reports that "at least 50 percent" of the world's population resides in urban environments.

2008: The Environmental Protection Agency (EPA) reports that approximately 65 percent of U.S. trash comes from households. The report also states that an estimated 12.7 percent of all trash found in landfills is food scraps.

2009: The Chinese city of Beijing produces over 6.5 million tons of garbage per year.

2009: *No Impact Man*, a documentary featuring a New York City resident who experiments with dropping out of the mainstream consumer culture, is released.

2010: Researchers estimate that 95 percent of the food consumed by residents of Alaska comes from outside the state.

2010: Researchers report that the country of Uruguay leads the world in per capita beef consumption.

2010: The United Nations reports that approximately 1.8 million children worldwide die each year due to toxins found in drinking water.

2010: Reports are released showing that residents of the state of Connecticut produce, on average, approximately five pounds of garbage per person per day.

2010: Researchers report that the Indian capital of Delhi produces approximately 7,000 tons of solid waste per day.

2010: The U.S. Environmental Protection Agency reports that Americans produce approximately 96 billion pounds of food waste annually, a result of discarding over 25 percent of prepared food.

2010: The World Health Organization reports that 1.6 million people die each year from waterborne diseases found in drinking water.

2010: Researchers estimate that the number of people worldwide who lack access to basic sanitation services is approximately 1 billion.

2010: American scientist Jared Diamond estimates that America's consumption of natural resources, combined with its output of wastes, is 32 times higher than in the developing world.

2011: Italy's Environmental Ministry enacts a ban on polythene shopping bags in an attempt to reduce plastic waste. The European Union's (EU) Commissioner for the Environment opens an online consultation to assess EU citizens' opinion of taxing or outright banning plastic shopping bags across the continent.

2011: The international terrorist Osama bin Laden is found and killed in a residential neighborhood in Abbottabad, Pakistan. Observers of his hiding place note the residents were unusual as they never put trash out for collection, opting instead to incinerate it within the walled compound.

2011: California announces more than 1 billion pounds of electronic waste collected have been recycled since the state enacted the United States' first e-waste law in 2005.

2011: India passes the E-Waste (Management and Handling) Rule, a law intended to make producers financially liable for the disposal management of e-wastes.

2012: The Netherlands' National Waste Management Plan (established in 2002) has a target goal of recovering 83 percent of the nation's waste this year.

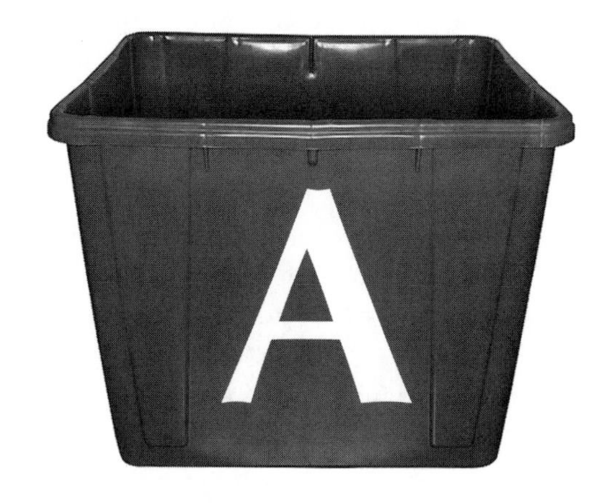

Acid Rain

The term *acid rain* describes any from of precipitation containing high levels of sulfuric and nitric acids. All precipitation, including snow, fog, hail, and sleet, can be acidic. The terms *acid deposition* or *wet disposition* are frequently used as synonyms. This form of pollution has caused severe environmental problems in many parts of North America, Europe, and Asia. It affects mainly heavily industrialized regions and densely populated urban areas, but substantive amounts of gases causing acid precipitation can travel hundreds of miles before falling.

Causes and Formation

The formation of acid rain generally starts with the emission of sulfur dioxide and nitrogen oxide gases into the atmosphere. These substances react with the water molecules in clouds and become sulfuric and nitric acid. By the time they reach the ground again in the form of precipitation, they are highly acidic. The degree of acidity is measured on a pH scale, ranging from 0 to 14.

The number indicates the hydrogen ion concentration in kilograms per cubic meter, thus lower numbers represent greater acidity. In severe cases, the pH value of rain can be as low as 2 or 3. Some parts of western Europe and the eastern part of the United States suffer from this highly acidic rain. In certain areas, such as Los Angeles, California, or Whiteface Mountain, New York, acid fog causes much more damage, showing about 10 times higher degrees of acidity than rain.

Acid rain is mainly caused by human activities. The largest share of these pollutant gases come from the combustion of fossil fuels, such as oil and coal, used by electric power plants as well as by cars and other motor vehicles. Livestock production is another important factor. A much smaller amount comes from wildfires. Natural phenomena can also cause acid precipitations. Emissions from active volcanoes can lead to fog and rain with high degrees of acidity, making human settlement and vegetation impossible in surrounding areas. A small amount of those gases causing acid precipitation occurs naturally in wetlands and oceans.

Effects and Damage to Nature

Acid precipitation contaminates streams and lakes and damages fish and other aquatic life. It is especially harmful in areas where soil is thin or mainly composed of granite rock, since there are few natural filters to buffer the acidic components of water. Acid precipitation further has the effect of dissolving aluminium from the soil, which aquatic organisms then absorb. Not all species can tolerate the same amount of acidity, and some are more vulnerable to increasing aluminium and lower pH values than others. Fish eggs are affected, as they will not hatch when the water is too polluted. Higher acidity can also kill adult aquatic organisms. As a result, there is reduced biodiversity in contaminated lakes and rivers, and some species have already been eliminated.

Acid precipitation causes great damage to many kinds of vegetation. It inhibits nitrogen fixation and leaches out nutrients from foliage. Trees can then become less tolerant to other outside influences, such as cold weather. High altitude forests are most affected, as they are more often immerged in contaminated clouds and fog, which are much more acidic than rain. Even though agricultural crops can be damaged, the harmful effects are counterbalanced by additional applications of lime and fertilizers that replace the nutrients. Some microorganisms are sensitive to changing degrees of acidity. Their enzymes are denatured, and they cannot survive. While toxic substances such as aluminium dissolve from the soil and enter the watershed (causing damage to marine life), useful and important nutrients, such as magnesium and calcium, are also leached away. This loss of nutrients especially affects sensitive plants. The addition of limestone helps stabilize the pH of soil—a method mainly used for agricultural soil.

Acid rain corrodes the surface of many buildings—mainly those made of limestone and marble. The acid rain reacts with calcium carbonate in the building materials and covers the surface with a white layer of gypsum. This light substance easily flakes off when it rains, uncovering the uneven and damaged surface below. It can affect either an entire structure or only small parts where the chemical reaction is more intense. The effect on sculptures can be very damaging, as the process mostly destroys the finely carved parts of the art piece. Acid precipitation erodes and damages many important historic buildings and cultural heritage sites. It can also increase the oxidation of certain materials, such as bronze and copper, which are often used for sculptures. Renovating damaged cultural heritage is very costly. The United Kingdom, for example, invested up to £10 million to repair the damage caused by acid rain to Westminster Abbey in London.

Solutions

There are two different approaches for dealing with the problem of acid rain. Some policy makers advocate the "cost-sharing" principle, meaning that all those countries affected by acid rain should contribute to the problem's solution collectively. Adherents of the "polluter pays" policy state that those who are responsible for pollution should pay for repairing the damage they caused.

Several new technical solutions such as flue gas desulfurization or the use of lime have been attempted in order to cut emissions. The car industry has also reacted by producing greater numbers of low-emission or high-fuel-efficiency vehicles in an attempt to emit less nitrogen oxide.

On the regulatory level, several national and international treaties and conventions attempt to regulate emissions and devise a system to handle the damage caused by toxic pollutants. Emissions trading is another possible solution that combines an institutional and a market-based approach, giving producers incentives to invest in new technologies to cut pollution. With the enactment of the 1990 Clean Air Act Amendment in the United States, the amount of pollution every producer is allowed to emit is regulated. A firm is given a certain number of emissions allowances. Those producers who decide to reduce pollution by installing environmentally friendly equipment can sell their extra units to other producers who are not able or not willing to cut emissions. This way, they recover the costs they previously invested in new environmentally friendly technologies.

Carmen De Michele
Ludwig-Maximilians University Munich

See Also: Clean Air Act; Emissions; Fuel; Gasoline; Pollution, Air.

Further Readings

Ellerman, A. Denny. *Markets for Clean Air*. Cambridge: Cambridge University Press, 2000.

Foster, Bruce A. *The Acid Rain Debate.* Ames: Iowa State University Press, 1993.
Krupa, S. V. et al., eds. *Air Pollution, Global Change and Forests in the New Millennium.* Developments in Environmental Science. Amsterdam, Netherlands: Elsevier, 2003.
Roglesfield, Lyman G., ed. *Acid Rain Research Focus.* New York: Nova Science Publishers, 2008.

Adhesives

Adhesives are substances that can hold at least two materials together by surface attachment. They have been part of human history since 4000 B.C.E. Early adhesives were made from natural materials, such as tree sap, vegetables, tar, beeswax, or animal parts. Synthetic polymer adhesives began to be used around the 1900s. Glue is made from organic compounds, while adhesives are chemical-based. Glue is a by-product of the animal processing industry, such as milk and meat processing. While glue used to be made from old horses (leading to the expression "ready for the glue factory"), 21st-century glue is made from synthetic material or cow hoofs and bones. Adhesives are used in almost every facet of modern human life, including applications in envelopes, cardboard containers, carpeting, automotive trim, film, laminates, and composite materials. The adhesive polymer industry has become an essential part of consumer society.

The first adhesive patent was granted in Great Britain in the late 1700s for a fish-based glue. Synthetic glues were developed in the early 20th century, and technological advances continue into the 21st century. There are two types of adhesives: structural adhesives and nonstructural adhesives. Structural adhesives, such as those used on bridges, are expected to last the life of the product; while nonstructural adhesives, such as those used for pressure sensitive envelopes, are less permanent.

Disposal and Recycling Challenges

Adhesives have always been a challenge for the recycling industry. Successful recycling depends on the quantity and quality of material collected for processing and sent to end markets. In the past, commodity processing required municipal residents to soak labels off jars, remove plastic film windows from envelopes, and other preparatory steps in order to prevent contamination or adverse impacts on the operation of machinery. Manufacturers had to use highly toxic solvents to remove industrial adhesives, such as those used in electronic, automobile, and carpet components. These requirements caused many people to view recycling as inefficient, cumbersome, and time consuming. However, technological advances in commodity processing have improved so that only a few collected commodities cause problems.

The bane of municipal and industrial recycling processes continues to be pressure-sensitive or press-and-seal adhesives, which are increasingly used in products because of cost and convenience factors. The challenge being approached by mills is designing machinery to better handle adhesives, while manufacturers continue to work on environmentally and machine-friendly adhesive formulas or detachable adhesives. Until processing technology catches up with consumption of adhesive products, the best way to prevent damage to machinery and contamination is through education.

Recycling Process

Often, the public is not educated about the specifics of the recycling industry because it is believed that if it is not "easy," people will not recycle. This leads

Technological advances in adhesives removal have made paper recycling less cumbersome and toxic. However, pressure-sensitive adhesives, which are cheap and convenient, continue to cause millions of dollars in damage to pulping and sieving machinery.

the public to believe that recycling rules are arbitrary and that putting materials such as self-adhesive notes in with paper has an inconsequential impact, while continuing to soak labels off of products and remove staples, despite the fact that technology has been developed to address one problem but not the latter. However, when people understand how their actions impact the recycling process, they are more likely to keep contaminants out of the waste stream.

Adhesives cause problems when bales of sorted paper, including those containing adhesives, are fed into a processing machine that blends the paper with water and chemicals, creating a pulp. This pulp goes through several processing steps and is sieved through different sized metal grating. Anywhere along this process, paper containing adhesives can damage the machinery by clumping, quite literally gumming up the works. This damage costs several million dollars per year.

Barbara L. MacLennan
West Virginia University
Laura J. Stiller
Monongalia County Solid Waste Authority

See Also: Culture, Values, and Garbage; Post-Consumer Waste; Pre-Consumer Waste; Recyclable Products; Recycling Behaviors.

Further Readings

Adhesive and Sealant Council (ASC). http://www.ascouncil.org (Accessed June 2011).

Cognard, Philipe. *Handbook of Adhesives and Sealants: Basic Concepts and High Tech Bonding*. Amsterdam, Netherlands: Elsevier, 2005.

Pizzi, A. and K. L. Mittal. *Handbook of Adhesive Technology, Revised and Expanded*. Boca Raton, FL: CRC Press, 2003.

Adorno, Theodor

The metaphors of garbage and waste exist throughout the postwar writings of German philosopher Theodor W. Adorno (1903–69), functioning in complex and often seemingly contradictory ways. A member of the Frankfurt school, Adorno's writings drew upon sociology, philosophy, and musicology and spanned the 20th century, including the modernism of the 1920s, the emergence of the cold war in the 1940s, and the student revolts of the 1960s. The majority of his work involved developing a critique of totality, the Enlightenment, capitalism, fascism, Stalinism, and the culture industry. For Adorno, the garbage of capitalist culture was pitched in opposition to the autonomous points of resistance he saw located in radical art, literature, and music.

Beliefs

The meanings Adorno allotted to both garbage and waste are neither straightforward nor interchangeable. Adorno argued that all culture after Auschwitz, including its critique, was garbage [*Müll*]. Or, in other words, that the sphere of "resurrected culture" (that which he saw as merely rehashing traditional values of truth, beauty, and goodness, as if the Holocaust had not happened) should be considered as mere refuse. At the same time, Adorno's critique of culture as garbage was also directly related to his condemnation of the trashy kitsch produced by capitalist culture.

Having fled Nazi Germany in 1938, Adorno lived in exile in the United States until 1953. The philosopher consequently experienced the janus-faced culture of the mid-20th century; the *Hochkultur* (high culture) of Germany versus the burgeoning U.S. culture industry and its soon-to-be all-consuming mass media. For him, the culture industry's production of lifestyle involved a problematic recycling of real culture, transforming aesthetics into commodities and muting any negative or critical potential the former might have had. According to Adorno, the culture industry dangerously sanctioned this demand for rubbish. He warned against capitalism's often ahistoric consciousness, which swept aside the past as garbage. He was critical of the logic of capitalist production, which he saw as relegating to the junk pile everything not in line with the most recent methods of industrial production. For Adorno, this also ultimately meant rubbishing what he thought of as life's valuable continuity. He argued that art had degenerated into culture by means of the entertainment industry, becoming confused with its own waste products as its aberrant influence grew. In correspondence with the German philosopher Walter

Benjamin, Adorno suggested that even rubbish could be subject to capitalist exchange value. At the same time, Adorno saw the communism of the Eastern Bloc as having transformed culture into rubbish as a means of control, and he accused Josef Stalin of throwing modernism's great heroes, such as Franz Kafka and Vincent Van Gogh, onto the rubbish heap.

On the other hand, and in keeping with this critique, Adorno suggested that people should address the "waste products" and "blind spots" of history so as to reclaim their radical potential. For him, the relationship between waste and memory was powerful. He believed that a history of waste and blindness was equivalent to a history of dissonant or negative elements. For Adorno, the modernist works of art that resisted being commodified and turned into capitalist garbage were those that managed to retain a negativity and autonomy. He suggested that it would be rewarding to investigate the piles of rubbish, detritus, and filth upon which the works of major artists appear to be erected and to which they still owed something of their character.

Analysis of Beckett

The complexity of waste as a figure in Adorno's thought is evident in his approach to Samuel Beckett's 1957 one-act play *Endgame*. For Adorno, *Endgame* presented culture after the destruction of culture with his protagonists literally living in history's trash cans. He argued that Beckett offered modernism as the obsolescence of the modern, presenting a philosophy simultaneously reduced to "culture-trash." Adorno's Beckett lays waste to culture, as the play depicts a world that, along with the continuity of tradition, has been thrown on the garbage heap. Beckett's play, like Adorno's postwar philosophy, pictures existence after catastrophe and the complete reification of the world. In the dialogue between Hamm and Clov (the play's two antiheroes), Adorno argues that thoughts become refuse—the day's leftovers—as humanity vegetates on a pile of ruins. For Adorno, Beckett's trash cans are the emblem of a culture restored after Auschwitz.

Adorno's perceptive reading of cultural production as a mode of recycling resonates with postmodernism's constant reappropriation of previous styles and ideas. Many scholars, such as Brian Neville and Johanne Villeneuve, who have explored the idea of cultural waste, have turned to Adorno's philosophy as a means of rethinking the semiological and epistemological conventions that underlie notions of waste, memory, and recycling.

Sarah Edith James
University College London

See Also: Capitalism; Commodification; Consumerism; Culture, Values, and Garbage; Garbage in Modern Thought; Germany.

Further Readings

Adorno, Theodor W. *Minima Moralia: Reflections on a Damaged Life*. London: Verso, 1984.
Adorno, Theodor W. *Negative Dialectics*. New York: Continuum, 2005.
Adorno, Theodor W. *Prisms*. Cambridge, MA: MIT Press, 1983.
Adorno, Theodor W. "Trying to Understand *Endgame*." Trans. by Michael T. Jones. *New German Critique*, v.26 (1982).
Harding, James Martin. *Adorno and 'A Writing of the Ruins': Essays on Modern Aesthetics and Anglo-American Literature*. New York: State University of New York Press, 1997.
Jenemann, David. *Adorno in America*. Minneapolis: University of Minnesota Press, 2007.
Neville, Brian and Johanne Villeneuve. *Waste-Site Stories: The Recycling of Memory*. Albany: State University of New York Press, 2002.

Aerosol Spray

Aerosol spray represents an inexpensive and convenient way to dispense a variety of liquids, including deodorants, insecticides, and paints. A dispensing system that creates an aerosol mist of liquid particles, aerosol spray uses a can or bottle that contains a liquid under pressure. Common in the United States since the 1940s, aerosol sprays use propellants to assist with the process of driving the payload out of the container. Chlorofluorocarbons (CFCs) were commonly used as propellants in aerosol sprays, but during the 1980s CFCs were found to have negative effects on the Earth's ozone layer. As a result, in 1989, the Montreal Protocol on Substances That Deplete the Ozone Layer (Montreal Protocol) sought to phase out the use of CFCs and has been largely successful in doing so. Replacements for CFCs have been found, although these are often flammable. This, along with the sometimes nonbiodegradable nature of aerosol spray packaging, has led to concerns about the waste these products generate.

History

Aerosol sprays were conceptualized as early as the late 18th century, although it took almost 150 years for the concept to be realized. Atomizers were used throughout the 19th century to dispense paint, perfume, and other liquids using a hand-operated pump rather than stored gas. In 1926, a Norwegian chemist, Erik Rotheim, invented the first aerosol spray can. Rotheim's invention forced liquid contained in a canister out of a small hole, emerging as an aerosol or mist when the container's valve was opened. Rotheim sold his patent to a U.S. company for 100,000 Norwegian kroner, but it was not until 1939, when Julian S. Kahn received a patent for a disposable spray can, that aerosol sprays became commercially viable.

During World War II, the U.S. government sought a convenient means to combat the malaria-carrying mosquitoes besieging troops in the Pacific. Lyle Goodhue and William Sullivan, working for the U.S. Department of Agriculture (USDA), designed a refillable spray can, dubbed the "bug bomb," which was patented in 1943. Goodhue and Sullivan's small, portable can used liquefied gas to provide it with propellant qualities, permitting soldiers to spray the inside of tents with aerosol pesticide. After the conclusion of World War II, the USDA granted licenses to three companies to manufacture aerosols for commercial use, including Chase Products Company and Claire Manufacturing, which continue to manufacture aerosols to this day. Robert Abplanalp, owner of the New York–based Precision Valve Corporation, invented the crimp-on valve in 1949; this controlled the aerosol spray, further expediting the commercial viability of aerosol sprays.

Packaging and Propellants

Aerosol sprays proved immensely popular with consumers as a means of dispensing liquids. The liquid to be dispensed is stored in a can or bottle that contains a gas under pressure. When the consumer opens the valve, the liquid is forced out through a small hole, emerging as an aerosol mist. The gas expands inside the container as it drives the liquid out, thereby maintaining an even pressure. Once outside the container, droplets of propellant evaporate rapidly, leaving the payload suspended as very fine droplets or particles.

Most aerosol-spray containers are comprised of three primary parts: the can, valve, and actuator. The can is usually lacquered tinplate or aluminum and is often two or three separate pieces of metal crimped together. The valve determines the spray rate of the liquid and is usually crimped to the structure of the can. The actuator is the button that the user depresses to open the valve. The nozzle's shape and size control the spread of the aerosol spray and can sometimes be adjusted by the user. Aerosol spray containers used with highly viscous products, such as food spreads, thick shaving creams, and sealants, often use a piston barrier system, which assures separation of the product from the propellant, ensuring purity of the product. Certain pet-care, pharmaceutical, sun-care, and other products use a bag on valve (BOV) system, where the product is separated from the propellant by means of a laminated, hermetically sealed, multilayer pouch. The BOV system is popular because it extends the product's shelf life. Regardless of the system used, the pressurization makes cans used for aerosol sprays difficult to recycle and, as a result, they often end up in landfills.

Inside the pressurized can, the propellant exists as gas in vapor form, existing in balance with its bulk liquid at a pressure that is higher than atmospheric levels, but not dangerously high. From the 1940s through the 1980s, CFCs were most commonly used for the propellant in aerosol sprays. After CFCs were identified as a major source of ozone depletion, however, a number of nations met to pass the Montreal Protocol, which was an international treaty designed to phase out the use of CFCs and other ozone-damaging substances. To date, the Montreal Protocol has been ratified by 196 nations and is considered one of the most successful international environmental agreements. As a result of the Montreal Protocol, the Earth's ozone layer is expected to recover fully by 2050. Aerosol sprays today use mixtures of volatile hydrocarbons, such as propane, n-butane, and isobutane, as well as dimethyl ether (DME) and methyl ethyl ether as propellants. While these do not harm Earth's ozone layer, all of these substances are flammable. This flammability requires additional safeguards during the production of aerosol sprays and limits the methods of disposal. Although aerosol sprays' packaging and propellants create waste, their convenience makes their continued popularity and prevalence likely.

Stephen T. Schroth
Jason A. Helfer
Knox College

See Also: Biodegradable; Clean Air Act; Cleaning Products; Packaging and Product Containers; Pollution, Air.

Further Readings

Glennon, John P. *Estimating Microorganism Densities in Aerosols From Spray Irrigation of Wastewater*. Washington, DC: Environmental Protection Agency (EPA), 1982.

Rappert, B. "The Distribution and Resolution of the Ambiguities of Science, or Why Bobby Can't Spray." *Social Studies of Science*, v.31/4 (2001).

Africa, North

North Africa includes Algeria, Egypt, Libya, Morocco, Sudan, Tunisia, and western Sahara. The geopolitical distinction between these countries and the rest of Africa (sub-Saharan Africa) is due to the barrier of the vast Sahara Desert and the proximity of the northern region of the continent to the trade routes and colonizing activities of Mediterranean civilizations like the Phoenicians, Greeks, Romans, and Vandals. Since the 7th century, north Africa has been a significant part of the Arab and Islamic world. Arabic continues to be one of the official languages of every country in the region, and North Africa has a strong Muslim presence. The entire region was at one point under the control of the Ottoman Empire (with the exception of Morocco), and during the height of European imperialism, various parts of the region were under the control of the United Kingdom, Spain, Italy, or France. The cultural, political, and economic influences on north Africa's history have thus been complex and varied, but have served to increase the commonalities among the nations of the region and the differences between it and sub-Saharan Africa.

In the 21st century, Algeria, Libya, and Sudan are dependent on their oil and natural gas reserves; Tunisia on tourism; Egypt on tourism and the strongest industrial sector in the region; and Morocco on phosphate mines and farming. North Africa has been one of the regions noticeably affected by global climate change, with record heat reported. Western Sahara is a disputed territory, claimed by both Morocco and the Sahrawi Arab Democratic Republic (SADR), the government of which operates in exile from Algeria, with that host country's support. SADR is recognized by the African Union, while Morocco's claim is recognized by most of the Arab world, again highlighting the transcontinental influences of the region.

Water Crisis

Sudan and Tunisia are among the countries most impacted by the worldwide freshwater crisis, with 12.3 million and 2.1 million people, respectively, whose sole water source is contaminated. Algeria and Egypt are among the countries with the most significant water deficits in the world, leading to overpumping and a lowering of the water table. Subsequent grain shortages require the importation of wheat and other basic staples. Morocco's water problems are exacerbated by a denser population, considerable population growth, and the seasonal influx of tourists in the summer months. The treatment of water and wastewater in north Africa is thus a critical issue germane not only to environmental concerns and long-term safety and welfare, but also to current ongoing crises of human survival.

Wastewater treatment removes contaminants from wastewater originating both from households and from effluents (sewers and other runoff). The by-products are environmentally safe fluids

People living along the canals in El-Gededa om El-Resh and El-Seds in Egypt's eastern El-Sharkia province used to throw their garbage and sewage into open canals. After USAID worked with Egypt's government to address environmental issues related to agricultural irrigation, canals passing through residential areas were covered and protected from this garbage and sewage problem. Municipal solid waste collection is problematic throughout north Africa, where urbanization is increasing and collection efficiency is poor.

and solids; the solids can be disposed of or turned into fertilizer, while the fluids are returned to the environment. Technological advances now permit treated wastewater to be purified back into drinking water. Wastewater treatment of any sort, however, is expensive to set up, with newer technologies even more so.

Rapid population expansion can result in an expensive arms race between the wastewater and its treatment facilities, particularly as—in many parts of the world—the per capita use of water is steadily increasing. In 2010, only Singapore produced drinking water from wastewater on a large scale. It may be decades before north Africa can follow, without outside financial intervention.

Further, water problems are exacerbated in the region because of a lack of sanitary landfills, insufficient municipal solid waste management, and a lax legal environment for overseeing industry and modern agriculture. Some rural areas do not have sewage networks or have failed to repair broken-down wastewater treatment plants. Urban wastewater and municipal solid waste are major sources of pollution in north Africa. Where sewage networks are lacking, wastewater is disposed of in septic tanks and may wind up either on the seashore, poured into valley courses, or deposited on open land. Most north African wastewater treatment plants were designed only for secondary wastewater treatment (treating it to a level of quality suitable for agricultural purposes, but still far from clean). When problems arise with these plants, the quality of the discharge drops to considerably below that level. Many households are improperly connected to their wastewater systems, leading to large amounts of above-ground runoff, especially near urban areas.

Lower-tech, lower-cost solutions include water treatment systems like solar-powered distillation to produce drinking water and desalination to harvest seawater and turn it into freshwater usable for either drinking or irrigation. While overexploitation of groundwater has dire consequences, it is politically unpopular to slow it down or reduce it because limiting the amount of groundwater available for crop irrigation can result in crop and farm failures in the short term, even though permitting it to continue will lead to crop failures in the long term.

Wastewater Irrigation

Because of the increasing demand for fresh water, there has been significant interest in reusing wastewater in agriculture in order to free up more clean freshwater for human consumption. Wastewater irrigation conserves freshwater, presents a low-cost method for sanitary disposal of urban wastewater, reduces the pollution of waterways, provides a water supply for farmers, and provides crops with nutrients, reducing the need for other fertilizers (especially artificial fertilizers), which can damage soil and water quality. Treated wastewater is used extensively for irrigation in Israel and has seen increased use in Morocco—one of the north African countries with the necessary capital for extensive wastewater treatment facilities. Wastewater for agricultural uses still needs to be treated to a reasonable microbiological standard, but can result in increased crop yields of as much as 25 percent.

Morocco operates 39 urban wastewater treatment plants and 23 in rural communities. Approximately 900,000 people of Morocco's population of about 13.5 million are served by a wastewater treatment plant. Reuse of wastewater for irrigation has been somewhat restricted by the fact that not all of the plants treat water to the necessary level of cleanliness, which could result in contamination of crops. Morocco's water pollution problems have been steadily increasing since the 1960s, and the present cost of damage caused by polluted water exceeds 1 percent of the country's gross domestic product (GDP), making it one of the most water-polluted countries in the world from a financial perspective. Most of north Africa has only ill-defined standards for wastewater treatment, which are often overlooked. Funding is available from the European Investment Bank, the World Bank, the United Nations, and the U.S. Agency for International Development for developing systems of wastewater reuse, including raising the standards of treatment plants.

The potential dangers of wastewater irrigation include the contamination of groundwater through nitrates; the build up in the soil of heavy metals and other chemical pollutants; the establishment of disease vectors; excess growth of algae and other vegetation in wastewater-carrying canals; and health risks to workers with prolonged contact with untreated wastewater or to consumers of fresh produce irrigated with disease-carrying water. These problems can be exacerbated by a bureaucracy and workforce unfamiliar with the risks and the methods for mitigating them, a lack of information on cost-effective wastewater treatment, a lack of local infrastructure for developing the proper technical specifications for such treatment, and farmers who are not sufficiently aware of the appropriate procedures to reduce the health risks associated with the use of wastewater. Theoretically, technical support in these areas could be offered by outside agencies. Morocco participates in the MEDAWARE project, a European-Mediterranean partnership for developing water management and wastewater policy solutions.

In many parts of north Africa, farmers without sufficient access to other water sources have resorted to using untreated wastewater for irrigation, which carries all of the above risks and a significantly elevated health risk both to farmworkers and consumers. Wastewater contains both industrial contaminants and biological contaminants (such as bacteria, eggs, and larvae) from sewage, both of which can be passed on to crops, especially dangerous in those consumed raw. Washing is not always sufficient to reduce this risk, particularly in countries where the water used for such washing cannot be depended upon for cleanliness.

Some farmers mitigate risk by using wastewater only for crops that will not be consumed (such as cotton) or that will be heavily processed before consumption (such as cereals, grains, and soybeans). This use can still pass contaminants into the soil and presents a health risk to workers. Algeria, for example, suffered a food poisoning epidemic in the summer of 2010 as a result of insufficiently treated (possibly untreated) wastewater used to irrigate watermelon.

In July 2010, the World Bank approved the Northern Tunis Wastewater Project, allocating $8.03 million to build a submarine outfall to reduce the environmental consequences of treated wastewater discharge into the Gulf of Tunis. The completed project will help farmers use treated wastewater for agricultural purposes. The remainder of the discharge will go through an underwater outfall into the Mediterranean Sea, and systems will be greatly enhanced to monitor discharge quality issues and increase Tunisia's wastewater treatment capacity, which—like the rest of the region—is far behind demand. Ideally, the project will see significant environmental, economic, and human health gains, and it will reduce Tunisia's dependence on foreign food imports as well as the possibility of a public health epidemic caused by irrigating with untreated wastewater.

Other Forms of Waste

Throughout north Africa, there are problems with municipal solid waste collection. While quality varies from area to area, efficiency is typically low, and often the waste is deposited in landfills without any sanitary measures taken. Some landfills are even located directly at the waterfront, polluting the sea through leaching. Sometimes, waste is merely deposited on a vacant lot or otherwise unused land by waste removal trucks. The increased urbanization of north Africa and general population growth have greatly increased the volume not only of household waste, but also of construction and demolition waste, and there are usually few controls or plans dealing with handling such wastes, leading to widespread uncontrolled and unmanaged dump sites. Remedying this situation is not a budget priority in any north African country, and the capital needed to begin a methodical waste collection management system is significant. Libya, for instance, did not even include municipal solid waste management in the National Physical Plan until 2000, despite addressing matters in and allocating monies for other urban utilities. Often, urban solid waste collection is more efficient than in rural areas, but this is not always the case.

The legal frameworks of north African countries often fail to address issues related to the disposal of poisonous, toxic, or household waste, as well as similar issues such as the correct storage and handling procedures for poisons, pesticides, and herbicides. The law lags behind the technology in these respects, since apart from Egypt, the area is relatively new to industrialization and has been plagued with social, political, and economic concerns that may have made such lawmaking seem a luxury easily put off. Egypt is rather more advanced in this area and initiated a 10-year program for Integrated Solid Waste Management in 2000, which addresses municipal, agricultural, healthcare, construction, and demolition wastes as the top priorities, followed by industrial waste, waste generated from clearing drainage canals, and municipal wastewater sludge as secondary priorities, as well as adopting the "polluter pays" principle. The program laid out a comprehensive legal framework for waste management in Egypt and initiated groundwater and soil condition studies to inform the subsequent 10-year plan. Egypt had been ahead of the region in the 20th century as well, finally adopting regulations specific to hazardous waste in the 1990s, while most other north African countries have failed to do so even in the 21st century.

Even in Egypt, as in the rest of north Africa, there is a considerable gap between the amount of waste generated and the amount of waste disposed of at disposal sites. The developed world considers it an aberration if these amounts do not approach identicality. In north Africa, scavengers pick over the waste for reusable materials, and some waste is simply never collected. It is difficult, however, for record keepers to know to what cause to ascribe the "missing" waste, which makes efficiency studies problematic.

Unlike in the more developed parts of the world, solid waste in north Africa is usually between one-half and two-thirds organic waste (two to three times as much as in the United States), with plastic, paper, glass, and metals together making up about one-quarter of the waste by weight. Rural areas dispose of much less solid waste because so much of what is disposed of in cities is likely to be reused in the country as fertilizer or animal feed. In Egypt, Morocco, and Tunisia, in particular, the food waste of tourists in urban areas is a significant contributor to municipal waste.

Sanitary landfills are in their infancy in north Africa. In most areas, there is insufficient regulation defining or overseeing proper landfill management. Laws may even be limited to their minimal demands of banning the burning of waste and the placement of organic waste too near to waterways. What regulations exist are frequently bent or broken, sometimes out of ignorance or financial inability to comply.

Energy Sources

Hydrocarbons are the primary source of energy for north Africa. Algeria and Libya are both members of the Organization of Petroleum Exporting Countries (OPEC). Tunisia and Morocco are less rich in oil, but still produce enough of it to depend upon it as a main source of energy. However, as oil reserves run low, the possibility of nuclear power in the future has been raised. Every north African country, barring Western Sahara, has worked on developing nuclear capabilities, often with Western support—especially from France. While the words "nuclear north Africa" have raised concerns about the development of nuclear weaponry in a historically volatile region, it seems that the region's interest in nuclear technology is purely related to the need for post-oil energy sources. Nuclear power plants could efficiently power seawater desalination plants to alleviate the water crisis, and the adoption of nuclear power would increase economic ties to Western corporations.

In 2007, Egypt announced plans to build nuclear reactors to meet its rising energy demands. Algeria, meanwhile, has pursued plans to build a nuclear power station by 2020 and has the advantage of considerable uranium deposits, though it had no capacity for enriching uranium as of 2010. Libya has negotiated with Russia for assistance in building a civilian nuclear power plant as well.

Bill Kte'pi
Independent Scholar

See Also: Africa, Sub-Saharan; Cairo, Egypt; Developing Countries; Iran; Middle East; Pollution, Water; Saudi Arabia; Sewage; Sewage Treatment.

Further Readings

Fatta, D., I. Arslan Alaton, C. Gokcay, M. M. Rusan, O. Assobhei, M. Mountadar, and A. Papadopoulos. "Wastewater Reuse: Problems and Challenges in Cyprus, Turkey, Jordan, and Morocco." *European Water,* v.11/12 (2005).

Pearce, Fred. *When the Rivers Run Dry.* Boston: Beacon Press, 2007.

Peterson, Mark Allen. *Connected in Cairo.* Bloomington: Indiana University Press, 2010.

World Bank. *Making the Most of Scarcity: Accountability for Better Water Management in East and North Africa.* New York: World Bank, 2007.

Zereini, Fathi and Wolfgang Jaeschke, eds. *Water in the Middle East and North Africa.* New York: Springer, 2010.

Africa, Sub-Saharan

Sub-Saharan Africa consists of four major regions in Africa: west, east, central, and southern Africa. About 48 independent countries exist within the region. The term *sub-Saharan Africa* is usually used to depict the geographical area of the African continent that lies south of the Sahara Desert, or those African countries that are located south of the Sahara. Some say it excludes the Sahel and the Horn of Africa in the north. The region is one of the poorest in the world and is characterized by low economic opportunities, unstable government and government policies, poorly developed social and technological structures, and unplanned and expanding urban centers.

The region is also characterized by high population growth rates, comparatively high fertility rates, and a youthful population. The region's urban population was 15 percent in 1950, 42 percent in 2000, and it was projected that by 2020, the majority of the population will live in urban areas. Sub-Saharan Africa has one of the world's fastest-growing populations, increasing at about 2.8 percent per year, and it is expected to be home to over 1 billion people by 2025. While in the first half of the 20th century urbanization was predominantly confined to countries that had the highest levels of per capita income, in the more recent past, the most visible changes in urbanization have occurred and will likely continue to occur in middle- and low-income countries like those in sub-Saharan Africa.

Consumption Patterns

The region has witnessed a decline in agricultural output since the 1970s and relies heavily on importation to feed its growing population. The patterns of consumption have tilted toward imported foods, including livestock (such as frozen foods and dairy) and staple foods (such as rice). This development has raised the consumption of mostly high-waste-generating imported food and has also led to increasing numbers of fast food restaurants. The volume of waste generation among fast food operators within the region is high, with very passive or no government regulation.

According to 2004 United Nations Economic Programme (UNEP) reports, countries in sub-Saharan Africa are facing serious problems related to natural resource management and environmental pollution owing to rapid growth in urbanization and industrialization. High consumption levels and a propensity to consume are putting a strain on the environment. However, excessive and indiscriminate consumption of resources is not limited to urban people; a disproportionately growing number of rural dwellers in sub-Saharan Africa engage in arbitrary consumption of resources such as deforestation and bush burnings. But the patterns and trends mark the difference. The contribution of urban and rural activities to harmful consumption of resources is a critical factor.

Consumption is clearly an essential property of human settlements. In industrialized countries, consumption of resources will mainly continue in the pursuit of a better of life, while in developing countries, such as sub-Saharan Africa, consumption is primarily driven by the desire to meet basic needs, increase economic growth, and sustain the development process. In economic terms, higher consumption levels lead to higher investment, which further leads to higher per capita income in the long run. Consumption will arguably contribute to human development when it enlarges the capabilities and enriches the lives of people without adversely affecting the well-being of others. Consumption becomes a positive factor in the development process when it is done in a sustainable manner. But the links between consumption and development are often broken. When they are, consumption becomes inimical to human and environmental development. The consumption pattern of the early 21st century is undermining the environmental resource base. If the patterns of consumption in sub-Saharan Africa are not changed, the problems of unsustainable and excessive consumption of resources and human development will worsen. Therefore, the real issue is not consumption itself, but rather its patterns and effects, including emissions and wastes that pollute the Earth and destroy ecosystems, and deplete and degrade renewable resources, undermining livelihoods and future development.

The current pattern of resource consumption in sub-Saharan Africa is overexploiting the natural resource base and stretching the carrying capacity of ecosystems that provide them to a point that they could collapse long before the world runs out of nonrenewable resources. Meanwhile, current consumption patterns do not show any prospect of abatement, threatening the health and livelihood of future generations. Pollution, resource depletion, energy consumption, increasing waste generation, and biodiversity and landscape destruction, which appear to be unsustainable by any standard, constitute the features of many consumption patterns in sub-Saharan Africa. Land conversion associated with both urbanization and tourism shoreline development is also noticeable.

The production and consumption of commodities requires the extraction and use of natural resources (such as wood, ore, fossil fuels, and water). It also requires the creation of factories and factory complexes whose operation creates toxic by-products, while the use of commodities themselves (such as automobiles) creates pollution. High consumption levels can never be separated from increasing volumes of waste. Excessive consumption in sub-Saharan Africa has brought about a deep concern for waste management as increasing volumes of waste are generated by urban areas.

The current conditions and trends of natural resources consumption in sub-Saharan Africa involve three critical areas: (1) energy and transport, (2) water, and (3) forest-based and mineral resources. These consumption activities provide clues and insights into the current consumption patterns of resources and prospects for the future.

Energy

Energy consumption in sub-Saharan Africa varies dramatically, with noncommercial fuels, such as wood and animal waste, dominating fuel consumption. The use of wood for fuel is predominant in both rural and urban locations, accounting for approximately 70 percent of total energy use and 90 percent of household energy use. Africa is the world's largest consumer of biomass energy (energy from sources like firewood, agricultural residues, animal wastes, and charcoal), calculated as a percentage of overall energy consumption. Nigeria consistently leads sub-Saharan Africa in commercial energy consumption. In 2001, Nigeria consumed 0.92 quadrillion Btu (quads), which represented 32 percent of all energy consumed in the region. Of the remaining countries, Zimbabwe registered the next highest level with 0.24 quads of energy consumed, and Ghana consumed 0.17 quads. These numbers are generally higher than most of sub-Saharan Africa, which averaged 0.06 quads in 2001. Although domestic demand for energy consumption in sub-Saharan Africa is growing rapidly, consumption levels remain well below world average.

The global primary energy consumption is rising by more than 3 percent annually, and most of this consumption takes place in human settlements. There are, however, wide variations in per capita energy use in different regions of the world; on average, the energy used by a person in a developed country is about nine times that of a person in a developing country. Large disparities also exist among developing countries; for example, the annual per capita consumption of modern forms of energy in sub-Saharan Africa is less than half the average of developed countries. Similar disparity exists within the countries of sub-Saharan Africa, as well as between the rural and urban poor and the higher income groups. Several factors contribute to this disparity in energy consumption: the level of urbanization, economic activity, and living standards. With rapid urbanization in low- and middle-income countries, this scenario is, however, changing fast. A World Bank study estimated that the energy demand in these countries is expected to rise to parity with Organisation for Economic Co-operation and Development (OECD) demand by 2015. Sub-Saharan Africa has a great potential for higher consumption of energy, with urbanization and motorization as the main contributing factors.

Transport

The rapid growth in demand in the energy sector in sub-Saharan Africa is taking place amid grossly inefficient energy use in the transport sector, households, and construction, which, together, account for most of the energy use in human population. The transport sector is the largest single consumer of commercial energy in human settlements, accounting for nearly half of all petroleum consumption. In sub-Saharan Africa, it accounts for nearly 80 percent of all commercial energy used. The key factors that contribute to inefficient energy use in the transport sector are the growing road congestion in urban areas resulting from urban growth, increasing dependence on cars, and vehicles with poor fuel efficiency coupled with poor vehicle and road maintenance in developing countries.

In households, considerable scope exists for efficiency improvements in common cooking, lighting, and appliance technologies. Vast amounts of energy are wasted in traditional cookstoves still in common use in rural areas of the developing world. The technology is already available for energy-efficient stoves, refrigerators, and lighting appliances, such as compact fluorescent lamps. Building insulation can reduce energy use by three-quarters, but its diffusion is slow, mainly because of higher initial cost to the consumer. Energy production and use, in almost every form, generates varying degrees of environmental externalities that affect human health, ecological stability, and economic development. The wide-scale use of fuel wood in Africa is contributing to deforestation and soil erosion. Also, the extensive use of biomass in traditional stoves exposes the users—mainly women and children—to high levels of indoor pollution. Transport-related air pollution is already threatening the health of some 300–400 million city dwellers in the region. If current trends continue, by 2015, transport-related air pollution in many large cities of the developing world will be much worse than projected levels for major cities in industrialized countries with a much higher level of motorization.

Another major characteristic of unsustainable energy consumption in sub-Saharan Africa is gas flaring. The flaring of natural gas in Nigeria, Angola, Cameroon, and Gabon has proven to be a significant source of carbon emissions in sub-Saharan Africa. In the process of oil production, natural gas is released. Because gas infrastructure in sub-Saharan Africa is extremely limited, this "associated" gas is often burned off, or "flared," rather than captured for use. Not only does this waste a potentially valuable energy source—the World Bank estimates that every day Africa flares gas equivalent to 12 times the energy that the continent uses—but it also releases carbon dioxide directly into the atmosphere. However, Nigeria—the major player in gas flaring—is working on developing a gas-to-liquids industry and expanding its liquefied natural gas trade to reduce

Margaret E. Heggan Public Library
606 Delsea Drive
Sewell, NJ 08080

flaring. The government has attempted to terminate all gas flaring, with heavy fines for companies that do not comply. However, this goal is far from being realized as high magnitudes of gas are wasted through flaring.

Water

Freshwater is a finite resource, which has to meet ever-increasing demands from competing users: agriculture, industry, and the domestic sector. The availability of freshwater is decreasing in many parts of Africa, notably in north Africa, the Sahel, and the Kalahari-Namib regions of sub-Saharan Africa. Some 80 countries with about 40 percent of the world's population are facing acute to chronic water scarcity. With rapid urbanization, many cities are also facing water scarcity. In sub-Saharan Africa, more than 300 cities face water shortages, with 100 in acute distress at an estimated cost of $14 billion in lost economic output each year. Excessive groundwater extraction and consumption in many cities in sub-Saharan Africa has caused saline intrusion and ground subsidence. According to the Nigeria Demographic Health Survey, 42.6 percent of its population in 2008 drew water from an unimproved source. In an attempt to have access to improved sources of water, many tube wells or boreholes have been dug. The implication of this activity is arbitrary degradation of the environment.

In developing countries, close to 85 percent of water is allocated to agriculture, 10 percent to industry, and only about 5 percent for domestic use. Less than 45 percent of the water for agriculture actually reaches the crops. In addition, cities in developing countries are losing between 30 and 60 percent of water—produced and treated at high cost for domestic use—through leakage from poorly maintained water supply systems before it reaches the intended consumers. There is also the need to put in place measures to promote more efficient use of water resources, reduce losses and wasteful use, and ensure equitable distribution among different uses and users for the present and future generations. There is also an important equity issue associated with water use, particularly in urban areas. The urban poor in developing countries almost always have to rely on private water vendors and end up paying a very high price for their water, often several times more than other affluent neighbors who have water piped into their homes. Thus the poor, in effect, subsidize water services for the more affluent and continue to suffer water scarcity and a heavy burden of healthcare. Inequitable pricing of water also leads to profligate use and wastage of this precious resource by the domestic sector and, to a lesser extent, the industrial sector.

Forest-Based and Mineral Resources

Sub-Saharan Africa has suffered the worst deforestation, with the forest cover reduced to 8 percent. The per capita availability of forests in developing countries is now about half of that in industrialized countries. Firewood accounts for more than half of the world's demand for wood. In African countries, biomass is the key source of energy, accounting for 50–90 percent of the total national energy supply. In Sudan, between 1962 and 1980, wood fuel consumption contributed to 92 percent of the 21,125 square miles of deforested areas. Zimbabwe's forests are reduced by 1.5 percent per year, with a growing fuel wood deficit and implications for women who gather fuel wood. In Nigeria, for instance, the livelihood of some 40 million of the country's indigenous people, many of them women, depends on access to fuel wood. Despite widespread deforestation in large parts of the developing world, the forest-products industry continues to thrive, with a global annual turnover of $330 billion. The annual demand for industrial wood is estimated to increase from 1.6 billion cubic meters to 1.9 billion by 2010, driven by increases in population and rising living standards. Industrial logging has also played an important role in the destruction of primary rain forests in central Africa.

In the case of mineral resources, central Africa is home to one of the world's largest rainforests and serves as one of the world's most important carbon sinks. Carbon sinks capture carbon dioxide from the atmosphere, thus reducing global carbon dioxide levels. Deforestation is one of the most pressing environmental problems faced by almost all sub-Saharan African nations, with one of the primary causes of deforestation being wood utilization for fuel. Many sub-Saharan countries have had over three-quarters of their forest cover depleted, and it is estimated that if current trends continue, many areas—especially that of the Sudan-Sahelian belt—will experience a severe shortage of fuel wood by 2025. Deforestation also has negative implications for the local environment, including increased erosion and loss of biodiversity. The highest rates of deforestation occur in areas with large growing populations, such as the East African Highlands and the Sahel.

The harvesting of wood for use as fuel also has contributed to the problem of desertification. *Desertification* is the term used to describe the loss of soil fertility and structure to the extent that its ability to support plant life is severely compromised. In sub-Saharan Africa, where desertification has its greatest impact, forest areas are often cleared in order to harvest fuel wood and for agricultural use. Traditional farming practices, which tend to be inefficient and land intensive, significantly degrade scarce arable land—the single most important natural resource in sub-Saharan Africa. Desertification can lead to downstream flooding, reduced water quality, and sedimentation in rivers and lakes. It also can lead to dust storms, air pollution, and health problems, such as respiratory illnesses and allergies.

At the World Summit for Sustainable Development, the United States joined forces with the governments of Cameroon, the Central African Republic, the Democratic Republic of the Congo, Equatorial Guinea, Gabon, and the Republic of Congo, along with multilateral donors to promote economic development in the Congo River basin through natural resource conservation programs. The program creates a network of parks and protected areas, well-managed forestry concessions, and communities that depend upon the conservation of the forest and wildlife resources in the Congo River basin. The United States alone planned to invest $53 million in this project. However, this plausible program on environmental sustainability has not yielded any appreciative change in people's attitude toward deforestation because of the complexities of the changing processes involved.

Urban Waste

Waste generation, both domestic and industrial, continues to increase worldwide in tandem with growth in consumption. In developed countries, per capita waste generation increased nearly threefold since the 1980s, reaching a level five to six times higher than that in developing countries. With increases in populations and living standards, waste generation in developing countries is also increasing rapidly and may double in volume within a decade. If current trends continue, according to some estimates, the world may see a fivefold increase in waste generation by 2025.

Urban areas not only concentrate consumption of resources, but also generate much of the domestic waste. Most city governments are facing mounting problems with the collection and disposal of wastes. In high-income countries, the problems usually center on the difficulties of disposing the large quantities of solid wastes generated by households and businesses. Land filling, the most widely practiced disposal method in these countries, is becoming increasingly costly because of land scarcity and rising transportation costs. In developing countries, many city authorities are unable to collect and transport more than one-third of the solid waste generated because of limited municipal budgets. The rising cost of land filling and the need to conserve resources have prompted most industrial countries to focus increasingly on recycling and reuse of both municipal and industrial wastes. In developing countries in general, and sub-Saharan Africa in particular, recycling is driven by the need for low-cost materials for industry, high levels of underemployment, and the low purchasing power of large segments of the population. Many cities in developing countries have developed extensive waste economies.

Agricultural Waste

In sub-Saharan Africa, increasing uses of agrochemicals in intensive farming practices have contaminated potable water resources to a level hazardous to health. Poor solid waste disposal affects all three environmental media. Air pollution from combustion products, illegally dumped hazardous wastes, and other gaseous products of waste decomposition have a significant effect not only on the local environment, but also on a global scale, contributing to the greenhouse effect and causing acidification of the environment. The pollution of water courses, both surface and subsurface, by decomposing wastes is also a problem, especially where it is ultimately used as a potable supply by communities. Increasing rates of waste generation in developing countries are threatening the environment. Countries in sub-Saharan Africa have not been able to develop effective and holistic approaches to waste management: mainly as a result of, a lack of infrastructure for waste management, lack of access to and adoption of appropriate technology to manage waste, widespread illegal dump sites, and limited presence of formal systems to sort and recycle waste. Developed countries are successfully managing waste through recycling systems and reuse of both domestic and industrial waste. The challenge for sub-Saharan Africa is turning the ever-increasing waste over to effective and adequate management.

In conclusion, countries in sub-Saharan Africa portend high consumption patterns of natural resources because of their growing population and poor waste management systems. The unprecedented increase in human population, with prospects for further increase, raises questions on the capability and potential of the region to maintain the current patterns in consumption levels without impinging on the sustainability of resources and environment. Consumption and production of waste will go on at an ever-increasing pace. Many countries in sub-Saharan Africa are finding it rigorously difficult to cope with the environmental challenges posed by unsustainable use of resources and mounting waste creation.

Akanni Akinyemi
Obafemi Awolowo University

See Also: Africa, North; Overconsumption; South Africa; Sustainable Development.

Further Readings

Bekele, Frehiwot. "Consumption Patterns Must Change." *Africa Recovery*, v.12/2 (1998).

Cointreau, S. *Declaration of Principles for Sustainable and Integrated Solid Waste Management.* Washington, DC: World Bank, 2001.

Economic Commission for Europe (ECE). "Workshop on Encouraging Local Initiatives Towards Sustainable Consumption Patterns in Africa." Vienna, Austria, February 2–4, 1998.

Montgomery, Mark R. "Urban Poverty and Health in Developing Countries." *Population Bulletin,* v.64/2 (June 2009).

Pimentel, David, Xuewen Huang, Ana Cordova, and Marcia Pimentel. "Impact of Population Growth on Food Supplies and Environment." http://jayhanson.us/america.htm (Accessed June 2011).

"Report of the Thematic Workshop on Population, Poverty and Environment: Highlights." *FAO Report on Sustainable Consumption Patterns.* Rome: October 26–30, 1998.

Sokona, Youba. "Energy in Sub-Sahara Africa." http://www.helio-international.org (Accessed June 2011).

United Nations Environment Programme. "Sustainable Consumption and Production in Africa." http://www.unep.org/roa/docs/pdf/sustainable_consumption.pdf (Accessed June 2011).

Air Filters

Particulate air filters remove solid particulates like mold, pollen, dust, and bacteria from the air. Typically they are composed of fibrous materials that are permeable by air but trap the particulates. Depending on the usage for which they are intended, filters may include a chemical component, which removes or neutralizes contaminants, or they may use a static electric charge to actively attract particles rather than simply passively catching them. Air filters have two types of environmental ramifications: they are used to improve air quality and their construction and disposal have environmental consequences.

Use and Production

Air filters are used in vehicles to both filter the air for the passenger compartment and prevent the entrance into an engine's cylinders of abrasive particulates that will contaminate the oil and contribute to mechanical wear and tear. Both types are usually made of pleated paper, sometimes felt. Ford introduced a new long-life filter system for engines in 2003 that uses foam instead of paper: these filters are placed in the bumper of the car with a stated

150,000-mile service interval. But apart from the use of these filters in the Ford Focus, nearly all engine air filters are made with pleated paper.

The paper used for filters is called "filter paper" and is similar to the paper used for coffee filters and tea bags. The material, usually made with high-cellulose bleached wood pulp, called "dissolving pulp," may be treated with chemical reagents and impregnated to increase moisture resistance before being pleated. It is semipermeable and placed perpendicular to the expected airflow, and many heavy-duty filters may be rinsed periodically in order to extend their life.

Dissolving pulp is made from pulpwood (timber grown specifically for paper production). Most pulpwood forests are reforested as standard practice, making them a renewable resource. Pulpwood can also be harvested from mixed forests, in which the better-quality trees are used for lumber production, while pulpwood is made from the four types of trees inferior for lumber: open-grown trees, which have heavy branches low on the trunk; dead or diseased trees; treetops; and trees too small for logging. Salvage cuts after natural disasters are often used for pulpwood. Typical tree varieties used for dissolving pulp include aspen, paper bird, red maple, balsam fir, jack pine, and white spruce, all of which have high cellulose content.

Particulate air filters are used to filter home and automotive systems from mold, bacteria, pollen, and dust. While reusable filters last longer and don't clog landfills, they are more expensive and difficult to clean properly, so disposable filters are more common.

Health Effects

On the face of it, the production of air filters is largely environmentally friendly. The bleaching, however, can be considered problematic. Whiteness is considered desirable in most filters because it makes it easy to visually ascertain how dirty the filter has become. Various chemicals are used to reduce the chromophores (color-causing substances) in the pulp, including chlorine dioxide, hydrogen peroxide, sodium dithionite, magnesium salts, and sodium silicate. Ozone is sometimes promoted as an oxidizing agent for bleaching because it can achieve results without any chlorine (in products designated "totally chlorine free") or without any elemental chlorine (in products designated "elemental chlorine free"). The main way in which bleaching pulp can damage the environment is through the release of materials into waterways, which are usually located near pulp mills because significant amounts of water are required for their operation. The substances released include chlorinated dioxins, which are recognized as a persistent environmental pollutant with significant human health effects. Most human exposure to dioxins is through food because they accumulate in the food chain through the fatty tissues of animals.

Recycling and Cleaning

Further, while the production of the filter uses a renewable resource, it is still a wasted resource if the filter is simply thrown away, particularly if it is not designed to be used for a long period of time. Remedies to this issue include recycling and cleaning. Cleanable air filters are more expensive than disposable ones, but they last longer. On the other hand, their maximum efficacy is considerably lower than disposable air filters because there is a tug-of-war between permeability and ease of cleaning; the smaller the particles a filter can effectively capture, the harder it is to clean it. It is particularly difficult to find an effective, reusable air filter for household antiallergen use, for instance, and claims of efficacy by manufacturers are often wildly exaggerated.

Recycling air filters, on the other hand, is often possible through specialized services. They can generally not be recycled as ordinary paper, depending on the use to which they were put, but services exist that exchange clean filters for used ones in order to recycle them.

Bill Kte'pi
Independent Scholar

See Also: Paper Products; Pollution, Air; Recyclable Products; Recycling.

Further Readings

Brower, Michael and Warren Leon. *The Consumer's Guide to Effective Environmental Choices: Practical Advice From the Union of Concerned Scientists.* New York: Three Rivers Press, 1999.

Garlough, Donna, Wendy Gordon, and Seth Bauer, eds. *The Green Guide: The Complete Reference for Consuming Wisely.* New York: National Geographic, 2008.

Rogers, Elizabeth and Thomas Kostigen. *The Green Book.* New York: Three Rivers Press, 2007.

Alabama

A Deep South state, Alabama is bordered by Tennessee, Georgia, Florida, Mississippi, and the Gulf of Mexico. The state has the second-largest inland waterway system in the country, with numerous rivers and creeks. The 2010 population was 4,779,736. The economy has become more diversified since World War II, with a gradual transition from agriculture to expanded mineral extraction, technology, and heavy manufacturing sectors, as well as the establishment of numerous military installations providing government jobs. Nevertheless, the state has often suffered from economic trouble, and in 2010, unemployment approached 9 percent. It was one of the states impacted by the 2010 BP oil catastrophe in the Gulf of Mexico.

Energy Consumption

Alabama ranks 15th in the country of carbon dioxide (CO_2)-polluting states, producing about 31 tons of CO_2 per resident per year, and it ranks highest in per capita energy usage. About half of the state's energy consumption is due to the industrial sector. It also ranks fourth-highest in per capita gasoline consumption. The state has considerable natural gas and coal resources, which encourage its dependence on fossil fuels, though the plentiful rivers also have untapped hydroelectric potential, and many of the state's regions are perfectly suited to growing switchgrass for biofuel.

Waste Disposal

The Land Division of the Alabama Department of Environmental Management administers the state's waste management and remediation programs, with primary jurisdiction over the disposal of solid and hazardous waste and the remediation of contaminated sites. Specific major programs within the Land Division include hazardous waste, solid waste, remediation, scrap tire, and brownfields/voluntary cleanup. The state's rules for hazardous waste disposal follow those of the federal government, and the delegable portion of the federal Resource Conservation and Recovery Act is administered by the Land Division. The Land Division designates permitted landfills in the construction/demolition, industrial, and municipal waste categories, and it investigates illegal solid waste disposal sites and complaints regarding waste disposal.

The brownfields/voluntary cleanup program is set up to redevelop brownfield sites in Alabama. Brownfields are properties whose redevelopment or reuse may be complicated by the presence of hazardous substances, pollutants, or contaminants. Their safe redevelopment and remediation is intended to create jobs or to increase public space without doing so at the expense of greenfields. The program assists local governments and nonprofit organizations with the assessment, cleanup, and redevelopment of such sites.

Emelle Landfill

Emelle, Alabama, in Sumter County (on the Mississippi border), is home to the country's largest hazardous-waste landfill. A young, tiny town, Emelle had a population of only 31 (79 percent African American) as of the 2000 census, and it was not incorporated until the 1980s, despite continuous population since the previous century. The landfill is owned by Chemical Waste Management, Inc. (CWM), which has been accused of and used as a textbook example of environmental racism (the practice of shunting waste, especially hazardous waste, to areas populated by African American, Native American, and other minority communities). The CWM landfill was purchased in 1978 and the site has become the depository for approximately 6 million tons of waste. It is located near the top of the Eutaw Aquifer, which supplies water to much of Alabama. This part of the state is poor—about one-third of the residents live below the poverty level—and almost entirely African American, being located in the Black Belt soil region. Political considerations may have been involved in the landfill's inception, as one of the original owners was the son-in-law of former Alabama governor and presidential candidate George Wallace, a noted racist. Resource Industries later sued CWM for fraud and misrepresentation, winning an award of $91 million, though the general public of Alabama and Sumter County received nothing.

The landfill was supported by local governments, as it was extremely profitable for the impoverished county (generating tens of millions of dollars in fees), but it attracted considerable opposition from activists, other Alabamans, and regulatory agencies. The company countered that it created local jobs. The landfill, originally 300 acres, expanded over time to 2,700 acres. Between 1984 and 1987, about 40 percent of the toxic waste in the entire country was brought to Emelle from the many businesses and other customers that CWM serves.

Federal regulations and a new state tax slowed the growth of the landfill in the 1990s, and as of 2010, only about 120,000 tons a year are added to the landfill, down from a peak of 800,000 tons. Problems at the site have included a cloud of acidic vapor released in 1984, a fire in 1985 requiring the evacuation of hundreds of workers, and at least one spill of liquid waste onto adjacent land without any contingency plans implemented or authorities notified. Activist groups opposing the landfill have included Greenpeace, the Sierra Club, Sumter Countians Organized for the Protection of the Environment (SCOPE), and Alabamians for a Clean Environment (ACE). ACE is primarily credited with attracting the attention of national organizations and activist groups. Meanwhile, the local Sumter County media portrayed landfill opponents as hippie holdovers out of touch with the economic realities of their hometowns and played up the fact that while most of Sumter County is black, most of the activists, especially those interviewed on television, were white. Emelle is an even smaller town than it once was, hundreds of landfill jobs having been eliminated since the early 1990s when operations were at their peak. Supporters of the landfill point out that Emelle now suffers from the worst of both worlds: the landfill is still there, still as much a danger as before, but now provides fewer jobs for the public and fewer taxes for the county.

Bill Kte'pi
Independent Scholar

See Also: Economics of Waste Collection and Disposal, U.S.; Landfills, Modern; Mississippi; Politics of Waste; Race and Garbage; Tennessee; Toxic Wastes; United States.

Further Readings

Canis, Wayne et al. *Living With the Alabama/Mississippi Shore*. Durham, NC: Duke University Press, 1985.
Gray, Aelred and David A. Johnson. *The TVA Regional Planning and Development Program*. London: Ashgate Publishing, 2005.
Williams, Randy, ed. *The Alabama Guide*. Tuscaloosa: University of Alabama Press, 2009.

Alaska

Alaska is often referred to as the "last frontier" of the United States. The largest of the United States by area, with 663,268 square miles, Alaska only has 710,231 residents, making it the fourth-smallest state in terms of population. Scale and distance shape both consumption and waste management practices in the state, with residents paying among the highest prices within the United States for commodities. Recyclables collected within the state are shipped through the Port of Tacoma for processing and sale, and the array of communities scattered across the state rely on landfills and incineration, depending on the size and remoteness of the settlement. Unique characteristics of Alaska include its marketing to outsiders, which has billed it as the "vacation of a lifetime." The influx of tourists, particularly to southeast Alaska during the summer season, presents challenges of waste management. Similarly, the resources available in Alaska, particularly oil and gas, continue to be central to key questions of climate change, science, public policy, consumption, and international affairs. The arctic and subarctic environments of Alaska present several challenges in terms of consumption, waste collection, and disposal.

While Alaska has a low population per square mile, the remoteness of many areas, including the state capital, presents limited options for local disposal of waste. In terms of consumption, Alaska is a state of ironies where, although there are many natural resources available and a tradition of independent living, there is heavy reliance upon outside sources of consumptive goods. This irony is perhaps no greater than in the context of fossil fuels, including the iconic Alaska pipeline and the controversial utilization of the Arctic National Wildlife Refuge. The isolation of Alaska brings with it high costs of gasoline and heating oil. This discrepancy is particularly marked between rural and urban areas of Alaska.

The higher price of oil in Alaska has contributed to increased prices in store-bought foods. Alaskans boast the highest consumption of subsistence resources in the United States. Subsistence resources utilization is much higher in rural than urban areas. There is, at the same time, a heavy dependence on imported foods. It is estimated that 95 percent of the food consumed by Alaskans, on average, comes from outside the state. Although urban Alaska residents pay a premium in relation to the "lower 48," the price in rural areas can be two or even three times the cost of goods in the "lower 48."

Transfer Stations

Transfer stations in Fairbanks have been a source of wealth for some. These transfer stations are places of exchange wherein individuals will dispose of items they no longer want, including building materials, functional electronics, or even a baby's crib. At these locations, individuals not only seek to get rid of their unwanted goods, but also very deliberately place some of their goods where they can easily be accessed by people scavenging for resources. Humans are not alone in scavenging for resources in waste as ravens, arctic foxes, and bears have all been reported as frequent users of waste in different parts of Alaska. Individuals will scavenge resources that they can reuse personally or to salvage for resale either in whole or for their component parts. For example, some individuals patrol different transfer stations to salvage the copper wire from working electronic devices in order to sell the copper.

Recycling and Toxic Waste

The relative isolation of Alaska presents challenges in developing programs to deal with different types of waste. Urban areas have different capacities to deal with recyclable materials. For example, although there are drop-off locations for recycling in Fairbanks, there was no pick-up program designed for recycling as of 2010; both Anchorage and Juneau have more thorough recycling programs. At the end of 2010, a community recycling center and education facility opened on the Palmer-Wasilla Highway. Built to LEED Gold Standard specifications, the facility serves the rapidly growing Matanuska-Susitna Valley north of Anchorage. There are very few locations for recycling and very little has been implemented in terms of recycling programs in rural Alaska.

Rural communities utilize a hub system wherein a centralized town with aviation capacity serves as the centralized distribution point for consumable goods from Anchorage. A 2010 initiative in Nome is beginning the process of using the plane's return trip to Anchorage to ship electronic, potentially toxic waste back from Nome and surrounding communities. Initiatives such as this have the potential to prevent the creation of new Superfund sites in rural Alaska. Previous disposal of toxic waste from households, military, and industry present serious health issues, particularly in rural Alaska.

Human Waste

In terms of human waste, the Alaskan outhouse is a primary—if not sole—option for many who live in "dry-cabins," or those without running water. Alaskan outhouses take a myriad of forms, including sanitary, closed-structure, or open-pit. This can have severe consequences during the spring when thawing occurs, resulting in large quantities of water on the landscape. In some areas, this water can lead to overflowing in outhouses and hence potential sanitation issues, including water pollution. This issue can be further compounded by the fact that much of this water can remain stagnant in certain areas. In some instances, rural and urban residents without access to indoor plumbing or limited access to indoor plumbing use "honey-buckets" in order to deal with human waste. This method of disposal offers the benefit of not having to venture outside, but involves the collection of human waste in an inside bucket, which is then later disposed of in garbage or placed outside in order to freeze during the wintertime.

Fuel

Several efforts are underway in Alaska to decrease the overall reliance on outside sources of fuel and food. In order to limit the amount of heating oil consumed in newly constructed housing, architects and organizations such as the Cold Climate Housing Research Center are working to incorporate energy efficient design elements in new construction and house retrofitting. In other areas, cities, towns, and organizations are exploring alternatives to fossil fuels, including wind and geothermal power.

Tourism

While there is heavy reliance in Alaska on natural resource exploitation, the tourist industry also plays a large role in Alaska's economy. The impacts of tourism are more noticeable in certain regions than others. On typical summer days in downtown Juneau, multiple cruise ships transport tourists through the famed Inside Passage and along the Northwest Coast.

Tourists from all over the world utilize these cruise ships and rush into ports of call en route to or from Anchorage. These cruise ships must also dispose of their waste products, and this has raised controversy as to how to best dispose of waste en route. The greater length of time these ships carry additional weight, the more fuel they burn.

Alaska is still aptly referred to as the "last frontier" as its unique positioning offers both challenges and opportunities for cost and environmentally effective means of dealing with challenges of consumption and waste production and utilization. The relative isolation of not only the state, but also rural communities and urban centers, may point to unique circumstances (such as high costs of transportation leading to price premiums on food and fuel) wherein the most cost-effective solutions involve tight cycling of energy and waste so that Alaskans may confront issues of waste and consumption in their own backyard. To describe Alaska as a "state removed" does not address the reality that the state will continue to grapple with its connections to not only the "lower 48," but also the global economy, as much-needed tourist dollars continue to flow into the state, bringing with them benefits and challenges.

David Fazzino
University of Alaska

See Also: Dumpster-Diving; Food Consumption; Gasoline; Grocery Stores; Human Waste; NIMBY (Not in My Backyard); Ocean Disposal; Public Health; Recycling; Recycling Behaviors; Supermarkets; Sustainable Development; Sustainable Waste Management; Toxic Wastes; Trash to Cash.

Further Readings

Manget, Sean. "Nome to Become Rural Trash Disposal Hub." *Alaska Journal of Commerce* (February 26, 2010).

University of Alaska Fairbanks Cooperative Extension Service. "Cooperative Extension Service: Alaska Food Cost Survey." http://www.uaf.edu/ces/fcs (Accessed June 2010).

Weiser, E. "Use of Anthropogenic Foods by Glaucous Gulls (*Larus hyperboreus*) in Northern Alaska." M.S. Thesis, University of Alaska Fairbanks, May 2010. http://users.iab.uaf.edu/~abby_powell/Weiser_Thesis_2010.pdf (Accessed June 2010).

Wolfe, R. J. "Subsistence in Alaska: A Year 2000 Update. Juneau: Division of Subsistence, Alaska Department of Fish and Game." http://www.subsistence.adfg.state.ak.us/download/subupd00.pdf (Accessed June 2010).

Alcohol Consumption Surveys

Alcohol consumption can be both beneficial and harmful to human health. In the 20th century, moderate drinking was associated with a lower risk of cardiovascular disease. However, recent trends have shown alcohol consumption to be one of the leading causes of death and poor health in the United States. According to the World Health Organization (WHO), the United States has the second-highest level of alcohol consumption after Europe. It had peaked in the 1980s, and after a slight reduction in the next decade, it has remained stable since 2000.

Statistics

Alcohol consumption surveys have been structured to collect and report information regarding alcohol consumption in the U.S. population. The National Survey on Drug Use and Health (NSDUH) conducts surveys on drug and alcohol consumption of U.S. consumers from different demographic (such as age, gender, race/ethnicity, education, and employment), geographic, and other variables. A 2008 survey report shows that 51.6 percent, or about 129 million Americans age 12 and older, consume alcohol. Approximately 23 percent of those groups are binge drinkers (with blood alcohol concentration 0.08 grams or above), and 7 percent are heavy drinkers (those who consume an average of more than two drinks per day). Of all ethnicities, white Americans have the highest (56 percent), and Asian Americans have the lowest (37 percent) percentage of self-reported alcohol use. These survey statistics also provide information about alcohol-related deaths, injuries, and other health problems.

Underage drinking continues to prevail in the United States. The percentage of male and female drinkers ages 12 to 17 is similar, at 14–15 percent. However, in the age group of 18–25, the percentage of male drinkers (64 percent) is higher than that of females (58 percent). A number of national surveys have confirmed that three-fourths of 12th graders, more than two-thirds of 10th graders, and about two in every five 8th graders have consumed alcohol. According to a survey by the National Institute on Alcohol Abuse and Alcoholism (NIAAA), each year, approximately 5,000 people under the age of 21 die of alcohol-related causes. This statistic includes approximately 1,900 deaths from car crashes, 1,600 deaths from homicide, 300 deaths from suicide, and hundreds from other injuries such as falls, burns, and drowning. Data from the National Epidemiologic Survey on Alcohol and Related Conditions (NESARC) show that teens who drank alcohol at an early age were at greater risk of developing alcohol dependence in adulthood.

Beyond the NIAAA's national-level surveys, many online and off-line surveys have been conducted in colleges and universities to learn about alcohol consumption among college-level students and the implications on campuses. These surveys include questions regarding campus culture, alcohol control policies, enforcement policies, availabilities, pricing, marketing, and special promotions of alcohol. They provide valuable information for the authorities to intervene to prevent drinking and alcohol-related consequences among students. The data further showed that the students who drank at "non-extreme" levels and not in the "high risk" group were harmed the most.

Some of the problems related to alcohol consumption as reported by the college students are missing classes, poor performance on tests, being injured, having arguments or fights, having trouble with authorities, getting arrested, damaging property, sexual assault, and drunk driving. The availability and effectiveness of appropriate programs and laws, such as "No drinking and driving," was shown to save the lives of underage drinkers, adults, and the people around them. The percentage of people 12 and older driving under the influence of alcohol dropped from 14 percent in 2002 to 12 percent in 2008.

Truth in Garbage

There is a considerable difference between the survey results and the true alcohol consumption in the adult population, as well as in underage drinkers on college campuses. According to William Rathje, an eminent garbologist, alcohol consumption is typically underreported by 40 to 60 percent. During surveys, people are not always truthful about how much alcohol they consume on a regular basis because they might be perceived wrongly or because

Alcohol consumption surveys, particularly among underage drinkers, are usually inaccurate by a wide margin. Counting beer or wine bottles in household garbage or campus dumpsters is the most reliable method of estimating actual alcohol consumption.

they are not aware of their drinking patterns. Looking through a household's garbage for beer cans and wine bottles is found to be a more reliable method in accounting for actual alcohol consumption. Similarly, true data for underage drinking in college campuses may be found by locating the number of beer cans discarded in trash bins in residence halls for freshmen and sophomores. Hence, alcohol consumption surveys and garbology are complementary tools in finding the true extent of alcohol consumption.

Mousumi Roy
Penn State University, Worthington Scranton

See Also: Consumption Patterns; Culture, Values, and Garbage; Garbage Project; Garbology; Public Health; Supermarkets; Surveys and Information Bias; Zero Waste.

Further Readings

Chaloupka, F. J., M. Grossman, and H. Saffer. "The Effects of Price on Alcohol Consumption and Alcohol-Related Problems." *Alcohol Research & Health*, National Institute on Alcohol Abuse and Alcoholism, v.26/1 (2002).

DeJong, W., M. E. Larimer, and M. D. Wood, eds. "College Drinking: New Research From the National Institute on Alcohol Abuse and Alcoholism's Rapid Response to College Drinking Problems Initiative." *Journal of Studies on Alcohol and Drugs* (July 2009).

Lilienfield, R. and W. Rathje. *Use Less Stuff: Environmental Solutions for Who We Really Are.* New York: Ballantine Books, 1998.

National Institute on Alcohol Abuse and Alcoholism (NIAAA). "Underage Drinking: Why Do Adolescents Drink, What Are the Risks, and How Can Underage Drinking Be Prevented?" http://pubs.niaaa.nih.gov/publications/AA67/AA67.htm (Accessed July 2011).

Rehm, Jurgen, T. K. Greenfield, and J. D. Rogers. "Average Volume of Alcohol Consumption, Pattern of Drinking, and All-Cause Mortality: Results From the U.S. National Alcohol Survey." *American Journal of Epidemiology*, v.153/1 (2001).

Ryan, B. E., T. Colthrust, and L. Segars. "College Alcohol Risk Assessment Guide, Environmental Approaches to Prevention." http://www.higheredcenter.org/files/product/cara.pdf (Accessed July 2011).

U.S. Department of Health and Human Services. "Results From the 2008 National Survey on Drug Use and Health: National Findings." http://www.oas.samhsa.gov/NSDUH (Accessed July 2011).

Wagenaar, A. C., T. L. Toomey, and K. M. Lenk. "Environmental Influences on Young Adult Drinking." *Alcohol Research & Health*, National Institute on Alcohol Abuse and Alcoholism, v.26/4 (2005).

Wechsler H. and T. F. Nelson. "What We Have Learned From the Harvard School of Public Health College Alcohol Study: Focusing Attention on College Student Alcohol Consumption and the Environmental Conditions That Promote It." *Journal of Studies on Alcohol and Drugs*, v.69/4 (2008).

Aluminum

A rare and precious metal just 150 years ago, aluminum is now a nearly ubiquitous element in consumer goods, cars, power lines, airplanes, and computers. Although aluminum is the most abundant metal in the Earth's crust, the process of extracting it from ore consumes extensive amounts of soil, water, and energy. Aluminum's physical qualities, particularly its shine, light weight, malleability, and versatility, make it attractive to artisans, industrialists, and artists, for whom it is a key ingredient in wiring, jet aircraft, cookware, sculpture, and antiperspirant, among other products. A great deal of the world's aluminum still comes from bauxite ore. Since the 1960s, however, recycling has become an equally important source. In the 21st century, while some people prize aluminum as a sustainable, reusable resource, others denigrate it for the high ecological impact of its production.

History

Throughout antiquity, people used alum as an astringent and as an ingredient in ceramics. It was not until 1807, however, that Sir Humphrey Davy proposed that alum contained a base metal, which he called "aluminum." In 1827, chemists Hans Orsted and Frederich Wohler isolated pure aluminum for the first time using electrolysis, but the process was slow and costly. In the mid-19th century, pure aluminum was considered as precious and rare as gold or silver. This notion changed in the 1880s, when chemists Charles Martin Hall and Paul Héroult, working independently in the United States and France, perfected a quicker and more efficient process of electrolyzing aluminum. The Hall-Héroult process ushered the transition of aluminum from precious metal to industrial element, and it remains a primary method in aluminum extraction.

Mining

Most aluminum in the Earth's crust resides in bauxite. Bauxite occurs naturally throughout most of the globe, but a great deal of the world's supply lies in developing areas, including Jamaica, Guinea, Brazil, China, and India. For people in these countries, bauxite mining presents a potential path to employment and economic development, but the benefits of mining have come with social and environmental costs.

Bauxite mining is an open-pit process, which means that topsoil and vegetation must be removed in order to obtain ore. The refining process requires the use of caustic sodium hydroxide, which yields valuable aluminum oxide and an alkaline toxic waste. This alkaline waste, known as "red mud," has some industrial applications, but it has also been a source of groundwater contamination in bauxite mining countries.

Both the refining process and the subsequent electrolysis by the Hall-Héroult process consume tremendous amounts of electricity. To provide this energy, governments and mining interests have built hydroelectric dams near bauxite sources. Even though they may provide local people with cheaper sources of power, dams are not always popular.

In the Indian state of Orissa, for example, ethnographers and activists contend that the aluminum industry and its dams have led to the displacement of thousands of *adivasi*, or "tribal," groups. They suggest that the forced relocation of people who live in bauxite-rich regions presents an economic and ecological burden that outweighs the benefits of aluminum extraction.

Recycling

Though bauxite mining remains a primary source of aluminum, industry leaders and environmentalists now recognize that it may be cheaper and more energy efficient to obtain aluminum by recycling. Aluminum is 100 percent recoverable, meaning it can be recast without any loss of its prized shine and malleability. Aluminum has a low melting point relative to other commonly used metals (about 600 degrees Celsius, or 1,200 degrees Fahrenheit). When heated, aluminum liquefies, leaving a small amount of waste product, called "dross." This dross can be further refined to recover pure aluminum.

The collection, recycling, and recasting of metals is not a new process, but the era of modern aluminum recycling did not begin until after the Hall-Héroult process made industrial aluminum production possible. The first large-scale aluminum recyclers sprang up in the metalworks of Chicago, Illinois, in 1904. During World War II, recycling in Chicago and other parts of the United States boomed as the military searched for cheap and quick sources of aluminum and other materials for weapons production.

A watershed moment in the history of recycling was the introduction, in 1964, of the aluminum beverage can. The aluminum can was a light, cheap, recyclable vessel that quickly became an industry standard. By the 1970s, can collecting had become commonplace, and by the 1990s, municipal recycling programs were established throughout the United States and Europe.

Aluminum remains one of the most commonly recycled materials in the world, and scrap aluminum has become a globally traded commodity. Although prices for aluminum are notoriously volatile, the recycling industry has grown consistently since the 1960s.

Economists and environmentalists alike celebrate the reductions in environmental impact and energy consumption that come from recycling aluminum instead of refining it from bauxite ore, but aluminum recycling also can have negative impacts on the environment and human health. As in mining, these impacts are not evenly distributed across the population.

Consider the example of Chicago, where residents of the low-income and largely African American neighborhoods located near aluminum smelting plants complain that local air quality has diminished. They point out that rates of asthma are higher in their neighborhoods than in others. Furthermore, the dominance of for-profit waste management firms in the recycling industry has meant that most of the economic benefits of aluminum recycling go to companies, rather than to people in the Chicago communities where homes and apartments are surrounded by can drops, aluminum smelters, and landfills. Chicago's recycling business has expanded over the past few decades, but low-income families still tend to bear the health burdens of recycling. Occupational hazards related to aluminum recycling include exposure to particulates and toxic fumes from plants and chronic and acute injury from sorting, hauling, and casting.

Informal and Alternative Uses

Around the globe, a robust informal economy for salvaged aluminum exists parallel to the industrial recycling industry. Scavengers and trash pickers recover aluminum in the form of cans, siding, wiring, alloy wheels, and other automobile parts, selling by the pound or kilogram to large brokers, who in turn sell in bulk to international traders. Since they deal in small quantities, informal scavengers frequently bear the economic impact of aluminum's price volatility. For example, many small-scale dealers and traders lost their livelihoods when aluminum prices plummeted after the 2008 global financial crisis.

An alternative to market-based aluminum recycling is artisanal recycling in which sculptors and cottage industrialists produce new items from found or purchased scrap aluminum. Aluminum's low melting point makes it an ideal metal for home foundries. Using homemade furnaces, artisanal recyclers design and build decorative and functional artwork, household utensils, and hardware.

Perhaps more than any other periodic element, aluminum encapsulates the promises and pitfalls of an age in which lightweight metals seem indispensable and when generalized concern with the problem of waste has led to the embrace of recycling. Aluminum's global circulation, moving from consumer products to dumps and back again, makes it a material bridge between poor and rich around the world.

Alex Nading
University of Wisconsin, Madison

See Also: Illinois; Mineral Waste; Packaging and Product Containers; Recyclable Products; Recycling; Street Scavenging and Trash Picking.

Further Readings

Padel, Felix and Samendra Das. "Double Death: Aluminum's Links With Genocide." *Social Scientist*, v.34/3 (2006).

Pellow, David. *Garbage Wars: The Struggle for Environmental Justice in Chicago*. Minneapolis: University of Minnesota Press, 2002.

Schlesinger, Mark. *Aluminum Recycling*. Boca Raton, FL: CRC Press, 2007.

Anaerobic Digestion

Anaerobic digestion (AD) is a form of waste treatment that involves the controlled decomposition of biodegradable and organic wastes. The main outputs of AD are methane, which is considered a renewable form of energy, and a possible fertilizer rich in nitrogen, known as "digestate." Consequently, AD has played a significant role throughout the world as an alternative to the open dump, modern landfill, or incinerator, all of which represent waste treatment systems with considerably greater environmental risks. More recently, European and American governments have developed policy initiatives to promote AD as both a replacement for landfills and a possible solution to global climate change. But there are differences of opinion concerning how AD is best designed and implemented, which ultimately reflect different ideologies of sustainable development.

Digestion Process

AD is also the name of the process of biological decomposition that occurs outside the presence of oxygen due to the activity of microorganisms. There are four stages to the process, as complex organic polymers are gradually broken down into their basic constituents; they include hydrolysis, acidogenesis, acetogenesis, and methanogenesis. Though this process occurs naturally, it is actually rather sensitive. It is ideal for anaerobic digestion to occur within landfills in order to increase decomposition and ultimately create more capacity, but there need to be sufficient conditions to promote microbe colonies, including the right pH, a digestible substrate, and an adequate temperature. Increasingly, landfills are tapped for their biogas, which also contains methane, albeit in smaller proportions than in naturally occurring biogas. However, landfills are not designed to provide easy access to the waste they contain. In this way, they resemble the simplest form of anaerobic digester, a batch system. In these contexts, waste material is sealed off and breaks down undisturbed. Other digesters are designed to promote more control over the process, although this is difficult to achieve.

Steel anaerobic digestion towers in Germany. The Clement plant, built in 1976, processes over 400 million cubic meters of purified wastewater per year. Heavy rain and floods risk the discharge of untreated wastewater into the Rhine, as occurred in 1981 and 1995.

Digesters can be designed in a variety of ways, depending on how the process is maintained, at what temperature it occurs, and the kind of feedstock it is meant for. When digesters are developed independently from landfills—typically contained within steel, cylindrical vessels—they can be harnessed continually and predictably. The benefit of continual mixing is better gas production in addition to the ability to add material to alter the pH to a more optimal level; for example, adding basic compounds to lower acidity. Similar advantages come from heating digesters, so that the microorganisms that power them are the more efficient thermophilic variety, rather than mesophilic. Overall, the digestion process must sustain a relative chemical balance between stages of fermentation so that none of the microbes overwhelm the others. One common problem involves the overproduction of volatile fatty acids, or VFAs, which are the product of the acidogenesis stage and can disrupt the pH necessary to maintain later stages. One way of dealing with such complications is through a multistage digester, which allows for a more sustained, controlled process by separating the four stages into particular tanks, each corresponding to the microbial populations involved in the process.

Uses

The inputs to AD are divided into three categories: high solids that are dry, high solids that are wet, and low solids. Of the three, high-solid feedstocks that are wet require considerably more energy to transport and process. Some of the earliest AD plants were used for sewage treatment in 19th-century England and India. Development organizations and aid workers have helped to promote the spread of basic AD techniques in rural areas of China, India, and other developing countries. In these contexts, AD is used primarily on farms, enabling households to make more productive and sanitary use of their animal wastes while providing a free source of fuel for heat and cooking. On an industrial scale, farmers increasingly turn to AD if they are "going organic" or simply attempting to reduce their reliance on the purchase of expensive, fossil-fuel-based fertilizers and energy sources. After the energy crisis of the 1970s, people returned to AD as a possible method of conserving and producing energy with wastes within and beyond the farm.

In the 21st century, AD has received increasing government support in the United States, Europe, and elsewhere as a sustainable alternative to more established waste disposal methods. In Germany, for example, AD has been promoted on a large scale through feed-in tariff programs, which guarantee developers a premium price for their energy. Controversially, these digesters typically codigest biological wastes with energy crops, such as corn, which are grown expressly to serve as catalysts for the digestion process and improve energy production overall. AD has been preferred, in particular, as a means of addressing directives from the European Union that require biodegradable waste materials to be diverted from landfills in proportionally greater amounts, leading ideally to a zero waste future. Food and green wastes, in particular, have been targeted as potential inputs for municipally run AD facilities. It is up for debate, however, to what extent large, centralized AD facilities can remain sustainable if they are drawing wastes from farther and farther distances and, presumably, requiring more fossil fuels to do so. Similar questions can be raised about whether electricity or heat is the more efficient application of AD-derived methane. As models of AD as a system of energy production draw more interest and investment, furthermore, smaller-scale, farm-based models may lose favor among regulators.

Josh Reno
Goldsmiths College

See Also: Landfills, Modern; Methane; Microorganisms.

Further Readings

Braun, R., B. Drosg, G. Bochmann, S. Weiß, and R. Kirchmayr. "Recent Developments in Bio-Energy Recovery Through Fermentation." In *Microbes at Work*, Heriber Insam, Ingrid Franke-Whittle, and Marta Goberna, eds. Berlin: Springer, 2009.

Chynoweth, David. "Environmental Impact of Biomethanogenesis." *Environmental Monitoring and Assessment*, v.42/1–2 (1996).

Appliances, Kitchen

The consumer durable goods used every day in food processing, preparation, and storage may often go overlooked in the story of waste and garbage. They often outlast their owners, transferring hand to hand several times over their life cycle. In many cases, it is not known how long they can last, given that working exemplars of several species have passed the century mark. However, in most cases, they are a central node in the production of waste at the household level.

Status

The abandoned, deadly refrigerator waiting in the dump to lure the unwary child is a familiar trope of urban legend. In fact, nearly 400 children died in old-style refrigerators, with mechanical latches, from roughly 1946 to 1984. This occurred despite legislation, community action, and the introduction of a technology (magnetic strips) designed to prevent the problem. The reason children continued to die in cast-off refrigerators and freezers for a quarter century after federal legislation was passed in 1956 is because of the unique status of kitchen appliances among waste items—in a very disposable society, they tend to last a very long time.

Kitchen appliances occupy a nebulous status between furniture and infrastructure, which is reflected in the variance of laws concerning occupancy of property and the rental or sales agreements accompanying transfer of property. They fall under the title of "amenities" and, often, landlords are not required to supply them, but are required to maintain them if present. This could mean that a 60-year-old stove is still in regular usage. When selling a home, the picture becomes even less clear, and appliance disposal or provision is dominated more by cultural convention and maintaining goodwill than by law. At times, sellers who wish to retain their appliances when they transfer the property are encouraged to write them into an exclusion rider just in case of controversy. Appliances, and the disposal thereof, can change status when affixed to real property; thus built-in appliances are generally considered fixtures and belong with the house during transfer, while a refrigerator that one simply unplugs remains personal property and leaves with the departing resident.

A good example of the limbo-like state many kitchen appliances inhabit is given by the garbage disposal, an item which, while an add-on to the basic utility of a sink, is nonetheless unlikely to be transferred from one home to another. It is a key player in the processing of food remains into waste, and it is also a controversial partner with sewage lines in the removal of waste from the home. The disposal of too much grease can clog the lines, while grinding bones or corn husks and other tough materials can render it inoperable. The use of potable water to transport food waste out of the household has been interpreted as a waste of a resource and as a possible danger to old, overwhelmed wastewater treatment plants. Many green information sources stress the alternative of composting or simply sending food waste to landfills via regular trash pickup. In these scenarios, the decomposing food wastes can be returned to utility; in the first through fertilizing household soil, and in the second as a possible source for biogas.

Life Cycles

In any given kitchen, there can be a mix of appliances from various periods, all with distinct life cycles. There are about 600 million appliances in existence, and the Association of Home Appliances Manufacturers states that a majority of them are not disposed of by their original owners, but are sold, traded in, left behind in a move, or given away two or three times over. According to the trade group, the actual lifetime of many appliances is difficult to estimate, but it can easily be 10–20 years, and maybe twice that. "Ask Heloise" is a newspaper and online helpful household hints column, and Heloise ran a contest for the oldest working appliance. The results were that among smaller appliances, like toasters and clothes irons, some were still working after 100 years! Larger categories included gas stoves from 1924 to 1929 that were still in use and a refrigerator from 1929 that was still in working order. A. P. Wagner, an appliance repair company that has been in business since 1928, held a video contest for oldest refrigerator, and the winner had a working ice box from the early 1900s, while a runner-up had a 1918 Kelvinator mechanical refrigerator converted from an ice box and still housed within the wooden cabinetry.

Hazardous Materials

One aspect many appliances have in common with respect to disposal is that they contain materials classified as hazardous waste. Older refrigerators contain chlorofluorocarbons (CFC) and hydrochlorofluorocarbons (HCFCs), both of which are ozone depleters and contribute to greenhouse gases and climate change. According to the Environmental Protection Agency, any refrigerator manufactured prior to 1995 contains CFCs in the refrigerant, and until 2005, they were manufactured using a foam containing HCFCs. Any appliance with switches and relays made before 2000 may contain mercury, and there are also polychlorinated biphenyls (PCBs) to be considered. Unfortunately, with respect to disposal, there are no perfect choices, as resale just transfers the problem to consumers and societies less able to bear the burden of expensive energy consumption and proper disposal. Recycling recovers many of the components, but in the process they generally release the foams containing HCFCs. Finally, simple disposal leaves them intact and nonbiodegradable, complete with dangerous components, in landfills.

Luckily, many working kitchen appliances buck the need for novelty to become cherished hand-me-downs. Nostalgia for their style can inspire retro recreations, like the Waring 60th Anniversary blender in chrome or the 1950s Easter-egg pastels of the KitchenAid mixer. This call and response of design across the decades helps to add luster to the concept of heirloom kitchen appliances, further ensuring their survival in homes. It also helps that the mechanics and design have changed little; KitchenAid attachments from the early 20th century still fit models from 100 years later. Being able to use a grandmother's kitchen appliance can also be a connection to the past, reaffirming foodways and family histories.

Susan Mazur-Stommen
Independent Scholar

See Also: Consumerism; Culture, Values, and Garbage; Home Appliances; Recycled Content.

Further Readings

Laumer, John. "Trash-Talking the Garbage Disposal: Examination of a Not-So-Green U.S. Export." http://www.treehugger.com/files/2008/02/trash-talking-the-garbage-disposal.php (Accessed September 2010).
Roberts, Jennifer. *Good Green Kitchen: The Ultimate Resource for Creating a Beautiful, Healthy, Eco-Friendly Kitchen*. Layton, UT: Gibbs Smith, 2006.

Archaeological Techniques, Modern Day

Archaeology is the study of past human behavior through the objects and sites left behind by ancient people. Archaeologists use a variety of techniques to study artifacts, the contexts in which they are found, and even the landscape itself.

Archaeology is most commonly associated with excavation. In ancient times, people often utilized parts of existing buildings for new construction so that a site would build up over time. Plant growth and windblown soil also would cover the underlying remains. Excavation is the process of removing those layers of accumulated soil and fallen debris to reveal sequences of ancient activity areas.

Tools and Techniques

Excavation tools include not only small handheld trowels and dental picks, but also backhoes and heavy equipment that—in skilled hands—can be very effective in removing tons of overlying debris. Three-dimensional recording of excavated features is achieved through electronic theodolites (also known as total stations) or global-positioning systems (GPS). Researchers frequently integrate these field measurements with Geographic Information Systems (GIS) to build databases of artifacts and their contexts.

Large-scale excavations are time consuming and generally expensive to carry out, so other techniques have been developed for assessing a site's age and contents. These include coring, which involves drilling a narrow-diameter shaft to recover a sample of an ancient site's layers, and "shovel-testing" in which archaeologists scoop up representative samples from the upper layers of a site.

Systematic reconnaissance techniques enable excavated areas to be placed into the larger landscape context. Pedestrian survey is a method in which teams of people walk over terrain to identify artifacts and features that are still visible on the ground. Buried remains also can be found through a variety of remote sensing techniques, including magnetic gradiometry, electrical resistivity, and ground-penetrating radar. Other advanced approaches include the use of satellite data to detect buried remains. The use of older, declassified military images, such as CORONA, can be particularly valuable for parts of the world where population growth or warfare has resulted in logistical constraints on site access.

Data and Analysis

Archaeological data management includes the recording of many types of information, from artifact types to landscape conditions. Pottery and stone tools are classified and organized in ways that will help them be compared with similar materials from neighboring sites to understand patterns of human activity over time. Ancient remains are dated by a variety of techniques. Radiocarbon dating is suitable for organic materials that are less than 50,000 years old, while dendrochronology is utilized to assess the age of wooden beams through the study of the patterns of tree rings. There are also techniques that measure changes in object energy over time, such as thermoluminescence for baked-clay objects and optically stimulated luminescence for sediments.

Studies of ancient food remains provide information about changes in human relationships to the environment, including the adoption of agriculture starting 10,000–12,000 years ago. Plant remains, such as seeds, often are preserved by charring in ancient cooking fires. These and other small remains can be recovered through flotation, which is a technique for washing dirt to release small organic fragments that float to the surface of the water.

In addition to studying the visible portions of plants, archaeologists evaluate nearly invisible plant residues such as pollen (often preserved in lake sediments as well as in archaeological sites) and phytoliths (silicate casts of plant cells). Archaeologists also study bones and other animal remains, both those species that were eaten and those that lived in the surrounding environment. The remains of commensal animals such as mice, rats, and dogs also can reveal patterns of migration as seen in the Pacific Islands and elsewhere.

Artifacts can be subjected to laboratory techniques to learn more about ancient technology. Chemical analysis and geological sourcing can pinpoint the origins of artifacts and raw materials to provide information about ancient trade routes. Vessels sometimes hold the preserved residues of contents such as wine and dairy products. Stone tools also can preserve the residue of blood along their cutting edges as well as marks indicative of cutting fibers or bone.

High-resolution scanning electron microscopy (SEM) can show detailed traces of use-wear as well as the analysis of production, such as the laborious perforation techniques utilized to make stone beads in ancient times. Other increasingly common techniques utilized for the study of durable artifacts include X-ray diffraction, X-ray fluorescence, instrumental neutron activation analysis (INAA), and inductively coupled plasma mass spectrometry (ICP-MS). These technologies are becoming increasingly portable, which enables researchers to analyze objects on the spot instead of having to export objects from their country of origin—an advantageous system because removal of archaeological remains from their home countries can be a politically sensitive issue.

Human Remains

The study of human burials and skeletal remains provides information about both biology and society. From bones, researchers can determine the sex, age of death, and general health of each ancient individual. Skeletons also sometimes preserve markers of the manner of death, whether through violence-associated trauma or through lengthy illnesses, such as tuberculosis, which leave distinctive physical scars. The skeleton also serves as a record of lifelong processes of energy expenditure as seen in the muscle attachments on limb bones, strain on extremities, and evidence of arthritis. Farmers, porters, and those who kneeled to grind grain for long hours all had different stresses on the body.

Teeth, a particularly durable part of the skeleton, offer a wealth of information about ancient activities, diet, and health as they preserve a record of what the individual ate over the lifetime. Strontium and oxygen-isotope analysis can show whether individuals moved from the coast inland, or even from one altitude zone to another. Teeth show that the earliest food-producing people (10,000–12,000 years ago) actually had worse dental health than their predecessors.

These early farmers suffered accelerated tooth wear from ingesting grit along with their foodgrains and had an increased number of dental caries because of their starchy, plant-heavy diet. When combined with skeletal markers of age and sex, archaeologists can evaluate associated social changes such as gender differences in tasks and in access to new foods.

The study of ancient DNA from human skeletal remains is being used to address questions of ancient migrations and social relationships within groups; for example, whether women or men moved from their ancestral homes after marriage. Whole human bodies are occasionally preserved either in exceptionally wet conditions (such as bog bodies of northern Europe), in glaciers and other frozen environments (such as the Italian Bronze Age man known as Otzi), or through processes of mummification as seen in Egypt, China, and South America. However, prior to conducting research on burial populations, archaeologists should be mindful of descendant populations who may object to studies of human remains and whose wishes are legally protected through legislation such as the 1990 U.S. Native American Graves Protection and Repatriation Act.

Specialized Fields

Several specialized fields of archaeology have emerged since the 1970s. Underwater archaeology focuses on the study of ancient ships as well as docks, bridges, and other submerged human-made features. Underwater archaeologists use SCUBA gear and—when the water is very deep—robots for the recovery of ancient items. Underwater archaeology is particularly useful for understanding ancient trade routes; for example, when shipwrecks are found with intact cargoes. High-altitude archaeology is another form of specialized inquiry, which has resulted in the recovery of ancient mummies from virtually inaccessible locations in the Himalayas and the Andes.

Ethnoarchaeology consists of interviewing and observing contemporary people who practice low-tech methods of manufacturing pottery, metal objects, baskets, and other crafts. Researchers examine the whole process of craft production, including the acquisition of raw materials and the social relations of the production process. Factors of apprenticeship and gendered task specialization are an integral component of production, which can also be modeled for the past.

Archaeologists also conduct experimental replications to learn about the skills and techniques required of ancient people. Experimental archaeology includes making stone tools and using them for woodworking and butchery; replicating the early trial-and-error processes of ancient metalworking; or building structures that mimic those found in the archaeological record. Sometimes, those experimental structures are left to decay naturally so that archaeologists can observe the slow degradation that eventually produces an archaeological site, and sometimes the structures are deliberately destroyed through fire or other processes to understand how the archaeological record is physically created.

Finally, conservation is an integral part of archaeological management. Sophisticated techniques of chemical analysis are used to simultaneously study and preserve ancient wall paintings; fragile objects, such as baskets; and unstable objects, such as metals that are subject to rapid oxidation and decay once they are removed from their buried environments.

Modern archaeological techniques are useful for not only building a picture of the past but also making projections about the future. Because of the long-term relationship of humans to the environment, archaeologists are now contributing information about the history of human-induced climate change that is increasingly featured in current debates about global warming.

Monica L. Smith
University of California, Los Angeles

See Also: Archaeology of Garbage; Dating of Garbage Deposition; Funerals/Corpses; Garbology; History of Consumption and Waste, Ancient World; Human Waste.

Further Readings

Miller, Heather M.-L. *Archaeological Approaches to Technology.* Amsterdam: Elsevier, 2007.

Renfrew, Colin and Paul Bahn. *Archaeology: Theories, Methods and Practice.* 5th ed. New York: Thames & Hudson, 2008.

Scarre, Chris, ed. *The Human Past.* 2nd ed. New York: Thames & Hudson, 2009.

Archaeology of Garbage

The field of archaeology involves the scientific study of the remains and artifacts left behind by past societies and cultures. These materials are recovered and interpreted to reconstruct the structure and behaviors of past societies. Contemporary archaeology is a subfield of archaeological research that focuses on current society or the very recent past. The archaeology of garbage, or "garbology," as it has become known, involves literally applying archaeological methods to analyzing modern society's waste.

Why Study Garbage?

Archaeologists have realized the value in studying refuse for decades. When uncovering information about a vanished society, especially one with few material goods, the garbage mounds, middens, or refuse pits often offer the richest information about the people who made them. The garbage people leave behind is not only proof humans once occupied a space but it also provides unbiased factual data on how people lived: what they ate, what they made, and how they acted. When reconstructing past civilizations, archaeologists often rely on subjective writings and records left behind. Carvings on temples and tombs glorify leaders and speak to how they want to be remembered (not necessarily how they actually behaved). Contemporary anthropologists and ethnographers often rely on recent interviews, surveys, and observations to analyze contemporary behavior. Unfortunately, human memory is unreliable. People tend to remember things the way they wish to, consciously or unconsciously applying personal bias to what is recorded. When surveyed about household behavior, individuals may overestimate good behaviors (such as eating vegetables) and underreport bad behaviors (such as drinking alcohol). There may also be a desire to please the person conducting the interview. Subjects may give the answers they think the interviewer is looking for, rather than answering factually.

History of Garbage Studies (Garbology)

There have been a number of attempts to systematically study contemporary garbage to answer specific questions. In the earliest decades of the 20th century, two civil engineers, Rudolph Hering and Samuel Greeley, evaluated the trash management methods used in a number of cities. They gathered enough information to publish *Collection and Disposal of Municipal Refuse* in 1921, the first known textbook about urban waste management.

During World War I, the War Food Administration collected data on food discarded across the United States. Researchers found that households threw away almost 30 percent of the food acquired for the home in 1918. In comparison, 21st-century citizens throw away less than 15 percent. The decrease is assumed to be related to the technology available for packaging and storing food to reduce spoilage.

The military also dabbled in garbology in 1941. Two enlisted men surveyed new recruits about U.S. Army life. They had to stop because it was against regulations to survey military personnel. They had already heard many complaints about the food, so they stationed observers in mess halls to record what the men threw away. After 2.4 million meals were observed, it was calculated that 20 percent of the food served was discarded. The report these soldiers generated listed many findings, from favored and not-so-favored foods to the simple observation that portions were simply too large. After the U.S. Army implemented a number of the suggestions made in the report, it saved about 2.5 million pounds of food per day.

A Peeping Tom version of garbology evolved in the 1970s. A. J. Weberman, a journalist fascinated with Bob Dylan, started ransacking his garbage cans and writing articles about what Dylan discarded. Weberman demonstrated once again that people say one thing while doing another. Dylan always said he had no interest in fan magazines, yet Weberman found many in his trash. He conducted similar raids on other celebrities, including Muhammad Ali, Neil Simon, and Abbie Hoffman, publishing exposés on his findings. Other reporters have followed Weberman's example by stealing garbage from the curbs of celebrities and politicians. Similar invasions of privacy still happen in the 21st century.

Another form of garbology is used in law enforcement. The Federal Bureau of Investigation and other police agencies have been known to delve into the discards of individuals suspected of illegal activity, often finding the proof needed to make an arrest. While these activities loosely fit the definition of garbology, they are conducted to stir up scandal and controversy or to find information on a specific individual. There is no application of the scientific process, and the data gathered are not used to analyze the behavior of the population as a whole.

Archaeologists, on the other hand, are not interested in the contents of the trash of one particular household. The true archaeology of garbage uses archaeological methodology to link artifacts (the garbage) to patterns of behavior in modern society. The seeds of this field were planted in 1971 when students in the archaeology program at the University of Arizona conducted small-scale research studies as class projects to evaluate common stereotypes versus reality. They collected the garbage from homes in the wealthy part of town and from a poor neighborhood to compare food consumed, household products used, and educational materials purchased. The study was too small to be valid as research, but the effort was promising. The idea that analyzing garbage would unobtrusively and objectively measure behavior was born. Similar projects followed, and, in 1973, the "Garbage Project" was officially under way.

The Garbage Project

In 1973, the archaeology department of the University of Arizona launched an in-depth study of modern society by analyzing the garbage of contemporary Tucson, Arizona. This Garbage Project was conceived by William L. Rathje, a Harvard-trained archaeologist who originally specialized in the Classic Maya. Rathje was dissatisfied with the research techniques applied to modern society. Cultural anthropologists and ethnographers frequently rely on interviews, observations, and questionnaires with members of the target population. Interviewees are very aware that they are being examined and may not answer completely truthfully. Traditional archaeologists primarily focus on societies that cannot be observed directly. As a result, they have developed methods to study the behaviors and attitudes of people that are long gone by carefully analyzing physical materials left behind and using them to interpret past practices. Rathje decided that these archaeological methods and theories could also be applied to modern society. He entered into an agreement with the Sanitation Department of the City of Tucson to deliver several randomly selected pickups from specific census tracts to a sorting area at a maintenance yard.

The Garbage Project, a prime example of ethnoarchaeological research, has been running continually since 1973 and has expanded from the analysis of the garbage of select households "fresh from the sanitation truck" to excavating samples from landfills across the country. Hundreds of (fully inoculated) students over the years have sorted more than two million pieces of garbage into about 150 categories of data that are recorded on computer forms and saved in a database. To protect the privacy of individuals, elaborate safeguards are employed to assure the public that the information is completely anonymous. All student workers sign a pledge to not look at or save personal items, and a senior supervisor is present at all sorting. Personal data, including paperwork or photos, are never recorded or analyzed. The sanitation department foreman responsible for delivering the sample to the processing area is not allowed to be present when the bags are opened and does not have access to the database. As soon as the contents are recorded, any aluminum found is recycled and the rest of the garbage is returned to a landfill.

The tremendous accumulation of data gathered over three decades has had a number of practical applications. A good example was demonstrated in the early years of the project. A primary research interest of the garbologists involved analyzing the behaviors leading to food waste. In 1973, there was a widespread beef shortage. From 1973 through 1974, beef waste was analyzed and compared to data after the shortage was over to see if the scarcity of beef led to changes in behavior. One would assume that in times where a product is scarce and expensive, there would be less waste, but the researchers found the exact opposite to be true. The results showed that in average years, beef waste accounted for 3 percent of all beef purchased, but during the shortage, beef waste jumped to 9 percent. A hypothesis was suggested that when individuals were confronted with a shortage, they went overboard to stockpile beef, purchasing any cuts they could afford, even if they were unfamiliar. As a result, people either bought more than they could store and eat before spoilage occurred or bought cuts they did not know how to prepare, creating more waste. This hypothesis was put to the test in 1975 with a sugar shortage. Citizens of Tucson, being close to the border, were bringing less-refined Mexican sugar home. The sugar hardened quickly, which resulted in many bricks of crystallized sugar in the trash. Other atypical forms of sweetener were also used, once again tripling the amount of sweet food waste during a time of scarcity. From studies like these, the Garbage Project developed the First Principle of Food Waste: "The more repetitive your diet—the more you eat the same things day after day—the less food you waste." Over the years, a large number of agencies and groups have approached the Garbage Project for waste disposal information to support their studies.

Garbage Collection

There are two general methods used by the Garbage Project to collect their data. The methodology and duration is determined by the nature of the project.

The first is called a "regular sort." Neighborhoods are divided into groups based on selected criteria: anything from family size to census tract to economic level. The Garbage Project arranges to have the refuse from randomly selected households within the group delivered to the sorting area. The data collected in this way can be used to compare different types of households or be used to get a snapshot of all garbage disposal patterns across the population. The regular sort of garbage can give clear information on a broad topic (like whether total food waste increases over time) or can be used to look at very specific cultural behaviors, such as alcohol consumption, condom use, or correct birth control use. Garbage does not lie.

The second methodology combines the regular sort with self-reports from the individuals being studied. This matched study method requires the agreement of everyone in the household. Despite assurances that all information is kept confidential, it is difficult to find households that will agree to keep a log of everything they throw away that can be compared to the actual refuse. A personal interview is also used in the matched study in which most people refuse to participate. Few of these studies have been done, but certain quirks remain consistent. What people say they discard rarely is even close to what is actually in their garbage bags. Unhealthy snacks and processed meats are repeatedly underreported, and, conversely, produce, dairy products, and high-fiber cereal are overreported. Over and over again, Garbage Project studies illustrate how the average person in the United States is completely unaware of their actual consumption and waste. It is possible that any government agency conducting a study based solely on surveys or interviews will be using fundamentally unsound data.

The Garbage Project does not limit itself to observing food product waste. In the late 1980s, regular sorts were conducted in Louisiana, Arizona, and California to evaluate the disposal of hazardous household waste. It was found that these materials made up about 1 percent of household garbage. This figure does not seem like much until it is multiplied by an 88,000 household community, resulting in almost 65,000 pounds of toxic chemicals being shipped off to the local landfill annually. While this percentage stayed the same across the board, the content of the hazardous waste varied by the socioeconomic status of the neighborhoods evaluated. Low-income households discarded mostly car maintenance fluids, middle-class homes disposed of home improvement products like paints and varnishes, and wealthy households primarily removed pesticides and herbicides.

Landfill Excavations

Shortly after World War II, the sanitary landfill became the most popular way to dispose of garbage in the United States. Previous generations relied on incinerators to burn refuse or open dumps where garbage was piled and left to contaminate the air and land. A sanitary landfill is one where garbage is piled and compressed, then covered with several inches of soil or some other inert material nightly to reduce problems with odor or pests. The Garbage Project began excavating landfills for two reasons. The first was to see if the data being gathered from sanitation trucks could be matched, or cross-validated, by samples collected from municipal landfills. The second was to determine exactly what happens to garbage over time after it is deposited. The Garbage Project to date has excavated in 15 landfills across the United States and Canada, analyzing about 12 metric tons of debris deposited since 1952.

Traditional archaeological techniques can be used to "excavate" a landfill. The layers, or "strata," of the site are easily dated by newspapers, magazines, and mail. Seasons are reflected in Christmas wrap, heart-shaped candy boxes, Easter candy wrappers, and other seasonal trash. Trenches up to 25 feet deep were created with backhoes, and bucket-auger wells were drilled down 100 feet for samples that were sorted and analyzed. One key component of this research was to determine what was actually there. The public has made assumptions about what is filling landfills without any measurements being taken. At different times, fast food packaging, Styrofoam (expanded polystyrene foam), disposable diapers, and plastics have been reviled in the media and blamed for the massive mounds of trash produced by society. The Garbage Project discovered that diapers, fast food containers, and Styrofoam combined comprised less than 3 percent of the content of the landfills. Plastics made up less than one-quarter of the contents. Building and construction debris claimed another 20–30 percent. By far, the largest and fastest-growing component is actually paper and paper products. While greater quantities of plastic are being discarded, the process of light weighting (making the same size container with less product) has kept the total volume of plastic from increasing. Despite the commitment to local recycling programs, the amount of paper found in landfills is rising. Packaging, phone books, magazines, catalogs, junk mail, and more are inundating the garbage.

Another myth commonly held about landfills is that the contents will biodegrade like a giant compost heap. Biodegredation does occur, but at a much slower rate than expected. A compost heap in a backyard works well because the materials are chopped up, water is added, and the pile is regularly churned up and turned over. This process does not occur in landfills. Materials are dumped, fluids are usually prohibited, and the garbage is regularly compacted so oxygen cannot reach the materials to begin the degradation process. Fifteen-year-old newspapers have been found that can still be read. The best way to reduce the amount of waste in landfills is to put less in them by reducing waste at the source—homes. A commitment to recycling and reuse programs would also be a pragmatic way to reduce garbage.

Garbology Applications

The archaeology of garbage has a number of useful applications. Excavations at landfills across the country by members of the Garbage Project have illustrated how little is known about what is actually discarded. These digs have had an impact on the perceptions of the public and on policy planners.

Various government agencies contract with the Garbage Project for information on U.S. citizens. Documented evidence on diet and nutrition, food consumption and waste, recycling, and the disposal of household materials will continue to be critical in the development of public policies.

Landfill archaeology can be used to measure the effects of these public policies after they are put in place. For example, to find out if a new recycling program established in a city reduces the amount of waste in its landfill. These research techniques could also be used to trace pollutants back to their source in the landfill and eventually back to the point of discard.

Garbology is used in the 21st century by scientists evaluating new processes for municipal waste management. A steadily growing population is generating an increasing amount of waste, despite the existence of recycling programs.

Decomposition in landfills generates methane gas, which has to be vented. Studies are being conducted to use this methane production to generate electricity.

Civilizations tend to cycle from a simple, efficient use of resources to a period of economic vitality and conspicuous consumption before returning to a "recycle and reuse" mentality. The United States is entrenched in that vital phase, and it has the mountains of trash to prove it. At the same time, the country has never had such a variety of resources and technology at its disposal to manage the waste stream. Rathje and Cullen Murphy have a number of suggestions to keep in mind:

- Americans have problems with garbage, but do not think of it as a crisis. Communities should steadily continue to make improvements to their waste management efforts, but nothing drastic.
- Be willing to pay for proper garbage disposal. The cheapest way to get garbage out of sight may not be the best option.
- Look at the big picture. Diapers and fast food packaging have taken big hits in the media, but the Garbage Project has proven that more than half of landfill waste is actually paper products and construction/demolition debris. Radically changing how these two categories of waste are handled will have a great impact on landfills.
- Modest attention to household behavior by average citizens will have a great impact. Rather than dealing with garbage after it is thrown away, individuals could easily put a little more effort into recycling and reusing products before they enter the waste stream. Deliberately purchasing goods made with recycled materials or things in recycled packaging would give a great boost to a growing industry.

The archaeology of garbage has proven that there is a great disparity between what people say they do and what they actually do. Widespread misconceptions about what is thrown out, the biodegradation of waste, recycling in communities, and other aspects of waste management have been clearly illustrated by the Garbage Project and its contemporaries. Researchers now have the knowledge and a great opportunity to use this information to improve society by raising awareness and increasing effectiveness in managing the waste stream.

Jill M. Church
D'Youville College

See Also: Archaeological Techniques, Modern Day; Archaeology of Modern Landfills; Construction and Demolition Waste; Garbology; Landfills, Modern; Paper and Landfills; Sociology of Waste.

Further Readings

Anonymous. "Prowlers Target Celebs' Trash." *New York Post* (October 27, 2008).
Buchli, Victor and Gavin Lucas. *Archaeologies of the Contemporary Past.* New York: Routledge, 2001.
Cote, Joseph A., James McCullough, and Michael Reilly. "Effects of Unexpected Situations on Behavior-Intention Differences: A Garbology Analysis." *Journal of Consumer Research*, v.12/2 (1985).
Hering, Rudolph and Samuel Greeley. *Collection and Disposal of Municipal Refuse.* New York: McGraw-Hill, 1921.
Katz, Jane. "What a Waste: The Generation and Disposal of Trash Imposes Costs on Society and the Environment: Should We Be Doing More?" *Regional Review*, v.12/1 (2002).
Rathje, William L. "The Garbage Decade." *American Behavioral Scientist*, v.28/1 (1984).
Rathje, William L. "Once and Future Landfills." *National Geographic*, v.179/5 (1991).
Rathje, William L. and Cullen Murphy. "Garbage Demographics." *American Demographics*, v.14/5 (1992).
Rathje, William L. and Cullen Murphy. *Rubbish! The Archaeology of Garbage.* New York: HarperCollins, 1992.
Reagin, Misty. "Turned on by Trash: Across the Country, Cities and Counties Are Benefiting From Methane-to-Energy Projects." *American City & County*, v.34/6 (2002).
Royte, Elizabeth. *Garbage Land: On the Secret Trail of Trash.* New York: Little, Brown, 2005.
Shanks, Michael, David Platt, and William L. Rathje. "The Perfume of Garbage: Modernity and the Archaeological." *Modernism/Modernity*, v.11/1 (2004).
Strasser, Susan. *Waste and Want: A Social History of Trash.* New York: Henry Holt, 1999.
Thomas, David Hurst. *Archaeology: Down to Earth.* Fort Worth, TX: Harcourt Brace College Publishers, 1999.

Archaeology of Modern Landfills

Archaeologists have long studied garbage (the remains of past societies) in order to both develop theories of behavior and gain overall insight into past cultures. The archaeology of modern landfills emerged from a desire to find out what was behind the media's portrayal of the "garbage crisis" faced by Americans and out of a curiosity about modern material culture. Defined broadly, material culture includes physical things modified by human behavior. Examples include plowed fields, polluted rivers, and modern landfills. In essence, the archaeology of landfills is the "archaeology of us." Studying contemporary garbage can provide useful insights into current behaviors that can be useful in the present. Learning about what society does now enables people to make changes in behavior now, instead of in the years to come.

The archaeology of landfills stems from the work of behavioral archaeologists. Specifically, the Garbage Project at the University of Arizona spent decades studying the garbage in modern landfills. Garbology (the study of trash and human societies) is carried out by garbologists—archaeologists who examine trash. This branch of archaeology posits that archaeology is the study of the relationships between human behavior and material culture. As such, behavioral archaeology is able to challenge societal myths about consumption patterns. The archaeology of modern landfills has led to myriad

Behavioral archaeologists study the contents of modern landfills to reveal societal practices regarding food waste and consumption. These discoveries can lead to more rapid implementation of practical changes in behavior rather than waiting years for habits to evolve. The Garbage Project has dispelled numerous misconceptions about landfill use that can help individuals and policymakers make better decisions about waste. For example, disposable diapers do not make up a significant percentage of landfill trash, as is commonly believed.

insights as well as the ability to debunk many modern garbage myths about what occurs in, and what makes up, landfill content. For example, the excavation of landfills has shown what the true composition of landfills is, and that organic matter does not decompose and biodegrade as many people believed it did. Important findings were discovered about food waste and consumption patterns and the accumulation of paper and construction debris in landfills.

Landfill Excavation

In 1987, the Garbage Project began excavating landfills throughout the United States and Canada. During these excavations, almost 30,000 pounds of garbage was removed, sorted, and analyzed. The process involved the use of auger machines that could dig down to 100 feet and remove undisturbed units of trash. Fully intact newspapers were used for dating the trash.

The excavation of landfills led to several insights regarding popular garbage myths and misconceptions. Among them were the myths of biodegradation, the realization of the myth of what landfills are actually composed of, and the misconception that society is running out of room for landfills. For example, while it is a popular notion that trash biodegrades inside landfills, in actuality most trash remains well preserved, even after decades of being buried. The main reasons for this are the lack of both oxygen and moisture inside landfills. The process of adding trash to landfills creates an environment that is not conducive to the breakdown of organic matter. When new material is added to a landfill, it is tightly compacted and covered with a layer of dirt. This process creates the ideal environment for garbologists to gather well-preserved trash, but not the conditions required for decomposition.

The idea that society is running out of landfill space has been prevalent since the late 1800s, but remains unfounded. Throughout the United States, there are plenty of open spaces that could accommodate new landfills. While there are certain areas with landfill shortages, this may be more a function of landfills getting bigger and bigger, but with fewer and fewer of them opening up as needed. This may, in part, be due to environmental justice advocacy in which Not In My Backyard (NIMBY) protesters oppose garbage dumps in their communities. A concern with this idea is that existing problematic landfills may receive less attention than new, well-planned landfills.

The excavation of modern landfills began, in part, as an attempt to find out if the popular media representations of the garbage situation were accurate. Archaeologists wanted to get an idea of the actual volume taken up by various kinds of garbage. For decades, citizens had been told that they were about to be buried by their trash. Garbologists wanted to find out if public opinion and media representation were accurate depictions of the reality of the garbage situation. Their findings led to some similarities with national estimates, but many findings were surprising. One was the low volume of items the public believed to be the most problematic, including disposable diapers, fast food containers, and plastics. The other surprise was the large volume of paper and construction and demolition debris found in landfills.

Landfill Composition

Surveys reveal that most Americans are concerned about the amounts of fast food packaging, polystyrene foam (Styrofoam), and diapers that are thrown away. Survey respondents believed that about 75 percent of landfill volume consisted of these three items. Garbage Project results found that, in actuality, fast food packaging, polystyrene, and diapers account for only about 3 percent of landfill volume. National estimates of landfill composition do not include construction and demolition debris, which, surprisingly, accounted for as much as 28 percent of the volume of mixed refuse according to Garbage Project calculations. These findings provided useful information that could be used as a basis for policy development as well as insights into the inaccurate imagination society has regarding its garbage behavior and the actual makeup of landfills.

The Garbage Project compared a poll of what people think is in landfills and data from landfill excavations. The results show the large discrepancies between the mental and material realities of what goes into landfills. For example, the survey indicated that people heavily overestimated the volume of disposable diapers (estimate: 41 percent; actual: <2 percent), plastic bottles (estimate: 29 percent; actual: <1 percent), and large appliances (estimate: 24 percent; actual: <2 percent) while they underestimated paper (estimate: 6 percent; actual: >40 percent), food and yard waste (estimate: 3 percent; actual: ~7 percent), and construction debris (estimate: 0 percent; actual: ~12 percent). These huge discrepancies between perceptions and reality can lead to policies and actions that are inappropriate or inadequate to address the actual nature of the behaviors leading to the garbage predicament.

Paper contributes to a large percentage of landfill space, but its impact is highly underestimated by the general public. Even with extensive recycling efforts, paper still makes up a large segment of landfill volume. The Garbage Project describes paper as the "invisible man" in landfills, with volumes increasing more rapidly than plastics, which people tend to think of as more of a problem than paper.

Plastics remain an issue of confusion for the general public because of the change in the production of plastics over the years. Lightweighting is a process by which the characteristics of plastics are maintained, but fewer resins are needed for the final product. Once these lightweighted plastics are crushed, they take up minimal volume in landfills. This translates into more pieces of plastic being added to landfills, but at a consistent volume to those in the past. While the number of plastic items has increased considerably since the 1970s, the amount of space taken up by plastics has remained constant.

Social Contexts and Implications

Garbage Project landfill excavations led to a call for a "two realities" approach to studying the relationship between behavioral-material patterns and societal self-perceptions. Such an approach, it is believed, could lead to an archaeological theory that integrates the interaction between the two. The hope is that this approach would lead to improved social policies that work as they are intended as opposed to falling victim to unrealistic expectations about behavior based on embedded and inaccurate societal self-perceptions that may serve to perpetuate behaviors, rather than reduce or eliminate them.

Several broad conclusions resulted from landfill excavations and may be useful in the development of more appropriate theory and policies based on the reality of behaviors and the perception of those behaviors. First, grand patterns of behavior can be overlooked by society, including policy makers and citizens. Social scientists, in this case archaeologists, can remove themselves from the entrenched social norms and perceptions and seek out and recognize the material realities of behavior. An example of this lies in the Parkinson's Law of Garbage. This law states that the rate of disposal increases with the size of refuse containers. Even though current waste management policies aim to reduce the amount of solid waste ending up in landfills, there has been an increase in items such as household hazardous waste ending up in landfills. The premise behind Parkinson's Law of Garbage states that garbage will expand to fill the available container. Even as policies have been implemented to reduce municipal solid waste (MSW), the container sizes available for trash have increased.

These contradictory messages may actually lead to an increase in MSW being landfilled as opposed to reduced. Second, systematic approaches to knowledge can lead to discoveries about the inaccurate self-perceptions that exist at a societal level. An example of this is the fact that construction and demolition debris are not included in national estimates on municipal solid waste. These estimates, on which many public policies are based, do not accurately depict the content of landfills. Third, it appears that there are certain conveniences that even self-proclaimed environmentalists will not do without. Among them is the disposable diaper, which according to Garbage Project findings, does not make up a significant amount of trash in landfills.

Why is it that modern culture, on the one hand, has such wild exaggerations about what is in landfills and, on the other, lacks insight into what trash is really made up of? The social, political, and economic realities of trash, what to do with it, and where to get information about the contents of garbage and landfills must be reconsidered if forward-thinking policies are to be developed that will actually help address the current garbage situation.

Major questions that emerge from landfill excavations have to do with why construction and demolition debris is excluded from national MSW estimates. With waste characterization studies being a popular source of MSW information, how can it be that such large amounts of actual landfill composition can be left out of the equation? Additionally, if paper is one of the largest categories occupying landfill space, why aren't there more or better policies and initiatives in place to reduce the amount of paper that ends up in landfills each year?

Moving forward, MSW policies may prove more effective at reducing the proportion of the waste stream that ends up in landfills each year. Landfill excavation reminds people to pay attention to where ideas come from and not assume that mainstream information is always correct. Through systematic analysis of landfills, garbologists have given society important insights into the reality of the garbage situation.

Maggie Ornstein
Graduate Center, City University of New York

See Also: Archaeological Techniques, Modern Day; Archaeology of Garbage; Garbage Project; Landfills, Modern; Paper and Landfills.

Further Readings

Rathje, W. L. "Once and Future Landfills." *National Geographic*, v.179/5 (May 1991).

Rathje, W. L., W. W. Hughes, D. C. Wilson, M. K. Tani, G. H. Archer, R. G. Hunt, and T. W. Jones. "The Archaeology of Contemporary Landfills." *American Antiquity*, v.57/3 (July 1992).

Rathje, W. L. and C. Murphy. *Rubbish! The Archaeology of Garbage*. New York: HarperCollins, 1992.

Argentina

Argentina is a vast, geographically diverse country. It includes the semitropical forests in the north, the vast agricultural bounty of the Pampas, the tallest stretch of the Andes, and the frigid Tierra del Fuego at the southern tip of South America. Argentina has the eighth-largest area in the world but contains the 31st-largest population. Over 90 percent of the population is urban, with about one-third sprawled in and around the capital, Buenos Aires. At the beginning of the 20th century, Argentina was one of the richest countries in the world; at the beginning of the 21st century, it has middling economic power compared globally—its overall gross domestic product (GDP) purchasing-power parity (PPP) is ranked only a little higher than its population, at 24th. Its GDP (PPP) per capita is about 80th worldwide, at about $13,400 in 2009. Other indicators of consumption are consonant with Argentina's population, providing a further indication that Argentina is, in many ways, about average on a worldwide scale. For example, it is 31st in electricity consumption, 27th in oil use, and 19th in natural gas consumption.

Brief History

Argentina's political and economic history since the start of the 20th century has been tumultuous. In recent decades, a conservative military dictatorship has ceded control to democratic institutions. Much of Argentina's contemporary political organization continues the legacy of Juan Perón (1895–1974), who, as president, favored state intervention in economic activity and co-optation of mass organizations, especially unions.

Since the 1990s, and especially before the economic crisis began in late 2001, Argentinian lawmakers have adopted a variety of progressive environmental measures. Most fundamentally, the Constitution of 1994 explicitly establishes environmental sustainability as the government's responsibility. The same document prohibits the entry of hazardous or radioactive wastes.

During the 21st century, Argentina has struggled to recover from a devastating financial crisis that began in late 2001 and led to massive unemployment, manifold increases in poverty, and a quick succession of governments. One way of examining the economic fortunes of Argentina is through garbage. Each day, Argentinians discarded on average about 0.9 kilograms (kg.) per person in 2000 and 2004–2005 (similar to the averages for Poland and Chile). However, in 2002, this figure fell to 0.67 kg and then rose to 0.8 kg in 2003, as the crisis provoked Argentinians into consuming less and increasing their efficiency with what they did consume. As a result of the global financial crisis of 2009, a similar drop-off occurred, to 0.86 kg.

Cultural Consumption

In addition to soccer, the tango, and gauchos, beef and *yerba mate* (a drink made from steeping leaves in hot water) are essential elements of a traditional Argentinian identity. Consumption of these two items is nearly ubiquitous. Over 90 percent of Argentine households consume yerba mate. Most of the concern regarding mate focuses on its consumption. Waste from its production has not provoked large-scale environmental concerns beyond those generally applicable to agriculture, such as the runoff of pesticides. Its consumption usually yields a relatively light amount of waste, as it typically involves reusable and shared items, such as kettles or even larger containers of hot water, along with a gourd and metal straw for drinking. Some evidence suggests that mate can cause serious health problems, but other lines of evidence point to its health benefits.

Argentina had been the long-running world leader in beef consumption, surpassed by Uruguay in 2010. Argentinians still far outstrip any other heavy consumers, such as the U.S. population. Overall rates of consumption in Argentina vary mostly because of fluctuations in cost rather than concerns about beef's effects on health or about the environmental effects of beef production. These effects in terms of waste are quite high, as beef production represents, in terms of gallons per pound, perhaps the most inefficient use of water of any major food; likewise, it arguably results in the greatest release of greenhouse gas emissions of any major food.

Greenhouse Gas Emissions

Argentina has garnered attention for its commitment to reducing greenhouse gas emissions. It is a non–Annex I signatory to the Kyoto Protocol, meaning that it can host Clean Development Mechanism projects and that it is not required to reduce its emissions. Nonetheless, in 1999, Argentina offered to voluntarily specify a legally binding limit on its emissions within the framework of the protocol. It proposed using a relatively stringent formula for calculating carbon intensity, which would allow its emissions to rise only slightly as overall GDP rose. However, some other developing countries opposed Argentina's proposal for fear that it would become a standard, and Argentina's economic collapse and political turmoil in the early 2000s, which actually caused its energy intensity to rise as its GDP dropped, caused it to withdraw this proposal. Renewable sources supply only about 1.5 percent of Argentina's electricity and about 7.5 percent of its total energy; both figures represent a decline compared to 2000.

Recycling

The financial crisis led to the precipitous expansion of *cartoneros* (Argentinians who filter and recycle other people's garbage). The root of this term is *cartón*, or "cardboard," one of the materials they collect. In Buenos Aires, perhaps as many as 40,000 people at one point—many of them newly unemployed because of the financial crisis that began in late 2001—stream into the city at night to pick recyclable materials out of garbage set out for officially designated collectors to retrieve. They then sell these materials to middlemen or directly to recycling plants. This process has become increasingly institutionalized. One early development was the provision of overnight train service between the city to its poorer outskirts, dedicated to cartoneros. One neighborhood in the city has worked with a cooperative of cartoneros to separate recyclables before the cartoneros arrive. The local government, as part of its implementation of an ambitious plan to reduce landfill deposits, has adopted aspects of this experience as it plans to further recognize and regulate cartoneros. The city is providing incentives for cartoneros to join cooperatives of collectors, giving them semiofficial status; new sorting centers might employ others.

The massive number of cartoneros in Buenos Aires has drawn considerable attention, but analogous phenomena occur in other cities. In Córdoba, the city has provided two cooperatives of cartoneros with uniforms, training, and specially designed motorized carts for transporting the recyclables to the various collection points for their loads. The city has also helped to arrange a contract with a recycling business and a secure method of payment for the collectors. Rosario's cartoneros have received permission to circulate in some areas with carts drawn by horses or humans. Some cities appear to leave cartoneros unregulated as evidenced by disputes among them over territory.

The resulting rate of recycling is difficult to determine. The federal government estimates that, nationwide, only 2.5 percent of solid waste was recycled; however, it is unclear whether this figure includes recycling performed by cartoneros before official collectors picked up garbage. Some have estimated that cartoneros recycle as much as 13 percent of the Buenos Aires' garbage. The potential exists for greater recycling. Half of the waste is organic; glass, paper, and cardboard compose almost half of the remainder.

Disposal

The largest percentage of trash ends up in modern, engineered landfills, especially in larger cities. A smaller percentage is placed in less-secure landfills. About one-fourth winds up in open dumps, especially in smaller towns and cities. As of 2009, none was incinerated, but this might change soon, although various proposals to incinerate trash have inspired considerable controversy. Few projects exist to convert trash or other waste, such as manure, into usable energy, but the great potential has generated considerable interest.

Water: Clean and Waste

Argentinians do not have universal access to running water in their homes. After considerable progress in the 1990s, rising from 66 to 78 percent of the population with connections, the 21st century has witnessed a slight dip followed by a slight rise, to an estimated 81 percent connected. The government's goal is to have 85 percent of the population connected by 2015. Access to sewage systems is more limited but growing more rapidly. As of 2010, only 55 percent of the population lives in homes with sewage connections, but this is an increase from 42 percent in 2001 and 34 percent in 1991.

A long-running dispute with Uruguay has centered on issues of waste and water quality. While political intrigue may explain part of the controversy, Argentina officially objected to the establishment of a pulp mill on Uruguay's side of the Río Uruguay, which flows between the two countries, on the grounds that the mill would pollute the river. Among other actions, Argentinians blocked a bridge over the river. The International Court of Justice eventually ruled in 2010 that Uruguay had failed to notify Argentina properly but that the plant did not present a danger. The countries then established an elaborate mechanism to monitor the river's water quality along its course, testing for pollution from either country.

Tracy Duvall
Universitas Indonesia

See Also: Buenos Aires, Argentina; Recycling; Street Scavenging and Trash Picking.

Further Readings

Brown, Jonathan C. *A Brief History of Argentina*. 2nd ed. New York: Lexington Associates, 2010.
"The Other Nightlife in Buenos Aires: The Story of Argentina's Cartoneros." *Verge* (Fall 2009).
Secretaría de Ambiente y Desarrollo Sustentable. http://www.ambiente.gov.ar (Accessed October 2010).

Arizona

The sixth-largest and largest landlocked state by population in the United States, Arizona is one of the Four Corners states of the southwest and shares an international border with the Mexican states of Sonora and Baja California. Phoenix, the capital and largest city, has the largest metropolitan area. The Garbage Project originated in Tucson, the second-largest city. The last contiguous state to join the Union, Arizona was granted statehood in 1912, on the 50th anniversary of its recognition as a territory of the Confederate states. Arizona is notable for its desert climate, but also holds pine forest and mountain ranges in the higher north and a cooler climate in the southern deserts. The Grand Canyon and many other sites of national importance are located in Arizona, and over one-quarter of the state is Federal Trust Land, home to the Navajo Nation, Hopi tribe, Tohono O'odham, the Apache, and Yavapai tribes.

The state's population grew exponentially after World War II, in part because advances in air conditioning allowed more people to tolerate the intense heat of Arizona summers. The population was 294,353 in 1910, 1,752,122 in 1970, and 6,392,017 in 2010, with most residents clustered in the Phoenix metropolitan area. In the early 1960s, retirement communities (then a relatively new concept) appeared in the state, catering to those who wished to escape the harsh Midwest and northeast winters.

Garbage

The 16th Nationwide Survey of MSW Management in the United States found the following: in 2004, Arizona had an estimated 8,197,591 tons of municipal solid waste (MSW) generation, placing it 19th in a survey of the 50 states and the capital district. Based on the 2004 population of 6,165,689, an estimated 1.33 tons of MSW were generated per person per year (ranking joint 19th). Some 7,172,000 tons were landfilled in the state's 40 landfills; 85,000 tons of MSW were exported, and 438,000 tons were imported. In 2006, Arizona was increasing its landfill capacity—it was ranked joint 14th out of 44 respondent states for number of landfills. Only whole tires and white goods were reported as being banned from Arizona landfills. Arizona has no waste-to-energy (WTE) facility, but 1,025,591 tons of MSW were recycled, placing Arizona 24th in the ranking of recycled MSW tonnage.

Garbology

In 1973, the Tucson Garbage Project originated at the University of Arizona as *Le Projet du Garbàge*. It initiated the anthropological study of contemporary rubbish often referred to as "garbology." Although the study has since expanded beyond Arizona, the archaeology of garbage has probably been studied more intensively there than in any other state.

Early History of Waste and Disposal

For the Spanish colonial and Mexican periods, most data on garbage disposal is from archaeological research, with only a few known historical ref-

The mechanical arm of a garbage truck in Scottsdale, Arizona, picks up 320-gallon trash containers and dumps the contents into a truck nicknamed "Godzilla," in 1972. Garbage pickup was captured as part of the U.S. Environmental Protection Agency's project to document environmental concerns via photographs in the 1970s, the same decade that William Rathje's Garbage Project, which originated in Arizona, began studying the archaeology of garbage. The latter project has been studied more intensively in Arizona than in any other state.

erences. Excavation suggests that midden deposits were dumped outside the eastern gate of the Tucson Presidio and that there are refuse pits and concentrations of gardens inside the north Presidio wall near an *horno*. Field surveys of the Barrio de Tubac identified a large refuse area in the site.

The 1849 Gold Rush, the Gadsden Purchase of 1854, Mormon colonization, and the railroads in the 1870s and 1880s all added to the increasing population and urban development of Arizona during the territorial period. The railroad also increased consumption of goods by allowing greater volume, variety, and availability of products to reach Arizona. When this population increase began, urban and rural refuse disposal was completely unregulated. Dumping was commonplace in arroyos, rundown areas, disused privies or wells, and vacant or abandoned property. Rubbish was usually simply thrown into the street for roaming animals to eat. The persistence of this practice led to streets filled with excrement, trash, and animal carcasses. Outbreaks of disease and the appearance of germ theory resulted in legislation passed by the newly established town and city councils to try to control garbage disposal.

Local governments began passing ordinances to increase community safety and quality of life; sanitary ordinances were usually first. These early ordinances placed the responsibility and cost of trash disposal with the landowner or tenant for the first time, and were reinforced by penalties up to $300 and up to three months imprisonment. Tucson passed the first ordinances for controlling garbage in 1871 and 1872, which included the earliest organized municipal trash collection, where the city marshall directed the emptying of rubbish pits every Saturday. Phoenix passed its first trash ordinances in 1881 when the city became incorporated, and soon after created the role of health officer to oversee public health. This office became common in most sizable communities, with Tombstone, then a village, gaining one in 1882. By the end of the century, however, the role was increasingly involved in infectious disease control.

Waste and Disposal in the 20th Century

Municipal government began taking more direct involvement in garbage collection and disposal in the early 1900s. This coincided with the public health director's remit becoming more focused on controlling infectious diseases, such as the influenza epidemic and tuberculosis. At this point, most incorporated population centers had ordinances in place to regulate garbage. Local government was directly involved in the removal of residential and business refuse, whether by agreement with an outside contractor or by their own dedicated department. Ordinances became more comprehensive and detailed, specifying containers for collection, separation of waste streams, and regulating garbage storage and pickup. Fees began to be charged for collection, and designated dumps appeared. Tucson opened an incinerator in the early 1930s, but (like many early incinerators) it was short-lived and was demolished in 1950. During World War II, recycling became a focus, with metal and rubber salvaged for the war effort; several metal and rubber drives took place in Tucson.

Post–World War II national environmental laws set countrywide standards for managing waste, which ceased to be under complete local control as the federal government targeted disease in the rapidly expanding population. Waste disposal sites were moved farther from populated areas, and landfills began to replace open dumping. Laws passed in the 1960s and 1970s regulated solid waste management, incineration, and discharge into waterways. The 1976 Resource Conservation and Recovery Act included the closure of all open dumps, which caused a crisis in some Arizona communities. Star Valley and Ponderosa dumps were closed by federal order but had to reopen briefly when a replacement landfill was not ready in time and illegal dumping skyrocketed. There were approximately 248 closed dumps and landfills in Arizona in 2010. The Silver Bell Golf Course in Tucson and Cave Creek Municipal Golf Course in Phoenix were both created to reuse space created by closure of old open dumps.

Water Consumption

The state's development in the desert would be impossible without extensive federal manipulation of waterways during the 20th century. Arizona's population grew substantially after President Lyndon B. Johnson signed the Colorado River Basin Project Act of 1968. That law created the Central Arizona Project, which redirected water from the Colorado River with dams and aqueducts to supply water and hydroelectric power to the growing Phoenix and Tucson areas. In the early 21st century, concerns over unsustainable water consumption patterns in the face of extended droughts led to conservation efforts throughout the state. Arizona entered into an agreement in 2007 with several other western states dependent upon the Colorado River to coordinate conservation efforts across the region, but sustainable water management remains a concern for the foreseeable future.

Garbage Archaeology

Beyond the Garbage Project, the relevance of garbage deposits to historical archaeology in general is highly appreciated in Arizonan archaeology. A key example of maximizing the value of garbage deposits is the work done at the Track Site. In the 1880s, with the Native Americans confined in the reservations, the federal government attempted a nonmilitary campaign to remove their presence by assimilating Native American children into Anglo-Saxon American lifeways. This was attempted by setting up Indian boarding schools, which set out to imprint a new identity on Native American children through education. The Track Site, a turn-of-the-century dump (use life 1891–1926) associated with the Phoenix Indian School, was excavated in the 1990s under federal government land exchange laws. Due to the size of the dump, approximately 9 percent of the deposit was sampled, yielding around 160,000 artifacts.

The methods used by the school to erode and replace Native American cultural identity were reflected in the excavated artifacts. The scale of the use of dolls to imprint socialization skills and reinforce U.S. gender stereotyping was demonstrated by the recovery of 136 fragments from 108 dolls. Indian school policy was to eradicate tribal identity by instilling a sense of the egotism of U.S. civilization, getting students to identity with "I and mine" instead of "we and ours." The success rate of this ploy is evidenced in excavated toothbrushes and combs that were meant to be marked with their owner's name, reinforcing ideas of personal rather than tribal property. Only 6 percent and 12 percent, respectively, had been marked. The marks made were often lines or dates rather than names, reflecting the survival of the Native American belief that a name is personal and not to be openly revealed.

Other artifacts symbolic of resistance included two clay objects that may be effigies and a possible fetish stone, which is bone shaped and of nonlocal geology. Items such as these were subject to confiscation by the school authorities. Shards of indigenous southwestern pottery were unexpected finds, as were knapped pieces of glass and ceramic, which showed that traditional lithic-working skills were still being practiced.

Jon Welsh
AAG Archaeology

See Also: Arizona Waste Characterization Study; Garbage Project; Garbology.

Further Readings

Arsova, Ljupka, Rob van Haaren, Nora Goldstein, Scott M. Kaufman, and Nickolas J. Themelis. "The State of Garbage in America: 16th Nationwide Survey of MSW Management in the U.S." *BioCycle*, v.49/12 (2008).

Diehl, Allison Cohen, Timothy W. Jones, and J. Homer Thiel. *Archaeological Investigations at El Dumpé, a Mid-Twentieth Century Dump, and the Embankment Site, Tucson, Arizona (Center for Desert Archaeology Technical Report 96-19)*. Tucson, AZ: Center for Desert Archaeology, 1997.

Lindauer, O. *Historical Archaeology of the United States Industrial Indian School at Phoenix: Investigation of the Turn of the Century Trash Dump (Anthropological Field Studies 42)*. Tempe: Arizona State University, 1996.

Reisner, Marc. *Cadillac Desert: The American West and Its Disappearing Water.* New York: Penguin Books, 1993.

Sullivan, Michael and Carol Griffith. *Down in the Dumps: Context Statement and Guidance on Historical-Period Waste Management and Refuse Deposits*. Phoenix: Arizona State Parks. 2005.

Arizona Waste Characterization Study

The Arizona waste characterization study was conducted by archaeologist William Rathje and his team in 1973, and it was one of the first major studies in modern garbology (the archaeology of garbage). Rathje (1945–) was then a new professor at the University of Arizona and was the director of the Cozumel Archaeological Project sponsored by National Geographic. He and his students studied the waste at Tucson's local landfill from an archaeological perspective to learn what it would say about the surrounding community. Quantitative data formed pictures of consumption patterns, as information from landfills and garbage bins was compared with what was known about individual households and residents. Among early findings, it was discovered that Tucson residents discarded 10 percent of their food, and middle-class households discarded the most food. Later studies found that food waste was likely closer to a 15 percent average. Landfill analyses cannot perfectly estimate the amount of food waste, due to the use of garbage disposals, which diverts a portion of food waste to wastewater.

Discrepancies With Self-Reporting

The study also tested people's self-reported behaviors, finding that alcohol consumption was significantly higher than residents reported on questionnaires. Recycling was lower than self-reported. The repeated studies of Rathje's team have discovered some interesting trends in questionnaire accuracy; for instance, a nondrinking member of a household is more likely to report the alcohol consumption of that household than a drinking member is, even though the drinking members would be presumed to have the most direct knowledge of it. Further, on average, female adults will overreport a household's usage of food and other goods by 10–30 percent, which Rathje calls "the good provider syndrome," reflecting an unconscious tendency on the part of mothers and housewives to assure the questioner that the household is well taken care of.

Disposal Habits

The studies have also highlighted dietary habits, such as the decline in purchases of raw red meat in the late 1980s and the tendency of eaters to trim away greater portions of fat (a decision informed by media campaigns warning of the dangers of heart disease). However, these same eaters ate greater amounts of processed red meat with less visible fat content, such as hot dogs, sausage, and salami, so that their fat intake actually stayed the same or rose. Similarly, during periods of specific food shortages, such as a beef shortage or sugar shortage, discarded amounts of the scarce food actually increase, presumably because of hoarding behavior.

The general trend with food waste seems to be that foods associated with food behavior that change little over time are the foods wasted the least: sliced bread, typically consumed frequently throughout the week, is wasted at a rate of less than 10 percent of purchased volume, which is below the average food waste level. Specialty breads, like hot dog rolls, muffins, and pita bread, are wasted at a rate of more than 35 percent. A similar principle is true of household hazardous waste: more of it is discarded when it is associated with one-off or infrequent tasks, such as reroofing or seasonal pool cleaning, than when it is part of regular periodic household maintenance.

A Marin County garbage investigation found that a household hazardous waste awareness campaign actually led to an increase in the amount of household hazardous waste being disposed of improperly, the exact opposite of its intent. The reason was that the campaign increased awareness of the hazards of certain materials, like paints and used motor oil, and households that missed the announced collection date still undertook to remove the materials from their homes using the easiest method available to them, disposing the wastes with their normal household trash. This discovery has led to Rathje's recommendation that such campaigns include and announce multiple, spaced-apart collection times for household hazardous waste.

Other Applications and Studies

The idea of using archaeological methods to study one's own modern civilization was a significant breakthrough in methodology and elicited interesting results across the board, leading to similar studies in other communities. University of Arizona teams continue to excavate landfills throughout North America. For instance, in the 1990s, their studies of 15 landfills found that the three types of items that were most commonly focuses of concern and emblematic of needless waste—grocery bags, disposable diapers, and fast food packaging—accounted for less than 2 percent of the landfill volume of the previous 10 years and that their absence would be barely noticeable if they were banned altogether.

What took up considerable volume and was rarely mentioned in public discussions of landfill issues was construction and demolition debris, which was not even included in the Environmental Protection Agency's national estimates of municipal solid waste landfills. Despite attempts to avoid this waste in its excavations because of the increased likelihood of equipment damage, the Arizona team found that construction and demolition debris accounted for at least 20 percent by volume of what it excavated.

Construction and demolition debris is thus the second-largest category of landfill material, next to paper. In most landfills, paper biodegraded very slowly and constituted the largest volume of landfill material not only in the 10-year-old landfills from the 1980s but also the 40-year-old landfills from the 1950s. By volume, nearly half of the excavated material consisted of paper products, like computer printouts, packaging paper, newspapers, magazines, and phone books. This discovery was an impetus behind the shift in the 1990s toward an emphasis on curbside recycling and the need to recycle paper products, whereas previously paper had been treated as a more or less acceptable landfill material relative to specters like plastics and diapers.

Bill Kte'pi
Independent Scholar

See Also: Archaeology of Garbage; Archaeology of Modern Landfills; Arizona; Garbology; Sociology of Waste.

Further Readings

Rathje, William and Cullen Murphy. *Rubbish! The Archaeology of Garbage*. New York: HarperCollins, 1992.

Royte, Elizabeth. *Garbage Land: On the Secret Trail of Trash*. New York: Back Bay Books, 2006.

Strasser, Susan. *Waste and Want: A Social History of Trash*. New York: Henry Holt, 2000.

Arkansas

One of the southern states, *Arkansas* is the Algonquin name for the Quapaw Indians, an exonym adopted by European settlers. Diverse in geography, the mountainous regions of Arkansas in the Ozarks and Ouachita Mountains are part of the U.S. interior highlands. The Delta and Grand Prairie are the state's lowlands, lying along the Mississippi River, which makes up most of the eastern border. Arkansas has around 150,000 acres of wilderness areas, which are reserved for hunting, angling, hiking, and basic camping; mechanized vehicles are banned from these areas. The state is known for extreme weather, including thunderstorms, tornadoes, and high rainfall; historically, high water coming down the White River has flooded the cities in its path.

With so much open wilderness, tourism is an important part of the Arkansan economy; the official nickname "The Natural State" was coined in the 1970s. State agricultural outputs are poultry and eggs, soybeans, sorghum, cattle, cotton, rice, hogs, and milk. Industrial output includes food processing, electrical equipment, fabricated metal, machinery, paper products, bromine, and vanadium. Several global corporations have based their headquarters in northwest Arkansas since the 1970s, creating an economic boom; these include Walmart, JB Hunt, and Tyson Foods. Automobile-part manufacturing has come to eastern Arkansas to support automobile plants in neighboring states.

Waste and Landfills

The 16th Nationwide Survey of MSW Management in the United States found the following: in 2006, Arkansas had an estimated 3,468,842-tons of municipal solid waste (MSW) generation, placing it 31st in a survey of the 50 states and the capital district. Based on the 2006 population of 2,809,111, an estimated 1.23 tons of MSW were generated per person per year (ranking joint 27th); 2,900,689 tons were landfilled in the state's 62 landfills and 160,937 tons of MSW were exported; the import tonnage was not reported. In 2006, Arkansas was increasing its landfill capacity—it was ranked joint 4th out of 44 respondent states for number of landfills. Yard waste, whole tires, lead-acid batteries, and electronics were reported as banned from Arkansas landfills. Arkansas has three waste-to-energy (WTE) facilities, which processed 35,464 tons of MSW (27th out of 32 respondents), and 532,689 tons of MSW were recycled, placing Arkansas 30th in the ranking of recycled MSW tonnage.

Waste to Energy

Arkansas has been at the forefront of waste-to-energy developments, a technology championed by Governor Mike Beebe and Senator Blanche Lincoln. Waste Management, Inc.'s Two Pine Landfill gas-to-energy plant is the state's first landfill gas (LFG)-to-energy plant. Operational since 2006, the 4.8-megawatt facility powers approximately 4,500 homes in North Little Rock. While the landfill has approval for a 144-acre extension, it plans to build another LFG-to-energy plant on the expansion. A partnership with Audubon Arkansas will turn 300 acres of the site's total 500 acres into a wildlife refuge. A $3 million investment was required to design and install an LFG collection system that met the Environmental Protection Agency's New Source Performance Standards. A 15-year contract to pipe and sell LFG to GEO Specialty Chemicals is being used to recoup this investment. Gas is piped to GEO to fire an industrial kiln, saving annual greenhouse gas emissions equivalent to taking 5,650 cars off the

In the Ozarks, Native American artifact collecting became such a major problem that anti-looting laws and government intervention were deemed necessary. In Yell County, 16th- and 17th-century Native American garbage pits have yielded significant deposits.

road. In Fort Smith, the city's landfill uses its LFG by injecting it straight into the natural gas pipeline. SouthTex gas-treating company partnered with Cambrian Energy to create a treatment facility that would remove carbon dioxide from the LFG, bringing it in line with Arkansas Oklahoma Gas Corporation specifications. This treatment and utilization of LFG, rather than combustion in a flare, reduces the Fort Smith landfill's overall carbon dioxide emissions by 60 percent. The city also receives royalties from the gas companies for the gas rights, providing a revenue stream from the operation that comes at no cost to the public as the gas companies carry out all of the work in return for ownership of the gas. Fort Smith is the largest landfill in Arkansas; a fifth cell was added in 2010, this 12-acre extension cost $1.9 million and should last for three to five years. Covering over 1,000 acres, the landfill serves Fort Smith, six Arkansas counties, and parts of two Oklahoma counties; the landfill is expected to serve until around 2075.

Historical Garbage

Carden Bottom, Yell County, is an alluvial floodplain created by the Arkansas River, known as a site of international importance for Native American sites dating back as far as the Archaic period. The site was extensively looted from the late 19th century for its renowned Late Mississippian pottery finds, to the extent that some looted pots were falsely claimed to be from the site or were forged. Test excavation in the early 1990s found Native American garbage pits in the area, and potentially more were discovered at one site targeted by geophysics in 2009. A midden deposit of major significance was discovered, dating to the 16th and 17th centuries, when trash was deposited into a gulley in the back of the village in an attempt to backfill it. This midden has produced the full range of Native American artifacts and also iron artifacts and beads of glass and brass that could only have come from European contact. Therefore, this garbage deposit has provided artifacts and environmental data from the climax of the Mississippian culture and the beginning of the contact period.

Picking up arrowheads in the Midwest is a generations-old activity, involving collecting surface finds that were regarded as expedient items by their Native American creators. However, prolific widespread looting, which has come to involve digging into archaeological deposits, has become a major problem, especially when the looters are acting at night, armed, and involved in illegal drugs. Experts have noted the connection between illicit drugs and looting of archaeological sites for several years.

The term *twigger* (a portmanteau of the words *tweaker* and *digger*) has been coined in the Ozark Mountains, where a methamphetamine epidemic has been building. Here, archaeological sites, including burial grounds, have been systematically attacked for artifacts and human remains and left littered by the twigger's refuse, typically cigarette packs and beer cans. It is thought that the collectible items are being traded directly for drugs and are ultimately sold on the Internet. Suggestions have been made that artifact collecting is carried out by addicts on a methamphetamine high, who can stay awake for days completely focused on a single task before collapsing. In northwest Arkansas, law enforcement officers have been shot and attacked with knives and digging tools when confronting twiggers.

Federal archaeologist Caven Clark estimated that more than 90 percent of the 130 cave sites and over 95 percent of rock shelters in the Buffalo National River unit of the National Park System have been looted. Special Agent Robert Still has stated that 70 percent of the archaeological crime cases he worked on are connected to drugs. Anti-looting laws were not well received in the Ozarks, where mound digging and arrowhead collecting are popular activities and government intervention on landowners' property is unpopular.

Garbage and Law Enforcement

In 2004, an Arkansas legal case saw a couple placed on trial for drug dealing (*Morris v. State*) seek to have prosecution evidence disallowed as the search warrant was issued based on items taken from their trash without a warrant. The judge refused this request, and (after being found guilty) the couple appealed that the evidence had been wrongfully presented. Their defense cited the Arkansas Constitution, claiming a reasonable expectation of privacy that was violated by police officers setting foot on their property to open their trash cans. This appeal was rejected, as Arkansas law upholds warrantless seizure and search of garbage on the same grounds as the federal courts.

Jon Welsh
AAG Archaeology

See Also: Archaeology of Garbage; Crime and Garbage; Landfills, Modern; Methane; Power Plants.

Further Readings

Ljupka Arsova, Rob van Haaren, Nora Goldstein, Scott M. Kaufman, and Nickolas J. Themelis. "The State of Garbage in America: 16th Nationwide Survey of MSW Management in the U.S." *BioCycle*, v.49/12 (2008).

Stewart-Abernethy, Leslie C. "The Carden Bottom Project , Yell County, Arkansas: From Dalton to Trade Beads, So Far." *Arkansas Archaeological Society Fieldnotes*, v.261 (November/December 1994).

Stewart-Abernethy, Leslie C. *Queensware From a Southern Store: Perspectives on the Antebellum Ceramics Trade From a Merchant Family's Trash in Washington, Arkansas*. Paper accompanying a poster session at 1988 Society for Historical Archaeology annual meeting, Reno, Nevada (1988).

Atomic Energy Commission

The U.S. Atomic Energy Commission (AEC) was established by Congress in 1946, transferring control of the development of atomic science from the military to civilian hands after World War II. The AEC was emblematic of the postwar American notion—best articulated in President Dwight Eisenhower's 1953 "Atoms for Peace" speech—that while atomic weapons had ended the largest war in the country's history, atomic energy was the key to the promotion of world peace, public welfare, and a free marketplace. While the AEC regulated nuclear technology, it did so with the aims of encouraging its development, at least in the United States. It was eventually supplanted by the Energy Research and Development Administration and the Nuclear Regulatory Commission (NRC), both of which were established by the 1974 Energy Reorganization Act.

Powers and Authority

The AEC was given broad regulatory powers over atomic science. The McMahon Act, which created it, explicitly prevented atomic technology transfer between the United States and foreign countries, and the act also required federal background investigations of any worker or researcher who sought access to AEC-controlled nuclear information. Employees of the AEC were exempted from the civil service system in order to give the commission greater independence, power, and flexibility in its hiring practices. All nuclear reactors and production facilities were required to be government owned, and the national laboratory system was created, building on facilities established during the war as part of the Manhattan Project.

Along with computers and the space race, nuclear technology was the most significant, prestigious, and exciting sector of scientific research in the early cold war era. The AEC's strict control over nuclear technology was vociferously criticized by free-enterprise proponents, who pointed out that the commission had exclusive jurisdiction over technology with significant social and public health implications—an important and potentially profitable technology that the private sector could only access by going through public-sector channels, which would seem to make competition difficult. Even one of the drafters of the McMahon Act referred to the AEC as "an island of socialism in the midst of a free-enterprise economy." But he and other supporters recognized that it was necessary; a technology had been created that was too dangerous to leave unregulated, and was too new to yet know what balance of power between the public and private sectors was safe. Rather than ban the technology, which has been done or at least proposed for other potentially dangerous technologies since (e.g., human cloning), its development was encouraged under strict government oversight intended to promote and protect the public interest.

Criticisms and Reorganization

Commercial nuclear power was legalized with the Atomic Energy Act Amendments of 1954, which tasked the AEC with regulating and encouraging the nuclear power industry, much as the Federal Communications Commission was meant to do with radio and television. However, the AEC was criticized for both erring too far on the side of encouragement at the expense of safety regulations and paying far too little attention to environmental concerns. As concerns about the environment grew throughout the 1960s and the 1970s, the AEC began to seem not like an obstacle to free enterprise so much as an agency in the pocket of industry, one with little vision for considering the long-term environmental consequences of nuclear power and nuclear technology. It therefore seemed like the AEC had little concern with regulations that would prevent those consequences. This was one of the factors leading to the 1974 Energy Reorganization Act, which divided the promotional and regulatory functions of the AEC into two bodies: the Energy Research and Development Administration, which was later absorbed into the newly created Department of Energy in 1977, and the Nuclear Regulatory Commission (NRC).

The NRC is tasked with the oversight of the licensing, safety, and security of nuclear reactors and radioactive material as well as the management of spent nuclear fuel. It is headed by five commissioners who serve five-year terms, appointed by the president and confirmed by the Senate. Every active nuclear power plant has resident inspectors who monitor day-to-day operations for the NRC, while special inspection teams made up of variously skilled specialists conduct regular site inspections.

Bill Kte'pi
Independent Scholar

See Also: Fusion; Hanford Nuclear Reservation; Hazardous Materials Transportation Act; High-Level Waste Disposal; Nuclear Reactors; Uranium.

Further Readings

Kragh, Helge. *Quantum Generations: A History of Physics in the Twentieth Century*. Princeton, NJ: Princeton University Press, 2002.

Mahaffey, James. *Atomic Awakening*. New York: Pegasus, 2010.

Vandenbosch, Robert, and Suzanne Vandenbosch. *Nuclear Waste Stalemate: Political and Scientific Controversies*. Salt Lake City: University of Utah Press, 2007.

Audio Equipment

Audio equipment includes the material devices involved in the processes of recording, processing, diffusing, and reproducing sounds and music. Developed after the invention of sound recording in 1877, 20th-century audio equipment has proliferated in many societal contexts in the form of a wide variety of devices such as microphones, recorders, players, mixers, amplifiers, loudspeakers, and sound editors. In the 21st century, audio equipment is used, and consequently discarded, in a wide range of social contexts, from professional fields (such as music production, medical diagnosis, or transport communication) to leisure activities (such as music and media consumption). A distinguishing feature of audio equipment waste is that the technology is subject to rapid and selective innovation cycles, making older equipment obsolete and in need of replacement.

Early History

Audio recording and reproduction are relatively recent activities in human history. It was only at the end of the 19th century that sound started to be recorded and reproduced. Media sociologist Jonathan Sterne noted that before that time, other kinds of audio equipment existed and were almost exclusively used in professions such as healthcare (where the binaural stethoscope had been in use since the middle of the century) and communication (where the telegraph mechanism was based on the Morse code). However, a larger diffusion of audio equipment took place only after Thomas A. Edison's invention of the phonograph, in 1877, and Isaac Berliner's development of the gramophone after 1888. Since then, the social diffusion of this equipment and its impact on culture and consumption has been highly relevant and subject to change following, as pointed out by media historian Patrice Flichy, the development of more general consumption patterns of the developing bourgeois family. Before the middle of the century, audio equipment became a common feature of leisure activities, contributing, as philosopher Walter Benjamin recognized, to the development of arts in the "age of mechanical reproduction."

The availability of sound recording equipment opened up new possibilities in many professional fields, communication, and art. Its impact on culture and consumer society became even more evident with further innovations introduced before World War II, especially with the diffusion of the radio and the introduction of electric sound recording devices. In the 1930s, the radio became the first great means of communication in modern society, establishing audio equipment as a common home appliance in modern houses. The use of electric reproducing devices became common in the 1940s, setting new sound quality standards, favoring the development of home high-fidelity systems, and becoming the basis for further innovations and transformations of technologies of sound recording.

Developments During and After World War II

During World War II, a variety of new equipment had been developed by military engineers, such as magnetic tape, which was mainly used in the coding and decoding of military communication and in SOund Navigation And Ranging (SONAR), which became an indispensable tool in naval transportation. In healthcare, audio equipment also had a crucial role in treating deafness and hearing impairment as well as in many diagnostic tests. Audio equipment became relevant in many fields: from automobile and air transport to industry and design, from work in organizations and commerce to political life. Composer Murray Schafer pointed out that the huge diffusion of audio equipment produced a radical change in the modern "soundscape" and an increase in the noise level in the contemporary world.

The use of the compact disc, which began to be marketed in 1982, is expected to be surpassed by newer technology in the 21st century. Audio materials have a rapid turnover rate as obsolete equipment is replaced and subsequently discarded.

One of the most relevant changes in audio equipment's consumption patterns consisted in its shift after World War II from a fixed and prevalently home-based use toward a mobile and individualized one. During the 1960s, audio equipment began to be installed in automobiles, and in the 1970s, it became miniaturized and battery powered, which made it possible to market devices such as the "boom box" and the personal Walkman. As a result, audio equipment became more and more personal, owned and used daily by an increasing number of people at any moment of their lives. Moreover, a further notable step in audio equipment development consisted in digitalization, a process that began with the digital compact disc and fully developed with the rise of the Internet and the commercial success of the iPod, personal computers, mobile phones, and digital personal devices integrating advanced audio features.

Disposal and Waste

Two phenomena related to the material presence of audio equipment in society can be identified: its accumulation and the consequent tendency to replace it periodically. Innovations occur rapidly with audio equipment, thus making old equipment rapidly obsolete. This fact produces a constant tendency to discard older generations of equipment in favor of the adoption of newer ones. For example, magnetic audio cassettes were introduced in the late 1960s and substantially disappeared in the late 1990s; the digital compact disc and its player started to be marketed in 1982 and is expected to be surpassed in terms of diffusion by other digital music formats in the 21st century. The continuous evolution of objects and formats is one of the main causes for the constant replacement of older audio equipment.

A second phenomenon at the basis of the structural turnover of audio equipment consists in the fact that innovation processes are often characterized by "format wars," which also represent a common feature of other technologies such as television and DVDs. Format wars consist of the competition between mutually incompatible equipment that tries to achieve a prominent position in the market and results in the exclusion of losing equipment, which has no other use than to be abandoned and discarded. Notable examples of these cases are the wars between cylinder records and disk records; the 8-track and 4-track cartridges and the compact cassette; and the DVD and the SACD.

A last phenomenon connected with the social presence of old audio equipment is renewed consumer interest in old and obsolete equipment, which is culturally reframed into "vintage equipment." This process of reconfiguration of the sociocultural value of audio equipment explains, for example, continued interest in turntable and vinyl records, tube amplifiers, and old-fashioned music instruments, such as analogue synthesizers.

Paolo Magaudda
University of Padova

See Also: Home Appliances; Household Consumption Patterns; Mobile Phones; Noise; Television and DVD Equipment.

Further Readings

Benjamin, W. *The Work of Art in the Age of Mechanical Reproduction.* New York: Classic Books America, 2009.

Flichy, P. *Dynamics of Modern Communication.* London: Sage, 1995.

Millard, D. *America on Record. A History of Recorded Sound.* Cambridge: Cambridge University Press, 1995.

Schafer, R. M. *Soundscape: Our Sonic Environment and the Tuning of the World.* Rochester, VT: Destiny Books, 1993.

Sterne, J. *The Audible Past: Cultural Origins of Sound Reproduction.* Durham, NC: Duke University Press, 2003.

Thompson, E. *The Soundscape of Modernity: Architectural Acoustics and the Culture of Listening in America, 1900–1933.* Cambridge, MA: MIT Press, 2004.

Australia

The tropes of *trash*, *rubbish*, or *waste* have saturated the Australian imaginary since white settlement, operating as a metonym for Australian culture at large and its relationship to European culture specifically. This is hardly surprising, given that contemporary Australia was founded as Britain's "dumping ground," functioning as an outpost for Britain's convicts for 65 years in most states since the first fleet of ships arrived in Botany Bay in 1788. Australia was henceforth understood as a colony of outcast populations, whose founding of the nation constituted the so-called convict stain right up until the 20th century.

Cultural Wasteland

Australia's identity as a cultural wasteland continued after Australia was federated in 1901, where it was repeatedly imagined as inferior in relation to its "mother country," the United Kingdom. This view continued to mire the cultural imagination for most of the 20th century, where Australia is repeatedly imagined to be devoid of the idealized versions of culture that characterize European civilization. A. D. Hope, for example, decried the nation as a vast cultural desert in his infamous poem "Australia"—a place where civilization has eroded, like a "sphinx demolished" or "stone lion worn away." This view of Australia as a cultural desert is expressed in numerous other national writings, including Robyn Boyd's treatise on the nation and its design sensibility in the 1960 work *The Australian Ugliness*. For Boyd, the nation is an empty cultural wasteland—a chaotic and accidental rubbish dump fashioned in the image of U.S. urban culture. Australia's design ugliness is categorically distinct from the United States', however, evident not only in the trash spawned by the greed of mass production but also by the dreary conformity, banality, and complacency of Australian culture at large. It is for this reason that Boyd regards the Australian ugliness as the worst of its kind, evident in the wasted potential of the nation's natural beauty: the cheaply produced cream-brick veneer houses that line the sprawling suburbs, the mess of electricity wires that blight every street, gaudy plastic decorative flowers placed on restaurant tables, old mattresses thrown over back fences, chair lifts harassing holiday areas, and soft drink signs marring the highways and service stations. "Most Australian children grow up on lots of steak, sugar, and depressing deformities of nature and architecture," he writes.

Ten years earlier, A. A. Phillips decried Australia's belief in itself as a cultural wasteland in his infamous 1950 essay "The Cultural Cringe." Phillips critically addressed the ingrained feelings of inferiority Australian intellectuals evidenced when evaluating the nation's theater, art, and music. He critiqued the perpetual deference to British and European definitions of culture in which anything produced by local artists, writers, and musicians was regarded to be derivative and deficient by comparison. This had a major impact upon the way Australians began to view their own culture, and by the 1960s, there was a growing interest in the distinctiveness of Australian culture, including a revalorization of the cultural products previously dismissed as inferior or secondary to European forms, including Australia's distinctive folk culture and convict ballads. By the 1970s, the convict stain had largely diminished, as Australia began to value its unique history as a colony of convict "underdogs." Australian history was embraced with pride, as numerous Australians began to research their cultural heritage in the hope of finding a convict ancestor.

Recent Cultural Trash

In more recent years, trash and waste have continued to gain prominence as a distinctive feature of the Australian culture and national identity. Australian cultural production has begun to celebrate the nation's purported lack of European cultural references as a uniquely Australian sensibility, with numerous films, television shows, and art =works embracing its identity as trash, refuse, and kitsch. This reached its apogee during the 1980s and 1990s, when a spate of films were funded by the Australian Film Commission in order to produce and market a unique sense of national identity.

In the aim of creating a national cinema, numerous filmmakers emphasized the absence of "high art" (central to European definitions of culture), while celebrating trash and kitsch aesthetics as one of the defining and marketable features of Australian national identity. Films like *Sweetie* (1989), *Strictly Ballroom* (1992), *Muriel's Wedding* (1994), *The Castle* (1997), *Love Serenade* (1996), *Priscilla, Queen of the Desert* (1994), and television series like *Kath and Kim* (2002) and *Bogan Pride* (2008) embrace suburban kitsch as the defining aesthetic of the national sensibility. Suburban "trash" has become a well-recognized and celebrated national aesthetic, embraced with the sense of laconic humor that is said to define the nation at large.

Clashes With Indigenous Culture

This celebration of the trashy, kitsch, and camp aesthetic is also a defining feature of the films and photographs produced by one of Australia's best-known indigenous artists, Tracey Moffatt, whose works self-consciously represent Australian culture and identity with a flagrant flair for the "distasteful." Many of Moffatt's works, including her films *Bedevil* (1993) and *Night Cries: A Rural Tragedy* (1989) represent primal scenes from Australian settler culture in which Indigenous and nonindigenous relations are evoked in a hyperbolic campy excess of color saturation and over-the-top caricature as if to highlight the fabrications of Australian history and the impermanence of settler attitudes and stereotypes of indigenous Australia—their staginess and lack of fixity.

Moffatt's celebration of trash as a defining Australian aesthetic also speaks back to the canon of "good taste" propagated by the European establishment, which allowed British colonists to celebrate European definitions of culture while denying that Australia had any culture at all. These definitions of Australia as a land devoid of culture perpetuated Australia's founding violent myth of *terra nullius*—the notion that Australia was cultureless by British standards—the "last" and the "emptiest" of lands, "without songs, architecture, history," as A. D. Hope puts it in "Australia." These reified views of European culture thus perpetuated Australia's founding erasure of indigenous cultural production that existed for 40,000–60,000 years in various forms, including storytelling, painting, and dance.

A Dumping Ground?

Despite recent attempts to embrace and revalorize Australia's unique cultural sensibility outside European standards and definitions of culture, traditional representations of Australia as an empty dumping ground continue to saturate the global imagination, with old imperial agendas being played out with new ecological imperatives. Since 1990, debate has ensued over Australia's suitability as a potential host for the world's increasing nuclear waste. These proposals have garnered support as nuclear power plants are increasingly touted as the solution to global warming and the reduction of coal emissions.

In 2005, Australia's prime minister, John Howard, introduced the Commonwealth Radioactive Waste Management Act to challenge indigenous property rights in the Northern Territory in a bid to secure parcels of land for Australia's nuclear waste dump. Amid much controversy, debate ensued about potential sites within Australia to locate a potential dumping ground, with Muckaty Station primed as the most suitable site for the dumping of nuclear waste in 2010. This proposal to dump nuclear waste at Muckaty Station—covering virtually 120,000 hectares in the geographical center of the Northern Territory—is part of a long history of imagining Australia as a dumping ground for the United Kingdom's waste. The United Kingdom's imperial view of Australia as a dumping ground was also evident in Australia's recent history as host to UK atomic bomb tests staged 1955–63. Seven major nuclear tests and hundreds of minor trials were staged at Maralinga in the Woomera Prohibited Area in South Australia. These tests had long-term health effects on some of the Maralinga Tjarutja people, to whom the government paid compensation money in 1994. The possibility of a new nuclear waste dumping ground has exacerbated ongoing issues relating to indigenous sovereignty in the Northern Territory, in addition to the future safety of the nation.

Waste Management

The primary method of waste management in Australia is landfill. In 2002–03, more than half of all the 32.4 million tons of solid waste generated in the nation (including 70 percent of municipal waste, 56 percent of commercial and industrial waste, and 43 percent of construction and demolition waste) went into landfills. The Department of the Environment and Heritage reported high rates of recycling, with 99 percent of Australian households reporting at least some sort of recycling or reuse activity in 2006. The overall recycling rate for all materials was estimated at 46 percent for 2002–03.

Conclusion

Waste has played a pivotal role in Australia's identity as a colonial outpost of the United Kingdom from 1788 into the 21st century. While Australia has been a dumping ground for numerous aspects of the United Kingdom's perceived rubbish—including some of its population and its material by-products—the usefulness of this waste, in and for the nation, is part of ongoing cultural, ecological, and economic debate. Therefore, given that trash rests at the very helm of Australian modernity—in the very foundations of the colonial culture—waste signifies the presence of colonial culture in addition to its demise. Waste continues to suture Australia's imagination of itself in ways that both corrode and confirm the colonial center.

Marita Bullock
Macquarie University

See Also: Culture, Values, and Garbage; Hazardous Materials Transportation Act; NIMBY (Not in My Backyard); Politics of Waste; Toxic Wastes.

Further Readings

Australian Bureau of Statistics. "4613.0—Australia's Environment: Issues and Trends," 2007. http://www.abs.gov.au/Ausstats/abs@.nsf/0/82D6EAD861A050C9CA2573C600103EA1?opendocument (Accessed December 2010).

Boyd, A. *The Australian Ugliness*. Melbourne, Australia: Penguin, 1960.

Hawkins, G. and S. Muecke, eds. *Culture and Waste: The Creation and Destruction of Value*. Lanham, MD: Rowman and Littlefield, 2003.

Hope, A. D. "Australia." *Meanjin Papers*, v.2/1 (1943).

O'Regan, Tom. *Australian National Cinema*. London: Routledge, 1996.

Phillips, A. A. "The Cultural Cringe." *Meanjin*, v.43 (1950).

"Submission to the Productivity Commission Inquiry Into Waste Generation and Resource Efficiency, Report no. 38." Canberra: Department of the Environment and Heritage, February 2006.

Summerhayes, Catherine. *The Moving Images of Tracey Moffatt*. Milan, Italy: Charta, 2007.

Automobiles

The automobile, pioneered by German engineer Karl Benz in the 1880s, became one of the great engines of consumption during the 20th century. Available to the masses after Henry Ford developed a system of mass production that produced millions of Model T cars, the automobile made possible suburban patterns of settlement, long-distance journeys, the development of the modern shopping mall, and the consumption of millions of gallons of petroleum every year. The automobile is one of the defining elements of modern consumption, and both its use and disposal have substantial effects on the environment.

Brief History

The automobile is a symbol of prosperity, freedom, and consumption. Within 20 years of Benz's first motorized vehicles, enthusiasts had founded the American Automobile Association (AAA) in the United States (1902) and the Automobile Association in the United Kingdom (1905) to promote driving. Enthusiasts developed high-end racing automobiles. Sport racing flourished in the 20th century and continues with the popularity of National Association for Stock Car Auto Racing (NASCAR), Indy Car, and Formula One racing circuits worldwide.

The automobile became popular because of its utility to people seeking affordable, convenient transportation. The Ford Motor Company's affordable Model T cars (introduced in 1908) allowed unprecedented mobility to the masses, a sentiment later captured by its competitor General Motors in its "see the USA in your Chevrolet" advertising campaign. If Ford mastered mass production, General Motors mastered market segmentation, offering a variety of models and colors as early as the 1920s under brands such as Cadillac, Buick, Oldsmobile, and Chevrolet. This strategy proved popular; after World War II, General Motors grew into the largest business on Earth. By the end of the 1950s, three American companies (General Motors, Ford, and Chrysler) dominated the U.S. automobile market, producing over 5 million cars each year between 1949 and 1965. All three were headquartered in Detroit, nicknamed the "Motor City."

Automobile purchases changed with 1950s patterns of conspicuous consumption. General Motors continued to alter the color, shape, and styling of its models with garish chrome bumpers and tail fins, encouraging customers to purchase new vehicles every year on grounds of stylistic obsolescence. Its competitors followed suit. Manufacturers introduced new amenities to make driving more comfortable, including air conditioners, radios, and 8-track tape players. Attempts to introduce vinyl record players were brief and unsuccessful, but the amenities proved popular for people spending more time in their cars. Models on the road by 1960 included two-person coupes, larger sedans, station wagons often sporting three rows of seats, and trucks. Public transit options shifted from mass rail to buses in many areas.

Social Effects

The mass consumption of the automobile has had several social effects. One has been a major role in the spatial redistribution of population. Within the United States, suburbanization had already begun (assisted by streetcars) in the late 19th century, but the automobile allowed suburbanites to conveniently live farther away from central cities. The U.S. landscape shifted rapidly between 1945 and 1970, reconfiguring in ways that made automobiles necessary for full enjoyment of the built environment. Population shifted from central cities to sprawling low-density suburbs, most lacking public transportation. Shopping malls with massive parking lots grew throughout suburbia, offering shopping and entertainment amenities that had previously only existed in downtown corridors of large cities. Drivers used automobiles to commute from suburban homes to urban jobs, as well as to shop and take children to school. The United States invested billions of dollars in an interstate highway system. This network of asphalt and concrete ostensibly provided security during the cold war, allowing the population to escape metropolitan areas in case of nuclear attack. The endless grey ribbon produced new road-based industries, including motels and fast food restaurants. Decentralized metropolises sprawled across the landscape of the west and south, forgoing densely spaced skyscrapers for endless swaths of parking lots and low, wide buildings. The billboard industry, on the rise since the introduction of the automobile, grew with the mass use of highways in the 1950s.

Emissions

Although the automobile has had a major effect on the spatial patterns of modern life, most attention to the environmental consequences of the automobile concerns the fuel used in the internal combustion engine. When the automobile first competed with the horse as a popular method of transport, observers found the lack of solid waste from "the horseless carriage" a marked improvement over the piles of manure left by animals. Tailpipe emissions seemingly disappeared into the air, leaving streets cleaner and safer.

Tailpipe emissions became atmospheric problems as the automobile grew in popularity. Although several inventors experimented with steam-powered engines, only automobiles powered by petroleum were mass-produced in the 20th century. Emissions leaving the tailpipe were not simply water, but a combination of carbon dioxide, carbon monoxide, nitrogen oxides, and particulate matter. Concern over smog produced by volatile organic compounds and particulate matter led to criticism of the automobile as a polluter of local atmospheres by 1970. Carbon dioxide emissions from the automobile were widely linked to global climate change by 1990. Despite international concern about atmospheric pollution, almost 1 billion oil-consuming automobiles were in use worldwide in 2010.

Some pollution concerns have been mitigated. Lead was added to gasoline for much of the 20th century to improve engine performance. Internal combustion vaporized the heavy metal and spread it over heavily trafficked areas, resulting in unsafe ambient lead levels as well as lead in the water table and soil. The United States began limiting lead as a fuel additive in 1973 after concerns about unsafe levels of lead found in children, eliminating leaded fuel use within 20 years. Lead reduction has since occurred in most of the industrialized world.

Oil Consumption

Although the automobile is not the only source of petroleum consumption in modern society, it is the primary one according to the U.S. Department of Energy, accounting for more that two-thirds of all petroleum consumption in the United States in 2008. The United States has increased imports of petroleum since World War II, complicating geopolitics because oil-rich nations often had disputes with the United States, leading to both diplomatic and armed conflicts. For consumers, the most visible geopolitical conflict was the Organization of the Petroleum Exporting Countries' (OPEC) oil embargo of 1973. OPEC members (including Iran, Iraq, Kuwait, and Saudi Arabia) opposed to the United States' alliance with Israel withheld petroleum from the United States. Since domestic oil production was insufficient for the needs of consumers, shortages led to rationing and long lines at gas stations. Rising energy costs contributed to economic decline in the mid-1970s.

In the wake of the 1973 embargo, several Japanese manufacturers (including Honda, Toyota, Mazda, and Nissan/Datsun) made inroads into the U.S. market with small models often getting more than twice the fuel efficiency as their larger U.S. counterparts. As the relative value of fuel lessened during the 1980s, automobiles gradually increased in size and amenities. At the dawn of the first Iraq

War in 1991, for example, a new Honda Civic was far larger than a 1977 Honda Civic.

Although fuel rationing had not been repeated in the United States as of 2011, concerns about the security of fuel supplies as well as criticism of the widespread environmental dangers of petroleum use have led to calls for alternative fuels for transportation. Automobile manufacturers have developed electric automobiles, and grassroots consumers have increased the use of biodiesel fuels, but the vast majority of automobiles on the road continue to be powered by petroleum.

Disposal

The environmental effects of the automobile also include the consequences of its disposal. Although millions of automobiles go out of service every year, most are not simply thrown into landfills. Mechanics scavenge automobile graveyards (specialized areas devoted to storing junked automobiles) for working parts to keep vehicles running. Reliable mass production meant that a transmission from a junked car might be used in a working car of the same make and model. Parts that cannot simply be transplanted into a working automobile still have value. The automobile, composed mostly of steel, became a major source of ferrous scrap in the 1920s as mass consumption of motor vehicles expanded. Scrap firms extracted the metal through manual labor, using torches and shears to harvest scrap in a slow, deliberate process usually requiring at least half a dozen laborers working on each automobile body. Scrap dealers classified the salvaged material as heavy No. 1 grade scrap steel, valuable in the making of structural steel skeletons for buildings, ships, and new automobiles. Automobile graveyards began dotting the U.S. landscape in the 1920s.

After the development of shredding automobiles and household appliances, scrap iron industry research estimated that by 1980, between 30 and 40 percent of U.S. ferrous scrap came from junked automobiles. Unfortunately, non-ferrous residue known as "fluff" has to be disposed of by incineration or dumping. The increase in automotive amenities, including airbags and their carcinogenic ingredient azide, has complicated the disassembly and shredding process and has contributed to the millions of tons of shredded material annually.

Although the manual process for harvesting ferrous scrap from an automobile was slow, it was sufficient for the demands of the 1920s. Military demand shaped World War II–era scrap sales, with much of the heavy scrap coming from salvaged ships and buildings. Automobiles were a useful source of scrap during the war, but not a main source.

By the end of the 1950s, when Americans disposed of over 9 million automobiles annually, more people were concerned with the problem of blight marring roadsides and the new suburban developments where millions escaped the dirt and noise of central cities for the promise of pristine, quiet communities. Like suburban subdivisions, scrapyards often were situated on the edges of urban settlements within plain view of major highways. More than 25,000 automobile graveyards lay scattered across the United States in 1951; as Americans migrated to suburbs, piles of automobiles once isolated from view became visible amid the malls and residential subdivisions springing up around them. State governments began attempts to control or remove junkyard blight in the mid-1950s, and federal attention increased once new First Lady Lady Bird Johnson began advocating the beautification of public space in 1964. One year later, her husband, President Lyndon B. Johnson, signed the Highway Beautification Act into law, requiring junkyards to remove or shield blight visible from the highways. Although the act did not eliminate automobile graveyards from the landscape, it was part of a widespread effort to limit the environmental effects of the automobile's use in the late 20th century.

Scrap firms adapted to the glut of junked automobiles by developing the automobile shredder, a device that automated the process of cutting up and separating an automobile's steel from its other materials, allowing more rapid and safer disassembly. Instead of having a team of 10–12 men work on an automobile with hammers, torches, and shears, one or two men fed the automobile body onto a belt that moved it into the shredder. Once inside, mechanized hammers and shears separated and shredded the entire vehicle. Once shredded, large magnets separated the ferrous material from the rest, allowing workers on the other end to quickly harvest the ferrous material. Instead of a dozen men, perhaps three or four were needed to supervise an automated disassembly process.

An analogy could be made between the disassembly of automobiles and the mechanization of the meat industry in the 19th century. The disassembly line transformed meat harvesting from a task performed by skilled butchers who honed their craft with long apprenticeships to a mechanized process involving little labor, processing a vast supply of material into sellable commodities. The analogy is not exact; the massive stockyards in Chicago fed giant killing floors in a handful of meatpacking plants, whereas automobile shredders spread to hundreds of scrapyards across the United States.

Non-Ferrous Scrap Disposal

The shredder was effective at quickly separating ferrous scrap from the other materials in an automobile, allowing recycling of steel; in 1980, scrap iron industry leaders estimated that between 30 and 40 percent of the United States' ferrous scrap came from junked automobiles. Unfortunately, that separation had consequences in the form of automobile shredder residue (ASR), also known as "fluff." All matter contained in an automobile was grist to be ground up during the shredding process. Once separated from the ferrous metal, the wastes might be incinerated, hauled to landfills, or left on the ground. However it was managed, ASR joined ferrous scrap as common products of the shredder. In 1960, the automobile typically contained steel, copper, lead, glass, several acids, rubber, several fabrics, asbestos (in brake pads), and a growing variety of plastics. As designs became more complex over the next half century, manufacturers added new materials to the automobile, including computer circuitry and airbags.

By the end of the 20th century, the amenities within automobiles increased, with cup holders being a particularly noticeable addition. Many models of sport utility vehicles (SUVs) feature DVD players and entertainment systems rivaling those in many homes. In the early 21st century, the automobile is as complex as ever, including computer circuitry and a variety of potentially hazardous materials. The airbag both increases safety and complicates disassembly. Since it reduces fatalities in accidents, it is a standard feature in automobiles. One ingredient in airbags is the carcinogen sodium azide. During disassembly, it adds to the dangerous materials making up ASR. As the hazards in ASR have increased, so has the volume of ASR produced. The Environmental Protection Agency estimated that shredders produced about 3 million tons of finely ground ASR in the United States annually by 1990, creating concerns over how to properly landfill or otherwise manage the waste.

Despite calls for an industrial ecology approach to automobile assembly and disassembly that would eliminate hazardous wastes, more attention is paid to performance of the machine, safety of the driver, and fuel efficiency of the engine than the life cycle of the product. The automobile has become more complex over time because of innovations that increase the enjoyment and safe use of the vehicle, but they also complicate disassembly. Over time, shredding and burning of junked automobiles has had environmental consequences, including the release of hazardous, corrosive, and carcinogenic substances into the ground, air, and water.

More than a century after its invention and half a century after the U.S. government first tried to limit the consequences of automobile use, the petroleum-consuming automobile remains a vital part of modern life, continuing to shape patterns of settlement, energy use, shopping, employment, waste disposal, and climate change. The consequences of its use will endure and grow for the foreseeable future.

Carl A. Zimring
Roosevelt University

See Also: Car Washing; Carbon Dioxide; Consumerism; Fuel; Iron; Junkyard.

Further Readings

Graedel, Thomas E. and Braden R. Allenby. *Industrial Ecology and the Automobile.* Upper Saddle River, NJ: Prentice Hall, 1998.

Mauch, Christof and Thomas Zeller, eds. *The World Beyond the Windshield: Roads and Landscapes in the United States and Europe.* Athens: University of Ohio Press, 2008.

McCarthy, Tom. *Auto-Mania! Cars, Consumers, and the Environment.* New Haven, CT: Yale University Press, 2007.

Avoided Cost

Avoided cost is a term used in waste management to represent monetary savings through the diversion of waste from disposal to a form of reprocessing, such as recycling or composting. Frequently used to justify either the initiation of new recycling or composting programs or the continuation of existing programs, avoided cost is an attempt to reflect how waste diversion may be economically beneficial and to assess alternatives to the disposal of material.

Motivations for Use

Municipal solid waste collection programs have existed in most communities since the early 20th century. Established as a response to public health, aesthetic, and commercial concerns, the collection of waste—or the provision of access to waste disposal locations—is considered a standard and necessary public service even in difficult budgetary times. Waste disposal, due to its necessity and long-term existence as a public service, is often financed by a complex and convoluted structure of fees and taxes.

Recycling and composting programs, which are forms of waste diversion, are newer additions to the waste management practices of most communities and are often viewed as optional services. When initiated, it is common that services for these programs are charged directly to citizens, making program costs easily visible. Presenting the full cost of waste diversion programs in this way can give the appearance that waste diversion methods are more expensive than traditional disposal of waste through landfilling or incineration. Avoided cost is a tool developed to provide a more accurate description and accounting of the comparison of the cost of waste disposal with potential savings through diversion activities.

Calculating Avoided Cost

Avoided cost is the amount of money not spent on disposing of waste because material has been diverted from the disposal stream. When calculating avoided cost, the net cost of waste disposal is compared with the net cost of a form of diversion, such as recycling or composting.

To determine avoided cost, the cost per ton for waste disposal must first be calculated. Several costs are associated with disposal of waste. First, waste must be collected from households, businesses, or other points of creation. Collection requires labor, equipment such as trucks and containers, and associated administrative or overhead costs. The cost of hauling this material to a disposal facility, also termed the *hauling fee,* may be included in collection costs or broken out as a separate cost. Finally, waste must be "tipped" or dumped into a landfill or incinerator, the cost of which is termed the *tipping fee*. Tipping fees include the cost for the management of disposal facilities. Disposal of waste in a landfill or through incineration is becoming increasingly expensive because of limited space available for landfilling, public outcry over placement of new landfills, and more stringent technological requirements regarding the operation and location of new landfills and incinerators.

The cost per ton of waste disposal is determined by calculating the cost of disposal (adding together collection and hauling fees, tipping fees, and other administrative overhead for overseeing the process) and dividing by the total number of tons of waste disposed. The equation may be described as: cost per ton = cost of disposal divided by total number of tons.

Next, the cost per ton of recycling is calculated. Costs for recycling include collection, which is similar to that of waste, and the processing of materials. The cost of processing recyclables differs depending on what material is being recycled and the market for use of the recycled product. In addition to costs for collection and processing, the sale of some recyclable materials also provides revenue. The amount of revenue gained is dependent upon local, regional, national, and global commodities markets.

The cost per ton of recycling is calculated by adding the cost of processing and collecting recyclable materials and subtracting any revenue gained from the sale of commodities. This number is then dividing by the total number of tons recycled. The equation may be described as: net cost per ton of recycling = (cost of processing minus commodity sale revenue) divided by total number of tons recycled.

Using these two figures, avoided cost is calculated by subtracting the net cost per ton of recycling from the cost per ton of waste. The result, if a positive number, is avoided cost. For example, if the cost for waste disposal is calculated as $100 per ton and the cost for the collection and processing recyclable materials is $100 per ton, but $25 in revenue per ton is gained from the sale of the recycled commodities, the cost benefit analysis is described as: $100 (cost per ton of waste) minus $75 (cost per ton of recycling) = $25 avoided cost. Recycling therefore results in not spending an additional $25. This formula does not always result in the calculation of a positive number. If the associated costs of recycling are greater than that of waste disposal, there will not be avoided cost or monetary savings.

Difficulties in Calculating Avoided Cost

Although the equations for calculating avoided cost use simple mathematics, determining appropriate numbers for use in the equation is often difficult. Waste disposal costs are often difficult to determine because of the nature of funding a long-standing public service. The calculation of overhead and administrative costs can also vary depending on the formula used for this determination.

Sarah M. Surak
Virginia Tech

See Also: Economics of Waste Collection and Disposal, International; Economics of Waste Collection and Disposal, U.S.; Recycling; Zero Waste.

Further Readings

Ackerman, Frank. *Why Do We Recycle? Markets, Values, and Public Policy.* Washington, DC: Island Press, 1997.

Williams, Paul T. *Waste Treatment and Disposal.* Chichester, UK: John Wiley & Sons, 2005.

Baby Products

Starting in the 1800s, the clothing, diapering, and feeding of infants—activities that once rested solely in the hands of a baby's primary caretakers—has transformed into a medicalized and commercialized multibillion-dollar industry that now lies in the hands of business. In the early 21st century, global expenditures on children's toys and video games alone are estimated at $86 billion annually. With these newly invented baby products have come a historically unprecedented amount of baby-related trash, including mass manufactured artificial feeding implements (such as bottles and nipples), infant clothing, and disposable diapers.

Brief History of Baby Feeding

In a little over 100 years, infant feeding has transformed from a practice that was primarily dependent on an infant's mother's ability to successfully breast-feed her child to a powerful industry that brings in $3 billion annually and that makes feeding an infant possible without the presence of a mother. Archaeological and historical documentation make it clear that feeding infants, especially those who do not have access to a lactating individual, has long been a concern of human civilization. Up until the early 1900s, a woman's inability to breast-feed her child during the first months of an infant's life usually determined the fate of her infant. The infant was thus entirely reliant upon her lactating mother, or other lactating women, for nourishment and survival.

That did not stop mothers and their families from trying to come up with other viable methods of nourishing their babies. Ancient attempts at artificial feeding (such as horns and ceramic vessels) have been found in Africa dating to the Neolithic period and from Bronze Age Europe. Wet nursing (the act of having a child nurse from a woman who is not its mother) or having an infant suckle an animal's teats were two alternatives for mothers who either chose not to or could not breast-feed their infants. Wet nursing is ancient in practice and is discussed in Egyptian hieroglyphs, medical texts written in India between 400 B.C.E. and the 7th century C.E., texts found in ancient Greece, and in the Old Testament. When a mother either could not breast-feed or secure a wet nurse, pap and panada were used as breast milk substitutes, which involved cooking flour, bread, or cereal in water, butter, or milk.

Condensed milk became available on the market in 1853, finally offering a reliable—though not fail-safe—option for mothers who could not nurse or afford a wet nurse. Gail Borden patented his milk evaporation technique in 1853 and then subsequently marketed Eagle Brand Condensed Milk starting in 1856. Because his formula contained sugar, pediatricians tended to shy away from recommending it to their patients. Unsweetened evaporated milk was introduced on the market in the late 1880s and 1890s by companies such as the Helvetia Milk Condensing Company (also known as the Pet Milk Company), Liebig, and the Pacific Condensed Milk Company (later known as the Carnation Company). With a safer condensed milk becoming available on a mass scale and advocated by pediatricians, the vehicle through which artificial milk could be fed to an infant also become available en masse. Still, pediatricians were hesitant to recommend formula over breast-feeding, as the medical benefits of the latter practice greatly outweighed the former.

It was at this time when baby feeding bottles were introduced on the market. Rubber bottle nipples were first distributed in 1830, yet consumers did not meet their introduction with delight. Lead, zinc, and arsenic were added to the rubber nipple, and bottles were labeled "instruments of death" in the mid-to-late 1800s. Sterilization, the widespread acceptance and application of germ theory, and pasteurization processes made drinking milk from a bottle much safer. Advertisements and articles promoting bottle-feeding appeared in women's magazines such as *Good Housekeeping* and *Ladies' Home Journal* as early as the 1880s. The surge in bottle and artificial feeding advertising and media helped usher in a new age of infant feeding. Rima D. Apple states that in the United States, breast-feeding dramatically decreased between 1917 and 1948, with more than one-third of mothers opting not to nurse their babies upon being discharged from their labor and delivery hospital stay. As Apple explains, this figure stands in stark contrast to the percentage of U.S. babies nursed upon their birth between 1917 and 1919, which ranged from 82 to 92 percent in a number of rural and metropolitan U.S. cities.

As of 2010, the Infant Formula Act of 1980 ensures that all formulas meet governmental standards for infant nutrition. Nonetheless, bottle-feeding remains such a contentious topic that the World Health Organization (WHO)/UNICEF Code of Marketing for Breastmilk Substitutes has requested that marketers not send out any products associated with bottle-feeding (such as samples of formula, nipples, or bottles) to pregnant women. The La Leche League also formed in 1956 to counter the recent tendency to opt for bottle-feeding and to highlight the benefits of breast-feeding for both mothers and their babies. The American Academy of Pediatrics and the WHO have taken a stand against bottle-feeding by recommending that babies be nursed for at least the entire first year of a their life, with 2 years of age being the optimal time for the mother's cessation of breast-feeding.

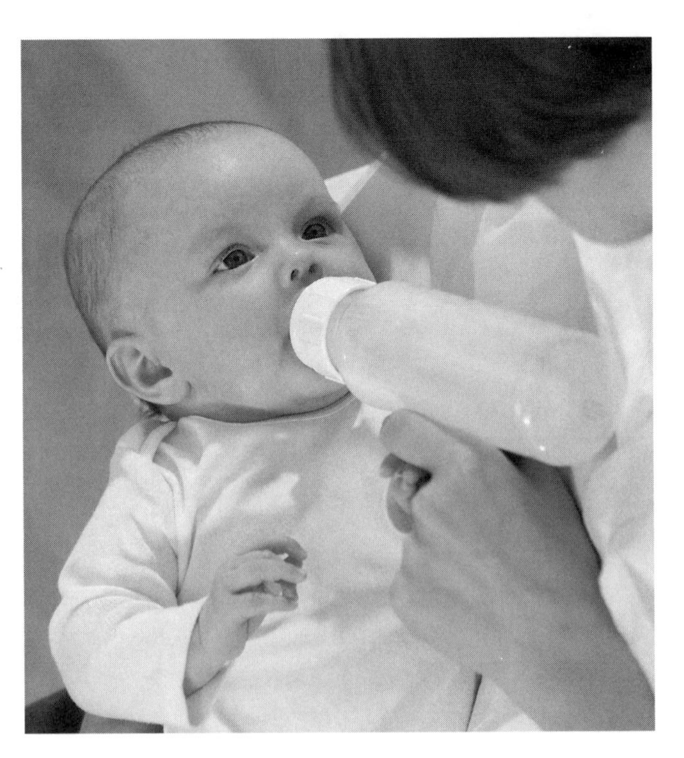

Formula must be made with food ingredients that are recognized as safe or approved as food additives for use in infant formula. Advances in sterilization and nutrition, along with marketing campaigns, created a surge in the use of commercial formula.

Solid Foods

The introduction and production of baby food has similarly been fraught with contention. Prior to the 1920s, it was common for families to wait to add solids until the first or second year of a child's life, with some cultures waiting until the age of 7 or 8 to introduce certain foods. Gerber, a well-known baby food company, unveiled its first line of canned baby food in 1927. Other companies, including Beech-Nut and Libby's, followed suit by unveiling their own lines of baby solids in the mid-to-late 1930s. Advances in canning technologies, sanitation standards, and new developments in nutritional sciences (including the concept of "vitamins") helped push back the age at which babies were fed their first solids and assisted with the popularity and success of these companies' foods.

Disposable Diapers

Perhaps the most controversial product associated with babies is the disposable diaper, which constitutes approximately 1–3 percent of all solid waste in the United States. Before the 1800s, diapers consisted of absorbent grass, moss, leaves, linen, or cotton, with animal skins, wool, flannel, or linen acting as a protective outer layer. Reusable diapers made out of knitted wool, linen, or cotton came into fashion with the advent of industrialized textile production in the 1800s. The safety pin, which was invented by Walter Hunt in 1849, allowed caretakers to secure cloth around a baby, and by the 1890s, rubber pants were invented to cover the cloth and linen diaper as well as to protect the outer layers of a baby's clothes. The soiled inner layer of the diaper, typically made of linen or cloth in the mid-to-late 1800s, had to be washed or boiled by hand in hot water and soap. This was not an institutionalized practice until the 1930s, when both rural and urban families finally understood the medicinal and hygienic advantages of boiling diapers and sterilizing them with a hot iron. The 1930s and 1940s witnessed additional changes in diapering. As of 2010, nearly all diapers in the Western world consisted of a cloth or latex outer layer, with disposable cotton or gauze acting as the absorbent inner layer. This provided a simple, hygienic solution in a time when soap was scarce and rationed during World War II.

The first completely disposable diaper was introduced by Procter & Gamble in 1959, but disposable diapers did not become widely used until the 1970s. This new invention had and continues to have its critics, however. The average infant goes through 10,000 diapers and produces 1 ton of trash in its lifetime of disposable diaper use. Environmentalists have protested and raised red flags about the manufacturing processes and impact of disposable diapers. To produce disposable diapers, oil must be harvested to make the plastic in the diaper and trees must be cut down. Dioxin, which is produced when diaper manufacturers bleach wood white, is admitted into the air and enters into the soil during the manufacturing process; some claim that dioxin causes, among other things, cancer, male and female infertility, and short-term memory loss. Air pollution, soil pollution, deforestation, relocation of indigenous populations, and the exposure of toxic chemicals to oil workers (which can result in asthma, infertility, birth defects in unborn children, and cancer) are all directly caused by the manufacturing of disposable diapers.

In addition, diapers can present health hazards for adults living near and working at landfills. Although disposing of human feces in household waste is illegal, most parents do not place infant feces in diapers into adult toilets. Once household waste is transported to a landfill, pathogens and microbes from soiled diapers become a cause for concern. According to pediatric doctors Marianne B. Sutton and Michael Weitzman, these microbes can potentially leach into the groundwater below as well as pose a health risk to landfill employees. Consequently, environmentalists have now begun to advocate the use of reusable cloth diapers, though there remains some debate as to the environmental impact of these diapers as well.

Together, these new products—bottles, diapers, and baby-food containers—have made parents' lives both easier and more complex. Parenting is easier in that infant food, diapers, and formula can be instantly purchased without going to great lengths to procure, make, and guarantee the safety of such items. However, it is more complex because with these recent innovations, society continues to face and produce insurmountable amounts of one-time-use disposable items (such as glass jars, plastic containers, and diapers) associated with babies.

Stacey Lynn Camp
University of Idaho

See Also: Children; Disposable Diapers; Industrial Revolution; Landfills, Modern; Packaging and Product Containers; Race and Garbage; Shopping.

Further Readings

Apple, Rima D. *"How Shall I Feed My Baby?" Infant Feeding in the United States, 1870–1940.* Ph.D. Diss. University of Wisconsin, Madison, 1981.

Bel Geddes, Joan. *Small World: A History of Baby Care From the Stone Age to the Spock Age.* New York: Macmillan, 1964.

Bentley, Amy. "Feeding Baby, Teaching Mother." In *From Betty Crocker to Feminist Food Studies,* Arlene Voski Avakian and Barbara Haber, eds. Amherst: University of Massachusetts Press, 2005.

Cone, Thomas E., Jr. *200 Years of Feeding Infants in America.* Columbus, OH: Ross Laboratories, 1976.

Fildes, Valerie A. *Breasts, Bottles, and Babies: A History of Infant Feeding.* Edinburgh, UK: Edinburgh University Press, 1986.

Prince, J. "Infant Feeding Through the Ages." *Midwives Chronicle & Nursing Notes,* v.89/1067 (1976).

Sutton, Marianne B., et al. "Baby Bottoms and Environmental Conundrums: Disposable Diapers and the Pediatrician." *Pediatrics,* v.88/2 (1991).

Van Esterik, Penny. *Beyond the Breast-Bottle Controversy.* New Brunswick, NJ: Rutgers University Press, 1989.

Barges

Garbage barges have had a place in U.S. waste disposal since the late 19th century. The garbage barge is a social and historical product whose notoriety loomed especially large after 1980. Currently, when U.S. methods of production and waste are under scrutiny, the garbage barge may offer an important lesson through which can more effectively wrestle with the implications of throwaway society. Waterways have served urban waste management as sinks as long as there have been cities. During the late 19th century, cities such as New York City employed long flat barges to transport the city's garbage away from land and into the Atlantic Ocean for dumping. Workers on board, known as "scow trimmers," rummaged through the dumped waste for rags, scrap, and anything with a resale value. Even after opposition to ocean dumping in the 20th century led to an emphasis on landfilling as the ultimate destination for wastes, barges continued to play an important role in transporting waste. This role was largely obscured from public attention until a pair of embarrassing incidents in the 1980s brought them notoriety.

The *Mobro 4000* and *Khian Sea*

Garbage barges have been involved in remarkable incidents involving attempted waste disposal. In March 1987, the *Mobro 4000* left Islip, New York, and carried 3,000 tons of trash to a landfill only to face bans by five states and three countries before it was forced to sail back to Long Island three months later. Its load was burned in an incinerator and then buried in the same place where it would have gone months earlier. More spectacular than the *Mobro 4000* is the *Khian Sea* episode in 1986. Whereas the *Mobro 4000* carried trash, the *Khian Sea* carried nearly 15 tons of Philadelphia incinerator ash in search of a landfill—a search that became a 16-year journey during which it faced repudiation by at least 11 different countries and five states (Florida, Georgia, Ohio, South Carolina, and the Cherokee Nation of Oklahoma) and committed the illegal acts of dumping thousands of tons of ash into the ocean and subsequently onto a Haitian beach. Activism brought to light the predatory nature of *Khian Sea*'s activities and forced both the Haitian ash cleanup and the return of approximately 2,000 tons of its ash to a landfill in south-central Pennsylvania in 2002.

As illuminating and disconcerting as the *Mobro 4000* and *Khian Sea* incidents may be, they were not isolated episodes. Waste could be legally exported by barge from the United States to many places on the planet if the carrier registered with the Environmental Protection Agency's (EPA's) prior notification program. Garbage barges during the 1980s registered with the program. Debate over the cause of the burgeoning waste trade and garbage barge boom ensued; notwithstanding academic disagreement over whether a "garbage crisis" can be considered the culprit, no one denies that the saga of the post-1980 garbage barge was not created out of thin air.

Predating the unpleasant garbage barge venture was the 1973 report "Cities and the Nation's Disposal Crisis," issued by the National League of Cities and U.S. Conference of Mayors, asserting that the culprits were the skyrocketing waste volume and sharp decline of available disposal sites. David Pellow, suggesting possible causality, asserts that during the 1970s and 1980s aggressive environmental justice and antitoxic movements challenged dirty and unsightly landfill and incinerator systems; their pressure helped prompt a series of stringent regulatory requirements derived from the framework of the 1975 Federal Resource Conservation and Recovery Act. Throughout the 1980s, landfills closed, landfill capacity dwindled, and waste disposal costs skyrocketed. Numbers representing the cost of disposal varied depending on location and waste type, but in a 1987 issue of *The Nation*, Andrew Porterfield and David Wier offer one instance of the trend: "In 1976, disposal cost $10 a ton; today the figure is between $60 and $140 a ton, in some cases even higher."

Waste haulers, who faced a creeping devaluation of their investments, employed the garbage barge in desperate searches for cheaper dump sites. In spite of objectionable episodes like the *Mobro 4000* and *Khian Sea*, the use of waste barges continued as an affordable way of transporting unwanted waste great distances. In the early 21st century, barges drew new notoriety due to changing composition of the waste stream. Barges carried proliferating forms of electronic waste—such as obsolete televisions, air conditioners, VCRs, cell phones, personal computers—from the United States, the world's leading user of electronic equipment, to a slew of developing countries, particularly in Asia and India, exacerbating environmental inequalities as impoverished workers were exposed to the hazards created by consumers thousands of miles away.

E-Waste Collection and Transportation

Superficially, the garbage barges that drew public scorn in the 1980s seem different from the wave of e-waste barges that draw attention today. The former barges came into use in part because of dwindling domestic landfill options, whereas the latter barges arose in part because of many disingenuous domestic recycling schemes. As much as 80 percent of the e-waste collected for recycling is not recycled domestically but rather exchanged with electronic recycling companies that sell it in third world countries. However, these older and newer barges have substantial commonality. Both waves of barges emerged because of the necessary and sufficient conditions like productive consumption and state powers of mediation. The garbage crisis of 1970–80 is consequential upon what sociologist Giovanni Arrighi calls the 1945–67 surge of productive consumption, the most monumental recurrent cycle of material expansion and commodity production and consumption in the history of the capitalist world economy.

The proliferation of obsolete e-waste is consequential upon a recent, recurrent global phase of financial and trade deregulation and what sociologist Manuel Castells calls the most dramatic growth in telecommunication and computer networks that, since 1990, has propelled the formation and swift diffusion of Internet-mediated communication networks for the whole spectrum of human activity. Furthermore, the power of law has yet to be brought to bear on the garbage barge industry. The United States refused to sign the 1989 Basel Convention, which was designed to reduce the transfer of hazardous waste to "less developed" countries and was ratified by 172 countries. The United States also refused to sign the Basel Convention's 1995 amendment, which strictly prohibits the exportation of hazardous waste from "developed countries" to "developing countries"—an amendment implemented by 32 of the 39 developing countries to which it applies.

U.S. Municipal Waste

The question now shifts to whether the reputation of garbage barges is unwarranted. U.S. municipal waste may constitute as little as 2 percent of the country's total production of waste, while its e-waste production comprises perhaps 1.5 percent of its municipal waste. According to Adam I. Davis, a former compost and fuel programs director for Waste Management of North America: "For the roughly 210 million tons of MSW [municipal solid waste] the United States generates each year, we generate an additional 7.6 billion tons of industrial waste, and 1.5 billion tons of mining waste, 3.2 billion tons of oil and gas, electric utility and cement kiln wastes, and 0.5 billion tons of metal processing waste." Davis's statement helps put the picture of waste production into perspective, and the military is another massive source of waste. When scholars

Barges carry electronic waste (e-waste)—obsolete televisions, air conditioners, VCRs, cell phones, and personal computers—from the United States, the world's leading user of electronic equipment. Up to 80 percent of e-waste recycling is not conducted domestically but, rather, exchanged with electronic recycling companies that sell it in underdeveloped nations. The law has not yet caught up with the proliferation of e-waste; the United States has refused to sign laws designed to reduce the transfer of hazardous waste to less developed countries.

concentrate too much on the barge and its contents and too little on the fuller picture of our production of waste, they forgo an important opportunity to understand the origins of that waste.

Consider the incongruity between the amount of waste carried on any single barge and what it took to produce it. Conceivably, we could have encountered and experienced directly *Mobro 4000*'s trash, the *Khian Sea*'s ash, and any one of the more recent waves of barges brimming with e-waste. The *Khian Sea* sailed at a time when the U.S. EPA estimated the nation's solid waste stream was nearly 160 million tons, and the e-waste barges were sailing during the first decade of the 21st century when the Basel Action Network estimated that 100 million of the nation's personal computers became obsolete each year. These productions and re-productions of solid waste and e-waste are neither small nor appear to be dwindling—particularly because the systemic scheme of planned obsolescence continues to propel streams of consumption and waste while large portions of the population continue to enjoy extravagant lifestyles. For a simple glimpse at the incongruity between the enormous accumulations of waste and the very wasteful endeavors it took to produce them, consider two remarkably suggestive statements. William McDonough says: "What most people see in their garbage cans is just the tip of the material iceberg: the product itself contains, on average, just 5 percent of the raw materials involved in making it." David Pellow, who quotes a report generated by the United Nations University, says:

> By the end of 2002, 1 billion personal computers had been sold worldwide. Even though computers are becoming smaller and more powerful, their ecological impacts are increasing. The average 53-pound desktop computer with a monitor requires 10 times its weight in fossil fuels and chemicals to manufacture . . . which makes it much more materials intensive than the automobile or refrigerator, which only require one to two times their weight in fossil fuels to make. The natural resources extracted during the manufacture of computers are so intense that is likely to have significant impact on climate change and depletion of fossil fuel resources.

The garbage barge is a fluid social and historical construct and embodiment of the present system of production and consumption. The demand to cease the processes of production and consumption with all the ensuing forms of waste, if taken literally, would likely prompt a public health calamity. The barges offer a significant lens through which one can either see yet ignore some of the most important roots of our socioecological woes or see and address them in ecologically and socially sensitive ways.

Blaise Farina
Rensselaer Polytechnic Institute

See Also: Archaeology of Garbage; Hazardous Materials Transportation Act; Industrial Waste; Politics of Waste.

Further Readings

Boroughs, D. L. "Dirty Job, Sweet Profits." *US News & World Report* (November 21, 1988).

Clapp, J. "Seeping Through the Regulatory Cracks." *SAIS Review*, v.XXII/1 (Winter–Spring 2002).

Clark, R. "City Still Struggling to Dispose of the Trash Problem." *Philadelphia Inquirer* (May 25, 1986).

Epstein, S. S., et al. *Hazardous Waste in America*. San Francisco: Sierra Club Books, 1983.

Handley, F. J. "Hazardous Waste Exports: A Leak in the System of International Legal Controls." *Environmental Law Reporter*, v.XIX (April 1989).

Komp, C. "Barely Regulated, E-Waste Piles Up in U.S., Abroad." http://www.ban.org/ Library/ Features/060511_barely_regulated.html (Accessed June 2011).

Ladou, J. and S. Lovegrove. "Export of Electronics Equipment Waste." http://www.ban.org/Library/ Features/080101_export_of_electronics_waste.pdf (Accessed June 2011).

Pellow, David N. *Resisting Global Toxics*. Cambridge, MA: MIT Press, 2007.

Third World Network. "Toxic Waste Dumping in Third World Countries." In *Toxic Terror: Dumping of Hazardous Wastes in the Third World*. Penang, Malaysia: Third World Network, 1989.

Uva, M. D. and J. Bloom. "Exporting Pollution: The International Waste Trade." *Environment*, v.31/5 (June 1989).

Valette, J. and H. Spaulding, eds. *The International Trade in Wastes: A Greenpeace Inventory*. 5th ed. Washington, DC: Greenpeace USA, 1990.

Beef Shortage, 1973

The 1970s are sometimes called the "decade of shortages." Perhaps best known is the oil crisis, but oil was not the only commodity in short supply. In the summer of 1973, the United States was faced with a beef shortage. Grocers had difficulty keeping their shelves stocked. Cuts of beef, such as better-quality steaks, began to be used as prizes at grand openings of retail stores. Beef had joined blenders and toasters as highly desired incentives and give-aways. Furthermore, concerned consumers began to stockpile beef. This eventually led to a greater-than-average waste of beef.

The cause of the beef shortage is multifaceted. U.S. President Richard Nixon froze the price at which beef could be sold. The cost of beef production, however, was rapidly rising. This rapid increase in production costs was tied to both natural and artificial causes.

Beef packers were unable to pass along the increased costs to their customers, which resulted in a greatly reduced supply of beef to retailers, such as grocery stores. A number of factors influenced the beef shortage of 1973 and researchers have learned a great deal about the public's response to the shortage through the study of garbage generated during the time period.

Causes

One factor in the shortage was the increased demand for meat due to a rise in the average income in the United States. For example, in 1958, the average U.S. consumer used approximately 80.5 pounds of beef a year. By 1972, consumption had risen to an average of 115.9 pounds per person. While the demand for beef was growing, so was the cost of feeding cattle. In the early 1970s, the southern corn leaf blight damaged corn crops across the country. The poor corn harvests resulted in higher prices on corn, which is commonly used in cattle feed.

Another common crop used for feeding cattle is wheat. Wheat prices also surged in the early 1970s. In the summer of 1972, the United States sold approximately 440 million bushels of wheat to the Soviet Union. Following this sale, the domestic price of wheat began to rise, further increasing the cost of feeding cattle. The increased cost of production was then passed on to the consumer in the form of higher beef prices.

By March 1973, the consumer price index for all beef, as well as other meats also affected by higher production costs, had risen by 15 percent. On March 29, 1973, President Nixon set a price ceiling on beef. The price of corn and wheat, however, were not capped. The price of live animals was also not capped. The price ceiling on beef meant that these rising costs of production could no longer be passed on to the customer. Meat packers could no longer afford to produce beef. Cattle production began to slow, and packing plants began to shut down. U.S. Department of Agriculture reports from August 1973 show a 17 percent drop in beef production from the previous year.

Reactions

Much of what is known about how people reacted to the beef shortage has been learned by studying garbage from the time period. The Tucson Garbage Project, which began in 1973, collected data on beef consumption and waste over a period of 15 months. The data collection coincided with the beginning of the beef shortage in the spring of 1973. Data collection ended in the spring of 1974. Using the labels from supermarket packages of beef, researchers were able to record the date the meat was packaged, as well as how much the package weighed. This information was then compared with the weight of the beef found in the landfill, and a rate of waste was calculated. This rate of waste included cooked and uncooked beef, but not fat or bone.

Once the data collection had been completed, the numbers were analyzed. The analysis revealed that after the beef shortage ended, the rate of waste was about 3 percent. During the beef shortage, however, the rate of beef waste was estimated at 9 percent. This means that during the beef shortage, 9 percent of all beef purchased ended up in a landfill, but after the beef shortage ended, only 3 percent ended up in a landfill.

Researchers offered several explanations for this increase in waste during a time of short supply. First, they suggested that consumers were concerned that stores would be unable to keep beef in stock, and so they stocked up on as much beef as they could find. Second, researchers suggested that consumers purchased unfamiliar cuts of beef, which they then did not know how to cook. Finally, researchers suggested that consumers did not know how to properly store the large amounts of beef they had purchased. These factors may have led to the increased amount of beef waste during the shortage.

Melissa Fletcher Pirkey
University of Notre Dame

See Also: Food Consumption; Food Waste Behavior; Garbage Project; Garbology; History of Consumption and Waste, U.S., 1950–Present; Meat; Sugar Shortage, 1975.

Further Readings:

Eckstin, A. and D. Heien. "The 1973 Food Price Inflation." *American Journal of Agricultural Economics,* v.60 (1978).
"Food: Yes, We Have No Beefsteaks." *Time Magazine* (August 13, 1973).
Rathje W. and C. Murphy. *Rubbish: The Archaeology of Garbage.* Tucson: University of Arizona Press, 2001.

Beijing, China

The People's Republic of China (PRC) is a compelling contemporary site for considering both the relationship of consumption and waste production to economic development and the cultural specificities of refuse in social life. Wealth per capita is generally positively correlated with the volume of refuse generated by each person in a country. This idea has been clearly evidenced in the PRC, where modernizing urban centers, like Beijing, have increasingly produced more household waste and environmental pollutants per capita as individual real income has rapidly increased during the decades of development since the 1970s. With economic development, however, the challenge of increased trash generation has also increased, with China currently producing about one-third of the world's garbage. Like many other leading urban locales in the PRC, Beijing has struggled with managing the growing amounts of waste materials produced by its populace. Infrastructure and public sector resources, ranging from landfill capacities to the number of sanitation workers, are strained by the unwanted byproducts of economic growth and consumption. In Beijing, there exists a simultaneous dilemma of how to judge and intervene in the social norms of behavior that lie behind the culturally particular conditions of waste. From a nascent consumerism to state-sponsored mass education campaigns on recycling, the questions raised regarding refuse in the Chinese capital are grounded in historically embedded practices.

Economic Expansion

The economy of Beijing has exploded since the 1970s along with garbage produced in this ever-expanding urban zone. Since 1978, the PRC has undergone *gaige kaifang* (reform and opening up) measures that have led to historically unprecedented levels of growth in personal wealth (though with a widening income gap). This has not been a uniform expansion with equally distributed benefits across the nation; large urban centers, such as Beijing, have disproportionately reaped the gains of these reforms. Led by state efforts at privatizing select sectors of the economy, attracting foreign direct investment, and allowing for an expanded personal employment market, not only has per capita gross domestic product (GDP) increased steadily in the capital, but a domestic consumer market has also taken hold.

Consumption Expansion

In addition to the PRC growing abstractly in terms of wealth, a structural change with economic and cultural ramifications has also occurred since the 1970s: the burgeoning consumer market. Before 1978, Beijing was barely a site of modern forms of consumption and consumer waste generation. A strictly limited range of choices in consumer products was readily available, while imported foreign items were almost impossible for the general public to obtain. Moreover, under the Maoist era before 1978, modern consumer values were not characteristic of the aspirations and dispositions of the citizenry of Beijing. Social norms and state ideology focused more on fostering a production-based economy and advocated the significance of consuming basic necessities, not consumer products believed to be characteristic of much-maligned Western capitalist cultures.

Beijing in the 21st century is a lesson in transformations brought about through economic expansion and rapidly changing consumer preferences. Shopping malls abound in the urban landscape as more and more wealth is spent in the capital on retail products. Within the capital, Beijing's residents saw an 8 percent increase in their disposable incomes in 2009, comparable to the average rate of yearly per capita income growth during the past decade. This is a provocative statistic, as garbage discharge in the city has also been growing in parallel at about 8 percent a year. Consumer marketing is in ascendancy as Beijing's citizens encounter billboards, print advertisements, social media, and television commercials on a regular basis. The combination of increased disposable income, the availability of a wide scope of goods, and a nascent cultural orientation toward consumptive practices has changed the scope of waste in Beijing.

Personal automobile ownership serves as a quick indicator of the rise of consumerism and its attendant waste production in Beijing. For the burgeoning middle class in Beijing, the automobile serves as a marker of socioeconomic status, much as it did in the United States in the 1950s. As over 15,000 new cars are bought in the capital every week, the Beijing government estimates that over one million new automobiles will be registered per year during the early decades of the 21st century. This explosive growth will lead to not only increased totals in environmental waste products, such as emissions, but will also mean increased garbage associated with automobile ownership and upkeep in the near future. Urban air pollution threatens the nation's status. In 2008, the government attempted to control air pollution during the Summer Olympic Games by temporarily restricting automobile use.

Garbage

The combination of these factors is leading to increased garbage production not only in the faraway factory sites where consumer goods originate but also within Beijing itself—the end point of purchase. While simple items associated with consumption, such as plastic shopping bags and item packaging, may seem negligible, the aggregate of such waste products disposed of by Beijing's over 17 million residents is substantial. Likewise, the turnover rate in product ownership is increasing, meaning that, more and more, Beijing residents are disposing of durable goods in favor of more up-to-date versions. Simultaneously, the private-sector food market, a minimal dimension of the economy in the 1970s, has led to an explosion in organic refuse as a surplus of food now reaches the capital.

The most immediate challenge facing Beijing is an aging waste-processing infrastructure that has not been able to keep up with the growth in Beijing's volume of household garbage. As of 2010, Beijing generates a little over 6.5 million tons of garbage per year (approximately 18,000 tons of garbage each day). By 2015, it is predicted that the capital will produce 30,000 tons per day.

Landfills

According to the Beijing Municipal Commission of City Administration, the stress on the city's 13 landfills is a critical issue, as their strained capacities are no longer a sustainable solution for future waste management. In 2010, Beijing depended on landfills to dispose of over 80 percent of its annual garbage. By comparison, Japan, a country considered to have a highly sustainable platform for waste processing, incinerates over 90 percent of its waste. With the majority of refuse generated in the urban region heading to landfill sites, Beijing's facilities are expected to reach maximum capacity by 2012. Moreover, existing landfills have side effects: toxic leakages into water sources and noxious odors for nearby neighbors have led to protests by residents against the construction of new landfills. For the city government, innovative solutions for addressing ever-greater levels of waste produced within the city limits are required. The problem is as much a matter of contentious municipal politics as it is of technological solutions.

Solutions

The city government of Beijing has recently initiated a three-part approach to solving these dilemmas. As landfills and their stored waste are a long-

Rising consumerism in Beijing has prompted more private automobile use, which in turn has contributed to more urban air pollution. Over 15,000 new cars are bought in the capital every week. Air pollution not only threatens the health of Beijing residents, as seen in the heavy smog surrounding the city above, but also its status: in 2008, the Chinese government attempted to control air pollution during the Summer Olympic Games by temporarily restricting automobile use.

term component of the urban landscape, the city has attempted to address local complaints about their side effects. In 2009, Beijing installed deodorant guns at many of its landfills in order to minimize the odors that permeate nearby neighborhoods. The scaling down of dependency on future landfills demonstrates increased state responsiveness to local citizens. These varied adjustments indicate the complex intersection of new technologies for addressing waste and the changing political scene in the PRC around matters such as government accountability and expressions of popular opinion.

As exclusive dependency on landfills is phased out, the municipal government is implementing a shift to a more diverse set of methods for processing waste. City officials intend to shift from landfills as the primary method of garbage disposal to incineration and biotreatment. Both of these more-novel methods of garbage processing are considered a step in waste processing sustainability, with incineration generating heat for energy production and biotreatment producing fertilizers. However, these changes are facing popular opposition. The construction of new incineration facilities has become particularly unpopular as urban residents in the capital fear the release of poisonous dioxins from the burning of trash in a city where air quality is already low.

The final approach by the state points to a desired transformation in the public's cultural response to waste, rather than a change in waste disposal technologies alone. Recalling the PRC's long history of mass mobilization efforts, the Beijing government has recently initiated a campaign to encourage residents to sort their garbage at the point of disposal. Likewise, public institutions, such as government offices and schools, mandate selective sorting of waste. While this may seem second nature to those in countries with a longer history of recycling, Beijing authorities have undertaken an education initiative in public areas and the media to raise awareness of this novel practice in the Beijing populace. The goal has been to actively foster a citizenship that holds itself responsible and adjusts its behavior in relation to large-scale environmental impacts.

Overall, the problems and practices associated with waste in Beijing are as much questions of cultural and political forms as they are of refuse disposal technology and infrastructure.

Andrew Hao
University of California, Berkeley

See Also: China; Consumerism; Household Consumption Patterns; Politics of Waste; Shanghai, China.

Further Readings

"Beijing Plans to Step up Recycling, Waste-Recovery." *China Daily* (December 13, 2006).

"Beijing to Have More Garbage Facilities." *China Daily* (July 24, 2009).

Hoornweg, Dan, et al. "Waste Management in China: Issues and Recommendations." http://www.worldbank.org/research/2005/05/6679942/waste-management-china-issues-recommendations (Accessed July 2011).

Jones, Samantha L. "A China Environmental Health Project Brief: Environmental and Health Challenges of Municipal Solid Waste in China." Washington, DC: Woodrow Wilson International Center for Scholars, 2007.

Beverages

Beverages are liquids that have been prepared by various means for human consumption. The term is broad and encompasses both alcoholic and nonalcoholic drinks, carbonated and noncarbonated drinks, and hot and cold drinks; ranging from beer, wines, liquors, coffee, tea, bottled water, soda, milk, juice, and more. Beverages have a history as long as humans have existed and have as much cultural significance as physical function.

Commercial beverages are one of the most ubiquitous manifestations of modern consumer society and generate waste both upstream and downstream in production and disposal. In the United States alone, consumers purchased more than 57 billion liters of carbonated soft drinks, 31 billion liters of bottled water, 24 billion liters of beer, and 21 billion liters of milk in 2006. Globally, more than 200 billion liters of bottled water are sold annually. Most of these beverages come in relatively small container sizes, up to one gallon. This means that there are billions of beverage containers (made of plastic, glass, aluminum, and paper) that require disposal or recycling every year.

Before the 20th century, beer and sodas were often consumed at the point of purchase (at the saloon or soda shop), and beverages like milk were delivered in refillable glass containers. Drinks—or "tonics" as they were often called—like sarsaparilla and root beer were popular in U.S. saloons in the 1800s. Historically, bottled beverages developed hand-in-hand with mobility and leisure.

In the post–World War II era, leisure time increased along with higher incomes in the entire Western world. Car ownership became more common in the United States and in Europe, which, combined with expanded and improved highways, encouraged both work and leisure on the go. Single-serving beverages became popular for consumption in the home and during recreational activities. No longer tied to the point of purchase, consumers could bring along their favorite drink wherever they went. Beverage container packaging also served as a branding and marketing surface in an increasingly competitive marketplace.

Container Disposal

Littering soon became a major problem. Because single-serving beverages were often consumed during outdoor recreational activities where trash cans were not available, consumers often simply threw the bottle or can along the highway, in nature, or in the city. During the 1960s, there was a worldwide backlash against littering, often targeting disposable containers, which had gained market share in place of refillable glass bottles. The iconic "Crying Indian" public service announcement from Keep America Beautiful in 1971 prominently featured empty bottles as litter and directed the awareness of an entire generation of Americans to the environmental damage done by littering.

While Keep America Beautiful promoted awareness campaigns as the most appropriate way to prevent littering, many states and countries introduced mandatory container deposit legislation (also known as "bottle bills") starting in the 1970s. Vermont had a brief period with a bottle bill in the 1950s, but the first lasting bottle bill was introduced in Oregon in 1972. By 2010, 11 U.S. states had some form of bottle bill, along with many European countries and others around the world. The most successful deposit-return systems reach recycling rates of over 99 percent, such as refillable glass bottles in Norway and Sweden.

Beer

Beer brewers led the way in the development of individual-sized containers and in the adoption of disposable containers. Until Prohibition, the United States had a multitude of local breweries serving a predominantly local market with beer in refillable glass bottles that often carried a deposit. Upon return, the bottles were washed, refilled, and resold, often with trippage numbers of up to 25–30. After Prohibition ended in 1933, the previously diverse U.S. brewery market was reduced to a few, larger companies. These large companies had the capital necessary to invest in new bottling and canning technologies for disposable containers (one-way glass and steel cans). These one-way containers allowed them to cut costs by not having to accept, handle, transport, or wash refillable glass bottles. Adolph Coors Company led the way in developing a recyclable can for beverages to replace the nonrecyclable steel can, introducing the first aluminum can in 1959. Although the U.S. beverage market continues in the 21st century to be controlled by a few companies (such as Anheuser-Busch Inbev and MillerCoors), there has been a large growth in the number of microbreweries.

Sodas

Sodas have seen a similar trajectory toward large multinational corporations like Coca-Cola and PepsiCo. Coca-Cola, which owns more than 400 brands worldwide, sells more than 1.6 billion drinks every day. While in the Western world one-way containers are the norm, refillable containers still dominate other markets like Latin America. Soda companies have been criticized about their use of scarce local water resources in the manufacture of bottled beverages, such as in the state of Kerala, India, where protesters first claimed in 2003 that Coca-Cola's manufacturing plant had depleted groundwater resources and emitted toxic sludge. In the United States, beverage companies often use high fructose corn syrup as a sweetener, while the rest of the world generally uses cane sugar. Diet sodas use artificial sweeteners.

Bottled Water

Bottled water is the fastest-growing beverage industry in the world. In the early 1970s, bottled water hardly showed up in the statistics on beverage consumption. By 2010, bottled water was the second most popular commercial beverage in the United States. Internationally, Italy, France, and Mexico lead in consumption of bottled water. Bottled water is highly sensitive to narratives of origin. Many of the well-established brand names have their roots in historical spa resorts around actual springs, such as Evian, San Pellegrino, Perrier, Poland Spring, and Vittel, though some of these brands have been criticized for having lost the connection to the original spring in the transition to industrial mass production. Poland Spring, for instance, taps water from boreholes far from the original spring. Most low-end brands are purified water from municipal water supplies.

Coffee and Tea

Coffee and tea have a ritual place in many societies, such as the morning cup of coffee and afternoon tea time. In 1999, 54 percent of Americans drank coffee every day. Scandinavians consume the most coffee per capita in the world, with Finland topping the list at 12 kilograms per person per year. Americans, in comparison, averaged 4.2 kilograms in 2008. Starbucks, founded in 1971 in Seattle, Washington, has become emblematic of modern coffee consumption, with more than 15,000 stores in 50 countries by 2010. These new coffee houses offer coffee primarily in takeaway paper or styrofoam cups, rather than asking the customers to consume their coffee onsite from washable ceramic cups. The waste production from these throwaway containers is considerable, which has prompted coffee companies like Starbucks to offer incentives to consumers to use their own refillable mugs as well as to recycle the paper cups. The newest technological development in coffee consumption is espresso capsule machines that use a small prepackaged coffee capsule to prepare one cup at a time, such as Nespresso, founded in 1986. While this technology allows the convenience of an instantly brewed cup of coffee, it also generates considerable packaging waste of plastic or aluminum capsules.

Other Beverages

Other beverages for household consumption, such as milk and juice, come in either plastic jugs or plastic-covered cardboard packaging. Although milk continues to be primarily a short-traveled commodity, juice products are marketed as place-based commodities, such as orange juice from Florida. The commercial orange juice industry developed in Florida in the 1920s and expanded dramatically after the discovery of pasteurization methods and improvements in canning technology. Tropicana Products, which was the first company to sell not-from-concentrate orange juice in 1954, now has a 65 percent market share in the U.S. orange juice market and sells its products worldwide.

Brand Recognition

Beverages as consumer products are hard to separate from their packaging, which makes them portable, protects them, and serves as a branding and marketing surface. The packaging comprises a majority of the waste produced by beverages, though transportation and water extraction can also be environmentally damaging. Beverage producers are often reluctant to change the packaging or contents of beverages because many beverage brands have developed a strong consumer following and brand recognition. In 1985, Coca-Cola faced massive protests from its customers when the company changed its recipe in an attempt to modernize the classic drink; this event is now remembered as one of the biggest failures in the history of marketing. Likewise, some consumers claim that the packaging material changes the flavor of the contents, saying Coca-Cola tastes differently in glass bottles, aluminum cans, and PET plastic bottles.

Finn Arne Jørgensen
Umeå University

See Also: Consumerism; Food Consumption; Packaging and Product Containers; Recycling; Water Consumption.

Further Readings

Hine, Thomas. *The Total Package: The Secret History and Hidden Meanings of Boxes, Bottles, Cans, and Other Persuasive Containers*. Toronto, Canada: Little, Brown, 1995.

Pendergrast, Mark. *For God, Country, and Coca-Cola: The Definitive History of the Great American Soft Drink and the Company That Makes It*. New York: Basic Books, 2000.

Royte, Elizabeth. *Bottlemania: Big Business, Local Springs, and the Battle Over America's Drinking Water*. New York: Bloomsbury, 2008.

Shiva, Vandana. *Water Wars: Privatization, Pollution, and Profit*. Cambridge, MA: South End Press, 2002.

Standage, Tom. *A History of the World in 6 Glasses*. New York: Walker Publishing, 2006.

BFI

See Browning-Ferris Industries

Biodegradable

Biodegradable matter is material capable of being decomposed by bacteria or other biological means. Biodegradable matter usually consists of organic materials such as plant, animal, and other substance matter originating from living organisms. It has the ability to be broken down into smaller, harmless products by way of the action of living organisms. The term is often used in relation to ecology, garbology, and waste management. Biodegradable products are often associated with perishables (products, food, and waste materials subject to death or decay). The term *biodegradability* of a product refers to its disposition to disintegrate as the result of natural processes.

Benefits

Biodegradable waste is usually regarded as less harmful to human health and the environment than nonbiodegradable waste. Environmental pollution caused by discarding biodegradable waste products, such as human, animal, and vegetable waste, is normally rendered harmless by natural processes and so causes no permanent harm. Most biodegradable products and waste products can be used for replenishing the natural food cycle; such products can provide soil nutrients necessary for the regeneration of life. In urban, industrial environments, biodegradable waste products, normally well suited for natural cycles such as fertilization, are used less frequently than in agricultural areas.

Complications

However, because of human population growth and the steadily increasing levels of consumption, biodegradable waste produced in large quantities cannot be easily disposed of and processed and can cause severe environmental and health problems. Biodegradable waste such as cattle manure is produced in increasingly large quantities because of large-scale commercial farming targeted to meet demands of an increasingly wealthy population. According to research in garbology and waste management literature in industrial countries, most biodegradable (waste) products are destroyed in waste incinerators and processed together with nonbiodegradable waste. Because of chemical and often harmful substances used for sewage treatment, human waste is also mostly wasted. Household refuse analysis shows that the populations of industrial countries discard as much as 40 percent of biodegradable products, especially food remains, into mixed garbage containers.

Collection and Use

Urban biodegradable waste collection varies across countries. In most countries in Europe and in the United States, biodegradable garbage is collected or separated at waste processing plants infrequently and often only in the largest metropolitan areas. While statistics differ in each country and by the type of research conducted in the industrialized countries, there are virtually no statistics available on the use of biodegradable waste in developing countries. In many developing countries, biodegradable waste is dumped together with nonbiodegradable waste (such as plastic and chemical waste) in mixed landfills or burned in mixed incinerators. Aside from a few countries, notably in northern Europe, that try to promote use for the vast quantities of daily discarded biodegradable waste, few global efforts have been made to address the issue of wasteful waste treatment practices.

Product Design Sustainability

In their influential 2002 book *Cradle to Cradle*, William McDonough and Michael Braungart pioneered the notion of product design sustainability that is primarily based on the importance of biodegradable products as a solution for modern industrial waste. In the cradle to cradle model, all materials used in industrial or commercial processes may be placed into two categories: technical or biological (biodegradable) nutrients. Biological nutrients are similar to biodegradable products and can be disposed of in any natural environment and decompose into the soil, re-entering the biological life cycle without affecting health or the natural environment. Technical nutrients, which are nontoxic, nonharmful synthetic materials that have no negative effects on the natural environment, can be used in continuous cycles without being "downcycled" into lesser products, ultimately becoming waste. In this view, both synthetic (but nonharmful) and biodegradable products can be either continuously "reused" or "re-entered into the (natural) cycle" without harming human health or the environment. While ecologically friendlier products, such as biodegradable washing detergents, gained in popularity among the industrial countries' environmentally conscious middle classes, these do not have a universal marketing appeal. Commercial companies have recently stepped up production of many types of biodegradable products, including cups, plates, and containers. Unfortunately, these products do not follow through the whole processing chain and get dumped in mixed garbage containers. In the early 21st century, such products are almost absent from developing countries, where few special processing facilities for such waste are present.

Biodegradable Plastics

One significant positive trend to counter the increased problem of waste is the introduction of biodegradable plastic and polymer technology. Biodegradable plastics and polymers are currently being developed, and, according to Emo Chiellini and Roberto Solaro, their use in industrial countries is on the rise. Originally, synthetic substances were produced to counteract biodegradable processes and were widely praised for their durability and inability to disintegrate for centuries. Even the simplest consumer products whose cycles were supposed to be stopped with one-time use, such as plastic water bottles or aluminum beer cans, were made of materials that would "live" much longer than the consumable product they contained.

The production of plastic materials has risen sharply in both industrialized and developing societies. This is due to expectations for the rising demand of polymeric production and the expected tripling of this demand in the first decade of the 21st century. The average annual consumption of plastic materials per capita in industrialized countries was over 100 kg (and rising) per capita per year in 2010, and developing countries were catching up. This created a serious plastic waste burden, which can be tackled by developing biodegradable plastic products.

Helen Kopnina
University of Amsterdam

See Also: Composting; Downcycling; Food Waste Behavior; Garbology; Human Waste; Incinerator Waste; Sewage Treatment.

Further Readings

Chiellini, Emo and Roberto Solaro. *Biodegradable Polymers and Plastics*. Boston: Kluwer Academic Publishers, 2003.

McDonough, William and Michael Braungart. *Cradle to Cradle. Remaking the Way We Make Things*. New York: North Point Press, 2002.

Rathje, William and Cullen Murphy. *Rubbish! The Archeology of Garbage*. New York: HarperCollins, 1992.

Reilly, M. D. and W. W. Hughes. *Household Garbage and the Role of Packaging: The United States/ Mexico City Household Refuse Comparison*. Tucson, AZ: Solid Waste Council of the Paper Industry, 1985.

Ritenbaugh, Cheryl K. *Household Refuse Analysis: Theory, Method, and Applications in Social Science*. Beverly Hills, CA: Sage, 1984.

Simmons, P., N. Goldstein, S. Kaufman, N. Themelis, and J. Thompson, Jr. "The State of Garbage in America." *BioCycle*, v.47 (2006).
Wilk, Richard R. "Household Archaeology." *American Behavioral Scientist*, v.25/6 (1982).

Books

In an industrial culture governed by the logic of planned obsolescence and disposability, books seem to enjoy—more than most commodities—a durable sense of value. Books are eminently collectible. Libraries, both personal and public, are dedicated to the preservation of durable hardback volumes, while even the flimsiest mass-market paperbacks can circulate for many years in a thriving resale economy. Nevertheless, this elect status among consumer goods is, for the most part, a matter of delaying the inevitable; the majority of books sooner or later end up in the landfill.

Production

Over the 20th century, the way books were both published and distributed changed substantially, contributing to a system based on the production—and, in many cases, the overproduction—of massive numbers of books every year. A booming paperback industry contributed greatly to this postwar abundance of reading material as did the rise of "big box" chain stores in the 1980s and 1990s, bringing the book market to consumers in smaller cities and suburbs.

Inexact and conflicting as they might be, figures on the number of books published each year are available, and these estimates suggest that the industrial-scale production of books is a massive potential source of material waste. Bowker, the company responsible for assigning ISBN numbers to published books, determined that 288,355 new U.S. book titles and editions were published in 2009. This means that approximately 794 books are published every day (around 33 new titles per hour). This figure does not, however, take into account the exponential growth of the "publication on demand" industry being pioneered by small presses. When the estimated 764,448 new titles and reprints produced in these nontraditional venues are accounted for, the United States published over a million new titles and reprints in 2009.

The total number of books published each year is likewise difficult to establish with any certainty. For trade books, the average "first run" of any new title is around 5,000 copies for a medium-sized publisher and 10,000 for the larger publishing houses. These numbers, however, can vary significantly; a first printing for a cultural phenomenon such as the Harry Potter series can run to several million copies, while the small presses catering to the "on demand" market can print as few as one copy at a time. Estimates that account for every major publication category (adult and juvenile trade, mass-market paperbacks, religious, professional/scholarly, university press, elementary, high school, and college texts) suggest that nearly 2.5 billion units were published in 2005, the largest share of those (nearly 1 billion) coming from trade books.

Disposal

Unfortunately, none of these figures help resolve the problem of determining how many of the books purchased in any given year end up in the trash. The findings of a landfill excavation project conducted by William Rathje of the University of Arizona suggest that paper is the greatest contributor to the municipal waste stream. The study attributes much of this paper waste to two sources: telephone directories and newspapers. However, considering the massive numbers of trade books published over the course of a year, it is reasonable to conclude that the multibillion dollar publishing industry makes a substantial contribution to the mass of waste paper consigned to the landfill every year.

Historical Recycling

Books have a complicated history with waste and recycling; the publishing industry at the beginning of the 19th century created a market for linen and cotton rags. It represented an important early market for postconsumer materials, allowing networks of rag dealers and scavengers to return discarded materials to paper mills. Changes in papermaking, including a transition to woodpulp, gradually lessened the market for rags and increased consumption of wood fibers by the beginning of the 20th century.

Waste

A startling amount of waste is actually built into the system of publication and distribution through the long-standing policy of buying back remainders (unsold titles) from booksellers. Estimates suggest that as many as 30 percent of hardbacks and 40 percent of paperbacks travel from the press to the landfill without being purchased or read. The economics of traditional publishing, which relies on lithographic printing technology, requires a set number of copies (usually a few thousand) to be "run" to allow for a sufficient return on the publisher's initial investment, and the more copies of a book produced in a run, the greater the return potentially becomes. Once all the preproduction labor required to produce a book has been completed, lithographic presses are extremely efficient in satisfying the demand that a bestseller can generate; a lithographic (or "offset") printing press can churn out more than a million books in a 24-hour period. It therefore makes better economic sense for a publisher to produce more copies of a novel than the amount that will actually be sold, since every subsequent run requires its own start-up costs that cut into the publisher's profits. Overproduction is the result of such a system, and warehouses (and landfills) full of remaindered books are the inevitable outcome when even a small percentage of the hundreds of thousands of titles produced every year fail to sell out their first edition.

Waste paper consigned to landfills, including discarded books and telephone directories, may be the biggest contributor to the municipal waste stream. Some estimates calculate that an average of 35 percent of printed books go directly to a landfill.

Print on Demand and E-Books

Print on demand (POD), which relies on digital production technologies that can publish individual texts to order, promises to alleviate some of the habitual excess attending mainstream publishing practices. While huge first runs for big name authors remain standard due to the efficiency of lithographic printing technology, POD is currently the most rapidly growing branch of the publishing industry.

Though e-books, which are electronic reading devices that allow customers to purchase and read portable document files (PDFs) of books, have made inroads into 21st century reading culture, they have yet to seize a substantial share of the publishing market. Estimates suggest that only around 10 percent of books purchased are electronic versions. While these devices, in conserving paper and trees, arguably offer an environmentally friendly alternative to the paper-based texts currently depleting forests; the rapid evolution of electronic reading platforms and programs suggests a calculated obsolescence to e-books that far exceeds the impermanence of their analog counterparts. In addition, these devices' status as toxic e-waste further underscores the fact that the publishing industry, even if it does successfully shift to electronic formats, can never completely eliminate waste from the technologies of reading.

Kevin Trumpeter
University of South Carolina

See Also: Consumption Patterns; Garbology; Magazines and Newspapers; Paper and Landfills.

Further Readings

Greco, Albert N., Clara E. Rodriguez, and Robert M. Wharton. *The Culture and Commerce of Publishing in the 21st Century*. Palo Alto, CA: Stanford University Press, 2007.

Hall, David D. *A History of the Book in America*. Chapel Hill: University of North Carolina Press, 2009.
McGaw, Judith A. *Most Wonderful Machine: Mechanization and Social Change in Berkshire Paper Making, 1801–1885*. Princeton, NJ: Princeton University Press, 1987.
Smith, M. *The U.S. Paper Industry and Sustainable Production*. Cambridge, MA: MIT Press, 1997.

Brazil

The Federative Republic of Brazil is the largest country in South America. It is the world's fifth-largest country, both by geographical area and by population, and the only Portuguese-speaking country in the Western Hemisphere. In 2010, its population was estimated at 193 million people by the Brazilian Institute of Geography and Statistics (IBGE), occupying an area of 5,290,899 square miles. The territory of the country is divided into five main regions (South, Southeast, Central-West, North, and Northeast), 26 states, one Federal District (where the capital city, Brasília, is located) and 5,565 municipalities. Some municipalities are considered megacities, with populations greater than 10 million inhabitants (for example, São Paulo and Rio de Janeiro), and some have less than 1,000 people, showing the level of diversity that represents the different regions of country.

The Brazilian economy was one of the fastest growing in the world by 2010, making Brazil one of the major emerging markets. The country occupies the eighth position when ranked in terms of nominal gross domestic product (GDP) ($1.571,979 trillion 2009) and the ninth world economy by purchasing power parity (PPP). This level of GDP was achieved despite significant parts of the population living at very low income levels and thus falling outside the consumer market for most significant sectors. This situation has been improving in the 21st century due to economic stability, public investments, and social programs.

History

Europeans first encountered Brazil on April 22, 1500, with a fleet of ships led by Pedro Álvares Cabral, a Portuguese noble, military commander, navigator, and explorer. Initially, the Portuguese called it *Terra de Vera Cruz* (Land of Vera Cruz) and the territory was claimed by Portugal after its discovery, although colonization was only effective in the early 1530s, when the first settlement was founded.

By the time of the Portuguese arrival, the territory was already occupied by different indigenous people divided into more than 2,000 nations and tribes. These people were traditionally semi-nomadic tribes who subsisted on hunting, fishing, and agriculture. Fossil records show evidence that the area had been inhabited by indigenous groups for at least 8,000 years. Some of these tribes were assimilated by the Portuguese, while others were enslaved or exterminated during various periods of war. The indigenous people were also affected by different types of European diseases (such as measles, smallpox, tuberculosis, and influenza) to which they had no immunity, killing tens of thousands. The Portuguese saw the natives as "noble savages," and miscegenation of the population was commonplace.

After years trying to enslave the native people in order to use them as labor to exploit the abundant natural resources and in sugar production, which had become Brazil's most important export and economic activity in the 16th century, the Portuguese started to import African slaves to work in the sugarcane plantations. Portugal controlled the African slave trade by the time of the sugar expansion. These processes are important to acknowledge as they have profoundly shaped the country's social structure and economy, and the cultural and ethnic diversity that still characterizes Brazil in the 21st century. Historical data show that about 3 million Africans were brought to Brazil from the 16th to the 19th century, when Brazil was officially a Portuguese colony. This process lasted until 1888, when slavery was finally abolished in the country by the *Lei Áurea* (the Golden Law), a legal act of Princess Isabel, Empress of Brazil. Brazil was the last nation in the Western world to abolish slavery.

The end of slavery was part of a broader process that contributed to the construction of Brazilian society in the same period of time that the country became independent from Portugal (September 7, 1822), when Dom Pedro I—the son of Dom João VI, King of Portugal—declared the independence of Brazil. After its independence from Portugal, Brazil became a monarchy called the Empire of Brazil, which lasted until the establishment of a republican government on November 15, 1889. The end of the empire in 1889 and the foundation of the republic was a direct consequence of the abolition of slavery the year before, which had created a serious threat to the interests of the economic and political oligarchy.

The period of the Old Republic (1889–1930) marked the replacement of sugar by coffee as the country's main export in the late 19th century. The economic prosperity brought by the coffee trade was the main attracting force for foreign immigrants, mainly from Italy, Japan, and Germany. This influx of labor allowed the country to develop an emerging industrial economy and expand its economic activities away from the coastlines. This period ended with a military coup that placed a civilian and dictatorial ruler, Getúlio Vargas, into power.

After 1930, the successive governments, with many periods of dictatorships (1930–34, 1937–45, and 1964–85) and democracy (1934–37, 1946–51, 1951–54), continued the expansion of industrial activities and the development of agriculture throughout the vast interior of Brazil. In 1960, the capital of Brazil moved from Rio de Janeiro to Brasília.

The period after the military dictatorship (1964–85) was marked by several financial crises, hyperinflation, and the beginning of the neoliberal agenda in the country. Although the consolidation of democracy and some economic and financial stability were achieved during the 1990s with the change of the currency to the Brazilian real (R$) before and during Fernando Henrique Cardoso's terms (1994–2002), it also deepened social inequalities and poverty. This socioeconomic contradiction opened the opportunity for the election of Luiz Inácio Lula da Silva, a former trade union leader, as the president of Brazil. Although Lula had social programs at the top of his political agenda during his electoral campaign and two administrations, he kept very strict macroeconomic policies that gained confidence from the market, providing a favorable environment for public and private investments. These investments, combined with policy improvements in the social area, have fostered a large-scale process of economic development and social inclusion.

Consumption

Since the 1980s, Brazil has undergone a notable and accelerated modernization process. From a rural and single-product economy, Brazil's economy has become dynamic. Brazilian industries have been growing not only in numbers but also in size and competitiveness, although major political and institutional reforms are still needed. The financial stability achieved, combined with more effective 21st-century social and economic policies, have paved the way for the recent economic growth that has also benefited from high commodities prices. These changes fostered an important internal market, as many historically poor families have been brought to the middle-class lifestyle that includes different patterns of consumption.

The new Brazilian middle class is an important part of the social, economic, and cultural change of the country. This group has the opportunity to buy a number of products that it was not able to in the past. According to official data, 97.8 percent of Brazilian households own an oven or gas cooker, 50.3 percent are equipped with water filters, 89.2 percent have a refrigerator, 16.4 percent have a freezer, 37.5 percent have a washing machine, 87.9 percent own an AM/FM stereo radio, 93 percent own a television, 22.1 percent have a personal computer, and 74.5 percent have at least one telephone (mobile or fixed).

The Brazilian population does not have a homogenous consumption habit, since different classes of consumers are encountered in the Brazilian internal market as a result of a historical inequality in income distribution. For instance, lower-income groups spend more on food and beverage items. On the other hand, new and traditional middle class consumption patterns show an increase not only in expenditures on transportation, communication, and durable goods but also on leisure activities, health, and education. Shopping is integrated into the lifestyle of the Brazilian urban population as a usual and necessary activity. Shopping malls, outlets, supermarkets, and convenience stores can be found in many cities.

Waste

As a direct consequence of the social inclusion process and economic growth, combined with an urbanization rate of more than 82 percent and a lack of adequate urban planning, solid waste in urban areas is considered one of the most pressing issues and a top priority for environment sustainability. It is also one of the most costly services provided by a Brazilian city to its residents, and is becoming even more expensive.

It is estimated that from the 170,000 tons of solid waste generated on a daily basis in the country (13 larger cities generate 32 percent of the total; 525 municipalities with more than 50,000 inhabitants generate around 80 percent), only 140,000 tons are collected. More than 60 percent of the total solid waste collected, however, does not have an adequate final destination. It is mainly disposed of in landfills or simply dumped into empty areas, the so-called *lixões* (big trash). The situation requires solutions for the final disposal of waste, with a view to increasing recycling participation and reducing the total volume of waste generated.

There is also a threat that the current model based on landfills will physically collapse. With an increasing urban population and higher consumption levels, the model to dispose solid waste in landfills is expected to be exhausted in the early or mid-21st century. With the upcoming end of many landfills in Brazil, some states and cities have been looking for more modern alternatives to dispose of their solid waste. This is especially true in the southeastern large cities, the region where the important areas of São Paulo, Rio de Janeiro, and Minas Gerais are located. They account for more than 40 percent of the total population and just over 50 percent of Brazil's GPD. Many disposal alternatives may economically value the waste, which in most cases is simply burned.

The Basic Sanitation Act is considered to be a landmark for the implementation of initiatives related to solid waste. At the same time, the National Solid Waste Policy was approved in August 2010, after 20 years of discussion in the national congress. It regulates the collection, final disposal, and treatment of urban, hazardous, and industrial waste. It also provides the guidelines to reduce the generation of waste and the level of pollution of the environment by instituting a waste management hierarchy. It has also introduced measures to facilitate the return of solid waste to original producers with emphasis on reverse logistics so that they can be treated and reused in new products.

The framework also provides the opportunity for the introduction of economic incentives for recycling activities at the same time that it establishes the principle of shared responsibility for the lifecycle of the products in which manufactures, importers, distributors, vendors, and consumers are responsible together with solid waste management agents. The law has also introduced the National Solid Waste Planning Program under responsibility of the federal government that will be updated every four years, taking into account new developments in the area. This planning process will also be able to establish recycling goals and provide financial support to improve solid waste models.

One of the opportunities that appears as an alternative to the old landfill model is solid waste thermal treatment with power generation systems, widely used in Europe and Japan. In Brazil, there are no specific instruments in place to encourage this type of thermal treatment or other possible existing technologies, although there are some experiments under way, especially in large cities such as São Paulo and Rio de Janeiro. Some of these alternatives still face a number of barriers, including a lack of adequate funding for new technologies and solid waste management models, technical and institutional barriers, and cultural resistance. However, alternatives to current models are not only desirable but also necessary. After a long history of low consumption patterns and inadequate solid waste management, Brazil will have to face this challenge to sustain its social development and economic growth.

Leonardo Freire de Mello
Rafael D'Almeida Martins
State University of Campinas

See Also: Developing Countries; Rio de Janeiro, Brazil; São Paulo, Brazil.

Further Readings

Bourne, R. *Lula of Brazil: The Story So Far*. Berkeley: University of California Press, 2008.

Fausto, B. *A Concise History of Brazil*. New York: Cambridge University Press, 1999.

Levine, R. M. and J. J. Crocitti, eds. *The Brazil Reader: History, Culture, Politics*. Durham, NC. Duke University Press, 1999.

Ribeiro, H. and G. R. Besen. "A Panorama of Selective Waste Collection in Brazil: Challenges and Prospects Taken From Three Case Studies." *INTERFACEHS: A Journal on Integrated Management of Occupational Health and the Environment*, v.2/4 (2007).

Waite, R. *Household Waste Recycling*. London: Earthscan, 1995.

Browning-Ferris Industries

Before its acquisition by Allied Waste, Browning-Ferris Industries (BFI) was the second-largest waste disposal company (second to Waste Management, Inc.), offering a full range of waste, recycling, and sanitation services. At one time, BFI operated in nearly 800 locations in Asia, Australia, Europe, the Middle East, New Zealand, and North America with BFI's North American operations including 104 solid waste landfill sites, 32 medical-waste treatment facilities, and 125 recycling facilities.

Birth and Rise of BFI

Tom Fatjo Jr. founded BFI in 1966. The birth of BFI was a result of national legislation, the Solid Waste Disposal Act, designed to regulate waste collection and disposal services, including higher standards of sanitation from collection trucks and increased use of landfill burial over incineration due to concerns about air pollution. Tighter government regulation necessitated larger capital investments by waste collectors, unobtainable by most small, family-owned businesses, which were the primary waste collectors of the time. This regulatory change created an opportunity for a larger, broader-reaching company that could professionalize the waste-management business.

Beginning by collecting residential waste with a single truck in Houston, Texas, Fatjo expanded his business to include collection of commercial and industrial waste. By 1968, Fatjo expanded into waste disposal, winning a large contract from the city of Houston. Partnering with corporate financier Louis A. Walters, Fatjo and Walters formulated a program of acquisitions of smaller collection and disposal companies across Texas and the southern United States. To raise capital for the acquisition program, in 1969, Fatjo and Walters acquired control of Browning-Ferris Machinery Company, a publicly traded manufacturer of, among other things, garbage trucks and landfill equipment.

In 10 short years, BFI had grown tremendously. Revenues were over $250 million, and the corporation accumulated over 60 landfills, becoming a critical asset. Regulation and increasing public anxiety made the creation of new landfills nearly unobtainable and maintenance of existing landfills increasingly expensive, thus helping create a natural monopoly. As early adopters, BFI and Waste Management were able to grow largely as a result of barriers to entry.

Congress passed the Resource Conservation and Recovery Act in 1976 (implemented by 1980), which was legislation designed to tighten control of potentially dangerous landfills, including chemical and toxic waste. With a similar effect as the Solid Waste Disposal Act, this legislation again helped create natural monopolies, as experienced companies such as BFI grew, while others were denied entry. By 1983, chemical waste provided 10 percent of BFI's revenue. However, by 1984, BFI was under investigation in seven states for charges of monopolistic practices, including price-fixing. These charges were denied, although BFI paid $15 million in out-of-court settlements by 1989. BFI's toxic landfills became a liability, ultimately resulting in BFI's withdrawal from the toxic-waste industry in 1990.

Recycling

Diversifying its interests, BFI experimented with paper waste recycling. However, declines in paper manufacturing proved detrimental to the paper recycling business for BFI, which created a spin-off company in 1975 for this segment of the business.

After instituting effective operational cost-cutting measures to resurrect the organization, BFI CEO and former Environmental Protection Agency director William D. Ruckelshaus was determined to grow the organization through recycling efforts despite wary naysayers. Focusing on paper waste, which accounted for 45 percent of the U.S. waste stream and soon would represent 85 percent of BFI's recycling business, BFI's recycling division grew astronomically, from less than $10 million in 1990 to $675 million annually by 1995.

With its recycling initiatives proven successful, BFI began to duplicate its earlier strategy of rapid acquisition. Also reflective of the past, government regulation in the waste industry ensured BFI's success by denying entry to smaller, would-be competitors.

Decline

In 1997, Browning-Ferris Industries Inc., retreating from an ambitious global expansion, agreed to sell its international operations to Sita, a unit of France's Suex Lyonnaise des Eaux SA, for $1.4 billion in order to focus on its core North American business. Sita's purchase of BFI's international operations made Sita Europe's largest waste-management company.

In 1998, BFI's internal growth averaged less than 1 percent, with Deutsche Bank Securities analyst Mari Bari downgrading the cost per share, maintaining a hold rating on BFI's stock, and not recommending BFI stock for purchase or sale.

By 1999, BFI had lost its acquisition power, selling its medical waste unit to Stericycle after a failed bid to acquire Texas-based MedServe, Inc., Stericycle's distant competitor. The remainder of BFI was acquired through a merger with Allied Waste Industries, making Allied Waste the second-largest waste management company in the United States (still second to Waste Management, Inc.). The U.S. Department of Justice's Antitrust Division approved the merger with the caveat that the merged corporation (Allied Waste) sell operations in 18 metropolitan areas in 13 states, from Massachusetts to California, to avoid monopolization.

Leslie Elrod
University of Cincinnati

See Also: Recycling; Recycling in History; Resource Conservation and Recovery Act; Solid Waste Disposal Act; Waste Management, Inc.

Further Readings

Funding Universe. "Browning-Ferris Industries, Inc." http://www.fundinguniverse.com/company-histories/BrowningFerris-Industries-Inc-Company-History.html (Accessed August 2010).

Heumann, Jenny. "Browning-Ferris Industries, Inc.: Facing New Challenges." *Waste Age*, v.29/12 (1998).

Treaster, Joseph B. "Justice Department Approves Merger of Allied Waste and Browning Ferris." *New York Times* (July 21, 1999).

Bubonic Plague

Bubonic plague is the most common and curable form of the bacterial disease plague and is caused by infection with the *Yersinia pestis* bacterium. The term *bubonic plague* comes from the classic symptom of swollen glands, or buboes, on the body. There are also pneumonic and septicemic forms of plague. Plague is a disease found naturally in rodent populations. While rats are the most common carriers, it occurs in mice, squirrels, prairie dogs, chipmunks, marmots, and dozens of other species. It is transmitted from rodent to rodent by fleas. Occasionally, humans enter the cycle, resulting in havoc.

Plague caused three of the most devastating pandemics in history, including the Black Death of medieval Europe. During a pandemic, spread of the disease is accelerated by specific environmental conditions. In the past, overcrowding in villages and cities, proximity to dirt and filth, poor personal hygiene, the absence of public sanitation, and limited medical knowledge all combined to create the ideal breeding ground for plague. Modern public health standards and antibiotics keep outbreaks of the plague contained in the 21st century, although cases do still occur where humans encounter affected animals in the wild.

Forms of Plague

Humans can display three different forms of plague. Bubonic plague is the most common, causing the vast majority of cases throughout history. It spreads by *Yersinia*-infected fleas transmitted from rodent to person. Two to six days after bites from infected fleas, individuals show signs of infection. The most characteristic symptom is the appearance of swollen lymph glands, called "buboes," in the groin, neck, and armpits. Other symptoms include headache, dizziness, vomiting, shivering, sleeplessness, apathy, delirium, and terrible pain in the extremities. Death typically occurs in untreated individuals within a week of the emergence of symptoms. The bubonic form of plague causes death in more than 50 percent of untreated cases.

Septicemic plague can occur when *Yersinia pestis* enters and quickly multiplies in the bloodstream. It can result from direct fleabites or as a complication of bubonic plague. The bloodstream can be invaded so quickly that a person can die from septicemic plague before the bubonic symptoms even appear. Symptoms of this form of plague include septic shock, meningitis, and coma. Uncontrolled bleeding under the skin can cause black patches. Untreated, the mortality rate is almost 100 percent.

Pneumonic plague is the least common but most fatal form of plague. It is highly infectious and is the only form that can be transmitted human-to-human. It can appear as a complication of bubonic plague. If *Y. pestis* invades the lungs, severe pneumonia follows. Weakness, shortness of breath, and coughing increase as the lungs fill with fluid. Coughing and spitting produce droplets filled with the highly infectious bacteria, which is easily inhaled by those nearby. Patients who do not receive treatment within hours of the onset of symptoms will not survive. Death can occur within 24 hours of exposure.

Plague in History

There have been three distinct and well-documented plague pandemics in the last 2,000 years, each having a huge impact on human society. The first, called the plague of Justinian, arrived in Constantinople in 542 from either northern India or central Africa. The plague followed trade routes as infected rats hitchhiked on ships of grain being delivered throughout the Mediterranean Sea. Waves of the plague reappeared throughout Europe for 200 years. It is estimated that over 100 million people died.

The second pandemic began in 1346. By the time it was over, the population of Europe and the Middle East was reduced from 100 million to 80 million people. At the height of the epidemic, thousands were dying daily in larger cities such as London and Marseille. The bodies could not be collected and buried fast enough, increasing the spread of sickness. At the time, the plague was called the Great Pestilence or the Great Dying. Centuries later, it was referred to as the Black Death. During the Great Pestilence, the infected fleas and their host rats traveled along the Silk Road from central Asia. From the caravan routes to ships that crossed the Caspian Sea, the rats traveled from port to port, spreading plague to humans living in filthy, crowded, rat-infested cities. By 1348, the plague was in England. It reached Scandinavia and Russia, making a complete circuit, by 1352.

No one in the medieval era knew that microbes cause disease. Public health measures were crude and often caused more harm than good. During the epidemic, ships were quarantined upon docking for 40 days, but rats left the ships by way of the ropes. Infected individuals were closed up in their homes with healthy family members, causing entire families to die. Human waste and dead rats littered the streets. People rarely bathed, changed clothes, or washed bedding, providing an ideal environment to harbor infected fleas. So many people were sick, dead, dying, or fleeing from the plague, there was no one left to maintain the few social services that existed—nursing the sick, cleaning waste from the streets, or collecting and burying the dead. Plague epidemics continued to strike Europe at irregular intervals through the 18th century.

The third pandemic began in the 1860s in the Yunnan Province of China. After killing millions again in China and India, the plague spread worldwide, thanks to modern rail and steamship trade. For the first time, bubonic plague spread to all inhabited continents. The pandemic was considered active by the World Health Organization until 1959, when worldwide casualties dropped to under 200 per year. Although the plague appears to be dormant, the disease still exists in rodent populations, and occasional cases of bubonic plague still appear where people come in contact with these animals.

Treatment and Prevention

Researchers working in Asia during the third pandemic finally isolated the plague bacterium and identified fleas as the plague vector (the delivery method). In the 21st century, vaccines are available, and antibiotics exist that can cure most cases of the plague when identified and treated quickly. Streptomycin, tetracycline, and doxycycline are all effective in curing bubonic plague. The mortality rate for plague has dropped to 5–15 percent, and the majority of those deaths are the result of a delay in diagnosis and treatment.

Modern medicine, well-developed public health and sanitation policies, and increased standards of personal health and hygiene have made outbreaks of plague very rare. However, it is still endemic in regions of Africa, Asia, and North and South America. Agencies like the Centers for Disease Control, the World Health Organization, and other state and local agencies perform active surveillance for outbreaks of the plague and other infectious diseases. Decreasing the rodent population through proper waste management, using insecticides to control the flea population, and educating the public about signs and symptoms of infection all help minimize the impact of the plague on 21st-century society.

Jill M. Church
D'Youville College

See Also: Funerals/Corpses; Germ Theory of Disease; Public Health; Slums.

Further Readings

Gage, Kenneth L., David T. Dennis, and Theodore F. Tsai. "Prevention of Plague: Recommendations of the Advisory Committee on Immunization Practices." *MMWR: Morbidity and Mortality Weekly Reports*, v.45 (1996).

Gottfried, Robert S. *The Black Death: Natural and Human Disaster in Medieval Europe.* New York: Free Press, 1985.

Kelly, John. *The Great Mortality: An Intimate History of the Black Death, the Most Devastating Plague of All Time.* New York: HarperCollins, 2005.

Sherman, Irwin W. *Twelve Diseases That Changed Our World.* Washington, DC: ASM Press, 2007.

Buenos Aires, Argentina

Buenos Aires is the capital of Argentina and the nation's center of economic activity since its establishment by Spanish merchants in the 16th century. As of the 2001 census, the greater metropolitan region is home to one-third of Argentina's population, with more than 13 million people.

The impoverished people of the shantytowns within the capital and in the larger Province of Buenos Aires carry out most of the recycling in the city. Since the 2001 economic crisis, a common figure roaming the capital's streets is the *cartonero* (a cardboard recycler) who retrieves the remnants of society's consumption, thereby contributing to the environmental cleanup and overall conditions within the metropolitan area. Argentine recycling thus is quite unlike efforts in more developed nations where recycling typically falls within the province of an educated, relatively affluent class of citizens. Recycling in Buenos Aires also is clearly decentralized and piecemeal; it is comprised of a series of individual efforts toward economic survival, rather than a coordinated campaign whose aim is ecological sustainability. Buenos Aires has no municipal department or organization devoted to recycling. Solid Waste Management Department is in charge of a recycling service.

Cartoneros

The cartonero first emerged on a widespread scale in Buenos Aires with the 2001 demise of the 1991 Convertibility Plan, whereby the Argentine peso was tied to the U.S. dollar. The collapse of the national economy resulted in the freezing of bank accounts and the biggest foreign debt default in world economic history. The percentage of people who were unemployed or underemployed increased dramatically in the wake of this crisis, and many of the newly unemployed were from the middle classes who encountered poverty for the first time. Confronted with this hardship, people turned to rummaging and sorting through trash to collect recyclable material—usually cardboard—in order to sell it to the recycling companies located in the suburbs of Buenos Aires. The term *cartonero* is somewhat of a misnomer because these informal circuits of recyclers also collect glass bottles, plastic, newspapers, and metal.

History

Although the cartonero is a recent phenomenon, the history of the disposal of material wastes and the circulation of garbage in Buenos Aires stretches back centuries, beginning with the foundation of the city, in 1580, by Juan de Garay. Governmental urban hygiene policies have shaped the living conditions of the poor who depend on others' trash to survive. In 1860, for example, the municipality established the system of the *Quema* (meaning "burn") in the southern part of the city where trash was sent for incineration. This was a new technique that illustrates a major change in how Buenos Aires treated and disposed of waste. Before the Quema, trash was typically thrown into streams, indoor wells, or fields. Accompanying the advent of the Quema was the emergence of a new neighborhood in the same area inhabited exclusively by people devoted to the collection of garbage. By the end of the 1920s, the lower Flores neighborhood dump in the capital became the largest waste disposal site in the city. It lasted until the 1960s, when it was cleaned up and urbanized with the construction of the Almirante Brown Park and the channeling of the Cildañez stream. For many years, men, women, and children had collected paper, used rags, cans, glass, and even bones, gathering them for use as a source of heat. Argentine painter Antonio Berni portrayed the misery of the garbage collectors in the lower Flores neighborhood by inventing a character, Juanito Laguna, a young man whose daily survival depends on scraping together all sorts of remnants from the city.

Between the 1930s and the 1970s, two different social actors emerged in the domain of informal trash collection: the *botellero* (the bottle collector) and the *ciruja* (the junkman), the former surviving by collecting trash that had been initially disposed of in both municipal and clandestine dumps. In 1977, with the creation of the company Cinturón Ecológico Área Metropolitana del Estado (CEAMSE or State Metropolitan Area Ecological Belt), the government forbade the use of industrial incinerators in the Buenos Aires area and replaced them with landfills. In this same year, through the enactment of law 8782/77, the government prohibited the act of collecting garbage and waste, which was thought to inevitably result in their incineration. This new policy put an end to burning trash and made garbage collection accessible to only a select group of companies. The economic crisis and unemployment of the 1990s once again led to the emergence of waste collection through new and very different modalities.

Effects and Culture of Cartoneros

Cultural anthropologists, sociologists, and historians do not agree on the exact date for the emergence of the cartoneros phenomenon. While some scholars directly connect it to the economic collapse in 2001, other scholars have suggested that the practice of collecting waste was already well under way by the middle of the 1990s. A major source of this scholarly disagreement stems from the fact that it is quite difficult to calculate the precise number of cartoneros circulating within Buenos Aires, as it is an ambulatory profession that is neither closely monitored nor legally sanctioned. For this reason, estimates have ranged from 25,000 to 100,000 cardboard collectors.

The cartonero has generated much interest for social scientists, writers, and documentary filmmakers. The 2001 novel *La Villa* (The Shantytown) by Argentine writer César Aira, and the 2006 documentary film *Cartoneros* by Eduardo Livón-Grosman are paradigmatic of the increased visibility of the cartonero as an object of scholarly inquiry. *La Villa* narrates the story of Maxi, a middle-class teenager, who offers every evening to help the cartoneros in the lower Flores neighborhood in collecting cardboard. The body of Maxi, which is portrayed as both athletic and healthy, is markedly contrasted to the cartoneros, who are represented as silent, feeble, hard-working, and frightened people. The Livón-Grosman documentary conveys the abrupt social changes that have been disrupting ordinary Argentines, emphasizing the transformations within the middle classes who have been reduced to rummaging through garbage in order to survive.

The cartoneros are to be credited with transforming the practice of waste collection in the city as their activities have made waste into a veritable commodity. What originally was an urgent effort to economically survive, rather than an ecologically informed gesture to preserve the environment, has become one of the most active waste management industries in Argentina. The cartoneros have led to the creation of cooperatives such as El Ceibo within the capital and Nuevo Rumbo, El Orejano and Alicia Moreau de Justo in the larger province of Buenos Aires.

Goals

In 2005, following a campaign by Greenpeace, the city council approved the first municipal zero waste plan in Latin America. The Integral Management Of Solid Urban Waste law set goals of 0 percent reduction of waste to landfill by 2010, a 50 percent reduction by 2012, and a 75 percent reduction by 2017. The law bans landfilling of recyclable and compostable waste by 2020 and bans incineration of solid waste until 75 percent diversion has been achieved. The baseline used in the law is the 1.5 million metric tons of solid waste disposed in 2004. While nongovernmental organizations have clashed with the city about adhering to its provisions (notably, the city attempted to incinerate waste in 2007), the law marks an important development in Latin American waste policy.

Gisela Heffes
Rice University

See Also: Argentina; Dumpster-Diving; Paper Products; Recycling Behaviors; Zero Waste.

Further Readings

Anguita, Eduardo. *Cartoneros: Recuperadores de Desechos y Causas Perdidas*. Buenos Aires, Argentina: Norma Editorial, 2003.

Escliar, Valeria. *Cartoneros: ¿Una Práctica Individual o Asociativa? Ciudad de Buenos Aires, año 2004–2005*. Buenos Aires, Argentina: Centro Cultural de la Cooperación Floreal Gorini, 2007.

Paiva, Verónica. *Cartoneros y Cooperativas de Recuperadores: Una Mirada Sobre la Recolección Informal de Residuos. Área Metropolitana de Buenos Aires, 1999–2007*. Buenos Aires, Argentina: Prometeo, 2008.

Schamber, Pablo J. *De los Desechos a Las Mercancías: Una Etnografía de los Eartoneros*. Buenos Aires, Argentina: Editorial SB, 2008.

Suárez, Francisco M., and Pablo J. Schamber. *Recicloscopio: Miradas Sobre Recuperadores Urbanos de Residuos de América Latina*. Buenos Aires, Argentina: Prometeo, 2007.

C&D Waste

See Construction and Demolition Waste

Cairo, Egypt

Cairo is the largest city in Egypt, with roots in settlements more than two millennia old. Its waste management challenges in the 21st century reveal longstanding social and economic tensions. The core of Cairo's solid waste management system is exemplary with its recycling rate of about 85 percent. Based on garbage collectors (called *zabaleen*), their families, and communities, this labor-intensive core is among the most sustainable systems in the world. The zabaleen have come under attack, however, by neoliberal privatization and gentrification. Metropolitan Cairo (including parts of Giza and Qaloubiya governorates) has about 17 million inhabitants. Authorities are overwhelmed with keeping up with urban services. Modern amenities abound in wealthier quarters; poor residents lack many such conveniences.

Colonial municipalities did not spend much time on waste. In the early 20th century, migrants from desert oases (called *wahiya*) started to collect waste from wealthier households. They sold waste as fuel to public bathhouses or makers of *fuul*, a bean dish. In the 1940s, as garbage increased, poor Christian migrants became subcontractors of the Muslim wahiya. These migrants, the zabaleen, bought organic waste (the largest part of the refuse) for their pigs, which the Muslim wahiya were prohibited by Islam to keep. Brokers (called *mu'allim*) set up new arrivals with pigs, shacks, and pigsties. Humans and pigs shared one yard. Wahiya oversaw the garbage business, and zabaleen worked the garbage routes. In the 1950s and 1960s, mu'allims set up zabaleen settlements on the urban fringes. With urban expansion, zabaleen were relocated to more distant locations.

As garbage proliferated, the zabaleen refined waste processing. Glass, metal, bones, or rags were sorted and sold to middlemen who resold them to workshops. Garbage was—and remains—a family business. Husbands left settlements in the early morning with donkey carts. Children traveled with them to guard carts, while fathers went into buildings to collect garbage. Upon returning home, men dumped the carts of waste into family yards, where women and children sorted the garbage and pigs and goats consumed organic waste. The wahiya-zabaleen system resembles those of 19th-century Western cities, where urban waste fed into rural-to-urban waste and agricultural circuits, except theirs is an urban-to-urban fringe circuit and has a higher recycling rate (New York's Barren Island recycled 60 percent in the 19th century).

In 1970, authorities relocated a zabaleen community to the Moqattam Mountain east of Cairo. The community received no municipal services. Regardless, the zabaleen worked and improved recycling. Driven by poverty and maintained by both family labor and an understanding of the value of resources, the system provided work for thousands. Leftover garbage was dumped on the street or dumping grounds outside the community and burned. Their smoldering clouded the community in pollution.

This efficient system had flaws. Because zabaleen sought valuable garbage, they did not service poor neighborhoods. However, no municipal waste management existed through the 1970s, and poor people simply threw their garbage on neighborhood dumps, which were regularly burned. The system's worst flaw was its human cost; in the 1970s, infant mortality among the zabaleen was 40 percent, as diseases were pervasive. The Moqattam community started organizing, and the first Garbage Collectors' Association was registered in the 1970s with the Ministry of Social Affairs.

In 1976, the Moqattam community burned down twice in one year. The fires and public attention they produced, combined with growing waste, generated a political understanding that waste management was necessary and that the zabaleen's conditions needed to be addressed. In 1980, local and international agencies started projects in the Moqattam village to establish cleaner ways of working and on-site recycling to add profitable jobs for the zabaleen. Electricity and water were installed and social services (such as clinics and schools) were added. A composting plant for manure was constructed, and plastic-processing facilities were built. Loan programs allowed families to set up recycling workshops. A rug-weaving workshop opened where women, upon completion of a course, could buy their own loom on loan. Combined with urban upgrading schemes, the projects transformed the community. Tin shacks were replaced by concrete buildings, where humans and animals had separate quarters. The village grew from 5,500 to over 15,000 residents in the 1980s. Similar interventions followed in the other six zabaleen communities.

Cairo and Giza, overwhelmed by waste problems in 1984, founded the Cairo (and Giza) Cleansing and Beautification Authorities. Solutions overlooked the existing system, and ecology and sustainability were secondary. In 1990, authorities banned donkey carts for garbage collection, and the zabaleen founded cooperatives to purchase vehicles. By the late 1990s, the zabaleen still handled 30 to 40 percent of the 14 million metropolitan residents' garbage (about 3,000 tons per day), recycling 85 percent. By the turn of the century, local companies were hired to take the waste of poorer quarters to public dump sites, where it was burned and produced pollution.

Decline of the Zabaleen

In 2003, with growing garbage mountains and pollution, Cairo and Giza contracted multinational waste management companies. The zabaleen were neither consulted nor were they given roles in the contracts. The new companies were to clean streets and collect household trash. Unlike the zabaleen, they did not collect waste from front doors; instead, containers were placed in streets for households to deposit their waste. Companies had to recycle only 20 percent of the waste. This system had various problems, including angry residents who had to pay more for less service; company vehicles were too large for small alleys, public garbage receptacles were stolen, and zabaleen and other scavengers mined containers and left a mess. Some zabaleen made informal arrangements with the companies and received garbage for fees or services. Wealthier households continued to pay zabaleen in addition to company fees. Companies contracted zabaleen for some routes.

Framed by neoliberal policies and real estate speculation, metropolitan Cairo mushroomed throughout the 1990s. Upscale Moqattam City expanded on the plateau above the zabaleen village. Gentrification and concerns about the city's aesthetics brought zabaleen villages under scrutiny. Authorities decided to provide land 15 miles outside the city for the Moqattam zabaleen's recycling operations. Other plans foresee the entire community's relocation to this site.

At the onset of the 2009 swine flu, the Egyptian government decided that pigs were to blame. In May, authorities slaughtered about 300,000 pigs, even though experts advised against it. This slaughter was a disaster for the 70,000 metropolitan zabaleen, and many saw it as another governmental measure to eliminate their jobs. Observers commented that the slaughter reflected the influence of Islamist groups on the government and its attempts to destroy the livelihood of Christian zabaleen. The zabaleen's response was straightforward: no pigs, no organic waste collection. Of the 6,000 tons of waste collected daily by the zabaleen before May, 60 percent had been organic waste, which was now left to rot in the streets. Simultaneously, the multinational waste company in central Giza went on strike that summer. In 2010, metropolitan Cairo was far from achieving viable waste management or any system that remotely approached the rate of recycling and sustainability that the zabaleen developed. Many zabaleen continue to work.

Petra Kuppinger
Monmouth College

See Also: Culture, Values, and Garbage; Developing Countries; Organic Waste; Street Scavenging and Trash Picking.

Further Readings

Assaad, M. and N. Garas. "Experiments in Community Development in a Zabbaleen Settlement." *Cairo Papers in Social Science*, v.16 (1993–1994).

Fahmi, W. and K. Sutton, "Cairo's Contested Garbage." *Sustainability*, v.2 (2010).

Haynes, K. E., et al. "Appropriate Technology and Public Policy: The Urban Waste Management System in Cairo." *Geographical Review*, v.69 (1979).

Iskander, M. *Garbage Dreams* (Film). CinemaGuild, 2009.

Kamel, L. R. I. *Moqattam Garbage Village*. Cairo, Egypt: Stallion, 1994.

Calcutta, India

See Kolkata, India

California

California and its waste practices are a study in contrasts, ranging from cities with some of the strictest laws and innovative ideas to some world-class, man-made environmental disasters. The Golden State is the third biggest state by geographic area and has both the largest population (37,253,956 as of 2010) and largest economy of any of the United States. California is not one culture but a myriad, each with a specific take on garbage and disposal. California is also a bellwether state, with legislative initiatives that are often the model for other states to adopt.

People have been creating, handling, and disposing of waste in California for thousands of years. The oldest archaeological site found as of 2010 in California is the Daisy Cave rock shelter in the Channel Islands off the coast of Ventura County, which has been dated to somewhere between 12,300 and 11,120 years old. While any archaeological site by definition traffics in garbage, a shell midden on San Miguel Island in the same group dating to 9,700 years old is evidence of collective, managed, and intentional disposal of food waste. Later, kitchen middens from Spanish colonial-period missions, like those in Santa Clara and San Juan Bautista, give evidence of the interaction between California natives and Spanish missionaries. In Riverside, two Chinatowns have been excavated, and the artifact-rich trash pits have helped tell the story of Chinese migrants to California in the latter half of the 19th century. Excavations in West Oakland have detailed changing practices with respect to waste disposal and the arrival of both sewage systems and indoor toilets in the 1880s. Collections of the California State Historical Parks contain artifacts recovered from residential refuse deposits in Monterey, which provide a detailed image of the material conditions in a mid-to-late 19th-century seaport on the west coast, while a collection of early-20th-century items from a site called the Spain Street Dump excavated at the Sonoma State Historical Park give insight into this period.

Modern Innovations

In the 21st century, the state of California and its cities have often been innovators in waste management, along with other environmental issues.

In 1986, the California Beverage Container and Litter Recycling Act was passed; three years later, in 1989, the California Integrated Waste Management Act mandated that municipalities achieve waste diversion from landfills of 25 percent by 1995 and 50 percent by 2000. Other acts, which include the Electronic Waste Recycling Act, Cell Phone Recycling Act, and Rechargeable Battery Recycling Act, require that either consumers pay an additional fee for managing the recycling of electronic devices, or that producers and retailers of products containing hazardous materials, such as batteries, are responsible for their collection when spent.

The Recycled Newsprint Act mandates that at least 50 percent of the newsprint used by printers and publishers in California contain a minimum of 40 percent post-consumer paper fiber. As of 2010, Californians were diverting 58 percent of waste, and no new landfills had been opened in over a decade, though California remains home to the largest active landfill in the United States. The Puente Hills Landfill is also using biogas generated by decomposition to fuel a power plant, called the Puente Hills Gas-To-Energy Facility, generating up to 50 megawatts of energy per day.

Disposal Diversity and Controversy

Across California, municipalities have been granted a wide degree of latitude in interpreting some of these acts. The form that waste diversion takes can vary in means and effectiveness from city to city, with some cities like San Francisco being consistently far in advance of the vanguard, while others, like Sacramento, can lag. San Francisco became the first city in the United States to ban plastic grocery bags in 2007 (a similar attempt to institute such a ban statewide failed to pass the California legislature in 2010). Already diverting an impressive 72 percent of waste into recycling and compost in 2009, San Francisco is aiming for zero waste by 2020. San Francisco has instituted the first large-scale urban program for collecting food waste for composting. Meanwhile, Sacramento as recently as 2010 has allowed residents to pile green waste in the streets for pickup, called "loose in the street pickup," by the Public Works department. A voluntary rollout of bins for green waste in March 2010 did not prevent a record number of complaints about street-side debris in April.

A famous example of the controversy that can arise over innovative plans is the "toilet to tap" process pushed by various Southern California municipalities at the beginning of the 21st century. Facing chronic water shortages and regular contamination of coastal waters by fecal coliform bacteria, "indirect potable reuse" was proposed as a method whereby local aquifers would be "refreshed" deep underground by injecting them with previously treated wastewater. The wastewater would, over time, pass through strata of rocks, sand, and minerals, rendering it safe. Unfortunately, the concept was christened "toilet to tap" by the media and seized upon by, among others, influential comedian Jay Leno, who used it as the basis for several monologues. The idea, miscommunicated as it was to the public in this disastrous public relations debacle, slunk away for the next few years. Within a decade, it had quietly been revived and put into practice in Orange County, politically a more conservative region than Los Angeles County to the north, but with similar issues concerning saltwater intrusion of groundwater and disposing of wastewater along the coast. Orange County's Groundwater Recharge Project is now the largest such project in the world.

Agriculture Operations

Further inland, concern over large-scale wastes turns from drainage issues to agricultural runoff and waste by-products. In Imperial County, the All American Canal brings in water from the Colorado River to irrigate fields, whereupon the runoff is discharged into the New River, commonly considered the most severely polluted waterway of its size in the United States. This channel originates in Mexico, crosses the border, and flows into the Salton Sea, thus becoming an international political issue—and has been so since at least the 1940s.

California's Central Valley is another agricultural center, with the Imperial Valley as the largest economic driver in the state. Enormous agricultural operations have been experimenting with the use of biosludge (a wastewater treatment by-product) as fertilizer. It comes from a variety of source material, ranging from human feces to slurry resulting from the papermaking process. The re-use of such material, which is sprayed onto fields supporting food crops, is not without controversy. Despite treatment, the biosolids can retain heavy metals and organo-chlorines, and there are limitations to how much can be spread upon a field at any given time.

Hazardous Waste

In addition to agriculture, one of California's other iconic industries, Silicon Valley, has had specific impact on the landscape. Silicon Valley, despite the aseptic appearance of high tech, is actually very toxic in terms of waste. Santa Clara County is home to the most Superfund (Comprehensive Environmental Response, Compensation, and Liability Act of 1980, or CERCLA) sites in the country—23 out of around 60. Semiconductor manufacturing has now moved largely offshore, but a toxic legacy remains. Until the 1976 Federal Resource Conservation and Recovery Act, hazardous waste disposal was largely unregulated, and companies could pump it into the ground at will, only later discovering that this threatens groundwater supplies. A prime example of this type of contamination is the Superfund site Newmark, in San Bernardino, where two underground plumes of contaminants, including chlorinated solvents, tetrachloroethylene (PCE), and trichloroethylene (TCE), resulted in the temporary closing of 20 wells within a six-mile radius. Eight of the 20 were never reopened, and the plume continues to threaten supplies down-gradient, including those of the neighboring city of Riverside, in Riverside County, which services approximately 300,000 customers. Across the state, many of the rural counties have also suffered from improper disposal of hazardous waste. Siskiyou County, on the border with Oregon, has a Superfund site at the former location of a lumber products facility. Groundwater contaminants discovered there include arsenic compounds, creosote, and heavy metals.

Due to their size and the concentration of population in urban areas along the coastline, many of these rural problems go underreported in the mainstream media. California is a study in contrasts, both social and environmental. Home to unique animal and plant life, gorgeous scenery, and diverse microclimates, it attracts tourists from around the world. Meanwhile, inland, where tourists less often go, the dirty work of maintaining a giant economic engine, the millions of people who labor in it, and the mountains of waste that they generate lie hidden.

Susan Mazur-Stommen
Independent Scholar

See Also: Archaeology of Garbage; Farms; Human Waste; Industrial Waste; Public Water Systems; Shopping Bags.

Further Readings

Baca, Marie C. "Toxic-Waste Sites Haunt Silicon Valley: Hazardous Materials Used in Early High-Tech Manufacturing Remain a Burden for Local Companies and Communities." *Wall Street Journal* (July 15, 2010). http://online.wsj.com/article/SB10001424052748704111704575355212354653420.html (Accessed September 2010).

Goldman, George and Aya Ogishi. "The Economic Impact of Waste Disposal and Diversion in California: A Report to the California Integrated Waste Management Board." Department of Agricultural and Resource Economics, University of California, Berkeley, April 4, 2001. http://are.berkeley.edu/extension/EconImpWaste.pdf (Accessed September 2010).

Kennett, Douglas J. *The Island Chumash: Behavioral Ecology of a Maritime Society*. Berkeley: University of California Press, 2005.

Praetzellis, Adrian and Mary Praetzellis. "Putting the 'There' There: Historical Archaeologies of West Oakland." Excavation Report by the Anthropological Studies Center at Sonoma State University. http://www.sonoma.edu/asc/cypress/finalreport (Accessed September 2010).

Campbell Soup Study (1930s)

The Campbell Soup Company was the largest marketer of soup cans in the 20th century and has a history that dates to the late 1800s. The development of Campbell represents an interesting case study regarding the field of consumption and waste research.

Innovations in Marketing and Advertising

Campbell is considered one of the pioneering companies regarding marketing, advertising, and brand development. Between 1910 and 1930, the Campbell Company (which changed its name to the Campbell Soup Company in the 1920s and specialized in soup production) was among the first to advertise in full color in women's magazines, explicitly addressing housewives as targets of its advertising campaigns.

In the same period, Campbell was part of the early stages of marketing research because of the study made about Campbell's consumption trends by Charles Coolidge Parlin, a forerunner in the field of marketing studies. Parlin started to work on Campbell in the early years of the 20th century, trying to convince the company to advertise in the *Saturday Evening Post*, a newspaper with readers who were typically in the lower classes. Campbell's marketers were uncertain about the potentials of a campaign like this, since they were persuaded that Campbell's typical consumers were higher-class people who could afford to buy soup cans. Lower-class people were believed to cook their own soup to save money and were therefore not interested in buying soup cans. Parlin, who used to work for the newspaper as an ad marketer, conducted a study in which he counted the number of soup cans collected from garbage in different neighborhoods across Philadelphia, Pennsylvania, and demonstrated that the lower classes consumed more soup cans than higher-class people, since the latter had cooks to cook soup for them at home. Campbell then started to advertise—successfully—in the *Post*.

Campbell was also one of the pioneers in the use of radio advertising. The company started to broadcast brief ads during radio shows in 1931, and the catchy slogan "Mm . . . Mm . . . Good!" (which used to characterize Campbell's radio ads) soon became very popular and contributed to the strengthening of the company's brand.

Art and Culture

Late in the 20th century, Campbell happened again to become part of another crucial development in consumption studies, this time regarding aesthetics. In the early 1960s, several examples of a

The columns of the Royal Scottish Academy in Edinburgh are wrapped with artist Andy Warhol's 1960s Campbell's soup can art in July 2007. Warhol's pop art image embraced mass consumerism while demonstrating its iconography and inherent contradictions.

Campbell's soup can became the iconic subject of one of the most famous serial prints by Andy Warhol, the founder of the art stream known as pop art, whose aesthetics enthusiastically embraced consumerism and mass culture while playing with their iconography and contradictions. The image of the Campbell's soup can printed by Warhol in so many different varieties has entered the collective imagination as the aesthetic icon of the age of culture industry and mass society, when consumption and mechanical reproduction met and conflated with culture, aesthetics, mass media, and advertising.

This is a crucial meeting, since it actually represents the core of what is referred to as postmodernism, or postmodernity. Postmodernism is the sensibility that epistemologically characterized the end of the 20th century, and pop art represents the aesthetic paradigm in terms of considering pop art as the "cultural crossroad." Postmodernism is usually contextualized as the epistemological evolution of modernism, whose aesthetics at the beginning of the 20th century were characterized by a strong cultural opposition to the emerging instances of mass society and mass culture, which were believed to be carrying the germs of a cultural impoverishment due to the spreading of standardized reproduction and mass consumerism.

The Campbell Study of the first half of the 1900s can therefore be read as running against the tide of modernist sensibility. It is a cultural forerunner in the sense that it perceived in advance the growing importance of marketing and consumption research. Campbell's cans portrayed in Warhol's prints can be considered as the iconographic display of consumer culture contradictions, showing the threat to originality that mass reproduction represented at the time, in the sense of a renewed relationship between the concepts of image and commodity in the era of mass society. Campbell's studies and activities, throughout the many different instances in which the company and its products have been involved, can therefore be read in a timeline perspective as a metaphor for the whole historical narrative of the 20th century.

Alessandro Gandini
Milan University

See Also: Capitalism; Commodification; Food Waste Behavior; History of Consumption and Waste, World, 1900s.

Further Readings

Huyssen, A. *After the Great Divide: Modernism, Mass Culture and Postmodernism*. Bloomington: Indiana University Press, 1986.
Jameson, F. *Postmodernism, or, The Cultural Logic of Late Capitalism*. London: Verso, 1991.
Kurtz, D. L. *Contemporary Marketing*. Mason, OH: South Western Cengage Learning, 2006.
Lippard, L. R. *Pop Art*. London: Thames & Hudson, 1966.
Parkin, K. "Campbell's Soup and The Long Shelf Life of Traditional Gender Roles." In *Kitchen Culture in America*, Sherrie A. Inness, ed. Philadelphia: University of Pennsylvania Press, 2001.

Canada

A nation of vast landholdings, yet fewer people than the state of California, Canada shares consumption and waste patterns similar to the United States. Canada generated 791 kilograms (kg) per capita of municipal waste in 2005, which is highest among all the Organisation for Economic Co-operation and Development (OECD) countries (the average of all OECD countries is 610 kg per capita) and is almost twice as much waste as Japan generated per capita in 2005. The waste generation in Canada—and, in fact, in all OECD countries—has been on the rise since the 1980s when the rate of urbanization, economic growth, income levels, revenues, and consumption started growing. With the increase in per capita income level and more disposal income, lifestyles and consumption levels increased and changed the waste stream pattern, increasing overall waste production. Waste generation in Canada has steadily increased from 510 kg per capita in 1980 to 737 kg per capita in 1995 to 894 kg per capita in 2007. Other OECD countries (which have experienced similar growth patterns) have, however, managed to keep the municipal waste generation under better control than Canada.

Ontario is the largest generator of waste. Waste generation data per province in 2002 is shown in Table 1.

Table 1 Waste Generation Data by Province, 2002

Province/ Territory	2000	2002	2000	2002
	tons		kg per capita	
Newfoundland and Labrador	X	231,291	X	445
Prince Edward Island	X	X	X	X
Nova Scotia	246,792	252,012	262	270
New Brunswick	243,300	256,190	322	342
Quebec	3,175,000	3,471,000	430	466
Ontario	4,191,337	4,388,239	359	363
Manitoba	501,921	494,535	438	428
Saskatchewan	305,901	321,069	299	323
Alberta	994,555	1,159,697	330	372
British Columbia	1,292,999	1,354,177	319	329
Yukon, Northwest Territories, and Nunavut	X	X	X	X
Canada	11,242,405	12,008,338	365	383

On average, the composition of Canadian municipal solid waste consists of 40 percent organics, 26 percent paper, 9 percent plastic, 3 percent glass,

4 percent metal, and 18 percent other (a mixture of animal waste, textiles, wood, and tires).

Disposal

In 2002, Canadian households generated over 12 million tons of municipal waste, of which only 2.5 million tons was diverted to either recycling, reuse, or composting. The remaining 9.5 million tons were disposed of mainly in landfills, and part was incinerated. According to the statistics for 2000, about 95 percent of the waste was disposed of at landfill sites, and the remaining 5 percent was incinerated. About 83 percent of the waste disposal facilities are publicly run, but they dispose of only 56 percent of waste, while the 17 percent of facilities that are privately run dispose of about 44 percent of the waste. Since landfills are the most commonly used method of disposal, they exist in every province and territory of Canada. The new landfill sites are sanitary, properly engineered landfills, but there are still a few landfill sites that are old and have issues with leachate and greenhouse gas emissions.

Most of the waste in Canada is landfilled, and municipalities are having trouble finding space for landfills, mainly because people do not want landfills in their neighborhoods. For example, in the 1990s, when Toronto's Keele Valley landfill reached near capacity, the municipality explored the possibility of shipping its waste 590 km north, by rail, to Kirkland Lake, Ontario. The 9,000 residents of Kirkland Lake protested the idea of having Toronto's municipal waste dumped into an abandoned mine in their community. In the end, the community won, and the decision to use the mine was overturned by the Toronto City Council. Instead of working on better solutions or waste management planning in Toronto, the decision was made to export the waste from Toronto to a landfill site in Michigan. Toronto paid $55 per ton for shipping its trash to Michigan ($35 for trucking and $20 for landfilling). There are additional environmental impacts associated with transporting the waste. Toronto is working on reducing the municipal household waste going to landfill sites; yet, instead of developing a landfill near the waste-generation source, it has begun trucking waste to London, Ontario, and its surroundings.

Other municipalities are also sending waste out of city boundaries; for example, the Greater Vancouver Regional District sends about 16 percent of the solid waste generated to Cache Creek Landfill (330 km northeast of the city). Kenora transports part of its waste out of province to the Winnipeg landfill site.

Recycling

Most of the waste management programs originate at the municipal level, hence waste management differs from city to city. Many Canadian municipalities have, however, initiated successful recycling programs. Paper, fibers, glass, cans, and organic materials are recycled and composted across Canada. Many municipalities have ambitious programs to divert as much as 60–70 percent of the waste away from landfills. In Hamilton, Ontario, some communities have a one-bag-per-week limit of waste going to landfill sites but can have recycling and compostable waste in addition to it. Toronto has set the ambitious goal of diverting 100 percent of all household waste—meaning everything will be somehow reused or recycled—by 2012.

In general, recycling in Canada has grown since the 1990s and has become an integral part of waste management practice. According to 2002 data, 6.6 million tons of nonhazardous waste material was collected by local waste management companies or organizations. The bulk of this material was mixed paper (23 percent) and organic waste (18 percent).

Composting has also become popular since the 1990s and is used at both household levels and on a larger scale. In 2002, about 1.2 million tons of organic waste was composted at 351 centralized composting facilities in Canada. There is no data available on composting performed in backyards or on-site by some institutions.

Incineration

In 2000, 5 percent of total municipal waste (1.1 million tons) was incinerated in about 21 incinerators across Canada. There is an ongoing debate about the use of incinerators in Canada. For example, in 1991, Ontario's New Democratic Party (NDP) government placed a ban on the construction of new incinerators to burn municipal garbage and the expansion of existing incinerators. One of the reasons given by the NDP government for placing in the ban was that "incineration is incompatible with the 3Rs." This ban was removed in 1995, when the Progressive Conservative government came into power.

Incineration opponents use two main arguments to support this contention. The first is that it is incompatible with the reuse, recycle, and reduce principles (the 3Rs) because incinerator operators and recycling operators compete for the same materials. Most of the time, the materials recycled, such as paper, plastic, and fiber, are the ones with high calorific value that are also needed for incineration. The only exceptions are glass and metals. When the Ontario government was considering a ban on new or expanded incinerators, the pulp and paper industry opposed the move because the industry had invested in recovery, recycling, and de-inking facilities, and it feared that it would lose the waste paper that it wanted back in production. The second argument given by opponents is the high cost of building an incinerator and its ability to move attention away from diversion activities. Incineration experts argue that for an economically viable operation, the capacity of an incinerator should accommodate at least 1,000 tons of garbage per day (which is massive if diversion activities are carried out). The cost to build such a facility is approximately $100 million. Operating costs to maintain the equipment, especially the pollution control equipment, is also high. Incineration opponents say that dramatic increases could be made in waste diversion programs if these resources were put into diversion programs instead of incinerators.

Changes made in 2007 at the behest of the municipalities to the Environment Protection Act (Ontario) made it easier to test emerging incineration and thus encourage municipalities to consider incineration as a means of disposing of garbage and probably cogenerate electricity. Although several municipalities in Ontario are poised to exploit the legislative changes, the chair of Toronto's works and infrastructure committee was against in 2010.

Waste Management Responsibilities

The responsibility of solid waste management in Canada is shared among all three levels of government: federal, provincial, and municipal. Collection, diversion (which includes both recycling and composting), and disposal fall under the jurisdiction of municipal governments, while the approval, licensing, and monitoring of operations come under the supervision of provincial governments. The federal government monitors issues related to toxic substances, the international movement and trade of waste, federal lands and operations, air emissions (including greenhouse gas emissions), and strategic planning to include sustainable development goals in waste management across Canada through federal funds.

Conclusion

Canadian municipalities are working toward ambitious goals of diverting waste away from landfill sites but more needs to be done to move toward sustainable and integrated waste management. New legislation is also in place to minimize waste; for example, user fees, tipping fees, and limitations of waste bags. Edmonton, Prince Edward Island, Guelph, and Halifax are some municipalities leading the programs to divert waste in Canada. Other instruments, such as deposits for bottles and producers' responsibility for packaging waste, have also been introduced.

Velma Grover
York University

See Also: Composting; Incinerators; NIMBY (Not in My Backyard); United States; Zero Waste.

Further Readings

Conference Board of Canada. "Environment: Municipal Waste Generation." October 2008. http://www.conferenceboard.ca/hcp/details/environment/municipal-waste-generation.aspx#way (Accessed November 2010).

Environment Canada. "Municipal Solid Waste." November 2010. http://www.ec.gc.ca/gdd-mw/default.asp?lang=En&n=EF0FC6A9-1 (Accessed November 2010).

Statistics Canada. Environment Accounts and Statistics Division. *Human Activity and the Environment: Annual Statistics 2005*. Ottawa, Canada: Ministry of Industry, 2005. http://www.statcan.gc.ca/pub/16-201-x/16-201-x2005000-eng.pdf (Accessed November 2010).

Candy

The package that candy comes in is a large part of its pleasure. From the tale of the five golden tickets in the Wonka chocolate bars in Roald Dahl's iconic novel *Charlie and the Chocolate Factory* to the heart-shaped gold and velvet embossed Valentine's Day candy boxes to Cracker Jack toys and comics or baseball cards bundled in with bubble gum, people love candy and the package it comes in.

Marketing

Even everyday candy sells itself through its packaging; a cursory look at the candy section of even the smallest convenience store will reveal an amazing display of marketing strategies in the form of package design, materials, color, and font. Novelty candy, like some breakfast cereals, have actually become part of the broader marketing strategy of other products. There are candies that are tied to the release of new movies, sporting events, and even concert tours as exemplified by the guitar- and microphone-shaped gummy candy sold to promote the 2010 concert tour of 15-year-old Hannah Montana.

Holidays

Candy is celebratory. Almost every culture celebrates with sweets, but in the United States, candy and its holiday packaging is central in the celebration of holidays such as Easter, Halloween, Valentine's Day, Christmas, and New Year's; a box of chocolates is often considered the perfect gift for the person who is "hard to buy for." Candy is also big business. Much of the candy industry's total sales of around $28 billion per year are sold around the holidays. According to the National Confectioners Association, of the $92.91 spent on candy per person in the United States, $20.39 is spent on Halloween candy alone.

Most candy packaging ends up in landfills and can also be found anywhere people litter, such as on streets or in waterways. The mixed polymers used in plastic-film candy packing do not recycle well; most of this waste eventually winds up in a landfill.

Consumer-Driven Packaging and Waste

All of this celebrating results in a punishing amount of material being sent into the waste stream. Waste and Resources Action Program (WRAP), the United Kingdom's waste management program, estimates that 3,000 tons of waste are generated by the packaging of Easter candy bought in the United Kingdom. In 2009, WRAP worked with confectionary manufacturers, who then significantly reduced the environmental impact of their Easter products.

Not all of candy's packaging is based on marketing strategy; much of it is driven by science. A candy dish is still seen as a form of hospitality, whether it is sitting on a grandmother's coffee table or on a hostess counter at a restaurant. Historically, people would simply reach in with fingers to grab a few unwrapped candies, but people now want after-dinner mints to be individually wrapped so that they are germ free. Keeping germs out of foodstuffs requires packaging.

Some design is concerned with keeping candy from breaking; people want to open candy and find it intact to have the pleasure of breaking it, dividing it up, unraveling or stretching it into bite-size pieces. Keeping candy intact during shipping and storing requires packaging.

People want candy to have beautiful colors, to be glossy, and to have a delightful aroma on its way to salivating mouths. In order to keep good odors in (and other odors out) and in order to minimize the exposure that will dull color and sheen, packaging is required. And all this cardboard, paper, aluminum foil, and plastic packaging that preserves the purity, appearance, and shape of the candy has to go somewhere.

Disposal

According to an Environmental Protection Agency 2008 report, containers and packaging generate 30.8 percent of municipal solid waste, representing the largest portion of the total amount. Food scraps made up 12.7 percent of the total municipal solid waste that year, but candy is really not a major offender in this category. Candy is not highly perishable, and for most people, candy is a prized food, meaning the candy is likely to be consumed. However, most candy packaging ends up in landfills. It is also present as litter: on the street, in waterways, and along roads. The packaging consists of some paper, aluminum foil, and plastic. The mixed polymers that compose the plastic film used in candy packing do not lend it to recycling; it can be incinerated, but most of this waste eventually finds its way into a landfill.

Bulk candy, some of which is not individually wrapped, is sold in some venues, and is the best choice for anyone concerned about unnecessary packaging entering the waste stream. As in other product categories, there has been some effort to use biodegradable packaging, and there is a small industry that uses recycled juice bags and some brands of candy wrappers to create tote bags, jewelry, and other accessories. But these efforts are small in proportion to the problem.

Maybe the worst offenders are products that are both candy and toys, such as lollipops that have flashing lights, lollipops attached to a battery-powered motor that makes it spin, and plastic covers that make them look like microphones.

Although it seems a large part of candy marketing is aimed at children, it may be more correct to say it is directed at childhood. Adults buy candy for children as a means of vicarious pleasure, remembering how much candy meant to them as children. Part of the pleasure of candy is nostalgia, not just of the candy itself, but also of the packaging. One development in the candy industry is marketing of nostalgia candy; consumers want the candy of their childhood, and this demand is being met primarily through Internet sales. The home delivery of products has some increased impact on the waste stream if consumers do not make the effort to recycle the cardboard boxes, plastic bags, and the other packaging required to get the product to them intact.

Production and Globalization

Candy consumers have been affected by both globalization and the difference in government standards for food quality. These factors have impacted the safety of the commodities used to make candy products. Lead is a toxin that disproportionately affects children, and starting in the 1990s, the California Department of Health and the U.S. Food and Drug Administration found that some Mexican candies had large amounts of lead in them because of the candies being stored in lead-glazed clay pots and candy wrappers printed with lead-based ink. Additionally, some Mexican chili powder (an ingredient in many kinds of Mexican candy) contains high lead levels. In 2007, the world learned about Chinese products that were laced with the toxin melamine; as the scandal unfolded, melamine was found in milk and milk products, which resulted in melamine being found in candy and other food products that were distributed inside China and exported to countries around the world, including the United States.

There are also some social justice issues tied to the consumption of chocolate, as the average person in the United States consumes 11 pounds of chocolate a year, but essentially all cacao beans, chocolate's primary ingredient, are grown outside the United States. The poor countries in west Africa produce 70 percent of the world's cocoa. In 2000, in response to reports of child labor exploitation, the Harkin-Engel Protocol was developed by two members of the U.S. Congress, nongovernmental organizations, labor experts, and the World Cocoa Foundation. There have also been some efforts to establish Fair Trade sources of cocoa, to improve the economic conditions of the farmers who grow this commodity in poor countries. Candy, as a processed food containing sweeteners such as cane sugar or high fructose corn syrup, is also a conspicuous symbol of caloric overconsumption in the industrialized world.

Fran Mentch
Cleveland State University

See Also: Children; Consumerism; Fast Food Packaging; Food Consumption; Food Waste Behavior; Gluttony; Packaging and Product Containers.

Further Readings

Jacobs, Andrew and Mark Mcdonald. "F.D.A. Opens Office in Beijing to Screen Food and Drug Exports: [Foreign Desk]." *New York Times* (November 20, 2008).

Lidsky, D. and K. Rockwood. "Numerology: The Business of Candy." *Fast Company*, v.139 (2009).

Medlin, J. "Sweet Candy, Bitter Poison." *Environmental Health Perspectives*, v.112/14 (2004).

National Confectioners Association. "Annual Confectionery Industry Review—2009 Industry Review." http://www.candyusa.com/Sales/content.cfm?ItemNumber=1440&navItemNumber=2685 (Accessed December 2010).

Parmar, Neil. "Idled Toy Inventors Find a Sweet Niche: 'Novelty' Candy." *Wall Street Journal* (Eastern Edition) (July 21, 2005).

"U.K. Confectioners Reduce Easter Egg Packaging." *Candy Industry*, v.174/3 (2009).

U.S. Department of Commerce. "U.S. Department of Commerce Industry Report Food Manufacturing NAICS 311." 2008. http://trade.gov/td/ocg/report08_processedfoods.pdf (Accessed December 2010).

World Cocoa Foundation. "Encouraging Sustainable, Responsible Cocoa Growing." http://www.worldcocoafoundation.org/addressing-child-labor/reports.asp (Accessed December 2010).

Capitalism

The word *capitalism* has become heavily politically loaded in contemporary usage; it is a term that nearly always arouses dispute, and there is no consensus on its exact definition. It has been identified with a variety of different phenomena across a range of historical contexts and is subject to very differing interpretations according to one's worldview. It has been used to describe a set of social and economic practices and relations, but it is also seen as an ideology or way of thinking. As a set of practices, capitalism refers generally to a free, competitive market economic system in which the means of production (such as materials, land, or tools) are privately owned, as opposed to cooperatively or state owned, and they are managed and operated for private profits. In addition, private individuals and organizations, as opposed to governments, make decisions regarding supply, demand, distribution, price, and investments. Most developed countries are regarded as mixed economies as opposed to purely capitalist economies as governments continue to play a role in regulating the economy through ownership and intervention. Countries and regions such as Europe, the United States, Canada, Japan, Singapore, and Australia all rely on a largely mixed economy, where many transactions occur outside the marketplace and are subject to public-sector regulation. These include, for example: public pensions, social security, unemployment benefits, healthcare, and education. In this sense, the operation of a completely free-market-driven economy remains largely elusive.

Ideology

As a way of thinking, capitalism embodies a series of ideologies. It embraces the ideology of individualism, wherein individuals are seen as free to make market choices based on self-interest. Along with competition, supply, and demand, self-interest is thought to be a key mechanism for allocating resources in society. Some scholars also observe that the amalgamation of individual interest will benefit society as a whole. A second key tenet of capitalism is the Enlightenment idea of progress. This idea of progress is seen as synonymous with economic growth achieved through the exploitation of productive labor and technological advancement. The idea of productive labor is central to capitalist thought, viewed as vital to economic growth and central to a meaningful life. Some also argue that capitalism is almost entirely future focused. In the view of humans as rational and self-interested, attention is placed on the future of economic decisions. Finally, in capitalist modes of thinking, the object takes on specific significance. The gap between consumers and producers widens, and consumers have little or no social relationship with the producers of the goods they consume. In the absence of this set of social relations, increased emphasis is placed on objects of consumption, and individuals become increasingly defined by—and take meaning from—their role as consumers, as opposed to producers, of goods.

Etymology

A useful inroad to understanding capitalism might be a consideration of the etymology of the term. Early use of the word *capital* is associated with the trade in livestock. It evolved from the word *capitale*, which is based on the proto-Indo-European word *caput*, meaning "head." It is also developed from the terms *chattel* and *cattle*, referring to movable property. While it is difficult to trace the first use of the term *capitalism* in English, novelist William Thackeray is widely cited as the first to refer to it in his 1854 text *The Newcomes*. In an essay published in 1851, a French socialist politician and one-time friend of Karl Marx, Pierre-Joseph Proudhon, argued that capitalism as an economic idea was analogous to government. In their earlier writings, Karl Marx and Friedrich Engels used the term *capitalism* only a few times, instead preferring to refer to the "capitalist mode of production," in the three volumes of *Das Kapital* (1867, 1885, and 1894). It was not until the beginning of the 20th century that the term came into widespread use and began to stimulate debate. Werner Stombart's 1902 *Modern Capitalism* introduced the term to the political arena. This was followed in 1904 by Max Weber's use of the term in *The Protestant Ethic and the Spirit of Capitalism*.

History

Historically, the roots of capitalism can be traced through a series of market forms: mercantilism, industrialism, Keynesianism, neoliberalism, and globalization. Commentators typically observe the roots of capitalism to lie in mercantilism, which originated in the Middle East and Rome in the Middle Ages. Mercantilism was underpinned by protectionist policies involving the restraint of imports and encouragement of exports in order to produce a favorable balance of trade. The differences between imports and exports were then paid by foreign countries in precious metals. One of the ideas central to mercantilism was bullionism, which stressed that the economic health of a nation could be measured by the amount of precious metal, gold, or silver it possessed. The mercantile system, however, was far from egalitarian; it largely served the interests of merchants and producers whose activities were protected or encouraged by the state. Organizations such as the British East India Company and the Dutch East India Company profited vastly from a system wherein their trade interests were protected and supported by nation-states. This era was characterized by sea travel and discovery, an increase in the capacity for trade, the establishment of a series of colonies beyond Europe, and significant growth of European industry relative to agriculture.

The period of mercantilism in the 16th to mid-18th centuries was followed by industrialism. Key thinkers such as Adam Smith challenged the central tenets of mercantilism, in particular, its focus on the idea that the world's wealth was finite and that gains could only be made by transferring capital between countries. Instead, he focused on creating efficiencies through the mechanisms of production. Smith is often cited as the "father of capitalism," and in his rationale for free trade, he promoted the idea that rational individual self-interest would promote the wider well-being of society. Therefore, along with supply and demand, self-interest ought to be a key mechanism for allocating resources in society. The mechanization that accompanied the Industrial Revolution marked the decline of small-scale artisan and handicraft production.

During this period, the idea of the factory and factory work emerged in the guise of Frederick Taylor's principles of scientific management, also popularly termed *Taylorism*. Here, the focus was on the application of scientific principles to management in order to improve economic productivity and labor efficiency. Work was divided into small, specialized functions involving a de-skilling of labor and the rise of monotony in the workplace. This period of industrialism also marked the heyday of a series of successful U.S. industrialists who were aggressively innovative and competitive, including Andrew Carnegie, the Scottish American entrepreneur who developed new methods to mass-produce steel rails for railway lines and John D. Rockefeller, who made his fortune in the oil industry by tightly controlling costs, using refineries' waste, and developing his own oil pipeline network.

Keynesianism and Neoliberalism

While mercantilism and industrialism both contributed to the capitalism commonly understood in the 21st century, the two economic systems most closely associated with its development—although in very different ways—are Keynesianism and neoliberalism. Keynesianism is so named after British economist and social liberalist John Maynard Keynes. Largely in response to the 1930s United Kingdom (UK) recession, Keynes developed a series of monetary and fiscal measures to alleviate the effects of economic recessions and stabilize the economy. These measures included greater government involvement in regulating the business cycle. These policies were largely successful in the 1950s and 1960s in the United Kingdom and other Western economies.

This period saw the establishment of the welfare state in the United Kingdom. By the 1970s, Keynesianism started to fall out of favor, spurred by the critiques of economists such as Milton Freidman, who argued forcefully for a laissez-faire economic system, which emphasized individual choice and reduced government intervention in markets. During the 1980s, the governments of Margaret Thatcher in the UK and Ronald Reagan in the United States were heavily influenced by the neoliberal school of thought, in particular, free markets, control over public expenditure, tax cuts, and privatization. Many of the received contemporary capitalist ways of thinking stem from the ideas of the neoliberal school of thought.

Globalization

The increasingly global nature of economic systems makes it important also to consider the impact of globalization on the development of capitalism. The term *globalization* embraces a wide range of cultural, political, social, and economic trends, but in general, globalization refers to the integration of national economies into the international economy. This is achieved in a variety of ways, but includes the reduction and removal of barriers between national borders, which has resulted in an increased flow of people, ideas, and capital. Another trend contributing to globalization includes improvement in communication technologies, such as the World Wide Web and satellite communications. In the West, it has also marked a shift away from the focus on manufacturing and production that formed the basis of industrialism. Trade liberalization and labor law reforms meant that companies could easily outsource production and labor to cheaper sources in developing countries as a way to increase cost efficiencies. Emphasis shifted away from manufacturing and instead toward marketing.

Whereas in the past, companies competed primarily on the strength of the goods they produced, 21st-century companies compete in the brand market for the most powerful brand image. Commentators such as Naomi Klein have termed this shedding of staff and manufacturing processes a "race toward weightlessness."

Major Social and Cultural Impacts

The idea of capitalism is not merely restricted to it as an economic system. Many of the critiques of capitalism are based on the social and cultural outcomes and implications of such a system. Critiques are diverse and varied, but it is possible to identify three key dimensions to them: alienation or separation, the exploitation of labor, and the profit motive.

Alienation. Karl Marx argued that alienation was a result of capitalism. His theory of alienation observed that workers become alien or foreign to the world they inhabit because workers in the capitalist system do not own the products of their labor. In addition, the means to define themselves as individuals resides entirely in a system of production that is privately—as opposed to collectively—owned. In such a system, the individual is not viewed as a social being, but rather as an instrument to produce surplus value through work. In individuals' bid to work to earn money for consumption, they are increasingly defined as consumers as opposed to producers, which brings with it critiques of consumerism as inauthentic and as ultimately unsatisfying. In a contemporary twist to the alienation perspective, sociologist Juliet Schor highlights the problems inherent in the work-to-spend cycle in the United States, observing spiraling levels of debt and increasing working hours among U.S. laborers.

Exploitation. Commentators observe that in a capitalist system, individuals are forced to sell their capacity to work in order to survive, and labor becomes a commodity. Here, work becomes the central, organizing motif of life and is seen as the only way to enjoy a meaningful existence.

Increasingly long working hours are required, and the gradual encroachment of work time on leisure time and a blurring of boundaries between working lives and private lives ensue. In a capitalist system, work therefore increasingly structures individuals' identity, sense of self, and sense of place within the world. In addition, the fundamentally unequal labor relation built into the capitalist system results in inequality on both local and global levels and stimulates the gap between the rich and the poor both within societies and between countries. Because capitalism is about the concentration of wealth, a concentration of power follows this flow of wealth.

Profit Motive. The third key criticism of capitalism is the limitless search for profit that is central to its survival. In its ceaseless search for growth, capitalism demonstrates a predatory disregard for both people and the environment, viewing them as resources for exploitation. Environmentally, capitalism's hunger for growth is depleting the Earth's finite natural resources. Capitalism has recently been heavily involved in harnessing global labor surpluses. Driven by the profit motive, corporations have looked to less-developed economies for cheaper sources of labor.

This process has been epitomized by the establishment of Export Processing Zones (EPZs) where trade barriers, including tariffs and quotas, are eliminated. Factories in these zones involve highly labor-intensive work to transform raw materials into products. However, as activist Naomi Klein observes, the lack of regulated labor laws in EPZs in parts of South America and Asia result in sweatshop-labor practices, with employees working long hours in very poor conditions for below-subsistence wages.

Effects on the Environment

The spatial reorganization of local, national, and global economies under capitalism has consequences for the environment. The wealth of Andrew Carnegie, for example, affected several regions. When Carnegie lived in New York City, his company extracted iron ore from rural Minnesota, leaving gaping holes and toxins in the soil. The ore was sent to Pennsylvania steel mills, producing pollution in the land, air, and water. Resulting steel was sold all over the world, with Carnegie amassing great profits. He subsequently engaged in philanthropy, redistributing wealth on his terms.

Twenty-first-century developments include the evolution of globalized waste markets, with exports of hazardous waste from wealthy nations to developing economies. These developments exacerbate concerns over growing environmental inequalities as the global economy grows.

Anticapitalism

In response to the many critiques, a series of anticapitalist groups have developed over the years. The anticapitalist position is a particularly difficult one to occupy given that capitalism is consistently promoted as the only possible—and only natural—means of organizing the economy. Therefore, much anticapitalist activity is about developing and promoting alternatives to a capitalist way of thinking. In doing so, anticapitalism is concerned with interrupting the ceaseless search for profit, reuniting individuals with the means and processes of production through collective activity, and placing values other than economic ones as central to meaning in life.

Elizabeth Parsons
Keele University

See Also: Commodification; Consumerism; Industrial Revolution; Materialist Values; Needs and Wants; Socialist Societies.

Further Readings

Braudel, Fernand. *Civilization and Capitalism: 15th–18th Century*. London: HarperCollins, 1982.

Harvey, David. *The Enigma of Capital and the Crisis of Capitalism*. London: Profile Books, 2010.

Klein, Naomi. *No Logo: Taking Aim at the Brand Bullies*. Toronto: Knopf Canada, 2000.

Schor, Juliet. *The Overspent American: Why We Want What We Don't Need*. New York: Harper Perennial, 1999.

Wallerstein, Immanuel. *Historical Capitalism ith Capitalist Civilization*. London: Verso, 2011.

Car Washing

Along with motels, auto cinemas, drive-ins, and other symbols of roadside architecture, car washing became an essential component of the U.S. car culture for a large part of the 20th century. Professional car washes thrived whenever the suburban Saturday morning dream of household car washing became a practical impossibility because of time constraints for car owners and new environmental restrictions and regulations.

Waste

There are some differences in the level of waste and resulting environmental damage derived from car washing, depending on the type of car wash (driveway versus professional). According to engineering studies, a 5/8-inch hose running at 50 psi uses approximately 14 gallons of water per minute. This means that a rather short domestic car wash can use between 116 and 180 gallons of water. Professional car washes consume approximately 60 percent less water than driveway washes, depending on the method they use (manual, conveyor, in-bay automatic, or self-service systems).

Pollution

Besides water-consumption issues, the dirt washed off vehicles as well as many of the materials used in the process of cleaning are generally harmful to the environment. Washing vehicles (both by hand or using a machine) on an impermeable surface can cause wash wastewater flow into storm drains. Soap from washing cars at home or charity car washes generally causes a lot of pollution, since soap and dirt flow freely through storm drains and ditches, ending up in streams untreated.

Contaminants in most of wash wastewater include the following:

- Petroleum hydrocarbons (such as gasoline, diesel fuel, motor oil, fluids, and lubricants) from automobile engines, leaks, and fuel combustion processes. Oil and grease contain hazardous materials such as benzene, lead, zinc, chromium, arsenic, and metals resulting from normal wear of auto brake linings (copper), tires, exhaust, and fluid leaks.
- Detergents, including biodegradable detergents. Surfactants that are present in detergents and cleaning formulations (both synthetic and organic agents) lower the surface tension of water, allowing dirt or grease to be washed off. Detergents can destroy the external mucus layers that protect fish from bacteria and parasites. Detergents can also damage the gills. Most fish will die when detergent concentrations approach 15 parts per million. Detergent concentrations as low as five parts per million can kill fish eggs.

  Surfactant detergents are implicated in decreasing the breeding ability of aquatic organisms. Organic chemicals such as pesticides and phenols are then much more easily absorbed by the fish. A detergent concentration of only two parts per million can cause fish to absorb double the amount of chemicals they would normally absorb, although that concentration itself is not high enough to affect fish directly.

  Phosphates and nitrogen-containing detergents (nitrates) that are plant nutrients could cause excessive growth of pest plants at or in nearby water bodies. Water naturally contains less than 1 milligram of nitrate-nitrogen per liter. State and federal laws in the United States set the maximum allowable level of nitrate-nitrogen in public drinking water at 10 milligrams per liter (10 parts per million). Infants who ingest water that is high in nitrate can develop methemoglobinemia.

  This condition is also called "blue baby syndrome" because the skin appears blue-gray or lavender. This color change is caused by a lack of oxygen in the blood. Common sources of nitrate contamination include fertilizers, animal wastes, septic tanks, municipal sewage treatment systems, and decaying plant debris.
- Chemicals harmful to living organisms, such as hydrofluoric acid and ammonium bifluoride products (ABF), along with solvent-based solutions. Also, chemicals and oils used for the maintenance of cleaning machinery (for automatic systems) can be discharged into sewers.

The U.S. Clean Water Act of 1972 requires professional car wash facilities to pipe their wastewater to water treatment facilities or to state-approved drainage facilities designed to protect the environment and avoid pollution. Contaminants in car wash wastewater include dirt, debris, and petroleum products from the automobile; biodegradable detergent products; surfactants; phosphates and nitrogen-containing detergents; and chemicals harmful to living organisms, such as hydrofluoric acid and ammonium bifluoride products.

- Debris that can clog storm sewer inlets and grates and thereby prevent storm water drainage to the sewer.

An empirical quantification of the environmental effect of driveway car washing was revealed by the Residential Car Washwater Monitoring Study published by the city of Federal Way, Washington, in July 2009. The study concluded that an estimated 18.85 washes per car per year sent a significant amount of pollutants directly to the community's water bodies. It was calculated that each car contributes each year with approximately 0.05 pounds of petroleum hydrocarbons (gasoline, #2 diesel, and motor oil), 0.093 pounds of surfactants (MBAS), 0.0042 pounds. of metal (chromium, copper, lead, zinc, and nickel), and 0.019 pounds of different inorganic substances (ammonia, nitrate-nitrite, and phosphorus) that are thrown to water bodies. Additionally, driveway car washing in the Federal Way sends 1.43 pounds of solids per car per year to those water bodies.

Treatment and Filtration

The U.S. Clean Water Act requires professional car washes to direct car wash wastewater to water treatment facilities or to state-approved drainage facilities designed to avoid pollution. Filtration of the wastewater is a recommended practice conducted before discharge to a sanitary sewer. Filtration leaves fewer solids present in the wash wastewater stream discharge to the sanitary sewer system. The sludge can be dried by removing it from the car wash system and allowing the water to evaporate. Since the transportation of sludge to be dried requires a special permit, some places have on-site drying facilities where water from sludge is evaporated before disposing of the resulting solids as nonspecial waste that can be disposed of with general refuse. Special waste derived from car washing must be handled and disposed of in accordance with specific Environmental Protection Agency (EPA) regulations. Special waste must be disposed of in a licensed, special waste disposal facility and must be transported by a licensed special waste hauler using a special waste manifest.

Most states require a National Pollutant Discharge Elimination System (NPDE) permit from the respective state EPA for businesses that discharge car wash wastewater directly into a surface water body or to a storm sewer that discharges to a surface water body. It is a common practice that if car wash wastewater is discharged directly to a sanitary sewer system, the business must apply for a state construction permit and may also need to apply for a state operating permit. Nonetheless, car wash regulations vary from city to city.

Mauricio Leandro
Graduate Center, City University of New York

See Also: Automobiles; Engine Oil; Pollution, Water; Sewers; Sociology of Waste.

Further Readings

Environmental Protection Agency. "Pollution Prevention and Toxins." http://www.epa.gov/oppt (Accessed November 2010).
Storm Water Center. "Pollution Prevention Fact Sheet: Car Washing." http://www.stormwatercenter.net/Pollution_Prevention_Factsheets/CarWashing.htm (Accessed March 2011).

Carbon Dioxide

Carbon dioxide (CO_2) is ubiquitous. It is a chemical compound that is commonly encountered, for example, in chemistry classes in high school. It also entered the global stage of climate change politics and economies as a currency of emissions to be traded on carbon markets. Thus, a definition of carbon dioxide must engage with the complexity of its status in society.

Definition

From the point of view of the scientific discipline of chemistry, the concept of carbon dioxide refers to a molecule in which two oxygen atoms are bonded to a central carbon atom. Technically speaking, at standard pressure and temperature (near mean sea level pressure and at 0 degrees Celsius) CO_2 exists as a gas. Only at lower temperatures, below minus 78 degrees Celsius, the gas deposits directly in its solid form: dry ice. Carbon dioxide exhibits a variety of traits, making it useful for contemporary societies. To illustrate, the beverage industry uses the acidic characteristic of CO_2 to produce carbonated drinks, like soda water. Another characteristic of the gas is its nonflammability. This is used, for example, in fire extinguishers.

Effects on Climate

In the second half of the 20th century, scientists increasingly alarmed the public about the relationship between the proportion of CO_2 in Earth's atmosphere and global warming. Svante Arrhenius provided an argument mentioning the possibility of such warming in 1896. This relates to another quality of the chemical compound: CO_2 in the atmosphere lets solar radiation pass onto Earth, but traps some of the radiation when reflected back toward space. In effect, the atmosphere is heating up—just like a greenhouse experiencing increasing temperatures under solar radiation. Thus, the more carbon dioxide is in the atmosphere, the more intense is the greenhouse effect. Based on this understanding, which is routinely elaborated on and updated by the Intergovernmental Panel on Climate Change (IPCC), many governments set up mechanisms to reduce the concentration of CO_2 in the atmosphere.

Reduction Initiatives

For setting up such mechanisms, the sources of the increase of atmospheric carbon dioxide had to be identified. The Environmental Protection Agency (EPA), for example, names as the five largest human-related sources of carbon dioxide emissions as the combustion of fossil fuels, nonenergy use of fuels, iron and steel production, cement manufacturing, and natural gas systems. All these processes are not primarily targeted to produce carbon dioxide. Rather, CO_2 is a by-product. Thus, carbon-as-waste is socially created by defining processes as designed to produce a specific output (for example, energy), while the by-product CO_2 emission is implied as something not wanted.

At Kyoto, Japan, in 1997, the international community devised a central mechanism through which the concentration of atmospheric carbon dioxide was to be reduced: carbon markets. At these markets, permits to emit CO_2 would be traded. Furthermore, the international community intended to reduce the amount of several other significant greenhouse gases in the atmosphere, rather than merely focusing on carbon dioxide. Other greenhouse gases have also been conceptualized in magnitudes of the global warming potential of CO_2. This allowed using the market mechanism to target CO_2 as well as equivalent greenhouse gases.

Carbon as a Signifier of Greenhouse Gases

The Kyoto Protocol referred to six greenhouse gases (and groups thereof) slated for reduction: carbon dioxide (CO_2), methane (CH_4), nitrous oxide (N_2O), hydrofluorocarbons (HFCs), perfluorocarbons (PFCs), and sulphur hexafluoride (SF_6). Within the hegemonic climate change discourse, these gases are measured in terms of their global warming potential. The global warming potential (GWP) of a gas is calculated by incorporating the lifetime of the gas and the degree by which it influences global warming as much as carbon dioxide. The GWP is defined relative to a time horizon, typically 100 years. The reason for translating greenhouse gases into carbon dioxide is its usability as a historical proxy of atmospheric warming due to being well distributed in the atmosphere. Therefore, scientists and policy makers co-constructed carbon dioxide as a currency in which the effect of greenhouse gases can be measured. For this to work, the global warming potential of CO_2 is defined as 1. For methane, for example, in 1995, the IPCC calculated methane's GWP for a time horizon of 100 years as 21 CO_2 equivalents (CO_2eq). Thus, over a period of 100 years, 1 unit of methane causes as much global warming as 21 units of CO_2. Later, the IPCC's 2007 analysis reported the GWP of methane as 25 CO_2eq over the same horizon. The reported GWP of the hydrofluorocarbon HFC-23 changed between 1995 and 2007 from 11,700 to 14,800 CO_2eq. Thus, GWPs are subject to change. The reports themselves recognize significant degrees of uncertainty.

To understand discussions about climate change and the significance of the amount of CO_2 emissions to be avoided, it is relevant to recognize that carbon dioxide is used as a currency representing GWPs and the unit is CO_2eq. When politicians discuss reducing the concentration of carbon dioxide, they are likely to refer to a variety of substances summed up as CO_2. The result of this has been the notion of carbon. The term *carbon*, then, refers to all of these six gases that have been made commensurable.

For the hegemonic discourse, the establishment of carbon as a signifier of greenhouse gases had positive effects. The fight against climate change was now addressable in a common currency. Different policy instruments and technologies can be assessed in terms of their carbon saving potentials. However, while carbon is used as a general token for those gases that have global warming potential, the scientific-legal implementation of the concept actually limits its range of representation to the six Kyoto gases, necessarily silencing other climate change–relevant dynamics.

Carbon Accounting

The regime to tackle climate change is fundamentally based on reducing carbon input into the atmosphere. The management of the atmospheric carbon load is based on quantifying (potential) carbon emissions. Such quantification can be achieved through a variety of direct and indirect paths of calculation. In principle, physical devices can be used to measure the amount of carbon emissions. This end-of-pipe measurement is, however, rarely used. Normally, actors use less direct quantification strategies:

To comply with the Kyoto Protocol, countries may determine their direct carbon emissions by including all the carbon emissions associated with, for example, fuels sold and cement produced in the given state. Then, the carbon emissions are counted as the country's emissions independently of where they are actually used. This approach requires knowing the quantities of those substances that are conceptualized as causing the country's carbon emissions. To know the amount of fuel sold in the country, accountants might use the tax on fuels. One can derive the amount of fuels sold via the sum of value-added taxes paid for fuels to the country.

A corporation, on the other hand, may want to establish the amount of carbon it emits. For that, it would use a number of indicators such as fuels and energy consumed or kilometers flown by airplane. These quantities are known indirectly. For companies often having to account for their expenses anyway, it is rational to use simple calculations to derive their carbon emissions. For example, suppose a company bought gasoline for $20,000 at a price of $2 per liter. Thus, it consumed 10,000 liters of gasoline. According to the United Kingdom (UK) Department for Environment (DEFRA), a liter of petrol causes 2,3307 kilograms (kg) of CO_2 emissions. Thus, this company's gasoline use caused 23.3 tons of carbon emissions.

Individuals may calculate their carbon emissions by using carbon footprint calculators. These typically ask for several lifestyle decisions, such as the kind of vehicle the individual uses for mobility, or the kind of nutrition consumed.

These examples show that carbon emissions are constructed. For managing emissions, it is necessary to count carbon emissions avoided by specific practices. Thus, to make decisions on how to reduce carbon emissions, actors have to assess the carbon emission–saving potential of an action. While it is simple to calculate the average amount of carbon emissions avoided when, for example, switching from a vegetarian to a vegan diet, an expert-based bureaucracy is involved with assessing the potential to avoid carbon emissions by reforestation projects.

The Kyoto Protocol envisaged such projects. These and other kinds of projects are intended to reduce the amount of carbon emissions that would occur if the project had not been implemented. This part of the Kyoto Protocol is called Clean Development Mechanism (CDM). By running a CDM project, the responsible operators are able to produce carbon reductions—Certified Emission Reductions (CERs). Thus, such CDM projects are built to produce negative carbon emissions. These negative counts of emissions can be summed up with positive counts of carbon emissions. Hence, a company that established 500,000 tons of carbon emissions can neutralize this amount on its carbon balance sheets by adding negative emissions—CERs. Thus, carbon emission counts may consist of both actual emissions and certified CERs.

Analytically, two processes are common to all these approaches to quantify carbon. First, accountants draw boundaries as to what emissions they will include in balance sheets. Thus, some sources of emissions are included and accounted for. Other sources, however, are silenced and will not show up in the balances. Second, carbon emissions counts are not obtained by direct chemical measurements, but rather are based on a variety of factors (for example, the GWP emission factor of natural gas). The emission factor of any specific good, service, or lifestyle is always an average of a number of measurements. Both social constructivist and statisticians' readings of measuring explain the reality of diverging measurements. Disparate organizations often come up with different carbon counts for the same activity.

Economic Implications

The hegemonic discourse and structure of nation-states installed carbon markets to trade permits to emit carbon. Following the economist Ronald Coase, the mechanism assumes that market structures allow for the optimal (the most cost efficient) allocation of emissions. In effect, the creation of carbon markets commodified the atmosphere. This has significant economic implications: the rights to emit can now be accumulated within the dominating economic dynamics of capitalism. This ties material carbon politics to the accumulation of capital by economic elites.

Conclusion

The management of carbon-as-waste uses market mechanisms with an underlying system of experts devising factors to translate selected emissions into carbon. Political discourse is able to hide its agency behind the scientific side. The politics is summed up by many observers as capitalist technocracy: rather than opening up a democratization of the politics of carbon-as-waste, climate change has been reduced to an engineering and accounting problem and an economic profit zone. Thus, the problem of getting rid of carbon has been transformed into a business challenge. Other possibilities to tackle climate change have been silenced in favor of commodifying the atmospheric commons.

Ingmar Lippert
University of Augsburg

See Also: Automobiles; Capitalism; Commodification; Emissions; Pollution, Air; Zero Waste.

Further Readings

Clark, Brett and Richard York. "Carbon Metabolism: Global Capitalism, Climate Change, and the Biospheric Rift." *Theory and Society*, v.34 (2005).

Lohmann, Larry. "Marketing and Making Carbon Dumps: Commodification, Calculation and Counterfactuals in Climate Change Mitigation." *Science as Culture*, v.14/3 (2005).

Mackenzie, Donald. "Making Things the Same: Gases, Emission Rights and the Politics of Carbon Markets." *Accounting, Organizations and Society*, v.34/3–4 (2009).

Celluloid

The first modern plastic, called celluloid, was initially developed as a synthetic replacement for diminishing natural resources, only later becoming closely associated with the development of motion pictures. Despite its extreme flammability, celluloid performed a crucial role, albeit in different ways, in the popularizing of both photography and the cinema. In addition, celluloid maintains throughout its history an intimate relation with both the conservation and consumption of animals, paradoxically promoted as preserving animals against environmental devastation while nonetheless depending upon a medium composed of the waste products of industrialized animal slaughter.

History

Marketed as a material for mass-producing cheap simulacra of scarce natural resources, principally ivory and tortoiseshell, celluloid was patented in 1870 by John Wesley Hyatt, who set up the Celluloid Manufacturing Company in the following year. By 1880, Hyatt's company issued licenses to a variety of firms, producing everything from dental plates and piano keys to jewelry, combs, and novelties, the latter being advertised as a luxury previously only available to the wealthy. In this way, Hyatt capitalized on the development of nitrocellulose by Christian Friedrich Schönbein in 1846. Two decades later, Alexander Parkes combined nitrocellulose with a plasticizer-solvent to produce the stable and fully formable forerunner of celluloid, which he named Parkesine. By varying the solvent employed, this new material was found to accurately mimic a wide range of naturally occurring substances.

This history of material mimicry is, however, largely forgotten today, eclipsed by the association of celluloid with the invention of film. By the early 1880s, photographers were already experimenting with transparent sheets of celluloid coated with a gelatin emulsion, seeking to replace the fragile and unwieldy glass plates that were in use at the time. Consistency remained a serious problem, however, and it was only with the introduction of John Carbutt's sheet film in 1888 that celluloid could finally be relied upon to provide a uniform thickness and unblemished surface, thereafter becoming widely available as a base for photographic plates. It was this base stock that Thomas Alva Edison (or rather,

Despite its extreme flammability, celluloid, the first modern plastic, was crucial in popularizing photography and films. The word film *originally referred to the coating of gelatin emulsion produced from connective tissue and other slaughterhouse byproducts.*

his chief engineer W. K. L. Dickson) used in the development of perforated 35mm celluloid film bands—Edison's major contribution to the invention of cinema—for use with his peep-show Kinetoscope. Thin and flexible, and thus allowing for its production in long continuous rolls, celluloid made photography available to amateur hobbyists for the first time. When, also in 1888, George Eastman introduced the Kodak system—a 100-image celluloid-backed roll film, which was to be returned to the manufacturers for developing—demand immediately outstripped supply.

Celluloid quickly established itself as the only suitable material base for "living pictures," used not only in the Edison Kinetoscope but also by the Lumiére brothers in their Cinématographe and by many others. Nevertheless, it was unable to shed its explosive origins, the nitrocellulose rendering it highly flammable and thus an unacceptable danger in the minds of many people. This danger was cemented in the popular imagination by a fire during the 1897 ball of the Société Charité Maternelle in which 143 people died (although celluloid was not in fact to blame). This hastened the imposition of safety restrictions upon the practitioners of early cinema, pushing it out of the domestic setting common to the magic lantern shows and into the less reputable theaters and fairgrounds. As a result, motion pictures became increasingly identified with popular public entertainment, as distinct from the domain of scientific research within which it originated.

Relationship With Animals

While often overlooked, celluloid maintains an intimate relation with animals throughout its history. As Nicole Shukin persuasively argues, the popular tours of the "disassembly" lines of the Chicago stockyards, insofar as spectators were treated to a view of time as a linear sequence of discrete moments, in fact constituted protocinematic technologies. More directly, celluloid and its constituents capitalized on the consumption of animals in a variety of ways, from the initial development of nitrocellulose as a gunpowder alternative to the marketing of celluloid as a material for mass-producing simulacra of commodities originally culled from the bodies of animals. Most notably, in the development of celluloid as a photographic and cinematic medium, animals again played a crucial role, with all of the three main figures associated with photographic motion in the late 19th century—Eadweard Muybridge, Étienne-Jules Marey, and Ottomer Anschütz—first employing the new technology to record the movement of animals. The promotion of early motion pictures was organized around a rhetoric of wildlife conservation, figured both by Marey's "Photographic Gun," and by the reemploying of such terms as *shooting* and *snapshot*.

While the words *celluloid* and *film* have long since become synonymous, the term *film* originally referred only to the coating of gelatin emulsion produced from connective tissue and other slaughterhouse "leavings." Film was thus the medium in which the movement of animals was first captured, permitting the reconnection of discrete images. With or without its film coating, celluloid was thus the first mimetic plasticity, at once replicating, breaking down, and reconstituting the lives of animals, while remaining materially dependent upon their slaughter. In other words, the strength and flexibility of celluloid, which serves to preserve animals—to "save" them by recording for public consumption that which might otherwise be lost forever—could nonetheless take place only through the "waste" of animals slaughtered for consumption.

Richard Iveson
Goldsmiths College

See Also: Commodification; Environmentalism; Television and DVD Equipment.

Further Readings

Rossell, Deac. *Living Pictures: The Origins of the Movies*. Albany: State University of New York Press, 1998.

Shukin, Nicole. *Animal Capital: Rendering Life in Biopolitical Times*. Minneapolis: University of Minnesota Press, 2009.

Central America

Central America is the geographic region that lies between Mexico and Colombia. Historically, it has included Guatemala, Honduras, El Salvador, Nicaragua, and Costa Rica, though contemporary geographers now also include Belize and Panama. Known for its biodiversity, rich archaeological heritage, and linguistic and ethnic diversity, the region is home to some 41 million people. The largest Central American country by geographical area is Nicaragua (80,778 square miles [sq mi]), and the smallest is El Salvador (13,000 sq mi). By population, Guatemala is the region's largest country, with 14 million inhabitants, and Belize is the smallest, with less than 400,000 people.

Colonized since the days of Christopher Columbus, Central America has been the scene of intense agricultural development, including cacao, coffee, bananas, and cotton. Military action during the leftist revolutions of the 1980s and 1990s (in El Salvador and Nicaragua), U.S. counterinsurgency projects in the same period (in Honduras), and decades of ethnic strife (in Guatemala) have led to intense deforestation in much of the region, along with extensive water and soil pollution. Central Americans have rolled back some of the ravages of colonialism and war through organic, fair trade, and cooperative forestry and agriculture, but consumption and waste problems persist. Notably, gold mining has been a source of fast wealth and ecological ruin in the central mountains that run from northern Panama to Guatemala. Although much of the population continues to depend on the production of a few key crops, namely, coffee, beans, rice, and corn, Central America is rapidly urbanizing in the 21st century. The region confronts five principal types of waste disposal issues: military, agricultural, mining, industrial, and municipal.

Military Waste

Often disregarded in discussions of waste, military refuse remains a pressing issue in Central America. Civil wars, revolutions, and narcotics-related violence have been particularly important in shaping the region's contemporary landscape. It was not until 2010, over 20 years after the cessation of hostilities between the leftist Sandinista government and U.S.-backed contra revolutionaries, that Nicaragua was able to declare its countryside land mine–free. Evidence of armed struggle, including unexploded ordnance, erosion, and unmarked graves, also pepper the countryside of El Salvador, Guatemala, Panama, and Honduras.

Agricultural Waste

Nicaragua's experience with industrial agriculture since the 1950s provides perhaps the most dramatic example of the region's struggle to govern industrial agriculture and its resulting wastes. In the mid-20th century, Nicaragua was Central America's breadbasket, with production of cotton, wheat, corn, and beans at record highs. By the 1970s, however, overuse of chemical pesticides had made Nicaragua a net importer of these products, with many fewer working farms in its once-fertile Pacific coastal plain. Malaria, a disease that was once under control in the country, made a comeback when mosquitoes became resistant to pesticide treatment. Across the region, pesticides applied on the banana plantations of the U.S.-based United Fruit Company have had serious health effects on workers and their families. Current and former plantation laborers continue to raise questions about the long-term environmental and health impacts of pesticide use, but they have had little luck in foreign courts.

Gold Mining

Gold and silver mining have been part of Central America's economy for centuries. Since colonization, the potential of a strike has drawn both large corporations and small prospectors to the mines. Though it presents opportunities for great wealth, gold extraction is not without its critics. For example, during the 2000s, indigenous Maya activists in the Western Highlands of Guatemala engaged in vocal and often tense clashes with their own government and the Glamis Corporation of Canada. The Maya leaders argued that although Glamis's open-pit gold mining project was initiated with the blessing of the United Nations Development Programme, indigenous people were not fully consulted about the mine's environmental impacts. The activists claimed that Glamis's use of cyanide for leaching gold out of ore was contaminating groundwater and ruining small farmers' chances of making a living.

As in the cases of military and agricultural waste, responsibility for mining and its attendant wastes often spans across national borders. Central American governments have been unable or unwilling to prosecute waste management oversights, and the home countries of multinational corporations often lack the legal structures to tackle these problems.

On a smaller scale, artisanal gold miners throughout the region use mercury to extract small amounts of gold from ore they find in streams, rivers, and abandoned mines. Activists and leaders in the gold-mining regions of Nicaragua, Guatemala, Costa Rica, and Panama have warned that mercury, a neurotoxin, will do long-term damage to brain and fetal development in infants and children as well as to the motor skills of adult and adolescent miners.

Industrial Waste

Though agriculture, forestry, and tourism comprise a great portion of Central America's economy, heavy industry has grown since the 1990s. In Guatemala, Honduras, El Salvador, Nicaragua, and Costa Rica, the Dominican Republic-Central America Free Trade Agreement (CAFTA-DR) has made free trade zones, or *zona francas*, important sources of low-income employment in the manufacture of goods for the global market, especially apparel. The CAFTA-DR provides some environmental and labor protections, but unions and community groups have found it difficult to organize workers and to confront the ecological damage caused by the chemical dyes, wastewater, and airborne emissions that apparel plants emit.

Before CAFTA-DR, industrial waste had already taken a toll on the region. In 1992, the presidents of Central America's countries voted to impose a ban on the importation of toxic waste to the region after activists successfully argued that waste importation was damaging the environment. Thanks to imported and domestic waste, Nicaraguans witnessed the nearly total destruction of the fishery and ecosystem of Lake Managua, Central America's second-largest body of freshwater. In the early 1970s, the U.S.-based Pennwalt Corporation opened a caustic soda and chlorine plant on the shores of the lake. By 1981, a U.S. research team found that 37 percent of Pennwalt's workers suffered from mercury poisoning. Mercury from the caustic soda production process continued leaking into Lake Managua through the 1980s, causing irrevocable damage to freshwater marine life and the local fishers who depended on it. When the Pennwalt plant opened in the 1970s, U.S. factories had already stopped using mercury in caustic soda and chlorine production. In Nicaragua, however, workers at Pennwalt grappled with long-term and often irreversible health problems caused by exposure to mercury—the same health problems that in the United States precipitated the first enforcement of the 1971 Occupational Safety and Health Act. Pennwalt continued operating until a series of chlorine leaks in the early 1990s sparked protests from local residents. In what became known locally as "El Caso Pennwalt," the company failed to rectify the chlorine leaks, and the plant closed.

Municipal Garbage

With its proximity to the larger economies of Mexico, the United States, and South America, Central America's access to disposable consumer products, appliances, and packaged foods outpaces its otherwise high levels of economic poverty. Rapid urbanization across the region has not come along with rapid development in urban infrastructure. The garbage dumps of Guatemala City, Tegucigalpa, and Managua are home to hundreds of families who make their living scavenging for waste, and the global recycling industry has sparked an informal economy for aluminum, plastic, and paper. Children comprise a large part of the workforce in this garbage economy, and all workers are exposed to the fumes of burning refuse, skin and eye diseases, malnutrition, and toxic chemicals.

No city in Central America has a sanitary landfill. Cities and towns deposit the majority of their waste in rudimentary dumps, from the massive pit in Guatemala City to Managua's "La Chureca," the region's oldest dump, sited on the shores of Lake Managua. International development organizations have proposed projects to construct sanitary landfills in the region, but these plans continue to face opposition from dump dwellers and their advocates, who fear the possibility that dwellers may lose both their homes and their livelihoods if dumps close.

Despite its mineral wealth, physical beauty, and unparalleled biodiversity, Central America remains the poorest region in the Western Hemisphere. In fact, it contains two of the hemisphere's three poorest countries (Nicaragua and Honduras). Agricultural transitions, deforestation, and the promise of work in free trade zones indicate that the region's cities will continue to grow. This urbanization, coupled with increased access to disposable consumer goods and packaged foods, indicates that Central Americans will likely struggle with political, economic, and environmental health challenges related to the rapid expansion of consumption and waste for years to come.

Alex M. Nading
University of Wisconsin, Madison

See Also: Developing Countries; Farms; Mineral Waste; Pollution, Land; Pollution, Water; Toxic Wastes.

Further Readings

Faber, Daniel. *Environment Under Fire: Imperialism and the Ecological Crisis in Central America.* New York: Monthly Review Press, 1993.
Imai, Shin, Ledan Mehranvar, and Jennifer Sander. "Breaching Indigenous Law: Canadian Mining in Guatemala." *Indigenous Law Journal*, v.6/1 (2007).
Iwerks, Leslie and Mike Glad. *Recycled Life* (Film). http://recycledlifedoc.com (Accessed June 2010).
Murray, Douglas. *Cultivating Crisis: The Human Costs of Pesticides in Central America.* Austin: University of Texas Press, 1994.

Certified Products (Fair Trade or Organic)

Claims made for certified products address issues surrounding human rights, fair wages, safe working conditions, animal welfare, and/or environmental stewardship. Products typically receiving certifications include food, fiber, and forest products. Timber products are certified through entities such as the Forest Stewardship Council (FSC) and are sold in lumberyards or home improvement stores. Not all FSC-certified products are labeled as such, and some require further research by the consumer. Fair Trade as well as Shade Grown certifications attest to the social justice and environmental health aspects of food production. In terms of Fair Trade coffee, consumers pay a price premium in exchange for the knowledge that their purchase is providing a price to coffee producers that is considered more equitable than they otherwise would get. Although this may yield greater revenues for some coffee growers, not all growers are situated, either physically or socially, to take advantage of these markets. Further, growers, usually through a cooperative, must meet certain standards of large-scale coffee marketers in order to have their coffee bought at a premium. As with all commodities, the premium price for Fair Trade coffee is subject to shifting market demand and competition for market share from other regions. Shade-grown products have been shown to be more expensive to maintain in some situations, suggesting scalar issues. That is, larger producers may be better situated to adopt more ecologically sustainable practices, eventually leading to certification, than small-scale producers would be able to take on. In this sense, some small-scale producers are effectively priced out of the market.

Certifiers make the claim that their certified products authentically represent a suite of socially and ecologically sustainable practices embodied in the product and its presentation. At the consumer level, these seals or marks and the claims they make concerning the process and/or product are scrutinized in relation to noncertified or alternatively certified products. These certifications and the subsequent labeling and marketing serve an educational purpose, informing the consumer about where a particular product came from, how it was produced, by whom it was produced, under what conditions it was produced, and what inputs were used to make the product and sometimes its package. This information is provided with the intent of weaving a narrative that calls the consumer to action, that is, to purchase. At the same time, the consumer must believe that the certifying group or agency has the power to uniformly enforce its standards upon all the producers of the product. In instances such as prepared organic foods, the consumer is called to trust a series of unknown producers, processors, and distributors over a complex supply chain that all the products meet the minimum standards.

These certified products have, on the one hand, been heralded as consumer activism, where consumers can vote for a better world through their wallets. On the other hand, they have been vilified as a means to further distinguish wealth disparities via consumption and a shallow sense of environmentalism, wherein nothing need be sacrificed, only consumed with greater prudence. In this respect, the slow food movement, organic foods, and community-supported agriculture have all been attacked as gustatory elitism.

The organic food market in the United States has been one of the fastest-growing sectors in U.S. agriculture over the past decade. This has been recognized as a lucrative market where businesses, including large corporations, can do well financially while also doing good for the environment. The majority of organic products in the United States reach the mouths of consumers in much the same way as their nonorganic counterparts: through highly centralized production, processing, and distribution facilities. In some instances, organic farms of today more resemble industrial agricultural operations than they do diversified farms, which attempt to balance economic and ethical considerations. Nevertheless, a shift to organic management regimes by definition means a prohibition on the use of chemicals that have historically been overused and misused in industrial agricultural production, resulting in pollution of land, water, and air. Further, chemical residues impact not only the physical environment but also the biological environment, including the resultant human health impacts.

The process of creating policies and practices concerning certification in terms of either Fair Trade or organic products can be very contentious as both multiple stakeholders vie to shape standards. As a case in point, the establishment of the National Organic Standards was marked by fierce debate as to what would be permitted to be considered organic. The controversy surrounded the product and process of organic food production. Notably, an outpouring of public comments in opposition to sludge as fertilizer, genetically modified organisms, and irradiation of foods were effective in eliminating the proposed elements as part of contemporary organic standards in the United States. Despite this victory, critics of the current organic certification and management regime point out that the legislation has served to consolidate the organic industry in the United States, although small farmers continue to work to develop innovative product, process, and marketing strategies using sustainable approaches and agroecological principles, regardless of whether or not they are certified organic. In some respects, farmers chose to forgo certification due to the expense involved, or adopted alternative labels that convey the high quality and ecological sustainability of their farming operations.

Those who have marketed certified products continue to advocate their social and ecological viability. Consumers purchase these goods for a variety of reasons that may or may not directly correspond with the intent of the producers. The ecological and social justice implications of certified product production, distribution, and consumption all deserve future study.

David Fazzino
University of Alaska

See Also: Capitalism; Carbon Dioxide; Commodification; Consumption Patterns; Farms; Food Consumption; Organic Waste; Personal Products; Pollution, Land; Pollution, Water; Recyclable Products; Slow Food.

Further Readings

Delind, L. "Transforming Organic Agriculture Into Industrial Organic Products: Reconsidering National Organic Standards." *Human Organization*, v.59 (2000).

Fisher, C. "Selling Coffee, or Selling Out? Evaluating Different Ways to Analyze the Fair-Trade System." *Culture & Agriculture*, v.29 (2007).

Romanoff, S. "Shade Coffee in Biological Corridors: Potential Results at the Landscape Level in El Salvador." *Culture & Agriculture*, v.32 (2010).

Children

Children ages 0–16 are one of the largest demographic categories of consumers in society. Their role is increasingly important, because their expenses have increased and their influence over parental spending has expanded. In the process of children's graduate consumer socialization, especially through ages 6–16, their independence as decision makers progressively grows, characterized by a dynamic cognitive and social development.

There is, however, an imbalance between children's needs and consumption in different parts of the world due to varied standards of living. In many countries, poverty prevents children from being satisfied even with minimal products of primary needs. In contrast, in more developed countries, consumption often exceeds the needs of children, although hunger remains a problem for some segments of society. For instance, in 2007, over 13 million children age 18 and younger lived in poverty in the United States.

Children as Consumers

One classification includes five stages of the development of children as consumers:

1. Ages 0–2: accompanying parents and observing
2. Ages 2–3: accompanying parents and requesting
3. Ages 3–4: accompanying parents and selecting with permission
4. Ages 4–5: accompanying parents and making independent purchases
5. Ages 5–6: going to the store alone and making independent purchases

P. M. Valkenburg and J. Cantor offer a four-stage development from a psychological point of view:

1. Infants and toddlers (ages 0–2): feeling wants and preferences
2. Preschoolers (ages 2–5): nagging and negotiating
3. Early elementary school (ages 5–8): adventure and the first purchase
4. Later elementary school (ages 8–12): conformity and fastidiousness

The categories of children's consumption include food, toys, games and other entertainment products, clothes, accessories and cosmetics, children's books, school and educational accessories, children's furniture and equipment, children's medicine, and other children's items. There are also adults' categories that have been consumed by some children at the expense of their health (such as tobacco).

Developing sound food habits for good health, recognized as a preventive health, has become a main task of modern parents, children, and society in general. One serious problem is the disproportionate consumption of condensed milk that makes children vulnerable to dehydration and death from diarrhea. There are many social programs throughout the world that help families and single parents' children get enough food, including milk.

There is a tendency toward overconsumption by children in some countries in the world, including the United States, which results in serious obesity problems. Plate-waste is a specific issue, since children leave a great deal of food on their plates, especially during school lunches. Many projects and programs try to both reduce the wasting of food through recycling and to decrease the obesity problem. The Robert Wood Johnson Foundation spends $100 million per year toward reversing childhood obesity—the single largest effort of this type in history.

Toys and games play an essential role in children's life and are also one of the most successful types of merchandise all over the world, with an annual profit of around $21 billion. Electronic games and other entertainment products have an ambiguous role, since electronic games are often thought to have ill effects. A successful marketing mechanism that develops a culture of extensive consumption is the promotion of collectible toys, including Barbie dolls, Dora, and numerous other brands. In contrast to toys and other small products, parks like Disneyland in California and Universal Orlando in Florida create an opportunity for creative entertainment that may lead to extensive consumption.

Clothes, accessories, and cosmetics are socially sensitive. This is the field of social life where children at their earliest ages learn about inequality of wealth in society and begin to construct their explanation models with age-related variations. The influence of fashion and pop stars has become typical for large segments of the global child community. China's children's industry, the largest in the world, is known as a fashion industry. The "sexualization of children" is also a reason for the development of

Children ages 0–16 are one of the largest demographic categories of consumers in society, and their influence continues to grow. The heavy marketing of collectible toys such as Barbie and Dora dolls has fostered a culture of extensive consumption among children.

a special attitude toward clothing and makeup. The prevalent marketing push toward profits attacks, in some cases, the world of children's innocence through sexual symbols and provocative imagery that in turn contributes to an underground culture of child pornography; this is one of the most devastating results of the consumer culture in the information age. Children's cosmetics vary from creative cosmetics for birthday parties to full imitations of adults' makeup.

Children's furniture and equipment include room furniture (such as beds, dressers, and shelves), garden equipment (such as shovels), baby carriages, miniature cars, bicycles, skis, and other sports equipment.

There are conventional and complementary and alternative medicines (CAM) for children. Data from the U.S. National Health Statistics Reports in 2008 show that children whose parents use CAM are almost five times as likely to use CAM compared to children whose parents do not use CAM.

Most children's consumption categories belong to two types of subcultures: primary consumers and secondhand consumers. Reusing children's and adult clothing, furniture, and recreational items is a policy developed through different commercial and not-for-profit organizations worldwide, and it results in billions of dollars in business.

There is a special category of children-consumers who belong to families with low incomes. Social programs worldwide help these children to grow healthy through a series of initiatives. For instance, in Utah and some other states, special programs provide milk to any child raised in a low-income family.

Children and Waste

Waste is a result of regular consumption or overconsumption. There is a difference between garbage and waste, although garbage is a part and an aspect of waste. "Reduce, reuse, and recycle" is a policy of the green culture that has been addressing children through a variety of programs. For many reasons, the children of the world not only increase the quantity of garbage, because of expanding consumer patterns, but are also exposed to greater amounts of human garbage that in some cases is deadly. In cities like Cairo, Egypt, children comprise up to 50 percent of the labor resources involved in waste cleanup in the streets. There is a gender division: boys pick up the garbage, while girls help the women in sorting the garbage. Children and adults who live or work near hazardous landfills are at greater risk for disease and suffer adverse health effects, including cancer and damage to the fetus in pregnancy.

Electronic waste (e-waste) is one of the greatest problems from a global perspective. For instance, recycled computers have been sent from rich to poor countries where children interact with them and sometimes throw the electronics onto fires. As a result, children breathe in highly carcinogenic fumes. Since the 1980s, cities like Guiyu, China, have been taking in e-waste from other countries for dismantling and safe processing. The lead poisoning level in children is 69 percent in such places. Items from the United States, the Netherlands, Germany, and South Korea were found at a dump site in Ghana's capital, Accra, where they also endanger children. Statistics of the United Nations Environment Programme estimate annual e-waste between 20 million and 50 million tons.

Another serious problem is poisoned water. According to a United Nations report, 1.8 million children younger than the age of 5 die every year because of poisoned water.

Cultural Diversity, Enculturation, Socialization, and Health

Cultural diversity among children's consumption and waste response depends on family, cultural traditions, and actual socialization. There are also differences in gender; female and male children have different needs and a varied psychology of consumption and waste response.

Consumer patterns influence the health of children. Overconsuming food or specific products may result in obesity. Poverty, hunger, or the use of drugs, alcohol, and tobacco may even cause the deaths of children. Children depend on an active system of consumption even before their birth. Many of them become victims of unhealthy pregnancy habits, including tobacco use by their mothers.

One of the traditional cultural patterns worldwide is shared-plate food and finger-feed. These create opportunities for the socialization of children, but they may also cause health problems because of deficits of important vitamins that benefit children or undeveloped personal hygiene.

Archaeological evidence indicates that at the beginning of sedentary economy and in the course of the advance of agriculture (the Neolithic Age and the Copper Age), plates with large food portions were popular, although small bowls also existed. It is very probable that shared-plate feeding was emblematic of these early human cultures. Accordingly, consumption patterns have been defining the style of human culture since the earliest stages of human civilization.

Childhood is an important period for preparing children for successful future economic, social, and cultural roles in the free market world. The process of socialization develops together with enculturation and reflects on the individual decision making of children. In particular, consumer behavior depends on the way in which children will be introduced to the problems of wealth. Although children's economic knowledge is socially differentiated, developing strong savings behaviors (including strategies and decisions) and saving beliefs (including attitudes and beliefs) is a cross-social ability. Such patterns of enculturation and socialization directly relate to children's health, since some new products of e-media, like video games, can be addictive.

Art is very powerful in enculturation. As an education tool, it helps children to access and develop their creativity. Many items that would normally end up in the trash can be turned into attractive, useful items through handcrafting. There is an excellent example of this in the Children's Garbage Museum in Stratford, Connecticut. Special artistic programs in the United States engage children in the problem of sustainable living.

Living in a diverse social environment, children's social sensibility as consumers depends on how they have been introduced to societal problems and how they develop as decision makers. The consumption models of children are in fact models of the future of humankind.

Lolita Nikolova
International Institute of Anthropology

See Also: Baby Products; Candy; Consumption Patterns; Cosmetics; Malls; Social Sensibility; Toys.

Further Readings

Aziz, H. "Improving the Livelihood of Child Waste Pickers: Experiences With the 'Zabbaleen' in Cairo, Egypt: Evaluative Field Study." *Waste* (2004).

Gunter, B., J. McAleer, and B. R. Clifford. *Children as Consumers: A Psychological Analysis of the Young People's Market*. London: Routledge, 2001.

Harris, J. L., J. L. Pomeranz, T. Lobstein, and K. D. Brownell. "A Crisis in the Marketplace: How Food Marketing Contributes to Childhood Obesity and What Can be Done." *Annual Review of Public Health*, v.30 (2009).

John, D. R. "Consumer Socialization of Children." In *Children: Consumption, Advertising and Media*, F. Hansen, J. Rasmussen, A. Martensen, and B. Tufte, eds. Copenhagen, Denmark: Copenhagen Business School Press, 2002.

Kirsh, S. J. *Children, Adolescents, and Media Violence: A Critical Look at the Research*. Thousand Oaks, CA: Sage, 2006.

Knox, E. G. "Childhood Cancers, Birthplaces, Incinerators and Landfill Sites." *International Journal of Epidemiology*, v.29 (2000).

Linn, S. *Consuming Kids*. New York: New Press, 2004.

Pliner, P., J. Freedman, R. Abramovich, and P. Darke. "Children as Consumers: In the Laboratory and Beyond." In *Economic Socialization. The Economic Beliefs and Behaviours of Young People*, P. Lunt and A. Furnham, eds. Cheltenham, UK: Edward Elgar, 1996.

Pole, C. "Researching Children and Fashion: An Embodied Ethnography." *Childhood: A Global Journal of Child Research*, v.14/3 (2006).

Shankar, A. V., J. Gittelsohn, K. P. West, Jr., R. Stallings, T. Gnywali, and F. Faruque. "Eating From a Shared Plate Affects Food Consumption in Vitamin A–Deficient Nepali Children." *Journal of Nutrition*, v.128/7 (1998).

Singer, D. G. and J. L. Singer, eds. *Handbook of Children and the Media*. Thousand Oaks, CA: Sage, 2001.

Valkenburg, P. M., and J. Cantor. "The Development of a Child Into a Consumer." *Journal of Applied Developmental Psychology*, v.22/1 (2001).

Ying, G. "Consumption Patterns of Chinese Children." *Journal of Family and Economic Issues*, v.24/4 (2003).

Chile

Chile has garnered international attention for its free market economic policies, devastating earthquakes, and decades of political change. Its neighbors remember Chile's history of territorial expansion. All of these factors have affected its changing patterns of consumption and waste.

Chile's long, thin profile along the Pacific Ocean contains a highly urbanized population of about 17 million. The country had a purchasing power parity (PPP) of about $15,000 per capita in the late-2000s, which ranked about 76th worldwide. Income inequality was relatively great.

Water

Chile has served as a paradigmatic case for studies of water privatization, and this process has affected patterns of water consumption and sewage treatment. In the early 1980s, the military government established water rights as a private commodity in the constitution. This created perhaps the world's most free market approach to water rights. The entities that delivered water and sewage services were still public, but democratic governments from the late 1980s into the early 2000s also gradually privatized this sector. Foreign economic interests, such as a pension fund for Canadian teachers, have invested heavily in Chilean water providers. Government agencies increasingly regulate water quality, and policies promote the extension of services to underserved areas.

This system has provoked considerable debate in Chile and internationally. Water is piped to nearly 100 percent of the urban population, and sewage treatment has risen from about 20 percent in 1999 to 84 percent in 2008, making Chile a leader in Latin America. While low-income Chileans can qualify for subsidies, the rise in water prices that accompanied privatization has led to a marked decrease in water use per capita. However, many observers complain that the system favors wealthy sectors of the country to the neglect of others. For example, the rates of water connection and sewage treatment in rural areas are much lower. And, contrary to expectation, Chile's level of water losses, already high, actually rose after privatization.

Exports

Much of Chile's wealth has come from the export of commodities. The War of the Pacific (1879–83) in which Chile took its northern extension in the Atacama Desert from Peru and Bolivia, allowed Chile to dominate the extraction and exportation of nitrates, which were valued abroad primarily as fertilizers. Wood used in this industry came from nearby forests of the tamarugo tree, which were severely depleted: 21st-century government efforts to replenish this tree must contend with falling groundwater tables to feed nearby towns' consumption.

In recent decades, copper has overshadowed nitrates as Chile's main export, and Chile remains the world's dominant producer of each commodity. Copper mining is centered in the Atacama Desert and the north-central region. Pollution controls have become progressively stricter since the return to democracy in the 1990s. Prior to that, legal environmental controls were weak or nonexistent, and mines' emissions of arsenic, cyanide, carbon monoxide, sulfur dioxide, and heavy metals into the air, ground, and water caused significant health problems among workers, nearby residents, and plants and animals. Despite marked reductions, waste from copper mining remains a significant environmental hazard.

Energy and Carbon

Chile's energy consumption has increased rapidly, along with its gross domestic product (GDP), since the 1980s. As with water, Chile has almost completely privatized its energy sector since then, and foreign investors control much of it. Access is almost universal. The vast majority of the supply has come from a mix of hydroelectric plants, coal, and natural gas. To reduce Chile's dependence on Argentina and on decreasingly predictable hydrology, the country is increasing its reliance on coal.

Chile has ratified the Kyoto Protocol, which does not require it to reduce emissions, and makes it eligible to host projects via the Clean Development Mechanism. In 2008, the government announced a plan to address climate change that focused on studies, organization, and planning, rather than on comprehensive provisions for reducing consumption or waste. Some of its more concrete provisions aimed at creating more secure supplies of water. Nonetheless, Chile has been the site of various greenhouse gas–reducing projects, including a campaign to promote compact fluorescent lamp (CFL) bulbs, methane capture at landfills, and wind farms.

To help account for an anticipated doubling of demand over the next decade, Chile has increased energy efficiency considerably, and overall electricity consumption has actually decreased in some years in the late 2000s. The government hopes that improvements in efficiency will account for 20 percent of the expected growth in consumption. These savings were achieved in part through building codes. Chile also has recently passed a requirement that large utilities—excluding hydropower plants—gradually increase their inclusion of renewable energy sources to equal or exceed 10 percent by 2024.

Overall, though, Chile's carbon footprint per capita grew rapidly over the 2000s, reaching about the worldwide average, in part because of industrial growth and greater dependence on coal. Calculations of its ecological footprint have ranked Chile anywhere from the most damaging country in Latin America—within the range of rich, European countries—to slightly lower on both scales.

Garbage

Each day, Chileans produce more than one kilogram of garbage per person. Most Chilean garbage goes to older sites that were not designed according to modern standards. In part, this is due to the low cost of dumping garbage. The federal government estimates that about 60 percent of garbage is deposited in landfills. Recycling is not widespread in Chile, and the federal government does not have a comprehensive recycling plan as of 2010, instead relying mostly on education and encouragement. Its goal for a public-private effort in Santiago is the recycling of 25 percent by 2020, up from about 10 percent nationwide. Programs such as methane capture have existed in scattered landfill sites, but the unfiltered trash reduces their effectiveness.

A massive earthquake in February 2010 created a set of challenges related to waste. Trash collection and recycling were interrupted, and environmentalists feared that leaks of sewage and industrial waste presented hazards for humans and other organisms. Most visibly, Chile had to remove vast amounts of rubble that remained piled along streets for weeks and—in some places—months after the main tremor. Ad hoc trash dumps also sprang up, replete with rats. Some cities created new collection fields for the rubble, without time to prepare them as they would a planned landfill. The cost of such a monumental endeavor, in the face of other costly priorities such as repairing roads and constructing temporary housing, kept local governments from tackling the rubble quickly. Some nongovernmental organizations proposed projects to create monuments out of rubble or to recycle parts from it.

Easter Island

Easter Island, ruled by Chile, has become the focus of a debate regarding consumption there in the few centuries before Europeans first landed on the island in 1722. All of the competing scenarios require considerable conjecture. Some observers, most notably Jared Diamond, argue that the inhabitants, or Rapanui, gradually deforested the island for a combination of reasons, including to build boats and to transport the island's famous statuary using logs. According to other researchers, deforestation derived mostly from environmental changes beyond the islanders' control. Some evidence suggests that a period of lower temperatures and drought in the early 1600s—coincident with the Little Ice Age and similar to a very lengthy El Niño pattern for this region—caused the simultaneous extinction of not only trees, but also many types of bushes and shrubs that islanders would not have used for major projects. This historical debate has attracted considerable attention, in part because some observers have explicitly likened the Easter Islanders' situation before 1650 to that of all present-day humanity.

Tracy Duvall
Universitas Indonesia

See Also: Capitalism; Developing Countries; Mineral Waste; South America; Sustainable Development.

Further Readings

Bauer, Carl. "Market Approaches to Water Allocation: Lessons From Latin America." *Journal of Contemporary Water Research & Education*, v.144/1 (2010).

Portal Web de la Comisión Nacional del Medio Ambiente. http://www.conama.cl (Accessed September 2010).

China

Over the past few decades, the People's Republic of China (PRC) has transformed from a poor, socialist country that was devoid of consumer culture in the 1970s to one of the world's largest consumer markets in the 21st century. With a booming economy and fast-paced development, China was the world's largest consumer in 2010, and is an exporter of many products. The enormous size of its market, rapidly rising levels of domestic consumption, and accompanying social changes have attracted the attention of commentators around the globe. China also plays a key role in consumption in the West, exemplified by the "Made in China" label, which dominates department stores and supermarkets. However, consumption is not without discontent; social ills once assumed to be exclusive to Western society now bring unhappiness to the Chinese. There has also been a heavy burden on the environment, such as desertification, drought, flooding, and pollution, and these problems threaten sustainable development. Another issue is the increasing volume of waste in response to industrialization and urbanization and inadequate waste management, which has social, financial, and environmental implications.

China's Consumer Revolution

Consumption in China in the 21st century is a complete contrast to its recent past. The PRC was established in 1949 by Mao Zedong's Chinese Communist Party, and it was essentially a socialist society with no commercial consumer culture. In the socialist Mao era, goods and services were rationed and limited in variety. The service sector consisted of state-operated shops, as private firms were banned. Workplaces (known as *Danwei*) consisted of state-owned enterprises and public organizations (or, in rural areas, communes and production teams), which provided all basic necessities and other goods to members, including food eaten in centralized canteens, clothing, education for children, housing, and medical services. The work units also controlled events such as travel, marriage, and family planning. Most families had only enough money to meet their basic needs, and the majority of household income was spent on food.

Consumption in China from the 1950s to 1980s is often described in the literature by the products people at that time dreamed about owning, such as a bicycle, transistor radio, watch, and a sewing or washing machine. However, these items were expensive and were not commonplace, as households needed to save up or pool their incomes. Socialist ideology also dictated consumer fashions; for example, it was patriotic to dress in standardized clothing, such as the blue and grey cotton suits of the late 1970s, but ideologically improper to wear items such as jewelry. The Mao era was also characterized by various social and economic movements such as the Cultural Revolution—the antithesis of a consumer culture—which brought momentous change and periods of turmoil.

Deng Xiaoping's rise to leadership in 1978, following Mao Zedong's death in 1976, was a major turning point in 20th-century China and completely changed the face of consumption. Deng abandoned socialism and a centrally planned economy in favor of a market economy. He instigated the "Open Door" policy that gave the country access to foreign goods, investment, and technology transfer. Reform policies included the liberalization of industry, which encouraged the growth of privately owned enterprises, and price liberalization, which replaced state-set prices with market forces. Deng's famous slogans included "To get rich is glorious," "poverty is not socialism," and "take the lead in getting rich." The government implemented the Four Modernizations Program that emphasized developments in agriculture, industry, education, science, technology, and defense. As a result of these changes, China, in the post-Mao era, is transitioning to a modern nation. The country's economic prowess is reflected in almost all measures of development, especially gross domestic product (GDP), which has grown at staggering rates.

One of the most visible features of China's development, particularly during the late 1980s to mid-1990s, has been the emergence and growth of a consumer culture and society. China's opening up to the world led to the increased availability and diversity of products. Rising incomes and living standards led to increasing consumer spending power and preferences. The consumption structure has changed; the Chinese now spend much less money on basic daily necessities such as food, clothing, and durables for the home and more on items such as education, entertainment, housing, tourism, technology products, and medical insurance. Chinese consumers follow the latest fashions and trends in ways similar to consumers in other countries. With fast urbanization, an increasing number of new consumption sites, such as large department stores, shopping malls, and skyscrapers, are appearing in cities across the country.

A food stall in China. Consumption in China in the 21st century is a complete contrast to its recent past. In the 1970s, China was a poor, socialist economy with no consumer culture; now, it is one of the world's largest consumer markets.

Consumer society is stratified. There is a rapidly expanding middle and upper class, with strong consumption power, especially in major cities. Many leading suppliers of luxury goods, such as automobiles, clothes, fashion accessories, jewelry, and cosmetics, have entered the Chinese market to cater to the affluent. China has several hundred thousand millionaires who consume high-end purchases, such as Bentley limousines. However, China's consumer revolution is mainly an urban phenomenon and is less evident in the poorer, rural areas. There are also differences across the country that reflect regional inequalities. Consumption is more conspicuous in regions along the southern east coast, especially in the special economic zones (SEZs), which have received most investment economic growth. In contrast, the inland rural areas (central and western China) have received relatively little investment and remain underdeveloped. Despite the glamour and glitz of major cities like Beijing and Shanghai, urban inequalities and poverty are rife, especially among rural-urban migrants working unofficially in urban areas.

Foreign products are popular in China, largely because of their perceived status and quality—a complete contrast to the Mao era when anything foreign was regarded as bourgeois. Chinese nationalism has, however, been reflected in some objections to foreign products. The country imports many goods and services from overseas, and Western-owned manufacturing companies have relocated to China, making their products and technology available.

The Chinese economy is dominated by overseas investment and trade, but the government hopes that domestic consumption will be a major source of economic growth in the future. This is in contrast to countries such as India, where domestic consumption has played a key role in growth—China is more dependent on exports. Numerous initiatives and policies have been introduced to entice domestic consumerism, such as higher income levels, holiday extensions, infrastructure development, and expansion of bank credit. China's accession to the World Trade Organization in 2001 opened up Chinese trade and services to the world market. In response to the global financial crisis, the government launched a 4 trillion yuan ($585 billion) economic stimulus package with the aim to stimulate the economy.

However, several factors in China tend to curb consumption, including competition repression in which state-owned enterprises and other government-sponsored firms lack competitors, which leads to higher prices for many goods and services; financial repression, such as when banks keep interest rates low, which reduces the income of consumers; underdeveloped social security and retirement schemes, which encourages the public to adopt high saving rates and consume less; and widening income disparities, particularly between urban and rural areas and between coastal regions and inland territories.

Consumption

China is a leading consumer and producer in a wide range of industries and services. It is also the world's largest exporter, and many products around the globe display the "Made in China" label.

China has the world's largest agricultural output, contributing 13 percent of GDP and employing over 300 million farmers. Rice is the dominant crop and is cultivated mostly in southern provinces such as Guizhou, Hebei, Yunnan, and Sichuan; wheat is also important, especially in the north. Other major food crops include beans, corn, millet, oats, peanuts, potatoes, tea, and various fruits and vegetables. Nonfood crops include cotton and tobacco. China's increasing affluence and growing population have led to increased demand for meat. Animal husbandry, livestock farming, and the Western intensive-farming model are increasing in popularity. China is the world's leading producer of pigs, chickens, and eggs; and it also has sizable herds of sheep and cattle. The country's aquaculture accounts for about one-third of the world's total fish production, concentrated in the middle and lower Yangtze Valley and the Zhu Jiang Delta. Since the early 1980s, the population's dietary patterns changed from staple foods to diverse meals with meat and dairy products.

Only about 15 percent of the country's total land area is suitable for farming. Yields are high because of intensive cultivation. Agriculture in China has always been very labor intensive, with limited agricultural machinery, although production has generally increased as a result of technological changes. Since the 1950s, and particularly since economic reforms in the 1980s, there has been a shift from an agricultural-based economy to industrialization. Large areas of farmland were converted for urban and industrial use; the proportion of the population working in agriculture declined; and a large number of farmers have pursued alternative livelihoods such as manufacturing and commerce.

China is the second-largest consumer of primary energy. Energy consumption has grown dramatically since 1980—largely in response to business and infrastructure development—and is likely to increase further over the long term. Most energy comes from fossil fuels such as coal and oil. Coal consumption is declining, and thousands of mines have closed. Output from China's oil fields is also in decline, replaced by an increasing number of imports. Hydroelectric resources account for about one-fifth of energy use, mostly in the southwest. The Three Gorges Dam, located across the Yangtze River in Hubei Province, is the world's largest hydroelectric power station by total capacity. As of 2010, natural gas production accounted for only 3 percent of energy production, and nuclear energy 2 percent. China is also developing alternative energy resources such as geothermal, tidal, wind, and solar power.

China is a major industrial and manufacturing base. Industrial development has been given considerable attention since the founding of the PRC. During the 1950s and 1960s, the Maoist regime emphasized heavy industry, especially machine-building and metallurgical industries, and the economy grew much faster than before 1949. Prior to 1978, most output was produced by state-owned enterprises. The transition era from the 1980s onward saw the development of light industry; new industries such as chemicals, electronics, and pharmaceuticals; and high-technology industries that produce computers, electronics, and telecommunications. This has contributed to changing consumption patterns as the general public now have wide availability of goods.

Overall, industrial output has grown at an average rate of more than 10 percent per year. Industry is concentrated in the coastal provinces of Guangdong, Jiangsu, Shandong, Shanghai, and Zhejiang. The automobile industry has developed rapidly since the early 1990s. China is the largest consumer of automobiles and has domestic and overseas companies.

Tourism is one of the fastest-growing areas of consumption. The hotel sector and travel agencies are expanding rapidly. China has become important as both a tourism destination and tourism-generating country. If the predicted growth of Chinese tourists materializes, there is likely to be strong demand for tourism services both domestically and internationally, although the domestic tourism market currently constitutes more than 90 percent of the country's tourism traffic.

Waste and Recycling

Alongside increasing consumption, China is the world's largest generator of household and industrial waste. Over 1 billion tons of industrial waste and 200 million tons of household solid waste are generated annually. Household waste in China accounts for one-third of the world's total and is increasing at annual rates of 8–10 percent, projected to be at least 480 million tons annually by 2030.

Formal waste management started in the late 1980s and remains at a preliminary stage. The Department of Urban Construction (within the Ministry of Construction) is the national authority responsible for urban waste management; municipal and district governments are responsible for management at the local level. There is enormous variation in waste collection and treatment services between and within cities, ranging from rudimentary collection systems in the poorer suburban areas to adequate services in modern, high-rise apartment blocks in major cities. Most household waste is buried in landfills, and only a small proportion is composted or incinerated. Although some cities operate good sanitary landfills, the majority are poorly operated and, in many cases, are simple, open dumps. Common constraints of landfills in China include inadequate equipment and technology, minimal or no landfill lining and chemical treatment, insufficient compaction and waste covering, little gas collection, and poor management. Since the introduction of incineration technology in the late 1980s, the number of incinerators is increasing, mostly in big cities, and the government has introduced policies to encourage investment in them. However, there is a lack of advanced technologies and integrated waste treatment systems, although several new treatment facilities such the Beishenshu landfill and Asuwei power plant offer integrated methods.

The composition of solid waste from Chinese households is dominated by kitchen waste because Chinese food is rich in fresh fruit and vegetables. Food sold in markets is generally unprocessed and unpackaged. However, the proportion of waste glass, paper, plastics, and other recyclables is rising. Coal ash is another major component in household waste because many homes use coal for cooking and heating, although coal is increasingly being replaced by natural gas. The waste stream is inhomogeneous because only a small proportion is separated and, during waste collection, domestic and industrial waste are often mixed.

Recycling is less prevalent in China than in the West. The recycling industry as a whole is underdeveloped, as there are few recycling companies, laws, policies, and regulations. Public recycling facilities are limited, although some cities such as Beijing are improving their recycling infrastructure. However, in China—where poverty is rife and waste is valuable—recycling is done by scavengers who sort through rubbish in streets, households, and even landfill sites to collect and sell recyclable products. The waste collectors are usually poor migrants from rural areas, and their work conditions are sometimes hazardous.

Every year, Western countries export millions of tons of waste to China, particularly plastics and paper, for treatment and recycling. It is transported by freight in empty shipping containers used for exporting Chinese goods. As disposal and landfill prices are high in the West, sending waste overseas, such as to China, is a cheaper option; another reason is to overcome stringent environmental and waste regulations. In some cases, Chinese companies set up offices overseas to buy waste, and they have been able to offer higher prices and accept higher quantities of waste than their counterparts in the origin country. However, the global recycling trade is not without criticism. Complaints include the environmental and social costs of shipping waste, no guarantee over environmental and other standards in China, reduced business and materials for recyclers in the country of origin, and claims that the West is dumping its rubbish on China. It also highlights failures in the recycling market in the West. Recently, China's recycling industry was adversely affected by the global economic crisis, as reduced consumer demand for products led to lower prices of recyclables; consequently, many Chinese recycling units closed.

Electronic waste (e-waste) consists of discarded electronic appliances such as computers, digital cameras, mobile phones, printers, refrigerators, televisions, washing machines, batteries, circuit boards, and electrical wiring. Most e-waste in China is from overseas, especially the United States, which exports more than 70 percent of its e-waste to China for recycling. The recycling of e-waste involves dismantling electrical appliances to extract valuable materials such as copper and gold. It is profitable, generating income for many urban and rural workers, and it dominates the economy and livelihood of entire villages and towns, such as Guiyu in south China, where 80 percent of families (more than 30,000 people) are involved. However, the processing of e-waste poses serious threats to public health and the environment. Pollutants and hazardous materials such as cadmium, lead, and mercury are discharged into the environment, and employees in China often work without any health and safety measures.

Society and Environment

China's consumer revolution is part of a broader social revolution that has transformed people's everyday lives. Since the 1980s, living standards and quality of life have improved substantially, and millions of people now have greatly improved lifestyles. However, China's development is not without drawbacks. Promoting economic growth, making money, and becoming rich have been the catchphrases in reform-era China—a mindset that has justified a host of unethical behaviors. Social development lags behind economic development, and there are widening disparities between different regions. A myriad of social ills hide behind the bright facade of data often used to glamorize the country's consumption and economic success. Social problems include corruption, crime, drugs, and prostitution. China's relentless pursuit of economic development has held greater precedence over other issues.

Consumer complaints are common and include fake products, poor services, product safety, and substandard goods. For example, in 2007, there was a series of scandals involving tainted food and products that led to recalls of Chinese exports. Well-publicized cases included tainted pet food imported by the United States and children's toys containing excessive lead levels. In 2008, there was a nationwide milk scandal in China involving milk and infant formula that had been adulterated with melamine.

Intellectual property piracy and counterfeiting is another major problem. The most common counterfeit and pirated goods are CDs, DVDs, cigarettes, and clothing; but counterfeiting is a problem in many industries, even the automotive and pharmaceutical industries. Counterfeit goods are appearing more and more in overseas markets. Counterfeit products tend to have substandard quality, damage the reputation of legitimate brands, and pose risks to public health and safety. There are also connections between counterfeiters and organized crime rings. Counterfeiting does, however, provide jobs to millions of people in China and, in some towns, sustains local economies.

China's consumption and development have placed a heavy burden on the environment. Air and water pollution are serious problems. China is a major emitter of carbon dioxide, sulfur dioxide, and greenhouse gases. Large amounts of household and industrial waste are untreated, causing pollution and threatening groundwater quality. Other environmental problems include desertification, drought, floods, soil erosion, and a declining water table. These problems extend to other countries, as China's huge consumption has led to imports of natural materials from elsewhere. For example, imports of timber from Indonesia have contributed to serious deforestation. Various measures have been taken to tackle environmental issues, including energy-saving practices, such as controlling waste discharge, financing research and development programs, and increased spending in environmental conservation and protection. Relevant laws and regulations have been enacted. Public awareness of environmental stewardship has been rising, and public surveys have reported widespread concern for environmental issues and sustainable development across a broad spectrum of issues. Further, the Chinese public has responded disapprovingly and emotionally to high-profile cases of environmental damage publicized in the media. The increasing popularity of vegetarianism and organic food—though still small in absolute numbers—can be explained to some extent by concern for the environment. In 2008, the government prohibited all supermarkets, department stores, and shops across China from giving out free plastic bags in response to concern for the environment.

Although environmental protection is receiving more attention, the development-first orientation dominates. The large-scale destruction of nature poses a real threat to sustainable development in the country and the world. China's increasing consumption of everything—from new construction, energy demand, and more cars on the roads—raises concerns about its impact on the world's resources and whether China's goal of achieving Western consumption levels for its entire people is achievable and sustainable.

Gareth Davey
Hong Kong Shue Yan University

See Also: Beijing, China; Consumerism; Developing Countries; Population Growth; Shanghai, China; Tianjin, China.

Further Readings

Davis, Deborah, ed. *The Consumer Revolution in Urban China*. Berkeley: University of California Press, 2000.

Economy, E. C. *The River Runs Black: The Environmental Challenge to China's Future*. Ithaca, NY: Cornell University Press, 2004.

Jaffrelot, Christophe and Peter van der Veer. *Patterns of Middle Class Consumption in India and China*. New Delhi: Sage, 2008.

Pong, D., ed. *Encyclopedia of Modern China*. New York: Charles Scribner's Sons, 2009.

Clean Air Act

The Clean Air Act (CAA) is a series of legislations passed between 1955 and 1990 for managing air quality by minimizing the air pollutants released from industries and motor vehicles so as to protect human health and the environment. The CAA is incorporated into the United States Code (USC) of Law as Title 42 (Public Health and Welfare), Chapter 85 (Air Pollution Prevention and Control). The CAA contains six separate titles (subchapters) for Air Pollution Prevention and Control, Emission Standards for Moving Sources, General Provisions, Acid Deposition Control, Permits, and Stratospheric Ozone Protection.

History of Air Pollution

Air pollution has been a concern in large cities for many years. As early as 900 B.C.E., air pollution was noted in Babylon in an asphalt mine as a "strange smell in the air." Air polluted with dark smoke, fog, stench, and soot was described as unbearable in Rome, Egypt, and, later, in England over the following centuries. Various rules and regulations were imposed to control air pollution. For example, in the beginning of the 14th century, the king of England banned the use of sea coal to reduce smoke in London.

The Industrial Revolution in the 18th century caused fast population growth and economic development in the Western world. A high level of coal combustion during this period led to the rise of pollutants in the air, which became hazardous to human health, especially in large cities. Chicago and Cincinnati were the first U.S. cities to attempt to reduce air pollution by means of legislation in 1881. Over the next half-century, other cities introduced their own regulations to control smoke emission.

Finally, because of public outcry and the continuous deterioration of living conditions in cities, the federal government decided to intervene and introduced the Air Pollution Control Act of 1955 under President Dwight D. Eisenhower. This act provided federal funding for the scientific research of air pollution. The Clean Air Act of 1963 was the first federal legislation to control air pollution. In 1967, the Air Quality Act was expanded to monitor and control interstate transport of air pollution. The Clean Air Act continued to be revised and amended, resulting in major amendments in 1970, 1977, and 1990.

Joint Effort

The CAA is the result of a continuous and joint venture in managing air quality, involving many parties in the United States. Three executive agencies—the Environmental Protection Agency (EPA), the Council on Environment Quality (CEQ), and the Office of Management and Budget (OMB)—worked on the legislation with other federal, state, and local government agencies, such as Congress, state governors, county officials, mayors, city council members, state legislators, state and local air agencies, the courts, and other federal agencies. Major industry and trade organizations, small businesses, farms, scientists, engineers, academia, research organizations, the Clean Air Scientific Advisory Committee (CASAC), environmental and public health groups, and the people of the United States contributed to the legislation and implementation of CAA and its amendments.

Standards

In a major amendment in 1990, the EPA established the goals for National Ambient Air Quality Standards (NAAQS). The acceptable limits of the six principal pollutants (carbon monoxide, lead, nitrogen oxides, sulfer dioxide, particulate matter, and ozone), measured in parts per million (ppm) or by volume, (mg/m3) of air, for primary and secondary standards are defined by the CAA. The Primary standards, are set to limit air pollutants to protect public health, including "sensitive" populations, such as asthmatics, children, and the elderly.

The secondary standards set limitations to air pollutants for protecting public welfare, including protection against decreased visibility as well as damage to animals, crops, vegetation, or buildings. Every state creates its own EPA-approved State Implementation Plan (SIP) to attain and maintain the NAAQS standards. The areas that are unable to meet the NAAQS standards are classified as Nonattainment Areas and are required to follow strict guidelines to achieve the design limits of the air pollutants by the deadline set up by the EPA. The worse the pollution problem, the more stringent the control requirements imposed by the EPA.

Effects

A committee on air quality management in the United States concluded that the implementation of CAA-specified regulations substantially reduced the emissions of several pollutants from both moving and stationary sources. However, emission control of many older and higher-emitting facilities was yet to be achieved as of 2010. The "cap and trade" policy has also helped to reduce pollution from major industries. Air quality monitoring networks have confirmed a decrease in air pollution, especially in urban areas. The direct cost of implementing air pollution control regulations has been more than $20–$30 billion per year. However, the economic value of the benefits to public health and welfare far exceeds the cost of implementation.

Future population and economic expansion will impose a challenge to the progress that has been made in improving ambient air quality. However, the CAA will continue to require strict vigil in improving and maintaining air quality with the help of new and innovative technologies, including production and implementation of alternative energy and zero waste technology.

Mousumi Roy
Penn State University Worthington Scranton

See Also: Acid Rain; Carbon Dioxide; Coal Ash; Emissions; Environmental Protection Agency (EPA); Pollution, Air.

Further Readings

Bachmann, J. "Will the Circle Be Unbroken: A History of the U.S. National Ambient Air Quality Standards." *Journal of the Air & Waste Management Association*, v.57 (2007).

Committee on Air Quality Management in the United States. National Research Council (U.S.). *Air Quality Management in the United States*. Washington, DC: National Academies Press, 2004.

Environmental Protection Agency. Office of Air and Radiation. http://www.epa.gov/air (Accessed July 2010).

Fleming, J. R. and R. Bethany. "History of the Clean Air Act: A Guide to Clean Air Legislation Past and Present. http://www.ametsoc.org/sloan/cleanair/index.html (Accessed July 2010).

Maritineau, R. J. and D. P. Novello, eds. *The Clean Air Act Handbook*. 2nd ed. Chicago: American Bar Association, 2004.

McCarthy, J. E. "CRS Issue Brief for Congress, Clean Air Act Issues in the 109th Congress, Congressional Research Service, The Library of Congress." 2006. http://ncseonline.org/NLE/CRSreports/06jun/IB10137.pdf (Accessed July 2010).

Clean Water Act

The Clean Water Act (CWA) is the basic structure that regulates the quality of groundwater in the United States. In particular, this act regulates the discharge of pollutants into water. This important piece of legislation has had a long history of political and public controversy, and CWA legislation has proven particularly difficult to enforce.

History

From the late 19th century, the U.S. federal government recognized the necessity of regulating national waterways, but it was not until the mid-20th century that the U.S. government really faced the threat that polluted water posed to public health. The pollution of U.S. waterways steadily increased because of the growth of cities and the expansion of heavy industries. In 1948, the first major U.S. law that addressed issues of water pollution was drafted by Congress. The Federal Water Pollution Control Act of 1948 (FWPCA) set out the legal authority of the federal government to regulate water quality. Under the FWPCA, the Office of the Surgeon General as well as other federal, state, and local entities were authorized to create programs to eliminate and reduce the pollution of interstate waters. The FWPCA provided funding to state and local programs limited to interstate waters. In 1956, the FWPCA underwent amendments that increased the power of the federal government to intervene when public safety was in question. In 1965, President Lyndon B. Johnson pledged that the nation's dirtiest rivers would be cleaned up by 1975. These powers were expanded further with the introduction of the Water Quality Act of 1965.

The main issue was that because of a lack of technology to monitor water quality, it was difficult for authorities to prove that a violator had caused a specific violation. The Federal Water Pollution Control Act and its many amendments were not effective enough in stopping water pollution; it was necessary to overhaul this hodgepodge of laws and create a unified program of regulations and fines that would cut water pollution off at the source. The fight against water pollution lagged behind the war on air pollution—the Clean Air Act was passed in 1970.

By the 1970s water pollution had reached a crisis level in the United States. On June 22, 1969, the Cuyahoga River in Cleveland, Ohio, burst into flames; oil and fuel from industrial waste that had been dumped into the waters was the cause.

This was not the first time—that particular body of water had suffered fire over the previous three decades—but it was the first time such a fire generated national outrage. What had changed by 1969 was the nation's diminished tolerance for environmental pollution, not only because of concerns raised by Rachel Carson in her 1962 book *Silent Spring,* but also because of a growing environmental movement in middle-class communities across the United States.

President Richard Nixon stated that the 1970s would have to be the era in which Americans would pay for their past debts by reclaiming the purity of air, water, and environment. Nixon's pledge gave Americans great hope that the federal government would take environmental issues more seriously. Nixon ended up vetoing the proposed Clean Water Act on economic grounds, but Congress was quick to overturn the veto. This political exchange between Congress and President Nixon begged the question of who was going to pay the hefty short-term price for the long-term improvement of America's waters?

Goals and Implementation

In 1972, the Clean Water Act (CWA) was introduced. The CWA differed in scope from previous legislation; it aimed not only to end the violations that had caused these extreme levels of pollution, but it also aimed to restore the quality of waters in the United States. The goals of the CWA were to have zero discharge of pollutants into navigable waters, have water quality that can sustain aquatic life and in which people can swim, and prohibit the discharge of toxic amounts of toxins. This law states that the nation's waterways are not to be used as dumping grounds.

The CWA proposed a permit system for regulating pollution at the source. These "point sources" included industrial facilities (such as manufacturing, mining, and oil and gas extraction, as well as service industries), municipal governments (including military bases), and agricultural facilities. These point sources needed to obtain a permit from the National Pollutant Discharge Elimination System (NPDES) managed by the Environmental Protection Agency (EPA) in conjunction with state agencies.

In 1977, Congress expanded the reach of the EPA, which was now asked to control the release of toxins into sewers and surface water. Frustrated by the slow progress in cleaning up U.S. waters, Congress made further changes to the Clean Water Act in 1987. This included programs to clean up site-specific areas, such as the Great Lakes via the Great Lakes Critical Programs Acts of 1990. In addition, nonpoint source pollution regulation programs were added, specifically addressing the pollution caused by stormwater runoff from industry and farms.

Overdue Revision

As of 2010, the Clean Water Act has not been revised since 1987, and many Americans feel that a large-scale reassessment is long overdue. The U.S. Supreme Court's uncertain definitions of which waterways are protected by the Clean Water Act have continuously undermined the EPA's attempts to impose penalties and fines. Many businesses have declared that CWA laws no longer apply to them, and pollution rates have actually risen. The Clean Water Act has largely failed to fulfill its mandate of ending pollution and protecting the nation's waterways. Regulators do not have the jurisdiction to prosecute some of the nation's biggest polluters, and it is estimated that more than half of major pollution cases in the 21st century have been shelved or discontinued.

The 2010 Gulf of Mexico oil spill brought the Clean Water Act to the public forefront once more as investigations were being opened against British Petroleum and their violations of the CWA, which carry criminal and civil penalties and fines. Many critics have claimed that the interests of big business and bipartisan politics have undermined a real overhaul of the CWA and any attempts to create strong federal policy that could effectively stem the pollution of national waterways and protect the nation's water supply.

Rachel Black
Boston University

See Also: Clean Air Act; Industrial Waste; Pollution, Water; Public Water Systems; Safe Drinking Water Act; Sewage Collection System; Sewage Treatment; Toxic

Substances Control Act; Waste Treatment Plants; Water Consumption; Water Treatment.

Further Readings

Adler, Robert, Jessica C. Landman, and Diane M. Cameron. *The Clean Water Act 20 Years Later*. Washington, DC: Island Press, 1993.

Duhigg, Charles and Janet Roberts. "Rulings Restrict Clean Water Act, Foiling E.P.A." *New York Times* (February 28, 2010). http://www.nytimes.com/2010/03/01/us/01water.html?scp=1&sq=clean+water+act&st=nyt (Accessed June 2010).

Environmental Protection Agency. "Summary of the Clean Water Act." http://www.epa.gov/lawsregs/laws/cwa.html (Accessed July 2010).

Hays, Samuel P. *Beauty, Health, and Permanence: Environmental Politics in the United States, 1955–1985*. New York: Cambridge University Press, 1987.

Ryan, Mark A. *The Clean Water Act Handbook*. 2nd ed. Chicago: Section of Environment, Energy, and Resources, American Bar Association, 2003.

Cleaning Products

There are cleaning products explicitly designed for toilet bowls, drains, showers, carpets, glass, tables, computers, toys, hard/soft surface floors, and appliances, just to mention a few. The range of uses for cleaning products is wide and their specificity is remarkable. They come in virtually any size and presentation as wipes, liquids, polishes, aerosols, and foams, and with or without disinfectant. Products meant to ease the daily duties of keeping households tidy are ubiquitous.

Most cleaning products are available for sale to the general public without any restrictions. As asserted by researcher Janice Hughes, detergents, degreasers, stain removers, and pesticides have made homes miniature chemical factories, since chemical levels can be up to 70 times higher inside the home than out. In fact, women who work only in the home have a 55 percent higher risk of getting cancer than do women working outside the home. Information regarding actual or potential risks to health or the environment is not necessarily on the label. Some manufacturers avoid publishing the risks by arguing that they have the right to keep their formulas a competitive secret. This commercial practice of not revealing some ingredients entails the possibility of concealing potential dangers to individual health and the environment. Concerned users can ask the manufacturers for Material Safety Data Sheets (MSDS) that contain detailed information about the ingredients used.

There are a number of polluting substances involved in the production, use, and disposal of cleaning products that imply some sort of environmental or health risk. For example, a recent study reported that about 200 xenobiotic organic compounds (chemicals not produced naturally by organisms) were found in grey wastewater from bathrooms. Those chemicals included fragrances, preservatives, phthalates, pharmaceuticals, and flame retardants. Several of those compounds could be associated with cleaning products consumed in the households.

Risks

Risks associated with substances included as ingredients of cleaning products are diverse. Some of the substances present in cleaning products that pose risks for users are not even active ingredients. For example, phthalates, which are suspected to have adverse hormonal effects, help to make dyes and fragrances more fluid. Other chemicals simply keep a product stable on the shelf, and others, such as glycols, act like antifreeze agents.

For most users, the risk of cancer or the ability of certain compounds to disturb the growth and development of an embryo or fetus (teratogenicity) are the biggest concerns, but some products are potentially harmful in different ways. Denatured alcohol is commonly used as a degreasing agent as an alternative to soap, detergent, or other degreasers. Denatured alcohol is also a popular disinfectant not listed as carcinogenic by regulatory bodies like the Occupational Safety and Health Administration (OSHA), the National Toxicology Program (NTP), or the International Agency for Research on Cancer (IARC). Nonetheless, it poses a fire hazard, particularly if poured onto electrical appliances. There is also a risk of toxicity if denatured alcohol is used in closed environments.

The range of uses for cleaning products is wide and their specificity is remarkable. They come in virtually any size and presentation as wipes, liquids, polishes, aerosols, and foams, and with or without disinfectant. However, this variety comes with risks. For example, phthalates, which are used to make dyes and fragrances more fluid, may have adverse hormonal effects. Many other household cleaning products contain harsh solvents and chemicals that destroy the natural processes involved in wastewater treatment.

Environmental and health risks for households necessarily include their associated goods like the family car, where members spend a substantial part of their lives. Cleaning products for cars are big sellers, but many of them involve well-known risks. Aliphatic hydrocarbons, naphthas, and petroleum distillates in general are used in some car waxes, furniture polishes, and general car cleaning products. These have been associated with neurotoxic reactions. Some of them may cause irritation to the skin, digestive system, throat, and lungs when inhaled. A common product used in car cleaning products (particularly in cleaners and waxes) is formaldehyde. This product is carcinogenic, neurotoxic, and poisonous.

Pet Care Products

Products for the care of pets are not exempt from risks. Allethrin, a synthetic pyrethroid (a form of a chemical found naturally in the chrysanthemum flower), is used in some pet flea-control products. It can cause damage to the immune system, with hay fever–like symptoms and also sudden swelling of the face, eyelids, lips, mouth, and throat tissues. Pyrethroids like Allethrin are generally harmless to human beings in low doses but they can be dangerous to sensitive individuals, fish, and other aquatic organisms. Studies show that in very small amounts (as low as two parts per trillion), pyrethroids are lethal to most beneficial insects such as bees and dragonflies. They are also toxic to invertebrates that constitute the base of many aquatic and terrestrial food webs. A particular matter of concern is that pyrethroids can pass through secondary treatment systems at municipal wastewater treatment facilities. Another product used in pet flea control is D-limonene, a known neurotoxin, which is also an eye and skin irritant. Studies have suggested that this substance is associated with cancer and is also teratogenic.

Flea collars are not necessarily safer. Diazinon, a common compound used in many of these devices, has been attributed with allergenic and neurotoxic characteristics. It has been found that diazinon is toxic to the human fetus and to birds.

Bathroom Cleaners

Products for bathroom and toilet cleaning are particularly dangerous to individual and environmental health. For example, bathroom cleaners often contain sodium hypochlorite, a corrosive that irritates or burns skin and eyes, and causes fluid in the lungs, which can lead to coma or death. Acid blue 9, also known as brilliant blue FCF or disodium salt, is used in some toilet bowl cleaners and deodorizers. It has been associated with cancer. Sodium bisulfate, a very common ingredient of toilet bowl cleaners and deodorizers, can cause asthma attacks, particularly in children. Some drain openers use aluminum, which has been reported as a cause of lung disease if inhaled and has also been related to Alzheimer's disease. Dichlorodifluoromethane, a neurotoxic and eye irritant substance, can also be found in the ingredients of some drain openers.

Some brands of the apparently innocuous toilet deodorizers contain paradichlorobenzene, a well-known carcinogen that causes liver and kidney damage. Ammonia and derived compounds like ammonium chloride or ammonium hydroxide are used in toilet bowl cleaners, deodorizers, and some air fresheners. They are generally safe when highly diluted but otherwise can cause eye irritation, cataracts, and corneal damage.

Combining Substances

One point of interest for researchers is the combined effect of toxic substances and in combination with substances already present in the environment. For example, diethanolamine, a mild skin and eye irritant used in a wide range of household products, reacts with nitrites (added as preservatives to some products or present in the environment as contaminants) to form highly carcinogenic nitrosamines. Air pollution resulting from the use of cleaning products is a matter of concern for researchers and policy makers. Research shows that several cleaning products and air fresheners contain substances that can react with other air contaminants to yield potentially harmful secondary products. For example, certain resins can react rapidly with ozone in indoor air, generating pollutants such as formaldehyde.

A good source for online information on potentially harmful products is the Household Products Database. It is administered by the National Library of Medicine and offers compiled data from the MSDS, product labels, and manufacturer Web sites for more than 4,000 products. A useful component of this database is its reference to HMIS® coding, the Hazardous Materials Identification System, a registered mark of the National Paint and Coatings Association (NPCA) but used by all manufacturers to comply with the requirements of OSHA's Hazard Communication Standard. The HMIS coding guide of colored bars, numbers, and symbols to convey the severity of hazards of chemicals is portrayed in every record of the Household Products Database. Ratings assigned to a product do not necessarily represent the outcome of an objective, third-party evaluation. Those ratings were determined for each brand by its manufacturer and documented in the Material Safety Data Sheet (MSDS) published by the manufacturer. Most of the ratings for health, fire, or reactivity dangers are based on short-term, acute reactions to the materials and not on long-term effects.

Alternatives to commonly used cleaning products are green products. Most manufacturers of traditional cleaning products now offer their own "green" brands. As a general practice, green cleaning products avoid the use of chlorine and phosphates. The latter are strongly associated with ocean pollution. Green products captivated a substantial part of the market in the first decade of this century. However, the market for most natural cleaning products has experienced a reduction as recession took hold, particularly in the United States. This means that, in terms of cleaning products, consumers are concerned about their own health and the environment only if they can buy products at the same price (or even less) than those that are harmful. Unless green products are guaranteed some sort of subsidy, it is going to be difficult to change traditional practices on the consumers' side.

In summary, small amounts of cleaning products seem to be harmless for both users and the environment. But real life behavior seems to favor large doses of several different products in a household, creating a harmful environment with unknown consequences for society and the environment as a whole.

Mauricio Leandro
Graduate Center, City University of New York

See Also: Aerosol Spray; Household Hazardous Waste; Pesticides; Pets; Toxic Wastes.

Further Readings

Carson, R. *Silent Spring*. Boston: Houghton Mifflin, 1962.

Hughes, J. *Hazards of Household Cleaning Products*. http://www.shareguide.com/hazard.html (Accessed June 2011).

National Institutes of Health. Household Products Database. http://hpd.nlm.nih.gov/index.htm (Accessed June 2011).

Cloaca Maxima

The Cloaca Maxima, literally meaning "greatest sewer," was the largest sewer system in the ancient world. According to the Roman historian Pliny, this wonder of the ancient world was constructed in the 6th century B.C.E. by the two Tarquin kings and was a permanent reminder to the Romans of their ancient history. Pliny emphasized the Cloaca's durability as well as its savage discipline of construction; sewer workers—mainly forced labor drawn from Rome's poorer residents—who committed suicide were said to have had their bodies crucified as an example to others.

It is probable that the Cloaca was originally an open drain, formed from streams from three of Rome's hills, which were channeled through the main Forum and then on to the river Tiber. This open drain would then have been gradually built over, as space within the city became more constricted. The sewer was originally meant to drain land only, with the city's other waste matter being flushed into the Tiber. However, as Rome's population grew, the sewer increasingly became a dumping ground for unwanted wastes and became choked with filth. In 33 B.C.E., the emperor Agrippa demonstrated how he had unblocked the Cloaca Maxima by riding through it on a boat. In the 21st century, this foundation layer of Rome's history remains open to curious tourists, and the original outfall of the sewer into the Tiber is still preserved.

Although widely documented in ancient writings, the Cloaca Maxima was only "discovered" more widely in the 18th century when Rome was undergoing renewed archaeological investigations. From 1748 to 1774, the artist Giovanni Battista Piranesi (1720–78) published an extensive series of views of Rome focusing on its ancient archaeological remains that included the Cloaca Maxima. Piranesi's son Francesco collected and preserved the views and published them in 29 volumes from 1835 to 1837. Thereafter, the Cloaca Maxima became widely known in Europe and was to attain enormous importance as a prototype for the new drainage systems being planned in major European cities, particularly London and Paris, both of which vied to become the "new Rome" in the 19th century.

Applications

In the 1860s, London's sewer system was being transformed by the engineer Joseph Bazalgette (1819–91), with new intercepting sewers constructed to prevent waste matter from polluting the river Thames. During its construction, comparisons to ancient Rome and its sewer were consistently evoked. The journalist Henry Mayhew (1812–87) described London's sewers as second only to the "giant works of sewerage in the eternal city," while Bazalgette used the Cloaca Maxima as a prototype for London's sewers in his lectures on his new drainage system.

When London's new sewers were completed, the press were almost ecstatic in their praise. In a ceremony held in 1865 to mark its formal opening, some newspapers compared the new sewers with the wonders of the ancient world. According to the *Daily Telegraph*, the main drainage system was a project alongside which even the pyramids of Egypt and the sewers of Rome "paled into comparison." The *Marylebone Mercury* made similar comparisons: the main drainage system was described as the "representation of a mighty civilisation," a civilization nobler than ancient Rome because it lacked its "despotic power." Underlying these comparisons was the view that London's new sewers were a permanent monument to the future when the city—especially compared with its main rival, Paris—would become the cleanest and most magnificent city the world had ever seen. If the content of London's new sewers was "not a bit better" than that in the sewers of Paris, their technological and political basis most certainly was. Under Napoleon III and Baron Haussmann in the 1860s, Paris was undergoing a more radical transformation than London. New boulevards were driven through the medieval city, and new sewers were constructed beneath them. While some criticized London's government for not "Haussmanizing" London enough, most celebrated the city's new sewers as making the city above comparison with any other European capital. If Bazalgette had done "what Tarquin did for Rome," he did it without the "despotic power" of the latter. The fact that Napoleon III was also self-consciously modeling his new Paris sewers on this Roman precedent also points to an implicit criticism of his despotic methods.

Paul Dobraszczyk
University of Reading

See Also: History of Consumption and Waste, Ancient World; Pollution, Water; Sewage; Sewers.

Further Readings

Gowers, E. "The Anatomy of Rome From Capitol to Cloaca." *Journal of Roman Studies,* v.85 (1995).

Hopkins, J. N. N. "The Cloaca Maxima and the Monumental Manipulation of Water in Ancient Rome." In *The Waters of Rome*, K. W. Rinne, ed. Charlottesville, VA: Institute for Advanced Technology in the Humanities, 2007.

Mayhew, H. *London Labour and the London Poor.* Vol. 2. London: Griffin, Bohn, 1862.

Pliny. *Natural History: A Selection.* London: Penguin, 1991.

Reid, D. *Paris Sewers and Sewermen: Realities and Representation.* Cambridge, MA: Harvard University Press, 1991.

Coal Ash

Coal ash is the noncombustible waste product left over from the burning of coal. It consists of airborne particles called "fly ash" as well as heavier particles called "bottom ash" that settle on the floors of coal-fired furnaces. Fly ash, which was once released into the atmosphere by coal-burning electric utilities, is now typically "scrubbed" from exhaust gases through a variety of mechanisms (such as bag filters, cyclone separators, or electrostatic precipitators) installed within smokestacks. Once the dry ash has been recovered, water is typically added to form a slurry, allowing for pipeline transport and reducing the potential for the finely grained ash to become airborne after capture. The coal ash or slurry can then be disposed of in a landfill; however, since such a solution entails not only transport costs but also the additional expense of paying the municipality in charge of the landfill, it is more frequently pumped into nearby impoundment ponds already owned by the utility. The coal-burning industries of the United States produced an estimated 131 million tons of coal ash annually as of 2010. Since the total weight of ash produced increases with each passing year—a phenomenon attributable to new coal-fired plants being brought into operation to meet increasing electricity demand as well as to improvements in scrubbing technology—the disposal of this waste material presents a serious and growing matter of ecological concern.

Composition

While precise percentages tend to vary depending on the composition of the coal that is fed into the boiler, coal ash consists primarily of silicon dioxide (SiO_2) and smaller portions of aluminum oxide (Al_2O_3) and iron oxide (Fe_2O_3). Although coal ash's principal compounds are largely benign, it also contains trace amounts of heavy metals and hazardous compounds such as arsenic, cobalt, lead, mercury, uranium, dioxins, and polycyclic aromatic hydrocarbon (PAH) compounds.

The disposal strategies for this waste, which include not only impoundment but also reuse in both industrial and commercial products, makes the risks posed by these trace elements a source of public controversy. Though the coal industry insists that the coal ash is perfectly safe and that traces of these hazardous elements exist in practically everything, environmental advocates are far less sanguine, pointing out that these naturally occurring toxic elements become concentrated in unregulated, unlined holding ponds, allowing hazardous pollutants to leach into public water supplies.

Health and Safety

Considering the incredible amount of coal that electric companies consume—estimates suggest that the average American's individual consumption of electricity requires the burning of a little more than 20 pounds of coal per day—the sheer volume of coal ash produced, apart from its potential toxicity, has become an equally pressing public safety concern. Potential problems with impound ponds were publicized in the wake of a massive coal ash spill in December 2008 at the Tennessee Valley Authority's Kingston Fossil Plant. A breach in the retaining wall of the pond ended up releasing over one billion gallons of coal ash slurry into a nearby river, killing fish and wildlife and damaging homes in the vicinity. Cleanup efforts are expected to cost more than $1 billion, making it one of the largest and most costly industrial accidents in U.S. history. The disaster has galvanized calls for the Environmental Protection Agency (EPA) to classify coal ash as a hazardous waste and subject its disposal to stringent federal control. Industry stakeholders, however, continue to emphasize the unnecessary economic consequences of such legislative action. The EPA has since responded with a "co-proposal" laying out a highly contingent and convoluted set of federal regulations that effectively hedges on the issue of coal ash's legal status.

Recycling

Also invested in this debate are industries that attempt to recycle coal ash into salable material. Ever since the 1930s, when coal ash began to be recovered in significant amounts, scientists and technicians have investigated ways to salvage the by-products of coal-fired electricity production for profitable use. Coal ash has been tried and rejected as a suitable substitute for a wide and incongruous variety of products, including pesticides and toothpaste. Similar in chemical structure to clay, it has been used with slightly better results as a filler material for rubber, paint, putty, roofing material, roads, and ceramics. Its greatest commercial success as of 2010 is as an ingredient in Portland cement. Coal ash, a variety of pozzolan, has been touted by industrial interest groups as an environmentally friendly substitute for cement in that the carbon footprint from producing new cement can be diminished substantially by mixing in portions of coal ash, which exists already in abundance. Nevertheless, coal ash's origin as the by-product of a carbon-intensive process of energy generation complicates claims about the carbon neutrality of industrial products that use coal ash as a filler material.

Other Criticisms

Concerns about the toxicity of materials made with coal ash also complicate its alleged environmental virtues. Opponents claim that products made from refashioned coal ash will follow the same cultural trajectory as asbestos—a product that has been vilified for its carcinogenic properties after being celebrated for so long as a miracle product. In spite of these critics, coal ash is becoming a more popular ingredient in civil and commercial construction efforts around the globe with concrete made from this coal ash mixture being used in high-profile construction ventures such as the Burj Khalifa in Dubai, the world's tallest building in 2010.

Coal ash thus remains, in both its material and legal aspects, a highly ambivalent substance. Environmentalists see its abundance and potential toxicity as further evidence for the ecological malfeasance of the energy industry, while industry stakeholders maintain that the benefits of reliable, affordable energy inevitably must outweigh public anxieties concerning what is arguably a relatively harmless and potentially useful by-product of modern existence. With the federal government reluctant to arbitrate this stalemate, coal ash promises to remain one of the more prominent and enduring waste concerns in the 21st century.

Kevin Trumpeter
University of South Carolina

See Also: Clean Air Act; Environmental Protection Agency (EPA); Power Plants; Scrubbers.

Further Readings

Environmental Protection Agency. "Proposed Rule for the Disposal of Coal Combustion Residuals From Electric Utilities." *Federal Register*, v.75/118 (2010).

Goodell, Jeff. *Big Coal: The Dirty Secret Behind America's Energy Future*. Boston: Houghton Mifflin, 2006.

U.S. National Research Council Committee on Coal Waste Impoundments. *Coal Waste Impoundments: Risks, Responses, and Alternatives*. Washington, DC: National Academies Press, 2002.

Colorado

Named by the Spanish for its red-colored earth, Colorado is a state of diverse geography and complex climate, consisting of mountains, foothills, high plains, and desert. Originally a mining economy in the mid-19th century, the development of irrigation brought agriculture to the fore. The federal government is a major part of the state's economy, with the North American Aerospace Defense Command (NORAD) and the United States Air Force (USAF) being just two of the federal agencies based in Colorado. This is partly due to the capital city, Denver, being equidistant both between Los Angeles and Chicago and between Seattle and New Orleans. As of 2009, 61.9 percent of the population live in the Denver-Aurora-Boulder Combined Statistical Area.

The 16th Nationwide Survey of MSW Management in the United States found that in 2006, Colorado had an estimated 8,690,005 tons of municipal solid waste (MSW) generation, placing it 16th in a survey of the 50 states and the capital district. Based on the 2006 population of 4,766,248, an estimated 1.82 tons of MSW were generated per person per year (ranking fifth) and 8,208,407 tons were landfilled (ranking 10th). Colorado did not report its number of waste-to-energy (WTE) facilities (presumably none) or any data regarding landfills. Colorado recycled 481,958 tons of MSW, placing the state 32nd in the ranking of recycled MSW tonnage.

In the early 21st century, Colorado faces several challenges relating to consumption and waste. Air pollution has been an issue in industrial Pueblo since the advent of the steel industry there in the 1880s. Air quality in the Denver metropolitan area is threatened by suburban sprawl, the high altitude, and the region's dependence upon automobile transportation. Water consumption in the state is divided between residential and agricultural uses; much state politics revolves around water rights. The city of Boulder, northwest of Denver, has adopted a policy of controlled growth expansion as well as initiatives to manage urban wildlife. Despite these efforts, sprawl in the region has been substantial since the 1980s, as communities such as Broomfield and Lafayette have become homes to commuters to Denver.

Cowboy Wash

One of the most controversial refuse finds ever found, and certainly the most notorious human coprolite, is the Cowboy Wash cannibal coprolite. The cannibalism argument is one of the most heated debates in southwest American archaeology. At a number of Anasazi sites in the Four Corners region, there are nonburial sites where human remains were excarnated, butchered, and burned between 1125 and 1175 C.E., including Aztec Wash, the Grinnel Site, and Hanson Pueblo. Cannibalism and extreme violence are interpreted at around 100 sites in the southwest since the discoveries at Cave 7, Utah, and Mesa Verde, Colorado, in the 19th century. While osteological studies show the bones from these sites were processed in a way consistent with food preparation, opponents of the cannibal theory have forwarded such interpretations as secondary interment, necrosadism, and the ritualized execution of witches. Processed and randomly discarded human remains pointed to cannibalism but there was no archaeological proof that it had actually taken place.

Cowboy Wash was excavated in 1996 by a mixed team of Anglo-American and Native American archaeologists. The excavation was part of a study of 17 Anasazi sites on the southern footslopes of Sleeping Ute Mountain, during which 105 structures, including 36 pithouses were excavated, dating from 450 to 1280 C.E. The site (5MT10010) at the southern base of Sleeping Ute Mountain consists of three pithouses (features 3, 5, and 15) and several other structures believed to be the settlement of an extended family of 15–20. The site was abandoned suddenly around 1150 C.E., with personal belongings left behind; the site was never reused, which is very unusual in an area where even building material is salvaged and reused. Valuable tools, pottery, and ornaments were found in direct association with bench and floor surfaces with no sediment between, indicating they were undisturbed.

Microstratigraphic evidence showed the roofs had decayed in place rather than being salvaged or burned as is the custom at almost all other sites in the area.

Bones and bone fragments were discovered scattered around floors and stacked on surfaces in two pithouses (features 3 and 13). Over 1,000 bone finds were recovered, representing a minimum number of individuals (MNI) of four adults, one adolescent, and two children. Almost all the bones were disarticulated, many bones were missing, including vertebrae, and all head, face, and long bones were broken. The bones had butchery marks where muscles had been cut, skulls and other bones were burned, and long bones were broken down to "cooking size." Some of the bones display "pot polish" on the ends where they have been rubbed smooth by friction, possibly with the inside of a cooking pot. Stone tools of types used in butchery were found around the hearth in feature 13, and in feature 15 there were fragments of cooking pots.

All of the evidence pointed to cannibalism but there was no actual evidence of the stage where the human remains were consumed. Human tissue can be butchered and cooked but proof of consumption, the final stage in an act of cannibalism, is the only way that cannibalism can be physically proven to have taken place and is irrefutable evidence of cannibalism.

Further Analysis

Three classes of find were tested using biomolecular techniques, cooking vessel fragments associated with the burned bones, stone tools from the apparent butchering area, and, crucially, a human coprolite. Seemingly contemptuously deposited in the remains of the hearth of feature 15, this is the only known coprolite from a structure hearth in the southwest. Protein residue analysis was used on the stone tools and the cooking-vessel shards; this technique examines residual blood and tissue adhering to pots and tools used to process animal remains. Two stone tools tested positive for human blood residue, and the cooking pot shards tested positive for human myoglobin. The last step of the cannibalism process was proven when the coprolite was analyzed and indicated that human remains had been eaten and excreted.

The coprolite that had been deposited into the cold ash of the hearth was of a size and shape consistent with human origin, and had a dry weight of 30g. Macroscopic analysis showed no apparent plant remains, very unusual for a human Anasazi coprolite. Microscopic analysis found it consisted entirely of meat, with virtually no starch crystals or plant evidence present except for windborne pollen. The proteins in the coprolite were then examined to find a human protein that was not found in blood and therefore possibly from any internal bleeding, or a protein that had been shed from the intestinal lining. The protein also had to be one that did not come from the human digestive system as this could also have come from the consumer. Myoglobin, a protein found only in skeletal and cardiac muscle, was tested for using immunoelectrophoresis and the enzyme-linked immunosorbent assay (ELISA) test.

Immunoelectrophoresis is a test used to identify proteins by their electrophoretic and immunological properties; it can also be used to extract DNA. Gel electrophoresis, around which the technique revolves, exploits the separation of molecules due to differences in size, charge, or other physical properties. When current is passed through a gel, the immunoglobin molecules move toward the oppositely charged electrode. Because the different molecules have different characteristics, they will migrate through the gel at different rates. Heavier, low-charged molecules will move slower than smaller, highly charged molecules. The migration therefore separates the different molecules. In immunoelectrophoresis, the molecules of individual immunoglobins, now separated, can be identified by immunological methods and their presence and quantity noted. The chemical composition of myoglobin differs between animal taxa, so the ELISA assay test can identify human myoglobin from the myoglobin of other species. The ELISA uses specific antigen-antibody reactions to test for antigens or antibodies by attaching an enzyme that produces a color to the antigen or antibody specific to the target substance. The sample is attached to small wells on an ELISA plate, the antigen or antibody is added, and then the plate is "washed." If the antigen-antibody reaction has occurred, this can now be demonstrated by the enzyme's changing color. In the Cowboy Wash case, an anti-human antibody was used, which reacted with the human tissue in the sample.

The coprolite tested positive for myoglobin, proving that cannibalism had occurred. As the results were so controversial, modern stools, ancient coprolites, blood-positive stool samples, and nonhuman myoglobin samples were tested as controls, and none tested positive for human myoglobin, giving conclusive proof. The interpretation was unpopular with some scholars and Native American groups and it was suggested that the coprolite was actually of coyote origin. However, the coprolite did not contain fur or bone fragments usually found in canine feces, and on rehydration did not have the characteristic lubricating coat that allows canines to safely digest pieces of bone.

The reason for acts of cannibalism at this time remains unknown. The 1125–75 C.E. period was one of change and turmoil in the southwest. One theory is that cannibalism was used as a terror tactic to establish religious or political control over a population. Other hypotheses include starvation brought on by drought and invasion by an outsider group.

Jon Welsh
AAG Archaeology

See Also: Archaeological Techniques, Modern Day; Funerals/Corpses; Human Waste.

Further Readings

Arsova, Ljupka, Rob van Haaren, Nora Goldstein, Scott M. Kaufman, and Nickolas J. Themelis. "The State of Garbage in America: 16th Nationwide Survey of MSW Management in the U.S." *BioCycle*, v.49/12 (2008).

Marlar, R. A., et al. "Biochemical Evidence of Cannibalism at a Prehistoric Puebloan Site in Southwestern Colorado." *Nature*, v.407/74–78 (2000).

Commodification

The term *commodification* is used in different ways in mainstream business theory, Marxist economics, and anthropology. All of the senses of the word are, however, relevant to the subject of consumption and waste. As a historical process, commodification has been associated with the rise of globalized industrial capitalism and consumer culture. In facilitating high levels of mobility and consumption, commodification has contributed to contemporary environmental challenges. It is a pervasive aspect of 21st-century society in which the market is increasingly prevalent in daily life and even waste itself is commodified.

In Business

In the language of business economics, the term *commodification* (or *commoditization*) has been in use since the early 1990s, referring to the standardization of formerly differentiated versions of a product so that consumers no longer perceive any differences in value among them. According to business theory, this transformation is associated with an increasingly competitive market for goods and services and normally leads to decreased prices. The result is a commodity that is relatively fungible in the sense that every unit of the commodity is considered to be equivalent to every other one for the purposes of economic transactions (as, for example, every dollar has the same value and exchangeability as every other dollar). An example from the arena of agriculture is that of corn (or maize). Dozens of cultivars of the *Zea mays* species were developed by indigenous American horticulturalists, suitable for various growing conditions and culinary uses. Seeds were saved from each crop to plant the next generation, and the product was produced, traded, and consumed within face-to-face social networks. The commodified version of corn designed for industrialized agriculture includes a small number of similar cultivars grown from commercially hybridized seeds that do not breed true but instead must be purchased anew each planting season. The product is highly standardized and fungible, forming a great stream of undifferentiated "commodity corn" severed from the identity of individual producers, which is mostly channeled into feedlots for commodified meat production and into the myriad intakes of the processed food industry.

In Marxism

The second meaning of commodification comes from Marxist economics, where the term has been used since the mid-1970s, referring to the transformation of goods that were not formerly considered in terms of their monetary value into products that are bought and sold on markets as commodities. This process of imbuing goods with exchange value has been associated with the expansion of market trade into areas with previously nonmarket economies and noncommercialized social relationships. For Karl Marx, the prime example was when the labor of an actual worker is abstracted into a manufacturing input that is measurable in hours.

Under conditions of "alienated" (commodified) labor, the worker exchanges for a wage the power to determine what they will produce, how and when they will produce it, as well as ownership and control over the product of their labor. The Industrial Revolution has been associated with commodification, both in terms of alienated labor (as with the transition from artisanal production to work on assembly lines) and in terms of the development of undifferentiated, depersonalized products. Commodification can be thought of as the difference between a landscape artist painting for pleasure and expression and one filling an order on a deadline for two dozen sofa-size pieces with stags, sunsets, and mountain lakes. The more-commodified paintings are designed to suit consumers' tastes and are sold through impersonal transactions to people the artist never meets, their value expressed as a money figure.

Building on the Marxian concept, anthropologists have developed an approach to commodification that emphasizes the ways incorporation into markets changes the meanings as well as the social relations surrounding objects (which may be material goods, services, or abstract items). In concrete human communities, objects are understood ethnographically in terms of the ways they are defined and used within specific social and cultural contexts. It is this context of social relations that structures the access persons and households have to particular items. When objects are commodified, they become detached from this immediate context and lose their original meaning. Traditional forms of access to them are replaced by generalized exchangeability. Disembedded from the local community and its ways of categorizing and evaluating things, the commodity is reconceived in terms of an unlimited sphere of exchange and universal measures of value. It may now be appropriated in other contexts and its meaning redefined for and by consumers.

This process can change both the meaning and the form of the commodified things, as they become products marketed to consumers. An example would be the case of a traditional community festival that becomes a tourist attraction. Marketing this item of living culture transforms it into an object performed for outsiders, frozen into a form attractive to consumers, its timing perhaps altered to coincide with tourism seasons. Cultural critics have identified the accelerating pace and scope of commodification in society (including the growth of markets for human embryos, organs, and other tissues) as a dehumanizing trend. Others have pointed out the positive and liberating results of commodification, as it creates new markets to generate new economic activity and frees individuals to sell their skills wherever they choose.

Marketing and Globalization

Commodification in all of these senses is historically linked both to the rise of modern industrialism and globalized consumer culture and to the resulting dramatic increases in waste of all sorts. In the 21st century, commodities may be mass-produced using nonlocal materials and marketed worldwide so that similar products may be found almost anywhere. Marketing designed to stimulate the desire to consume commodities has accelerated the pace at which populations have embraced the lifestyle of consumption offered by industrial capitalism. In this sense, commodification has been a force for global cultural homogenization. On the other hand, ethnographic studies have underscored the ways people may reformulate the meanings of such products in the process of consuming them, appropriating commodities as resources for constructing identities and recontextualizing them in heterogeneous ways.

Effects on the Environment

The worldwide growth of industrialized commodity consumerism has generated environmental challenges. The expansion of commodification in order to profit from continuously stimulated consumer

Hybridized corn varieties bred for high-production agriculture must be repurchased each planting season. The standardized final product forms a stream of commodified corn that is nearly indistinguishable from producer to producer.

markets has occurred with insufficient regard to the resulting wastes from the extraction of raw resources; the transportation of inputs, products, and shoppers; industrial manufacturing processes; and nonbiodegradable packaging materials and material products themselves that are sometimes designed to be used briefly, then disposed of and replaced. Commodification characterizes the social relations and products at every stage along the way. The lifestyle of commodity consumption is filled with material possessions, most of which end up in landfills and garbage dumps. Even these may be privately owned properties, and the services they provide are also commodified. Waste itself is recommodified when it is sold for manufacture into products made from recycled materials. Such is the pervasive and systemic process of commodification in modern industrial society that even the legal right to pollute is being commodified, as a market develops for buying and selling carbon credits.

Caroline Tauxe
Le Moyne College

See Also: Capitalism; Consumerism; Post-Consumer Waste; Socialist Societies.

Further Readings

Ertman, Martha and Joan C. Williams. *Rethinking Commodification: Cases and Readings in Law and Culture*. New York: New York University Press, 2005.

van Binsbergen, Wim and Peter Geschiere, eds. *Commodification: Things, Agency and Identities*. Berlin: LIT, 2005.

Composting

Composting in solid waste management systems is the decomposition and stabilization of the organic fraction of municipal solid waste (MSW) carried out by a microbial community under controlled, aerobic conditions. Though composting has been practiced by people since they first settled in agricultural communities, it is now emerging as a centralized waste management method that both reduces the volume of waste that must be disposed and creates useful products. Most biogenic matter can be composted; its degradation occurs in four stages and is carried out by an ecological succession of microbial communities. The resulting product (compost) can be used as a soil conditioner, fertilizer, mulch, or a replacement for peat.

Overall Process

Composting is a naturally occurring biological process undertaken by a succession of microbial communities, and it is used by people to manage the organic fraction of waste. The aerobic degradation process can be written as

$$C_aH_bO_cN_d + 0.25\,(4a+b-2c-3d)\,O_2 \rightarrow aCO_2 + 0.5(b-3d)\,H_2O + dNH_3.$$

Overall, a consortium of microorganisms breaks down organic matter in the presence of oxygen, reducing the volume and mass of waste by approximately 50 percent (on a dry-weight basis); the other 50 percent of the mass is released as carbon dioxide (CO_2), water (H_2O), and ammonia (NH_3). Composting is an exothermic process; it releases heat and raises the temperature of the substances being degraded. The process is considered complete when only stabilized organic matter—matter that has low biologic activity that can be stored without giving rise to health or nuisance problems—is left over.

Objectives and Uses for Compost

There are four main objectives to composting as a waste management method: to reduce the volume of waste, stabilize waste, sterilize waste, and produce a valuable product from the waste. Composting greatly decreases the mass and volume of waste to be managed, which translates into a reduction of costs for managing that waste. Stabilization of waste allows for safe storage of the waste; if not composted, stored organic waste is likely to emit odors and to contain pathogens. The high temperatures reached in the composting process (upward of 65 degrees Celsius) destroy most pathogens and weed seeds contained in the organic waste. Finally, the main objective of composting is to create a valuable resource. Compost can be used as a soil conditioner, fertilizer, mulch, or a replacement for peat. Though compost's nutrient content is lower than that of commercial fertilizer, its nutrient release is slow and sustained.

Most plant matter, animal tissue, and microbial components can be degraded in an aerobic composting process. Cellulose is the most abundant component of plants and is found in most organic wastes; under aerobic conditions, many fungi and microbes are involved in cellulose degradation. Lignin, a structural component of plants and a major component of wood, is degraded much more slowly. Organic matter such as plastic and leather are difficult to break down, while inorganic substances such as glass or metal are relatively inert.

Biology and Chemistry of Composting

Composting involves an ecological succession of microorganisms with a presence depending on environmental conditions. Bacteria, archaea, fungi, protozoa, and worms are all involved in the aerobic degradation of organic waste. Microorganisms use the carbon in organic waste to produce energy and to synthesize cellular components. In addition to carbon, microorganisms need other macronutrients (N, P, and K) to thrive as well as several other micronutrients (Co, Mn, Mg, Cu, and Ca). A ratio of carbon to nitrogen of 20–25:1 is considered ideal for composting. An effective substrate for composting will balance carbonaceous wastes (such as dry leaves, hay, and paper) with nitrogenous wastes (such as grass, food waste, and sludge). To create an environment in which microorganisms are able to degrade organic waste, the moisture content must be between 12 percent (the minimum required for biological activity) and about 65 percent, the level at which oxygen availability becomes too low. The ideal moisture content for composting is 45–50 percent. The pH levels must hover around neutrality; bacteria prefer 6–7.5, and fungi prefer pH levels between 5.5 and 8. Temperature naturally varies throughout the composting process, but fluctuates between the mesophilic (25–40 degrees Celsius) and thermophilic (45–60 degrees Celsius) ranges. These ranges are shown in Table 1.

Table 1 Key Variables for Composting Material

Substrate Variable	Ideal Range
Moisture Content	45%–50%
C:N	20–25:1
pH	6–7.5

Composting is defined by four main stages, each with distinct microbial populations that carry out the degradation of waste under the reigning conditions of each stage:

1. *Mesophilic stage.* As soon as composting conditions are established, microbes begin to proliferate. Bacteria and fungi break down the easily degradable, energy-rich substances, and their activity causes the temperature to rise. Worms and millipedes may act as catalysts.
2. *Thermophilic stage.* Rising temperatures promote the persistence of organisms adapted to hotter conditions; these thermophilic

organisms continue to consume both the organic substrate and the mesophilic organisms. Most of the remaining biodegradable organic matter is consumed in this stage.

3. *Cooling phase.* When easily decomposed material becomes exhausted, thermophilic organisms begin to decline in number. As their activity slows, the temperature of the compost decreases, and mesophilic organisms recolonize the composting mass. This phase is dominated by the presence of fungi and bacteria that can degrade materials that are more difficult to break down, such as starches and cellulose. The time it takes for a compost pile to progress through the first three phases is a few weeks.
4. *Maturation phase.* All of the easily degradable material has disappeared by this stage. Microbial populations decline, and the compost pile begins to stabilize. The maturation phase can take between several weeks and a year.

Compost Systems

In practice, compost systems may be either closed or open and may occur on small or large scales. Open composting can take place in windrows (heaps that are turned either manually or mechanically) or can take place in static piles through which air is blown. Closed composting occurs in a vessel, which may take many forms, but must be aerated through tumbling or rotation. Open composting requires a large land area. Small-scale composting can occur at the household-level, and large-scale composting occurs at the municipal or regional scale.

Proliferation of Composting

Composting is growing in importance as an MSW management strategy. In the United States, the number of composting programs has increased from 700 in 1988 to 3,800 in 1998, and more than 20 states have banned the disposal of yard waste in landfills. The European Union's Landfill Directive 1999/31/EC requires member states to reduce the amount of biodegradable waste that is landfilled to 35 percent of their 1995 levels by 2016. Centralized composting programs are less common in developing nations, though household-scale organic waste management strategies may be more common. Despite the widespread availability of substrate in the organic fraction of MSW and the growth of centralized composting programs in industrialized nations, the percentage of MSW that is composted is small for most nations, with values ranging from 1 percent in the United Kingdom to about 9 percent in the United States to 22 percent in the Netherlands.

In the mesophilic stage, fungi and bacteria aid in breaking down plant materials that are difficult to degrade, such as starches and cellulose. The three other stages are thermophilic, which raises temperature; cooling; and maturation, when the compost stabilizes.

Constraints on the Use of Compost

Producing good compost from MSW faces some challenges. What people throw away is heterogeneous, but the separation of waste into organic and inorganic fractions is necessary for the production of compost. If this separation happens after collection of mixed solid waste, there is a risk that heavy metals, toxic organic compounds, or inorganic materials present in the MSW will contaminate the organic waste. Inclusion of these substances in a compost pile threatens the biodegradation process and makes the use of the resulting compost difficult and hazardous. Compost faces large market barriers; while the production of compost requires energetic and monetary resources, the market value of the product tends to be low. This discrepancy discourages widespread composting.

Sintana E. Vergara
University of California, Berkeley

See Also: Biodegradable; Food Waste Behavior; Microorganisms; Organic Waste.

Further Readings

Diaz, Luis. F., et al. *Compost Science and Technology*. Boston: Elsevier, 2007.

Diaz, L. F., G. M. Savage, and C. G. Golueke. "Composting of Municipal Solid Wastes." In *Handbook of Solid Waste Management*, 2nd ed, G. Tchobanoglous and F. Kreith, eds. New York: McGraw-Hill, 2002.

Gajalakshmi, S. and S. A. Abbasi. "Solid Waste Management by Composting: State of the Art." *Critical Reviews in Environmental Science and Technology*, v.38/5 (2008).

McDougall, Forbes, Peter White, Marina Franke, and Peter Hindle. *Integrated Solid Waste Management: A Life Cycle Inventory*. 2nd ed. Malden, MA: Blackwell Science, 2001.

Rhyner, Charles R., Leander J. Schwartz, Robert B. Wenger, and Mary G. Kohrell. *Waste Management and Resource Recovery*. Boca Raton, FL: Lewis Publishers, 1995.

Comprehensive Environmental Response, Compensation, and Liability Act (CERCLA/Superfund)

More than a century after the Industrial Revolution, economic growth and wealth creation in the United States remained closely tied to industrial output. As recently as the 1960s, fully one-third of the country's workforce was engaged in manufacturing activity. In subsequent decades, however, the United States underwent a transition to a post-industrial, or service-based, economy. In the early 21st century, manufacturing provides employment for only about one in ten workers, and the country has become littered with literally hundreds of thousands of abandoned industrial sites. Including machine shops, steel mills, automobile factories, old mines, and others, the variation of these contaminating sites is likewise extensive. Although the doors to these sites may have closed, many continue to pose significant environmental and health risks to the surrounding community through the risk of contamination. The most notorious chemical contamination occurred in the Love Canal neighborhood of Niagara Falls, New York. Receiving national attention in 1978, this incident spurred the federal government to enact the 1980 Comprehensive Environmental Response, Compensation, and Liability Act (CERCLA, or Superfund). This act stands as the most important federal response to the dangers of the uncontrolled release of hazardous chemicals into the environment.

Goals, Priorities, and Classifications

The goals of the CERCLA legislation are to identify those sites associated with hazardous contamination and to assign the costs of cleanup to the responsible parties. Eligibility for federal funds for cleanup requires that a site be designated as a "national priority." Congress initially required the Environmental Protection Agency (EPA) to identify 400 sites for listing on the National Priorities List (NPL). The sites listed on the NPL are only those of the absolute highest priority and do not include the many thousands of sites associated with far lower levels of contamination. As of July 2010, the number of sites on the NPL had subsequently ballooned to 1,277, more than tripling the original list. Furthermore, only 343 sites had been identified as needing no further response and were deleted from the NPL. While the NPL has received the most attention, CERCLA empowers the EPA to pursue remediation action at any contaminating site. Perhaps the greatest difficulty in satisfactorily addressing these potentially hazardous sites is confusion over the requirement that the site be cleaned, with the question arising: just exactly how clean is "clean?"

This confusion led to the 1986 Superfund Amendments and Reauthorization Act (SARA), which established criteria that the EPA was to consider when determining an appropriate course of action for the remediation of each site. While providing some clarity, SARA did not, however, establish a national set of standards for the cleanup of polluting sites. Thus, there remains heightened concern over a lack of clarity regarding CERCLA standards and enforcement policy. The reason for this concern is that liability is both strict and retroactive.

In other words, parties may be held responsible for the costs of cleanup even if the contamination occurred before 1980 and was legal at the time. Likewise, parties that had nothing to do with the original use of the property and assume ownership of the property only after it has been abandoned may also be held liable for the cleanup. Apprehension over the possibility of being forced to pay untold dollar amounts in remediation has served as a deterrent to potential buyers of contaminated sites. While various amendments to CERCLA have been implemented in order to address this concern, estimates still place the number of abandoned Superfund sites in communities across the United States at more than 500,000.

Brownfields

People often notice abandoned properties in desirable locations, (for example, a gas station on a corner lot at a busy intersection) and wonder why no one has invested in that potentially prime piece of real estate. The reason could very well be that it is a Superfund site. Investors are frequently unwilling to incur the risk of being exposed to liabilities associated with the remediation of that site. The uncertainties and costs associated with the redevelopment of these brownfields are costing many older industrial cities potential tax revenue and job opportunities. Seeking to minimize costs, developers frequently look to open space at the fringe of urban areas as ripe for development. However, building on these greenfields contributes to urban sprawl and takes jobs and dollars out of the central city. Furthermore, there is also significant concern that these abandoned and potentially hazardous sites are disproportionately located in neighborhoods or communities with a high minority population. Associated with the concept of environmental justice, evidence suggests that the poor—and, in some cases, minorities—may not receive equality when it comes to environmental cleanup of Superfund sites.

Effects

Three decades after its passage, the relative merit of CERCLA remains uneven. While CERCLA has spurred investigation by the EPA into tens of thousands of sites thought to be releasing hazardous chemicals, and while billions of dollars have been spent on cleanup, there have also been a number of unintended consequences. Most acute are those concerns about liability that have brought to a halt untold numbers of projects aimed at the redevelopment of brownfields in central cities. While CERCLA has received legislative attention over the years, questions also remain regarding the appropriate level of federal funding. Superfund remediation did receive additional funding through the American Reinvestment and Recovery Act of 2009; however, only through a long-term and holistic strategy for the application of CERCLA will all parties involved in the remediation of Superfund sites truly benefit.

Christopher Cusack
Keene State College

See Also: Environmental Justice; Love Canal; Politics of Waste; Pollution, Land; Pollution, Water; Resource Conservation and Recovery Act; Toxic Wastes.

Further Readings

Bartsch, Charles and Elizabeth Collaton. *Brownfields: Cleaning and Reusing Contaminated Properties.* Westport, CT: Praeger, 1997.

Judy, M. L., et al. "Superfund at 30." *Vermont Journal of Environmental Law*, v.11/2 (2009).

Probst, Katherine N., David M. Konisky, Robert Hersh, Michael B. Batz, and Katherine D. Walker. *Superfund's Future: What Will It Cost?* Washington, DC: Resources for the Future, 2001.

Computers and Printers, Business Waste

Evolving from tabulation machines used to handle growing volumes of information in the early 20th century, computers became integral to the function of businesses large and small by the 21st century. Beginning in the 1950s, gigantic mainframe computers—primarily manufactured by International Business Machines (IBM)—handled information at large corporations; a decade later, advances in microprocessor technology allowed smaller businesses to purchase microcomputers or even desktop models to handle records. Today, businesses in both developed and developing nations rely on mass-produced computers to handle their needs. Though increasingly large volumes of computers and printers are disposed of annually, they do not necessarily follow a linear trajectory from consumption to disposal and waste. Instead, business disposal practices increasingly involve further rounds of consumption through reuse and refurbishment prior to eventual disposal. Like other forms of cast-off electronics, the consumption and wasting of computers and printers used in business environments is a culturally distinctive practice that varies from place to place.

Disposal Options

In North America and Europe, business concerns about data security and liability arising from breaches tend to trump concerns about the environmental consequences of information technology (IT) disposal. These concerns have spawned an industry known as information technology asset disposition (ITAD). The ITAD industry offers a number of disposal options including reuse, refurbishing, re-marketing, data sanitization, and recycling of computers, printers, and other electronics. In 2010, nine leading ITAD firms in the United States had combined revenues of between $265 and $345 million annually. Fueling the growth of this industry are a number of information privacy laws that carry stiff economic penalties if companies are found in breach.

ITAD firms actively encourage reuse, remarketing, refurbishing, and the like for their business clients. ITAD company Websites market these services in a way that suggests that recycling should occur only when equipment cannot be repurposed or redeployed. The emphasis in the business sector on reuse and refurbishment thus marks an important contrast with the household consumer sector, which is encouraged to replace older computers and printers with new ones, even when they could be reused or repaired. Meanwhile, in Asia, Africa, and South America, computers, printers, and other business IT assets circulate within complex informal recovery economies that refurbish, repair, and remanufacture this equipment as well as disassemble them into their constituent components and materials, which are then fed back into the production economy. However, the disposal and recovery of computers and printers from business waste is far from being a closed-loop production system. Patterns of disposal and recovery raise a number of controversial issues that link economic questions with moral ones about poverty, survival, economic production, health, and the environment.

Criticisms

While there are key differences between the business sector and the personal consumption sector in terms of consumption and disposal practices, the disposition of electronics from either sector has similar effects. Attempts to mitigate the health and environmental effects of electronics disposal focus on formalizing recycling. These strategies include product take-back programs and industrial-scale material and energy recovery systems. In North America, Europe, and parts of Asia and Africa, these strategies are increasingly mandated by law. While such efforts may appear beneficial, there is a debate about their efficacy. Formal industrial-scale recycling can recover substantial amounts of material and energy. It can also reduce the need for mining new raw materials. However, it also leads to what some argue is a wasteful destruction of working computers and printers that could, under the right conditions, be fit for reuse by people and businesses otherwise unable to afford them. Moreover, relying on recycling to manage waste computers and printers from business environments cannot escape the problems of materiality. Recycling at an industrial scale typically requires transportation of equipment over long distances to recycling facilities, thus adding to the environmental footprint of disposed electronics. Industrial-scale recycling machinery requires substantial amounts of energy that must be generated in some manner, thus raising the likelihood of CO_2 and toxic emissions. Smelting for material and energy recovery from electronics can release toxic substances such as lead. Some argue that formalized recycling risks merely shifting the loci of toxic burdens of cast-off electronics rather than truly eliminating them. Others contend that the emphasis on recycling is a trap that risks foreclosing on options for the cleaner production of original products.

Josh Lepawsky
Chris McNabb
Memorial University of Newfoundland

aste;

High-
nain/

ur
ntent/
d.pdf

; of
d

lian

Ya[illegible] Eric Williams. "Logistic Model-Based Forecast of Sales and Generation of Obsolete Computers in the U.S." *Technological Forecasting & Social Change*, v.76/8 (2009).

Computers and Printers, Personal Waste

The personal computer developed later than the mainframes used by corporations in the 1960s but the introduction of the microprocessor in the 1970s made this technology compact and affordable to sell to millions of individuals. International Business Machines (IBM) introduced the 5100 as a desktop model in 1975; two years later, Apple entered the desktop market with the Apple II. IBM responded in 1981 with the PC, which ran on an operating system designed by Microsoft. The PC quickly became a mass consumer product in the developed world as manufacturers built machines using Microsoft's operating system. Apple followed with the Macintosh in 1984, serving a smaller but devoted market.

Desktop computers became commonplace in businesses, schools, and households over the next 20 years. Microsoft founder Bill Gates became one of the wealthiest people on Earth due to the omnipresence of the Windows operating system, and Apple became associated with an elite design aesthetic that the firm shifted from computers to telephones, music listening devices and, by 2010, the largest music-selling store in the world (iTunes). In August 2011, shares of Apple surpassed those of Exxon to make it the most valuable company traded on the NASDAQ stock market.

The broad market of personal computers running either Microsoft or Apple operating systems continues into the 21st century, but the machines that run this software are far more powerful than their 1980s counterparts. Under Moore's law (named after Intel founder Gordon Moore, who came up with the concept in 1965), processor capacity doubles roughly every decade, accelerating functional obsolescence. This affects all computer markets, but once the computer became a mass-consumer product marketed to households in the 1980s, functional obsolescence had profound effects on the waste stream. Beyond processors, innovations in design allowed the personal computer to evolve from desktop machines to laptops in the 1990s, and began blurring the line between computers, telephones, and television equipment in the early 21st century.

Studies suggest that North America and Europe will reach their peak production of obsolete computer equipment by 2020. In developing regions, on the other hand, volumes of obsolete computer equipment will reach a peak by 2030. However, as with the consumption of computers, printers, and other electronics, disposal practices also vary widely. Estimates put the useful life of computers and printers at two to three years in North America, down from four to five years in the 1990s. Elsewhere in Asia and Africa, however, the same equipment remains in use for a longer time period. The differences in lifespans of electronic equipment over time and between regions shows that there is more to obsolescence than purely technical change. At least as important is status-seeking behavior on the part of consumers, the latest styles.

Recycling

In North America and Europe, consumer attitudes toward obsolete computer equipment suggest that consumers are grappling with a complex set of moral questions that belie stereotypes about a throwaway society. Though increasingly large volumes of computers and printers are disposed of annually, they do not necessarily follow a linear trajectory from consumption to disposal and waste. Instead, like other forms of cast-off electronics, they may continue to circulate in substantial recovery economies that include international commodity networks of trade and traffic.

North Americans and Europeans increasingly have access to formal recycling systems for computers, printers, and other electronics. Elsewhere in Asia, Africa, and South America, computers, printers, and other electronics circulate within complex informal recovery economies that refurbish, repair, and remanufacture the equipment as well as disassemble it. Their constituent components and materials are then fed back into the production economy. However, the disposal and recovery of personal computer and printer waste is far from a closed-loop production system. Patterns of disposal and recovery raise a number of controversial issues that inherently link economic questions with moral ones about poverty, survival, economic production, health, and the environment. For example, the burgeoning electronics recycling industry, a multibillion dollar activity, will continue to expand. At the same time, those parts of the industry operating in North America and Europe will see increasing competition from the developing world, in particular Asia, which is also the dominant location for the industrial production of new electronics. Regulatory regimes that seek to control international movements of obsolete electronics from developed to developing countries will need to evolve or risk irrelevance, since the highest volumes of obsolete electronic equipment will be produced in developing countries.

Effects on Environment and Health

The toxic consequences of disposing of computers, printers, and other electronics is an issue of increasing concern. Images of poor and marginalized people in various Asian and African countries picking through waste dumps in search of recoverable electronics or disassembling them by hand have provided a major impetus for legislation to control the international trade and traffic of cast-off electronics, including personal computers and printers. In many ways, these images have framed the public debate in North America and Europe about this particular slice of the waste stream and legislative attempts to control it. Such images have also spurred research by environmental nongovernmental organizations and academics concerned about the environmental consequences of electronics disposal. Such research has documented significant health and environmental consequences for people and places associated with the informal processing of these objects and materials. At the same time, studies of the material and chemical behavior of electronics disposed of in well-designed landfills suggest that they remain largely inert under those conditions and are not major sources of toxic releases.

Mitigation Strategies

More than a billion personal computers were in use worldwide by 2009, with China and the United States the largest markets for new machines. Entering the second decade of the 21st century, the definition of the personal computer is increasingly fluid, as desktop machines are sold not only along portable laptops but also smart phones, tablets, and other devices that rely upon microprocessors. All of these devices consume electricity and allow users to engage in consumption through on-line shopping. All are mass consumer objects subject to Moore's law and their inevitable obsolescence generates increasing volumes of electronic waste

For the most part, attempts to mitigate the health and environmental effects of electronics disposal focus on formalizing recycling. These strategies include product take-back programs and industrial-scale material and energy recovery systems. In North America, Europe, and parts of Asia and Africa, these strategies are increasingly mandated by law. While many laud these efforts, they are also subject to serious debate about their efficacy. Formal, industrial-scale recycling can recover substantial amounts of material and energy. It can also reduce the need for mining new raw materials. But industrial recycling in general, and of electronics specifically, involves energy and materially intensive processes. It often requires transportation over substantial distances from collection to processing facilities, which adds to its overall environmental impact. The industrial machinery used to shred and sort computers, printers, and other electronics at an industrial scale requires substantial amounts of energy to run. Smelters involved in material and energy recovery from electronics can release toxic substances, such as lead, from their operations.

Some argue that formalized recycling risks merely shift the loci of toxic burdens of cast-off electronics rather than truly eliminating them. Others contend that the emphasis on recycling leads to a recycling trap—a situation where a multitude of options for the cleaner production of original products are bypassed in favor of recycling them as an alternative to disposal. Changes to production processes and product design that would reduce the environmental and health risks of electronics are thus delayed or deferred by an emphasis on recycling. As a consequence, a vast range of opportunities may be missed for implementing greater material and energy efficiencies to the production process; altering product design to make repair and disassembly easier; and using less- or no-toxic components in the manufacturing of computers, printers, and other electronics. Meanwhile, a recycling system is instituted to deal with personal computer and printer waste only after such waste has already been created and only tries to reduce or eliminate that waste at the last possible moment, before the products enter the waste stream.

Josh Lepawsky
Memorial University

See Also: Computers and Printers, Business Waste; Mobile Phones; Recycling Behaviors; Toxic Wastes.

Further Readings

Barba-Gutiérrez, Y., et al. "An Analysis of Some Environmental Consequences of European Electrical and Electronic Waste Regulation." *Resources, Conservation and Recycling*, v.52/3 (2008).

Gregson, Nicky. "Identity, Mobility, and the Throwaway Society." *Environment and Planning D-Society & Space*, v.25 (2007).

Saphores, J. D. M., H. Nixon, O. A. Ogunseitan, and A. A. Shapiro. "How Much E-waste Is There in U.S. Basements and Attics? Results From a National Survey." *Journal of Environmental Management*, v.90/11 (2009).

Toxic Link India. *Scrapping the High-Tech Myth: Computer Waste in India*. New Delhi: Toxic Link India, 2003.

Yu, Jinglei, et al. "Forecasting Global Generation of Obsolete Personal Computers." *Environmental Science & Technology*, v.44/9 (2010).

Connecticut

The state of Connecticut is a leading innovator in waste disposal. Connecticut burns the highest percentage of garbage in the United States, such that over 60 percent of all waste collected is incinerated in an environmentally friendly manner. According to the Connecticut Department of Environmental Protection, state residents dispose of 2.7 million tons of trash annually. With a population of slightly over 3.5 million people, that figure equates to 1,500 pounds of waste per person per year, or nearly 5 pounds of garbage generated by each person every day. In order to responsibly deal with the collection and disposal of this garbage, the Connecticut Resources Recovery Authority (CRRA) was established to develop environmentally sound solutions to solid waste disposal and recycling management. The CRRA has promoted waste management standards that have made the "town dump" method obsolete in Connecticut. Moreover, through the foundation of two museums dedicated to garbage and its impact on the state as well as the nation, the CRRA continues to educate and raise awareness among a new generation of constituents. Contrary to these positive aspects of waste management in Connecticut, the relationship between organized crime and trash disposal was an example of an undesirable development of the garbage business in the state. Regardless of any negative attention garnered from such accusations and eventual convictions, energy consumption, waste collection, and disposal continue to be important issues to the state of Connecticut and for its residents.

Disposal Methods

During the 2008 fiscal year, 3,401,085 tons of solid waste were collected and disposed of in the state of Connecticut. From this total, only 163,543 tons

(4.81 percent) were deposited in local landfills. The Connecticut Department of Environmental Protection burned 62.06 percent, or 2,110,855 tons of trash. Of this portion, 544,709 tons of ash were deposited in landfills, and 48,070 tons of metal were recycled. Nearly 8 percent (261,255 tons) of the waste produced and collected in Connecticut was disposed out of state, and only about 25.45 percent (865,432 tons) was recycled. In comparison, according to an Environmental Protection Agency (EPA) report issued in November 2009, the national recycling rate—the amount of trash generated divided by the volume of material recycled and composted—is slightly over 33 percent. This apparent underperformance is surprising, since recycling is mandatory in the state of Connecticut. In accordance with the Connecticut General Statutes and the Regulations of the Connecticut State Agencies, items to be recycled include glass and metal food and beverage containers, corrugated cardboard, newspaper, white office paper, scrap metal, and waste oil, among other items. This means that residents, all public and private institutions, as well as every business including nonprofits are required to recycle. In addition, some municipalities have established ordinances that specify the recycling of other items, such as old magazines and junk mail. In fact, under state law, garbage haulers must report individuals who do not separate their trash or put out their locally provided recycling bins. The total amount of garbage for Connecticut's disposal statistics does not, however, include bottle deposits or the amount of auto scrap metal and storage batteries recycled annually.

Incineration

Until 1973, Connecticut waste management consisted of the "town dump" method wherein each municipality in the state had its own landfill to handle its locally produced garbage. The acting governor at the time, Thomas J. Meskill, established the CRRA to tackle the solid waste management issues afflicting the state. He also created the institution in order to develop and implement environmentally responsible methods of garbage collection and disposal. This quasi-public agency was designed to assist individual municipalities and serve their best interests in terms of waste management. In accordance with these objectives, the CRRA has built trash-to-energy facilities to burn garbage. The heat generated from this process is used to boil water and create steam that, in turn, spins turbines and produces electricity. Furthermore, the ash yielded by this process contains 75–80 percent less volume than the same trash if it were disposed of in a landfill. Therefore, the amount of space saved through burning trash is considerable when compared to more traditional methods of garbage disposal (such as landfills).

The CRRA serves 96 out of 169 cities and towns in the state, or about two out of every three residents. The agency oversees four large waste management projects across Connecticut, including ones in Bridgeport, Wallingford, the central counties, as well as the southeastern portion of the state. Four trash-to-energy plants have been built, and they process 2 million tons of trash per year. Collectively, the reduction of trash to ash produces 630 million kilowatt hours per year. This amount is enough electricity to satisfy the energy needs of 170,000 homes. The gases burned from the Hartford landfill alone can provide energy for 1,500 homes. In sum, Connecticut recovers 80 percent of its energy from nonrecyclable waste, thus saving the state 32 million barrels of oil since 1992.

Recycling and Hazardous Materials

In addition to burning the state's garbage, the CRRA manages the recycling facilities in those areas it serves. They process 130,000 tons of recyclables per year, including glass and plastic containers, steel and aluminum cans, and newspapers. Five hundred tons of containers and paper are recycled daily. Moreover, the CRRA deals with hazardous materials. Since 1999, 1.7 million pounds of recovered electronics have kept massive amounts of cadmium and other heavy metals out of the environment.

Education

In conjunction with the efforts of the CRRA, the state has founded two museums: the Trash Museum in Hartford and the Garbage Museum in Stratford. These institutions provide fun, educational, hands-on programs, activities, and exhibits for 50,000 visitors per year. The smaller of the two, the Trash Museum, occupies 6,500 square feet of space. Here, visitors can learn about the old "town dump" model and the solutions that Connecticut has implemented since the 1970s, including plans to reduce and recycle, the construction of trash-to-energy plants, and more environmentally friendly landfills. The Garbage Museum in Stratford, on the other hand, features 15 interactive exhibits, including a 125-foot glass wall that overlooks an adjacent recycling plant. According to the Garbage Museum's Website, its mission is "to help teach youngsters how to reduce, reuse, recycle, and rethink throw-away lifestyles." The museum attempts to accomplish this goal in a variety of ways. Visitors are encouraged to not only observe the waste management process, but also participate through several learning labs, one of which allows visitors to sort trash. A featured exhibit called the Worm Tunnel is also located at the Garbage Museum. This giant simulated compost pile containing five-foot worms and other types of crawling insects simultaneously entertains children and teaches visitors about the compost cycle. The highlight of the museum, the "Trash-o-saurus," is a 2,000 pound, 24-foot-long, 11.5-foot-high dinosaur sculpture. It is designed to impress visitors and to draw attention to the fact that the entire statue is made of the average amount of trash a Connecticut resident disposes of per year. Although the Garbage Museum has attracted 306,000 visitors since it opened in 1993 and over 32,000 in 2008 alone, economic hardships have afflicted the institution. Until 2009, the Garbage Museum was funded by revenues from recyclables sales in the area cities that contribute to the regional recycling facility. However, with economic problems and the departure of six towns from the CRRA in late 2008, the museum administration is struggling in 2010 to balance operating costs with what little revenue it receives from its modest admission rate, tours, and program fees.

Crime

Although museums dedicated to garbage are one means to stimulate interest and to raise awareness of waste management, the trash business in Connecticut is probably recognized more in the national media for its involvement in organized crime. A 2006 federal investigation filed against the Genovese crime family of New York City implicated several Connecticut waste disposal companies in attempts to fix prices and rig government contracts. Crime syndicates are attracted to garbage collection because it is a legitimate service that is necessary to the public and easy to infiltrate. Members of the Genovese crime families were accused of monopolizing businesses by asserting "property rights" to particular locations and stops in the trash collection market. When competing companies attempted to offer lower rates, the crime syndicate responded with threats, violence, or arson.

Nevertheless, this negative publicity does not negate the positive steps the state of Connecticut has taken to optimize its waste collection policies and disposal processes. The efforts of the CRRA in particular have shaped the way Connecticut residents perceive and deal with garbage in their own lives. As waste and energy conservation becomes an important issue in society, Connecticut is a forerunner in waste management on the national stage through its unique methods of dealing with its trash. For example, in addition to the high percentage of garbage that is reduced to ash in incinerators, a handful of Connecticut municipalities have implemented SMART programs, also known as Pay-As-You-Throw (PAYT). This method of charging for trash collection, based on the amount disposed, is designed to provide incentives for residents to not only increase the amount they recycle but also envision ways to produce less waste. SMART is a low-cost strategy that can help achieve Connecticut's target of increasing the state's recycling rate to 58 percent by 2024. Therefore, Connecticut is part of a larger, nationwide drive to provide waste management solutions that are economical, fair to residents, and beneficial to the environment.

Matthew Piscitelli
University of Illinois, Chicago

See Also: Incinerator Waste; Incinerators; Recycling; Sustainable Waste Management.

Further Readings

Connecticut Resources Recovery Authority. "Where Does Your Garbage Go?" http://www.crra.org/index.html (Accessed September 2010).

"Connecticut Trash Hauler Pleads Guilty in Mob Case." *New York Times* (June 4, 2008). http://www.nytimes.com/2008/06/04/nyregion/04trash.html (Accessed September 2010).

Department of Environmental Protection. "State of Connecticut Solid Waste Management Plan Amended December 2006." http://www.ct.gov/dep/site/default.asp (Accessed September 2010).

Construction and Demolition Waste

Construction and demolition waste or debris (C&D) is waste that results from the activities of construction, renovation, demolition, and (under some definitions) excavation. The objects of these activities include all features of the built environment, such as houses, buildings, industrial facilities, roads, and bridges The C&D waste stream is generally urban, bringing it into proximity with municipal solid waste (MSW) in terms of collection, processing, and disposal infrastructure. However, unlike MSW, C&D is nonputrescible, meaning it does not contain food or other organic constituents that would lead it to rot. For this reason, the collection and disposition of C&D does not have the same history or implications for public health that MSW does, and the regulatory structures for C&D collection and facility permitting are somewhat different from those for MSW. Like MSW, however, C&D is generally considered a nonhazardous waste, although it may contain certain hazardous components in small quantities from time to time, particularly asbestos and treated wood. Because of its generally inert status, C&D therefore falls outside the regulatory structures governing hazardous industrial wastes in terms of collection, transport, and facility siting.

C&D waste has existed for as long as human civilizations have been building permanent structures—a practice dating to the late Neolithic era (three to five millennia B.C.E.). Historians have documented the reuse of stone, bricks, and wood in deconstruction and construction over millennia as a matter of course, depending on material availability.

Quantities

Unlike MSW and some industrial processes, wastes that are generated as by-products of relatively steady, year-round activity, C&D waste generation fluctuates considerably with the economic conditions in a particular country or region. There is a close association between C&D generation and economic activity. Periods of building boom, moreover, presage periods of demolition decades into the future as structures age and require repair or demolition. Thus, increases in C&D in a particular year may represent echoes of a prior period of prosperity many years before, a phenomenon that has been studied and documented in European countries in particular. In rapidly developing countries, such as China, industrialization, increased consumption, and the expansion of the middle class has meant rapid escalation of C&D generation in the 2000s. Finally, C&D generation is influenced by disasters (which are by nature unpredictable and of varying magnitude) that result in destruction of the built environment.

For this reason, estimates of world C&D generation are not possible and national-scale estimates are highly specific year-by-year. According to the Environmental Protection Agency (EPA), the C&D waste stream in the United States as of 2003 was 170 million tons annually. The European Union (EU) reports a roughly estimated 450 million tons for all member countries, but the data quality, inclusiveness, and consistency varies greatly among nations and generally dates to the late 1990s. The Organisation for Economic Co-operation and Development (OECD) adds that Japan generates 123 million tons and Korea 38 million tons annually. Independent reports from China indicate that overall C&D generation is averaging 120 million tons annually and is growing every year, causing the Chinese government to regulate demolition permit issuance to stem the transport and disposal burdens facing Chinese cities. The Indian government estimates a rapidly growing C&D stream of roughly 20 million tons annually.

Composition

In contrast to widely fluctuating C&D quantities, the range of constituents of C&D remain similar across nations and time, although their relative fractions may differ. Inert materials, including concrete, stone, brick, and tile make up a great deal of C&D, which also may contain structural steel, wood, plaster, ceramic, carpet, sheet or plate glass, heavy duty cardboard, asphaltic roofing tiles, and, in some cases,

Inert materials, including concrete, stone, brick, and tile make up a great deal of construction and demolition waste, which also may contain structural steel, wood, plaster, ceramic, carpet, and plastic components. The percentage of demolition debris in the waste stream is difficult to calculate, but in 2003, the U.S. Environmental Protection Agency estimated America's annual volume at 170 million tons. The whole of the European Union creates an estimated 450 tons annually, and Japan's generation is estimated at 123 million tons per year.

plastic components. Asbestos (naturally occurring filamentous minerals that were, in the 19th and 20th centuries, used in Europe and North America for fireproofing and insulation) is a relatively small quantity of waste generated by renovation and demolition projects and in the 21st century is generally removed under highly-controlled conditions prior to any renovation and demolition activities. Other potentially hazardous constituents of C&D waste include wood treated with arsenic or other heavy-metal laden preservatives; plasterboard, which produces hydrogen sulfide gas when landfilled; and PVC piping, which may produce dioxin if incinerated.

Disposal

Disposal (including landfilling of inert constituents and combustion of wood, plastic, cardboard, and asphaltic roofing) is one frequent end for C&D. In the United States, it is estimated that 52 percent of all C&D generated is disposed of, largely through landfilling. Rates in northern Europe are markedly lower, with between 10 and 20 percent of all C&D generated disposed of. Rapidly developing nations such as China and India, in contrast, struggle with a lack of developed infrastructure for C&D disposal and dispose of the majority of C&D in landfills or waste piles.

Materials Recovery

The reuse of the inert quantities of C&D, ground into aggregate, is a well-established and widespread practice in the developed world, as is the recovery of structural steel from C&D, manually or by use of magnets, for recycling in most countries. In most cases, crushed concrete, stone, and bricks are used in place of gravel or sand in road building and drainage applications. Unlike steel recycling, a highly efficient process in which scrap metal is reintroduced into mill production, the crushing and reuse of inert C&D as aggregate is considered downcycling in that the end product does not conserve much of the energy or use value that went into making the substance in the first place.

Two 21st-century movements are under way in many countries of the world to address C&D waste in a more sustainable manner. The building deconstruction movement, which is active in North America, Europe, and Australasia and is closely tied with the zero waste and green building movements, is an effort by nonprofit and for-profit social enterprises to change the way demolition and renovation are carried out. Building on long-standing practices such as the removal of copper piping, mahogany, and other highly valuable materials prior to demolition, this movement has organized labor forces, storage spaces, and markets around a wider array of salvageable elements, deconstructing fixtures, moldings, doors and windows, tiles, structural wood, and other elements before the wrecking stage begins. Such items that may be of higher quality than contemporary substitutes, depending on their vintage, are then resold to builders for integration into new construction.

A second set of developments are organizing around the constituents that make up concrete, a large fraction of C&D that is growing rapidly in a wide range of building applications. The production of Portland cement, the agent for binding aggregates into concrete, has considerable impacts in terms of carbon dioxide emissions. The fortuitous discovery that fly ash from coal combustion can substitute for large portions of Portland cement in concrete production has led to research and project development around the use of this industrial waste in the construction industries of nations the world over in which coal is used for energy production. Utilization of fly ash for this purpose has a dual benefit of diverting it from disposal while reducing the need to fabricate cement in kilns. Another set of technologies, termed *closed cycle construction*, is being developed to thermally and mechanically release aggregate from crushed concrete debris so that it is of a high-enough quality to be reintroduced into new concrete production. As opposed to use as low-grade aggregate, C&D concrete processed using this technology, which is often fueled by the combustion of wood, cardboard, plastic, and asphaltic components of C&D waste, closes the loop in the fabrication of concrete.

Samantha MacBride
New York University

See Also: Coal Ash; Downcycling; Economics of Consumption, International; Economics of Consumption, U.S., Landfills, Modern.

Further Readings

Environmental Protection Agency. *Estimating Building-Related Construction and Demolition Materials Amounts.*" Washington, DC: U.S. Government Printing Office, 2009.

Laefer, Debra and Jonathan Manke. "Building Reuse Assessment for Sustainable Urban Reconstruction." *Journal of Construction Engineering and Management*, v.134/3 (2008).

Mulder, Evert, et al. "Closed Cycle Construction: An Integrated Process for the Separation and Reuse of C&D Waste." *Waste Management*, v.27 (2007).

Rao, Akash, et al. "Use of Aggregates From Recycled Construction and Demolition Waste in Concrete." *Resources Conservation & Recycling*, v.50 (2007).

Consumerism

Consumerism is a way of life rooted in mass production and the marketing industry. It includes a practice where social identity and prestige are constructed, experienced, and signaled through the purchase and possession of consumer goods and services. Consumerism is fueled by easy credit and by advertising designed to create desire for commodities by associating their acquisition with valued states such as happiness, peace of mind, attractiveness, gratification, affluence, and success. Consumerism is central to an economy in which people are preoccupied with material consumption to the point where the amount of goods acquired may be far in excess of actual need. Producers of commodities in industrialized societies profit by ever-expanding consumption, but meeting this growing demand has been using up natural resources at an unsustainable rate.

At the other end of the product life cycle, consumerism includes the practice of discarding broken, out-of-fashion, and even slightly used products, making room for new acquisitions. This has resulted in a huge and rapidly moving waste stream that itself has become problematic. Most social theorists agree that contemporary consumerism began at the dawn of the 20th century and gathered momentum with an expanding middle class in Europe and North America after World War II.

With modern globalization it has spread worldwide, wherever consumer products and associated images and narratives have penetrated societies whose traditional cultural economies have been disrupted by colonialist or neoliberal restructuring.

The term *consumerism* also has a number of more specialized meanings. In economic theory, it refers to the idea that continuously expanding mass consumption is beneficial to an economy. In a related usage, consumerism is the idea that the choices made by consumers should shape production and, by extension, the structure of the economic system as a whole. In economic policy terms, it is used to characterize policies that promote consumption. Finally, consumerism is a synonym for the modern consumer protection movement, which advocates the rights and interests of those who purchase products. In this context, consumerism promotes policies to ensure product safety, quality guarantees, and truthful advertising.

History and Significance of Consumerism

The origins of modern consumerism are linked to the invention of factory assembly lines at the turn of the 20th century. Mass production increased worker productivity to the point at which prices of consumer goods became affordable to most workers in industrially developed societies. It also created workplace and profit conditions conducive to labor union organizing. Industrial wages rose, boosting many into the new middle-class lifestyle of consumerism. In 1899, economist Thorstein Veblen identified a pattern of conspicuous consumption among middle-class people seeking to cement their social status through their consumptive behaviors.The expanded world of goods available in the industrialized world made such behavior possible.

Early in the mass consumption era, engineers began designing products in such a way that consumers would need to replace them periodically, either because they are used up or have become obsolete before wearing out. Consumerist policies promoted demand as a growth strategy in the 1950s, especially in the United States, where members of a growing middle class viewed their upward mobility in material terms, such as owning many household appliances, a single-family home, and an automobile. "Store-bought" clothing, processed foods, and manufactured chemical products acquired the glamour of modernity and high status. Attributing to such commodities the power to gratify, modernize, and uplift consumers became a common form of commodity fetishism. The novel luxury of buying such items off the shelf imbued them with special significance in the 1950s and early 1960s. This was a time when admiration for the scientifically up-to-date was a cultural norm, along with a rationale in which "second hand," "home grown," and "homemade" came to signify either the lingering frugality of Depression-era poverty or nostalgia for a romanticized pre-industrial past. The implicitly competitive character of consumerist status display was captured in the imperative of "keeping up with the Joneses."

Growing consumerism has been linked to other systemic changes, including the decline of mass transit, increased industrial pollution, suburban sprawl, an industrialized food system, and growing dependence on cheap fossil fuels and easily available consumer credit. By the 1970s and 1980s, rising lifestyle costs and personal debt made a two-wage-earner household common for middle-class American families.

With the rise of globalization, industrial corporations have managed to increase profits while keeping prices affordable by moving production to economically underdeveloped areas of the world where labor costs are low and regulatory regimes are lax. This has led to the decline of industrial employment in the most developed parts of the world, where nonunion service and professional jobs now predominate. Globalized communications have promoted consumerist values and lifestyles to a receptive worldwide audience aspiring to middle-class status understood in terms of material consumption. Products that facilitate media connectivity such as televisions, computers, cell phones, and other handheld devices are often among the first major purchases of such households. Using these products exposes a global population of would-be consumers to a stream of ideas, values, and images of affluent Western consumerist lifestyles, together with a steady barrage of commercial advertising specifically designed to stoke consumerist desires. This is one reason for today's high level of labor migration (both legal and illegal), as households deploy young adult members to work in parts of the world where higher wages make it possible for them to remit money and goods to increase the material living standard and social prestige of the sending family. Studies tracking worldwide buying trends have shown that by the close of the first decade of the 21st century, more than two billion people have adopted consumerist lifestyles of exchanging labor for money to buy nonessential goods such as large houses, late-model cars, and processed foods.

Critiques of Consumerism

Economic efficiencies have made mass consumption possible, but their real purpose is generating profits rather than social good. Consumerism has been widely critiqued as having detrimental effects on personal autonomy, family and community values, psychological health, household finances, social capital investment, and the natural environment. These impacts are among the hidden costs of consumerism, along with waste that is not accounted for as such. According to its critics, the consumer culture that began with providing basic conveniences and well-paying jobs has grown into an oppressive hegemony that degrades human well-being as well as the planet's ecosystems.

Critics charge that consumerism transforms the natural motivation to acquire a sufficient supply of necessities into an artificially generated and insatiable quest for things and the money to buy them, and that this transformation distorts healthy social relationships. This is said to occur, for example, when people come to judge their own and others' personal value in terms of their buying power, a calculation that renders the poor worthless. Conspicuous consumption has become central to claiming a respectable level of social worth. This presents the marginally middle class with the difficult choice of either working extra hours in order to buy more and more esteemed goods, going into crippling debt, or doing without the status-conferring goods and suffering the consequent social inferiority. In consumerist culture, even close social relations are often mediated through spending money on non-necessities as a way of demonstrating love and acknowledging connection.

Consumers identify with their possessions and perceive them in part as representing who they are and who they aspire to be. Such displays communicate identity through richly nuanced symbolic associations of style and brand, so that consumers may quickly assess the socioeconomic and cultural status of others. The construction of one's persona through shopping choices can be thought of as a creative process. Thus, people may actively "consume" products in the sense of assimilating them to their own uses and reworking their advertised meanings into new and personally significant ones. This sort of consumption is limited by the available choices (those that producers deem most profitable) and social pressure to conform to the norms of popular culture shaped by advertisers. Some argue that keeping in step with changing fashions and sharing media-based experiences have come to occupy the place of socially integrating rituals and community engagement in less-consumerist societies.

Money cannot buy love, nor can it guarantee happiness. Psychological research shows that a strong consumerist orientation can promote unhappiness, because striving for money takes time away from the close relationships that are a foundation of happiness for most people. A number of studies have also confirmed that those who organize their lives around consumption goals that exceed their financial reach report significantly lower overall well-being as measured in quality of relationships, mood, self-esteem, and psychological problems. With real wages in decline, consumers in the early 21st century also face the frustration of exchanging precious time for money in order to buy things they no longer have time to enjoy.

The psychological critique of consumerism also argues that it interferes with the process of individuation by connecting it to media-based fantasies and advertising ideals. Many children now spend more time engaged with digital media than they do interacting directly with family members. As young consumers become more alienated from close social relationships, the self is increasingly constructed in relation to the personas portrayed in the media, rather than the real persons around them. For adults, the habit of buying things without knowing where, or how, or by whom they were made supports an infantile understanding of the significance of products they use every day. Similarly, when used products are routinely removed by garbage disposal services, consumers may just as easily shed their sense of responsibility for the consequences. To the extent that affluent consumers no longer perform for themselves activities of domestic reproduction such as cooking, laundry, repair, and child minding, they surrender self-sufficiency and competence in these basic life skills.

Critical analyses of consumerism within the field of political economy also claim profoundly negative consequences. According to them, continuous restructuring of the world economy to produce goods for those with the money to buy them at sufficiently profitable prices has resulted in greater social inequality. A disproportionate amount of the resources that are essential to all people, such as fertile land, have been diverted into the production of commonplace luxuries such as tropical fruits for temperate-zone consumers, while peasant farmers dispossessed by agro-industry go hungry. The privatization of clean-water systems in underdeveloped societies keeps the consumer class healthy, while those who can't pay are exposed to devastating waterborne illnesses. Workers with few options suffer oppressively low wages and dangerous conditions in the duty-free industrial zones of hundreds of countries to manufacture clothing, toys, and electronics products that they cannot afford to buy.

Through these kinds of processes, consumerism increases social inequalities, making the lives of some very comfortable, while making it harder for others to meet their basic needs. Social equity also suffers when resources that could be used to support the public education, nutrition, health, and housing programs essential for marginalized citizens to have a chance at a good life are used instead to produce consumer items for the affluent. A final line of political economic critique characterizes consumerism as a hegemonic form of social control by which middle-class masses are pacified and rendered less prone to resisting the disproportionate power of commercial interests in public policy and the operations of the state.

An economic system predicated on continuous expansion implies ever-growing consumption, but environmentalists argue that the current rate and methods of production, use, and disposal of consumer products are degrading the natural systems upon which all life depends. Consumerism has been linked to most of the environmental problems seen today, including water and air pollution, depletion of nonrenewable resources, unsafe and costly waste disposal, and climate change. Critics argue that the high level of fossil-fuel-powered consumerism today ensures a lower standard of living for future generations.

Anticonsumerist Alternatives

Since consumerism began, some have resisted its power by pursuing alternative lifestyles. The "hippie" youth culture of the late 1960s and 1970s rejected consumerism by embracing the values of living simply and communally and of making things by hand rather than buying them. This ethical response to the perceived excesses of consumerism encouraged people to seek gratification in caring relationships, community involvement, and closeness to nature instead of material possessions.

More recent formulations of anticonsumerism are often phrased in terms of "sustainability," a term most often used to mean living in such a way that the living standards of future generations are not compromised. A standard adopted from Native American cultures into popular anticonsumerism is to consider the impact of decisions on others seven generations into the future. Living an anticonsumerist lifestyle today also entails practicing voluntary simplicity, such as cycling or using public transportation, even though one could afford the convenience of a personal automobile. Anticonsumerist activists today promote "community sustainability," an ideal understood as having three interlinked components: social equity, economic security, and environmental stewardship. The politically radical version of sustainability-based thinking argues that uncritical consumerism supports oppressive political economic forces that are antithetical to an ideal of living in sustainable communities. In its green-business version, sustainability thinking emphasizes cost savings through more efficient resource use and long-term, rather than short-term, management planning.

Most 21st-century environmentalism calls for reforming consumerist practices to reduce the harmful and wasteful use of natural resources through energy conservation, product reuse and recycling, and switching from fossil fuels to renewable energy sources. It differs from earlier anticonsumerist movements in the gravity of its stated purpose, which is urgent action to avoid catastrophic disruption of the natural systems underpinning human civilization and all life on Earth. "Green consumerism" does not challenge the profit motive of business, but seeks to use ecoconscious consumption strategies to influence how products are made and the sort of consumer choices that are available. It requires purchasers to consider the environmental impacts of all aspects of product life cycles—the supply chains, working conditions, and waste involved in its manufacturing process, packaging, transportation, use, and eventual disposal.

The practice of buying more environmentally benign products has already gained broad acceptance and become a significant market force, for example, in the case of organically produced foods that have become much more available and affordable as a result. Buy-local and eat-local movements seek not only to reduce the fossil fuel pollution associated with transportation, but also to support the economic viability of local communities. The success of such movements indicates that cultural change in consumer habits is occurring. Such change may be crucial, since identifying conventional consumerism as a threat of great magnitude leads to the conclusion that the human future depends on reshaping human's way of life into one that is sustainable within the natural limits of the Earth.

Caroline Tauxe
Le Moyne College

See Also: Commodification; Home Shopping; Household Consumption Patterns; Malls; Materialist Values; Overconsumption; Post-Consumer Waste; Shopping.

Further Readings

Cross, Gary. *Time and Money: The Making of Consumer Culture*. New York: Routledge, 1993.

Kanner, Tim and Allen D. Kanner, eds. *Psychology and Consumer Culture: The Struggle for a Good Life in a Materialistic World*. Washington, DC: American Psychological Association, 2004.

Shell, Ellen Ruppel. *Cheap: The High Cost of Discount Culture*. New York: Penguin Press, 2009.

Veblen, Thorstein (1899). *The Theory of the Leisure Class: An Economic Study of Institutions*. Mineola, NY: Dover, 1994.

Worldwatch Institute. *State of the World 2010: Transforming Cultures: From Consumerism to Sustainability*. New York: W. W. Norton, 2010.

Consumption Patterns

Consumption patterns characterize individual, group, and community use of the means and wealth for subsistence, enculturation, comfort, and enjoyment. They have quality and/or quantity characteristics anchored to different lifestyles. Specific factors characterize consumption patterns during different historical periods and in different regions. Increasing consumption has been a general trend in the evolution of human civilization. This has resulted in the modern and postmodern consuming society of the industrial and technological era, when consumption of many products has been higher than needed. The reason is the changing type of consumer: from a traditional consumer of fixed needs, toward a consumer of endless needs and wants.

The general classification of consumption patterns includes nine clusters with eight subclusters in the first group: food, beverages, and tobacco (bread and cereals, meats, fish, dairy products, fats and oils, fruit, and vegetables, beverages and tobacco, and other food products); clothing and footwear; housing; household furnishings and operations; medical care and health; recreation; transportation and communications; education; and other items. A variant is the classification based on nutrition, mobility, housing, clothing, health, and education as functional components of lifestyle.

Globalization and worldwide income growth have been increasing the similarities across countries in what consumers eat and where they shop. A crucial factor for change in the 21st century is the evolving of the sustainable style of life with attention to the environment and the increasing role of sustainable products.

Analysis suggests that low-, middle-, and high-income countries all respond differently to changes in income and food prices. As a result of the global economic crisis, poor families in Asia, for instance, spend more than half their household income on food and are bearing the brunt of the economic burden. In many countries of the world, including in eastern Europe, the incomes of a significant percentage of the population are not enough for subsistence and people experience hunger. In this case, the consumption pattern of using the entire income for necessities can be observed. Another characteristic of the early 21st century is the changing social profile of consumers according to changing incomes. The Internet and the knowledge economy have been creating new wealthy segments while others, whose incomes were based on property and traditional sectors of the economy, have experienced a limitation in the consumption of luxuries.

Age preferences are well established in some patterns of consumption. For instance, pizza restaurants and burger bars are more popular among the younger generation, while older people prefer pubs and restaurants. However, children and adults of different ages are both consumers of new products such as video games.

Consumption patterns depend on traditions, cultures, a changing economic base, and many other factors including even ideology (e.g., religious beliefs). They are socially determined and abrupt shifts in attitude and practice that can be observed. Settlements of tobacco lawsuits followed by an increase in cigarette prices is a good example. Increased knowledge of nutrition changes food consumption patterns toward healthier food worldwide, especially in the United States. The turn toward preventive health as a leading social agenda expands the role of alternative medicine that, in many cases, can become the preferred service for many people. This creates a conflicting consumption pattern between traditional and conventional medicine. An example of a complementary pattern of consumption would be the combination of slow and fast food based on health criteria and on the limits of income. Low-income segments of society have often consumed substitutes for luxuries that create a pattern of consumption that includes secondhand products.

Fashion and tourism are among the leading factors in the development of consumption patterns as a communication of wealth, culture, and status (after Veblen's model). These patterns include not only materialism, but also the consumption of experiences, relationships, symbolic expressions, and many other aspects of the social environment related to novelty, enculturation, socialization, and the desire for pleasure.

From consumer and sociological perspectives, the term *consumption* has a negative nuance because of the period of overconsumption in the late 20th century, although consumption is inherent in human society since human culture started with pure consumption of the products of nature (hunter-gatherer societies). Accordingly, early 21st-century consumption is accepted as a main trait of human culture along with production, a coin with two sides. Most consumption patterns have positive values because of the increasing role of the sustainable culture ideology worldwide.

Lolita Nikolova
International Institute of Anthropology

See Also: Consumerism; Economics of Consumption, International; Economics of Consumption, U.S.; Overconsumption; Underconsumption.

Further Readings

Botterill, J. *Consumer Culture and Personal Finance: Money Goes to Market.* New York: Palgrave Macmillan, 2010.

Daunton, M. J. *The Politics of Consumption: Material Culture and Citizenship in Europe and America.* Oxford: Berg, 2001.

David, M. C. *John Ruskin and the Ethics of Consumption.* Charlottesville: University of Virginia Press, 2006.

Francks, P. *The Japanese Consumer: An Alternative Economic History of Modern Japan.* Cambridge: Cambridge University Press, 2009.

Cohen, S. and R. L. Rutsky. *Consumption in an Age of Information.* Oxford: Berg, 2005.

Garon, Sh. M. and P. L. Maclachlan. *The Ambivalent Consumer: Questioning Consumption in East Asia and the West.* Ithaca, NY: Cornell University Press, 2006.

Halkier, B. *Consumption Challenged: Food in Medialised Everyday Lives.* Burlington, VT: Ashgate, 2010.

International Food Consumption Patterns. *International Food Consumption Patterns* (2011). http://www.ers.usda.gov/Data/InternationalFoodDemand (Accessed June 2011).

Reusswig, F., H. Lotze-Campen, and K. Gerlinger. *Changing Global Lifestyle and Consumption Patterns: The Case of Energy and Food*. http://www.populationenvironmentresearch.org/papers/Lotze-Campen_Reusswig_Paper.pdf (Accessed June 2011).

Seyfang, G. *The New Economics of Sustainable Consumption: Seeds of Change*. Basingstoke, UK: Palgrave Macmillan, 2009.

Shove, E., F. Trentmann, and R. R. Wilk. *Time, Consumption and Everyday Life: Practice, Materiality and Culture*. New York: Berg, 2009.

Copper

Copper, from the Latin word *cyprium*, is an orange-red ore that has been used by people for nearly 10,000 years. A penny may contain may be recycled ore as old as the first Egyptian pharaoh. Copper is an essential trace element necessary for healthy development in most living things, but it is also one of the most toxic heavy metals for living organisms.

Recycling

Copper is one of the most used and recycled metals, appearing in everyday products including cookware, electric wire, pipes, tubes, coins, automobile radiators, home fixtures, jewelry, as a pigment in paint, and as a preservative in paper, wood, and other products. Copper is infinitely recyclable, and there is as much copper recovered from scrap and recycled material as from ore that is newly mined. Recycled copper has a high value, with approximately 95 percent of the value of newly mined ore. This is beneficial, because mineral deposits like copper form so slowly that they are considered nonrenewable. Since global copper use is expected to continue to rise, the copper industry is dependent on the economic recycling of this heavy metal. In fact, copper's recycling rate is higher than any other engineering metal, according to the Copper Development Association.

Theft

The extensive utility of copper in electronics and machinery at times combines with its value on the global market to cause problems of theft and vandalism in the industrialized world. For example, as the value of copper rose between 2006 and 2008, communities across the United States experienced elevated rates of theft of copper pipes and downspouts from houses, abandoned buildings, and construction sites as well as copper wiring from traffic signals, industrial machinery, and agricultural machinery. While theft reports declined after a market collapse in autumn 2008, law enforcement officials in urban and rural communities alike have attempted to impose controls to keep dealers from purchasing stolen copper.

Production

The United States is second only to Chile as the largest producer of copper in the world. Copper towns began to appear in the United States around the same time in history as coal towns in the early 1900s as a result of mining advances that made possible the recovery of low-grade ore. The copper mining industry, like most mining industries, has a long history of worker inequality and community environmental injustice. There was little government oversight until the passage of the Clean Air Act in the 1970s. Even after the passage of this legislation, the mining industry received preferential treatment by most state governments because of the economic power they wielded. Since the 1970s, the combination of increased government regulation, industry improvements, and greater public awareness of environmental and health hazards has resulted in improved waste disposal and air quality processes.

Disposal

Disposal of copper waste varies with the form of the ore and the processes employed. There has been some debate about the most effective ways of disposing of the ore with the least amount of damage to humans, animals, and the physical environment. Chromated copper arsenate (CCA) has become a common wood preservative, which research has shown to leach over time when exposed to water. As increasing amounts of CCA-treated wood require disposal, this leaching process, which can be harmful to human health, soil, underground water sources, and other aspects of nature, is of specific concern for landfill operators and environmentalists. The Environmental Protection Agency deems copper slags and copper flotation waste as hazardous waste. About 24.6 million tons of copper flotation waste are created annually from world copper production.

There are also opposing viewpoints on the best method to reduce leaching from copper slags and flotation waste, with some scientists recommending the use of palm kernel fiber, while others advance a process called "vitrification." In addition to copper slags and flotation waste, copper ore mining can generate waste rock piles, which are disposed of in mine dumps. These waste rock piles can create acid draining, depending on their interaction with sulfide, sulfur, oxygen, and moisture. Despite the hazards this ore can create if improperly disposed of, it is a needed element that has a long and varied history of use by humanity. The need for ore and the high level of recycling in the industry make it a green product that has a long life.

Rosalind Fisher
University of West Florida

See Also: Construction and Demolition Waste; Pollution, Water; Water Consumption.

Further Readings

Copper Development Association Inc. "Recycling of Copper." www.copper.org/environment/uk/ukrecyc.html (Accessed June 2010).

Coruh, Semra and Osman Nuri Ergun. "Leaching Characteristics of Copper Flotation Waste Before and After Vitrification." *Journal of Environmental Management*, v.81 (2006).

Hyde, Charles K. *Copper for America: The United States Copper Industry From Colonial Times to the 1990s*. Tucson: University of Arizona Press, 1998.

Kapur, Amit. "The Future of the Red Metal: Scenario Analysis." *Futures*, v.37 (2005).

Mastny, Lisa. "World's Copper Use Strains Nonrenewable Resource Base." *World Watch*, v.21 (2008).

Ofomaja, Augustine E. "Equilibrium Studies of Copper Ion Adsorption Onto Palm Kernel Fibre." *Journal of Environmental Management*, v.91 (2010).

Sicotte, Diane. "Power, Profit and Pollution: The Persistence of Environmental Injustice in a Company Town." *Human Ecology Review*, v.16 (2009).

Townsend, Timothy, Brajesh Dubey, Thabet Tolaymat, and Helena Solo-Gabriele. "Preservative Leaching From Weathered CCA-Treated Wood." *Journal of Environmental Management*, v.75 (2005).

Cosmetics

Cosmetics reveal tensions in human society between the quest for aesthetic beauty and health and environmental risks that arise from their uses. The cosmetics industry has evolved into a practical knowledge and science for health and beautification. Cosmetics are a somewhat socially sensitive aspect of human civilization, and their use is influenced by a host of complex factors: economical, technological, cultural, ethnical, ideological, and biological. There are several categories of cosmetics: solids, semisolids, powders, and liquids. Functional classification of cosmetics includes antiperspirants and deodorants; bath and shower products; dental and mouthwash preparations; depilatories, masks, scrubs, bleaching kits, and other skin preparations or corrections; and face powders and creams, lipstick, rouge, blusher, and eye cosmetics. Soap is not considered a cosmetic product according to official standards.

Consumption

People have used natural pigments for beautification since the beginning of human civilization. The earliest and most impressive instances of using plant pigments as cosmetics come from Paleolithic cave paintings and decorated prehistoric ceramics. The ornaments depicted on the pottery can be seen also on the bodies of miniature ceramic figurines and on the walls of houses; for instance, in the Copper Age cultures of the Balkans. These early examples are evidence that people also applied decoration as beautification or with ideological meanings on their bodies. Ethnographic instances of tattoos indirectly confirm that human ideals of beauty include body decoration. The Sitagroi, a female head of a figurine with linear painted ornamentation, is a possible replication of 5th-millennium-B.C.E. fashions of face decoration, and it is among the most beautiful examples of how people looked in the ancient past. Learning more is a question of increasing access to archaeological records and learning about early uses of natural fragrances or more natural means of caring for hair, for example. Generally, the development of different styles of prehistoric female figurines during the Neolithic and the Copper Ages (later 7th–early 4th millennium B.C.E.) reflects an evolution in women's beauty ideals, and these, in turn, indicate not only the existence of, but also the development of cosmetics and fashion styles.

Cosmetics became a hallmark of state civilizations during the origin of state societies in the later 4th millennium B.C.E. Egypt, Mesopotamia, Crete, and Mycenae offer exclusive instances of ancient makeup and fashion related to wealthy and ruling segments of society. Cosmetic jars excavated in the tomb of Tutankhamun (1341–1323 B.C.E.) contained skin cream composed of approximately nine parts of animal fat to one part of perfumed resin.

Cosmetics in the Middle Ages and Renaissance were used for beautification, but poisonous components, such as mercury and white lead (and likely unknown ingredients), often damaged the face. There was no toothpaste, although some herbs were recommended for maintaining white and pearly teeth. Because bathing was not considered of importance, perfume became increasingly necessary. Queen Elizabeth I introduced "sweet coffers" containing paint, power, and patches. Cosmetics were socially sensitive through the ages, promoted by some or blamed as a sign of lower social status or amorality by others. In the 21st century, the worldwide cosmetics market totals more than $140 billion (32 percent for skin care, 27 percent for hair care, 12 percent for toiletries, 17 percent for makeup, and 12 percent for fragrances).

There are three lines of development in this industry that influence the consumer pattern: the megamultinational businesses of global chains (such as L'Oréal, Elizabeth Arden, Revlon, and Cover Girl), growing new international companies, and relatively small national or international businesses. The Internet stimulates this sector's growth by offering an ever-increasing variety of standard cosmetics and green or all-natural cosmetics. Cosmetics are also a large part of home-based businesses of distributors that sell essential oils, such as Avon, Mary Kay, Yves Rocher, and Oriflame.

Cosmetics in the 21st century are permanent and temporary. Organic cosmetics have become popular, although chemical components still dominate the worldwide distribution of cosmetics, even from the most popular companies. The quality of cosmetics depends on brand recognition, price, availability, and popularity. Cosmetic products are socially distinct and reflect social differences. A specific field of cosmetics use is in the artistic world. In some countries, body painting and tattooing have developed from a fashion to a tradition and a specific subculture.

Health and Cosmetics

Cosmetics are anchored to ideas of people's physical and mental health. Thanks to cosmetics, cleanliness has become a hallmark of contemporary human civilization. Traditional medicine saves people's life, and some essential oils have claimed to dissolve certain types of cancer in the human body. Cosmetic surgeries reshape human faces, increase self-esteem, and even affect personalities. Cosmetics that have a therapeutic component, typical of pharmaceutical products, are named "cosmeceuticals"; for example, antiaging creams and moisturizers.

A specific category includes products with ingredients from the Dead Sea. Many skin care products have been produced with Dead Sea minerals because of their rejuvenating effect. During the early 20th century, skin bleachers, lead makeup, and burning hair straighteners were widely sold to and within the African American community. Beauty became a substantial industry, with Madam C. J. Walker of Indianapolis an important pioneer in developing products and salons between 1910 and her death in 1919. In the century since Walker built her empire, beauty standards within African American communities have, alternately, valued light or dark skin and straightened or "natural" hair, depending on sociocultural conditions.

Allergic reactions to cosmetic skin products are common and may cause serious health problems.

Bacterial and fungal infections and skin viruses, such as warts, are the most common nail infections as a result of use of acrylic nails. Some cosmetics may be toxic to children if swallowed.

Cosmetics Refill, Recycling, and Waste

Refillable cosmetic containers reduce the amount of energy for their production and shipment, while the refillable containers from recyclable goods, such as aluminum and postconsumer plastics, reduce waste. M.A.C. and other companies promote recycling programs and offer free cosmetics for a certain number of returned empty containers. Aveda, Stila, and the Body Shop make empty compacts that can be customized and reused with different eye shadow shades, blush, or powder.

Cosmetics usually do not list an expiration date, although opened makeup can be harmful if it develops bacteria. Cleaning makeup accessories and knowing the expiration date for different types of makeup are excellent ways to ensure the healthy use of cosmetics. The cosmetics recycling movement was not well defined or available worldwide by 2010. In 2009, the Origins company initiated a recycling program and began to collect all brands of packaging (such as empty bottles, jars, and tubes).

A number of the ingredients in cosmetics are considered hazardous waste. According to the Environmental Protection Agency, trash with cosmetics waste could cause contamination and become harmful to people and the environment.

The best way to reduce the negative consequences of waste is to try to reduce use both during the production process and as consumer products. Oriflame has introduced site treatment at two facilities to decrease their total effluent waste. Yves Rocher has installed linerless versions of Domino's M-Series print and apply labeler that "virtually eliminates" the wastage generated from their label printing.

Not completely used and unfit cosmetics are considered hazardous waste. These are cosmetics that are improperly sealed; damaged, expired, and improperly stored; improperly labeled; counterfeit, substandard, and adulterated; and prohibited or unauthorized. Solids, semisolids, and powders require landfill, incineration, or waste immobilization; liquids require sewer, high-temperature incineration, or treated waste; antineoplastics require treated waste and landfill, high-temperature incineration, or return to manufacturer; aerosols require landfill without waste inertization; PVC plastics and glass containers require landfill and recycling; paper and cardboard require recycling, incineration, or landfill.

Cosmetics and Social Sensibility

Cosmetics have economical, biological, and age implications. Economically, cosmetics are part of pop culture, offering a relatively inexpensive means of integrating people of different ages with fashion and traditions, on the one hand. On the other hand, cosmetics can become a symbol of wealth, prosperity, and elitism through the monied symbolization of specific brands that offer products with extremely rare ingredients and high prices. Remarkably, the men's and children's cosmetics markets have been increasing as consumer needs increase. As such, the use of cosmetics is not necessarily anchored to women alone.

Cosmetics and Culture Diversity

Cosmetics connect and divide people. In the process of globalization, leading cosmetics companies stimulated a process of the unification of cosmetics subculture. At the same time, new (especially green) culture-oriented companies have been trying to make a change in society by promoting more innovative and planet-saving cosmetic products. There are also ethnical and religious peculiarities, since cosmetics include elements of social identity that directly interact with diverse cultures.

Lolita Nikolova
International Institute of Anthropology

See Also: Children; Consumption Patterns; Social Sensibility.

Further Readings

Butler, H. *Poucher's Perfumes, Cosmetics and Soaps.* Vol. 3. London: Chapman & Hall, 1993.

Corson, R. *Fashions in Makeup. From Ancient to Modern Times.* London: Peter Owen, 2003.

Peiss, Kathy. *Hope in a Jar: The Making of America's Beauty Culture.* New York: Henry Holt, 1998.

Radd, B. L. "Cosmeceuticals: Combing Cosmetic and Pharmaceutical Functionality." In *Multifunctional Cosmetics*, R. Schuller and P. Romanowski, eds. New York: Marcel Dekker, 2003.

Renfrew, C., M. Gimbutas, and E. S. Elster, eds. *Excavations at Sitagroi: A Prehistoric Village in Northeast Greece*. Los Angeles: University of California Press, 1986.

Scranton, Philip, ed. *Beauty and Business: Commerce, Gender and Culture in Modern America*. New York: Routledge, 2001.

Crime and Garbage

Crime and garbage refers to the myriad associations of crime and garbage, including crime-related materials, ownership, organized crime, not following procedures, illegal dumping, recycling scams, and environmental crimes. There is also the issue of crime and deviance. While crime is the violation of a written statute, deviance is a violation of social norms. In such cases of violation, there is no official penalty; rather, condemnation from a group or individuals may be the result. There is differing social opinion as to the deviance of acts related to garbage. In some communities, the response to negligent disposal of garbage is severe; while in others, the practice may be the norm and thus be more accepted. There is also a large degree of variance based on individuals. Their education about the environment, personality, and range of social associations can be factors in determining whether they view acts related to garbage as deviant.

In popular culture, one of the most common contexts of crime and garbage is the incident in which a material related to a crime—such as a weapon, some form of evidence, or even a body—is found in a trash bin, garbage bag, or dumpster. U.S. crime serial shows and newspaper headlines like "Baby Found in Trash" point to the macabre interest that people often have in crime and garbage.

Garbage, because it represents the way in which most people dispose of their personal possessions, is subject to the question of control: who owns garbage? Freegans (people who seek free food) and dumpster divers are known for their interest in obtaining disposed food items and making their meals from them. Some are concerned that this practice could lead to lawsuits if a freegan gets sick from such a meal. Others are concerned with the disruptions and potential invasions of private property caused by such dumpster diving.

Collecting Information

Trash trawling, *dumpster diving*, and *waste archaeology* are all terms given to the practice of culling garbage for potentially valuable information. This information can be of interest to private detectives, law enforcement officials, and identity thieves. Businesses that wish to understand trade secrets or interpret corporate planning documents may engage in forms of dumpster diving. Individuals are even more commonly targeted for such crimes. Information gleaned from garbage—including personal identification numbers and other personal information—is used to commit various forms of identity theft. This form of crime stems from the convenience of people throwing their garbage out without regard for what is thrown out. In response, in the United States and other nations, there has been a growth in personal shredders as well as businesses that pick up documents and shred them on the spot. There is also the issue of privacy and due process of the law related to garbage.

In 1988, in overturning a California court's decision in *California v. Greenwood*, the U.S. Supreme Court ruled that the Fourth Amendment's protection against unwarranted search and seizure does not extend to trash. Criminal defense attorneys have attempted to argue that police searches of garbage should occur only after obtaining warrants. Some U.S. cities have passed legislation making it illegal to go through garbage. The laws are focused on reducing identity theft resulting from trash pilfering. Canadian officials, responding to cases of their own (including the case against a man who was discovered to have marijuana paraphernalia in his trash), have expressed that laws related to garbage are especially not well defined. In 1995, a British Columbia justice wrote that, "putting material in the garbage signifies that the material is no longer something of value or importance to the person disposing of it . . . when trash is abandoned, there is no longer a reasonable expectation of privacy in respect to it." This ruling is similar to the U.S. Supreme Court's decision in 1988 that defined garbage as *bona vacantia*, or "ownerless goods." The reasoning behind this view of garbage as no longer belonging to the individual who disposed of it relates to interpretations that the individual exercised free will in disposing of the garbage. Some advocates complain that such laws result in violations of the individual's privacy, regardless of what is contained in the garbage.

Cities have also weighed in on who has access to garbage. In 2008, 28 Redding, California, waste management workers were disciplined for crimes considered theft, including the pilfering of old tools, metal, and other items that had been thrown away. The probe that netted these workers began in response to the mismanagement of waste by the city's former solid waste manager. In response to the scandal, a Redding city official expressed, "When people put their trash out they expect it to be buried." Scavenging has long been a problem at the city's transfer station and, while some believe that people should be able to scavenge through the garbage, city officials state that this is a crime since it robs residents of potential funds generated by recycling the garbage.

Organized Crime

Because of garbage's ubiquity, the control of garbage disposal can be profitable. Throughout the world, the Mafia and organized crime have played a major role in hauling garbage. In some cases, Mafia involvement is in the legal trade of hauling trash. Organized crime groups take control of the waste disposal businesses in a city, set the prices, fix bids, and control the organization of trash hauling such that people have no choice in the matter. In New York City, the garbage industry was controlled by La Cosa Nostra until the city took control of it in the 1990s. In Taiwan, there have been cases of organized crime groups collecting gravel, selling the material to companies, and then filling the holes with waste. In the United Kingdom, there is also a prevalence of organized crime involvement in illegal waste practices. In many societies, businesses handling waste are socially marginalized. In the postbellum United States, waste hauling was left to eastern and southern European immigrants because the work was seen as dirty. The resulting businesses were suspected of unscrupulous behavior in part because of criminal activity and partly because of xenophobic fears.

Illegal Disposal

In 2007, New Orleans imposed a new law that requires that all city trash be disposed of in specific legitimate trash bins that allow for easier collection by robotic-arm trucks. In the aftermath of Hurricane Katrina, the city was beset with massive amounts of dumping, and the new law was aimed at curbing such offenses. Penalties range from a $5,000 fine to six months in jail for each offense. Also, for individuals who refuse to obtain a city-approved bin, the city plans to stop collecting trash at that residence. The law also specifies rules about how garbage may be disposed of in carts, making it illegal for residents to fill carts excessively such that the lids will not close. Residents in cities throughout the United States and the United Kingdom express discontent that overly bureaucratic disposal rules will result in criminal justice agencies targeting individuals when, in fact, such agencies should target more wide-scale forms of garbage crime, such as illegal dumping.

In the United Kingdom, the Environment Agency has cracked down on a range of forms of illegal dumping. Such illegal dumping is referred to as "fly-tipping." Between 2005 and 2006, some 2.5 million cases of illegal dumping were recorded in the United Kingdom, and it is estimated that 63 percent of all fly-tipping is in the form of black bags or garbage bags that are thrown out by consumers, often in alleyways. There are also organized gangs who profit from the fly-tipping business. Offenders are prosecuted at a rate of one out of 100. Fly-tipping may be tied to a technique of neutralization known as "denial of a victim." The assumption that there is not a victim, however, is contradicted by the fact that private property owners have to deal with the cleanup of such garbage and the fact that the environment is a victim. In 2008, the UK's Environment Agency created a National Environmental Crime Team composed of former detectives, intelligence officers, and forensics individuals. According to the agency, the team is charged with investigating organized waste crime. Examples of successful investigations include a metal recovery company that was attempting to export hazardous waste cable to China, a serial fly-tipper who was collecting household and business garbage for a fee and then disposing of it around the city of Bristol, a car company caught dumping numerous types of garbage at a fly-tipping spot in the city of Bristol, and an individual found accepting illegal dumping on his property. The agency's work has not been without its detractors. In 2007, a Boston, Lincolnshire, man was fined for what the city called "illegal fly-tipping." The city had tracked a bag that the man placed in a public bin.

In the United States, though less organized than in the UK, there have been numerous attempts to crack down on illegal dumping. In Midland, Texas, officials have attempted to educate the public about the crime of illegal dumping. Many Texans are familiar with dumping trash on other people's properties and dumping used oil or old car batteries. For some, this is culturally accepted. City officials have responded by illustrating the effects of the damage that such crimes pose to the environment and emphasizing penalties that include fines up to $50,000 and five years in prison for each day of the offense. In explaining the severity of the punishments, officials explained that one quart of motor oil can contaminate 250,000 gallons of water, that illegal litter creates germs and contains sharp objects that can harm people, and that disposed car batteries can leach lead into the environment. A major form of illegal dumping is known as "cocktailing." This involves the mixing of hazardous waste with nonhazardous waste, with the hopes of the material passing as nonhazardous waste.

A man rummages through a dumpster at the back of an office building in Central London. Personal information such as bank account numbers and other identification extracted from garbage is often used to commit various forms of identity theft.

In 2010, 31 individuals were charged with recycling fraud in the state of California. The alleged perpetrators trucked millions of cans and bottles into the state, taking advantage of state redemption money. In some cases, cans were filled with sand so that they would weigh more. In the United Kingdom, similar crimes of recycling have taken place. Police and officials from the Environment Agency have discovered the illegal exporting of computers and other e-waste. The materials are often shipped to China, countries in Africa, and India where small children are exposed to hazardous materials in the process of recovering precious metals like gold, copper, and steel. Law requires that such materials be recycled only in the United Kingdom. Recycling illustrates one example of the attempt to harness the productive potential of garbage (in terms of the money generated) for legal means.

Pay as You Go

In the United Kingdom, one response to fly-tipping has been to pilot a new pay-as-you-go system in which individuals' garbage bins are fitted with computer chips. The bins are then weighed, with the amount of each household's garbage tracked. Households with less excess waste could receive reward vouchers. Some residents, however, are concerned that this attempt to better deal with refuse could result in crime also. Some worry that computer systems could be hacked to determine if residents are not home and then burglaries could be committed. Others express that the system may result in an invasion of privacy, and some worry that households that might be eventually charged for excessive amounts of waste could resort to illegal dumping or fly-tipping in order to avoid taxes. Policies and plans like these, while originating in the efforts to address crimes like illegal dumping, may also affect the growing environmental concerns that result from waste and overconsumption. The overall issue of garbage's impact on the environment is the subject of green and environmental criminology. These specializations of criminology include a number of arenas in which environmental harm has resulted from human action. Air pollution, deforestation, water pollution, despeciation, and the dumping of various forms of hazardous waste are some of the emergent concerns. These many environmental issues may be connected to the issue of risk society, identified by Ulrich Beck.

Environmental law aims at risk reduction, deals with social issues that are inevitable and pervasive, and is aspirational—meaning that it attempts to radically improve environmental quality. While some environmental laws may not directly apply to garbage, most do. The many harms caused by illegal dumping and other forms of waste disposal include immediate and future injuries, emotional distress, disruption of economic and social activities, remediation costs, property damage, and ecological damage.

Two emerging issues with crime and garbage include sentencing and mens rea. In the first area, there are concerns that sentences handed down against environmental criminals are especially light. In the second, criminologists have questioned whether a "guilty mind" is present in cases in which the scope of garbage crimes may not be known beforehand. Since the 1970s, there have been a number of international efforts to ban the dumping of hazardous materials sea, into the air, and less-developed nations. The last of these identifies a growing concern with environmental racism, which may include the dumping of hazardous garbage onto the lands of peoples considered less important than those doing the dumping. Crimes have also occurred in response to the many environmental groups that have attempted to draw attention to these issues. In 1985, the Greenpeace ship *Rainbow Warrior* was sunk by explosives of French commandoes. The associations of crime and garbage will, no doubt, continue to capture the imagination of criminologists, politicians, environmentalists and laypeople alike.

Scott Lukas
Lake Tahoe Community College

See Also: Dumpster Diving; Environmental Justice; Fly-Tipping; Freeganism.

Further Readings

Gibbs, C., M. Gore, E. McGarrell, and L. Rivers. "Introducing Conservation Criminology: Towards Interdisciplinary Scholarship on Environmental Crimes and Risks." *British Journal of Criminology*, v.50 (2010).

Jablon, R. "31 Arrested in $3.5 Million California Recycling Fraud." *Associated Press* (May 5, 2010).

O'Hear, M. "Sentencing the Green-Collar Offender: Punishment, Culpability, and Environmental Crime." *Journal of Criminal Law and Criminology*, v.91/1 (2004).

Rebovich, D. *Dangerous Ground: The World of Hazardous Waste Crime*. Piscataway, NJ: Transaction, 1992.

South, N. and P. Beirne, eds. *Issues in Green Criminology: Confronting Harms Against Environments, Humanity and Other Animals*. Cullompton, UK: Willan, 2007.

Sullivan, M. "New Jersey Justices Ponder the Privacy of Garbage." *New York Times* (February 18, 1990).

Sykes, G., and D. Matza. "Techniques of Neutralization: A Theory of Delinquency." *American Sociological Review*, v.22 (1957).

"UK Fly-Tipping 'On Massive Scale.'" *BBC News* (March 19, 2007).

Warren, P. "Organised Crime Targets Waste Recycling." *The Guardian* (July 8, 2009).

White, Rob. *Crimes Against Nature: Environmental Criminology and Ecological Justice*. Cullompton, UK: Willan, 2008.

Zimring, Carl A. "Dirty Work: How Hygiene and Xenophobia Marginalized the American Waste Trades 1870–1930." *Environmental History*, v.9/1 (January 2004).

Culture, Values, and Garbage

All societies are connected to garbage. Early human societies, while much smaller in population and less impactful in terms of resource use and waste, produced garbage. Mobile groups of early humans produced single-use midden sites, but as human inhabitation became more sedentary in nature, the physical relationship of humans and waste changed.

Population increase, the development of complex systems of production, social differentiation, and changing lifestyle patterns all contributed to differing amounts, distribution, and human relationships to garbage. All humans have relationships with garbage, therefore it is important to focus on how values impact these relationships. Values refer to abstract cultural understandings of what is considered right and desirable in society. All cultures maintain values, yet cultures express values in different ways and, thus, cultures display different relationships to garbage.

Trash Talk Project

In 2009, the Massachusetts Institute of Technology Trash Talk project was developed in order to better understand the lifecycle of garbage. The project uses small location tags that frequently update the position of a given item of trash and sends this information to researchers. Through this system, researchers have tracked over 3,000 pieces of individual garbage in New York, Seattle, and London. In many Western societies, including the United States, a value association of "out of sight, out of mind" governs many people's relationships to trash.

The goals of Trash Talk include the reworking of refuse infrastructures and behavioral change in consumers. What is most prescient about the project is how it illustrates the complex ways in which culture, values, and garbage are intertwined. While a person viewing the project data might become more aware of the problems with the attitude that once an item leaves one's possession it is no longer of consequence, actually motivating that person—and millions of others—to act on the data and to change practices is a much more complex matter.

Human Connections to Garbage

The Trash Talk project emphasizes the complex, yet overlooked, relationships that garbage and people share. In terms of their relationship to garbage, all people interact with it on two levels. One is a material connection, indicative of the physical and sensory contacts that people have with garbage. In some households, this connection begins with an individual removing an item from packaging, disposing of that item in the kitchen receptacle, placing that item and others into a larger bin, taking that bin to the curbside, and then the material connection ends. Others, including workers in sanitation plants and recycling centers, then continue a material connection with the garbage, but the material connection of the consumer and the garbage ends with the bin on the curbside.

The second connection that people maintain with garbage is an ideational one. Unlike the material one, which is manifested in things that can be touched, moved, and sensed, the ideational connection operates on the level of cognition. The differentiation of an item of value from an item of trash, for example, has nothing to do with the material principles of the object. Instead, humans determine whether the object is of value or whether it is considered trash. The decision of whether an individual decides to dispose of a broken radio or to consider it an heirloom to be kept is highly subjective and rooted in the value systems of a culture.

Values and Behaviors

As the thought picture indicates, any everyday human behavior is connected to a myriad of material and ideational contexts. These contexts impact the specific ways in which individuals think about garbage. For example, a person needs to eat. This individual might choose to consume a packaged product for dinner, an item with production contexts that impact the environment and produce waste but also supplies the body with value in terms of calories and pleasure. The media, worldview, specific values, and their associated norms and sanctions all impact the behavior in question. After the item is eaten, the individual has to decide what to do with the remainder, such as the leftover package. The package might be reused, repurposed, or recycled but, most likely, will be disposed

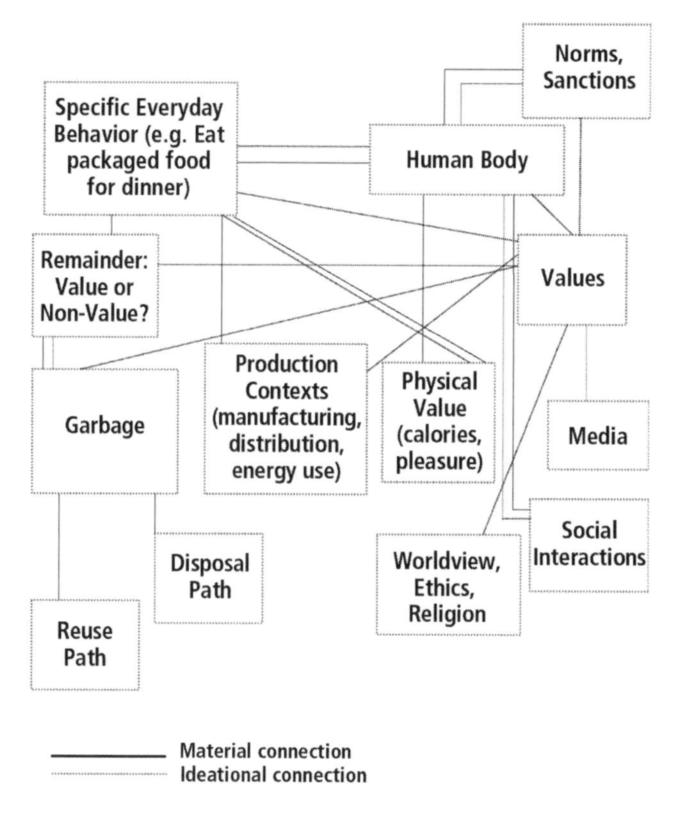

As this thought picture indicates, any everyday human behavior is connected to many material and ideational contexts. Note: while the diagram indicates meaningful connections, it does not posit the amplitude or direction of the effect of one thing on another.

of in the trash. Values are particularly significant in this chain of connections. Values help determine whether an individual will choose a certain product in the first place, how it will be used, and how any remainder will be reused or disposed of. The functional relationship of these material and ideational connections is especially relevant. If, for example, the media emphasizes the happiness that can accompany the purchase of high-environmental-cost products and if a particular religion shared by the consumer does not emphasize the value of natural theology, it becomes more likely that because of the preponderance of like values, the consumer will maintain current relationships with garbage.

Investigators, such as those in the documentary *No Impact Man*, have discovered that it is difficult, if not impossible, to simply alter material lifestyle patterns. One could decide to pursue a path of less material waste, but this decision is not left to the individual alone. Society, through its value structure, in large part determines how individuals interact with the material environment. In the case of garbage, values are impactful in three ways.

First, values condition people to think about the material world in specific ways. For a member of a Native American culture, the relationship to the material world might be conceived of in terms of mutual respect and stewardship, while for a North American consumer, the same relationship might be thought of through the economics of production. Such values impact how individuals go about a particular behavior and how people will react to certain behaviors that might result in greater waste than others. Second, values impact people's perceptions of "remainder." In cultures in which people do not think about how garbage will impact the environment, disposing of an item appears to be of no cost. In cultures in which people perceive a close connection between themselves and the natural world, reuse of an item might be done in lieu of disposal. Finally, values—even when seemingly ingrained in cultures—may impact people in indirect, as opposed to direct, ways. Sociologists speak of ideal culture as including those values that people purport to uphold, as compared with real culture, which refers to the values that people actually express. In the case of garbage, some people purport to be environmentally aware, yet they dispose of items that have large environmental impacts. William Rathje and Cullen Murphy indicate that a common facet of people's connections to their garbage is the disconnect that exists between the mental (people's perceptions about waste) and the material (what they actually waste). In this sense, people could, conceivably, consider themselves to be environmentally friendly, yet dispose of items, such as newspapers, that could be detrimental.

Sociologists have also pointed to the presence of value contradictions—situations in which competing values coexist in a society. For example, people might believe in both the ideas that abundance is good and frugality is good. A person might be inclined to save old items and, at the same time, purchase new ones. Perhaps this individual ends up having less of an impact at the end of throwing out garbage, but the impact at the production end of the products being purchased is significant. In some cases, apparent contradictions between values may be the result of structural requirements of the culture. In the United States, when CDs were sold in stores, a common practice was to sell them in large plastic cases. The cases were needed because storekeepers desired to address loss prevention—a larger CD case means that the item will be more difficult for shoplifters to steal. While this packaging was wasteful, it did prevent the material losses that storekeepers experience from shoplifters. In cases such as these, nations or local communities may have to decide how to balance the costs and benefits of various forms of consumption and associated waste and environmental costs.

Social Norms

As compared to values, norms, while dependent on a culture's value structure, are more specific codes that dictate social behavior. Some norms, including mores, are addressed more stringently by a society, while others, including folkways, are met with few, if any, sanctions for violation. The specific ways in which norms are played out in society determine how strongly people will be impacted by values. Sociologist Robin Williams indicated that norms represent those things that people should follow in society, but they often do not due to distribution (social knowledge about the norms), enforcement (how norms are punished or not), transmission (how people learn norms), and conformity (the extent to which deviance occurs in society or not). A reaction, such as a scowl following a person's disposing of garbage alongside a highway, may have little impact on the violator, depending on distribution, enforcement, transmission, and conformity present in the culture.

Norms reflect the specific ways in which values are acted upon in a local context. In terms of garbage, norms will vary highly if a person is living in a large metropolitan area that tolerates forms of excessive consumption—such as driving cars, permitting open dumping, or not recycling—as compared to a person living in a small town that has established consumption codes of conduct for both businesses and individuals. Were one to move between these fictional communities, one would discover that throwing out a plastic cup could be a folkway in the one place and a more in the other.

Table 1 American Values and Garbage

Value	Garbage Impact
Achievement and Success	If success is defined in terms of material items, it could result in use and disposal of more material products.
Efficiency and Practicality	Desire for up-to-date technology could create more waste in production and disposal.
Material Comfort	The desire for material comfort may involve wasteful forms of living.
Freedom	The belief in absolute freedom may result in individuals not following governmental recommendations on the environment and waste.
External Conformity	Individuals may imitate others, including celebrities, who display wasteful behaviors.
Individual Personality	By focusing only on the self, an individual might be inclined to think less about the impacts of consumption.
Live for Today	People's behavior patterns with respect to material culture will result in more waste and less conservation.
Obsolescence	Instead of fixing a broken item or living with an old but still useful one, people are inclined to buy new replacement items.

As well, one would begin to see specific associations between norms and, in terms of the connections discussed in the thought picture on the previous page, it would be possible to see that certain norms correlate with others, such as "One should be kind to passersby" and "One shouldn't litter."

When cultures express norms, they are reflecting on the specific values that are cherished in their cultures. As the world pushes toward more integrated and global forms of connection, it becomes clear that certain value systems are at odds with others. This is particularly clear when the relationships of culture, values, and garbage are addressed in a cross-cultural manner.

A number of the values presented in Table 1 are based on Robin Williams's study of U.S. values. Williams identified a series of core U.S. values, many of which can be tied to consumption patterns and resultant environmental impacts and garbage effects. In a global context, a number of these values are taking root in cultures that have adopted new values or are in the process of shedding old

Table 2 Non-Western Values and Garbage

Value	Garbage Impact
Community Spirit	Because people feel connected to a community, they may be less likely to produce excessive garbage, especially of the form that could impinge on other's rights or well-being.
Bricolage	People may feel inclined to make do with what is available, to act as jacks of all trades, and to remake rather than buy anew.
Environmentalism	People may live in a manner that expresses deeper respect for the environment and thus may deliberately alter wasteful practices.
Sustainability	People focus on the idea that they should preserve the environment for future generations.
Responsibility	As opposed to feeling free to do anything possible, people live with the idea of sharing responsibility with others.
Nonmaterialism	While people in the Western world sometimes view money as the end of all things, other people focus on things beyond material ones that could bring happiness and value to living.

ones. Table 2 illustrates some of the indigenous values that are being replaced by U.S. values.

Bricolage

Values are abstractions and approximations of cultures, and therefore it is possible to find non-Western values in the West and vice versa. The comparison of Western and non-Western values shows that there are radical differences in terms of how people relate to their garbage. The *zabbaleen* of Cairo, Egypt, practice a form of recycling that has roots in the ways in which they relate to garbage and consumption.

Instead of seeing items that in the Western world would be detritus, the zabbaleen look at the item as a resource—as something that can be reshaped in a new, useful form. They express what anthropologist Claude Levi-Strauss notes: the *bricoleur* works not from the principle of making things only if natural resources are available but makes things according to those things at hand, making do with what is available. The spirit of bricolage is not simply a utilitarian one. While the practice is expressed in cultures including those in Papua New Guinea, Mexico, Morocco, Senegal, Trinidad, and Kenya and can be considered practical, it is much more than that. It is an expression that, like the natural cycles of the Earth, attempts to make something new from something old.

As a practice and value, bricolage is dependent on the other values that correlate with it. In a society that remakes, people are less inclined to think of consumption as a process of buying an item from a store, disposing of its packaging, eventually throwing the item out, and buying a new one. Instead, people are likely to look at the entirety of how they are living.

One example of this approach to living on a larger scale is present in the small nation of Bhutan. In the 1970s, King Jigme Singye Wangchuck established a system known as gross national happiness (GNH). The Buddhist-based GNH focuses on four pillars: sustainable development, preservation of cultural values, environmentalism, and governance. Bhutan has an economy that establishes both material and ideational connections between people, resources, and garbage, but what is unique with GNH is how it revalues and reevaluates these connections.

While tourism and logging could be immensely profitable, the nation has agreed to limit the former and outlaw the latter because sustainability is a value considered to be of higher worth than materialism. Cultures like those of Bhutan express a model that, while favorable in terms of its lessened impact on the planet, is perhaps not practical for cultures based on large-scale systems of production and consumption.

Garbage Informs Culture

Martin O'Brien indicates that garbage is not the ephemera of society—rather, it is at the heart of a given society's productive activity. How a culture decides to act on its material objects—such as in whether to reuse or dispose of an item—speaks to the nature of the culture. All cultures are integrated, and thus the ways in which a culture deals with its garbage is connected to issues independent of garbage. Susan Strasser expresses that the difference between how the developed world and the underdeveloped world relates to garbage is that the developed world throws out garbage simply due to the fact that people do not want something any longer. It is a choice and a form of power. Thus, quite independent of values, the myriad relationships of cultures and garbage in the world are tied to forms of economic and cultural power. If members of a culture are living in poverty, they have little choice but to remake or reuse discarded items. If other people are living lives of affluence, they will likely not even consider reusing something. As globalization continues and as cultures increasingly connect (or collide) with one another, the issue of the power of value systems will necessarily come to the forefront. The world is a marketplace of products, not of values and, as such, people will not likely find themselves in the position of choosing between the values of gross national happiness and those of global capitalism.

Values and their associated connections with garbage are not immutable. Like any form of culture, the relationships that people share with their trash changes over time. A case in point is the early U.S. colonies. Susan Strasser points out that forms of bricolage were present and prevalent in this historical period. The remaking of animal by-products into soaps and candles, quilting, reworking clothes, metal recycling, and the fact that early manufacturers encouraged reuse of their products (such as tobacco tins as lunch boxes) paint a picture of U.S. culture and consumption quite different from that in the 21st century.

In 21st-century U.S. society, freecycling, freeganism, and dumpster diving represent a social expression that harkens back to the spirit of remaking in the U.S. colonies. What is curious about the ideology of freeganism is that many of its practitioners, while clearly enjoying the utilitarian benefits of the practice, express that what they are doing is re-valuing garbage. They are establishing a practice that is both useful and, in terms of values, something different than the mainstream. In studying examples like those of freeganism and dumpster diving, researchers have discovered that by reducing garbage and by changing lifestyles, new values could be expressed. Consuming and throwing away is socially significant, and perhaps what is most significant is the realization that not only could garbage be meaningfully reduced in such re-envisioning but also that new social forms could take hold in the aftermath.

Scott Lukas
Lake Tahoe Community College

See Also: *Garbage Dreams*; Garbage in Modern Thought; Sociology of Waste.

Further Readings

Cerny, C. and S. Seriff. *Recycled Re-Seen: Art From the Global Scrap Heap*. New York: Harry N. Abrams, 1996.

Gould, S. J. "Tires to Sandals." *Natural History*, v.98 (April 1989).

Hawkins, G. *The Ethics of Waste: How We Relate to Rubbish*. Lanham, MD: Rowman & Littlefield, 2005.

Laporte, D. *History of Shit*. Cambridge, MA: MIT Press, 2000.

Le, P. "Researchers Track 3,000 Pieces of Seattle Trash." Associated Press (September 14, 2009).

Levi-Strauss, C. *The Savage Mind*. Chicago: University of Chicago Press, 1966.

McDonough, W. and M. Braungart. *Cradle to Cradle: Remaking the Way We Make Things*. New York: North Point Press, 2002.

Massachusetts Institute of Technology. "Trash Talk." http://senseable.mit.edu/trashtrack (Accessed July 2010).

O'Brien, M. "Rubbish-Power: Towards a Sociology of the Rubbish Society." In *Consuming Cultures: Power and Resistance*, J. Hearn and S. Roseneil, eds. Houndsmill, UK: Macmillan, 1999.

Packard, Vance. *The Waste Makers*. New York: David McKay, 1960.

Rathje, W. and C. Murphy. "Five Major Myths About Garbage, and Why They're Wrong." *Smithsonian*, v.23/4 (1992).

Rathje, W. and C. Murphy. *Rubbish! The Archaeology of Garbage*. New York: HarperCollins, 1992.

Scanlon, J. *On Garbage*. London: Reaktion, 2005.

Strasser, S. *Waste and Want: A Social History of Trash*. New York: Metropolitan Books, 1999.

Thompson, M. *Rubbish Theory: The Creation and Destruction of Value*. Oxford: Oxford University Press, 1979.

Williams, R. *American Society: A Sociological Interpretation*. 2nd ed. New York: Knopf, 1964.

Dairy Products

A popular advertising campaign in U.S. print media asks people if they "got milk?" The ads depict popular celebrities with an exaggerated white "moustache" that one may have above their upper lip after a big gulp of milk. Regardless of the grammatical faux pas, this campaign poses a very interesting question about human food consumption. To some adults, the answer to this question is "yes." Roughly one-quarter of the nearly 7 billion people across the globe regularly consume milk and other dairy products, which refers to any food produced from mammalian milk. While typically coming from cattle, dairy products can also be produced from the milk of several other animals, including goats, water buffalo, sheep, camels, donkeys, reindeer, yaks, and horses. Raw milk from these animals can be processed into a variety of high-energy-yielding food products. The human species, which drinks and eat dairy into adulthood, is an exception to the rule, as other mammals do not consume milk beyond infancy. A look at dairy products not only provides insight into human food consumption patterns but also illuminates the reasons why some humans can digest dairy as adults while others cannot.

Nutrition

Mammals have highly specialized sweat glands, called mammary glands, which produce milk. Evolutionary researchers suggest that the production of this opaque white liquid evolved as a means to keep eggs moist and point to the egg-laying monotremes as evidence for this claim. Milk provides a highly effective means of delivering nutrients to mammalian infants, who cannot yet digest many types of foods that older members of the species consume. Milk also supplies infants with antibodies that boost the immune system. The composition and components of mammalian milk vary from one species to another, but all contain large amounts of protein, calcium, saturated fat, and vitamin C.

Digestion and Lactase

In order to consume dairy products, mammals must produce lactase. This enzyme breaks down the milk sugar lactose and enables digestion. Lactase is essential for infants of all mammalian species, as they receive key nutrients from their mother's milk. In humans, breast milk contains the proper balance of water, protein, fat, and sugar necessary for growth and development. Additionally, a 2007 U.S. Agency for Healthcare Research and Quality–compiled report for the World Health Organization showed that breastfeeding provides the infant greater immune health, lower chance to develop allergies, higher intelligence later in life, less chance to develop diabetes, decreased risk of obesity, and reduced risk of sudden infant death syndrome. While milk consumption is necessary for mammal infant growth and development, adult mammals, with the exception of humans, do not consume milk. Humans are unusual mammals because they not only consume milk as adults but they also consume the milk of other animal species.

Adult Consumption

For much of human history, the ability to digest dairy products was unnecessary beyond the age of weaning. As food foragers, ancient humans consumed vegetation and animal protein and procured food by gathering, hunting, scavenging, fishing, and beachcombing. With the rise of agriculture during the Neolithic revolution, which occurred independently in as many as seven or eight separate regions beginning around 10,000 B.C.E., human food consumption patterns forever changed. Groups that adopted agriculture replaced nomadic, food-foraging methods with sedentary, food-producing methods. Some of these groups also brought animals permanently into their settlements. Beginning in the Middle East and then spreading out both east and west from there, this domestication of animals is what led to adult human dairy consumption.

Initially, it appears that animals were domesticated only as an easily accessible source of meat. Archaeological evidence shows that humans began to use milk for food about 7,000 years ago during what is known as the Secondary Products Revolution. In terms of economy, consuming milk is much more efficient than killing an animal for meat. Like all other mammals, ancient humans were not

Cow's milk is produced on an industrial scale in the United States and is the most widely consumed dairy product. As of 2010, there were 62,500 U.S. dairy farms, predominantly in California, Wisconsin, Idaho, New York, and Pennsylvania. U.S. per-capita milk consumption was 22 gallons a year in 2008, compared to the high of 45 gallons a year in 1945; competition from soft drinks caused the decline. Meanwhile, Between 1970 and 2008, annual per-capita cheese consumption tripled from 11 pounds per person to nearly 33 pounds per person.

biologically equipped to consume milk beyond infancy because the ability to produce the lactase essentially switches off after weaning, resulting in drastically reduced levels of the enzyme and the condition known as lactose intolerance. Humans with this condition cannot metabolize lactose in their stomach, so dairy products remain intact as they pass into the colon. Here, dairy begins to ferment and produces gas, resulting in symptoms such as loose stools, bloating, flatulence, nausea, and stomach cramps. Gastroenterologists estimate that roughly 75 percent of the world's population is lactose intolerant.

Though researchers debate the origins of adult lactase production, some suggest that a single mutation event that occurred in the Middle East during the Secondary Products Revolution is responsible. This trait then radiated outward among both pastoral and agrarian groups. One commonly finds adult lactase in populations from northern and central Europe, the Middle East, northern India, and among several groups indigenous to Africa. Conversely, one is much less likely to find adult lactase among groups native to east Asia, Australia, and the Americas. The key evolutionary factor seems to be whether these ancestral populations used large domesticated animals. Though animal domestication occurred both in the Old World and New World (but not in Australia), one only finds large domesticated animals in the Old World. The one notable exception is the llama, which was domesticated in Peru around 4000 B.C.E. However, this camelid species was geographically restricted by the Andes Mountains and thus was not widespread. Natural selection acted on Old World populations that used large domesticated animals because possessors of the adult lactase trait were able to consume dairy in times of food scarcity. This increased their reproductive fitness and led to greater concentrations of the trait in future generations.

Modern Production

In the 21st century, cow's milk is the most widely consumed dairy product in the United States and is produced on an industrial scale. Before reaching the consumer, milk and other dairy products must first undergo processing in dairy facilities. Raw milk destined to be whole milk is first clarified to remove impurities. Instead of being clarified, skim or partly skimmed milk is separated, which eliminates butterfat and also removes impurities. Next, both whole and skim milk are heated for a short time in order to kill any harmful microorganisms in a process called pasteurization. Pasteurized milk is then stored cold to prevent spoilage. After sitting for a day, whole milk has a natural tendency to separate into a higher-fat cream layer and a skim milk layer. To prevent this separation, dairies use homogenization, which is a process that breaks up fat globules by pumping milk through narrow tubes at high pressure.

Several grades and types of milk can be produced from raw milk. Some milk products, such as cultured buttermilk, or kefir, use fermentation. Other products, like powdered milk or condensed milk, convert milk into a more concentrated form with a much longer shelf life. Dried milk can also be made into infant formula. Another common application for milk is a variety of thick creams that include sour cream, the slightly fermented crème fraîche, and clotted cream. Milk can also be used to produce a wide range of other dairy products, which include butter, ghee, cheese, curds, whey, casein, yogurt, and ice cream. Regardless of the grade, type, or product, dairy products are widely consumed by specific segments of the global population. Those who can eat and drink dairy have natural selection to thank for this ability.

Jeffrey Ratcliffe
Temple University

See Also: Baby Products; Beverages; Farms; Food Consumption; India.

Further Readings

Diamond, Jared. *Guns, Germs, and Steel: The Fates of Human Societies*. New York: W. W. Norton, 2005.

Sherratt, Andrew. "The Secondary Exploitation of Animals in the Old World." *World Archaeology*, v.15/1 (1983).

Wiley, Andrea. "Drink Milk for Fitness: The Cultural Politics of Human Biological Variation and Milk Consumption in the United States." *American Anthropologist*, v.106/3 (2004).

Dating of Garbage Deposition

Garbage leaves a fossil record, the accumulation and degradation of which can de dated to bring insights into patterns of consumption and waste. The origins and methods behind the dating of garbage have an interesting history. Cultural behaviors relative to the formation of trash deposits can be analyzed with relative and absolute dating methods.

Significance

Why date garbage? The answer lies deep in the history of scientific thought and the study of the past. In late-18th-century Western Europe, scientists were looking for ways to understand the origins of geological deposits, challenging the assumed recent origin of the planet. While their efforts ultimately led to the recognition of the Earth's antiquity, they also cleared the way for an empirical study of the past. As field research intensified in the 19th century, the discovery of ancient stone tools alongside the remains of extinct animals showed the utility of stratified deposits to date ancient artifacts. More importantly, it became apparent that the bulk of past cultural remains is composed of refuse—objects and other remains that have been produced or modified by humans but which have lost their usefulness, becoming garbage.

Methods

The discard of trash and its accumulation are different than geological depositions. Consequently, garbologists (archaeologists who specialize in the study of rubbish and refuse deposits) have developed methods to order, sequence, and date trash deposits, from fresh rubbish to ancient archaeological contexts. Broadly, garbologists rely on two categories of dating strategies.

First, patterns of accumulation and superimposition of buried deposits can serve to create relative sequences of events. These patterns can bring insights into the conditions and the time elapsed throughout periods of trash deposition. The number of distinct strata as a proxy to the number of discard events, coupled with the thickness and compaction of the deposits, can all point toward frequency, intensity, and length of the use of middens (heaps of rubbish that represent the accumulation of daily refuse). Meanwhile, comparisons of changes through time can shed light on historical events and broader social changes. More broadly, stratified middens can be organized into broad periods by looking at the characteristics of broken artifacts and their change over time.

Second, the analysis of the physical and chemical properties of garbage deposits can provide chronometric and absolute dates. Chronometric measurements from stratified deposits give archaeologists the opportunity to place sequences of deposition in an absolute time frame. The most commonly used chronometric techniques for dating ancient garbage are radiocarbon and, more recently, accelerator mass spectometry (AMS). Radiocarbon methods measure the decay of a radioactive isotope (C-14) to evaluate the time elapsed since the death of a living organism. These techniques can be advantageously applied to many types of rubbish, including animal bones, charcoal, plant remains, shells, and paper. Precautions are needed when interpreting radiocarbon results, including archaeological associations to contamination, accuracy, and precision. Nevertheless, coupled with relative sequencing, chronometric dating is a powerful tool to date middens and their formation.

Challenges

The challenges of interpreting the chronology of garbage deposition are manifold. At the most basic level, a distinction has to be made between primary and secondary refuse. Primary discard typically happens at the place where trash is produced, whereas secondary depositions imply a certain delay in discard, such as the storage or transportation of trash away from the original activity area. The context of a hearth located in the vicinity of a habitation area where there is evidence for food processing and consumption provides a viable example. In this case, the discovery and radiocarbon dating of burned organic trash can provide a range of absolute dates for activities associated with that same household. Here, it can be assumed that the dating of a garbage deposition is somehow representative of the time when that trash was produced and discarded. In opposition, in the context of secondary trash deposits such as construction and landfills, garbage may have been accumulated and stored for an unknown amount of time before it was discarded. In this particular instance, absolute chronometric measurements provide a range of dates associated with the production of waste, rather than its discard. Moreover, secondary deposits might possibly contain remains from various places and primary activity areas, leaving the chronological interpretation of mixed trash uncertain.

Study and Cultural Interpretation

Middens are the materialization of complex actions and meanings reflected in heavy garbage deposits. The study of garbage deposition thus requires a consideration of cultural variables, including curative and recycling practices, the impact of trash on immediate human life conditions, cultural habits, symbolic meanings, and the ritual use of waste. Trash represents the sum of discrete events situated at unique space-time intersections, and—although it is subject to similar and uniform postdepositional processes—its formation is as culturally laden as any other human behavior.

Archaeologists and other trash specialists must differentiate and characterize the types of garbage deposition to be dated and their respective implications. Modern garbologists might be interested in assessing garbage deposition on a daily or weekly basis, while prehistorians might focus on seasonal patterns of hunting and fishing. The Garbage Project of the University of Arizona, for example, keeps track of individual trash pickup events with detailed time and day data. In contrast, the study of prehistoric state societies often deals with mixed deposits and multiple discard events. Here, the objective might be to evaluate amounts of trash produced and discarded over specific periods of time as proxies for past demographics and patterns of consumption through time.

Processes of decomposition and preservation are also key to date garbages, as the rates of material decay vary whether refuse is deposited in open air dumps, housepits, or landfills. Modern garbologists monitor the production of biogas such as methane to evaluate the rate of biodegradation in different landfill contexts. In contrast, research at ancient cave sites might focus on stone tool debitage and fossilized bones.

The dating of garbage deposition must be considered from a situational perspective in relation to specific preservation conditions, types, and life cycles of refuse, as well as corresponding research questions. The complexity of garbage deposition points toward the importance of developing adequate schemes to date, sequence, and understand discard events.

David Chicoine
Louisiana State University

See Also: Archaeological Techniques, Modern Day; Archaeology of Garbage; Archaeology of Modern Landfills; Garbology.

Further Readings

Maschner, H. D. G. and C. Chippindale, eds. *Handbook of Archaeological Methods*. Lanham, MD: Altamira Press, 2005.
Rathje W. and C. Murphy. *Rubbish: The Archaeology of Garbage*. Tucson: University of Arizona Press, 2001.

Definition of Waste

The term *waste*, as a noun used colloquially, means some form of devastation such as wasteland; on a smaller scale, it is synonymous with rubbish, garbage, and trash. Waste also has specific meanings, defined for a particular purpose. In economics, waste implies the unproductive use of resources. For example: labor used to move work-in-progress around a factory is defined as waste because the cost incurred in doing so adds no value to the finished product. To protect the environment, governments in many developed countries define waste to include, in effect, every conceivable substance in any conceivable form. Comprehensive definitions enable regulators such as the Environmental Protection Agency to control as tightly as possible the disposal of anything that might potentially injure the environment. Such definitions, however, may not be conducive to achieving optimum results.

Implications for Industry

Historically, substances discarded as waste have been dumped, typically in dedicated landfills, waterways, or the ocean, even though those substances might have been useful. When regulators progressively restrict land available for (legal) dumping and prohibit discharge to waterways or the atmosphere, pressure to use waste increases. Industry is particularly sensitive to such pressure as the range of options to dispose of waste narrows and the costs increase. However, industry faces a dilemma: it generates many types of waste that are not readily recycled in the way that various resources such as glass, paper, and plastic are routinely recovered from municipal and commercial waste. The dilemma is that in abiding by the immutable definitions of waste prescribed to protect the environment, the regulator acts to inhibit rather than facilitate the use of waste. A substance discarded by its generator (such as the manufacturer that created it) is defined as waste by the authorities even though it may be regarded by others as usable. This implies the transition of a substance from one state to the other, notwithstanding the fact that it remains physically unchanged. The dilemma may be resolved by a definition of waste that accommodates such a transition and incorporates a notion of where it is deemed to occur.

To this end, industrial waste is defined as any substance not required by a generator and is therefore to be discarded from the site at which the substance is located. Transiting the site boundary represents the point at which the waste may become usable. This definition includes any form of matter that is accidentally or otherwise discharged to the environment.

Want and Custody

The notion of want as expressed by "not required" is the defining criterion for what constitutes waste. A substance remains waste for as long as it remains unwanted for a purpose (other than preparation for dumping). If, after a substance has left the generator's site, it is wanted by a user, then it may be accorded a different status. Some part of the original waste may become waste again, if that part ceases to be wanted by one or more subsequent users.

The issue of custody is relevant to this definition of waste, although its legal meaning may be construed differently from one jurisdiction to another. In the context of industrial waste, the general meaning of custody implies physical possession of the waste or a responsibility for its care and its fate. The encumbered entity (a person or a corporation) is the custodian who, in the first instance, would be the generator of the waste. In some jurisdictions, custody—responsibility for its fate—resides with the generator until the waste in question is "unrecognizably transformed," even though it may have been in the possession of others before that point.

Robin Branson
University of Sydney

See Also: Culture, Values, and Garbage; Garbage in Modern Thought; Politics of Waste.

Further Readings

LaBreque, Mort. "Garbage: Refuse of Resource?" *Popular Science*, v.210/6 (1977).

Scanlan, John. *On Garbage*. London: Reaktion Books, 2005.

Strasser, Susan. *Waste and Want: A Social History of Trash*. New York: Henry Holt, 2000.

Delaware

The second-smallest U.S. state by area, Delaware occupies the northeast part of the Delmarva Peninsula. While the population is not large (the 2010 census estimated 897,934 residents, 45th in the United States), Delaware has the 6th-highest population density, with over 60 percent of the population residing in New Castle County, the northernmost of the state's three counties. Although statewide development has recently increased, New Castle County is more industrialized, while the two southern counties—Kent and Sussex—have historically been more agricultural. The state has the second-highest number of civilian scientists and engineers by percentage of workforce, mostly because of the prominence of the Du Pont family (owners of DuPont, the world's second-largest chemical company) in Delaware's economic and industrial history. Delaware's economy is generally larger than the national average; its main industrial outputs are chemical products, cars, processed food, and paper, rubber, and plastic products. Agriculture is based on poultry, nursery stock, soybeans, dairy produce, and corn.

Before European colonization, Delaware was home to the Lenni Lenape (the Delaware), Susquehanna, and other Native American tribes. The Dutch were the first to establish trading posts in the early 17th century; however, in 1664, England asserted its claim on the area and expelled the Dutch. Settlement at this point was predominantly rural and situated along waterways, which were the main transportation routes. In the mid-18th century, more towns appeared and large numbers of immigrants arrived, bringing industry and commerce with them. Over-farming saw many farmers move west, and after the Revolution, industry manufacturing grew steadily.

Statistics and Rankings

The 16th Nationwide Survey of MSW Management in the United States found that, in 2006, Delaware had an estimated 988,433-tons of municipal solid waste (MSW) generation, placing it 46th in a survey of the 50 states and the capital district. Based on the 2006 population of 852,747, an estimated 1.16 tons of MSW were generated per person per year (ranking 31st). Delaware landfilled 885,283 tons in the state's three landfills (only Connecticut and Rhode Island report having fewer landfills, at two each). Delaware exported 12,617 tons of MSW (third-lowest export in the country), and 650 tons were imported (the lowest reported import in the country). In 2006, Delaware had 5 million cubic yards of landfill remaining and was increasing its capacity; it was ranked joint 32nd out of 44 respondent states for the number of landfills. Only whole tires were reported as being banned from Delaware landfills, with a ban on yard trimmings introduced at Cherry Island Landfill in 2008. Landfill tipping fees across Delaware were an average $58.90 per ton, where the cheapest and most expensive average landfill fees in the United States were $15 and $96, respectively. Delaware has no waste-to-energy (WTE) facility, but 103,150 tons of MSW were recycled, placing Delaware 44th in the ranking of recycled MSW tonnage.

Archaeology

As a coastal state, the Delaware archaeological record has numerous shell midden sites, mounds of discarded shell created by the processing of gathered shellfish and incidental food debris from the foods eaten alongside them. These sites are important as they are markers of small Woodland I period base camps, which are otherwise difficult to identify. The Wolfe Neck Site provided the finds that enabled the type description of Wolfe Neck Complex ceramics. Similarly, settlement sites of the Delmarva Adena Complex are little known. The Wilgus Site is a micro-band base camp containing material from the Delmarva Adena Complex and the later Carey Complex. While the living area of the site was on a low knoll that was raised and exposed to plowing and weathering, the nearby middens (both earth and shell middens) had survived relatively intact. As the Wilgus Site rubbish had been thrown down a slope to take it away from the living area, it had subsequently been buried by soil being washed down the slope by rainwater (slope wash or hill wash). The middens were large, around eight meters in diameter, and such a large mass of organics resists biodegradation, particularly shell middens. The calcium carbonate leaching from the clam and oyster shells created an alkaline burial environment, which buffered the acidity of the soil and thus preserved other organics in the deposit bone—antlers, fish scales, and charred seeds. Growth-ring patterns in the oyster shell hinges indicates they were gathered in late winter–early spring.

The Wilgus middens identified food sources such as freshwater fish (indicating the Indian River had not yet become brackish), deer, snake, turtle, and birds. The deer bones were mainly from the skull and legs. Trunk meat was probably consumed at the kill site, or fed to dogs, as it is difficult to carry. The presence of skulls was perplexing as they are of low meat value, although the tongue and brain are desirable foods among hunter-gatherers. Brains can also be used in tanning and processing hinds. The pattern of tooth eruption in the deer skulls indicated winter kills. The wide range of turtle species present was typical of a hibernating turtle population, a late-fall and winter food source. Amaranthus and Chenopodium seeds were found in both earth and shell middens. In the earth middens, they were associated with large pots thought to be related to processing seed plant foods, available late-summer and fall in these species. The seeds in the shell middens were associated with the winter faunal material and were probably stored for winter consumption. When all of the seasonality evidence is considered, it seems Wilgus was occupied from late summer into the following spring, with shellfish comprising a large part of the winter diet.

In Delawarean archaeology, there has been reconsideration of what were thought to be Native American rubbish pit features in which human remains had been deposited. These had been interpreted as refuse pits based on their shape, the presence of potsherds, flakes, fire-cracked rock, and absence of perceived grave goods. However, the use of ethnohistory and insights of Nanticoke tribe representatives have shown that Native American burial in a trash pit is extremely unlikely and that deposited items that were usually seen as trash can have ritual and symbolic meaning. Careful study of many of the recovered artifacts had attributes that differed from those of superficially similar items from nongrave settings. These findings meshed with data from ethnohistoric, ethnographic, and linguistic studies to create a consistent cultural context for the ritual and symbolic significance of the artifacts.

Reefs

Like the other mid-Atlantic states, Delaware has an artificial reef program. Since 1995, it has created thriving ecosystems on the otherwise featureless sandy ocean floor. There are 14 reefs in Delaware and along the coast, five of which are in federal waters. The eight reefs in Delaware Bay are primarily made from donated concrete culvert pipe and other concrete products augmented with scuttled tugboats and other material. Ballasted tire units have been used at three ocean sites. In 2000 alone, 24,500 tons of concrete products, 8,000 tons of ballasted tire units, and 86 decommissioned military vehicles were recycled into Delaware's artificial reef program.

The Redbird Reef, 16 nautical miles east of the Indian River Inlet, is the state's most popular reef site, with over 10,000 angler visits every year. Over 1.3 square nautical miles of reef have been created from 890 subway cars, 11 large vessels, 86 military vehicles, and 6,000 tons of ballasted tire units. It was begun in 1996 and is named after the 619 decommissioned New York Metropolitan Transportation Authority subway cars donated to create it. Delaware received the subway cars after several states had rejected them because of environmental-safety concerns. Environmental groups had campaigned against the use of subway cars because of the small amount of asbestos used in their glue and insulation. The groups also considered subway cars unsuitable for use in artificial reefs based on the survey of a single car that had been hit and destroyed by a dragger. These concerns proved unfounded and several states have competed for more of the subway cars. The reef became so popular that overcrowding and traffic became problematic, as did crime when the theft and sabotage of fishing traps and pots escalated. The problems reached such a level that the state applied to federal marine officials to make the area off-limits to large commercial fishermen who were clashing with anglers over space and tangled lines.

Jon Welsh
AAG Archaeology

See Also: Definition of Waste; Ocean Disposal; South Carolina.

Further Readings

Arsova, Ljupka, Rob van Haaren, Nora Goldstein, Scott M. Kaufman, and Nickolas J. Themelis. "The State of Garbage in America: 16th Nationwide Survey of MSW Management in the U.S." *BioCycle*, v.49/12 (2008).

Clark, Charles C. and Jay F. Custer. "Rethinking Delaware Archaeology: A Beginning." *North American Archaeologist*, v.24/1 (2003).

Custer, Jay F. *Prehistoric Cultures of the Delmarva Peninsula: An Archaeological Study*. Cranbury, NJ: Associated University Presses, 1989.

Delhi, India

Delhi is a megacity (population 10 million or more) and the second most populous city in India. Since the 1970s, Delhi has undergone rapid and unplanned growth in terms of urbanization, industrialization, and population increase. This tremendous growth in population and focus to improve the standard

In Delhi, India, the changing lifestyle, increased standard of living, and differences in the types of goods consumed have affected the amount and type of waste generated by citizens, including increases in plastic waste and multimaterial packaging. The 500 grams of waste per person daily is five times the national average. Landfilling is the formal disposal option. There is also less separation of waste, which has supplied the informal recycling industry conducted by ragpickers, scavengers, junk dealers, middlemen, and big merchants.

of living has resulted in land development, growing slums, industry, and automobiles, leading to degradation of the environment. Delhi faces the problem of supplying potable water, basic sanitation, and shelter on one hand, while battling with environmental problems from development activities, like pollution from solid waste, vehicular traffic, industrial emissions, deforestation, and loss of biodiversity, on the other hand. The changing lifestyle (which results in differences in types of goods consumed) and changing standards of living have also affected the amount and type of waste generated in terms of increases in plastic waste, paper packaging, multimaterial packaging items, consumer products, and other related types of waste associated with affluence. The increase in standards of living has also decreased the separation of waste as it happens in rural areas (mainly for economic reasons).

Quality and Quantity of Waste

With a total area of slightly more than 900 square miles (sq mi) and a population of more than 14 million, Delhi generates 7,000 tons of solid waste every day. The generation rate is about 500 grams per person per day, which is almost five times the national average. Moreover, garbage generation is likely to increase to 18,000 tons by 2021.

For solid waste management purposes, Delhi has been divided into three areas. Municipal Corporation of Delhi (MCD) has jurisdiction over 868 sq mi, or 94.22 percent of the total area. New Delhi Municipal Committee (NDMC) has jurisdiction over only approximately 26 sq mi, and Delhi Cantonment looks after another approximately 26 sq mi. MCD is the largest municipal body among the three bodies responsible for waste management in Delhi and has to provide services both to rural areas and a very large urban area. But this does not cover the entire city. There are virtually no arrangements for waste management in squatter settlements, slums, and illegal colonies, which comprise around 50 percent of the urban population in Delhi. As a result, waste is littered in open spaces and drains.

Waste generated from a household depends on the income level and the lifestyle of the people. Delhi households can be divided into a high-income group (HIG), with an income greater than Rs. 8,000 per month; a middle-income group (MIG), with an income between Rs. 8,000 and Rs. 5,000; a low-income group (LIG), with an income less than Rs. 5,000; and slum dwellers.

The quantity of MSW generated per capita varies for the LIG at 0.35 kilogram (kg) per day to 0.48 kg per day for the HIG. Affluent communities generate more recyclable waste than lower-strata dwellings.

Urban solid waste is normally a complex mixture of household, construction, commercial, toxic industrial elements, and hospital wastes. A physical analysis reveals that it consists of about 32 percent compostable matter. The recyclable components include paper (6.6 percent), plastics (1.5 percent), and metals (2.5 percent). Because of economic growth, the composition of waste is expected to change. Plastic is expected to increase to 6 percent, metal to 4 percent, and glass to 3 percent. Paper content is expected to increase to 15 percent, while organic waste will go up from 40 percent to 60 percent. It is projected that other materials (such as ash, sand, and grit) will decrease from 47 percent to 12 percent.

Waste Management

Solid waste management in MCD comes under the jurisdiction of the deputy municipal commissioner (Conservancy and Sanitation Engineering). This person is assisted by two directors: one is responsible for the Trans Yamuna area and the other for the rest of the MCD. These directors are helped by three joint directors, who are supported by subordinate staff. Each zone has a sanitary superintendent (SS) who, with the help of subordinate staff, looks after the collection and transport of solid waste: sweeping of roads, lanes, and by-lanes; removal of garbage from dustbins and community bins; and reports to the appropriate joint director. Administratively, the SS works under the zonal assistant commissioners and additional deputy commissioners.

The director is entrusted with solid waste management, along with sewerage and drainage activities. The engineering staff responsible for allowing the director to devote more time to solving sewage problems, such as choked drains, and in turn solid waste management, does not get the attention it deserves. The operation of the solid waste management is thus left to conservancy staff.

Waste transportation vehicles are maintained in zone workshops. The person in charge of the workshop is responsible for maintaining the vehicles, while the responsibility of loading, transport, and unloading of the vehicles lies with the assistant sanitary inspectors (ASIs), who are often not technically qualified for these operations.

Collection Systems

The system expects each household or commercial premise to deposit its waste at a local waste accumulation area (also called a community bin). These functions are normally carried out by private individuals called *zamadars* who are employed by householders or businessmen for a small monthly fee. MCD employees (called *safaikaramcharies*) are responsible for street sweeping, and the wastes they collect are also placed in the waste accumulation area.

Wastes are collected from these locations by truck, usually on a daily basis, and taken to disposal areas. Safaikaramcharies are involved in loading, transport, and dumping operations. Mechanized curbside collection systems such as those that are standard in Western cities are not used in Delhi. Trials with such systems have failed.

Localized systems, mainly organized by nongovernmental organizations (NGOs), attempt to collect wastes that can be recycled or composted rather than entering the waste deposition, collection, and dumping system. Key problems associated with waste collection systems have been identified as the following:

- Poor civic responsibility (some people dump waste indiscriminately)
- Zamadars are poorly paid, morale is low, and the work is often badly done
- Safaikaramcharies are paid better, but they are poorly organized and equipped
- Health and safety issues for workers are largely ignored

- The number and distribution of waste accumulation areas appears to follow no rational pattern
- The design of built structures at waste accumulation areas is poor, the size and design is often incompatible with the loading and transportation equipment, and inefficient manual loading is usually required
- Maintenance of built structures at waste accumulation points is poor
- The use of steel trolleys (similar to the mechanized bin system used widely for commercial waste in the United Kingdom) has proved unsatisfactory because of equipment costs and lack of maintenance of the complicated machinery
- The timing of waste collection (early in the day) is not ideal since most of the waste generation takes place later in the day

Waste Transport

There is a large fleet of assorted vehicles for transporting wastes from collection points to disposal sites. However, at any given time, a large proportion is out of action because of the age the fleet, overcomplicated types of vehicles, and poor standards of care and maintenance:

- The available fleet is not used efficiently
- The shift system for the drivers and safaikaramcharies is badly organized

Specific problems also include the following:

- Waste handling is often manual, meaning that loading and unloading are time consuming, the round trip time increases, and the labor and vehicle productivity is reduced
- Waste is handled several times because the existing transport and loading vehicles do not match with communal storage bin design
- There is a potential health hazard for workers as all types of wastes, including infectious hospital wastes, are disposed of in the same bins
- Workers are not given protective devices and those who are given any seldom use them
- There is no separate arrangement for transport of infectious wastes from hospitals or nursing homes
- The service level is very low as several bins are not cleared on a day-to-day basis and waste is not transported on Sundays or holidays
- Vehicle capacity is not fully utilized
- There are no standby vehicles for deployment during periodic maintenance or breakdown of vehicles in service

Waste Disposal by Landfilling

Landfilling is the main system for waste disposal in Delhi, and there were three main landfill sites in use as of 2010. Although the composition of waste in Delhi is conducive to composting, this method of disposal is not widely used. There are a few local initiatives organized by NGOs to compost waste in certain localities.

Delhi does have an incinerator, the Timapur incinerator, which was built in 1989. It has not been a success because of the low calorific value and high moisture content of the waste used as the feedstock. It requires the use of additional fuel oil to get the waste to burn, making this a very expensive disposal option. Pelletization to produce refuse derived fuel (RDF) is a possibility. However, this is not considered as a viable option because of the low calorific value and high noncombustible (ash) content of the waste feedstock, the high capital and operating costs of a pelletization plant, and potential atmospheric pollution impacts on combustion.

The informal recycling industry consists of different kinds of people: ragpickers, scavengers, junk dealers (*kabariwalla*), middlemen, and big merchants. There is a set hierarchy among these operators, with scavengers forming the lowest level and big merchants forming the top level of the hierarchy. Ragpickers pick up things from litter, segregate these, and sell them to either junk dealers or middlemen, while junk dealers buy things from households and sell them to the middleman, who sells the same to primary industry. In this system, ragpickers get the least pay, and middlemen make the maximum profit. Junk dealers go around the city on bicycles or with wheelbarrows, buying waste materials such as newspaper, plastic, glass, metallic containers, and utensils. Another type of dealer collects old textiles, which are usually exchanged for new metallic utensils used for domestic purposes.

The segregation system at the source is very efficient because of the economic incentive. Discarded paper (not crumpled), plastic bags, and other packaging material are collected by the ragpickers from the streets, local garbage collection centers, and landfill sites. According to Sristiti estimates, there were over 75,000 ragpickers in Delhi as of 2010, with their number increasing. It is a profitable business in the early 21st century, where it is often difficult to find a job. Ragpickers are mostly women and children, while junk dealers are generally males. The main reason for the existence of the recycling sector is the economic incentive. The households get extra income (though it is not substantial), but it is a major source of income for the people involved in the recycling business.

Privatization

There have been a few litigations against the local government for mismanagement of waste management in Delhi. This has led to privatization of waste collection and transportation in six zones of Delhi. Some of the terms of this privatization include the following:

- Ownership and Control of Recyclable Wastes (Article 5.15 of the Contract)
- Control and Rights over the Dhalao Space

According to some reports, this has led to loss of jobs for ragpickers and others in the informal waste recycling system.

Velma Grover
York University

See Also: Developing Countries; India; Mumbai, India; Recycling; Street Scavenging and Trash Picking.

Further Readings

Fernandes, Boniface. *Making Delhi a Better Place: Promoting a Vision of Urban Renaissance*. Delhi: Kalpaz, 2006.

United Nations Human Settlements Programme. *Solid Waste Management in the World's Cities: Water and Sanitation in the World's Cities 2010*. London: Earthscan, 2010.

Developing Countries

The term *developing countries* generally refers to nations with low standards of living, although high levels of variability in the societies and economies of developing countries exist. Many such nations, including ones in the Americas, Africa, and Asia, were colonies of the wealthiest nations of the 19th and 20th centuries. Evidence of this history is found in trade and development relationships in the 21st century, as extraction of mineral resources and agriculture endure even as new industrial activities and forms of waste management develop.

A feature of many developing countries is the megacity (a city with a population of 10 million people or more). As many nations won independence from colonial powers between 1945 and 1980, economic development spurred creation of megacities. Whereas there was only one megacity in 1945 (New York City), there were at least 14 such cities in 1995, and 10 of these cities were located in Latin America, Africa, and Asia. It is expected that 22 of the 26 megacities by 2015—also known as the city-region—will be located in developing countries. But this development comes with some consequences, such as environmental degradation, pollution, and growing social inequality.

Developing countries are usually defined by income group, which means their consumption patterns are comparable. Waste management is closely related to resource consumption patterns, urban lifestyle, jobs and income levels, and other socioeconomic and cultural factors. There are marked differences in urban and rural areas, and this is also reflected in the type of waste they generate. There are also differences among countries, but some common elements exist.

There is a significant difference between the waste from industrialized (developed) and developing countries. There are notable variations of waste and managerial problems from country to country and even from city to city. On average, however, low-income countries generate waste of 0.4–0.6

kilograms (kg) per person per day while the waste generation rate in developed countries is 0.7–1.8 kg per person per day. This implies that, generally, the rich produce larger quantities of waste. Waste in developing countries differs from that of developed countries:

- Developing countries have a much larger percentage of organic waste (such as vegetables and fruits) as compared to developed countries. Since the temperatures are usually warmer in developing countries and so is the rainfall (and humidity), there is a higher rate of decomposition and calls for more regular collection of waste.
- Amounts of dust or dirt in street sweepings are greater in developing countries compared to developed nations
- In terms of moisture content, developed countries have two or three times greater moisture in waste
- Waste density in developing countries is also two or three times greater than in developed nations
- Wastes generated in developed countries have much higher calorific value

Waste Management

Since most cities in developing countries are growing at a fast rate, most of the municipal governments are stretched beyond their capacities to deal with the change in waste quality and quantity. The vast majority of countries are struggling with ensuring sufficient collection services and implementing some sort of control as disposal sites are simultaneously facing increasing waste amounts due to the trend of urbanization. It is interesting to note that the problems faced by the cities of low economies in the 21st century and those of 19th-century North America and Europe are comparable. In both cases, the pace of population growth outstripped the capacity to manage urban services.

Just like common characteristics, there are some common problems with solid waste management across developing nations. The issues related to solid waste management can be grouped into a few factors: inability of the local municipal authorities to collect waste; lack of data, data collection, or monitoring; lack of financial resources; lack of capacity and required skills; improper disposal facilities; either lack of legislation or outdated legislation; improper organization structure of the authority responsible for waste management; and improper attitudes of people. The behavior of people is important to reduce or manage waste properly. In addition to income, there are other factors that influence waste management, including the sociocultural factor.

In Asia, for example, organic waste is seen as a resource and has been used for animal feed, fertilization of land—and even, in some, cases cooking—while in Africa waste is seen as dirty and people are often averse to coming in close contact with it. Another factor influencing waste management is religion. Literature and research has shown that in some religions, such as Hinduism, waste is seen as a resource and traditionally handled by certain castes. In some religions, such as Islam, people are averse to handling waste. Based on these factors, some of the common threads for waste management among developing countries are the following:

1. Lack of collection of waste (a large percentage of waste goes uncollected). It has been reported that in some cities, uncollected waste is as high as 70 percent.
2. There is a lack of proper collection bins either in household collection (in most cases, plastic grocery bags are just left outside the houses, attracting stray animals, rodents, flies, and mosquitoes) or even for collection at neighborhood levels (even when these bins or collection points are present, they are not properly maintained and cleaned). In India, for example, in most major cities, municipal solid waste accumulates in communal bins until waste collectors pick it up. Since there is no reliable pickup time from these communities, there are piles of garbage around these bins, attracting rodents and animals. Most of the uncollected waste then gets dumped on streets (usually side streets) or in neighboring areas, leading to environmental and health issues.
3. Transfer stations are available in only a few metropolitan cities where hauling distance

to disposal sites is too long. Waste is usually transported from collection points to transfer stations or to disposal sites through different ways (depending on how big the city is). Waste is carried on head by staff; on handcarts; on bicycles; in some cases animals such as donkeys, horses, or ox-carts are used to transport waste; and in larger urban areas waste is transported via trucks. Most waste handling is done manually, exposing waste workers to serious health hazards, especially in countries where wastes are usually not segregated, forcing workers to handle medical and hazardous wastes.

4. Transport trucks used for collection and transfer trucks are (in most cases) too mechanized for developing countries. In some cases, the collection and transfer trucks do not function to maximum capacity. According to United Nations Environment Programme (UNEP) estimates, in some cities in west Africa, up to 70 percent of collection and transfer vehicles may be out of service at any given time.
5. At times, the machinery used for waste management is not fit for developing countries either. However, changes in consumption patterns affect the quality of waste, requiring different technologies. For example, when the development level is low, there is not much packaging or post-consumer waste, and the waste going to the landfill does not need a compactor. However, as consumption and consumerism increase with increases in income, packaging and other waste that needs compaction increases, justifying the need for a compactor.
6. There are no controlled disposal systems (landfills are not engineered with lining and there is no collection system for leachate), and most of the disposal sites are uncontrolled, open dumps. Leachate can—and does—contaminate groundwater and run off into water systems if disposal sites are based near water systems. Traditionally, waste disposal consisted of hauling waste beyond the city boundaries and dumping it there. Open dumps often become breeding grounds for mosquitoes, thus contributing to the spread of diseases. In some countries, waste is burned or dumped in water systems, leading to air or water pollution.
7. There are inadequate guidelines and lack of legislation for proper siting, design, and operation of landfill sites. This causes problems when landfills are not in appropriate places. Another problem is the proper enforcement of legislation.
8. Most of the dumpsites have scavengers who recycle waste, but this can be hazardous since the waste is mixed with hospital waste, human waste, and hazardous industrial waste, which can lead to serious injuries and illnesses. For example, the Smoky Mountain dump, in Manila, the Philippines, had as many as 10,000 families living in shacks on or adjacent to the dump site. In addition to the health problems, these concentrations of people further complicate transport and unloading procedures and present numerous safety and logistical concerns. There have been cases of landslides and fires leading to many casualties, since it cannot be controlled easily. UNEP estimates also state that about 100,000 people currently scavenge wastes at dumpsites in the Latin American region alone. Even in Phnom Penh, Cambodia, garbage collection trucks are rarely seen on the streets, but female waste pickers are a common sight. Because of the availability of cheap labor, there is a high potential of mechanical recycling in developing countries compared to developed countries.
9. There exist operational inefficiencies of services.
10. There is inadequate management and separation of hazardous and healthcare waste. All waste collected should be separated so that it does not contain any medical, industrial, or construction waste before transferring it to community bins (the primary waste collection point). All biodegradable waste collected from slaughterhouses and markets should be collected and handled separately and in an appropriate manner. Absolutely no waste should be burned under normal circumstances.

Normally, no stray animals should be permitted to move freely around waste storage facilities by walling and gating the storage area.

11. While municipal authorities can choose their own method of collection, storage, and final disposal, it is recommended that collection of waste be carried out on a regular basis and must include the collection of waste from squatter colonies and slums. Normally, squatter colonies and slum areas are excluded from municipal services, but they are located within the city boundaries and leaving them unclean creates health hazards.
12. One of the challenges in solid waste management is when authorities under the influence of developed countries want to use technology, even when it is not applicable or suitable for local conditions. For example, calorific value of solid waste in developing countries is low, and when all the recyclables are sorted, there is hardly anything left to recover energy from in the waste cycle. Still, some municipalities want to own an incinerator or an energy recovery plant.
13. Composting is another option that is used mainly in rural areas in developing countries but can be used in urban areas as well, based on the high amount of organic matter present in the waste in these countries. On average, the organic content is as high as 50 percent. In cases such as Bandung (Indonesia) and Colombo (Sri Lanka), the residential waste stream has been found to have about 78 percent and 81 percent organic waste, respectively.
14. Lack of capacity and financial reasons are generally cited as major causes for lack of waste management in these countries.
15. Lack of data collection affects waste management. Information on local conditions and waste composition helps in selecting the frequency of collection and disposal technology. However, most developing countries lack this data and depend on consultants to come up with data and recommendations for disposal technologies, which can then be biased.
16. In order to prevent a lot of waste going to landfill sites, the highest priority is to achieve reduction in the amount of waste generated irrespective of economic growth and to raise public and political interest in sustainable waste management. This may be accomplished by educating and changing people's attitudes toward waste and increased participation in minimization of the waste generated by them. This minimization can occur through recycling as well as reuse and composting, even though informal recycling communities are very active in some areas—and hence almost all material that can be recycled for profit is siphoned off. Future progress is also required in improving the markets for recyclables. This is further illustrated through an example from Delhi, India, in which this activity is carried out through a chain of self-employed individuals or groups of dealers for whom this work is a source of income. The informal recycling industry consists of ragpickers, scavengers, junk dealers, middlemen, and big merchants.

Legislative Examples

The United Nations Conference on the Human Environment in 1972 prompted India to amend its constitution in 1975 to include provisions for improvements to the environment by amending its constitution. This was done so that sustainable development occurs, which is a balance and harmony between the economic, social, and environmental needs of the country.

Besides the center, every state has its own Department of Environment and Pollution Control Board for planning, promoting, and coordinating the entire country's environmental programs. The Ministry of Environment and Forests (MOEF) has identified the Central Pollution Control Board (CPCB) as the chief monitoring agency. In 2004, the MOEF enacted a National Environment Policy (NEP) for all agencies and civic bodies responsible for environmental management.

This forced the government's commitment to environmental protection and to review all regulatory reforms and legislations of the central, state, and local governments and to infuse a sense of commonality into the various environmental sectors, such as pollution control, waste management, and water resource management. The key objectives of the environmental agenda of India include the integration of environment concerns into development and economic policies, and the application of the principles of good governance—transparency, accountability, efficiency, and participation—to their system of environmental management.

No policy is going to meet all principles and dimensions of Integrated Solid Waste Management (ISWM), but the Indian policy meets the majority of the requirements. Besides, the policy makers seem cognizant of the importance of addressing the political situation in that they clearly lay out roles and responsibilities for different levels of authority. Some aspects of reality in India are left out, for example, there is no mention of huge populations of waste pickers prominent in every major city throughout the country.

Various different pieces of legislation regarding solid wasted management (SWM) and environmental protection were passed between 1939 and 1999 in the Philippines. For example, the 1939 Anti-dumping Law prohibited dumping of refuse, waste matter, or other substances into bodies of water. The 1999 Republic Act—the Clean Air Act was the first piece of legislation to outlaw incineration, including that of medical waste. The 2000 Ecological Solid Waste Management Act (ESWMA) was a comprehensive piece of legislation that was enacted after the tragedy at the Payatas landfill in the Philippines. The act was written in order to replace the piecemeal legislation that previously covered SWM in the country. Besides India, the Philippines is another country that decided to enact strong legislation to deal with SWM in an integrated fashion. The Philippine government has repeatedly stressed its commitment to good governance and a healthy environment through its emphasis on the decentralization of power for environmental management to local authorities. This ESWMA is almost perfect on paper—decision makers have put a lot of thought into its drafting. The main issue is that it is very hard to tell what aspects of the act have been implemented and what aspects have remained on paper.

Private Sector

Even before the 1980s, there were some private-sector companies involved in waste collection. However, the national governments and development agencies actively began to promote the private sector as a provider of municipal services during the 1980s. However, the private sector had hardly any experience in the provision of solid waste services. As such, they tended to copy government services and made the same mistakes, especially if burdened with using the same old equipment that was being used by the public sector. Local governments in developing countries have access to interest-free grants or transfers, which may not be the case for the private sector, resulting in higher costs, even though productivity might be better.

Most of the activities undertaken by the public sector could be done by the private sector as well. However, only those activities that are performed by government employees most inefficiently should be privatized—activities like solid waste collection. Maintenance of vehicles could be another area of interest for privatization because of labor restrictions on hours of work and associated high costs relating to overtime work in the public sector. The aim of privatization should be to reduce governmental control, ownership, and activity within a service like solid waste collection and disposal.

A significant number of African cities have been implementing pre-collection systems since the 1990s. In private subscription, residents are concerned about the removal of the waste but do not pay much attention to its disposal, meaning a strong regulatory and enforcement framework is required. In that case, a clause could be added that the contract could be terminated if the waste is disposed of illegally.

It is relatively easy to improve solid waste collection and disposal systems by getting the private sector involved. Private-sector participation is not considered successful if it just means improvement in service. On the contrary, such a service should be financially sustainable and cost effective. Before privatizing, a developing country should ensure that the private sector is well established. Otherwise, it is strongly recommended that the public sector should retain services to at least one-third of the service area. After continuous five years of successful operation by the private sector, the control area could be reduced to 20 percent for another five years.

Another area of concern of developing countries when privatizing is to minimize the termination of employees. One of the ways of handling such a situation is to freeze further hiring and not to replace the retiring employees. Thus, natural attrition creates significant flexibility in transition toward privatization. Standards need to be created for the whole waste industry that include new and emerging technologies and also the management of specific waste types, such as agricultural waste.

Conclusion

There should be a move toward integrated solid waste management and improving the legislations (and their enforcements) in regard to waste collection, transportation, and disposal. There is also a need for education, public information, and public participation.

If human society has to endure but for thousands and thousands of years, people need to learn a way of life that can be sustained. Human society must learn to control population and develop more efficient green technologies that produce as little harmful waste as possible (or no waste at all as the new theme of zero waste or cradle-to-cradle concepts are evolving). People must learn to rely on resources that are renewable.

Velma Grover
York University

See Also: Delhi, India; Recycling; Slums; Street Scavenging and Trash Picking; Sustainable Development; Zero Waste.

Further Readings

McNeill, J. R. *Something New Under the Sun: An Environmental History of the Twentieth-Century World*. New York: W. W. Norton, 2000.

Thomas-Hope, Elizabeth, ed. *Solid Waste Management: Critical Issues for Developing Countries*. Kingston, Jamaica: Canoe Press, 1998.

Wilson, D. C., C. Velis and C. Cheeseman. "Role of Informal Sector Recycling in Waste Management in Developing Countries." *Habitat International* (December 2006).

Dhaka, Bangladesh

The south Asian country of Bangladesh and its capital Dhaka (a megacity of over 12 million people), show similar consumption characteristics and face comparable challenges to other developing countries in the region.

Consumption Patterns

The population of Bangladesh spends more than half of its income on food and beverages (54 percent), while only 12 percent is spent on housing and rent. Clothes and footwear, as well as fuel and electricity, both take about 6 percent of income. There has been a trend showing that the amount spent on housing is increasing sharply, both in urban and rural areas, while less money is spent on consumption goods.

With higher income, people also change their consumption patterns. With more money, the population consumes more expensive food, such as fish, chicken, and mutton, while their demand for wheat, potatoes, and eggs decreases. More consumption also means more waste. The use of packaging made out of plastic, such as plastic bags and water bottles, has become very popular among consumers, drastically increasing the amount of solid waste.

Waste Management

The capital of Dhaka is not capable of coping with all the garbage caused by its inhabitants. Solid waste management runs smoothly in rich and middle-class neighborhoods, as they have enough political influence to guarantee regular governmental service. Poor slum areas can hardly rely on organized waste management. The government fails to provide most of its citizens with this public good. The agency in charge, Dhaka City Corporations (DCC), covers about 224 square miles, but not all parts are serviced equally. Since the city is not able to provide all its inhabitants with reliable service, some neighborhoods have established their own form of solid waste management, which is based on mutual trust and reciprocity among neighbors. Usually, the garbage collection system in Dhaka involves regular pickups by municipal workers and the deposit of waste in large, centrally located dumpsters. Individual households

A fruit seller in Dhaka, Bangladesh. The population of Bangladesh spends more than half of its income on food and beverages. They are also spending their higher incomes on more expensive food as well as food with plastic packaging, increasing solid waste.

dump their garbage in small dumpsters in the side streets of their neighborhoods. The municipal workers are responsible for collecting the garbage from these alleys within the neighborhoods and bringing it to the main dumpsters. This service has proven very unreliable, and DCC employees often do not come to collect the trash for weeks. Some neighborhoods have managed to find an alternative to the public system. They hired private contractors to regularly collect the garbage from the neighborhood, and they are paid by voluntary contributions from community members.

Since waste management must be organized collectively, neighborhoods established their own trash disposal committees. While this voluntary solid waste management works very well in some neighborhoods, other areas are not capable of organizing themselves. The ability of self-organization depends on the homogeneity of the community (same ethnic or religious background) and its social capital. Reciprocity is the key to functioning voluntary solid waste management, as it excludes free riders.

The government uses these success stories in order to encourage other neighborhoods to copy this self-help scheme. The authorities further discuss public–private partnerships as another possible solution for coping with waste.

Water and Sewage

Providing safe drinking water to the population has been another big challenge. The rural population gets its water from tube wells. About 11 million of these wells exist in the country, but almost half of them are contaminated. Bangladesh faces a mass arsenic contamination of its water, and the water cannot be used for drinking or cooking. In the city, many slum inhabitants do not have running water at all. Water supply is also frequently cut.

As in many large cities in developing countries, sewage is a major problem. A functioning sewage system needs governmental planning, so the inhabitants of poor areas cannot cope with the problem without the authorities' help. Most of the inhabitants of slums do not have private toilets in their households and have to use public ones, which they share with their neighbors. Open defecation poses large hygienic problems and is a severe health threat. Women in particular have problems using public toilets. They dislike walking across the neighborhood alone in order to use dirty community toilets. Sexual harassment is not unheard of, especially at night. Therefore, many women try to avoid using these toilets as long as possible by not drinking enough. Infections, constipation, and nephritic stones are frequent effects. In some cases, parents do not sent their daughters to schools because they do not have separate toilets for girls.

In many Muslim countries, such as Bangladesh, dealing with feces is taboo, so the problem has long been ignored. In 2010, a promising project was tested in the slum of Maimansingh. The inhabitants were encouraged to use special biodegradable plastic bags as toilets. They can use these bags at their homes and clean themselves with water in their backyards. The inside of these bags is covered with urea, which kills harmful germs. After use, these bags are collected separately from the ordinary household trash and disposed of on an empty field. It is also possible to bury the bags in the backyard garden. After six months, the feces become fertilizer. The Bangladesh Agricultural University is experimenting with this type of fertilizer and is successfully using it for growing lemon trees. In densely populated regions such as Bangladesh, agricultural areas are often close to urban settlements and slums, so both can profit from the symbiotic effects. As of 2010, the project showed positive results and the inhabitants of Maimansingh accepted the biodegradable plastic bag as an alternative toilet. Regardless, this can only be a temporary solution. Even when produced on a large scale, each bag costs about $0.03, which is unaffordable for most inhabitants. The Sustainable Sanitation Alliance is working on an improved, cheaper version of the plastic bags.

Ship Breaking

While recycling consumption goods can help to ameliorate waste problems in the city, large-scale ship breaking has become an important national industry. At the end of their sailing lives, vessels are bought by steel companies and then dismantled. In Bangladesh, the scrapping of ships is the main source of steel. Since the government does not have sufficient financial means to buy steel on the world market, most of the eight million tons of building materials used in the country each year comes from recycled ships. Many industries profit from ship breaking, as almost all parts of the vessel can be reused or resold. Steel mills remanufacture the scrapping of ships. A large number of businesses specialize in reselling the technical equipment, paint, lubricants, and furniture removed from the ships.

The government supports the profitable ship breaking industry, as it generates a large amount of tax revenues through import taxes and yard taxes. The industry mainly concentrates in the poor and underdeveloped coastal zone of Chittagong. About 20,000 people in northwestern Bangladesh are directly employed in the industry. Almost all are unskilled laborers who must work for low wages under precarious conditions without any labor standards. They do not have written contracts and therefore cannot enforce their rights. Still, economic benefits for the region are great, as those people would otherwise be unemployed.

Despite the economic advantages, ship breaking has many negative side effects. Hazardous waste contaminates the environment, while poisonous fumes and chemicals threaten workers' health. The vessels are cut up and dismantled by hand on the open beach. Usually, they are only cleaned superficially and still contain large amounts of dangerous material. Oil tanks always carry many liters of residual oil in addition to the oil used for engines or as lubricant. It mixes with the seawater and pollutes most of the coastal area from Fauzdarhat to Kumira and Chittagong. Older ships contain asbestos, which was formerly used as a heat insulator. The workers have no protection against it. On the ship breaking beaches, flocks of asbestos fiber fly around in the air, and workers constantly inhale them. Exposure to asbestos causes a wide range of pulmonary diseases and is lethal in the long run. During scrapping, high concentrations of highly toxic persistent organic pollutants (POPs) are released into the environment. They remain intact over long periods of time and accumulate in the fatty tissue of organisms. POPs have become geographically widely distributed, causing problems even in Bangladesh's hinterland. For humans, these toxic substances cause serious health problems by increasing the probability of cancers and by disrupting the hormonal system.

Carmen De Michele
Ludwig-Maximilians University Munich

See Also: Culture, Values, and Garbage; Developing Countries; India; Sewage.

Further Readings

Alauddin, Mohammad. *Environment and Agriculture in a Developing Economy: Problems and Prospects for Bangladesh*. Cheltenham, UK: Elgar, 2001.

Harpham, Trudy and Marcel Tanner. *Urban Health in Developing Countries: Progress and Prospects*. New York: St. Martin's Press, 1995.

Pargal, Sheoli, Mainul Huq, and Daniel Gilligan. *Social Capital in Solid Waste Management: Evidence From Dhaka, Bangladesh*. Social Capital Initiative Working Paper No. 16. Washington, DC: The World Bank, 1999.

United Nations Human Settlements Programme. *Water and Sanitation in the World's Cities. Local Action for Global Goals*. London: Earthscan, 2003.

Diet and Nutrition Surveys

Numerous diet and nutrition surveys worldwide attempt to assist preventive health programs through analyzing people's diets and nutrition. Four main types can be distinguished: national nutrition monitoring, national nutrition and health, international nutrition and health, and individual with online analysis of the results. The methodology includes all or some of the following basic steps: selecting samples sizes, interviews, food diaries, measurement and medical examination, and monitoring. There are government agencies, nonprofit organizations, and social firms (e.g., Casals & Associates) that provide the surveys.

In the United States, the basic surveys are part of the National Nutrition Monitoring and Related Research Program (NNMRRP), which was strengthened with the passage of the National Nutrition Monitoring and Related Research Act of 1990. The National Health and Nutrition Examination Survey (NHANES) is a program of studies designed to assess the health and nutritional status of adults and children in the United States. In the United Kingdom, there is the National Diet and Nutrition Survey (NDNS), a rolling program that provides a continuous cross-sectional survey of the food consumption, nutrient intakes, and nutritional status of people aged 18 months and older living in private households in the United Kingdom. Demographic and Health Surveys (DHS) are nationally representative household surveys that collect data for a wide range of monitoring and impact evaluation indicators in the areas of population, health, and nutrition. There are two main types of DHS surveys: Standard DHS Surveys (large samples, typically conducted every five years) and Interim DHS Surveys (smaller samples, focused on key performance monitoring indicators).

A series of specific surveys have been executed by nonprofit organizations. For instance, every two years the American Dietetic Association (ADA) surveys a large sample of American adults to identify changes in people's attitudes, knowledge, beliefs and behaviors related to nutrition, dietary habits, and eating.

International collaborative projects may include two or more countries as researchers and participants, such as China Health and Nutrition Survey, International Nutrition Survey at the Intensive Care Units (ICU), Healthy Lifestyle in Europe by Nutrition in Adolescence (HELENA), The European Prospective Investigation Into Cancer and Nutrition (EPIC), and Vitamin D Status Reports by the International Osteoporosis Foundation (IOF).

There are also individual surveys that provide online results for participants. The diet and nutrition survey results are comprised of data about food consumption, the interrelation between nutrition and health, nutrition therapy, and other goals.

Among the data of NHANES is the percent of kilocalories (kcal) from protein and the mean daily caloric intake (in kcals). More than 50 percent of the participants in the recent ADA surveys report they carefully select foods to achieve a healthful diet.

EPIC research had reinforced the hypothesis that a diet high in fiber reduces colorectal cancer risk. It also strongly supported the hypothesis that consumption of red and processed meat increases colorectal cancer risk while intake of fish decreases risk, and that being overweight and having low physical activity increases breast cancer risk after menopause.

Lolita Nikolova
International Institute of Anthropology

See Also: Alcohol Consumption Surveys; Candy; Dairy Products; Food Consumption; Grocery Stores; Meat; Slow Food; Supermarkets.

Further Readings

Bates, B., A. Lennox, and G. Swan, eds. *National Diet and Nutrition Survey. Headline Results From Year 1 of the Rolling Programme (2008/2009).* Food Standards Agency and the Department of Health. http://www.food.gov.uk/multimedia/pdfs/publication/ndnsreport0809.pdf (Accessed June 2011).

Church, S. M. "Diet and Nutrition in Low-Income Households—Key Findings of a National Survey." *Nutrition Bulletin,* v.32/3 (2007).

McKevith, B. and A. Jarzebowska. "The Role of Breakfast Cereals in the UK Diet: Headline Results From the National Diet and Nutrition Survey (NDNS) Year 1." *Nutrition Bulletin,* v.35/4 (2010).

O'Donnell, O., E. van Doorslaer, A. Wagstaff, and M. Lindelow. *Analyzing Health Equity Using Household Survey Data. A Guide to Techniques and Their Implementation.* Washington, DC: International Bank for Reconstruction and Development and World Bank, 2008.

O'Neil, C. E., Th. A. Nicklas, M. Zanovec, and S. Cho. "Whole-Grain Consumption Is Associated With Diet Quality and Nutrient Intake in Adults: The National Health and Nutrition Examination Survey, 1999–2004." *Journal of the American Dietetic Association*, v.110/10 (2010).

Disposable Diapers

As a universally identifiable consumer product, the disposable diaper emerged in the United States with the introduction of the Pampers brand by Procter & Gamble in 1961. Infant disposable diapers are now a mainstay of U.S. childrearing. A Mass Market Retailers and Information Resources, Inc., report states that U.S. sales of infant disposable diapers reached nearly $1.9 billion in 2009. Other estimates suggest that as of 2010, 96 percent of U.S. parents prefer disposable diapers. Globally, Euromonitor reveals that infant disposable diapers are a $20 billion industry. Aging populations worldwide have also resulted in an increase of adult disposable diaper sales. What is most interesting about the history of disposable diapers is what the industry's target markets reveal about patterns of consumption amid shifting global demographics and cultural norms in developed and developing countries. Also of special interest is what a waste product as ubiquitous as disposable diapers means for the natural environment, and what—if any—alternatives exist.

Developed Nations

In developed nations, the disposable diaper market has reached a point of saturation, so major disposable diaper manufacturers must increase profits by innovating brands with new technology and styles. For example, the company Kimberly-Clark updated its Huggies brand diaper to make them look like blue jeans; the brand's slogan is "it's the coolest you'll look pooping your pants." However, with the rise of the Internet and social media technology, especially blogs and Facebook, innovations are sometimes met with resistance from consumers empowered by social media networking features. In 2010, Procter & Gamble faced a social media storm of criticism in the United States from "mommy bloggers" when the company's new Pampers Dry Max diapers were reportedly giving babies chemical burns. A swift response by Procter & Gamble to consumer concerns was required to defend the purported safety of the product.

Developing Nations

Developing nations have thus far been more indifferent than resistant to disposable diapers. But with global birthrates and consumer purchasing power expected to increase markedly in the 21st century, markets for disposable diapers in developing nations are beginning to burgeon. China and India in particular are major targets of large diaper manufacturers. As of 2010, it was estimated that only about 6 percent of China's population used disposable diapers. To increase usage in that country, Procter & Gamble began its "Golden Sleep" campaign in 2007, which promised Chinese consumers that their babies would sleep better while wearing Pampers.

As a result, Pampers is now the top-selling diaper brand in China, and the company expects to add one billion new customers over the next five years. In India, disposable diaper usage is estimated to be only around 2 percent. Following the success of its Chinese campaign, Procter & Gamble has already begun implementing it in India. With birthrates in India projected to be double those of China in the early 21st century, India may soon become the world's largest disposable diaper market.

Environmental Effects

This huge increase in disposable diaper consumption has some worried about the effect it will have on the natural environment. Data from the U.S. Environmental Protection Agency (EPA) reveal that 3.7 million tons of disposable diapers were introduced to the U.S. municipal waste stream in 2007, or about 2.2 percent of total discards. Constructed of plastic, wood pulp, polyester, and special superabsorbent synthetic polymers, estimates suggest that the average disposable diaper will biodegrade in about 500 years.

Even biodegradable varieties are not much better, because they do not receive sufficient airflow in landfills to break down properly. Disease is another concern. Since most parents do not empty waste from diapers before throwing them away, most go to the dump filled with excrement, which is potentially introduced into groundwater or spreads disease to sanitation workers. Others have expressed worry about the toxicity of the chemical dioxin used in the bleaching of disposable diapers and the introduction of this chemical into the environment, and the skin of infants.

Cloth Diapers

As a response to these environmental concerns, some Western consumers now prefer reusable cloth diapers. However, some studies suggest that the benefit to the environment may be minimal, because of the added need of washing the diapers as well as the production of cotton to make them. One definite benefit for the average consumer, however, is that of cost: reusable diapers will cost $200–$600 for the time a child is wearing them, versus $2,000 or more for disposables.

The ultimate in diaper chic, however, is no diaper at all. Elimination communication, a process whereby parents learn to predict their child's bodily functions, is on the rise among some Western consumers. Ironically, cloth diapers and elimination communication are methods that have long been popular among parents in developing countries with ripening disposable diaper markets.

Conclusion

In the midst of fears of environmental degradation, disposable diaper manufacturers are ultimately high-tech innovators. For instance, consider that in the mid-1990s, the size of disposable diapers shrank by one-third and continue to shrink in the early 21st century while becoming more effective. Efforts are underway to make disposable diapers compostable and recyclable.

It is also worth considering the benefits disposable diapers give to women living in poverty globally. In the United States, because of the need for constant washing, reusable diapers are outside the realm of possibility for women without easy access to a washer and dryer. In developing nations, disposable diapers may save women living in poverty time and energy and allow them to pursue other opportunities.

In the end, it seems safe to predict that, whether in landfills or nurseries, disposable diapers will be around for a long time.

Wesley W. Roberts
University of Pittsburgh

See Also: Baby Products; Biodegradable; Developing Countries; Human Waste; Landfills, Modern.

Further Readings

Mortimer, Ruth. "Absorbing Market Share." *Brand Strategy*, v.192 (2005).
Onion, A. "The Diaper Debate: Are Disposables as Green as Cloth?" ABC News (May 26, 2005). http://abcnews.go.com/Technology/Story?id=789465&page=1#.TtsT8Ea0S0Y (Accessed July 2011).
Rothman, Barbara Katz, ed. "Diapers: Environmental Concerns." *Encyclopedia of Childbearing: Critical Perspectives*. Phoenix, AZ: Oryx Press, 1993.

Disposable Plates and Plastic Implements

The use of disposable plates and plastic implements is having a harmful effect upon the environment. The increase in their use can be largely attributed to three social factors: cultural changes in the perception of meals and family time, the capitalist economic system, and an increased pace of human tasks. Early human food implements were made of more durable materials. Knives were made of various materials, including rock and obsidian. Vessels for food storage, cooking, and serving were made of clay and pottery as early as 2000 B.C.E. Forks in the Chinese Qijia culture were made of bone (ca. 2000 B.C.E.). These materials made the implement durable enough for repeated use by many generations.

In modern society, different materials have been used to construct utensils and other cookware. In the 21st century, many human food implements are made of ceramics, glass, or melamine (plates and bowls) and stainless steel (cutlery). These inexpensive, durable, and reusable materials have become less desirable as people seek alternatives to accelerate food preparation, consumption, and cleanup.

Dining Culture

Food preparation and dining is a universal practice of family and friendship, but this activity is changing. Almost one-quarter of Americans eat fast food every day, according to a 2005 CBS poll. In some cases, gathering to share meals is the exception in family life. Group meals can be difficult for 21st-century families because of an increase in dual-earner households, single-parent families, and a greater push for children to participate in extracurricular activities away from the home. Approximately 24 percent of families eat together three or fewer nights per week.

A desire for convenience supports a greater reliance upon disposable plates and utensils. Plastics, first developed in the mid-1800s, have resulted in products that are durable, lightweight, fairly cheap to produce, and easy to discard. These implements are an example of planned obsolescence, which is a profit-making strategy used by companies that involves producing goods to be used once and then discarded by the consumer. In the United States, an estimated 60 billion paper cups and plates, 70 billion Styrofoam plates and cups, 190 billion plastic containers and bottles, and 40 billion plastic utensils are used annually, and many of these disposable items are not used again. Culture may also contribute to the use of disposable utensils. For example, in Japan, people refuse to reuse chopsticks served in restaurants, contributing to an increased reliance on disposable wooden chopsticks. One company reportedly produces 8 million pairs of chopsticks per day, a consumption rate that is unsustainable and contributes to deforestation.

A swan uses plastic garbage to build her nest. Sunlight breaks plastic down into small pieces rather than into its basic elements, allowing it to easily enter waterways and float into seas and oceans. In a landfill, plastics can take 1,000 years to degrade.

Environmental Effects

The popularity of disposable utensils and plates is counterbalanced by their negative environmental impacts. Plastics and polystyrene can leach toxic chemicals and do not biodegrade. Plastics can break down into smaller pieces when exposed to sunlight, but when buried in landfills, they can take 1,000 years to degrade. Photodegraded plastic and lightweight polystyrene can easily enter waterways and float out to sea. In the oceans, the Great Pacific and North Atlantic Garbage Patches are areas concentrated with plastic particulates and other debris in the upper water column. Organisms that ingest plastics and polystyrene, whether in the oceans or on land, risk death by choking or disease from consumption of toxic chemicals.

In an effort to divert disposable items from the waste stream, consumers, industry, and nonprofit and government organizations are seeking ways to reduce and recycle plastics and polystyrene. For example, members of the Kokua Hawaii Foundation aim to educate the public on the environmental and health benefits of a plastic-free lifestyle. Many municipalities collect recyclable waste from businesses and residential communities. Recycled plastics conserve the amount of oil used to produce new plastics (approximately 8 percent of the world's oil supply). Additionally, some Japanese companies are finding other uses for recycled chopsticks by remaking them into particleboard, paper, and tissue.

Laura Chambers
Nisse Goldberg
Jacksonville University

See Also: Fast Food Packaging; Food Waste Behavior; Pollution, Land; Recycling.

Further Readings

CBS Poll. "How and Where America Eats." http://www.cbsnews.com/stories/2005/11/20/opinion/polls/main1060315.shtml (Accessed September 2010).

Ferrant, Joan. *Sociology: A Global Perspective.* Belmont, CA: Thomson Higher Education, 2007.

Fox, Robin. "Food and Eating: An Anthropological Perspective." Social Issues Research Center. 2003. http://www.sirc.org/publik/food_and_eating_9.html (Accessed September 2010).

Gallup Poll. "Empty Seats: Fewer Families Eat Together." http://www.gallup.com/poll/10336/empty-seats-fewer-families-eat-together.aspx (Accessed September 2010).

LifeCycle Studies. "Chopsticks." *World Watch* (January 1, 2006).

District of Columbia

Neither a state nor a territory, the District of Columbia holds a unique role as the United States of America's capital city and one of the more elaborately planned cities in the nation. President George Washington commissioned French planner Pierre Charles L'Enfant to design a capital city in 1791. Article One, Section Eight of the U.S. Constitution grants Congress ultimate authority over Washington, D.C., and while the city has elected a municipal government since 1973, Congress retains the right to intervene in municipal affairs.

Waste Statistics

With an estimated population of 599,657, the District of Columbia disposed of 899,608.96 tons of municipal solid waste (MSW) and 169,831.67 tons of construction and demolition debris for a total of 1,069,440.63 tons in 2009. However, since D.C. has a ban on private landfill sites in the district, its MSW is collected by truck, taken to transfer stations and then hauled to Virginia, accounting for 7 percent of Virginia's annual MSW disposal. The residential recycling program run by the Department of Public Works (DPW) diverts 33,414 tons of material, or 24 percent of MSW to recycling from the 103,000 residential dwellings from which it collects, matching the national recycling rate. However, residents of the district produce 1.78 tons per capita of refuse, well over the national average.

Population and Labor

The district grew substantially after World War II, and between 1940 and 1970, the African American population became a majority, growing from 30 to 70 percent of the total population. The 1,700-member waste management workforce employed by the city was largely African American by 1970. Two years after the Memphis Public Works Department went on strike in 1968, the district's sanitation workers struck for the right to a comprehensive contract, which was won after a five-day strike.

Recycling and Composting

The district mandates that 45 percent of the total waste stream be diverted to recycling under the D.C. Solid Waste Management and Multi-Material Recycling Act of 1988. To aid in achieving such ambitious goals, the DPW adopted single stream (commingled) recycling and issued larger recycling bins for residential and commercial properties starting in 2005. The core government buildings for the District of Columbia already surpass the mandate by 10 percent, while the overall city rate for recycling is estimated to be only 34 percent. One incentive used by the DPW to increase compliance is the issuance of citations. The 3,114 commercial inspections conducted in 2009 resulted in 1,409 violations.

The district's DPW has expanded collection sites and materials in an attempt to stimulate greater participation in the recycling program. Residents dropped off 173 tons of electronic waste, 1,155 tons of scrap metal, and 93 tons of shredded paper at district transfer stations in 2009. On-site composting is strongly encouraged in the district for residences and commercial buildings, but no large-scale composting facilities are yet available. During 2009, the DPW composted 4,577 tons of leaves as part of the seasonal leaf collection program.

Environmental and financial incentives are driving the recycling and waste diversion program in the district. Hauling and processing fees for MSW cost D.C. $60 per ton, where a ton of recycled materials comes in at $25 per ton. In an effort to better understand the types and quantities of materials being recycled after the introduction of single stream collection, the DPW conducted a study across the city analyzing data from all of the city's eight wards. The result showed that D.C. residents were much more likely to recycle newspapers and green and brown glass than the rest of the nation, but much less likely to divert steel cans, corrugated cardboard, or clear glass.

All federal buildings and foreign missions are responsible for their own MSW and recycling. No data is available for the embassies and consulates housed in the district. As with the U.S. government though, any potentially sensitive material leaving as waste must be incinerated.

The District of Columbia has set ambitious goals of reaching a 45-percent recycling rate. Perhaps this goal is too ambitious, as a recent study conducted by the DPW concluded that only 36.2 percent of the refuse produced by D.C. residents could be recycled. If composting becomes a viable option for commercial and residential sites, then a 45 percent diversion rate can be attained. The district will remain dependent upon Virginia for meeting its landfill needs for the foreseeable future.

John Cook
Perishable Produce

See Also: Definition of Waste; History of Consumption and Waste, U.S., 1950–Present; Recycling; Virginia; Waste Disposal Authority.

Further Readings

District of Columbia Department of Public Works. "Residential Waste Sort." October–November, 2007. http://dc.gov/DC/DPW/Education+and+Outreach/Brochures+and+Fact+Sheets/2008+Residential+Waste+Study (Accessed July 2011).

Government of the District of Columbia Department of the Environment. "Public Report on Recycling Fiscal Year 2009." http://green.dc.gov/green/lib/green/fy09_recycling_report.pdf (Accessed June 2011).

Haaren, R., N. Goldstein, and N. Themelis. "The State of Garbage in America." *BioCycle*, v.51 (2010).

Jordan, Chris. *Running the Numbers: An American Self-Portrait*. New York: Prestel Verlag, 2009.

Wurf, Jerry, "The Revolution in Government Employment." *Proceedings of the Academy of Political Science*, v.30/2 (December 1970).

Downcycling

Downcycling is the reprocessing of material into a new product of reduced quality or value. The reduction in value of the new material results in the use of the material for an alternative, but potentially similar, purpose. Downcycling occurs when recycling most types of materials because of technical limitations of the recycling process, composition of the products recycled, and the state of remanufactured product collection processes and commodities markets. Downcycling is not a method for indefinitely maintaining the use of a material through recycling, but rather serves to delay inevitable, eventual disposal. The concept of downcycling, although not invented by William McDonough and Michael Braungart, was popularized in their 2002 book *Cradle to Cradle: Remaking the Way We Make Things*.

Motivations for Recycling

The act and process of recycling holds great importance in the early development of the modern environmental movement. Recycling served as both a catalyst for organizing and a symbol of change from what was considered the environmental status quo of degradation. Citizen concern resulted in the institution of municipal waste recycling programs in Western nations beginning in the 1970s, and recycling continues to serve as an outlet for the expression of environmental values today.

The symbolic position of recycling has played an important role in increasing consumer demand for collection and reprocessing of waste material and the availability to purchase products produced with recycled content. Recycling of materials, however, is not limited to products encountered by consumers in the municipal waste stream. Recycling waste from the industrial production process along with building and infrastructure construction and demolition, though less visible to consumers, is also an important and long-standing component of the global recycling market.

Recycling provides both environmental and economic benefits. Reprocessing of material generally utilizes fewer resources and creates less pollution than initial processing of virgin materials. Recycling also reduces the need for the extraction, mining, or harvesting of virgin material.

Limitations of Recycling

Because of technical limitations of the recycling process and to the composition of products, almost all materials lose value from changes in material structure when recycled. Most plastics, when broken down, result in a lower-quality polymer. This material cannot be recycled back into the original form (for example, a bottle) but instead can only be used to produce more durable products, such as carpet or clothing. These new products cannot be recycled again. Another example of the limitation of materials and the recycling process is that of paper fibers. Each time paper is recycled, paper fibers decrease in length, resulting in the production of a lower-quality product. Eventually, the fibers become too short to be reprocessed into any form of paper. Whereas the product for recycling may have been first produced as white office paper, after subsequent recycling, the fibers are only adequate for the production of paperboard or paper towels, a lower-quality product with a lower material value.

Limitations of Recycling Collection and Markets

The process through which recyclable material is collected and the state of the market in which this material is sold also impacts product downcycling. In commingled collection of materials, several types or grades of materials are collected together in one stream or container. Commingled collection is popular because it requires a fewer number of separate recycling bins. For example, in commingled glass bottle collection, all colors of glass bottles (brown, green, and clear) are collected in the same container. Recycling bottles back into their original form requires the separation of each color because the mixing of colors in the recycling process causes clouding of glass in the final product. In order to recycle commingled glass bottles into their original form, processing companies must separate each color. This separation requires additional labor and other associated costs. It therefore may be more efficient and cost effective for a recycler to reprocess the commingled glass into a lower-grade and lower-value product, such as fill material for construction projects.

Recycling markets also affect how material is recycled or downcycled. As consumers increasingly request municipal collection of material, more and more "recycled content" is available for use. Recycling markets, however, are not always available or easily accessible for the remanufacture of products from collected material that may be recycled into a new form of lesser value.

Sarah Surak
Virginia Tech

See Also: Economics of Waste Collection and Disposal, International; Economics of Waste Collection and Disposal, U.S.; Recycling.

Further Readings

Lambert, A. J. D. and Surendra Gupta. *Disassembly Modeling for Assembly, Maintenance, Reuse, and Recycling*. Boca Raton, FL: CRC Press, 2005.
McDonough, W. and M. Braungart. *Cradle to Cradle: Remaking the Way We Make Things*. New York: North Point Press, 2002.

Dump Digging

Dump digging refers to the practice of excavating old garbage sites. From beachcombing with metal detectors to digging through 18th-century outhouse contents, whatever the location, dump diggers share a common pursuit—unearthing items that people threw away in the past. Dump diggers also refer to themselves as "historical diggers" to highlight the differences between their work and professional archaeology, particularly garbology. Unlike archaeological garbologists, dump diggers normally search for intact items of historical interest, particularly household artifacts like glass and ceramics, leaving shards or scraps behind. Other dump diggers search for military relics, buttons, stoneware, and coins. Motivations for dump digging range from recreation to improving private collections and commercial sale.

Dump diggers are usually hobbyists, though some supplement their incomes by selling excavated materials at antique fairs and on the Internet. Dump diggers measure the value of the artifacts they find based on the objects' "collectiblity." To measure collectibility, diggers visit antiquarian shows and public gatherings, circulating photographs and stories about what they find. For example, glass bottles from the mid-1800s are exciting dig finds and popular collectors' items because diggers know that during this period, the United States saw a boom in the production of commercial medicines and home remedies, many of which were sold in glass bottles. During the same era, glassblowing and bottle manufacture evolved rapidly. Dump diggers value medicine bottles not only for the unique advertisements and labels they bear but also because 19th-century glassblowers constructed them using a range of ingredients and blowing techniques.

Techniques

To locate potential dig sites, dump diggers use spring steel probes and metal detectors as well as historical maps and photographs maintained in local libraries and historical societies. Unlike archaeologists, dump diggers generally do not obtain special permits. Most dump digging occurs on private property or on sites marked for development or demolition. Earth-moving machinery often reveals potential new dump sites. According to most veteran diggers, gaining permission from landowners or construction supervisors is as important in the digging process as digging test pits and consulting archives.

While most dump diggers search through abandoned town dumps they locate with archival maps and city plans, disused latrine wells are some of the most publicized dump-digging sites. As archaeologists and diggers know, privies doubled as household trash receptacles in the era before widespread household plumbing and curbside garbage removal. Digging organizations like the Manhattan Well-Diggers, who excavate wells on construction sites, have popularized privy digging, particularly in urban areas, which contain a surprising number of old latrine pits. Diggers look for undisturbed "night soil," or the privy's human waste layer, where they hope to find glass or ceramic bottles, pots, pitchers, plates, and bowls. After the advent of household plumbing, many privies were "dipped." Their human waste contents were removed and used as fertilizer. This process did not necessarily strip the wells of all their contents, and dump diggers continue to excavate these pits in search of artifact-filled night soil.

Controversy

Dump digging is controversial, and digging artifacts out of privy wells elicits particularly active debate between professional archaeologists and dump diggers. Archaeologists have clashed with diggers over how best to conserve and document the contents of dump sites. In particular, archaeologists critique amateur diggers' focus on the bottom of the wells. Archaeologists stress that in order to understand a site, they must excavate and document all of its contents in a meticulous and thorough fashion. Archaeologists of 19th- and early-20th-century U.S. cities point out that a great deal of meaningful evidence, including clues about social and cultural context, lies in pieces along the sides of privy wells. They claim that dump digging hinders their understanding of historical consumption patterns because diggers, in their haste to extract the most intact pieces, disturb or destroy the shards and organic material that are crucial to providing details about social life around dump sites. For this reason, some professional archaeologists have derided dump digging as "looting." Others have recommended that cities place stricter limits on digging at construction and demolition sites. They have also warned journalists and historical societies against overpublicizing or glamorizing the practice.

Looting is hard to define, but it is conventionally thought of as excavation without permission. There are many examples of dump diggers who criminally trespass to excavate sites, but most dump diggers obtain permission from property owners or developers. In the United States, private property laws protect digging on private sites. While digging on private property with the permission of the owner is not illegal, archaeologists still categorize the practice as unethical because artifacts are unscientifically removed from their context. They claim that little about the past can be understood from examining only diggers' high-profile, high-value collectibles.

Dump diggers see themselves as historical preservationists. Dump-digging handbooks, organizational documents, and Websites make a clear distinction between dump digging and haphazard "treasure hunting." They contend that communities of collectors are just as capable of preserving artifacts as professional archaeologists. Dump diggers see themselves as reclaiming, preserving, and even rescuing historical materials that would otherwise be lost to the hands of development and demolition. Diggers often justify their work as stewardship, and many feel that the trade in collectibles helps raise their value and profile, eventually moving them toward a place in a museum.

Furthermore, they point out, most sites they visit will be permanently altered or destroyed before archaeologists have a chance to excavate them, as small towns and rural areas become overrun by new housing and industry, or as urban historic districts are cleared for skyscrapers. If the context of these archaeological sites will be destroyed, they argue, dump diggers can help preserve something of the historical record.

For many individuals and families, dump digging is a valued recreational pastime. Dump diggers usually begin with little or no formal training, but in some areas, museums and public archaeologists are attempting to engage them more directly in the hope that academic and recreational diggers can aid one another in evaluating the material cultures of the past.

Sarah Besky
University of Wisconsin, Madison

See also: Archaeology of Garbage; Archaeology of Modern Landfills; Dumpster Diving; Garbology; Landfills, Modern; Residential Urban Refuse; Street Scavenging and Trash Picking; Trash to Cash.

Further Readings

Botha, Ted. *Mongo: Adventures in Trash*. New York: Bloomsbury, 2004.

Boxer, Sarah. "Threat to Archaeology: The Privy Diggers." *New York Times* (July 28, 2001).

Hollowell-Zimmer, Julie. "Digging in the Dirt: Ethics and 'Low-End Looting.'" In *Ethical Issues in Archaeology*, Larry Zimmerman, Karen Vitelli, and Julie Hollowell-Zimmer, eds. Walnut Creek, CA: Alta Mira Press, 2003.

Dumpster Diving

The large metal trash containers called "dumpsters" in the United States were developed in the 1930s to allow mechanical lifting and upending into a dump truck. Typically, they contain the refuse of several households or one or more businesses. In urban areas, they are usually placed outside apartment complexes, dormitories, or businesses. Temporary dumpsters are set up near construction and demolition sites. In rural areas, a series of dumpsters might be placed outside towns where individuals without garbage service dump their refuse. The volume of refuse that dumpsters contain makes them attractive spots for people looking to intercept the waste stream. These people participate in what is known as "dumpster diving," "skip dipping," or "skipping" in Australia and the United Kingdom.

Purposes

People dumpster dive for different purposes. For some, dumpster diving is an adventurous treasure hunt for choice items that they can use or give away. Certain artists collect items from dumpsters in order to create new and interesting art from old and discarded items. Other people regularly and systematically go through dumpsters, methodically collecting recyclable materials. Probably the most common form of dumpster diving in the United States as of 2010 was the collection of cans. Many low-income people collect and sell cans from dumpsters in order to purchase items that are not covered by a Supplemental Nutrition Assistance Program (SNAP). Some poor people, often mentally-disturbed or homeless, forage in dumpsters for food, while other people derive a decent living from recuperating items from dumpsters to sell. Others glean from dumpsters to provide charitable contributions, or simply to hoard. Freegans resist society's hyperconsumerism by withdrawing as much as possible from money exchange and living off capitalism's waste. They protest societal norms that define consumption as good and waste as filthy.

Food

The aversion to using the waste of others is particularly strong when considering food, and the enormity of edible food discarded in dumpsters is mind-boggling. The same people who might brag to a friend about finding a particular treasure in a dumpster will often blanch at the suggestion of eating food from the same dumpster. However, for some people, dumpsters provide their primary source of nutrition. Food not Bombs is a "dis-organization" that retrieves food from the waste stream in order to prepare free meals for people. Stores discard food that is past its expiration date, and restaurants and caterers discard food that has been served. Most people do not adhere to such strict regulations in their own kitchens, so at one level people know that this food will generally not make them sick. Dumpster divers often have stories of feasts prepared from free food. They complain of a preponderance of breads and pastries, which have a shorter shelf life than canned foods and seem to be regularly overproduced.

Stories of feasts from dumpsters are paralleled by the long lists that nonfood dumpster divers produce when talking about the activity. The tone is one of disbelief, often tinged with self-righteousness, when recording items like flat screen TVs and boxed sets of china. The very new and very old hold special places in the inventory. Rural dumpster sites are sometimes called "the general store," indexing the abundance of usable items there.

Dumpster diving is often seen as a legitimate form of recycling: U.S. Navy volunteers on the Naval Air Station in Whidbey Island, Washington, participate in the base's annual Dumpster Dive in 2005. The teams compete to recycle as many treasures as possible.

Ideal Times

Timing is important to dumpster divers. Urban dumpster diving is often done in the evening after businesses close and people put out their trash. Some dumpster divers get to know the trash disposal rhythms of particular businesses and particular neighborhoods and adjust their schedules accordingly. Collecting cans after large public events, such as concerts and football games, is particularly rewarding. The calendar is also important. *Hippy Christmas* is a term that has been applied to specific days during the year when residents are allowed to dispose of large items, but more commonly to the end of university terms when students move out of apartments and residence halls, leaving piles of usable objects behind. A 2010 trash inventory at an Oregon State University residence hall midterm found that only one-third of the contents of a residence hall dumpster was trash. Everything else was usable or recyclable.

Restrictions and Hazards

Some businesses discourage dumpster diving by locking their dumpsters or putting them behind tall fences. This brings up the question of when an item ceases to be private property. Does ownership transfer from the individual or business to the waste management company the instant an item is thrown in the dumpster, or is there a period of non-ownership from the time it is disposed to the time it is picked up? Some cities have passed ordinances making dumpster diving illegal.

Dumpster diving may be hazardous as well as illegal. Some businesses pour noxious materials over edible food to discourage dumpster divers. Other hazardous materials are disposed of in dumpsters with no thought as to who might be handling the contents later. A recent concern in rural areas has been the disposal of waste from methamphetamine labs including drain cleaners, white gas, and large bags of matches from which they scrape the red phosphorous. The fire hazard and poisonous gases that noxious chemicals can produce has caused some waste management companies to remove remote dumpsters.

Lists of "best practices" exist on the Internet for those wanting to try dumpster diving. These include the type of clothing that should be worn and useful equipment to take along. They tell dumpster divers not to climb fences or trespass and not to take confidential records. Dumpster-diver etiquette instructs people not to leave a mess and to take only what one can use, leaving the rest for subsequent divers. However, most dumpster divers learn the practice through oral transmission.

Joan Gross
Oregon State University

See Also: Consumerism; Food Waste Behavior; Freeganism; Garbage Art; Overconsumption; Trash to Cash.

Further Readings

Clark, Dylan. "The Raw and the Rotten: Punk Cuisine." In *Food and Culture: A Reader*, 2nd ed., C. Counihan and P. Ven Esterik, eds. New York: Routledge, 2008.

Edwards, Ferne and David Mercer. "Gleaning From Gluttony: An Australian Youth Subculture Confronts the Ethics of Waste." *Australian Geographer*, v.38/3 (2007).

Eighner, Lars. "On Dumpster Diving." *Threepenny Review* (Fall 1991).

Ferrel, Jeff. *Empire of Scrounge: Inside the Urban Underground of Dumpster Diving, Trash Picking, and Street Scavenging*. New York: New York University Press, 2006.

Earth Day

April 22 is the day when more than 180 nations celebrate Earth Day. Earth Day is the world's most widely observed secular holiday. It is supported by the international Earth Day Network and characterized by activities focused on environmental issues.

History

Earth Day was founded by Gaylord Nelson on April 22, 1970, turning April 22 into a national day of observance of environmental problems. According to the founding father himself, a U.S. senator at the time of establishment, the idea for Earth Day started with the realization that the state of the natural environment was not an issue in the politics of the United States.

This was the case despite the growing concern of U.S. citizens in the 1960s, which was stimulated in part by the 1962 publication *Silent Spring*, an influential book by U.S. marine biologist Rachel Carson (1907–64) about the use of chemical pesticides and their influence on health and the environment. The book raised awareness of the dangers of chemicals, led to the formation of environmental protection movements in many European countries, and put environmental issues on the U.S. political agenda.

Nelson managed to persuade U.S. President John F. Kennedy to give visibility to this issue by going on a five-day, 11-state conservation tour in September 1963. The tour became the germ of the idea that ultimately flowered into Earth Day. In 1970, a grassroots protest was organized by Denis Hayes, who was a professor of engineering at Stanford, the director of a national laboratory (SERI), and president of the Bullitt Foundation—an environmental foundation in Seattle. He was an environmental activist and proponent of alternative energy. This demonstration of popular protest tapped into the environmental concerns of the general public and infused the student anti–Vietnam War energy into the environmental cause. The demonstration of approximately 20 million people, many of whom were students from over 2,000 colleges and universities and about 10,000 primary and secondary schools, as well as other citizens, established a forum to express their concern with the environment.

Effects and Programs

Earth Day has been gaining in international popularity since the last decade of the 20th century.

The Earth Day Network (EDN) was founded on the premise that all people have a moral right to a healthy, sustainable environment. The Earth Day Network's mission is to broaden and diversify the environmental movement worldwide and to mobilize it as the most effective vehicle for promoting a healthy, sustainable environment. The EDN provides a combination of education, public policy, and activism campaigns and has, as of May 2010, more than 20,000 partners and organizations in 190 countries. The EDN coordinates more than 1 billion people who participate in the Earth Day activities, making it the largest secular civic event in the world. EDN's programs and activities are guided by goals including promoting civic engagement, broadening the meaning of "environment," mobilizing communities, implementing groundbreaking environmental education programs, and inspiring college students to become environmental leaders.

EDN's Environmental Education Program is one of the most successful in the United States, providing tools to educators and students for integrating environmental issues into core curriculum across disciplines and grade levels, in and out of the classroom. This program is also becoming increasingly known in Europe. In her book *The Morning After Earth Day*, Mary Graham described the sociopolitical effects of Earth Day on the international community in general and U.S. society in particular. The great increase in political efforts to address environmental problems following the establishment of Earth Day was unprecedented in the history of the U.S. (and, partially, European) environmental movement. Earth Day has also raised questions about settled habits, having to do with existing technology, dependency on energy and transportation, and the role that different political and social stakeholders play in environmental degradation. Despite its international success, Earth Day is less known in developing countries.

Earth Day activities range from greening schools in post-Katrina New Orleans to improving water and sanitation services in a refugee community in Ghana. Earth Day activities are widely publicized through social network sites such as Facebook, MySpace, Twitter, and LinkedIn. There were a number of children's books and online activity Websites containing Earth Day in the title published in the last five years.

Helen Kopnina
University of Amsterdam

See Also: Environmentalism; European Union; United States.

Further Readings

Carson, Rachel. *Silent Spring*. New York: Houghton Mifflin, 1962.

Earth Day Project. "Earth Day Network 2010." http://www.earthday.org/about-us (Accessed June 2011).

"Earth Day 2010 Events." http://action.earthday.org/events (Accessed May 2010).

Freeman, A. "Environmental Policy Since Earth Day I: What Have We Gained?" *Journal of Economic Perspectives*, v.16/1 (2002).

Graham, Mary. *The Morning After Earth Day*. Washington, DC: Brookings Institution, 1999.

Nelson, Gaylord. "How the First Earth Day Came About." http://earthday.envirolink.org/history.html (Accessed May 2010).

East Asia (Excluding China)

East Asia, because of its ancient and fascinating civilizations and economic rise, is an important part of the world. It holds a large part of the world's population, plays an ever-increasing role in the global economy, and is thus pivotal for global sustainability—or its failure.

Definition and Location

Geographically, east Asia is the eastern part of the Eurasian landmass. Politics and culture interfere, however, with this simple geographical definition. For example, Siberia is included in this geographic definition, but it has a distinct culture and history. When thinking geographically, it also makes sense to aggregate the diverse countries, regions, and cultures of the area, allowing one to compare the relative impact and importance of different world regions. This is occurring, in particular, in the context of organizations such as the Asia-Pacific Economic Cooperation (APEC) and the Association of South-East Asian Nations, together with China, Japan, and the Republic of Korea (ASEAN Plus Three). Even an east Asian community ultimately meant to be similar to the European Union has also been proposed.

These organizations are moving east Asia and its neighbors toward regional integration (or at least cooperation). Trade and security concerns are at the forefront of these talks. Here, however, "Asia-Pacific" and even "east Asia" are commonly meant to include southeast Asia and potentially even Australia.

The term *east Asia* is used in various ways, and it hides considerable diversity within any of those definitions. Even regarding ecosystems, east Asia's ecosystem ranges from boreal forests to subtropical jungles, as well as from coastal urban conglomerates to remote, inland deserts. Parts of Russia, Mongolia, but most importantly China, the Koreas, and Japan all belong to this area. Most commonly, historical and cultural ties that connect these latter three countries are given the greatest importance in defining east Asia.

Taking China's influence as a common thread, east Asia is the region that has been influenced by Chinese thought, language, and literature. China's historical sphere of influence would allow the area to be expanded even further, however. Singapore, for example, is often included in the cultural sphere of east Asia because of its strong ties with China. By the same count, Vietnam could be included, but it is usually seen as part of southeast Asia. Then again, east Asia is sometimes used to refer to China.

Through those historical-cultural links between China, Korea, and Japan, some elements of culture—and also of consumption—are similar across this otherwise diverse area. Thus, not only in terms of geography (as an abstract region to compare with others), but also in terms of culture, it is possible to consider an east Asia that is internally linked and different from other cultural areas. In fact, the very distinctness of east Asian cultures, coupled with their economic successes, led to some resurgence of interest in possible causal connections between the values that are particularly pronounced in a culture (such as education and work ethic in east Asia's Confucian tradition) and the economic success of countries belonging to that culture.

Historical Consumption

Consumption in east Asia is mainly discussed in the context of its economic growth since the second half of the 20th century. Much of the talk is about the (assumed) shift of global power and influence toward this region in an "Asian Century." Historically, however, east Asia long used to be a center of world population, urbanization, and economic activity. As the original home of some of the world's first widely traded luxury goods (such as silk, porcelain, and tea) as well as ancient civilizations with great economic activity, east Asia played a large role in the world economy long before it ever became global.

The region accounted for some 30 percent of global gross domestic product (GDP) between 1500 and 1850 (according to the data compiled by Angus Maddison), when the rise of the Industrial Revolution in Europe and then the United States shifted world economic activity toward the West (colonialism caused further decline in Asia). Even in these early times, there was widespread trade. This occurred along the Silk Road connecting east and south Asia with Europe, the Tea-Horse Route connecting southwestern China and Tibet with continental southeast Asia (what is now Burma/Myanmar, Vietnam, Laos, and Thailand), and far-ranging marine trade. The role of consumption is most apparent in descriptions of trade, not least those by Marco Polo, as well as in the archaeological remains of high-ranking persons, who were buried with a wide range of goods, both quotidian and luxury. Thus, a certain status of consumption and trade in valued products have a long history, although even Confucianism (let alone Daoism and Buddhism) would tend to hold scholars in high esteem and traders in rather low esteem. Moreover, the history of east Asia is also one of shifts between east Asian empires' openness to trade (and cultural exchanges) and closing off from the outside world.

Modern Consumption Patterns

In the 21st century, east Asia is spanning similar chasms: between tradition and modernity, between rampant consumerism and high savings rates, and

The teachings of Confucius introduced traditions that stress the need for humility and moderation in daily life in east Asia. Still, the region has long been a center of world population, urbanization, and economic activity, where luxury goods were first widely traded.

between a fast pace of development in recent history and a future that may be very different. Encompassing the world's most populous country (China) and two of the world's largest economies (China and Japan), east Asia has great importance. It holds 1.5 to 2 billion people and, according to World Bank data, the Asia-Pacific has a share of the world economy that amounts to 7 percent (based on GDP) or even 13 percent (by GDP based on purchasing power parity). Thus, it has a high impact regarding consumption and waste.

Even in 2009, when global primary energy consumption declined by 1.1 percent, energy consumption in the Asia-Pacific increased (according to the BP Statistical Review of Energy). The three largest economies in Asia (China, Japan, and South Korea) consumed 14.78 million barrels of oil per day, 17.6 percent of the total world consumption of 83.62 million barrels per day. U.S. consumption, however, is still higher, at 18.69 million barrels per day, or 22.4 percent of the world total daily consumption; these are absolute numbers and percentages and do not consider the respective population numbers.

Similar contributions to CO_2 emissions are found: eastern Asia produced 7,165 million metric tons of CO_2 emissions in 2007 (a 24 percent share of the world's 29,595 million tons). A particular problem in east Asia is the wide availability of coal, the use of which is greatly contributing to regional carbon emissions as well as atmospheric pollution. Per capita, however, emissions are only slightly above the world average of 4.4 tons, at 5.1 metric tons, and half of the recent growth in emissions results from production of goods for export.

Considering consumption of resources and colonization of nature, east Asia's challenges are apparent. Taking the entire Asia-Pacific together, the region holds 60 percent of the world's population. Even considering only China, Mongolia, the Koreas, and Japan, east Asia is home to 22 percent of the world's population. With 133 inhabitants per square kilometer (sq km), this region is populated three times more densely than the world average (45 persons per sq km). Availability of freshwater in the Asia-Pacific is only 36 percent of global availability, availability of biologically productive area per capita is less than 60 percent of the global average, and availability of arable land per capita is less than 80 percent of global average.

Thus, even while many of the economies are still developing, the ecological footprints of east Asian (and Asian-Pacific) populations exceed their locally available resources. Also, using the work of the Global Footprint Network, it can be shown that the Asia-Pacific region's ecological footprint, in terms of pollution and the capacity of regional ecosystems to absorb it, in particular as it is exacerbated by the high population numbers, is already the most negative in the world. Both the need for resources and the production of waste are set to increase still further into the 21st century. Economic growth in the area (high in the majority of its economies) is driven by improving living standards and by exports (so that consumption of resources in east Asia is not only for the local population, but also for consumers in other parts of the world).

Waste

East Asia's role in the global problem of waste is peculiar. On the one hand, development and urbanization work together to increase the production of waste. East Asia is a main source of pollution not only in terms of CO_2 emissions but also in terms of atmospheric pollution—one of the most visible and spreading forms of waste. Solid waste production has also been on the rise, and recycling systems for consumers are hardly existent in most east Asian countries (with the exception of Japan and Korea). Therefore, a big challenge is the increasing amounts (and mixing) of organics and plastics, which are mostly deposited in landfills. As one can observe in many rural parts of developing east Asia, attitudes toward waste are anything but environmentally conscious.

Waste, from organics that should be composted to electronics that need special treatment, is simply thrown out. Residential waste, which tends to receive the greatest attention, is still less of a problem than industrial waste. However, industrial pollution of air, soil, and water in China, in particular, is increasingly troubling. According to attempts at calculating a green GDP for China, most if not all of the country's economic growth over the last decades is cancelled out by the increasing cost incurred through pollution. On the other hand, there is a major recycling industry, especially for plastics and metals, utilizing waste as primary material and employing people from local trash collectors up to specialized recycling companies.

It is not only in the production of waste that an attitude shaped by agricultural, local traditions comes to the fore. In the 21st century, consumption behavior in east Asia follows a seeming union of opposites (as was encountered through its history), between materialism and a disdain for it. East Asia has not only shown rising consumption of raw materials because of export-led economic growth in general, but also because of growth in private consumption. One particularly noticeable sector, maybe unexpectedly, is that of luxury goods. Buying for quality—and more importantly for the status that can be shown by it—is a widespread behavior. Japanese consumers used to show this behavior very strongly at the height of their economy but have since become very thrifty and reluctant spenders. The Chinese, on the other hand, have increasingly turned into voracious consumers. Among those who can (or want to) afford it, luxury products are very popular. The reason is less of a Western materialism than a Confucian conspicuous consumption in which possession of luxury goods denotes that one has succeeded, elicits the respect and deference desired from others, shows that one can care for one's dependents, and shows that one is a productive member of the relevant social groups. The wider wish to participate in the pleasures of developed lifestyles is also a reason underlying the wide availability of cheap goods, both for the Chinese consumer eager to purchase and the Japanese consumer wishing to act thriftily.

In the role of thrift, one encounters the other side of consumption in east Asia. The region is also well known for its notoriously low private consumption and the widespread focus on savings. Savings rates in households are typically very high; thrift and frugality are highly prized. It is not just traditional culture that accounts for the savings, however. Research suggests that high household savings in China result from the necessity to save up for a son's wedding; in order for a bachelor to be seen as eligible, he must be able to afford a house. Moreover, social safety systems are insecure (or barely extant) across east Asia, so that the aging population also hold savings as insurance for their own old age. Filial piety and care for elderly parents is a main tenet of the Confucian tradition, but one that has become rather weakened by declining family size, whether resulting from socioeconomic factors or (in the case of China) government policy.

Future

The future development of east Asia's impact is hard to gauge. On the one hand, continuing development is needed and likely. With economic growth and improving lifestyles, there is a trend toward increasing consumption. Waste produced is also increasing and changing, for example, toward a higher percentage of plastics. On the other hand, aging alone implies that growth will slow as there are fewer people in the workforce—in particular, providing the pool of cheaper labor that much of east Asia's recent, rapid growth was based on. Additionally, younger people may try to limit their consumption as they need to make a living both for themselves and their children, save up for their own retirement, and support their elders (whether directly or in the form of tax money paid into pension systems). At the same time, even if elderly people hold a large share of east Asia's private wealth, they are wont to spend this money because it serves them as insurance in case of health problems or should the need for assistance arise. Predictions are further complicated by the possibility of ecosystem collapse or negative impacts of climate change, which would significantly impact east Asia. Pollution of air, water, and soils was already having a negative impact on population and ecosystem health by 2010). However, in the area of economic-industrial development, there is an increasing drive (not least for their economic potential) toward alternative energy technologies and green production methods. Concern over security and health has a further influence in promoting more sustainable lifestyles.

Gerald Zhang-Schmidt
Independent Scholar

See Also: Beijing, China; China; Culture, Values, and Garbage; Japan; Seóul, South Korea.

Further Readings

Fukuyama, Francis. "Culture and Economic Development: Cultural Concerns." In *International Encyclopedia of the Social & Behavioral Sciences*. New York: Elsevier, 2001.

Garon, Sheldon and Patricia L. Maclachlan, eds. *The Ambivalent Consumer: Questioning Consumption in East Asia and the West*. Ithaca, NY: Cornell University Press, 2006.

Guan, D., G. P. Peters, C. L. Weber, and K. Hubacek. "Journey to World Top Emitter: An Analysis of the Driving Forces of China's Recent CO_2 Emissions Surge." *Geophysical Research Letters*, v.36 (2009).

Lane, Jan-Erik and Reinert Maeland. "The Ecological Deficit of the Asia-Pacific Region: A Research Note." *Pacific Journal of Public Administration*, v.30/2 (December 2008).

United Nations Statistics Division. "Millennium Development Goals Indicators." 2010. http://mdgs.un.org/unsd/mdg (Accessed November 2010).

Wong, Nancy Y. and Aaron C. Ahuvia. "Personal Taste and Family Face: Luxury Consumption in Confucian and Western Societies." *Psychology and Marketing*, v.15/5 (1998).

Economics of Consumption, International

The forms and meanings of the term *consumption* in human society have varied between time and place. For much of human history, consumption has meant "using things up" produced within households. This is still the case in many parts of the world, and even in advanced societies, such consumption survives on the margins and is especially associated with women's labor. But as societies have developed, the economic meaning of consumption has come to be associated with the consumption of an increasing range of goods and services produced for exchange as commodities. In macroeconomics, consumption is the part of national output that is not saved. Consumption shares vary considerably between countries, as does the balance of private and public consumption. Private consumption is that of the individual. Public consumption arises where the state deter-

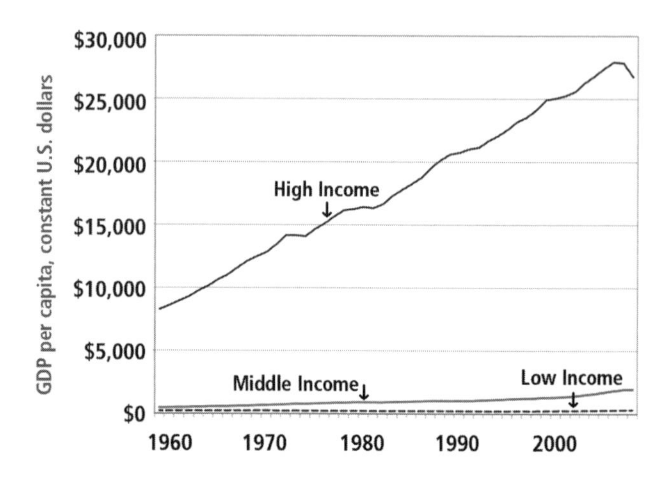

Figure 1 Growth in per capita GDP by country income group
Source: World Bank Development Indicators

mines the collective supply of public goods, such as education, health, and defense.

National Consumption and Income

In 1900, public and private consumption was estimated at a total of $1.5 billion worth of goods and services. By 1975, the figure was $12 trillion; in 2010, the total amount consumed worldwide is $31 trillion (at constant prices). The usual focus of the analysis of global consumption patterns is the nation-state, although some claim that there is a transnational ruling class defined both by its control of production and its consumption pattern. The world's population is over 6.5 billion spread over 194 officially recognized countries with different average income levels, which determine the overall patterns of consumption. The World Bank categorizes these countries as low income, middle income, and high income. Figure 1 shows that the gulf between high-, middle-, and low-income countries has grown substantially since the mid-20th century. gross domestic product (GDP) per capita in Switzerland was 191 times higher than that in Malawi in 1960; in 2010, it was 229 times higher.

Seventeen percent of the world's population lives in high income countries, yet they account for 80 percent of the world's consumption. The United States, with just 4.5 percent of the world's population, accounts for 32 percent of its total consumption. Despite their huge populations and booming economies, China and India only account for 4 and 2 percent of world consumption, respectively. In other words, these two countries, with 37 percent of the global population, consume only 6 percent of global output. For those in less-developed countries, consumption is strongly linked to the fulfillment of basic human needs.

Consumption is also a function of inequality—an aspect that has been increasing since the mid-20th century—within countries. The way in which the number of billionaires and millionaires has risen on all continents is one visible aspect of this. Although the exact figures are disputed, there are over 1 billion people in the world living in extreme poverty (surviving on less than $1.25 per day) and an additional 1 billion people living on less than $2 per day. Such individuals have limited access or hope of access to sophisticated mass consumption goods like mobile phones, computers, or even a regular electricity supply. The volatile prices of basic commodities compound the problem of fulfilling consumption needs for those in extreme poverty. Increasing global demand for food, with supply hampered by uneven distribution and environmental disasters, has driven up global food prices, leading to fears of a global food crisis. Food riots in low-income countries reflect a moral economy of consumption that goes beyond market forces.

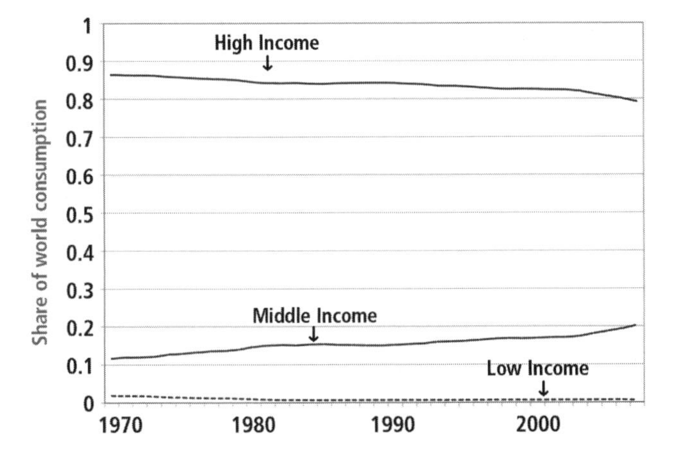

Figure 2 Share of global consumption by country income group
Source: World Bank Development Indicators

Pattern of Consumption and Economic Development

Human needs are different to human wants. For thousands of years, humans have lived lives structured around the fulfillment of basic needs: food, clothing, and shelter. The most basic human needs are independent of patterns of economic development. As long as average income per person allows individuals to satisfy only their basic necessities, no other types of consumption can develop.

Historically, the production of food has been the dominant use of labor, land, and capital. Efficiency growth in agriculture, with the development of new tools, labor-saving machinery, and plant and animal breeding, has led to a dramatic increase in yields and a relative fall in the inputs required.

Efficiency growth created the spare resources to drive the process of urbanization and industrialization. As average income per person increased, the pattern of consumption broadened, but the gains are

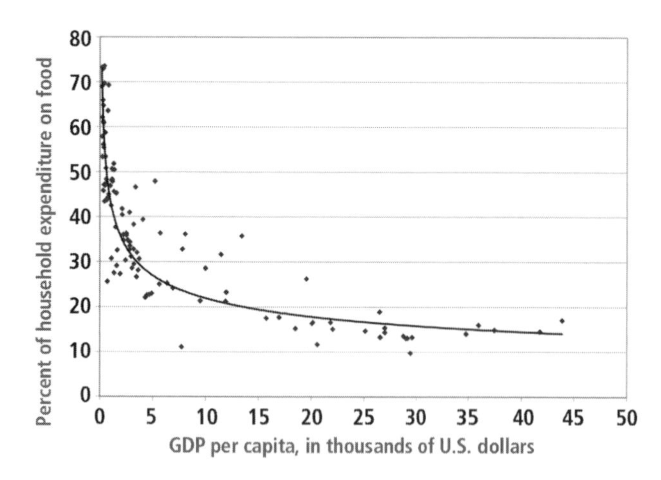

Figure 3 Engel Curve of relationship of household food expenditure to average national per capita GDP in 1996
Sources: GDP—IMF, Food Budget—U.S. Department of Agriculture

unequally distributed, which contributes to increased social differentiation. As society transformed from hunter-gatherers to settled agriculture, it allowed some individuals to become paid for functions other than food production. Warriors, priests, and administrators became full-time professions. It has only been in the last few centuries that (in the West) the mass population felt more of the benefits of this change. In many other parts of the world, consumption remains low for the mass of the population, despite the world becoming 50 percent urbanized by the turn of the 21st century. With increasing average income per person, households have an increased portion of their budget available to consume nonnecessities and buy goods previously seen as only available to the rich. This pattern was first quantified by German economist and statistician Ernst Engel into an Engel Curve, which shows the negative correlation between average income per person and the share of household budget spent on food. Over time, as countries developed, the balance of consumption shifted toward manufactured goods and then began to give way to the consumption of service industry goods, including leisure goods, as the length of the working week declined in richer countries.

Theories of Consumption and Global Consumption Patterns

There are several theories regarding how this changing pattern of international consumption can be explained. Neoclassical economic consumption analysis is based on three assumptions. The first assumption, called "asocial individualism," states that individual preferences are not affected by either social and economic institutions or the behavior of others. The second, called "insatiability," states that it is human nature to want more of everything. The third, called "commodity orientation," posits that consumers make rational choices based on full perfect knowledge of the market and all possible alternatives.

Thorstein Veblen, on the other hand, in his 1989 work *Theory of the Leisure Class*, argued that consumption can be driven by a desire for intangibles, such as status. This was not only to be found in primitive tribal behavior but also in consumer behavior in both advanced and developing societies. Consumption goods yield both status and use value to consumers. Consumption patterns "trickle down" as status-good consumption of the wealthy is mimicked by the lower ranks of the wealth distribution. New luxury products are developed to replace the goods that become objects of mass consumption. Veblen saw individuals as intentional status maximizers who live in a social environment of shared values and who consume so as to raise their social status, rather than for any intrinsic product benefit.

With the development of a more integrated world economy and a globalized media and culture, Veblen's ideas can help to explain the ways in which aspirational-status consumption patterns develop across widely different levels of national development. They can also contribute to an explanation of the patterns of overconsumption identified in more-advanced economies.

In poorer countries, consumption by the rich can take place at the expense of investment. Ragnar Nurske argued that if consumers in developing countries attempt to emulate the consumption patterns of those in the developed world, then a lower household savings rate will result and expenditure will leak out, hindering the process of development. At any income, an increased level of consumption can only result in a lower level of saving. The luxury jeweler Cartier, for example, now has "boutiques" in over 125 countries, including very low-income countries, such as Angola, Myanmar, and Yemen.

Globalization and Consumption

For those who can afford it, one is never far away from the opportunity to eat a burger and fries, drive a luxury German car, or watch the latest Hollywood movie. According to Theodore Levitt, "the products and methods of the industrialized world play a single tune for all the world, and all the world eagerly dances to it. Ancient differences in national tastes or modes of doing business disappear." But is a global consumption pattern desirable?

George Ritzer coined the phrase "McDonaldization" to describe the situation where the principles of fast food restaurants are coming to dominate other aspects of society. In McWorld, the term *efficiency* means using the most cost-effective method to achieve a given end. Processes are streamlined in a rationalized setting. Calculability is the organizational emphasis on a calculable advantage. Quantifiable returns are more important than quality. "Bigger" is seen as "better," when in reality it merely provides an illusion of quality. Food portion sizes increase; yesterday's "large" size is today's "regular." The World Health Organization estimates that there are over 1 billion overweight people globally. The problem is not restricted to industrial societies; the fastest growth often occurs in developing countries adopting more advanced patterns of consumption—including the spread of high-calorie processed foods.

A third dimension of McWorld is predictability. A Big Mac is a Big Mac whether it is eaten in Manila or Montreal. Television shows like *The Simpsons* are increasingly seen throughout the world. Control is the fourth dimension of McWorld, the substitution of nonhuman technology for human technology, as result of which both consumers and workers are subject to greater levels of supervision.

These arguments have given rise to critiques of emerging global consumption patterns. One approach has been nationalist. Consumption culture is closely related to national identity, which some argue needs protection from potential homogenization as globalized goods and services sweep the world. The French government, for example, spends billions of euros and employs thousands of people each year to protect and promote French culture. Spain, South Korea, and Brazil all place domestic content requirements on their cinemas.

Writers and political activists like Naomi Klein have challenged branding and logos and the ways in which they are used to hide chains of globally exploitative relationships. At the bottom of global supply chains, those who produce Nike trainers or Apple computers cannot afford to buy the goods they make and may not even understand the uses to which their work is put.

More theoretical critiques of global consumption have been made from several quarters. J. K. Galbraith argued that consumption is not a product of "consumer sovereignty"; rather, it is producers who create consumer desires through market control and advertising. Global firms such as Vodafone and Marlboro turn to sales and marketing in order to create demand for more. The United Nations has suggested that aggressive global advertising, estimated at $450 billion in 2009, is creating a global consumption space. Producer control shifts the balance of support toward private consumption at the expense of collective consumption goods, despite the weak link between privatized consumption and well being.

However, in the late 20th century, neoliberal economic regimes developed in many countries with policies centered ostensibly on the idea of consumer independence and sovereignty. Consumption was often based less on income and savings as the encouragement of growing consumer debt eventually contributed to financial instability in some parts of the world.

The Frankfurt school developed a related critique of global consumption patterns. This group had a pessimistic view of culture and the global culture industries. Drawing from Karl Marx's concept of worker alienation in capitalism, they argued that leisure consumption becomes an afterlife structured by the dehumanized workplace. The centrality of profit leads to a cultural scene that is formulaic, soothing, and banal but maximizes audience share. Global media conglomerates such as Disney, Time Warner, and News Corporation control film, television, news gathering, music, and the publishing industry. Rather than have a revolutionary edge, art becomes increasingly passive, impoverishing existence outside the workplace and allowing individuals to return to the mindlessness of work.

These arguments have, however, been contested. Supporters of globalization argue that although common patterns of consumption may diminish differences between societies, they can create more diversity within, giving individuals a wider choice from which to consume. Cultural homogenization and increased heterogeneity are not mutually exclusive. Culture has always been a process of creative destruction.

Even if one accepts the power of global producers to try to determine any emerging consumption patterns, it does not follow that they will be successful. Whether the consumer or producer is "sovereign" has been discussed as the "agent" versus "dupe" question. Dupes are believed to have ceded their agency to producers. Others argue that individual agency is central to the operation of the market and consumer society. Consumers may have lost the power to reject consumption as a way of life, but firms increasingly use and construct consumer agency as a way to develop products. Those firms who best use consumer agency will profit from it. The failure of Coca-Cola to create a "New Coke" is an example of a global firm falling foul of its market.

Agency can also exist in terms of critique and opposition. With culture always creating a passive citizenry, there is no possibility of a mass revolt. But consumer resistance is always present in the market, although it can be contained, as when anticonsumerism is itself a marketed lifestyle and the antiglobalization movement and its key players become brands in their own right.

Are Global Products Really Global?

It is important to leave open the question of how far a homogenous global consumption pattern is squeezing out local firms. McDonald's is the organization most identified with a global product, but on closer inspection, even they adapt their product to sell in local markets. In India, they do not sell beef; they use halal or kosher meat in certain restaurants, have different policies on the sale of alcohol, and have different interior designs of premises. Films, books, and music all rely on local understanding and knowledge, hence, the existence of national industries despite the threat of Hollywood. The Netherlands, a country of 10 million Dutch speakers, produces most of the books on its best seller list. Organizations are far from overriding local consumption patterns, having to adapt and create products to fill these patterns, which have been described as a process of "glocalization." Moreover, for those living in poverty, access to branded global goods is limited. Although these products may be offered in the global market, they are not in reach of many consumers.

Global Consumption and Economic Sustainability

Some economists have defined the end point of development as the creation of an age of high mass consumption. Whether or not economic development is leading to equally high levels of consumption around the world remains controversial given the huge inequalities that exist between countries. Were such an age of equalized global high mass consumption possible, it could not be at the level of the most advanced countries given the finite limits on the planet's resources. This raises the question of how future allocation of consumption will be made. The political and economic challenge is to allow all humans in all countries to have good consumption levels without threatening the sustainability of the planet.

Joseph Haynes
UK Government Economic Service

See Also: Consumerism; Consumption Patterns; Developing Countries; Economics of Consumption, U.S.; Externalities; Household Consumption Patterns; Sustainable Development.

Further Readings

Fine, B. and E. Leopold. *The World of Consumption (Economics as Social Theory)*. New York: Routledge, 1993.

Klein, N. *No Logo*. New York: Picador, 2002.

Levitt, T. "The Globalization of Markets." *Harvard Business Review* (May/June 1983).

Ritzer, G. *The McDonaldization of Society*. Thousand Oaks, CA: Pine Forge Press, 2004.

Schor, J. "In Defense of Consumer Critique: Revisiting the Consumption Debates of the Twentieth Century." *Annals of the American Academy of Political and Social Science*, v.5/10 (2007).

United Nations Development Programme. *Human Development Report, 2009*. New York: Palgrave Macmillan, 2009.

Economics of Consumption, U.S.

U.S. consumption has provided markets for the world, enhancing and sustaining economic opportunities for many worldwide. Technological advances have created new methods of communication and access to information that have opened up a world of consumption opportunities. U.S. consumers can—if they have the money—have anything. The U.S. consumer can be anything with an identity in continual flux based on and around the possession and ownership of what the consumer buys. On the other hand, many Americans live in abject poverty or are barely surviving above the poverty level. Without access to markets in everything, they cannot be what they want based on what they buy. Competitive consumerist capitalism has led to a host of ills, including a gulf between the haves and have-nots, its existence dependent on continually luring both haves and have-nots into spending themselves into unsustainable levels of debt.

Poverty

Despite the ability of many Americans to access markets in everything with a click of the mouse or the swipe of a card, the U.S. Census Bureau reported the 2008 poverty rate at 13.2 percent of the U.S. population. Many U.S. families live just above the income level considered minimally sufficient to sustain a family with basic housing, food, clothing, and medical needs, and they move in and out of poverty from year to year. Forty percent of the U.S. population has been in poverty since 2000. The ability to meet basic consumption needs in the United States is unevenly distributed, with minorities and urban residents at higher rates of poverty. Both the absolute levels of poverty and the fact that the burden falls on the poor is underreported in media, almost making the problem invisible. In the United States, one can avoid poverty altogether by simply avoiding pockets of poverty and pretending that they do not exist. Media, advertising, and marketing dollars are directed toward the hyperconsuming, sovereign U.S. consumer.

Sovereign U.S. Consumer

The sovereign U.S. consumer is potentially the most sought-after customer on the planet; relied upon for world economic security; pampered and abused by marketers; seduced and plundered by banks and financial institutions; prodded, poked, and analyzed by researchers; and treated simultaneously as brilliant diviner of taste and unwary sucker. The U.S. consumer makes up approximately 18 percent of the spending in the world. Consequently, changes in U.S. consumption are important to the world economy. Nowhere was this more evident than during the worldwide financial crisis of 2008–09. The financial crisis was due in large part to the housing crisis and subsequent credit crisis in the United States. U.S. consumers who had been living on money borrowed from inflated home values had gone on a consumptive binge for nearly seven years, then suddenly recoiled and left the marketplace. Consumers, fearful of the future, stopped buying and started saving, businesses cut production and stopped borrowing, and banks ceased lending. With decreased borrowing, lending, and investment, consumption continued to decline globally.

Marketplace ramifications for employment around the globe are dependent upon the U.S. consumer. The types of jobs dependent on consumption have changed. Information technology and communications advances have eliminated many desk jobs and have made existing workers more productive. This means that unless new jobs are created or population growth reverses direction, the decline in U.S. consumption implies a global glut of goods with no one to buy them and, simultaneously, a surplus of workers. The irony of the economic situation of the early 21st century is that the Keynesian solution to replace missing consumer spending with government spending has met with substantial resistance. In order to keep or replace the U.S. consumer, marketers resorted to the most extreme forms of exploitation, via invasive entreaties into personal privacy, hidden fees, deception, courting and entrapping younger and younger children, and infantilizing all forms of public and private discourse in order to sell more goods.

Fundamental Question of Economics

Economic historian Robert Heilbronner describes economics as how individuals and societies materially provide for themselves, suggesting at a minimum that the fundamental question of economics is how people meet basic consumption needs. How does society produce, allocate, and distribute what it needs for consumption? The answer to the question of what economics is, and specifically what consumption is, has changed over time, as has the meaning of economics. Economics no longer directly addresses meeting society's basic needs by allocating scarce resources but instead studies behavior, individual choice, and well-being. Lost is the idea of society's provisioning and even the idea that there are basic needs. In the land of plenty, even the idea of scarcity has lost its central place in economics.

1776–1880

The history of consumption and study of consumption in the United States can be divided in four periods. From the founding of the country to approximately 1880, consumption met everyday needs and behaved with respect to the economic theory of the day—a static, simplified model where consumers had fixed preferences and were maximizing some measure of utility or happiness subject to the prices of goods and their budget constraint. In mainstream economics, the sole reason for production is consumption. Assumptions underlying the standard economic model of consumption used almost exclusively in undergraduate texts are commonsense.

Consumers are supposed to know their preferences, these preferences are transitive, the consumer is supposed to be rational in the economic meaning of rationality, more is preferred to less, and consumers are supposed to be able to ignore irrelevant alternatives. Consumers are bound by their budget constraint and prices—if they know them. These models served economists well before the late 19th century, but they were crude and based on primitive, unrealistic assumptions. The modelers refused—or could not incorporate—history, social relations, culture, or institutions.

1880–1950

Thorstein Veblen seized on classical economists' oversimplified descriptions of human economic behavior. He then proceeded to verbally eviscerate them. Veblen's work summarizes the second period of U.S. consumption from 1880 to 1950. Veblen, an economic heretic, shocked and embarrassed his profession while amusing the general public with his sometimes cryptic writings about how institutions, laws, history, and social relations impact economic behavior. Veblen coined the phrase *conspicuous consumption* to describe the idea that consumption serves not only as a tool to meet basic needs but instead as a status symbol, and *conspicuous waste*—the idea that one could consume something of no value at all to oneself or society as a show of status. Veblen fingered the vacuous nature of consumer theory: it entirely ignored the vital needs of individuals for social relations and status. His notion of pecuniary emulation described the competitive nature of conspicuous consumption and waste. He opened up a new way of thinking about consumption that provided the fodder for marketers and advertisers to use in providing consumers the symbolic meaning they craved, while simultaneously exploiting these cravings.

1950–1980

The growth of advertising and marketing post–World War II marks the third period of the U.S. consumer. Producers had earlier recognized in the 1920s that they needed to create new products and demand and build some obsolescence into their products, otherwise profits would be short-lived. Marketers recognized that consumers were so fixated on keeping up appearances that they would even forgo necessities to do so. The critical assumption underlying the still-dominant, unchanged, but renamed, neoclassical theory of the consumer—that tastes were immutable and not the domain of economic theory—was challenged vigorously in both practice and the ivory tower. Combined with the assumption that more is better (the gluttony assumption), advertisers could, theoretically, create wants and desires. Consumers could then be victimized and fleeced of their money; in effect, they could be coerced into making transactions that they otherwise would not. The idea of coercion hit at the core tenet of neoclassical economics that mutual gains could be had from voluntary exchange. If the exchange was coerced—and hence not voluntary—

the theory of the consumer and of markets themselves falls apart. The chief critic of advertising and its potentially devastating impacts on especially the poor was John Kenneth Galbraith, the Harvard professor, U.S. ambassador to India, and adviser to U.S. President John F. Kennedy. Galbraith challenged the idea that tastes are exogenous to the system and echoed many institutional economists' critique of capitalism in general that satisfying wants through the competitive capitalist markets neither leads to greater individual nor societal well-being.

While Galbraith did not necessarily call for bans on advertising, like others had in the 1920s, his cautionary tales were prescient. The average 21st-century American is exposed to over 3,000 advertisements daily. From eggs to bathroom doors to helmets to planes flying overhead at the beach to tattoos on college students—media messages are nearly impossible to avoid. Advertising dollars spent in the United States annually grew from $150 billion in 1997 to $280 billion in 2006. In 2001, over $40 billion (of $230 billion) spent on advertising was spent on children. Advertising spending in the United States is comparable to the amount of money spent on education.

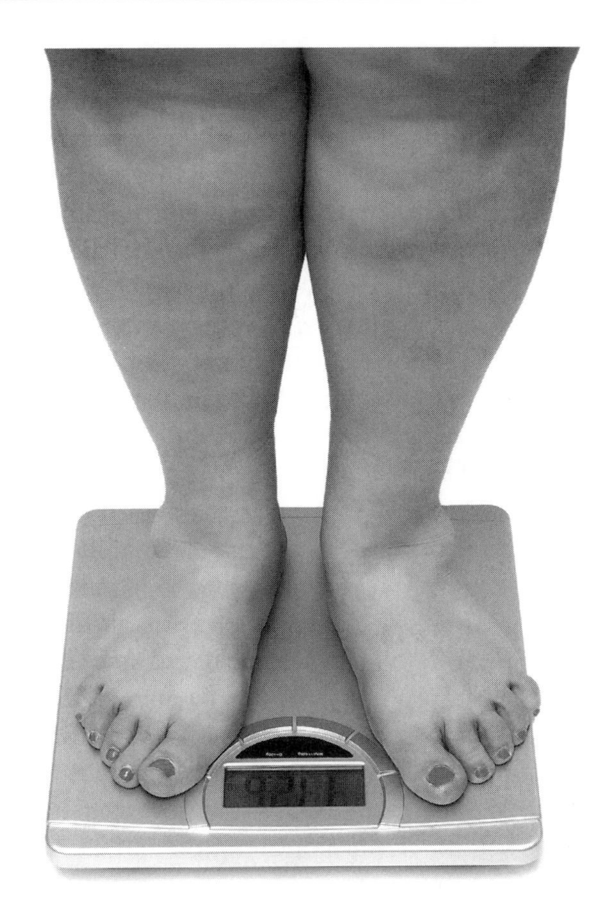

Food is abundant and relatively affordable in the United States, which is one of the world's most efficient producers. However, the nation has an obesity problem bordering on an epidemic, blamed on increased consumption, poor nutritional values, and lack of exercise.

1980–Present

The fourth (and current, as of 2011) distinctive period of U.S. consumption began in the 1970s and 1980s with a significant increase in status-motivated consumption. Juliet Schor provides an extensive body of work documenting the inability of consumers to escape the work-and-spend cycle as consumers are locked in an unending gerbil wheel of a battle to keep up with, not the Joneses or their neighbors, but with reference groups they could never hope to emulate—TV heroes like the characters in *Friends* or *Frazier* or other sitcoms. This run on the gerbil wheel results in more spending, frustration, and reduced joys of consuming precisely because the status-conferring qualities of competitive consumption never materialize.

Overconsumption

From an individual standpoint, competitive consumption, consumerism, and commodification may be disappointing in terms of utility maximization. From a public standpoint, concerns about overconsumption are manifold. Scientific consensus has identified increasing consumption levels around the world and especially in the United States as increasing carbon dioxide levels and other greenhouse gases exacerbate climate change and threaten the overall quality of life. Materialism in the United States is envied and mimicked worldwide. Mimicking competitive consumption globally, combined with U.S. unwillingness to act unilaterally to curb consumption and political gamesmanship, may support the sovereign world consumer and individual consumptive utility maximization, but it may also end up tragically magnifying a global crisis.

The United States has an obesity problem bordering on an epidemic. Food production in the United States is a source of pride and revenue. The United States is a net exporter of food and one of the most efficient world producers of a number of commodities. Food is, for most consumers, abundant and relatively affordable. Cheap, inexpensive food and profit-seeking marketers have combined with several generations who spend hours a day in front of a monitor, small or large, with little exercise. Questions about whether it is the amount or type of food, the lack of exercise, or a combination of the two has created increases in obesity (particularly childhood obesity). Childhood obesity increases health costs later in life and shortens life spans. Resources devoted to obesity-related costs could be used elsewhere.

While advances in measuring consumption preferences for public and nonmarket goods and services like those that the environment and ecosystems provide for all practical purposes, acknowledgement of this type of nonmarket consumption in practice is rare. Public goods, like infrastructure, public safety, or consumer protection, took the backseat to marketing, politics, and society's focus on the individual. Consequently, public consumption lags behind private consumption in the United States, and infrastructure is crumbling. All levels of government are struggling to raise revenue to cover long-term investments such as bridges, roadways, drinking-water systems, and sewage treatment plants.

Will Delavan
Lebanon Valley College

See Also: Carbon Dioxide; Consumerism; Food Consumption; History of Consumption and Waste, U.S., Colonial Period; History of Consumption and Waste, U.S., 1800–1850; History of Consumption and Waste, U.S., 1850–1900; History of Consumption and Waste, U.S., 1900–1950; History of Consumption and Waste, U.S., 1950–Present; Marketing, Consumer Behavior, and Garbage; Materialist Values; Sustainable Development.

Further Readings

Barber, Benjamin R. *Consumed: How Markets Corrupt Children, Infantilize Adults, and Swallow Citizens Whole*. New York: W. W. Norton, 2007.

Durning, Alan. *How Much Is Enough? The Consumer Society and the Future of the Earth*. New York: W. W. Norton, 1992.

Goodwin, Neva R., Frank Ackerman, and David Kiron eds. *The Consumer Society*. Washington, DC: Island Press, 1999.

Goodwin, Neva R. et al. *Microeconomics in Context*. 2nd ed. Armonk, NY: M.E. Sharpe, 2009.

Heilbronner, Robert L. and William Milberg. *The Making of Economic Society*. 12th ed. New Jersey: Pearson Prentice Hall, 2008.

Manning, Thomas D. *Credit Card Nation: The Consequences of America's Addiction to Credit*. New York: Basic Books, 2000.

Mason, Roger. *The Economics of Conspicuous Consumption*. Northampton, MA: Edward Elgar, 1998.

Sackrey, Charles, Geoffrey Schneider, and Janet Knoedler. *Introduction to Political Economy*. 5th ed. Boston: Dollars and Sense, 2009.

Schor, Juliet B. *The Overspent American*. New York: Basic Books, 1998.

Economics of Waste Collection and Disposal, International

Economics and waste have a complicated relationship; as national economies grow, the waste stream from that country also grows. That seems simple, but if two concepts—sustainability and globalization—are added to the picture, the complexity becomes more apparent. Ever since the Brundtland Commission published its 1987 report *Our Common Future*, the notion of sustainability (defined as "meeting the needs of the present without compromising the ability of future generations to meet their own needs") has had an impact on how people regard waste. Images of overflowing landfills, accidents with hazardous wastes, and widespread awareness of the potential impacts of climate change have changed attitudes toward the environment. The environment is no longer a free good, in economic terms, but a valued scarcity with an existence that is threatened by the growing waste production of growing economies. Globalization acts with sustainability to exacerbate human awareness of waste, for these two key concepts point to a closely connected world. Local and international economies intersect, and environmental policy decisions in one country may affect economic conditions in other parts of the world. In this context, waste is even more of a conundrum. A traditional definition of waste is of material that is no longer wanted and has no value for its owner; however, worldwide environmental regulation has enlarged that definition. Waste now can be viewed both as negative (a threat to the environment and the health of the planet) and positive (a resource or commodity that may be fed back into the production or commercial process to benefit local and national economies).

At the same time, most governments and economists in the developed world, with the developing world following close behind, view the production and consumption of goods and services as necessary to foster economic growth. However, increased production rates, along with increased use of natural resources and raw materials, lead to increased waste. Can increased waste, economic growth, and environmental concerns be reconciled?

Perhaps not, but German theorist Martin Jänicke suggests that local and international markets and environmental protection need not be mutually exclusive. To understand how this might be the case, one needs to examine what is meant by the term waste, how regulation and economic policy affect waste collection and disposal, and how businesses are responding to dramatically increasing international waste management opportunities.

Types of Waste and Waste Management Systems

All countries generate waste. But the kind of waste a country generates and how each country decides to manage its waste stream will, to some degree, depend on many intersecting factors. Geography, systems of government and law, manufacturing capabilities, public attitudes, and even climate will affect what waste is produced and the economic costs and benefits of managing its collection and disposal. Waste has a tendency to slip between categories, but five areas can give a basic image of the range of waste types: municipal solid waste (MSW), commercial, industrial, agricultural, and hazardous wastes. Because these categories are not completely separate from one another, the techniques for collection, disposal, or recycling of each type of waste may also cross categories. Various institutions—such as the Organisation for Economic Co-operation and Development (OECD), the World Bank, and the United Nations (UN)—as well as private consulting firms and some nongovernmental organizations—collect statistics on global production and management of different waste streams, but these figures are not easily comparable because collection methods and definitions of waste vary.

Worldwide, the most common form of rubbish disposal is the solid waste landfill, although amounts of waste disposed of in this way vary between and within continents. In Asian countries with large landmasses, such as India and China, almost 90 percent of MSW is disposed of in landfills. Countries in the Middle East have also traditionally relied on landfills, burying most waste in the desert. In many countries, availability of land for landfill is decreasing (because of population growth, urban sprawl, and resistance by communities to noisome dumps) and incineration provides an alternative. Despite the high costs of constructing and maintaining incineration plants—especially where there are strict emission regulations—these are a popular primary or secondary choice for waste disposal in many countries. The practice is most common in northern Europe, where landfill costs are high, and in Japan, where 73–78 percent of MSW is incinerated.

Other countries, such as the Philippines and Malaysia, have no incineration, in part because of the dangers of emissions from biomedical and hazardous wastes. Other waste streams come from commercial, manufacturing, and industrial sources. As with MSW, disposal practices vary from country to country, and each sector presents its own problems in disposal. Biomedical waste, for example, frequently cannot be recycled, and may contaminate other waste if not appropriately separated at the collection source.

Similarly, electronic waste (e-waste) causes significant disposal problems because of the high content of toxic materials. Other sources of waste that are often overlooked—yet have a direct relationship to economic life—are the waste created by natural disasters and the waste streams produced by the tourist and travel industries. Airports, for example, may generate amounts of waste similar to small cities, while cruise ships have been condemned for their impact on local environments and economies. Climate change experts suggest that the number and intensity of natural disasters will increase worldwide, and these forecasts have been borne out by the amount of destruction and the number of casualties in recent disasters. The impact of these events on waste production occurs both at the time of the event and in the process of recovery, placing great additional strains on local and national waste practices and budgets.

Economic Growth

While waste is generated by every country, no matter how rich or poor, the amount of waste a society generates is closely related to a country's gross domestic product (GDP). In general, the higher the income in a society, the higher its waste generation will be. Within OECD countries, municipal waste increased an astounding 59 percent from 1980 to 2002 and further increases, though at a lower rate, are expected up to 2020. The pressure on developed nations to continue production and consumption as a means of growing their economies has reinforced this pattern of waste generation.

The implications for land use and the environment are enormous. A similarly disturbing pattern holds true in other parts of the world, especially as economies develop. Some scholars suggest that if developing countries aim for a Western consumer lifestyle, the detrimental impact of human activity on the environment will increase tenfold by 2050. The building boom in the United Arab Emirates (UAE) provides one such example. Not only does the construction industry generate waste but also other technologies—such as desalination plants—that support its cities generate both waste and emissions at rates that give the UAE one of the world's highest per capita carbon footprints. Disposal rates are so high that Dubai's single waste treatment plant is often overwhelmed, and a similar plant in nearby Sharjah has a landfill that processes up to 39,000 tons per day, an amount that is three times the amount processed at Puente Hills landfill (east of Los Angeles)—the largest operation in the United States.

This pattern is duplicated in other parts of the world. Recent economic development in Asia has shown that growth is accompanied by increased solid waste generation, with the management of this sector becoming a major social, economic, and environmental issue. Economic growth and urbanization in Malaysia, for example, is correlated with a rapid increase in the generation of MSW. The World Bank identified MSW management as a major environmental concern for Malaysia, where some areas have a similar solid waste generation to many OECD countries. China, the world's fastest-growing economy as of 2010, has been the subject of even greater scrutiny. With its increasing population, urban development, and rising consumption patterns, China experienced an annual increase in urban domestic refuse of 8.2 percent between 1986 and 1991; by 2005, China had surpassed the United States as the world's largest MSW generator. As a World Bank report from that year noted, "No country has ever experienced as large, or as rapid, an increase in waste generation." The same report suggests that China will have to deal with an eightfold increase in its municipal budgets by 2020 as it struggles to deal with the increased waste stream across the country.

It is a particularly pressing problem to understand how to decouple economic growth from increased waste generation. It is not only MSW that is produced as developing countries scramble to reach the level of consumer comfort that most citizens of developed economies take for granted; as economies grow, so do their energy needs. In many places where efficient and clean energy sources are sought to bolster growing economies, countries are turning to nuclear power—most of nuclear plants now being built are in Asia and in central and eastern Europe. While many commentators laud nuclear power as a green energy because of its relatively insignificant emissions, critics point to the difficulties of safely storing nuclear waste as the more relevant point in the debate.

Waste Management and Regulation

With all of these issues—growing waste streams and problems posed by hazardous technologies and materials—how do governments around the world approach disposal practices, and how do these approaches affect business opportunities? Approaches include both regulatory practice and economic incentives and penalties to control the way goods are produced, consumed, and disposed of. Just as practices of disposal vary widely, regulatory approaches also differ around the world; however, common standards are increasingly being acknowledged worldwide. The work of the European Union (EU), through its executive body, the European Commission (EC), is particularly notable because it has actively pursued regulation of waste management since the 1990s. Aiming for sustainability, competitiveness, and cohesion, the EC promotes four key principles of waste management, all of which have an economic impact. These principles, as paraphrased by Mark Boyle in his 2003 article on waste management in the Republic of Ireland, are proximity (each member state looks after its own waste and should not need to export waste), the waste management hierarchy (prevention and minimization of waste production, then recycling and reuse over disposal), the precautionary principle (erring on the side of safety if there are any questions of hazard), and the polluter pays principle (PPP: the producer bears the costs of collection, treatment, and disposal). Of these key principles, the waste management hierarchy and PPP have been the most significant in both the EU and in influencing the development of policy and economic instruments in waste regulation around the world.

The EU requires its members to develop waste management plans (based on the hierarchy) with specific goals. A series of directives, such as those focusing on packaging and plastics, mining and extraction wastes, and hazardous wastes, have been promulgated since 1989, and member countries are meant to comply with these standards so that the environmental impact of waste will be halved by 2015. Within this broad framework, European countries develop their own strategies to meet EU targets. Germany, for example, introduced its Green Dot program in 1991, a program that directed producers to collect and recycle their own packaging. Countries seeking membership in the EU are also obliged to develop waste management plans as a condition of entry.

Since the beginning of the 21st century, waste management regulation has also had a high profile outside of Europe. Japan introduced its Fundamental Law for Establishing a Sound Material Cycle Society in 2000, with Korea following a similar approach with its Resources Conservation and Recycling Promotion Law in 2003. Waste management regulation, specifically directed at MSW, has also been enacted in other parts of Asia, including India, China, the Philippines (the Ecological Solid Waste Management Act was the first bill signed by incoming President Gloria Macapagal-Arroyo in 2001), and Taiwan. Other countries, including Indonesia, Malaysia, and Thailand, also have programs in place to control hazardous wastes and to reduce and recycle MSW. Australia has also developed a series of waste minimizing and recycling guidelines and targets ("Less Waste: More Resources") heavily influenced by the EU directives.

Regulation aimed at reducing packaging and e-waste has also been enacted across Asia and some parts of Africa. E-waste presents particular problems for developing nations: although it is banned from export from developed to less-developed countries through the Basel Convention, computers, cell phones, and other electronic goods still find their way to countries such as India, Nigeria, and Ghana, either illegally or through loopholes such as donations of goods. While India is attempting to address this issue with active recycling programs and regulation, the problem (often referred to as toxic colonialism) had not been solved as of 2010. Plastic bags and packaging are also often restricted or banned altogether by countries as diverse as Rwanda, Bangladesh, India, and the Republic of Korea, while recycling programs have been actively pursued in the Middle East (the United Arab Emirates and Israel, for example) and increasingly across Asia. Regulation, however, is not always the first step in encouraging waste management; in the UAE, cooperation and education (with a remarkable success rate of 85 percent usage of recycling receptacles, similar to rates in countries such as Canada) have been the foundation of recent programs, with compliance through regulation and economic incentives still to follow.

The climate for regulation and cooperation between government and industry has been slower in Latin America, but policy development indicates that this region is paying close attention to waste management practices. A short and incomplete list of examples includes Brazil, which has national regulation of pesticide packaging, with state laws on recycling, plastic takebacks, battery and medicine disposal; Mexico, which has similar covenants on pesticides and packaging and classifies expired medicine as hazardous waste; and Peru, which has national legislation that promotes producer responsibility and recycling.

Regulation Compliance: Incentives and Penalties

In all cases of policy development, the costs involved—to consumers, producers, and governments—are carefully scrutinized and actively debated. Regulation is generally accompanied by some form of economic instrument, whether in the form of incentive or penalty, but the underlying question of who will bear the cost can make or break the success of waste management programs. There is no uniform answer to this question: regions and countries decide on the best approach depending on their particular mix of industry, legislation, communal wealth, waste type, and attitudes toward the waste management hierarchy. Japan's sound-material-cycle approach expects both producers and consumers to take an active role in contributing toward the safe disposal of goods, while policy in India suggests that responsibility for disposal and recycling should be shifted away from the end user. These varying approaches underscore the local interpretation of international standards. Despite these variations, economic instruments are needed along with standards and effective monitoring and enforcement.

Incentives and penalties often work together. In recycling programs worldwide, households and businesses are encouraged to increase recycling by having deposit refund programs, while use-based waste generation fees and disposal taxes discourage larger waste streams, and outright penalties encourage compliance with environmental laws. In Japan, for example, the Home Appliances Recycling Law mandates that producers actively recover and recycle products, while consumers actively contribute to recycling costs. Similar programs are being tested in Australia. Japan has a similar approach to its end-of-life vehicle laws, where manufacturers are obliged to accept CFCs, air bags, and automobile shredder residue to recycle, but consumers contribute to the cost of recycling.

In areas where municipal waste management practices are still being introduced, such as some towns in the Philippines, surveys showed that over 80 percent of residents and businesses agreed on the principle of collection fees; 49 percent of households and 67 percent of businesses agreed that costs should be correlated to the volume of waste generated. Increased organization of waste management in developing areas, though, has implications for government budgets, with local and national expenditures increasing. Budget gaps are frequently addressed by strengthening private-sector participation, developing markets for sale of recyclables, and encouraging regional cooperation to develop economies of scale.

In many cases, compliance with regulation is often seen as an economic burden. It costs consumers more to cover disposal or recycling costs; it costs producers more to design and implement preventive or end-of-pipe measures, as well as bear partial or full responsibility for recovery and disposal; and it costs governments greater portions of waste management budgets to properly collect and dispose of waste. In particular, producers in Europe, where regulation has been active, see the cost of compliance as a negative; if legislation is not uniformly observed around the world, competitors are able to produce goods at lower cost without regulatory penalties.

Economic Benefits

However, that negative picture is incomplete. Despite free market rhetoric around the world, regulation and economic instruments may in some cases act as protectionist policies rather than always disadvantaging local producers. In addition, stricter environmental and waste regulation may offer many economic opportunities at local, regional, and international levels. These opportunities exist for providers of goods and services and for workers across the spectrum of informal and formal sectors. Recycling programs are an example of the complexity and the economic potential for waste management. Successful recycling programs (which target both supply and demand) are cheaper than new landfills, provide used materials that may substitute for virgin resources, and create jobs in the informal and formal sectors. As efforts become more organized, recycled goods become market commodities, rather than waste. This effort, while generally positive, does have its downside: larger quantities are collected, causing oversaturated markets and volatile pricing. For example, the post-consumer commodities market fluctuated wildly in 2008, with prices for old corrugated cardboard and old newsprint falling by 79 percent and 72 percent, respectively. Such fluctuations cause local stockpiling of recyclables and may also lead to temporary, improper disposal of collected goods.

Recycling

Despite these pitfalls, new businesses abound in the greening of waste management around the world. In the formal sector, start-ups are entering markets to promote recycling. In the Middle East, for example, where recycling has been minimal, new joint ventures in the UAE (between municipal governments and local and international private companies) encourage voluntary recycling through collection points at Abu Dhabi's National Oil Company stations. Start-ups in India are also trying to address the safe disposal and reuse of Waste Electrical and Electronic Equipment (WEEE) goods, while China has emerged as a leader in recycling, with the largest plastics recycling plant in the world in 2007. In the informal sector, waste pickers collect and recycle material at the local level. According to a World Bank report, approximately 2.5 million waste pickers worked in China in 2005, earning up to $150 per month (about half the earnings of cab drivers). This kind of informal collection, present in both developing and developed countries, connects local activity to world economic cycles, since goods circulate both in the pre- and post-consumer phases before making their way to disposal sites.

Waste to Energy

Waste management may also contribute to economic opportunities in other ways. Waste-to-energy (WTE) incineration plants are attractive to many municipalities because they simultaneously deal with two pressing issues: the disposal of waste and the need for energy. Although WTE plants need to be larger than typical incinerators to be cost effective and to meet emission regulations, they are utilized in Asia (notably China and Japan), in many parts of Europe (including the United Kingdom, Italy, and Scandinavia), and in North America (Canada and the United States) and are under consideration in other countries, such as Israel.

The proposal for the WTE plant in Jerusalem, undertaken with the assistance of U.S. companies, illustrates another area of economic growth within waste management: the provision of expert advice and services. Many countries, especially those with developing economies, have a dearth of waste management experts and must turn to international firms and government experts for advice, equipment, and services. International firms look for opportunities around the globe. North American firms, for example, may provide expertise to clients in the Middle East and Asia on waste management, while seeking expertise from countries such as Denmark on WTE projects. The EU, as a leader in waste management regulation, is also a key provider of expertise, especially in the developing world. The global market for environmental and facilities services stood at $227.1 billion in 2008, with the leading revenue source in this industry being solid waste management services (nearly 52 percent of the total). With ongoing increases of over 14 percent, industry watchers estimate that global waste services will be worth $260.3 billion by 2013, making it a growth industry for investors.

Small Business

It is not only large firms that benefit from the greening of waste management. In China, an estimated 90 percent of enterprises working in the environmental protection section are small firms. In addition, migrant workers have joined the stream of economic opportunities, traveling within countries (e.g., from the south to the north of India), or between countries (e.g., from India and Pakistan to the UAE), to work in recycling projects.

International Institutions

International institutions, such as the World Bank, also support the economic flows of waste management projects. Funding from the World Bank has gone to projects in locations as diverse as the West Bank, Jordan, Belarus, Vietnam, North Africa, and China. The EC also plays a significant role in the provision of funds and advice, working in conjunction with institutions such as the World

Waste-to-energy (WTE) incineration plants, such as this Teesside WTE power station at Haverton Hill, United Kingdom, address the pressing issues of waste disposal and energy needs. The European Union is a leader in waste management regulation. The European Commission promotes principles of waste management that aim for sustainability, competitiveness, and cohesion; all have an economic impact. Member states look after their own waste, prevent and minimize waste production, err on the side of safety, and use the "polluter pays" rule.

Bank or providing technical expertise to projects in India, China, and elsewhere.

Conclusion

Waste management services is one of the fastest-growing industries in the world, yet its existence presents an economic and environmental conundrum. If the industry continues to grow, it could well mean that one of the primary goals of current waste management practice—to reduce the amount of goods disposed of in landfills and by incineration—has not been met. A thriving industry—a success from an economic standpoint—will mean that the environmental mission identified by government regulation worldwide has failed. Conversely, if environmental targets to encourage less consumption and less waste are met, the waste management industry will cease to be an economic powerhouse. Clearly, this conundrum will continue to pose challenges for governments and businesses worldwide, but perhaps the biggest challenge of all is to the dominant vision of incessant growth as positive, necessary, and inevitable.

Deborah Breen
Boston University

See Also: Consumption Patterns; Developing Countries; Economics of Waste Collection and Disposal, U.S.; European Union; Incinerators; Sustainable Development; Zero Waste.

Further Readings

Felke, Reinhard. *European Environmental Regulations and International Competitiveness*. Baden-Baden, Germany: Nomos Verlagsgesellschaft, 1998.

Hoornweg Dan, Philip Lam, and Manisha Chaudhry. *Waste Management in China: Issues and Recommendations*. World Bank, Urban Development Working Paper, East Asia Infrastructure Department, Working Paper No. 9 (May 2005). http://siteresources.worldbank.org/

INTEAPREGTOPURBDEV/Resources/China-Waste-Management1.pdf (Accessed July 2010).
Lisney, Robert, Keith Riley, and Charles Banks. "From Waste to Resource Management." *Management Services* (December 2003).
Loughlin, Daniel H. and Morton A. Barlaz. "Policies for Strengthening Markets for Recyclables: A Worldwide Perspective." *Critical Reviews in Environmental Science and Technology*, v.36 (2006).
O'Neill, Kate. "The Changing Nature of Global Waste Management for the 21st Century" *Global Environmental Politics*, v.1/1 (2001).
Porter, Richard C. *The Economics of Waste*. Washington, DC: Resources for the Future, 2002.

Economics of Waste Collection and Disposal, U.S.

The economics of solid waste management in the United States have a wide range of variability by region. Many factors coalesce to generate this unevenness, such as availability of land for the placement of landfills, the political views of residents, the degree of urbanization and industrialization of the region, the type of waste and the amount generated, the prevailing wages, the choices of solid waste managers, and the degree of interaction with their communities. The modern approach to solid waste management in the United States and other advanced countries is labeled integrated solid waste management (ISWM) because it makes use of a variety of procedures to handle the solid waste that is collected. The chief strategies pursued through the ISWM system are landfill disposal, incineration, recycling, and composting.

About one-third of every dollar spent on municipal solid waste (MSW) management goes into the collection of waste. This basically accounts for the capital costs of the equipment (and its maintenance), for the labor costs (which constitute a large fraction of the collection costs), liability, and insurance costs. Unless the final disposal or processing site is within a short distance of the collection area, the waste is typically transported to a strategically located transfer station where it can be redirected to its destination: a landfill, an incinerator, a composting facility, or a recycling center—also known as a materials recovery facility (MRF). In the case of landfills, the waste may be transported hundreds and even over one thousand miles, adding significantly to transportation costs.

Landfills

Modern landfills have to meet strict design and operation criteria imposed by federal and state governments. Title D of the amended (in 1984) federal Resource Conservation and Recovery Act (RCRA) requires that landfills be lined at the bottom and sides by several layers of plastic and clay to prevent rainwater that percolates through the landfill to reach the water table of the site where the landfill is located. This percolated water, known as the leachate, is highly contaminated with organic chemicals and heavy metals. The RCRA also mandates that this leachate be monitored and treated. Furthermore, these modern sanitary landfills slowly decompose organic matter buried deep within them in an anaerobic biological process that generates methane. This methane must also be extracted and either burned on the spot or used to generate electricity through controlled combustion. All these requirements make the construction and operation of a sanitary landfill a relatively expensive undertaking.

For example, a landfill that processes 300,000 tons of waste per year with a life span of 20 years can have a fixed cost of $86.7 million, including post-closure costs. According to the RCRA, the owner of the landfill is liable for methane gas recovery and leachate monitoring and treatment for up to 30 years after the closure of the landfill. These costs vary with location. In the northeast United States, land availability for landfills is scarce. Also, the closer a landfill is to a metropolitan area, the more expensive it is to build and operate. This is reflected in the tipping fees that landfills charge for waste. The average tipping fee for the 100 largest cities in the United States is $47 per ton of waste. However, if one takes New York City as a reference point, the tipping fee for a landfill located 100 miles away from the city is $94 per ton, while that of a landfill 500 miles away is about $38 per ton.

The siting of a new landfill is also a politically arduous process with many communities not willing to allow proximate placement because the landfill is expected to create air pollution, possibly water contamination, a decline in the quality of life due to increased truck traffic and odors, and a concomitant decrease in property values. Within two miles of a landfill, housing values drop 6.2 percent for every mile that the home is closer to the landfill. This reluctance by citizens to allow the placement of a landfill (or, for that matter, an incinerator) near their community is what is known as the NIMBY (Not in My Backyard) phenomenon. As a result of both NIMBY and the RCRA regulations, the number of landfills has declined steadily, their size has increased substantially, and their locations have become more remote. The current trend is in the regionalization of landfills that accept waste from many towns, villages, and cities. These landfills have become larger, doubling their expected life span from 10 to 20 years. Many of the closed landfills were small waste dumps that did not meet state and federal regulations. Between 1988 and 1997, 500 landfills closed every year, decreasing from 8,000 in 1988 to 2,500 in 1997. By 2009, the total had declined to 1,800 landfills. Despite these reductions, the 10th-largest cities in the United States use a land area larger than the state of Indiana to discard their solid waste. In fact, the importation of solid waste from one state to another became a source of revenue for some states. By 1999, 8 percent of all waste generated was disposed of in another state. Pennsylvania went from burying 3.8 million tons of solid waste in 1993 to 7.9 million tons in 1996. Since it closed its huge Fresh Kills landfill in 2001, New York City sends 12,000 tons of solid waste every day to 26 different landfills and incinerators in New York, New Jersey, Virginia, Pennsylvania, and Ohio.

Incinerators

Incinerators have always been a controversial choice for the disposal of solid waste. They require a substantial capital outlay for their construction, their operating costs are high, and they have no flexibility regarding the throughput of waste—they must always operate at near their maximum capacity for economic reasons. Furthermore, their solid and gas outputs are considered hazardous, which results in their rejection by many communities with a more vigorous NIMBY response than that expressed against landfills.

Incinerators are designed to handle on the low end as much as 300 tons of waste per day, while the largest ones can handle up to 3,000 tons per day. Their corresponding initial costs range from $39 million to $780 million, respectively. Thus a mid-range-capacity incinerator burning about 1,200 tons per day, 365 days a year, for 20 years will need to charge a minimum of $39 per ton just to cover the capital investment. Even before considering operating and maintenance costs, this figure comes close to the tipping fees of many landfills in the United States. A reflection of this economic reality is the fact that the national average tipping fee for incinerators is about twice the national average of the tipping fee of landfills.

Most incinerators, as of 2010, are privately owned and generate and sell electricity to offset the operating costs. Incinerators operate under the principle of economies of scale and thus need to operate at or near maximum capacity in order to cover their expenses and guarantee a profit. This can be achieved only if the municipality agrees to deliver sufficient burnable solid waste to the facility. However, incinerator operators suffered a setback in 1994 when the U.S. Supreme Court issued a decision that made invalid all the "flow control" ordinances that municipalities had enacted in order to supply incinerators with this trash. This, in turn, forced incinerators to increase their tipping fees, which made them even less attractive as a waste disposal alternative.

Incinerators make economic sense either when they have a guaranteed supply of trash or when landfill operations are too expensive due to the scarcity of land, as in the northeast of the United States. The higher population density in this region makes it prohibitively expensive to find suitable sites for new landfills. Incinerators produce a variety of pollutants in their flue gas. Despite strict regulations that compel them to operate sophisticated air pollution control equipment, including the capture of fly ash (very fine and hazardous ash captured in filters), a very small fraction of the initial contaminant generated (such as carcinogenic dioxins) is released into the atmosphere. Furthermore, incinerators do not get rid of 100 percent of their burnable inputs; they produce a significant amount of ash collected at the bottom of the incineration segment of the unit. Original burnable trash is only reduced in volume by 70–80 percent. The remaining ash contains heavy metals and other toxic substances and was also declared hazardous by a U.S. Supreme Court decision, meaning that it has to be discarded in special landfills reserved for hazardous waste. Incinerators have to send this waste to a landfill, but a hazardous-waste landfill is more expensive than a regular landfill. It costs approximately $260 per ton to landfill ash. Comparing this to about $65 per ton to landfill the original (preburned) waste makes incineration significantly uncompetitive with direct preburn landfilling. In addition, the hazardous chemicals in this ash are more concentrated and more bioavailable, making it a substantially riskier mode of waste disposal.

Taking all these factors into consideration, communities have massively rejected the construction and operation of incinerators in their midst. By 1985, more incinerator capacity was being cancelled than proposed. Nationally, the number of incinerators peaked in 1991 at 170; by 1997, this number was down to 130. In addition, communities all over the United States successfully forced hospitals to close down the incinerators in which they combusted their medical waste.

Recycling

The other two components of modern ISWM systems are recycling and composting. Frequently, composting is treated as a special case of recycling. Organic waste is processed just like any other recyclable waste and the final product is put to use by a third party. However, recycling (of discarded manufactured items) and composting are often addressed separately.

Recycling has been the subject of intense controversy since the mid-1990s, when some economists began to question its costs and benefits and estimated that the costs of most recycling programs exceeded the economic benefits. Recycling was contrasted to landfill disposal and found to be significantly more expensive. Ever since, the debate about acceptable rates of recycling and about what is appropriate to recycle has continued. Many local governments initially resisted the implementation of recycling programs, and the efficiency of these programs is fairly diverse from city to city and between the states. Recycling has three major components: collection and processing from the waste generator by municipalities and businesses, use of the recycled material by manufacturers to produce new products, and the purchase by consumers of products made from recycled materials. The first two components have been subjected to deep scrutiny and to laws and regulations to propitiate an environment in which the recycled material makes it into a product. Recycling involves the cooperation of households in segregating recyclables from the rest of the waste, making them available to the collection entity, which then takes them to a transfer station, much like with trash destined to the landfill. From here, it is transported to the processing center—the MRF. The bulk of what is recycled as of 2010 falls under the categories of white paper, newspaper, mixed paper, cardboard, plastics, metals, glass, batteries, electrics, and electronics.

Recycling: Supply

According to 2006 Environmental Protection Agency (EPA) estimates, 32.5 percent of solid waste was recovered for composting and recycling as a national average, with 24.3 percent being recycling alone. However, there are cities like Portland, Oregon, in which the recycling rate is 50 percent, with a target rate of 60 percent. Nearly 9,000 curbside recycling programs have been operating nationwide since 1997, serving almost half of the population. Recyclables are also collected at drop-off centers. An analysis of the composition of solid waste, according to 2006 EPA estimates, shows a distribution of paper (33.9 percent); yard waste (12.9 percent); food waste (12.4 percent); metals (7.6 percent); wood (5.5 percent); glass (5.3 percent); rubber, leather, and textiles (7.3 percent); and other (3.3 percent). According to these figures, about 89 percent of the solid waste stream is either recyclable or compostable.

There is room for improvement. Such is the view of the EPA's Office of Solid Waste that considers the costs of efficiently run curbside recycling programs to be comparable to those of the collection and disposal of solid waste in landfills ($50–$150 per ton versus $70–$200 per ton, respectively, in

1997 dollars). In many programs, the cost of collection and processing of recyclables exceeds the revenue generated by the sale of recycled materials. Moreover, depending on the location, the cost of collection and landfilling is also lower than that of recycling. However, the more densely populated an area is and the more recyclables a collection vehicle collects per run, the lower the recycling costs because of economies of scale. Besides, in the northeast, where land is scarce and the NIMBY phenomenon is strong, recycling costs for an efficiently run program can have an advantage over landfill disposal.

Solid waste managers have improved the efficiency of recycling programs simply by educating the population about the benefits of recycling, compelling more households to participate in recycling or increasing their recyclables. Other strategies have involved providing containers for recyclables or increasing the size of the available containers. Further efficiencies can come from the redesign of the collection vehicles so that they can collect more recyclables per run or to have multiple-use trucks that can collect recyclables simultaneously with the regular trash destined for landfills. The fine-tuning of collection routes has been attempted with the idea of maximizing the number of collection stops per hour while minimizing the driving time. In addition, as recyclables collection increases, the number of routes assigned to collect landfill-bound trash can be reduced and the number of recyclable collection routes can be increased with the same personnel.

Another way that local governments have found to increase the relative amount of recyclables collected is by charging the households directly per bag of landfill-bound trash put at the curbside. The reasoning is that if the households want to cut their trash disposal expenses by cutting down the number of bags they produce, they will recycle more items. For example, Grand Rapids, Michigan, charges $1.25 per bag. By 1999, 4,000 communities in the United States priced their garbage directly. Nevertheless, this approach has its drawbacks because of relatively high administrative costs incurred by local governments trying to determine how much garbage each household has generated and to deter the illegal dumping of those trying to avoid paying their per-bag fees.

Recycling: Demand

Municipal solid waste, in principle, is collected from households and institutions like schools and government offices. This accounts for 250 million tons of solid waste per year for the whole country. However, commercial and industrial waste that is nonhazardous is also allowed to join the municipal waste stream, approximately doubling this figure. The most significant role that industry plays in solid waste economics is in the purchasing of recycled materials to use as raw material inputs into their products. This is the demand side of the recycling question. Although the situation has progressed since the 1990s, many manufacturers are still reluctant to accept recycled materials into the processing streams, either because they claim it is too expensive to refit the equipment for this purpose, or because they allege that the supply of recycled materials is inconsistent.

In order to compel manufacturers to use recycled materials, a series of laws and regulations have been imposed by local, state, and federal governments. Governments require that products that they purchase have a minimum amount of recycled material in them. Some are considering instituting a program similar to Germany's Green Dot program in which manufacturers are made responsible for the recycling of their packaging. About 30 percent by volume of landfilled waste is made up of packaging materials, and for every dollar spent by a consumer on disposable goods (such as food), approximately 10 percent goes to pay for the packaging. Another way to compel manufacturers to accept more recycled material is through landfill taxation, which can be used to subsidize recycling programs or to tax virgin materials. Business-oriented economists decry most of these measures. But virgin materials benefit from subsidies in the form of federal tax breaks. Income generated by the timber industry is taxed at the capital gains rate and not at the corporate income tax rate. Minerals are depleted from their mining region, and this is used by the mining companies as a deduction from their income in the form of a depreciation; and much of mineral exploration and extraction is carried out on public lands at no extra cost to the companies. In addition, freight rates paid for recycled materials are often higher than those paid for virgin materials.

Composting is another solid waste treatment method that avoids either landfill disposal or incineration. Most types of organic materials can be composted, including yard waste, household food scraps, commercial organic waste, and paper. At the end of the process, the original waste has been reduced by as much as two-thirds. The humus byproduct has an economic value as topsoil, mulch, fertilizer, or landfill cover. In 2010, quality mulch could be sold for $39 per cubic yard. Composting realizes a $14 per ton cost savings over landfilling.

The argument that recycling programs lose money and therefore should be cut back in favor of the cheaper disposal of trash in landfills goes against the grain of the thinking of large segments of the population. It baffles economists that the public is willing to pay even more money for their solid waste disposal if it benefits recycling programs. Surveys indicate that households are willing to pay an extra $3.27–$4.91 per month for access to yard waste collection and $6.44–$9.66 to have access to both yard waste and recyclables collection. Deposit-refund systems (for example, for bottles and containers) are popular, with the state of Michigan having a 100 percent return rate. It is estimated that a 10 percent reduction in solid waste destined for the landfill can be achieved through the institution of a system that collects and refunds $59 per ton of returnable material.

Proponents of recycling argue that the fact that a number of recycling programs do not break even—they lose money because the costs exceed the revenue from the sale of recycled materials—is a red herring because most government programs "lose money." This is particularly true of solid waste disposal in a landfill or through incineration. People have to dispose of their solid waste in some way; when it is done by these two latter means, this is money spent without the expectation of getting any of it back. Meanwhile, there are a variety of benefits when people recycle that cannot immediately be easily allocated a value. Benefits include the recovery and reuse of materials (for example, the trees that can be spared by recycling paper), the reduction in the disposal of waste (and the land that is conserved with the concomitant suppression of pollution generation), and the reduced efforts at collecting waste (and the reduction in greenhouse gas emissions that this entails). Moreover, there is significant disparity between the number of jobs generated by landfill disposal in contrast to that generated by recycling programs. According to some estimates, recycling 10,000 tons of material generates 36 jobs, while landfilling the same amount creates six jobs.

Composting

Composting is another solid waste treatment method that avoids either landfill disposal or incineration. All sorts of organic materials can be composted, including yard waste, household food scraps, commercial organic waste, and paper. Up to 15 percent of the solid waste stream is food waste, while yard waste constitutes 15 to 20 percent, either by weight or volume of municipal solid waste. Composting is a relatively simple process in which the shredded organic waste is combined with municipal sludge and then aerated properly (through a variety of technologies) for a period of weeks or months. At the end of the process, the original waste has been reduced by as much as two-thirds of the original waste, producing carbon dioxide, water, and humus (decayed organic mater). Humus has a variety of uses, such as topsoil, mulch, fertilizer, or as landfill cover. Initially, the "mulch" was given away for free to farmers and horticulturalists, especially when the quality of the mulch was inconsistent. In 2010, a properly processed mulch could commanded a revenue of $39 per cubic yard. Composting can be performed year-round on the ground in warmer climates like those of the Gulf Coast states, but it can also be carried out indoors with the proper technology and process design. For example, the city of Edmonton, Canada, has indoor composting facilities the size of 14 professional hockey rinks. Composting is economically superior to landfilling, with a net cost savings of $14 per ton.

Conclusion

The data presented describing the economics of solid waste management in the United Sates is hard to gather because of inconsistent, unavailable, or faulty records maintained by local governments. In spite of this, the economic picture shows that solid waste managers have a variety of options through the ISWM approach. Which methods they choose depends on a variety of factors. These factors depend on the population density of the locality and the particular idiosyncrasies of local government bureaucracies. Often, local residents form coalitions to force changes in the solid waste management system, for example, increasing recycling efforts or closing down incinerators. When solid waste is not directly handled by municipal governments, the job is done by private companies with the oversight of the government. By 2010, two companies, Waste Management, Inc., and Republic Services, Inc., handled more than half of the solid waste generated in the United States. These companies reduced costs by using fewer personnel. In general, government collection and disposal of solid waste is more costly than when private companies perform these services. According to some studies, whether collection is by private companies or by municipalities depends on the ideological bent of the local population. In conservative municipalities, waste collection and disposal is carried out by private companies, while in more liberal towns and cities, it is the municipality that takes care of solid waste.

Héctor R. Reyes
Harold Washington College

See Also: Composting; Incinerator Construction Trends; Incinerators; Industrial Waste; Landfills, Modern; NIMBY (Not in My Backyard); Post-Consumer Waste; Recycling; Waste Management, Inc.

Further Readings

Barlaz, Morton, et al. "Integrated Solid Waste Management in the United States." *Journal of Environmental Engineering* (July 2003).

Glenn, Jim. "Efficiencies and Economics of Curbside Recycling." *BioCycle*, v.33/7 (1992).

Kinnaman, Thomas C. "The Economics Of Municipal Solid Waste Management." *Waste Management*, v.29 (2009).

Kinnaman, Thomas C. and Don Fullerton. "The Economics of Residential Solid Waste Management." *NBER Working Paper no. 7326* (August 1999).

Porter, Richard C. *The Economics of Waste*. Washington, DC: Resources for the Future Press, 2002.

Tawil, Natalie. "Flow Control and Rent Capture in Solid Waste Management." *Journal of Environmental Economics and Management*, v.37/2 (1999).

Emissions

The generic term *emissions* is used to portray gases or particles pumped into the air by various sources. Emissions are not necessarily pollutant or harmful for living creatures. The Environmental Protection Agency (EPA) is mainly concerned with emissions that are or could be harmful to people. The EPA calls this set of principal air pollutants "criteria pollutants." The identified criteria pollutants are carbon monoxide (CO), lead (Pb), nitrogen dioxide (NO_2), ozone (O_3), particulate matter (PM), and sulfur dioxide (SO_2).

Carbon Monoxide

Carbon monoxide is an odorless, colorless, and toxic gas. Known sources for carbon monoxide are unvented kerosene and gas space heaters; leaking chimneys and furnaces; back-drafting from furnaces, gas water heaters, wood stoves, and fireplaces; gas stoves; generators and other gasoline-powered equipment; and tobacco smoke. Incomplete oxidation during combustion in gas ranges and unvented gas or kerosene heaters may cause high concentrations of CO in indoor air. Worn or poorly adjusted and maintained combustion devices (such as boilers or furnaces) can be significant sources, as can the flue if it is improperly sized, blocked, disconnected, or leaking. Automobile, truck, or bus exhaust from attached garages, nearby roads, or parking areas can also be a source. The effects of CO exposure can vary greatly from person to person depending on age, overall health, and the concentration and length of exposure.

Lead

Lead's toxicity has been recognized for more than a century; the metal is associated with the impairment of neural development in infants and young children and with cardiovascular disease and premature death in older people. Governments have long tried to reduce exposure by controlling industrial emissions, removing lead from gasoline, and mounting campaigns to remove lead-based paint from homes. In the United States, some of the highest lead levels in blood can be found in children in older cities like Philadelphia, Pennsylvania; Providence, Rhode Island; and Cleveland, Ohio. In 2008, the EPA set new limits for exposure at 0.15 micrograms per cubic meter of air, down from 1.5 micrograms.

Nitrogen Dioxide

Nitrogen dioxide is a highly reactive oxidant and corrosive gas. The primary sources indoors are combustion processes, such as unvented combustion appliances like gas stoves, vented appliances with defective installations, welding, and tobacco smoke. Effective measures to reduce exposures to this gas are venting the NO_2 sources to the outdoors and assuring that combustion appliances are correctly installed, used, and maintained.

Ozone

Ozone is a molecule consisting of three oxygen atoms. Ozone in the lower atmosphere is an air pollutant with harmful effects on sensitive plants and the respiratory systems of animals. The ozone layer in the upper atmosphere is beneficial, preventing potentially damaging ultraviolet light from reaching the Earth's surface. Exposure to ozone has been associated with premature death, asthma, bronchitis, heart attack, and other cardiopulmonary problems. The EPA lowered its ozone standard from 80 parts per billion (ppb) to 75 ppb in 2008. Nevertheless, the World Health Organization recommends 51 ppb.

Particulate Matter

Particulate matter is a mixture of very small particles and liquid droplets. Particle pollution is made up of a number of components, including acids (such as nitrates and sulfates), organic chemicals, metals, and soil or dust particles. According to the EPA, the size of particles is directly linked to their potential for causing health problems. Particles that are 10 micrometers in diameter or smaller can pass through the throat and nose, invading the lungs. Once inhaled, these particles can affect the heart and lungs and cause serious health effects. The EPA groups particle pollution into inhalable coarse particles (those found near roadways and dusty industries, larger than 2.5 micrometers and smaller than 10 micrometers in diameter) and fine particles (those found in smoke and haze that are 2.5 micrometers in diameter and smaller). These particles can be directly emitted from sources such as forest fires or they can form when gases emitted from power plants, industries, and automobiles react with other substances in the air.

Sulfur Dioxide

Sulfur dioxide is a highly reactive gas. According to the EPA, the largest sources of SO_2 emissions are from fossil-fuel combustion (especially coal) at power plants (73 percent) and other industrial facilities (20 percent). Smaller sources of SO_2 emissions include industrial processes such as extracting metal from ore and the burning of high-sulfur-containing fuels by locomotives, large ships, and non-road equipment. Sulfer dioxide is linked with a number of adverse effects on the respiratory system.

Other Pollutants

There are also a large number of compounds that have been determined to be hazardous and are called "air toxics." They are pollutants known or suspected to cause cancer or other serious health effects, such as reproductive effects or birth defects, or adverse environmental effects.

Categories and Measurement

The EPA has grouped emissions into two main categories: point and mobile emissions. Point source emissions come from clearly identifiable and relatively fixed sources like factories and electric power plants. Mobile sources include cars, trucks, lawn mowers, airplanes, and other moving items that put gases or particles into the air. Additional classifications can include biogenic (produced by living organisms or biological processes) and area emissions (spread over a spatial extent, like a county or air district).

Measurement of emissions is a complex process involving the detection of thousands of different substances in a variegated set of environments. In the United States, there is a central repository for measurement data from emissions. The Clearinghouse for Inventories and Emissions Factors (CHIEF) is a centralized resource for emissions data. The emissions data that is stored by CHIEF is normally used to create models, which can help to predict what air quality will be like in the future and what effect new regulations might have on air quality.

Greenhouse Gases

Greenhouse gases (GHGs) are released into the atmosphere by several sources. One of the main areas of interest regarding emissions has to do with global climate change due to anthropogenic sources of GHG emissions. The main GHGs in the Earth's atmosphere are water vapor, CO_2, methane, nitrous oxide, and ozone. Once in the atmosphere, those gases absorb and emit radiation within the thermal infrared range. This process is the cause of the greenhouse effect. Controlling for emissions coming from GHGs is a major environmental challenge for the United Nations. The Kyoto Protocol is an international agreement linked to the United Nations Framework Convention on Climate Change. The major feature of the Kyoto Protocol is that it sets binding targets for 37 industrialized countries and the European community for reducing GHG emissions. These amount to an average of 5 percent against 1990 levels over the five-year period 2008–12.

Natural gas is mainly composed of methane, which produces 20 times more radiation than carbon dioxide. Carbon dioxide receives more attention in the media over other greenhouse gases because it is released in much larger amounts. Still, it is inevitable when natural gas is used on a large scale that some of it will leak into the atmosphere. Current estimates by the EPA place global emissions of methane at 3 trillion cubic feet annually, or 3.2 percent of global production. Direct emissions of methane represented 14.3 percent of all global anthropogenic greenhouse gas emissions in 2004. According to the United Nations' Intergovernmental Panel on Climate Change (IPCC), from 1750 (the dawn of the Industrial Revolution) to 2005, the concentration of atmospheric CO_2 has risen from 280 to 379 ppm. Emissions have mounted in the 21st century. In 2009, total global CO_2 emissions had increased 25 percent since 2000 to 31.3 billion tons and almost 40 percent since 1990, the base year of the Kyoto Protocol.

Current trends in GHG emissions have impacted global weather. The amounts and types of emissions change every year in different countries. Those changes are caused by changes in global and regional economies, industrial activity, technology improvements, traffic, and many other factors. Air pollution regulations and emission controls also have an effect.

In 1970, the U.S. Congress passed the Clean Air Act (CAA) Amendments (the CAA was passed in 1963), which set into motion a nationwide effort to improve the country's air quality. Since then, additional laws and regulations have been added, including the 1990 amendments to the Clean Air Act. The global economy is a powerful driver for emissions. For example, in 2009, for the first time since 1992, there was no growth in global CO_2 emissions from fossil fuel use, cement production, and chemicals production. In 2009, a strong world economic recession led to a dramatic decrease in CO_2 emissions of approximately 7 percent in most industrialized countries. This drop of 800 million tons in emissions compensated for the continued strong increase in CO_2 emissions in China and India of 9 percent and 6 percent, respectively. This would have been the largest drop in more than 40 years because the global recession froze economic activity and slashed energy use around the world.

Since 2000, carbon emissions in China have more than doubled; in India, emissions have increased by more than half. Since the end of 2008, China has been implementing a large economic stimulus package over a two-year period. This package includes investment in transport infrastructure and in rebuilding Sichuan communities devastated by the 2008 earthquake. In 2009, CO_2 emissions jumped by 9 percent to 8.1 billion tons, even though China has doubled its installed wind and solar power capacity for the fifth year in a row. India, where domestic demand makes up three-quarters of the national economy, remained relatively unaffected by the credit crunch. Emissions continued to increase in 2009 by 6 percent to 1.7 billion tons of CO_2. India surpassed Russia as the fifth-largest CO_2 emitter.

A recent global estimate of the share of energy use attributed to urban areas (approximately 66 percent) suggested that the GHG emission share reached 70 percent. A global, geographically disaggregated estimate of GHG emissions from urban areas, however, had yet to be conducted as of 2010. Nonetheless, most of the measurements are performed in industrialized countries. The assessment excludes CO_2 emissions from deforestation and logging, forest and peat fires, postburn decay of remaining above-ground biomass, and decomposition of organic carbon in drained peat soils. The latter mostly affects developing countries. These sources could add as much as a further 20 percent to global CO_2 emissions.

Mauricio Leandro
Graduate Center, City University of New York

See Also: Automobiles; Car Washing; Engine Oil; Fuel; Gasoline; Tires.

Further Readings

Environmental Protection Agency. "Air Pollution Emissions Overview." January 2011. http://epa.gov/airquality/emissns.html (Accessed January 2011).

Hill, Marquita K. *Understanding Environmental Pollution*. Cambridge: Cambridge University Press, 2010.

Tiwary, Abhishek and Jeremy Colls. *Air Pollution: Measurement, Modelling and Mitigation*. 3rd ed. New York: Routledge, 2010.

Engine Oil

Engine oil is used to both prevent friction between internal metal surfaces of an engine and transfer heat away from the combustion cycle. Rubbing of metal engine parts produces microscopic metallic particles from the tiring and wearing of surfaces. Such particles could circulate in the oil and grind against moving parts, causing wear. To prevent this, oil is typically circulated through a filter to remove damaging particles. An oil pump powered by the engine pumps the oil throughout the engine, including the oil filter, catching the waste gathered by the oil along the cycle. Engine oil also cleans the engine of waste chemicals and buildups by grabbing the hurtful by-products of combustion, such as silicon oxide and acids, in suspension. Many engine oils have detergents and dispersants added to help keep the engine clean and minimize oil sludge buildup.

Modern varieties of engine oil are designed to work for extended periods under tremendous heat and pressure without losing mechanical or chemical characteristics. Engine oil is supposed to reduce oxidation that occurs most frequently at higher temperatures. Corrosion inhibitors are sometimes added to the engine oil. The specific composition of any given corrosion inhibitor depends on many factors, including the material of the system it has to act in, the nature of the substances it is added into, and the operating temperature. Some of the most common inhibitors are nitrites, chromates, and phosphates.

Used Oil and Oil Filters

An average car utilizes up to five quarts of engine oil, which has to be replaced approximately three or four times per year. Most of these changes are performed at service stations, where worn oil and filters are generally collected for different uses. A primary use for used engine oil is to re-refine it into a base stock for lubricating oil. This process is very similar to the refining of crude oil. According to the American Petroleum Institute, lubricating oil does not wear out, it simply becomes dirty as it does its job. Once water and contaminants are removed from used oil, it is returned to a full and useful life as re-refined base oil. The claimed result is that the re-refined oil is of as high a quality as a virgin oil product. Other sources assert that only 2.5 quarts of re-refined lubricating oil can be produced from one gallon of used oil.

A secondary use of the used oil is to burn it for energy production. Some used oil is sent to power plants or cement kilns to be burned as fuel. On a smaller scale, small quantities of used oil are burned in specially designed heaters to provide heating for small businesses.

Each year, the United States generates 425 million used automotive oil filters containing 160,000 tons of iron units and 18 million gallons of oil, since used oil filters can contain more than 45 percent used motor oil in weight when removed from the vehicle. Recycling all the filters sold annually in the United States would result in the recovery of about 160,000 tons of steel, or enough steel to make 16 new stadiums the size of Atlanta's Turner Stadium. Nonetheless, in the United States, less than 60 percent of used oil and filters are recycled.

An average car requires up to five quarts of engine oil to run properly, which should be replaced three or four times a year. Engine oil can be repeatedly refined as a base stock for lubricating oil, in a process similar to crude oil refinement.

Runoff

A significant amount of car owners change engine oil routinely at home. In California alone, one in five households have a do-it-yourself (DIY) oil changer. Those DIY devices can collect used oil for recycling, but some of the oil is not recovered. Cars leaking oil and oil spilled directly into driveways are an important cause of what is called *nonpoint source* (NPS) *pollution* or *polluted runoff*. This phenomenon is created when rain, snowmelt, irrigation water, and other water sources run over the land, picking up pollutants and transporting them into local water bodies. NPS pollution is also called "people pollution" because much of it is the result of activities that people do every day. Because each individual contributes to NPS pollution by performing daily activities, NPS pollution is the biggest threat to ponds, creeks, lakes, streams, rivers, bays, estuaries, and oceans. With each rainfall, pollutants are washed from surface and land areas into storm drains that flow into nearby waterways. Oil dumped on land reduces soil production. In the United States, more than 60 percent of water pollution comes from NPS, such as runoff from agricultural areas draining into rivers, fertilizers from gardens, or failing septic tanks. The oil from a single oil change can ruin the taste of a million gallons of drinking water (1 part per million), the supply of 18,250 people for 24 hours. It also has been demonstrated that concentrations of 50 to 100 parts per million (ppm) of used oil can pollute sewage treatment processes. According to the Environmental Protection Agency

(EPA), one quart of motor oil can pollute 250,000 gallons of water.

Oil and petroleum products are toxic to people, wildlife, and plants. Used motor oil can contain toxic substances such as benzene, lead, zinc, and cadmium. Since oil does not dissolve in water, it lasts a long time. Oil sticks to everything from beach sand to bird feathers. Films of oil on the surface of water prevent the replenishment of dissolved oxygen, impair photosynthetic processes, and block sunlight. The EPA has calculated that used motor oil is the largest single source of oil pollution in lakes, streams, and rivers. For Christine Todd Whitman, former administrator of the EPA, countless small acts, such as changing cars' oil in driveways without cleaning up leaks, can add up to big problems. According to the EPA, Americans spill approximately 180 million gallons of used oil each year into the nation's waters. This is approximately 16 times the amount spilled by the *Exxon Valdez* in Alaska in 1989.

Mauricio Leandro
Graduate Center, City University of New York

See Also: Car Washing; Emissions; Fuel; Gasoline; Tires.

Further Readings

Agency for Toxic Substances and Disease Registry. "Case Studies in Environmental Medicine: Taking an Exposure History." Atlanta: U.S. Department of Health and Human Services, 2001.

State of California's Used Oil Recycling Program." http://www.calrecycle.ca.gov/usedoil (Accessed July 2010).

U.S. Department of Health and Human Services. "Toxicological Profile for Used Mineral-Based Crankcase Oil," September 1997.

Environmental Defense Fund

Nonprofit organizations have played a crucial role in both forming state and national policy on waste and environmental issues and shaping public behavior. One of the oldest and most influential nonprofit organizations, with more scientists and economists on staff than any other similar organization, is the Environmental Defense Fund (EDF). EDF primarily focuses its resources on four key areas: global warming; land, water, and wildlife; oceans; and health. It is a critical source of information and advocacy, and its efforts have provided four decades of significant results for environmental and human health.

Beginnings

EDF was founded by middle-class New York suburbanites in 1967 as a result of focused attention on environmental issues such as those exposed in Rachel Carson's 1962 book *Silent Spring*. Carson revealed the damaging effects of DDT, a commonly used pesticide, on wildlife and humans. The EDF founders, resolved to halt the use of DDT, initiated a successful lawsuit to ban the dangerous chemical. After their success, they incorporated as EDF and grew from a small, volunteer organization operating out of an attic in the Stony Brook post office to the leading environmental organization in the country, employing around 350 individuals, enjoying a membership of over 700,000, and working on environmental issues worldwide. The majority of EDF's funds come from individual donations and memberships. Foundation grants provide another large portion of funding, while less than 15 percent of operating support and revenue comes from bequests, investment income, and government grants. Charity Navigator, an independent organization that evaluates the financial responsibility of charities, gives EDF its highest rating.

Activities and Successes

EDF's activities are governed by four key principles: a commitment to sound science, the power of partnerships, the power of the market, and a commitment to getting the law right. Nonpartisan and rigorous, EDF's scientific pursuits help guide its efforts and are used by state and national authorities in setting policy. The first major success of the fledgling EDF, the ban on DDT, was based on sound science. Basing its advocacy on careful studies provides EDF a trusted and respected voice. As a result, scientific studies linking chemicals in drinking water to high rates of cancer helped secure 1974 passage of the Safe Drinking Water Act, which, with amendments, is still used in the 21st century to ensure that Americans have safe water to drink. In 1985, EDF scientists played a crucial role in banning lead from gasoline, resulting in a significant decrease in childhood lead poisoning. EDF builds on past successes in 21st-century initiatives. One project promotes public understanding of the science behind climate change. Another initiative utilizes partnerships with leading universities to investigate the science of ocean mining and the implications of nanotechnology.

Corporate Collaboration

When EDF initiated a waste-reduction partnership with McDonald's in 1990, it was the first environmental group to work with a corporation. Up to that point, corporations were largely regarded as the enemy of environmental organizations, and working with a for-profit company engendered controversy within the environmental movement. Critics wondered if EDF was selling out to corporate interests or appealing to the lowest common denominator. The EDF-McDonald's partnership, however, was successful. It led to the elimination of 150,000 pounds of waste through new product packaging, and other fast food companies adopted the best practices developed by the EDF-McDonald's task force. A 2000 partnership with FedEx to create a more energy-efficient delivery truck enhanced fuel economy and decreased emissions and soot. In 2007, EDF opened an office in Bentonville, Arkansas, home of Walmart's headquarters, to help Walmart develop environmentally friendly policies and to assist the behemoth retailer in setting guidelines for its suppliers. EDF has partnered with local youth groups as well as rural communities to promote sustainable growth. Through partnerships with businesses and other groups, EDF aims to promote best practices within industries and to show the economic value of environmental stewardship.

Other Successes

The focus on improving the bottom line through environmentally friendly practices is also evident in EDF's principle of the power of markets. It is a proponent, along with corporations such as Ford Motor Company, Caterpillar, General Electric, Shell, and others, of a national cap-and-trade program to reduce global warming. As a result of scientific studies showing the far-reaching effects of acid rain, EDF successfully lobbied for a provision in the Clean Air Act of 1990 to include a cap on sulfur dioxide emissions. A 1995 plan allowed landowners compensation for protection of wildlife covered under the Endangered Species Act. EDF hopes to use market-based incentives to reduce deforestation in the Amazon, to reduce wasteful and dangerous fishing practices, and to limit air pollution, congestion, and overdevelopment in U.S. cities.

EDF prefers to enter alliances to negotiate for tough laws to protect the environment. However, when alliances fail to achieve their goals, EDF will initiate or participate in lawsuits to ensure the law is tough, fair, and based on sound science. It has been instrumental in creating guidelines for commercial vehicles, eliminating loopholes that led to significant pollution from diesel engines. EDF played a key role in the U.S. Supreme Court case that gave the Environmental Protection Agency the authority to regulate pollution from global warming.

Aimee Dars Ellis
Ithaca College

See Also: Environmentalism; Greenpeace; Packaging and Product Containers; Safe Drinking Water Act.

Further Readings

Gaudiani, C. *The Greater Good: How Philanthropy Drives the American Economy and Can Save Capitalism*. New York: Times Books/Henry Holt, 2003.

Graff, T. J. "Development of a Leading Environmental NGO: Thirty Years of Experience." *The Role of Environmental NGOs: Russian Challenges, American Lessons*. National Research Council, ed. Washington, DC: National Academies Press, 2001.

Krupp, F. and M. Horn. *Earth, The Sequel: The Race to Reinvent Energy and Stop Global Warming*. New York: W. W. Norton, 2008.

Environmental Justice

Environmental justice refers to both a protest movement that emerged in response to unequal exposure to waste and a set of principles developed out of that movement. Growing out of a realization of social inequities relating to waste, the environmental justice movement has spurred consciousness throughout the United States and around the world about the racial and economic dimensions of waste production and management.

The term *environmental justice* has roots in the U.S. civil rights movement in both the rhetoric and organizational strategies used. Of particular note is the Memphis Sanitation Workers strike of 1968, which attracted international attention when Dr. Martin Luther King Jr. supported the strike (and was ultimately assassinated while participating in it). In 1982, an African American community in rural Warren County, North Carolina, protested that state's siting of a toxic waste dump. The residents were joined by national civil rights figures, including the Southern Christian Leadership Conference's (SCLC) Joseph Lowery.

While the protest was unable to stop the site's development, it inspired similar protests as well as research into the relationship between waste and race. Five years later, the United Church of Christ's Commission for Racial Justice published a study, "Toxic Wastes and Race in the United States: A National Report on the Racial and Social Economic Characteristics of Communities of Hazardous Waste Sites" that concluded that race was the dominant factor in the siting of waste in the United States.

Scholars and activists across the nation mobilized during this period, with sociologist Robert Bullard (who had worked on these issues in Houston as early as 1979) emerging as a leading advocate and scholar. Bullard advised the Clinton administration as it took office, upon which the Environmental Protection Agency established environmental justice guidelines.

Legislative History

Enacted by the U.S. Congress in 1969 and signed into law by President Richard Nixon on January 1, 1970, the National Environmental Policy Act of 1969 (NEPA) requires all federal agencies and programs to consider the impact of their actions on the environment. Provisions of the act require that federal programs must be administered in an environmentally sound manner. Of particular historical significance is Section 102(2)(c), which established the requirement of a detailed statement of the environmental impact of the proposed action, any adverse effects, and alternatives. This review process results in an Environmental Impact Statement that solicits expert testimony and data that is circulated to affected parties, is reported in the Federal Register, and includes public hearings. The review and corresponding comments are then used to mitigate any actions that are taken. Shortly after this landmark act was instituted, the Environmental Protection Agency (EPA) was established in 1970 to oversee this and other environmental protection legislation.

The earlier Civil Rights Act of 1964 and the provisions of Title VI-Nondiscrimination in Federally Assisted Programs, specifically Section 601, stated that "no person in the United States shall be excluded from participation in or otherwise discriminated against on the ground of race, color, or national origin under any program or activity receiving Federal financial assistance."

In theory, this should have by default included a greater involvement of minority and underrepresented segments of the population when the NEPA process was enacted. However, this was often not the case when stakeholders most impacted by a proposed project were unaware of or unable to attend public hearings because of scheduling and location.

In addition, growing concerns were raised by grassroots organizations, which felt that their communities were being targeted for actions that had potentially negative environmental consequences because they were not politically or economically empowered. They believed that their views were not being fully considered as part of the environmental review process. Concerns were heightened by projects such as the placement of environmentally sensitive waste transfer stations and trash incinerators in mostly impoverished and minority communities. This led to the growth of Not in My Backyard (NIMBY) resistance movements.

As a response to many of these public concerns, in 1992, the EPA created an Office of Environmental Justice. In 1993, the EPA established the National Environmental Justice Advisory Council (NEJAC) to gather independent advice and recommendations from a variety of stakeholders, including academia, community groups, industry, nongovernmental organizations and environmental organizations, state and local governments, and tribal governments and indigenous groups.

To involve a wider array of government agencies, President Bill Clinton issued in 1994 Executive Order 12898, Federal Actions to Address Environmental Justice in Minority Populations and Low-Income Populations. The executive order refocused attention on Title VI of the Civil Rights Act of 1964 wherein ". . . each Federal agency shall make achieving environmental justice part of its mission by identifying and addressing, as appropriate, disproportionately high and adverse human health or environmental effects of its programs, policies, and activities on minority populations and low-income populations. . . ."

Goals and Effects

Environmental justice must involve the fair treatment of people regardless of their racial, ethnic, or socioeconomic status. No community or subpopulation should have to carry a disproportionate share of negative environmental outcomes. There must also be meaningful involvement of the potentially affected community. This requires seeking out those groups who will bear the greatest risk of a proposed action by scheduling public meetings during times and in locations that are fully accessible to members of the community and must consider the participants' concerns in the decision-making process.

Environmentalists and grassroots organizations involved in promoting environmental justice as a civil rights issue have developed additional terms specific to actions that have been viewed as creating a negative environmental impact on minority and indigenous populations, including the concepts of environmental racism and environmental colonialism.

The history of environmental protection has been one of balancing risk to human health, the preservation and conservation of natural resources, and economic development. Environmental crime resulting from noncompliance with environmental protection legislation raises public apprehension about the regulatory oversight of these policies. This is of particular concern when specific populations and locations are disproportionately affected.

Although a number of policies are in place to promote environmental justice for government-related projects, they do not necessarily cover actions by private industry. In response to these public fears, both the Emergency Planning and Community Right-to-Know Act of 1986 and the Pollution Prevention Act of 1990 were enacted to protect public health and the environment by improving community access to information about chemical hazards in the air, land, and water and to increase the capacity of emergency response and risk management. These were further clarified in 1995 through President Clinton's Executive Order 12969, Federal Acquisition and Community Right-To-Know Federal Actions.

Furthermore, to enhance public notification of environmental activities that may affect air, water, and land in certain neighborhoods, the EPA has created in support of its environmental justice initiatives EnviroMapper, a Web-based database that maps all facilities that have to report their use of controlled materials to the EPA. This can range from locations that produce and release air pollutants to Superfund sites.

Barrett Brenton
Anne Galvin
St. John's University

See Also: NIMBY (Not in My Backyard); Politics of Waste; Race and Garbage; Social Sensibility.

Further Readings

Bullard, Robert D. *Dumping in Dixie: Race, Class, and Environmental Quality*. Boulder, CO: Westview Press, 1990.

Environmental Protection Agency. "EnviroMapper for Ecofacts." http://www.epa.gov/emefdata/em4ef.home (Accessed December 2010).

Environmental Protection Agency. "Environmental Justice." http://www.epa.gov/compliance/environmentaljustice/index.html (Accessed December 2010).

Environmental Protection Agency. "National Environmental Justice Advisory Council." http://www.epa.gov/compliance/environmentaljustice/nejac/index.html (Accessed December 2010).

McGurty, Eileen. *Transforming Environmentalism: Warren County, PCBs, and the Origins of Environmental Justice*. New Brunswick, NJ: Rutgers University Press, 2007.
National Environmental Policy Act of 1969. http://ceq.hss.doe.gov/nepa/regs/nepa/nepaeqia.htm (Accessed December 2010).
Pellow, David N. *Garbage Wars: The Struggle for Environmental Justice in Chicago*. Cambridge, MA. MIT Press, 2002.
Presidential Documents. "Executive Order 12898 of February 11, 1994. Federal Actions to Address Environmental Justice in Minority Populations and Low-Income Populations." http://ceq.hss.doe.gov/nepa/regs/eos/ii-5.pdf (Accessed December 2010).
Presidential Documents. "Executive Order 12969 of August 8, 1995. Federal Acquisition and Community Right-To-Know Federal Actions." http://ceq.hss.doe.gov/nepa/regs/eos/eo12969.html (Accessed December 2010).
Rechtschaffen, Clifford, et al., eds. *Environmental Justice: Law, Policy & Regulation*. 2nd ed. Durham, NC: Carolina Academic Press, 2009.
Schlosberg, David. *Defining Environmental Justice: Theories, Movements, and Nature*. New York: Oxford University Press, 2009.

Environmental Protection Agency (EPA)

The Environmental Protection Agency (EPA) is a government agency of the United States. Its overall aim is to safeguard the environment and human health in the United States and to cooperate with other countries to protect the global environment. Specifically, the EPA's duties include developing national regulations and standards based on environmental laws passed by the U.S. Congress; monitoring and enforcing regulations via fines, legal sanctions, and other measures; collaborating with government and industry in pollution prevention and energy conservation programs; conducting environmental assessment, research, and education; and providing grants and support to external environmental programs and scientists. In the 21st century, the EPA has focused on priorities such as climate change, air and water quality, safety of chemicals, environmentalism, and building partnerships with state and tribal governments. An up-to-date listing of the EPA's current projects and programs is available on its Website.

History

The EPA was established on December 2, 1970, under the direction of President Richard Nixon. It was formed following a merger of smaller government agencies and departments involved in environmental monitoring and enforcement. The 1960s and 1970s in the United States were characterized by increased public concern for the natural environment. Rachel Carson's 1962 book *Silent Spring*, which discussed the widespread pesticide poisoning of nature, and other literature and media revealed the scale of the country's environmental problems to a largely unaware general public. There was a gradual cultural shift in people's attitudes toward the environment, including a public outcry for direct government action to protect the natural world, which marked the beginning of the modern era of U.S. environmentalism.

Administrators and Structure

The EPA is led by an administrator, appointed by the U.S. president. The first administrator was William Ruckelshaus (1971–73), followed by Russell Train (1973–77), Douglas Costle (1977–81), Anne Gorsuch (1981–83), William Ruckelshaus (1983–85), Lee Thomas (1985–89), William Reilly (1989–93), Carol Browner (1993–2001), Christine Whitman (2001–03), Michael Leavitt (2003–05), Stephen Johnson (2005–d09), and Lisa Jackson (2009–). The administrator heads a senior management team, including a deputy administrator, several assistant administrators, and regional administrators. The EPA has approximately 18,000 employees and several thousand contractors located across the country; most are employed as engineers, environmental protection specialists, scientists, support staff, and legal professionals. The agency's headquarters is located in Washington, D.C., and is made up of several offices, such as the Office of Air and Radiation, the Office of Environmental Information, the Office of Research and Development, the Office of Water, and the Office of Solid Waste and Emergency

Response. In addition, there are 10 regional offices located around the United States; each is responsible for implementing the EPA's programs within several states, except for programs delegated to U.S. states and Native American tribes. In 2010, the EPA's annual budget was about $10.486 billion.

Powers and Responsibilities

As a regulatory agency, the EPA is authorized by the U.S. Congress to create and enforce regulations to implement environmental laws and presidential executive orders. Approximately 130 substantive regulations are issued every year. Regulations written by the EPA help businesses, government, individuals, public authorities, and others to abide by environmental laws as they explain the necessary technical, operational, and legal details for compliance. The laws that are regulated by the EPA cover a range of environmental and public health protection issues; examples include the Clean Air Act, the Clean Water Act, the Energy Policy Act, the Pollution Prevention Act, and the Safe Drinking Water Act.

There are also a number of laws that guide the processes by which the EPA develops regulations. The EPA provides compliance assistance activities such as explaining regulatory requirements and ways to abide by them (such as providing specific compliance information); offering support services and resources, such as counseling, training, fact sheets, and guides; and providing technical assistance. This is accompanied by monitoring and assessment of compliance through inspection and evaluation at facilities and sites, detailed investigations, providing self-evaluation compliance tools for organizations, detailed review of information submitted to the EPA, and incentives for organizations and the public to report violations of environmental laws.

Despite the Environmental Protection Agency's strong reputation, a number of its decisions have been controversial and criticized. Scientists and environmentalists call for more action on health and environmental threats, while industry wants fewer regulations.

Enforcement

The EPA uses three main enforcement programs to ensure compliance with regulations. Criminal enforcement is used against serious violations and intentional disregard of the law and can result in stringent sanctions such as jail sentences. Civil enforcement uses legal means, but without criminal sanctions. Cleanup enforcement ensures the cleanup of pollution and waste or arranges reimbursement to the EPA for cleanup. The Comprehensive Environmental Response, Compensation, and Liability Act (CERCLA, or Superfund) is often used by the EPA as a cleanup enforcement mechanism. The EPA's enforcement programs work closely with the Department of Justice and local governments to take legal action.

Research and Planning

The EPA's Office of Research and Development conducts scientific research about the environment and awards grants and fellowships to scientists not employed by the EPA. It also provides expertise and technical support to address environmental issues. Research topics are wide-ranging and include air, computational toxicology, drinking water, ecology, human health, land, nanotechnology, pesticides and toxics, sustainability, and water quality. The office's work is coordinated across three national laboratories, four national centers, and various research centers and facilities. The national laboratories are the National Exposure Research Laboratory, National Health and Environmental Effects Research Laboratory, and the National Risk Management Research Laboratory.

Similar to other federal agencies, the EPA is required (under the 1993 Government Performance and Results Act) to performance manage its activities and to communicate information about its performance to the U.S. Congress and the public. Accordingly, the EPA conducts performance activities such as strategic and annual planning, detailed analysis and reporting, sets performance targets, and seeks input from internal/external stakeholders and external reviewers. The EPA writes documents about its performance, such as its Strategic Plan Annual Performance Plan and Budget, Performance and Accountability Report, and program evaluations, all of which are published on its Website.

Controversy

On the one hand, scientists and medical experts argue that not enough is being done to combat many environmental and health threats, whereas on the other hand, business leaders and others often argue for fewer regulations. As a government organization, the EPA follows the political agenda of the majority party, yet Democrats and Republicans disagree on a number of issues concerning environmental stewardship; and there have been reports of political interference in the work of the EPA.

Other criticisms include not enough reliance and power within the EPA among nonscientists, research activities that are not in tune with the agency's regulatory goals, failure to acknowledge some environmental and health risks, and lack of progress toward environmental justice.

Gareth Davey
Hong Kong Shue Yan University

See Also: Clean Air Act; Clean Water Act; Comprehensive Environmental Response, Compensation, and Liability Act (CERCLA/Superfund); Environmentalism; Public Health Service, U.S.; United States.

Further Readings

Collin, Robert William. *The Environmental Protection Agency: Cleaning Up America's Act.* Westport, CT: Greenwood Press, 2006.

Landy, Marc, Marc Roberts, and Stephen Thomas. *The Environmental Protection Agency: Asking the Wrong Questions.* New York: Oxford University Press, 1994.

Environmental Tobacco Smoke

Environmental tobacco smoke (ETS) is smoke that has originated from the use of tobacco products and remains in the environment. It is usually a combination of smoke emitted from a burning cigarette or cigar (known as sidestream smoke) and smoke exhaled by the smoker (known as mainstream smoke). In contrast to direct smoking, environmental tobacco smoke is inhaled by people not using the tobacco product, and, therefore, exposure to it is often labeled as involuntary smoking, passive smoking, secondhand smoke, and tobacco smoke pollution. The phrase *thirdhand smoke* is used to describe by-products from tobacco smoke remaining on smoke-exposed materials like clothing, skin, and upholstery.

Composition

Environmental tobacco smoke contains more than 4,000 chemicals, including at least 250 harmful to health and 50 carcinogens. Carcinogens in tobacco smoke include arsenic (a heavy-metal toxin), benzene (a chemical found in gasoline), beryllium (a toxic metal), cadmium (a metal used in batteries), chromium, ethylene oxide, nickel, and vinyl chloride. The chemical composition of smoke is highly variable and dependent on factors such as the tobacco type, chemical additives, wrapping paper, the way the product is smoked (such as smoking rate), contact with indoor surfaces, and environmental conditions. Research shows that environmental tobacco smoke can have a different composition than mainstream smoke, including more toxic substances, and it is not simply diluted smoke. The presence of tobacco smoke in indoor air is determined by testing for smoke constituents; and exposure of a nonsmoker to smoke can be tested by measuring cotinine (a nicotine by-product in body-fluid levels) in blood, saliva, or urine.

Risks

There is a scientific consensus that environmental tobacco smoke is a risk to human health, although the causal link is not always clear. Exposure to smoke in the environment is associated with the same health problems as direct smoking, particularly, a significant increased risk of cardiovascular diseases, cancers, respiratory problems, and cognitive impairment and dementia.

The Environmental Protection Agency (EPA) classifies environmental tobacco smoke as a known human carcinogen and estimates that each year it is responsible for several thousand lung cancer deaths among nonsmokers. Environmental tobacco smoke can also trigger asthma, allergies, and other respiratory problems; cause breathing and other problems among young children and infants, such as asthma, bronchitis, colds, infant death, ear infections, and pneumonia; and be a risk for pregnant women and babies, such as low birth weight, premature birth, and foetal death.

The health risks of exposure to environmental tobacco smoke are substantially lower than those associated with direct smoking. However, there is no safe exposure level, as studies have shown that even low amounts in the environment can be harmful, although the magnitude of risk remains debated by scientists. The overall risk is higher when nonsmokers spend many hours in an environment where environmental tobacco smoke is widespread, such as in the home or workplace. As a consequence of potential health risks, many state and local governments around the world have implemented smoking bans in indoor public facilities such as airports, bus terminals, cafés, hospitals, nightclubs, residential care facilities, restaurants, and schools. At the national level, some countries have adopted legislation to protect citizens from exposure to environmental tobacco smoke; for example, in March 2004, the Republic of Ireland became the first country in the world to impose an outright smoking ban in all indoor workplaces. Since then, many others have followed suit, including Italy, New Zealand, Norway, and Sweden. In countries without bans, an increasing number of private work and other places have taken voluntary measures to be smokefree. Smoking bans tend to be more effective when complemented by additional initiatives such as health awareness campaigns concerning health risks of smoke and programs to help people quit smoking. Alternatives to smoking bans include ventilation and air cleaning systems to reduce tobacco smoke pollutants and separating smokers from nonsmokers, but a ban is the only way to fully protect against smoke exposure.

Effects of Smoking Bans

Scientific research shows that smoking bans eliminate exposure to tobacco smoke in the indoor environment, thereby dramatically improving air quality, while exposure remains high in venues with no smoke-free regulations. Some studies have reported health benefits following policy implementation, such as a considerable reduction in the incidence of heart attacks and significant improvement in the respiratory health of workers, but more research is needed to substantiate these claims. Smoking bans also contribute to a reduction in tobacco consumption among employees and, therefore, lower the prevalence of tobacco-related diseases and their economic burden and social costs.

In the long term, the potential health improvements could have major economic benefits. Opinion polls reveal that most people are aware of the health risks of environmental tobacco smoke and that smoking bans have considerable public support. Criticisms of smoking bans include difficulty enforcing bans and possible harm to the economy (especially the hospitality and leisure industries), although there is no evidence of an overall negative impact. Some scientists disagree on the risks associated with passive smoking and its causal link to ill health, and they also criticize policy proposals to restrict smoking in communal places. The tobacco industry is also known for criticizing, and funding scientific research that casts doubt on, these issues.

Gareth Davey
Hong Kong Shue Yan University

See Also: Air Filters; Pollution, Air; Ventilation and Air-Conditioning.

Further Readings

Parker, Philip. *Environmental Tobacco Smoke: Webster's Timeline History, 1974–2007*. San Diego, CA: Icon Group International, 2009.

Watson, Ronald and Mark Witten. *Environmental Tobacco Smoke*. Boca Raton, FL: CRC Press, 2001.

Environmentalism

Environmentalism (as a philosophy) calls into question the impact of human actions on the environment and (as a movement) seeks to alter human-environment interactions to lessen the human footprint on the environment and foster environmental health. As a philosophy, environmentalism is usually associated with human values for conservation, ecosystem protection, restoration, and a deep concern for the natural environment.

As a social movement, environmentalism is associated with local, regional, national, and international political involvement in environmental policy development and reform, green technology development, advocacy for more protected areas, and improved citizen opportunities to shape and influence policies and development activities that affect the environment.

History

Sociologist Riley Dunlap and several others have documented an increase in environmentalism, in the form of increased concern about the environment in North America and a number of other countries. In the United States, long standing concern about conservation of natural resources and local concerns about public health gave way to rising concern about environmental health and quality after World War II. This process intensified in the 1960s, as organizations such as the Environmental Defense Fund grew, the popularity of Aldo Leopold's 1949 book *A Sand County Almanac* extolled the notion of humans having an ethic to protect the land rather than simply produce wealth, and Rachel Carson's 1962 book *Silent Spring* galvanized worries about toxins. Public pressure produced political action, including the advent of Earth Day in 1970 and a myriad of environmental protection policies that (for a time) enjoyed bipartisan support in the United States.

Worldwide, environmental politics spread in industrialized nations, aided in part by concern over nuclear weapons and nuclear power. Those concerns spurred the growth of West Germany's Green Party in the 1970s (a party that continues in the early 21st century as a substantial presence in parliament), and green parties exist, with less influence, in several other nations.

Greater public interest in the environment in the United States was heightened by the widely read book *Silent Spring*, which detailed the destructive role of pesticides and other agricultural applications on bird life; the first Earth Day in 1970; and the energy crisis in the 1970s (due to an oil embargo), where then U.S. President Jimmy Carter urged Americans to consume less energy. The environment began to hold greater sway over public media and interest in North America and elsewhere, and suggested a move away from a dominant social paradigm that viewed humans as exceptional creatures who are able to overcome environmental limits, particularly through technological advancements.

The stewardship ethic under the dominant paradigm is based on utility for humans, where land is primarily useful for harvesting natural resources, and it is reasonable to use land and water for waste disposal. Under the new environmental paradigm, humans are one among many creatures and increasingly learn to recognize their interdependence with the physical world. The new environmental paradigm is associated with environmental protection, population control, and constraints on industrial activity.

Approaches

Since the 1960s, surveys measuring adherence to the new environmental paradigm have found increasing levels of support, as support for the dominant social paradigm lessens. The work of environmentalism is largely undertaken by environmental organizations (typically termed *environmental nongovernmental organizations* or *ENGOs*. The approaches of various ENGOs reflect distinct views on how to best address environmental crises. Such views include the following:

- Degrowth, or scaling down the amount of raw material taken from the Earth and placed back into the Earth via wastes or pollution. There is considerable advocacy involved in this approach, often to stop a development, to build a hazardous waste facility, for example, or to increase the safeguards taken to assure that environmental harm is minimized (such as the one might expect now for offshore exploratory oil rigs, especially after the Big Horizon Gulf oil spill off the coast of Louisiana).
- Ecological modernization, or the increasing role of institutional reform to channel economic growth and industrial development in a direction that minimizes harm to the environment and incorporates environmental science into the design of new infrastructure, technology, and decision-making systems.
- Environmental conservation, or the ways in which wildlife, habitat, natural resources, and ecosystems can be protected at the individual, organizational, and governmental level.
- Environmental economic instruments, or the tools that government, industry, and other organizations can use, such as taxes or traceable pollution permits, to provide both incentives and disincentives for improved environmental behavioral outcomes.
- Public education, or the provision of learning opportunities to enhance public knowledge, attitudes, and behaviors about how natural environments function and how human beings can manage their behavior and ecosystems to live sustainably.

While each approach may adhere to a new environmental paradigm, there are often differences in the level of focus (e.g., individual, private sector, public sector) and in the acceptance or rejection of industrial capitalism (with its assumptions of constant economic growth). The listed approaches simplify the terrain considerably, but provide illustrative representations of the state of environmentalism in the 21st century.

Degrowth

The cornerstone of the degrowth approach to addressing environmental concerns is the notion that more economic growth can never be a solution to the environment, as it is the primary cause of the problem. Taking a critical stance of "sustainable development," adherents to degrowth feel that the aim of making economic growth environmentally sustainable is untenable: economic growth can never coexist with environmental sustainability. The *Limits to Growth* report published by the Club of Rome in 1972, which predicted an ecological collapse as a result of accelerated economic growth, is an influential document for the degrowth movement as was the work of Nicholas Georgescu-Roegen (a Romanian economist), who established the study of the bioeconomy in the 1970s.

Numerous ENGOs adopt a degrowth perspective on at least some of their work. For example, Greenpeace has been active in attempting to halt industrial production in Canada's oil sands (a large oil extraction project in Alberta). The premise for these organizations is that no level of industrial production can justify the damage inflicted to the natural environment and local people, including water pollution, toxic waste, and massive mining projects. A quotation from Greenpeace Canada's Website depicts a degrowth orientation: "Greenpeace is calling on oil companies and the government to stop the tar sands and end the industrialization of a vast area of Indigenous territories, forests and wetlands in northern Alberta." To achieve such ends, ENGOs involved in degrowth often engage players at multiple levels, including government, industry, community organization, and individual. However, unlike those supporting ecological modernization and environmental economics, those working from a degrowth perspective would be critical of working with government and industry, instead choosing to place pressure on them from the outside and trying to gain adherents from the broader public.

Ecological Modernization

Ecological modernization is a sociological theory of environmental reform. Growing out of Western Europe in the 1980s, early ecological modernization theorists sought to make room for advances in addressing environmental problems, rather than focusing solely on negative human-environment interactions. Initially, much attention was placed on the role of technological innovation in environmental reform. While this remains a feature of ecological modernization in the early 21st century, its prominence is less than in early conceptualizations. Ecological modernization theorists think little of approaches that concern only the public sector, maintaining a favorable attitude toward the marketplace as a driver of positive environmental change. The distinct features of ecological modernization are a view of environmental problems as challenges for social, technical, and economic reform; an emphasis on transformations in these arenas undertaken by the public and private sectors; and a rejection of degrowth perspectives that see economic growth as a threat to environmental sustainability. Rather, ecological modernizations suggest that economic growth and improved environmental reform can go hand in hand, as the innovators in ecological modernizing set new standards for others to follow, and those successful at reducing environmental harm find that it is also more profitable given the reduced costs they face for wastes and energy inefficiency.

In contrast to the focus on degrowth, ecological modernization is focused largely at the national or industrial level, rather than at the individual or community level. For example, the Pembina Institute (a sustainable energy thinktank in Alberta, Canada) approaches environmental action from an ecological modernizationist mindset. The group has a corporate consulting branch that, according to its Website, "provide[s] solutions for select industry leaders in Canada seeking to make their businesses more sustainable and climate-friendly." By supporting technology that is ostensibly beneficial to the environment and involving a range of stakeholders (such as communities, governments, and industry), the Pembina Institute works within the capitalist system to achieve results that improve environmental outcomes. Such results include life cycle assessments, alternative measures of progress, and ecological fiscal reform.

Environmental Conservation

Environmental conservation (or protection) is the form of environmentalism with which most people are familiar. This approach also has the strongest historical precedent. The Fabian Society of the late 19th century supported environmental protection as a means of creating a more civil society; John Muir acted to save tracts of wilderness (and cofounded the Sierra Club) in the western United States at the same time; and Henry David Thoreau expounded on the restorative qualities of pristine environments in the mid-19th century. Most governments have federal bodies that oversee the establishment of national parks and a mandate to protect at least some percentage of the country's land mass.

An example of a conservation ENGO is the Canadian Parks and Wilderness Society (CPAWS), with the primary aim to establish more protected areas in Canada. On its Website, its mission is "to keep at least half of Canada's public land and water wild—forever." To do so, CPAWS works with community groups and lobbies government. Those ENGOs seeking environmental protection are not necessarily for or against increasing industrial development. Conservation groups have varied stances on industrial growth; their goals tend to focus on policies that can further protect landscapes that are rare, hold endangered species, or that are of high social value given their proximity and potential for human recreational use.

Environmental Economic Instruments

Environmental (or ecological) economics focuses on the relationship between the economy and the ecosystems that support it. An environmental economics approach to environmentalism is at the state or industrial level and typically seeks to "fix" the current economic system rather than replace it—as a degrowth environmentalist might wish to do. Thus, efforts in this arena have emphasized the importance of modeling ecosystem functions in order to better understand the natural world as well as improving measures of progress—such as gross domestic product (GDP)—to account for decreases in human and environmental health as a result of economic activity. Additionally, this field of study addresses the valuation of ecosystem services, recognizing that in a system where the worth of a good is determined by its price, the environment will continue to be overlooked by most players in the economy. Finally, ecological economists seek to establish policy that uses fiscal incentives and disincentives to better the relationship between humans and the environment. For example, some of the economic instruments that are debated among environmentalists, industry, and government as possible solutions to environmental problems include taxes, pollution credits, mitigation banks, and carbon credits.

In addition to the Pembina Institute, other ENGOs that take an environmental economics approach to their work include the World Resources Institute (WRI). For example, in its Markets & Enterprise division, WRI's stated goal "is to harness markets and enterprise to expand economic opportunity and protect the environment." There are overlaps with ecological modernization, and the two approaches are quite similar. But while ecological modernizationists seek to involve stakeholders in the environmental policy arena, ecological economists are more closely focused on the use of economic instruments to achieve change.

Public Education

The final major approach is public education. Focused on the individual or community, the aim of public environmental education is to provide information to the general public on how the environment functions and how to live in an environmentally sustainable manner. Such education need not be in a school but can rely on other media as well (e.g., labeling, television shows, and advertising campaigns). Approaches to environmental education range from outdoor education (where individuals are taken into the wilderness to experience the natural world) to public information campaigns (e.g., on topics such as littering or idling a vehicle). In general, the aim of environmental education is to create a more aware and conscientious citizenry that cares about the environment and thus seeks to increasingly behave in informed and deliberate ways to minimize environmental harm. Environmental educators seek to equip citizens with the skills and information to make wise choices informed by environmental information.

The Evergreen Foundation is an ENGO with a mission, according to its Website, "to support and inspire community action [that] promotes the connection between a sustainable natural environment and healthy communities." As with most environmental educators, its focus is not on industry or government but on teaching individuals and community leaders about how to make cities more liveable by creating more outdoor spaces. The foundation works on community gardens, native plant identification, and local restoration projects. Like environmental conservationists, environmental educators do not necessarily subscribe to a capitalist or anticapitalist doctrine and often leave the motivations for social change from political activism to those they educate.

Conclusion

Environmentalism is constantly changing itself and the world. Despite differences in opinion about how best to address environmental problems, there exists agreement among environmentalists that nonhuman species deserve protection and similar moral rights as humans. Taken together, the various approaches described are affecting change in policy, in public attitudes, and on the landscape.

Emily Huddart Kennedy
Naomi Krogman
University of Alberta

See Also: Capitalism; Consumption Patterns; Greenpeace; Sustainable Development; Zero Waste.

Further Readings

Canadian Parks and Wilderness Society. "CPAWS Is Canada's Voice for Wilderness." http://www.cpaws.org (Accessed October 2010).

Costanza, R. *Ecological Economics: The Science and Management of Sustainability.* New York: Columbia University Press, 2001.

Dunlap, R. E. "The Impact of Political Orientation on Environmental Attitudes and Actions." *Environment and Behavior*, v.7/4 (1975).

Dunlap, R. E. and K. D. van Liere. "Commitment to the Dominant Social Paradigm and Concern for Environmental Quality." *Social Science Quarterly*, v.65/4 (1984).

Evergreen Foundation. http://www.evergreen.ca (Accessed October 2010).

Gottlieb, Robert. *Forcing the Spring: The Transformation of the American Environmental Movement.* Washington, DC: Island Press, 1993.

Hays, Samuel P. *Beauty, Health, and Permanence: Environmental Politics in the United States, 1955–1985*. New York: Cambridge University Press, 1987.
Pembina Institute. "Pembina Corporate Consulting: Sustainable Energy Solutions for Leaders." http://corporate.pembina.org (Accessed October 2010).
Rauwald, K. S. and C. F. Moore. "Environmental Attitudes as Predictors of Policy Support Across Three Countries." *Environment & Behavior*, v.34/6 (2002).
World Resources Institute. "Working at the Intersection of Environment and Human Needs: Markets and Enterprise." http://www.wri.org/markets (Accessed October 2010).

European Union

The European Union (EU) is an economic and political union comprising most of the countries in Europe. With roots in several economic and diplomatic alliances forged in Europe after World War II (including the European Economic Community), the EU was established by the Treaty on European Union (signed in Maastricht on February 7, 1992, and which entered into force on November 1, 1993). In 2011, member states included Austria, Belgium, Bulgaria, Cyprus, the Czech Republic, Denmark, Estonia, Finland, France, Germany, Greece, Hungary, Ireland, Italy, Latvia, Lithuania, Luxembourg, Malta, the Netherlands, Poland, Portugal, Romania, Slovakia, Slovenia, Spain, Sweden, and the United Kingdom. Candidate states included Croatia, Macedonia, Iceland, and Turkey.

The EU maintains common policies on trade among its member states, including a common currency (the euro, used in most member states, although not in the United Kingdom), as well as the movement of people, goods, and services. While its policies generally promote free movement of goods and services within its borders, the EU has developed policies aimed at controlling the production and distribution of wastes.

The 21st-century EU legal framework on waste is based on Directive 2008/98/EC. This directive includes a preamble of 49 points indicating the general orientation of the EU toward waste policy, management, recovery, and disposal. The directive is structured into seven chapters, which contain a sum of 41 articles. The chapters respectively include subject, scope, and definition; general requirements; waste management; permits and registrations; plans and programs; inspections and records; and final provisions. According to the EU summary, the most significant articles are those concerning waste hierarchy, waste management, permits and registrations, and plans and programs.

Waste Hierarchy

The complex legislation of waste management has been developed not only in relation to the geopolitical transformation of the EU legislative domain, but also with regard to technological progress in the field of environmental sciences. The numerous amendments to the laws reflect the need to cope with these two factors and are expressions of the awareness by the European Commission (EC) of the need to maintain a certain degree of flexibility over time. The 2008 regulation enlarged the scope of the previous one, introducing clearer definitions not only of the types and qualities of waste, but also of the hierarchy of waste treatment procedures. This step underlines the importance of differentiating possible ways of waste management within a general priority for the safeguard of human and animal health and of environmental protection. Thus, according to the 2008 directive, waste hierarchy is, in order of priority: prevention, preparing for reuse, recycling, recovery (such as energy recovery), and disposal. This hierarchy replaces the one formulated in the 2006 directive, which prioritized prevention, followed by recovery and reuse. In particular, the 2008 directive specifies that waste prevention should be the first priority and that reuse and recycling should be preferred to energy recovery from waste.

Waste Management

The section related to waste management refers to the responsibility of waste management, which is in principle that of the original producer or holder. Even in case of transfer of the responsibility for recovery or disposal, this responsibility is not completely discharged from the original producer. Also, the principles of self-sufficiency and proximity are sanctioned. The former concerns the idea that the community as a whole should reach self-sufficiency in waste management. The latter refers to the cooperation needed by (especially bordering) member states to establish an "integrated and adequate network of waste disposal and recovery installations." Articles 17–22 regulate the control and labeling of hazardous waste and oils.

Permits and Registrations

The section on permits and registrations underlines that any member states carrying out waste treatment shall obtain permits that specify the types and quantities of waste, technicalities, safety measures, method of treatment, monitoring operations, and other provisions when necessary.

Plans and Programs

The section on plans and programs defines the types of management plans that have to be issued periodically by each member state following the objectives and provisions of the directive. One relevant part of these plans concerns waste prevention programs, according to which each member state is expected to establish its own program by December 2013, indicating existing prevention measures as well as planned quantitative and qualitative benchmarks for assessing the progress of prevention measures.

Orientation and Coordination

Concerning the general orientation of the EU legislative framework, attention is paid to harmonizing EU directives with the freedom of each member state to amend or adapt its own legislation. This is done under the aim to "set general environmental objectives for the management of waste within the Community," or "to enable the Community as a whole to countries' self-sufficiency in waste disposal and recovery." Again, a strong emphasis is added to "help the EU move closer to a 'recycling society'" in order to allow each state to become aware of the need to reduce generation of waste and to learn to use waste as a resource. These general orientations somehow juxtapose with pluralistic indications over the freedom of member states "to limit incoming shipments to incinerators classified as recovery," to stress individual states' decisions over the use of waste as an economic resource, and to develop states' individual waste prevention programs. Even the ambitious call for a "European recycling society" is modulated through awareness that member states maintain different approaches to the collection of household waste and to waste composition; therefore it is considered appropriate that general targets "take account of the different collection systems in the Member States." One particular and extremely significant case of pluralism in the recent directive concerns the list of waste. This section specifies that a member state is allowed to consider waste as hazardous even though this does not appear in the annexed list, and an item present in the annex as nonhazardous. This tension between the EU and national freedom of interpretation of the EU legislation has often provided space to nontransparent and even illegal practices, such as the case of waste shipment.

The introduction of the single market in the EU in 1993 stimulated the development of trans-boundary shipments of goods—including waste—requiring more attentive regulations. These followed the expansion of EU boundaries, especially since the mid-1990s. There are four main principles regulating the shipment of waste. First, since waste for disposal is considered a heavy environmental burden, the general EU policy orientation is to promote self-sufficiency in the state members, thus reducing as much as possible the shipment of waste among them. Second, in principle, waste for disposal should be treated in the nearest installations, thus avoiding long shipments. Third, regulations concerning shipments of waste for recovery are less restrictive than those of waste for disposal, allowing a higher degree of mobility of the former within the EU. Fourth, export of hazardous waste within EU boundaries is allowed only if following special procedures. Hazardous waste shipment to non-Organisation for Economic Co-operation and Development (OECD) countries is prohibited, since these countries are not expected to possess proper treatment capacities. Trans-boundary shipments of hazardous waste are regulated by the Basel Convention, according to which the ratifying countries are expected to inform the other members of hazardous waste typologies itemized by national standards, designate and communicate the lists of competent authorities to control and authorize the shipment, and set the relevant duties and rights in the light of cooperation agreements among countries. The Basel Convention had been ratified by 172 countries as of 2010.

In 2007, EU countries exported a total of about 5.4 million tons of hazardous waste. It is calculated that only 80 percent of this amount is notified through the annual country reports; the remainder is assessed. This number shows a constant increase, whereas the percentage of hazardous waste exported compared to the overall generated waste remains stable (around 12 percent). This seems to prove that the EU is achieving self-sufficiency in handling its waste disposal activities. The main items of the hazardous waste exported are demolition waste, including mainly polluted soil and asbestos; waste from thermal processes; waste from organic chemical processes; and oil wastes of liquid fuels.

Illegal Waste Shipment

In spite of the regulations issued by the Basel Convention, the monitoring process of the European Environment Agency (EEA), and several transnational and national environmental organizations, the number of illegal waste shipments from EU countries is increasing. Official figures on this type of trade are still scarce and their relevance is often overlooked. A large portion of illegal waste exported to non-OECD countries is made by waste from electric and electronic equipment (WEEE), a portion of municipal waste. It is thus not directly classified as hazardous waste, in constant increase in the EU countries. Most of this waste is shipped, legally and illegally, to non-OECD countries such as China, Africa, and the Middle East.

A similar case is that of end-of-life cars, another significant item of illegal waste shipment from EU countries. Cars are shipped as secondhand items, hence avoiding the controls and documentation necessary for waste. The most common destinations are eastern Europe, western Africa, and the Middle East. The great economic incentives generated by the illegal waste trade make this traffic difficult to track, especially in the case of ordinary items such as electronic devices and cars. In the case of the most common types of hazardous waste generated by building and manufacturing, it is expected that illegal practices will continue to increase thanks to the interest of criminal organizations, which in some EU countries have been reported as participating in the waste trade business.

Household Waste Production and Recycling

The quantity of household waste generated by EU countries shows several degrees of variations. In most EU countries, industrial waste is from eight to 10 times larger than household waste, even though the latter has gradually increased. From 1995 to 2008, household waste generated in the EU increased from 474 to 524 kilograms per capita. This increase was not linear but showed a degree of stability in growth from 2006 onward. Generally, Western European countries produce quantities of household waste that exceed eastern European countries by 75–240 percent. In 2008, the four major producers per capita were Denmark, Cyprus, Ireland, and Luxembourg. The five minor producers were Poland, Slovakia, Latvia, Romania, and Lithuania.

Figures about national outcomes of recycling policies are even more telling of the differences among EU countries. The overall tendency is toward an increase in recycling among all countries, although with significant differences. The main body of the "EU recycling society" project is comprised of paper and cardboard, biowaste (kitchen and garden waste), glass, plastic, and metal. Recycling of bulky waste is significant only in a few of the old member states, whereas construction and demolition material is recycled in over 50 percent of the reporting countries as of 2010. Considering the national figures from 2005, the countries that produced less household waste roughly correspond to those that recycled the smallest percentage of the produced waste. The least recycling countries were Poland, Lithuania, and Slovakia. In the case of the larger producers, only Denmark scores among the top five in recycling, led by Germany, Belgium, and the Netherlands. The gap between old and new members in recycling levels is not alone, telling of the difference between older and new EU members since the data are to be read against the significant difference in generated waste. Also, a confrontation between generated waste per capita and population points out this imbalance. Confronting equivalent countries in population, such as, respectively, Belgium–Czech Republic, Denmark–Slovakia, and Austria–Hungary, it is possible to obtain a more complete picture of the difference between old and new member states. The 2008 data of kilograms per capita generated waste are 493–306 for Belgium–Czech Republic, 802–328 for Denmark–Slovakia, and 601–453 for Austria–Hungary. This confrontation reveals that in principle eastern European countries recycle less than Western European state members, but the former also generate comparatively much less household waste. The gap between old and new member states is telling of the difficulty of implementing a common policy of prevention and recycling in the EU. Although the complex body of community regulations (over 60) concerning waste management have set up the legislative and organizational framework for preventing harmful treatment of waste, several questions and problems remain unsolved. The two pristine EU projects of self-sufficiency and free circulation of goods have proven particularly porous to the spread of scarcely accountable and even illegal practices.

Davide Torsello
University of Bergamo, Italy

See Also: Economics of Waste Collection and Disposal, International; Politics of Waste; Recycling.

Further Readings

Dorn, N., S. Van Daele, and T. Vander Beken. "Reducing Vulnerabilities to Crime of the European Waste Management Industry: The Research Base and the Prospects for Policy." *European Journal of Crime, Criminal Law and Criminal Justice*, v.15/1 (2007).

O'Neill, Kate. *The Environment and International Relations*. New York: Cambridge University Press, 2009.

O'Neill, Kate. "Globalization and Hazardous Waste Management: From Brown to Green?" *University of California International and Area Studies Digital Collection*, v.1/4 (2002).

Torsello, Davide. *The New Environmentalism? Corruption and Civil Society in the Enlarged EU*. London: Ashgate, 2011.

Van Calster, Geert. *Handbook of EU Waste Law*. New York: Oxford University Press, 2008.

Williams, Paul T. *Waste Treatment and Disposal*. West Sussex, UK: John Wiley & Sons, 2005.

Externalities

Externalities are uncompensated costs or benefits, also known as interdependencies, central to environmental economics. Arthur Pigou, a British economist, is credited with initiating discussions about the theory of costs and benefits in the early 20th century. As such, the benefits of production and profit are weighed with the costs of production (assuming perfect knowledge).

A lack of perfect information means that there are outcomes outside the production and market environment that remain uncalculated. Pigovian economists suggest that indirect taxation should be used to correct for losses associated with these uncalculated outcomes, thereby enabling competitive allocation of resources in a market economy. This tax, levied on the externality producer, is then passed on to the consumer but not necessarily or typically distributed to those affected most directly by detrimental externalities. Environmental externalities, generally considered at the societal rather than the individual level of analysis, occur when natural resources are affected by nonnatural occurrences such as emissions from vehicles or industrial waste. Such emissions are considered external to the polluter's pricing and production decisions.

Detrimental Externalities

Most often considered when studying the impact of environmental externalities are negative, or detrimental, effects. Scarce natural and public resources, such as air and water as well as public and private lands, may be negatively impacted by the by-product pollutants of industrialization. For example, logging not only extracts wood for profit to produce salable products such as furniture and building lumber but also may create loss of habitat for wildlife, soil erosion, and an increase in pollutants from logging machinery, including air pollution from exhaust emissions, ground and water pollution from cast-off oil and gasoline, and noise pollution from the engines of logging equipment.

While logging produces products with economic value such as furniture and lumber, the uncompensated losses from techniques such as clear-cutting can include a loss of habitat for wildlife, soil erosion, and an increase in pollutants from logging machinery. Environmental externalities such as these, according to Pigovian economics, are generally considered at the societal rather than individual level of analysis. Indirect taxation on the externality producer is a suggested method of correcting for losses associated with these uncalculated outcomes.

Similarly, while increased retail development brings new jobs and purchasing power, detrimental externalities often emerge. From the increase in consumerism comes increased consumer traffic, bringing noise and light pollution, increased carbon dioxide output, increased crime, and decreased property values. There may, however, be beneficial or positive externalities that occur as a result of increased industrialization.

Environmental Racism

Environmental racism—the imposing of a disproportionate amount of pollutants on minority and low-income communities—represents a great injustice: having to choose between health and of economic stability or the acceptance of hazardous pollutants. Denials about the impacts of environmental discrimination continue largely due to the effects of NIMBYism (Not in My Backyard) in more economically advantaged communities, a structural lack of education and resources, and economic reality. Rather than merely convenient and easy, it has become economically expedient to dump in disenfranchised areas. The reality is in the economic correlation between dumping fees and profits. If manufacturers were to pass on the true cost of waste disposal, including the cost of externalities, the loss of profit would result in an increased retail cost that most consumers would not willingly accept.

Beneficial Externalities

Less often considered but viable issues are beneficial externalities. These occur when a nonmarket benefit affects the public but remains uncalculated in the traditional cost-benefit model. Extrapolating from the above examples, new plantings from reforestation might be enjoyed as both a wildlife area and a recreation area not suitable for logging due to premature growth. Light pollution, artificial light that brightens the night sky, disrupts stargazers and the natural activities of nocturnal creatures. While light pollution may be aesthetically displeasing to humans and endangering to wildlife, the extra traffic and lighting in the area may serve to deter criminal activity or bring new home buyers to an area not previously considered. Once accounted for and compensated, the emissions become internalized and cease to exist as externalities, leading to Pareto efficiency.

Internalization and Pareto Efficiency

Pareto efficiency occurs when allocation alternatives have been exhausted, with no individual better or worse off. This is not, however, the same as equity. Pareto efficiency merely indicates that there is no way to further benefit one party without further harming another. To reach Pareto efficiency, environmental externalities may be internalized through the implementation of any of four commonly used policy instruments, such as emissions fees, tradable emissions allowances, offset requirements, or source-specific standards.

With emissions fees, polluters such as industrial manufacturers pay for specified measurable pollutants. These manufacturers may elect to produce fewer pollutants, either through decreased output or through use of technological advancements that decrease emissions, thus receiving a lower abatement fee. Alternately, polluters who opt to not decrease emissions will pay relatively higher emissions fees to compensate for their abundant emissions.

The Environmental Protection Agency's (EPA) Clean Air Act of 1990 established a tradable allowance system by limiting levels of pollutants while creating a market for polluter rights. Those manufacturers who were low polluters, using less than their allotted amount of emission credits, could sell allowances to high polluters. In this way, high polluters could continue producing emissions above the governmental threshold without governmental penalty and the environment would not be polluted above the governmentally established limits. Similar program initiatives, such as the EPA's Federal Water Pollution Prevention and Control Act (Clean Water Act), sought to limit waterborne pollutants by enforcing water-quality standards and encouraging technological innovation to limit emissions. Similarly, offset requirements place a maximum threshold on emissions, but the penalty predominantly falls on new entrants to the market. Existing polluters consume a larger portion of the available emissions threshold, decreasing the proportional level of pollutants permissible to newcomers.

Although similar in intent to the Clean Air and Water Acts in its attempts to limit excessive noise pollution, the Noise Control Act of 1972, while not abolished, lost funding and so does not currently serve to limit auditory externalities.

Finally, source-specific standards are based on specific technologies that are expected to be able to control emissions as a means of forcibly diminishing pollution and, thus, externalities. For example, the utilization of smokestack scrubbers designed to minimize sulfur dioxide emissions decreases the detrimental externalities by decreasing emissions.

Conclusion

Environmental externalities, generally considered at the societal rather than individual level of analysis, occur when natural resources are affected by nonnatural occurrences. Such emissions are considered external to the polluter's pricing and production decisions. While the use of policies embracing Pareto efficiency enable the public to recapture compensation for some of the external, previously uncalculated costs, this is less likely to occur in areas in which people are disenfranchised, furthering environmental racism.

Leslie Elrod
University of Cincinnati

See Also: Air Filters; Carbon Dioxide; Clean Air Act; Clean Water Act; Emissions; Environmental Justice; NIMBY (Not in My Backyard); Noise Control Act of 1972; Race and Garbage; Scrubbers.

Further Readings

Ayers, Robert U. and Allen V. Kneese. "Production, Consumption, and Externalities." *American Economic Review*, v.59/3 (1969).

Cornes, Richard and Todd Sandler. *The Theory of Externalities, Public Goods, and Club Goods*. New York: Cambridge University Press, 1999.
Osborne, Martin J. "Pareto Efficiency." http://www.economics.utoronto.ca/osborne/2x3/tutorial/PE.HTM (Accessed April 2011).
Owen, Anthony D. "Environmental Externalities, Market Distortions and the Economics of Renewable Energy Technologies." *Energy Journal*, v.25/3 (2004).
Sandmo, Agnar. "Optimal Taxation in the Presence of Externalities." *Swedish Journal of Economics*, v.77/1 (1975).

Farms

Farm operations in the 21st century have vast potential to not only produce waste but also serve as sinks for wastes. In this way, farm operations can contribute to environmental and human health or illness, depending on the scale and the technologies utilized to grow food. Farms are typically located long distances from where the food will ultimately be consumed, leading to an increase in the inputs consumed in order to get food distributed. These inputs include the transportation infrastructure and fossil fuels.

As food crises have all too clearly demonstrated, as the price of oil increases, so too does the price of food, which leads to increased instability and food insecurity around the globe. The scale of farms and the use of a variety of technologies lead to disparate outcomes in terms of consumption of resources in farming operations and their associated waste products.

Farms also compete with alternative uses for limited and necessary resources. Water resources are of particular concern in arid regions as agriculture use often conflicts with other potential uses (such as residential and industrial) of water resources.

Centralization and Intensification

Scale and use of technology are quite evident in confined animal feeding operations (CAFOs), which centralize the production of animal foods like never before in the history of humanity. In CAFOs, intensive methods rely heavily—if not exclusively—upon the consumption of outside resources in order to grow animals in the most cost-effective manner possible. In essence, this heavy reliance upon outside sources as well as mechanized facilities allow for even less labor investment. Labor is a key component in the cost of food production; with increased intensification in mechanized, industrial agriculture, there are fewer farmers who are responsible for the majority of foods produced. These CAFOs are ideal sites for the spread of diseases, as animals live in very close proximity to one another, often living in a number of waste products including feces, bodily fluids, and sometimes corpses. CAFOs raise several ethical questions about the consumption of their products. In CAFOs, animals themselves become waste, sometimes as mistakes are made (downer animals), and sometimes as a part of the normal course of operations. An extreme example of the animals as waste mentality is evident in examination of egg production facilities where live male chicks have literally been disposed of in plastic garbage bags.

The intensification of agriculture in mechanized industrial agriculture farms brings with it several challenges in terms of waste and inefficiencies. Large monocultures of crops are incredibly vulnerable to disease and pest infestations. Oil, fertilizers, and pesticides are necessary inputs into these mechanized, industrial systems of food production. Increasingly, genetic modification of plants as well as animals also contributes to these systems. Most of the research and development in terms of genetic modification has been toward developing herbicide-resistant plants.

In this strategy, broad-spectrum pesticides can be utilized to control for undesired plant species, ideally wiping out all competition for water, sunlight, and nutrient resources. In reality, these interventions continue the technology treadmill as weeds and other agriculture pests adapt to these chemical control regimes, leading to a continuous cycle wherein further development of agricultural technologies is needed in an attempt to combat pests and increase food production. There is a particular philosophy of separation from and control over nature that goes along with these industrial systems of production: its (bio-)technologies and techniques are not neutral; rather, they are the result of a particular way of looking at the world.

Commodity Foods

Since former secretary of agriculture Earl Butz encouraged farmers to plant from fencerow to fencerow and "get big or get out," U.S. agriculture has remained nearly singularly focused on commodity food production. This emphasis on commodity food production has not only led to an overproduction of commodity foods in the United States but also overconsumption of a number of foods. This overproduction of commodity foods, particularly grains in the United States, has contributed to a number of human health impacts, including high rates of diabetes and obesity, both in the United States and globally. At the same time, overproduction of grains in the United States has led to grain dumping, wherein imported and subsidized U.S. commodities undercut the price and agricultural sectors of many other states.

Pollution

In terms of environmental health, farming creates pollution of the land, water, and air. A large part of this pollution is the result of excessive nutrients in the agroecosystem. Artificial fertilizers are particularly vulnerable to nutrient loss. Excessive nutrients can cause a number of issues, both locally and regionally. This manifests locally with high levels of nitrogen in local sources of drinking water and regionally with the zone of hypoxia caused by excessive nitrogen in the Gulf of Mexico. On-farm nutrient management strategies include building organic matter in the soil, conservation tillage, and maintenance of riparian buffer zones, which can prevent the loss of these nutrients and lead to greater on-farm nutrient capture. Smaller, labor-intensive farms utilize a number of tight nutrient cycling strategies, including composting or using worms to eat organic materials.

Subsidies

Farm subsidies structure the fundamental shape of 21st-century U.S. agriculture. These supports are necessary to many U.S. farmers in order to maintain a viable business and have led to vast areas being planted in single-species monocultures. The issues of scale in farming and increased food production cannot be overstated. In popular imagination, farms remain idyllic locales filled with the abundance and diversity of nature.

Biodiversity

Plant genetic resources play a major role in supporting industrial systems of food production. Globally, humans rely on a very narrow range of species to provide for 90 percent of the food supply. Well over half of the world's food supply comes from corn, rice, and wheat. These grains have been bred to produce high yields, with breeders selecting the traits that ensure the greatest increase in yields under a specific set of agronomic conditions, usually high-input conditions, including chemical fertilizers. The overemphasis on yield has in the past resulted in a degree of genetic uniformity and, hence, susceptibility to diseases and pests. There have been several documented crop failures due to genetic uniformity from the 1800s through a failure in 1984 in Florida, where 18 million citrus trees were destroyed. Crop failures, coupled with market structures of food dis-

U.S. agriculture is primarily focused on commodity food production like corn, soybeans, and wheat, which has led to an overproduction of commodity foods in the United States and overconsumption of a number of foods, particularly grains. Monocropping also relies heavily on commercial fertilizers, which leach into water sources. Excess nitrogen causes hypoxia in water bodies such as the Gulf of Mexico. On-farm nutrient management strategies can focus on building organic material in the soil and increase its water holding capacity.

tribution, can result in large loss of both income and human life, as happened during the Irish potato famine. Pesticides have been utilized in order to compensate for the lack of genetic diversity: however, their application has been largely ineffective in increasing yields because, in part, of pesticide resistance of insects as well as the negative impacts of pesticide application on soil biodiversity and beneficial insect populations.

On-farm biodiversity is conceptualized at the species, ecosystem, and genetic levels. On the farm, humans are the cultivators and destroyers of biodiversity. At each level of biodiversity, benefits accrue, regardless of whether these benefits are recognized (named) or acknowledged (valued) in cost-benefit analyses. There has been increasing international interest in the genetic level of biodiversity, particularly since the 1970s, because of the rise in utilitarian applications of genes in agricultural, industrial, and medical biotechnology product development. The value of these genes in agriculture is in the tens of billions of dollars in the United States alone.

Alternatives

Alternatives to the neoliberal model of food production continue to emerge. These challenge the dominant paradigm of food production, hence mitigating much of the consumption and pollution caused by larger-scale, mechanized approaches to food production. Many of these are community-based initiatives wherein participants are, to a varying extent, actively participating in food production through arrangements such as farmers markets and community-supported agriculture. Although small, the impact of these arrangements continues to be explored by researchers in both the social and natural sciences.

David Fazzino
University of Alaska

See Also: Carbon Dioxide; Certified Products (Fair Trade or Organic); Dairy Products; Food Consumption; Organic Waste; Pollution, Air; Pollution, Land; Pollution, Water; Slow Food; Sustainable Development; Water Consumption; Worms.

Further Readings

Altieri, Miguel. *Agroecology: The Science of Sustainable Agriculture*. Boulder, CO: Westview Press, 1995.

Jackson, Dana L., Laura L. Jackson, and Nina Bradley. *The Farm as Natural Habitat: Reconnecting Food Systems With Ecosystems*. Washington, DC: Island Press, 2002.

Posey, D. A., ed. *Cultural and Spiritual Values of Biodiversity*. London: United Nations Environment Programme, Intermediate Technology Publications, 1999.

Fast Food Packaging

Fast food packaging generally refers to an assortment of portable, disposable wrappers, containers, bags, boxes, and cartons of various shapes and sizes that are used to transport ready-cooked foods purchased from a quick-service restaurant (QSR). The terms *takeout*, *take-away*, *carryout*, and *to go* all reflect this intended portability. Fast food packaging has evolved over time to include more nuanced adaptations for functionality, product branding, and ease of use. Foil-lined wrappers treated with special coatings ensure that hot food items not only remain at temperature but also prevent grease from seeping through. Insulated paper cups keep coffee and other hot beverages hot; reusable plastic cups are designed to hold icy drinks without leaking. Cup lids prevent spillage in transit, while cup bottoms are sized to nestle comfortably in standard automobile cup holders. The fast food experience is expedited by the fact that these packages can be embossed with company logos, product information, or other promotional details; quickly filled with the assembly-line-produced item; and placed in a bag, box, or carrier when purchased. Fast food packaging has changed over time to better accommodate how, where, and why people eat out.

Early History

In premodern, rural societies, the average person did not regularly venture far from home. When people did travel, they usually stayed with relatives and acquaintances and consumed their meals there. As people began to trek farther from home and were unable to dine with people they knew, eating out transformed into a commercial venture. The restaurant, as it is experienced in the 21st century, is a modern, urban phenomenon that evolved as a result of the concentration of people, money, and commercial activity in urban hubs. Unlike fine dining restaurants, fast food restaurants responded to the perceived need for inexpensive, hot meals that could be quickly prepared, served, and eaten on the go; fast food packaging became the transport mechanism for these meals.

Like most innovations, the fast food wrapper was born out of necessity. While people tend to associate the term *fast food* with any number of U.S. QSR chains like McDonald's, Taco Bell, and Pizza Hut, the earliest fast food was street food, peddled to pedestrians in urban cities throughout the world by independent vendors. Some fast food packaging was—and remains—edible. In the preindustrial world, a fast food wrapper may have been a type of leaf, gourd, bread product, dumpling, or seafood shell.

The Industrial Revolution brought about major changes in fast food packaging. The same wood-pulp technology that made it possible for newspapers to be more widely circulated throughout Europe and the United States was eventually adapted to make disposable paper bags, boxes, food wrappers, and containers. The same workers who labored in the factories that produced these products likely ate the fast foods that they wrapped; they worked grueling hours and often did not have enough time to return home to eat at mealtimes. Small fish stands selling quick-fried whitefish and sliced potatoes served in a rolled-up piece of newspaper (later inserted into a paper bag) became a mainstay in London factory districts. Pushcart vendors, lunch wagons, and temporary food counters fed factory workers and railroad travelers with foods that could be wrapped in paper and quickly eaten without utensils. Leisure and recreational destinations like the World's Fairs and Coney Island also provided opportunities for carryout "novelty" foods. Hot dogs, sandwiches, hamburgers, and ice cream could also be conveniently contained in edible packages—buns, waffles, and the like.

Modern Fast Food

The production of paper goods increased throughout the 19th and early 20th centuries. In 1921, the White Castle System of Eating Houses Corporation was founded in Wichita, Kansas. White Castle is credited with revolutionizing the fast food industry. In addition to establishing the first operational model for the chain restaurant, White Castle developed the unique idea of packaging its hamburgers in individual boxes designed to keep them hot. One of the founders, Edgar Waldo "Billy" Ingram, is also credited with developing the technology to create the company's paper packaging products. Learning from White Castle's success, a number of entrepreneurs seized the opportunity to sell carryout food to the U.S. public, and the hamburger was their star product. Eventually, all manner of packaging—butcher paper, wax paper, small boxes, aluminum foil, paper trays, and inserts—were manufactured to keep takeout foods hot and convenient to carry and eat.

Fast food business continued to flourish during the Great Depression. However, during World War II, fast food restaurants were especially hard hit, experiencing an overall decline in sales because of food and material shortages, low unemployment rates, and the refocusing of industrial manufacturing technologies to the war effort. Wartime did yield the development of expanded polystyrene foam (EPS), the nonbiodegradable material from which styrofoam peanuts, coffee cups, and "clamshell" containers are made. Global food shortages continued to haunt the industry at the war's end, and the U.S. fast food industry struggled to regain its former glory. A cultural shift accompanied postwar prosperity in the United States. By the 1950s, more families were able to afford cars and suburban homes. Successful fast food franchises relocated to suburbia or established highway outposts. Roadside drive-ins and drive-thru restaurants became popular, and food packages were eventually designed so that drivers could eat, drink, and drive at the same time.

Waste

Despite the popularity and convenience of takeout foods, much criticism has been levied against the fast food industry for the quality of its food, low wages, and hiring practices. Critical attention has also been directed toward the health risks and environmental impact of fast food packaging. The coated wrappers that many chains use because they are resistant to oil and water are called polyfluoroalkyl phosphoric acid diesters (diPAPs). According to some researchers, these chemicals break down into perfluoroctanoic acid (PFOA) or perfluorcoctane sulfonate (PFOS), members of a class of perflurochemicals that may be carcinogenic.

Fast food packaging also generates a tremendous amount of litter and waste. Significant numbers of restaurants have attempted to counter the garbage problem on several fronts; most make a point of keeping their properties especially clean. Some also provide additional trash receptacles outside their locations, hand out certain items (such as napkins or ketchup packets) only by request, and have switched to more environmentally friendly packaging materials, eschewing styrofoam containers for BPA-free plastics and biodegradable paper. A number of leading fast food chains—most notably, McDonald's and Starbucks—are credited with adopting sustainable paper-purchasing policies and making some efforts to protect forests.

For better or for worse—much like fast food itself—fast food packaging is an iconic marker of U.S. popular culture.

Lori Barcliff Baptista
University of Chicago

See Also: Culture, Values, and Garbage; Food Waste Behavior; Restaurants.

Further Readings

Brody, A. L. and K. S. Marsh. *Encyclopedia of Packaging Technology*. Hoboken, NJ: John Wiley & Sons, 1997.

Chritser, G. *Fat Land: How Americans Became the Fattest People in the World*. New York: Houghton Mifflin, 2003.

Jakle, J. A. and K. A. Sculle. *Fast Food: Roadside Restaurants in the Automobile Age*. Baltimore, MD: Johns Hopkins University Press, 1999.

Leidner, R. *Fast Food, Fast Talk: Service Work and the Routinization of Everyday Life*. Berkeley: University of California Press, 1993.

Renner, R. "First Commercial Perfluorochemicals Found in Human Blood." *Environmental Science and Technology*, v.43/12 (April 29, 2009).
Schlosser, E. *Fast Food Nation*. New York: Houghton Mifflin, 2001.
Spurlock, M. *Don't Eat This Book: Fast Food and the Supersizing of America*. Berkeley, CA: Berkeley Trade, 2006.

Federal Insecticide, Fungicide, and Rodenticide Act

The Federal Insecticide Act of 1910, the first pesticide control law, was created primarily to protect consumers from ineffective products and deceptive labeling. On June 25, 1947, the U.S. Department of Agriculture (USDA) enacted the Federal Insecticide, Fungicide, and Rodenticide Act (FIFRA). This statute was created to regulate the use and intentional release of pesticides into the environment with a focus on the efficacy, not use, of pesticides.

Section 2(u) of FIFRA broadly defines pesticides as chemicals and other products used to kill, repel, or control pests that harm crops and reduce yield. This includes insecticides, herbicides, fungicides, and other pest eradicators. Pesticide products used in food production are also regulated under the United States Federal Food, Drug, and Cosmetic Act (FDCA).

FIFRA was amended regularly (1959, 1961, 1964, 1972, 1973, 1975, 1978, 1983, 1984, 1988, 1990, 1991, and 1996) to contend with potentially harmful environmental effects and to improve precautionary labeling. FIFRA was functionally rewritten in 1972 after having been moved to the jurisdiction of the newly created Environmental Protection Agency (EPA), when it was amended by the Federal Environmental Pesticide Control Act (FEPCA). Further additional amendments, such as the 1996 Food Quality Protection Act (FQPA), required manufacturers to provide information about all potential environmental and health impacts.

Review Process

Through FIFRA, the EPA is specifically authorized to (1) strengthen the registration process by shifting the burden of proof to the chemical manufacturer, (2) enforce compliance against banned and unregistered products, and (3) promulgate the regulatory framework missing from the original law.

A 1996 amendment created an expedited review process in cases where risk to human health, nontarget organisms, and environmental contamination was reduced. The 1996 amendment also required chemical registration every 15 years. If a pesticide is believed to cause potentially unreasonable environmental risks, a review process is initiated to both consider general ecotoxicological and environmental test data and determine whether the continued use of the pesticide presents unreasonable environmental risks. The burden of proof is on the manufacturer to demonstrate whether the pesticide can be used as regulated without unreasonable, adverse environmental effects. FIFRA enables the EPA administrator to issue one-year experimental use permits if needed for data acquisition unless the pesticide contains a previously unregistered chemical, which may call for additional testing.

Before the EPA registers a pesticide to be used on raw agricultural products, it either grants an exemption or establishes a tolerance (the maximum amount of safely consumed pesticide). In conjunction with the FDCA, a raw agricultural product is deemed unsafe if it contains pesticide residue unless the residue is exempt or is within the limits of a tolerance established by EPA.

Enforcement

The EPA is authorized to use a range of enforcement actions to address violations of FIFRA, including violations relating to the initial and annual production reports. Enforcement actions include notices of warning (NOWs), civil administrative penalties, termination of establishment registrations, and criminal sanctions.

It is unlawful under FIFRA to distribute or sell unregistered pesticides, use a registered pesticide in a manner inconsistent with its labeling, submit false information, or violate an order of the administrator issued under the act (with certain exemptions). Penalties include issuance of stop sale, use, or removal orders, seizure of the pesticide, and assessment of civil penalties. Intentional violations of FIFRA also subject the violator to criminal penalties.

When determining the amount of a civil penalty, several factors are assessed, including the appropriateness of the penalty to the size of the business, the effect on the ability to continue in business, and the gravity of the violation. The maximum allowable penalty in 2010 was $7,500 per violation.

Worker Protection Standard (WPS)

Generally, restricted-use pesticides must be applied by or under the direction of a certified applicator. The EPA's Worker Protection Standard for Agricultural Pesticides (WPS) is a regulation created to reduce the risk of pesticide poisonings and injuries among approximately 2.5 million agricultural and pesticide handlers working at over 600,000 agricultural establishments. The WPS contains requirements for pesticide safety training, notification of pesticide applications, use of personal protective equipment, restricted-entry intervals after pesticide application, decontamination supplies, and emergency medical assistance.

In 2004, the EPA amended the WPS glove requirements by adding the provisions for worker safety. First, the provisions permit all agricultural pesticide handlers and early-entry workers covered by the WPS to wear separate glove liners beneath chemical-resistant gloves (previously disallowed), reducing the discomfort of unlined chemical-resistant gloves, especially during hot or cold periods. Second, agricultural pilots are not required to wear chemical-resistant gloves when entering or exiting aircraft, as chemically resistant gloves were not found to add any significant protection against minimal pesticide residues found around the cockpit of aircraft.

Conclusion

Since its humble beginnings focusing on labeling, FIFRA and the EPA have come a long way in instituting environmental and health protections not only for consumers but also for agricultural workers and pesticide handlers.

Leslie Elrod
University of Cincinnati

See Also: Environmental Protection Agency (EPA); Farms; Pesticides; Toxic Wastes.

Further Readings

Edwards, Clive A. "Federal Insecticide, Fungicide, Rodenticide Act." *Pollution Issues*. http://www.pollutionissues.com/Ec-Fi/Federal-Insecticide-Fungicide-and-Rodenticide-Act.html (Accessed September 2010).

Environmental Protection Agency. "Federal Insecticide, Fungicide, and Rodenticide Act (Federal Environmental Pesticide Control Act) 7 U.S.C. §§ 136–136y, June 25, 1947, as amended 1972, 1973, 1975, 1978, 1983, 1984, 1988, 1990, 1991 and 1996." http://www.epa.gov/oecaagct/lfra.html#Summary%20of%20the%20Federal%20Insecticide,%20Fungicide,%20and%20Rodenticide%20Act (Accessed September 2010).

First Principle of Waste

According to official Environmental Protection Agency (EPA) and U.S. Department of Agriculture (USDA) estimates, approximately 25 percent of prepared food is wasted in the United States. Other estimates of total food waste from food produced in the United States are as high as 50 percent. Such discrepancies may be the result of varied measurement tools and methods as well as the large differences between what people report they waste and the actual amount of food waste produced.

In the 1970s, archaeologist William Rathje, along with colleagues and students at the University of Arizona, founded the Garbage Project. The project sought to use archaeological methods of inquiry in order to understand current society by studying the garbage people produce. While archaeologists often delve deeply into the waste of past cultures, this project was different in that it was looking for insights into the present nature of contemporary society by sorting through households' freshly discarded garbage.

Through this process, they came upon a number of interesting and surprising discoveries that led them to develop the concept of the First Principle of Food Waste (FPFW). The FPFW states that food products that are used on a regular basis are wasted at a much lower rate than specialty ingredients that are used less frequently. The more repetitive one's diet—the more one eats the same foods day after day—the less food waste will be produced.

The FPFW explains why sliced bread is hardly wasted at all, while specialty bread items, such as hot dog buns and bagels, are wasted at a rate of 30–60 percent. The FPFW also helps to explain why garbage collected from Mexican American census tracts had as much as 20 percent less food waste than garbage from Anglo census tracts. Mexican cooking offers a diverse array of dishes, but there are few ingredients in them. Tortillas, beans, beef, chicken, pork, avocados, tomatoes, lettuce, onions, chili sauces, and salsa are all used in many dishes. These staple ingredients are used constantly, with leftovers easily incorporated into new dishes.

Through continued and diligent recording of fresh household garbage, the Garbage Project fully investigated household maintenance waste. The intimate knowledge acquired about everyday products like detergents and cleansers, as well as more episodically used products like paint and varnish, led them to expand the FPFW to include other consumer products such as household maintenance products and hazardous waste. This led to the development of the more general First Principle of Waste.

Social Contexts

Insights from the Garbage Project led to realizations that seemed counterintuitive to general human behavior. During times of shortages, food is wasted at a higher level than usual. This seems to be due to crisis buying and hoarding behaviors that lead to increased food waste. The waste may be the result of people buying either foods they would otherwise not buy (for example, certain cuts of meat or sugar substitutes) and, therefore, do not know how to prepare, or foods that they simply do not like. These insights are helpful in better understanding consumption and waste patterns that may be contrary to common knowledge about food waste. Additionally, these archaeological methods are useful in understanding actual behavior, since peoples' self-reported behaviors are quite different from reality.

There has been an increasing awareness in the 21st century of waste and its related social and environmental impacts. As part of the rise in attention toward waste, the EPA and USDA have developed recommendations for households and institutions to reduce food waste that include a Food Waste Recovery Hierarchy, which advises certain ordered actions, including the following:

- *Source separation*: reducing the amount of food waste being generated
- *Feed people*: donating excess food to emergency food organizations
- *Feed animals*: providing food scraps to farmers
- *Industrial use*: processing waste food into fuel or industrial animal feeds
- *Composting*: recycling food into nutrient-rich soil amendments
- *Landfill and incineration*: disposing of food is the least desirable option

In 2008, according to the EPA, at least 12.7 percent of total municipal solid waste in the United States was food waste, with less than 3 percent of food waste being recovered or recycled. Thirty-one million tons of food waste were disposed of in landfills or incinerators. Landfilling food waste or any organic matter is problematic due to the anaerobic conditions in landfills. These conditions lead to the production of methane, a greenhouse gas more problematic than carbon dioxide.

Implications

The FPFW conflicts with current dietary advice from nutritionists and public health officials who encourage diverse dietary intakes. At the same time, recommendations for the FPFW parallel food movements, which encourage food consumption that focuses on local and seasonal diets. As of 2010, there has been limited research in this area beyond the Garbage Project. Further research into the specifics of food waste may lead to a more-nuanced understanding of it and may help in its overall reduction. While the Garbage Project focused on North America, this field would benefit from additional study that includes other regions of the world and incorporates various cultural contexts.

Maggie Ornstein
Christine Caruso
Graduate Center, City University of New York

See Also: Beef Shortage, 1973; Food Waste Behavior; Garbage Project; Garbology; Sugar Shortage, 1975.

Further Readings

Jones, T. W. "Food Loss and the American Household." 2006. http://www.redorbit.com (Accessed July 2010).
Rathje, W. L. "From Tucson to Armenia: The First Principle of Waste." *Garbage*, v.4/2 (1992).
Rathje, W. L. and C. Murphy. *Rubbish! The Archaeology of Garbage*. New York: HarperCollins, 1992.

Fish

Fish remains are readily identifiable components of garbage worldwide, if for no other reason than their smell, yet the contribution they make to the amount of garbage that accumulates is difficult to calculate. Fish bones are relatively delicate compared to those of land animals and do not stand up as well to post-depositional chemical and mechanical processes.

If one were to excavate, for example, a five-year-old garbage dump and tally up the number of fish bones recovered, they would be underrepresentative of the original number. In addition, an unknown—but large—percentage of fish never make it to land but are dumped dead back into the waters from which they were caught. Likewise, fish are often processed on or near the water, with the by-products never reaching land.

Calculating Waste

Perhaps the only method available for calculating the contribution of fish remains to garbage is to use catch figures as a baseline and then examine those figures against known or suspected patterns of human consumption. Annual worldwide fish production, including wild capture and aquaculture (farm) production (excluding shellfish) averages around 140–150 million tons. This figure is susceptible to the vagaries of any number of variables, including price and ecosystem disruptions, especially climate anomalies such as El Niño. Although precise production figures from China are unreliable, that country accounts for approximately 30 percent of the world's fish production, followed distantly by Peru, the United States, and Japan. Of the total tonnage of fish processed each year, 70–75 percent are for human consumption and 25–30 percent are for reduction into fishmeal and fish oil. Fishmeal, which is a coarsely ground powder made from cooked fish, was once important as a fertilizer but is used primarily in pet food in the 21st century. Peru is the leader in fishmeal production, using the Peruvian anchoveta, which is the leading fish caught in the world in terms of tonnage.

Worldwide, annual per capita fish consumption is about 35–40 pounds. In some Asian and African countries, where fish provide 50 percent or more of the animal protein, annual per capita consumption can run as high as 45–50 pounds. In some small island states, such as Seychelles, the annual per capita consumption can exceed 165 pounds. In other countries, especially meat-exporting countries such as the United States and Argentina, annual per capita consumption of fish is under 20 pounds. Based on these figures, one can roughly estimate that annual per capita fish waste worldwide is about 10 pounds.

Fish Smell and Its Effects on Garbage

Rotting fish produce one of the most noxious smells imaginable, in large part a result of adaptations that fish have made to aquatic environments over millions of years—adaptations that are vastly different than those made by land-living animals. Fish contain an odorless chemical called trimethylamine oxide (C_3H_9NO). Once a fish dies, bacteria in the body begin breaking down the compound into two foul-smelling compounds, putrescine ($C_4H_{12}N_2$) and cadaverine ($C_5H_{14}N_2$). The meat of land animals contains far less trimethylamine oxide than that of fish and hence does not smell as putrid when decaying. There is considerable variation in trimethylamine oxide levels even in fish, with cold-water species, especially those that feed near the surface, having the highest levels.

The presence of noxious compounds plays a role in the contribution of fish remains to garbage regarding two variables: (1) changes in human dietary preference and (2) how fish are handled immediately upon and after death. It is becoming increasingly clear that omega-3 fatty acids, found

Fishmongers are selective in determining the quality of their catch and discarding fish that lack freshness. Improper disposal, such as tossing on beaches, can negatively impact tourism. For this reason, the United Kingdom requires complete incineration of fish waste.

in fish oil, have medical benefits, including prevention of psychotic disorders in high-risk children and heart disease in adults through a reduction in blood-triglyceride levels. What is not clear is the range of health problems that omega-3 fatty acids prevent or ameliorate—although if one were to believe the hundreds of claims made in television and magazine ads, they prevent everything from male-pattern baldness to impotence. Not surprisingly, given Westerners newly found emphasis on eating supposedly healthy foods, cold-water fatty fish such as herring, sardines, and salmon have assumed a more important part of their diet. Salmon, for example, contains roughly five times the amount of omega-3 fatty acids as either cod or catfish and only slightly less than that amount compared to either mahi-mahi or canned light tuna. Oily fish also often contain heavy metals, such as mercury, and fat-soluble compounds, such as dioxin and PCBs, but Westerners tend to accept the risk in the quest for the much-heralded omega-3 fatty acids.

But Westerners, for the most part, also have a natural avoidance of any "fishy" smell and taste. Freshly killed fish lack any odor other than what normally is thought of as a "sea" smell. If prepared and consumed immediately—or if kept on ice or refrigerated for a short period—bacteria do not have time to break down the trimethylamine oxide into putrescine and cadaverine. Fishmongers tend to be highly selective in determining the quality of the fish they sell and are constantly culling their offerings and tossing out fish that are beginning to decay. In many nonindustrialized areas, the remains are either collected for disposal in a garbage dump or simply cast aside. Some Caribbean countries are becoming concerned over the negative impact on tourism caused by improper disposal of fish remains on beaches.

In contrast, many developed countries prohibit meat and fish from being dumped, even in landfills. In the United Kingdom, for example, which has stringent disposal laws, all raw meat and fish trimmings, including bones, heads, scales, and skin, must be disposed of in accordance with what are called "Category 3" requirements, which means complete incineration within an approved incinerator or processing for pet food at an approved plant.

Archaeology

Fish remains might be problematic for modern health concerns and aesthetic reasons, but they are excellent tools for discovering more about what went on in the past. Fish bones from archaeological sites, for example, indicate not only what past peoples ate but also what kinds of fish were common in a region at a particular time. Ecologists can use archaeological fish bones to examine how the diversity, distribution, and abundance of marine life has changed over time. For example, fish bones from the Mesolithic levels of Franchthi Cave in southern Greece show that the bluefin caught in the Aegean Sea in the 21st century pale in size compared to those caught some 8,000 years ago—a result of modern overharvesting of larger fish.

Conclusion

Fish will always be an integral part of the human diet and their remains will remain a component of garbage. The goal is to minimize the impact they have on the social and natural environment by disposing of them in ecofriendly ways. One area with considerable potential is the biodiesel industry, which is beginning to use fish remains as a renewable energy source. With some fish-processing plants in southeast Asia producing over 250,000 pounds of waste daily, there is no present shortage of raw material.

Michael J. O'Brien
University of Missouri

See Also: Dating of Garbage Deposition; Food Waste Behavior; Meat; Ocean Disposal.

Further Readings

Casteel, Richard W. "A Method for Estimation of Live Weight of Fish From the Size of Skeletal Elements." *American Antiquity*, v.39 (1974).

Delgado, Christopher L., Nikolas Wada, Mark W. Rosegrant, Siet Meijer, and Mahfuzuddin Ahmed. *Fish to 2020: Supply and Demand in Changing Global Markets*. Penang, Malaysia: WorldFish Center, 2003.

Food and Agriculture Organization of the United Nations (FAO). *The State of World Fisheries and Aquaculture*. Rome: FAO, 2000.

Floor and Wall Coverings

The first hand-knotted pile carpets were woven in prehistoric central Asia, probably for protecting nomadic tribes from the cold, temperate-zone nights. Following the Crusades, carpets and rugs entered Europe as luxury items and were only used as floor coverings in royal and ecclesiastical settings, often being hung or covering furniture. By the end of the medieval period, carpets were woven in Europe "in the Turkish manner." Being woollen, these carpets generally remained in use for around 200 years and are mainly studied via depictions in paintings.

In the 21st century, carpets are usually nylon or polypropylene. The vacuum cleaner was first marketed to clean carpets in the late 19th century. This was a luxury item until after World War II; they are ubiquitous in the 21st-century Western world.

Floorcloth

Stone and timber floors in larger houses were artisan-created works, intended to be seen. Painted and stencilled floors were augmented by floorcloths, hard-wearing painted canvas. Floorcloth (also known as floor oilcloth or painted floorcloth) first appeared in the early 18th century, but there are 15th- and 16th-century references to a similar material.

Floorcloth became popular despite its expensiveness and tendency for patterns to wear off. It was used in high-traffic areas, such as entrance halls and stairways, where easy cleaning was important. It was also favored in ground-level rooms in the summer, although it became malodorous and tacky when hot. Manufacturing was labor intensive and dangerous, using painting frames six storys high on which workers balanced to coat the canvas. The processes of drying and curing meant that production took several months. By the close of the 19th century, floorcloth had fallen in price enough to be available to the working classes.

Linoleum

The next significant development in flooring was Linoleum, which developed from a product called Kamptulicon. In 1855, Frederick Walton peeled the skin from the top of a can of oil paint and used this skin, produced by oxidation of linseed oil, to develop a product he called Kampticon, but later rebranded as Linoleum. Walton started the Linoleum Manufacturing Co., Ltd., in 1864. At first, the company ran at a loss as the fierce competition between floorcloth and Kamptulicon continued. A large advertising campaign and the opening of two Linoleum shops in London reversed the situation and sales skyrocketed.

The success of Linoleum spawned its imitation by floorcloth companies. Walton began legal action against Nairns of Kirkcaldy in 1877, but lost, having never registered the trade name. The word *linoleum* became the first product name to be ruled a generic trademark in court.

Linoleum's heyday was from the 1860s until after World War II, when the do it yourself (DIY) boom saw other hard floorings overtake linoleum, which was seen as old-fashioned and vulnerable to damage from the fashionable high-heeled shoes of the time. The "poor man's carpet" was undermined by the appearance of cheap carpeting. By the 1960s, linoleum was almost completely replaced by cheaper vinyl flooring and struggled throughout the 1970s when environmental change increased the price of linseed oil. A recovery in the late 1970s and 1980s credits linoleum's natural ingredients, durability, and nontoxic breakdown and combustion products. Linoleum has enjoyed resurgence as a green product since the 1980s, when it became possible to recycle linoleum for linseed oil, making it 100 percent recyclable. Natural lino production also recycles all remnants back into the production process.

Linoleum remains popular in the 21st century, but it suffers from a preconceived image of being déclassé and has a bad environmental reputation because of the term *lino* being applied to other flooring such as vinyl. It is actually one of the modern world's oldest natural floorings, being made from renewable materials. Lasting up to 40 years, linoleum is biodegradable and can be used as fuel since it is similar to coal in terms of energy and releasing CO_2 roughly equivalent to the uptake of its ingredients. Linoleum is therefore a closed loop system, where energy from incineration equals or exceeds energy used in production.

Polyvinyl chloride (PVC), or vinyl flooring, still holds most of the market because of its brightness and fire-retardant properties. However, vinyl's monomers are associated with occupational cancers, combustion products are highly toxic, and phthalate additives are a suspected health risk. Post-consumer PVC recycling is possible but was still in its infancy as of 2010 and not widespread.

Having few obvious reuses, sheet flooring is often fly-tipped. In Dresden in 1905–13, Die Brücke created the Linocut printmaking technique, applying the woodcut technique to linoleum. In Hip-Hop culture of the late 1970s and early 1980s, linoleum was used to provide a dancing surface for b-boying, or breaking (the preferred terms for breakdancing).

Wall Coverings

Woodcut-printed wallpaper became popular with the elite of Renaissance Europe, replacing tapestry and paneling as a means of decoration and insulation. The expense and import availability of tapestries led to the lower echelons of the elite adopting wallpaper, which featured similar designs to tapestry and was hung loose like tapestry. Prints, which were collectible artworks at the time, were often pasted to walls instead of being framed. They became common during the 15th century, and their price fell dramatically. The largest prints by prominent artists covered several sheets and were pasted on walls and hand colored. The Triumphal Arch, commissioned by Holy Roman Emperor Maximilian I, measures 2.95 by 3.57 meters and was printed using 192 separate woodblocks. The 1517–18 first edition was gifted to princes and cities of the empire to hang in palaces and city halls, a classic example of Habsburg propaganda.

England and France led European wallpaper production. One of the earliest known examples is printed on the back of a recycled 1509 London proclamation. Wallpaper became widespread in English homes of status when Henry VIII's excommunication curtailed tapestry imports from Catholic Europe. The Puritan government of Cromwell's Protectorate halted wallpaper production, but the Restoration saw demand rise after the dull material culture imposed by Puritan luxury-item bans. Britain became the European leader in wallpaper production, both supplying the large British middle class and exporting.

A wallpaper tax was introduced to Britain in 1712 and abolished in 1836. This tax was avoided by buying plain paper and having it hand-stencilled or decorated. During this period, flock wallpaper was developed, the same technique (using adhesive to coat a surface with fiber particles) having been used on walls in medieval Germany. A cheaper "mock flock" was developed by printing a deeper shade over a lighter shade of the same color. Jean-Baptiste Réveillon imported flock wallpapers into France from 1753. When the Seven Years War stopped this trade, Réveillon began manufacturing and created some of the most successful wallpapers ever.

The Napoleonic Wars caused Britain's wallpaper industry to decline. Their conclusion, in 1815, caused a massive European demand for British goods again, including cheap wallpaper. The development of steam printing presses in the interim made wallpaper cheap enough for the working classes.

Post World War II, the DIY boom made wallpapering accessible to ordinary householders; prior to this, it had been a luxury item that had to be hung by skilled tradesmen. Old wallpaper in great houses seems to have been simply papered over as sequences going back to the 18th century have been recovered. In some houses, wallpaper cannot be papered over and has to be removed, initiating a new range of hardware products designed to remove wallpaper and culminating in strippable wallpaper, which can be simply peeled off.

Jon Welsh
AAG Archaeology

See Also: Fly-Tipping; Industrial Revolution; Paint; Paper Products.

Further Readings

Beard, Geoffrey. *The National Trust Book of the English House Interior*. London: Viking in association with the National Trust, 1990.
Hardie, K. and P. Langdown. "Flockage: The Flock Phenomenon." *TEXT: For the Study of Textile Art, Design and History*, v.35 (2007–08).
Kosuda-Warner, Joanne and Elisabeth Johnson. *Landscape Wallcoverings*. Washington, DC: Scala Publishers, National Design Museum, Smithsonian Institution, 2001.
McClelland, Nancy. *Historic Wall-Papers: From Their Inception to the Introduction of Machinery*. Philadelphia: Lippincott, 1924.
Powell, Jane and Linda Svendsen. *Linoleum*. Salt Lake City, UT: Gibbs Smith, 2003.
Simpson, Pamela H. *Cheap, Quick, & Easy: Imitative Architectural Materials 1879–1930*. Knoxville: University of Tennessee Press, 1999.
Simpson, Pamela H. "Linoleum and Lincrusta: The Democratic Coverings for Floors and Walls." *Perspectives in Vernacular Architecture*, v.7 (1997).
Simpson, Pamela H. "Substitute Gimcrackery: Ornamental Architectural Materials, 1870–1930." *Ideas*, v.5/1 (1997).

Florida

Florida is one of the largest states in the United States by population and by size. It provides a glimpse into the nature of consumption and waste in the United States. Florida also has unique issues associated with waste management.

Perhaps what sets Florida apart in the cultural landscape of the United States is its appellation as the "Sunshine State." As the southernmost of the conterminous states and the one with the most coastline (over 1,200 miles), it is widely known as an international vacation spot with world-class destinations such as Orlando, Miami, Key West, and the Gulf Coast beaches. However, it is not only a vacationer's paradise and is also home to nearly 20 million people who live and work in the state. As the fourth-largest state by population, Florida is part of the broader U.S. consumer culture; however, its subtropical climate sets it apart from all other U.S. states.

Consumer Activity in Florida

The size of the state has a significant impact on the overall national economy. The gross domestic product (GDP) in Florida is the fourth-largest in the country. The most important economic activity in the state is tourism, which brings millions of visitors each year from throughout North America and abroad. Several major attractions include the Walt Disney World Resort, Busch Gardens, SeaWorld, and the Universal Orlando Resort. However, the beaches and warm weather are the main natural attractions that draw visitors from cooler climates. Tourism produces a tremendous waste burden on communities. Visitors' garbage, sewage, and their voluminous presence must be managed by local governments so as to ease encumbrances placed on the local environment.

The second most important economic activity in the state is agriculture. Florida is the largest producer of citrus in the United States and is an important source of winter fruits and vegetables. Copious volumes of Florida tomatoes, strawberries, blueberries, and a variety of other seasonal crops are shipped all over the world. In addition, unique crops, such as sugarcane, mangos, orchids, tropical fish, and landscape plants add to the state's agricultural diversity.

The greatest problem with agricultural waste in Florida is the runoff of agricultural nitrogen and phosphorus from fields. As most ecosystems in the state are low-nutrient environments, nutrient-rich agricultural runoff poses particular problems for these bionetworks. One particular ecosystem, the Everglades, has received particular attention because of these issues. Water used in agricultural processing is also an important waste product that must be managed with great care. By-products of agricultural processing, such as orange rinds, dead tomato vines, and leftovers from sugarcane processing, are sometimes taken to landfills, left to rot or burn, or enter a waste-to-energy stream.

The third most important economic activity in the state is phosphate mining. The Bone Valley Formation, the source of the phosphate, is present in many areas throughout central Florida. The phosphate is mined in complex strip-mining operations, where voluminous piles of waste known as "gypsum stacks" are left behind. In addition, many acres of acidic and poisonous wastewater ponds are present. During a 2004 hurricane, water from one of these ponds breached its banks and flowed into the Gulf of Mexico, where untold damage impacted the environment.

Florida is also home to some of the nation's most important ports and interstate highway systems that bring in consumer goods destined for Florida's homes. Although consumerism remains strong in the state, double-digit unemployment rates and plummeting real estate values have dampened consumer confidence, and many businesses have been forced to downsize or close.

Waste Management

Waste in Florida is managed by the state's Department of Environmental Protection, which ensures that state and federal laws are enforced, although local governments may have more specific rules that local entities manage. Each day, tons of solid waste from homes and commercial operations are picked up by local workers or by waste collection operators and transported to local and regional landfills. Most large communities in Florida have curbside recycling pickup or access to recycling drop-off centers. Specialized waste, such as medical and hazardous waste, requires specialized handling and disposal.

Prior to state and federal involvement in waste management, hundreds of unregulated garbage dumps existed throughout the state. Waste from these facilities often leaks into the environment, vexing environmental managers. Most problematic are old industrial dumps, such as those associated with battery manufacturing, which leak hazardous materials into the soil and groundwater. Some abandoned garbage dumps also lead to community redevelopment problems. These sites can be classified as brownfields, which are sites that have real or perceived contamination and lead to lower property values in communities.

Florida's unique geology requires special attention from waste managers. The underlying bedrock in most of the state is a highly porous and permeable limestone. The holes in the rocks are highly interconnected and allow easy flow of water from the subsurface into the aquifer. Therefore, a contaminant released at the surface can easily find itself miles away from its source very quickly. There are numerous examples of waste products entering the drinking-water system in Florida. These wastes can also leak into freshwater or saltwater wetlands.

Special Issues With Florida Waste

There are several issues associated with waste management in Florida. These include waste-to-energy facilities, Superfund sites, and waste associated with storms.

Located throughout Florida, there are several waste-to-energy facilities that provide a significant amount of energy to the state. These plants typically take in household garbage and convert it into electricity. The remaining ash is separated magnetically so that metals can be recycled. There are also several biodiesel facilities in Florida that convert plant material to a biodiesel product or ethanol. Originally conceived of as a scheme to deal with waste vegetation and agricultural by-products, biodiesel has created a new agricultural niche in Florida.

The state of Florida contains several Superfund sites. These are sites with complex environmental contamination problems associated with poor handling of waste products. Many of the sites are associated with old or abandoned factories that produced a variety of products, including chemicals and pressure-treated wood. For example, the Stauffer Chemical Company site in Tarpon Springs, Florida, produced a variety of products from raw phosphate ore, including elemental phosphorus for weapons, between 1950 and 1980. During this time period, nearly half a million tons of wastes were disposed of in unregulated landfills and unlined ponds on the 160-acre site. The site is particularly dangerous because it contains a variety of hazardous chemicals in the subsurface that can leak directly into the aquifer, nearby rivers, and wetlands.

Another waste challenge for Florida is debris associated with tropical storms and hurricanes. A tremendous amount of debris, including tree limbs, furniture, construction waste, and hazardous waste, is produced when hurricanes damage communities. Lacking adequate forewarning, managers are faced with finding ways to deal with thousands of tons of excess debris. Safety issues associated with downed power lines and trees and purification of refrigerated items due to loss of electricity compound these challenges. During such events, waste managers work hard to educate impacted communities on how to separate their waste into tree limbs and yard waste, construction debris, household waste, and hazardous waste. It may take months for waste to be entirely cleaned from some of the most impacted communities.

The 2010 Deepwater Horizon oil spill in the Gulf of Mexico is of concern to many associated with waste management. Cleaning a large oil spill produces a significant amount of oily contaminated waste that includes contaminated sediment, cleaning supplies, and protective clothing.

Conclusion

Florida's diverse economy produces different wastes in varied amounts. Federal and state rules that govern the management of waste are overseen at the state level by the Florida Department of Environmental Protection. Florida has a number of waste-to-energy facilities that provide electricity to consumers. The state must manage a number of special wastes, including hurricane debris, phosphate mining waste, and agriculture waste.

Robert Brinkmann
University of South Florida

See Also: Alabama; Comprehensive Environmental Response, Compensation, and Liability Act (CERCLA/Superfund); Georgia; Pollution, Land; Pollution, Water; Weather and Waste.

Further Readings

Brown, M. "Landscape Restoration Following Phosphate Mining: 30 Years of Co-Evolution of Science, Industry and Regulation." *Ecological Engineering*, v.24/4 (2005).

Florida Department of Environmental Protection. "Florida National Priority List (Superfund) Sites." http://www.dep.state.fl.us/waste/categories/wc/pages/cleanup/pages/nplsites.htm (Accessed July 2010).

Perry W. "Everglades Restoration and Water Quality Challenges in South Florida." *Ecotoxicology*, v.17/7 (2008).

Fly-Tipping

Fly-tipping, originally a British term, refers to the illegal dumping of waste anywhere other than an officially licensed site such as a landfill or municipal tip. Material may be dumped onto public or private land, in the city, or in the countryside. In the United Kingdom, those who permit, rather than carry out, the dumping of waste onto ground that

Fly-tipping (or illegal dumping of waste) in a private parking lot in Tottenham, United Kingdom. The vast majority of fly-tipping material is domestic waste. Limited hours and accessibility at public dump sites are cited as reasons for the prevalence of this practice.

does not have a waste management license are also culpable. That both offenses are potentially punishable by a fine and imprisonment is an indication of the seriousness with which the practice is viewed. However, the proportion of successful prosecutions to reported instances of fly-tipping is minuscule. Fly-tipping constitutes an eyesore, may have serious environmental impacts, and can often be a danger to health and safety.

Less obvious effects occur when areas prone to repeat fly-tipping start to suffer declining property prices or decreased economic activity. Moreover, the costs of clearing fly-tipped material can be high, ongoing, and divert much-needed resources from elsewhere. To fully explore the phenomenon of fly-tipping, it helps to know who tips, the nature of the material dumped, the various sites favored for fly-tipping, and varied rationales invoked to justify the practice.

Fly-Tippers

A varied range of fly-tippers can be established. A significant majority, around three-quarters in the United Kingdom, of all the material dumped is identified as domestic waste, and the presumption is that individual householders dump general domestic waste and unwanted items that they are not willing—or able—to dispose of in other ways. Fly-tippers also include small businesses that illegally tip in order to circumvent official, highly legislated, and costly methods and sites of disposal. Larger business fly-tippers comes in two varieties: those who simply illegally dump larger quantities of their own material, such as building waste and hazardous chemicals, and those operators who make money by illegally dumping waste collected from third parties. Such third parties may or may not know that the waste they pay to have removed ends up illegally disposed of.

Composition and Purposes

While the majority of the material dumped is domestic waste, what is surprising is that over half of this constituent comprises plastic sacks containing general household waste, which is the kind of waste that would also be disposed of via municipal waste collection. That it is dumped instead brings to attention various waste collection issues that remain unresolved for many householders. These include both lower levels of domestic waste collection than that of waste generation and a lack of proper disposal facilities such as local recycling points. The next most common items comprising fly-tipped domestic waste include old mattresses, broken refrigerators, washing machines, and vehicle tires. While much of the motivation for such illegal dumping may well be based on ignorance, convenience, or the avoidance of specific charges related to the disposal of such items, this material too draws attention to some oft-noted issues, which may be seen with some sympathy. These include limited opening hours at civic amenity sites, the strictly enforced banning of vans at those sites, and unaffordable charges for bulky, dangerous, and specific items of refuse collection. While much fly-tipped waste may occur as the result of actions designed to avoid charges for trade waste collection or to avoid paying landfill taxes, this is not necessarily the case for much dumped domestic waste.

Urban and Rural Issues

Fly-tipping is predominantly an urban phenomenon, although instances of rural fly-tipping are often more highlighted in the news media. This may be because the countryside is seen to be more pristine in principle than the city, such that any despoiling is judged to be more offensive and antisocial. It may also be the case that a greater material contrast can be usefully exploited, that between the domestic waste of the city and the supposedly natural materiality of the countryside. Moreover, the contrast between discarded, disordered, and defunct waste and green, managed, and productive nature is powerful. The costs of removing dumped waste and of identifying those responsible for fly-tipping also pose greater challenges in the countryside.

Within urban centers, fly-tipping can be viewed as a quite-complex social phenomenon. For example, it may be regarded as one of a series of commonly designated social problems, particularly those practices that despoil the urban environment. These include fly-posting, graffiti, noise pollution, littering, and dog fouling; the latter two generate far more public complaints to local authorities than fly-tipping does. Other forms of disposal of domestic waste may also be confused with fly-tipping. For example, legislation specifically excludes from its definition of "illegal tipping" waste dumped on domestic land with the owner's permission. However, a good deal of domestic waste dumped in the urban and suburban environment remains on domestic private property, often rear and front gardens. While not technically illegal fly-tipping, such practices, nevertheless, give rise to the same problems and safety issues with the added potential to cause acrimonious neighbor and neighborhood disputes. The difference between semipermanent, visible storage and tipping is often tenuous and contested.

Neil Maycroft
University of Lincoln

See Also: Culture, Values, and Garbage; Household Consumption Patterns; Midnight Dumping; Sociology of Waste.

Further Readings

Hawke, Neil, Brian Jones, Neil Parpworth, and Katharine Thompson. *Pollution Control: The Powers and Duties of Local Authorities*. Crayford, UK: Shaw & Sons, 2006.

House of Commons Environmental Audit Committee. *Environmental Crime: Fly-Tipping, Fly-Posting, Litter, Graffiti, and Noise*. London: Stationery Office, 2004.

Food Consumption

Humans eat food for a variety of reasons. On the most basic level, food provides the macro- and micronutrients necessary for life. Humans also experience food through a multitude of means, consuming food not only as it is ingested but also in the variety of meanings humans attach to ingredients, meals, cuisines, sharing, celebrations, and even being seen with foods. Humans experience food beyond the meal not only while consuming it but also in the selection of certain products over others in meal planning and preparation. Humans experience foods as both a biological necessity (nutrition) and within a social context through which people ascribe meaning and value to foods as cuisine, as tradition, as identity, for creating and maintaining relationships, through food's preparation, presentation, packaging, cost, and within and throughout its social context. There are a variety of ways of reflecting taste at the individual, group, social, and cultural levels. Meals and the way they are shared both physically and ideologically inform people of who they have been, who they are, and where they might be going.

The meaning of foods, even at the most base level of what constitutes proper food, varies through space and time. Foods that at many times of the year might be considered inferior have historically been readily consumed in times of resource scarcity. Food is not only a matter of individual, household, or community preferences but is also rather intertwined in overlapping and often conflicting values and ideologies. As such, the production, distribution, preparation, and consumption of foods is increasingly legislated and contested in a variety of political and economic contexts. The notion of what is good to eat and what is possible to eat is a product of these processes.

Food Variety and Domestication

Most humans eat a rather narrow variety of foods. This may be a product of several factors, including cultural preferences, cost of food acquisition in terms of money and time, and overall availability of a diversity of foods. For the most part, humans believe—depending on the cultural context—there are substances that are good to eat, those that will be eaten in less than ideal circumstances, and substances that should never be ingested. This can include utilization of certain parts of an animal or plant. In addition, how food makes its way to the mouth and the process surrounding this have a tremendous amount of influence upon how much waste is created in the process of making food.

The innovation of domestication dramatically altered human relationships to other species. With the adoption and utilization of these technologies, humans came to rely on fewer and fewer species with greater and greater intensity. This reliance created anatomical and physiological changes in other species, which, in turn, made them more reliant upon humans for their propagation and continuance. The overall impact of these innovations has been a decrease in the variety of foods readily available to large segments of the population. These processes also led to the continuing separation of humans from both the plants and animals consumed and the suite of physical, geological, ecological, and biological processes that allow for the production of food.

This is not to say that other food consumption patterns did not lead to the landscape modifications. Foraging activities, including hunting and gathering, have historically resulted in landscape modifications, although the scale and intensity of modification increased with the introduction of agriculture. The overall decrease in the variety of foods has been explored to determine the nutritional consequences of agriculture.

Food Rituals

Food consumption has become a distinguishing factor in identity, particularly along the lines of cultural and national identity. Historically, certain foods were ritualistically consumed in a variety of cultural events. In some instances, the ritualized consumption of certain foods marks a change in season. In the context of indigenous peoples of North America, these foods have been examined by scholars and activists as part of cultural revitalization movements, which encourage a return to remembered ways of doing and being. These foods have, in some instances, been viewed as a means to perpetuate health and confront disease in a number of cultural contexts. They are symbolically important in that they represent a return to traditional ways of eating and thinking in an effort to decolonize diets.

The consumption of traditional—sometimes conceptualized as precontact foods—is hence viewed as a response to contemporary health epidemics, most notably the type II diabetes that is prevalent in many 21st-century Native American communities. The call to consume more traditional foods is for some a resistance to the colonization of both food systems and the body. In other instances, indigenous peoples readily use traditional foods and methods of healing in unison with biomedical and other approaches to health and healing. At the same time, conceptions of traditional foods vary intra-culturally as the meaning of traditional foods can vary dramatically between and within households, communities, and nations.

Fusion

In some instances, these foods are twinned with other nontraditional or postcontact foods to produce fusion cuisine. Contemporary chefs continue to expand the offering of fusion cuisine by combining regional cuisines in new and interesting ways and integrating "new" foods in familiar dishes. This occurs in both high-end restaurants and resorts as well as in local communities in order to make the foods more exciting for younger generations who may not have grown up consuming traditional foods. One notable example of this is on the Tohono O'odham Nation of southern Arizona, where cholla cactus buds are now combined with spinach, pineapple, and dressing to produce a healthy and nutritious example. Other popular combinations among those promoting traditional foods on the Tohono O'odham Nation include tepary bean hummus and raspberry chia seed smoothies. These serve to entice non-O'odham to eat traditional O'odham foods and offer a new and unique way to experience traditional foods for O'odham peoples.

The mixture of traditional with nontraditional foods has appeal for many Tohono O'odham. At the same time, there has been a marked interest in these foods among outsiders who wish to try new foods. Depending on the context, this can lead to an increased interest in and capacity to grow traditional foods at the local level or to a siphoning off of traditional foods to fulfill the desires of consumers in other regions. Hence, traditional food enthusiasts and organizations such as Slow Food USA must also consider the impacts that their making traditional foods readily available to the general public might have on indigenous peoples who are working to revitalize traditional food systems.

International Exchange and Markets

This contemporary example of the circulation of plants and animals and their subsequent utilization in different locations, each with their own suite of judgments regarding the value of the food, mirrors earlier exchanges. The term *Columbian Exchange* (coined by historian Alfred Crosby), or the exchange of plants and animals for a variety of uses, including food, beginning with the voyage of Christopher Columbus in 1492, has had tremendous impacts on cuisine throughout the world. The most readily recognizable 21st-century foods are themselves products of interaction and exchange of plants, animals, and ways of preparing these for human consumption. The Columbian Exchange had dramatic impacts on the diets of those in the New World and those in the Old World. One of the starkest examples is the overreliance upon one type of potato for subsistence utilization by the Irish people. This crop failure, coupled with British control over Ireland, led to the Great Hunger, or Great Famine, which led to the starvation or emigration of over 1 million people. The Columbian Exchange was a product of colonialism and can be viewed as an early intensification of global flows of goods and people, a precursor to globalization.

In the 21st century, goods, people, and ideas flow through and across borders with varying ease via instantaneous and enveloping communications. Products come with their own biographies that are coproduced by both employees and, increasingly, in conversation with consumers. In the consumption of food—or any other consumables for that matter—humans are consuming the physical products based on sensory taste as well as perceptions of taste. Hence, the consumption of foods is also necessarily social in nature.

Food and the meanings that humans attach to it vary over space and time. Some of this can be related to the cost of products, which fluctuate over time. In its early years as a commodity, sugar was so expensive that only the elite could afford it. As sugar plantations expanded in the New World, aided with the forced labor of the transatlantic slave trade, the price of sugar dropped. This price drop allowed for consumption by the masses. Market fluctuations of commodities can dramatically impact farms of various scales. This can translate into increased vulnerability in terms of economic and food security.

Consumption Levels: Overnutrition and Undernutrition

Quantity of food consumption can be problematic at multiple levels if too much or too little food is consumed. The question of too little food being consumed in its most dramatic form occurs in large-scale collapse of the food system's ability to meet the dietary requirements of large segments of the population. In order to prevent large-scale famines, international monitoring systems coupled with news media coverage can mitigate the impacts of caloric shortfalls by assessing food insecurity. Food security, according to the 1996 World Food Summit, "exists when all people, at all times, have physical and economic access to sufficient, safe, and nutritious food to meet their dietary needs and food preferences for an active and healthy life." In order for this model to work, enough food has to be available for either direct assistance or for purchase to vulnerable populations. Often, foods that are utilized to confront food insecurity on a large scale are commodity foods.

The United States has focused on commodity production and has historically made claims of feeding the world through agricultural development in the Great Plains, or what has historically been referred to as "the breadbasket" of the United States. Despite this optimism and the tremendous historical growth in U.S. agriculture production, there are still numerous food security issues globally and in the United States. Both urban and rural populations suffer from a lack of food security as defined in international law and policy.

There are a number of health and nutritional disparities in the United States surrounding both undernutrition and overnutrition. One phenomenon has been the occurrence of both over- and undernutrition in the same individual, such that the person consumes an overabundance of calorie-dense, nutrient-poor foods. Hence, there is overnutrition in terms of calories and fat consumed and undernutrition in terms of key vitamins and minerals. These health disparities are particularly apparent when examining ethnic and poor communities in the United States and England, where there have been reports of food deserts. Food deserts, variously defined in the literature, refer generally to those communities that do not have ready access to fresh foods, notably fruits and vegetables. Neighborhoods with lower per capita income tend to have less access to these nutrient-dense foods and more access to foods that are relatively calorie dense and nutrient poor. A policy response to this has been to promote the construction of large supermarkets in food deserts in an effort to increase consumption of fruits and vegetables. According to preliminary reports, this has been successful by encouraging those in former food deserts to consume more fruits and vegetables and greater quantities of all types of foods.

In some instances, deficiencies in key nutrients have been viewed as government responsibility and in other instances corporate opportunity, although the two are not necessarily mutually exclusive. Government-permitted and industry-sponsored fortification of foods has helped to alleviate some health concerns caused by nutrient deficiencies. In 1969, the U.S. government allowed fortification on a wide scale, such that nutrients could be added to foods of any kind—whatever their preprocessing nutrient levels—if it was thought that the diets of a significant number of people were deficient in them. Attempts have also been made to increase the health potential of staple foods, hence meeting nutrient shortfalls. In one case, genetically modified rice has been promoted as a means to target vitamin A deficiency in children. Rather than attempt to ameliorate the conditions that led to a consumption of a narrow range of foods, technical expertise is seen as the "magic bullet" to alleviate suffering in an attempt to legitimate gross disparities that may arise in commodity production.

Caloric shortfall has been a major concern throughout the world, even in the United States, particularly during times of economic crisis, such as the 2008 "heat or eat" crisis in western Alaska, where some residents of rural communities were paying very high prices for fuel oil relative to those in urban areas of Alaska and those in the "lower 48." These can be mitigated through community-based safety nets such as food banks, as well as government-based food assistance programs.

Consuming too many calories can also lead to a number of health concerns. The increase in the number of calories consumed in relation to caloric expenditure leads to weight gain. Weight gain over time can lead people to become overweight or obese. This increase in weight is a risk factor for a number of diseases that have been collectively referred to as "diseases of affluence." Films, including *Supersize Me!*, *Fast Food Nation*, and *Food Inc.*, have done much to increase public understanding of the health risks inherent in contemporary food systems. These films are a means through which greater segments of the population are able to consider the impact of the food system. According to one recent study, fast-food consumption has strong positive associations with weight gain and insulin resistance, suggesting that fast food increases the risk of obesity. While these accounts may raise awareness among specific segments of the population, mere recognition of potential health consequences does not necessarily translate into actual dietary modification, particularly for those with neither the time nor monetary resources to make dietary modifications.

Obesity and Diabetes

The majority of people over 30 in the United States are either overweight or obese and the prevalence of overweight or obese individuals is likely to increase. The vast majority of deaths in the United States are due to chronic illnesses. These conditions are a product of lifestyle choices in terms of diet and exercise and environmental or structural factors that limit access to ready means of exercise and fresh food options. Narratives that emphasize personal responsibility in lifestyle decisions, or a "you've got it coming mentality," and minimize factors that are outside individual control do a great disservice to the promotion of health and wellness, as they ignore the multiple actors and entities that attempt to capture bodies for their own political, economic, and ideological ends.

These narratives become internalized at the individual, community, and societal levels with a cultural logic that emphasizes personal responsibility and freedom. Since individuals have the choice whether to consume foods that are deemed by a number of experts to be "bad" for them, individuals must have the strength to resist consuming these foods out of care for the self. The ramifications of dietary consumption patterns are staggering in terms of their cost to life expectancy and quality of life. Among the illnesses that result from lifestyle patterns, type II diabetes may present the most serious challenge to community health professionals worldwide. The worldwide incidence rate of diabetes continues to rise, likely leading to an increase in the overall number of people living with type II diabetes. In the 21st century, the highest incidence rates are in more affluent countries such as the United States, which had over 20 million cases of diabetes in 2005. Diseases of affluence, such as diabetes, increase premature mortality of the population, medical costs, disability, and individual and family suffering. Medical interventions are made to both save someone's life and affirm poor individual choices and the redemptive power of medical interventions. In this manner, environmental and structural elements of health outcomes are made invisible, yet they are ever present.

Production

Food consumption impacts personal health and the amount of resources that are consumed in order to bring foods to consumers. Food choices at the individual, household, community, and national levels regarding law and policy surrounding food production produce demands on net primary productivity, water, and soil resources. These choices have dramatic environmental impacts at the local, regional, national, and global levels. Included among these is local infiltration of nitrates into groundwater, aquifer depletion, soil erosion, the dead zone in the Gulf of Mexico, and human-induced climate change. Given the nutrition transition in several regions of the world where consumers are demanding greater quantities of resource-intensive foods, these impacts are likely to continue and expand unless there are dramatic shifts in policy and economics.

Despite these impacts—and potentially because of them—continuing efforts by multiple actors are being made to promote myths of agriculture and food production in the United States. The dominant myth of agricultural history in the United States is that it was built by family farmers, intrepid pioneers wresting sustenance from the soil. Alternatively, farmers are portrayed as working in unison with the natural environment to produce food in idyllic, pastoral scenes. These images are frequently employed in the marketing of farm products from a variety of production systems and scales. Effective advertising and marketing has led to the close affiliation people have with particular products as well as brands. These efforts dwarf media campaigns of both public health professionals and agencies that stress the importance of fresh-food consumption and environmental groups who call into question the ecological costs of food production.

Idyllic modes of production are quite separate from the realities of commodity-oriented production systems, which concentrate singularly upon yield maximization of one species to the detriment of others. This disconnect between the idyllic representations of the U.S. family farm and reality in food production continues to be contested among a variety of actors who agitate for continued change in food production systems. Consumers in grocery stores are thus faced with a number of claims made on the packaging—or rather the representation of reality—as idyllic and pastoral. Hence, the imagining of the farm as holistic and integrated with natural systems may be thought of as the first level of consumption. This imagery is not lost on advertising executives and public relations firms that work to maintain scenes of plenitude, the wealth of nature, and harmonious human–environment interactions. This obfuscation of farm practices is in part a product of geographic distance. This physical distance leads to a decrease in an understanding of the processes that go into the making of food. Since foods are consumed hundreds or thousands of miles away from where they are grown, consumers rely upon a number of third parties to certify food quality. Consumers concerned with quality have led a number of movements for more government and independent oversight of food. It is assumed through a series of labeling regimes—some state regulated, while others are promoted by nongovernment or industry organizations—that consumers will have an indication of the process that went into the production of the foods they are eating. This information serves as the basis for consumer decisions regarding whether foods will be purchased, how they will be prepared, and to whom and how they will be served. Hence, consuming foods that are labeled in a particular manner confer value to the consumer. This is not to say that various regimes of food labeling are entirely aligned with one another, as product manufacturers in the United States may make health claims concerning their food production with the caveat that a statement appear on that same label noting that the U.S. Food and Drug Administration has not evaluated the claims made. This includes the products themselves and the processes behind the products, such as the injection of recombinant bovine growth hormone (rBGH) in dairy cows.

In addition to health claims, food labeling serves to confer other domains of knowledge to the consumer. Presumably, questions regarding how the food was grown, how the animals were treated, and how the workers were treated can all be answered through reference to product labeling. As an example, the label *Fair Trade* confers to consumers a sense that they are reconciling their consumer wants and needs with their personal or household beliefs regarding equity and social justice. The U.S. Department of Agriculture's "Organic" certification also indicates to the consumer that the product was grown using organic methods. This label is increasingly questioned by some scholars and consumer watch groups who note the shifting nature of organic agriculture practices and industrial organic food producers and processors.

The amount of food available remains a pressing concern. Food is consumed through a web of political and economic relations. Consumer agency regarding food choices is always contingent upon a number of overlapping factors, including cultural preferences, accessibility, availability, convenience, and price. Increased industrialization of food production leads to increasing disparities in consumption and allows those affluent enough the luxury of not having to be concerned with where food comes from. However, concern is ever present in grocery stores whose burgeoning shelves hold a plethora of heavily labeled products claiming to represent a myriad of just and equitable relations. As anthropologists have shown, these products are never static and never fully consumed but, instead, are contingent and aspired to in a myriad of consumption practices.

David Fazzino
University of Alaska

See Also: Carbon Dioxide; Certified Products (Fair Trade or Organic); Consumerism; Dairy Products; Environmentalism; Farms; Food Waste Behavior; Grocery Stores; Meat; Organic Waste; Pollution, Air; Pollution, Land; Pollution, Water; Population Growth; Slow Food; Sustainable Development; Underconsumption; Water Consumption; Weather and Waste.

Further Readings

Foster, R. "The Work of the New Economy: Consumers, Brands, and Value Creation." *Cultural Anthropology*, v.22 (2007).

Glanz, K., M. Basil, E. Maibach, J. Goldberg, and D. Snyder. "Why Americans Eat What They Do: Taste, Nutrition, Cost, Convenience, and Weight Control Concerns as Influences on Food Consumption." *Journal of the American Dietetic Association,* v.98 (1998).

Kaplan, Martha. "Fijian Water in Fiji and New York: Local Politics and a Global Commodity." *Cultural Anthropology*, v.22 (2007).

Levenstein, H. *Paradox of Plenty: A Social History of Eating in Modern America.* Berkeley: University of California Press, 2003.

Food Waste Behavior

Food waste can occur during the production, storage, distribution, consumption, and nonconsumption of foods. This occurs from both human action and inaction. In terms of food production, foods can be wasted in the actual harvest process where food is not harvested through the initial harvest process or otherwise gleaned. In terms of storage, inadequately designed storage systems and storage for too long a time period can lead to waste. From the food distribution aspect, food can be wasted through accidents, too long a time period for distribution, or overdistribution, when too much food is distributed at the same time.

Food consumption depends upon a number of factors and varies from country to country and culture to culture. Decisions are made at multiple levels as to whether or not foods are good to eat and what portion of the animals and plants will be consumed. Hence, some parts may be deemed culturally inappropriate for human consumption. At the same time, individual preferences regarding food consumption also impact the amount of food waste produced. Foods that are not consumed may either be shared with other people, fed to animals, composted, sorted, or disposed of in trash bags and eventually landfills. Some municipalities concerned with shrinking landfill space and subsequent costs of landfill development or waste shipment are making efforts to reduce the overall waste stream by combinations of recycling programs and organic waste diversion.

Food Production Waste

In terms of food production, foods can be wasted throughout the growing process; in particular, food losses are readily apparent in the actual harvest process. Food waste during food production can occur from a number of factors and is often the result of farm management choices, including tools and techniques used in growing food, geology, or weather-related factors. Specifically, crops can be lost to pest infestations, climatic variations in temperature and water access, and soil fertility issues (which decrease the ability to plants of attain the water and nutrient sources needed for growth, thus leading to greater vulnerability). Mechanical harvesters may be inefficient in harvesting food in its prime, hence leading to waste of crops that are not yet mature. Crops that are missed through mechanical harvesting methods can be utilized through gleaning efforts and hence do not contribute to overall food waste. In addition, foods that are unfit for human consumption may be utilized to feed animals.

Postharvest events can also create food waste. This can occur in transit or in the multiple areas where food is stored. In terms of storage, inadequately designed and managed storage systems can lead to waste. Spoilage can occur in regions with high temperature and humidity. Another consideration that leads to food waste is loss to pests, including insects and microorganisms. These are a concern primarily in high-temperature and high-humidity regions.

Industrial systems of production are further vulnerable to weather and geologic events. Perhaps the most easily recognized vulnerability is created through confined animal feed operations (CAFOs). CAFOs centralize production using industrial methods in an attempt to control every aspect of an animal's life toward the goal of maximizing yields. This centralization of production creates serious health vulnerabilities for animals, which are often subject to large doses of antibiotics. Further, CAFOs are intensive in terms of their management, inputs, and waste. There is a heavy concentration of animals and, hence, animal waste in one area. Flooding events, as well as everyday operations can lead to damage to marine life as well as to water quality and safety. Even where these waste products are properly collected, there are potential water pollution issues associated with the utilization of liquid manure from CAFOs.

Regulation

In addition to the food waste associated with the actual production of foods, food waste can also occur due to government regulation and policies. The scale of agricultural systems can also produce vulnerabilities in terms of loss. Food safety regulations act to protect the public from potential negative health outcomes. Recalls are widespread in the food system. This leads to the waste of massive quantities of foods on an annual basis. To illustrate the scope of the food recalls, in November 2010 alone, there were food recalls on potato chips, chocolate, cheese, chicken pasta salad, dried taro, eggs, turkey, cookie dough, apple cider, meat and poultry canned products, ready-to-eat pork products, beef sticks, and tuna. During a food recall, attempts are made to ensure that all of the improperly labeled, contaminated, or potentially contaminated food is collected and destroyed. Food safety is not the only instance where governments become involved in regulating the distribution of foods. Government policy may in some instances be counterproductive in protecting the health of its population. This was particularly apparent when the U.S. government, in an effort to control the price of food during the Great Depression, deliberately destroyed hogs, produce, and milk despite chronic malnutrition of the population. This example highlights the political and economic relations through which food distribution and food waste occur.

Food Consumption Waste

After food makes its way without incident from producers through various distributors and means of transport to the grocery stores and then to consumers, there is still vast potential for loss. Food waste during consumption can occur through the selection of foods that members of the household may not eat, storage of foods in the household both before and after cooking (including temperature, humidity, and presence of pests), and in the cooking process. Some of the loss can be attributed to consumer food preferences. Food preferences are a result of a number of factors. These overlapping factors include cultural preferences, accessibility, availability, convenience, and price. Convenience in terms of time spent preparing meals was a theme that dominated advertising for food in the 1950s. Hence, Americans as a whole were spending less time dealing with food, either its production or processing, around and in their homes. But ready-made meals and convenience foods can hide the waste produced in making meals.

Cooking times are generally shorter than in the past, and as a whole, those who consume processed foods spend less time dealing with food, either its production or processing, around and in their homes. These ready-made foods contribute in other ways to the overall waste stream. In particular, packaging of heat-and-eat meals, snack foods, and prepared fruits and vegetables contributes plastics and paper to the waste stream. Efforts continue to be made to reduce packaging of food products in order to mitigate food-related waste related to consumption practices.

Convenience is not always a motivating factor in decisions regarding food consumption and food waste. According to one study done by Ann Allison, food consumption in Japan is a measure by which the state can assess the performance of women. Allison describes obento boxes as an ideological state apparatus, noting that children must quickly consume all of the obento that their mother has prepared for them. Food waste post-mealtime is viewed as a failure of mothers to properly care for their children.

Uneaten Food

The absence of strict codes of compliance for finishing a meal, coupled with increasingly larger portions of meals, often leads to excess prepared foods in the United States. A number of approaches have arisen in attempts to mitigate food waste at the household level. One of the simplest approaches is managing foods that are not utilized immediately after they are cooked for later consumption. There are recipe books to assist those who, out of necessity, desire, or a combination of both, seek to create additional meals from leftover foods. These cookbooks include *Waste Not Want Not: A Cookbook of Delicious Foods from Leftovers* by Helen McCully, *The Use-It-Up Cookbook, A Guide for Minimizing Food Waste* by Lois Carlson Willand, *The Use it Up Cookbook: Creative Recipes for the Frugal Cook* by Catherine Kitcho, and *The Thrifty Cookbook: 476 Ways to Eat Well With Leftovers* by Kate Colquhoun. The use of leftovers can lead to real savings for households that need to maximize their utilization of resources. However, some purchased foods are not utilized by many households. According to one study, a majority of consumers in the United States reported purchasing foods that they never use. This was matched by a study in the United Kingdom, which found that unused foods cost households substantial sums of money.

Several studies have been conducted to measure waste as well as waste behaviors in institutional settings, such as schools. These studies utilize a number of methods to examine not only the amount of food being wasted but also which foods are being wasted. One measure is the amount of overall waste generated. Institutions can establish a baseline by which to quantitatively measure, in terms of food weight, efficacy of waste reduction strategies. This has been utilized in several universities to show that the total amount of food waste produced in cafeterias can be dramatically reduced by simply eliminating the use of trays. Since students are limited to taking only what they can carry in their hands, acquiring the same amount of food requires more trips than it would otherwise with a tray. Further, it gives someone the opportunity to consider whether or not they would really like an additional plate or drink.

A more qualitative and nuanced approach is conducting plate, or plate-waste, studies. These studies specifically address that foods are being taken but not consumed. Plate waste studies are of three primary types, each yielding particular data sets: weighed, visual, and recall. In general, data from these studies are valuable not only in terms of eliminating waste at the institutional level but also in targeting nutritional interventions at both the individual and institutional levels. Through utilization of this information, it is possible for cafeterias to reconfigure their menus in order to

A number of approaches have been developed to attempt to lessen household food waste. One of the simplest is to encourage proper storage and use of foods for later consumption that are not utilized immediately after they are cooked. Cookbooks aimed specifically at leftovers cite the cost savings from using up these items. However, most U.S. consumers reported in a survey that they never use some of the foods they purchase. A United Kingdom survey had similar results; in addition, wasted food added up to a great deal of lost money.

reduce waste and contribute to better nutritional outcomes for children.

According to the U.S. Department of Agriculture's Economic Research Service, nutrient-dense foods such as fruits and vegetables comprise 20 percent of food waste. Among U.S. primary and secondary students, waste of salads and vegetables can range from one-third to half of the foods served to them. Taken together, this work indicates a pressing problem for nutritionists and other professionals dealing with health outcomes. Further, children, as well as adults and institutions, are likely to waste some of the best sources of nutrients and hence increase vulnerability in both the long run and short term to illness and disease.

Food Archaeology

Archaeologists have also attempted to address food waste, as food waste behavior can be assessed utilizing the archaeological record. Waste sites can serve as rich sites for data collection for archaeologists. These prehistoric and historic dumps or kitchen middens preserve organic remains for analysis by archaeologists who can determine historic consumption patterns. Data such as these can assist archaeologists in their attempts to reconstruct past diets as well as give clues to subsistence systems and trade networks. These techniques also lend themselves to analysis of contemporary waste. Archaeological studies have been conducted to analyze contemporary food waste behaviors; this study of garbage is known as garbology. The information gathered from landfills can be valuable in determining the composition of household waste as well as overall waste behaviors. This information is crucial to policy intervention to decrease the overall burden on landfills and the pollution potential of landfills, as well as to capture economic and environmental benefits.

One example of garbology is a waste characterization study in Phoenix, Arizona. This study noted that most of the residential waste stream is comprised of near-equal proportions of organic waste and recyclable materials. Further, 16.8 percent of the waste was determined to be food waste. Although the percentage of the residential waste stream comprised of food waste varies from municipality to municipality, it is fair to say that food waste comprises a significant percentage of residential waste. The amount of the food waste produced in the preparation of foods may not necessarily be captured in studies that examine only trash, because garbage disposal units in many households are the means by which people dispose of their unutilized portions of food. Some studies suggest that this figure may be much higher for some households.

Food Reuse

While archaeologists utilize waste in academic and practical applications, it is viewed as a resource in order to learn about specific behaviors regarding waste. Quite distinct is the utilization of waste as a food and livelihood resource. Scavengers position themselves to make valuable use of resources. Some do so out of economic necessity but they are not necessarily the most impoverished people in society. Scavenging of resources can provide high economic returns. Some scavengers or dumpster divers will utilize waste for their own purposes, including food. The act of dumpster diving varies in legality among countries and jurisdictions. In some areas there are laws, but these are infrequently enforced. In some instances, supermarkets, which contribute large amounts of food waste, will gate and lock their dumpsters to prevent dumpster diving. In some municipalities, work continues to capture as much of this food as possible through institutional arrangements, such as food banks, homeless shelters, and religious organizations. In many of these settings, the wasting of food is seen as an ethical as well as a moral issue.

Food waste can be captured and utilized in a number of ways to produce both economic and environmental benefits. Properly sorted food waste and other organic materials not only can be utilized on the smaller scale to enrich the soil but can also be utilized post-collection by either institutions or municipalities. University campuses and citywide programs have adopted separation of food from other waste products in order to minimize the amount of food waste that needs to be transported to landfills. Some residents may choose to bypass these systems of collections by handling their food wastes at home. At the household level, food waste can be converted through composting or vermiculture to produce soil amendments. Further food waste can be fed to animals on a farm in order to recycle nutrients. On a larger scale, it can be collected and utilized to feed animals in agricultural operations. Regardless of the scale, this reduces the amount of overall waste, decreasing economic and environmental costs, including water pollution, as well as recapturing nutrients that might otherwise be lost.

Collection of food waste may be problematic. Municipal collection programs may face severe challenges from residents, as food and other organic wastes may produce a nuisance. In order to prevent this, municipalities need to pick up organic waste materials frequently. This is particularly true in regions that experience hot temperatures and humidity, which may lead to nuisances such as a rotting smell or attract animals and insects. This can lead to decreased support of separate collection programs for organic and nonorganic materials.

As food waste is typically organic in nature, during its decomposition, methane, an extremely potent greenhouse gas, is produced. This presents both a challenge as well as an opportunity. Food waste in landfills can produce methane, which contributes to global warming; but food waste can also be utilized as an energy source. Anaerobic digestion plants burn methane produced from organic waste materials. Another energy source derived from spent fryer oil or yellow grease is biodiesel. Efforts are also being made to utilize brown grease, which includes oils and other residues, as biodiesel. This conversion of food waste into a usable fuel for transportation not only decreases the amount of waste that the food system produces but also allows for a smaller carbon footprint.

David Fazzino
University of Alaska

See Also: Archaeology of Garbage; Archaeology of Modern Landfills; Arizona Waste Characterization Study; Biodegradable; Carbon Dioxide; Consumption

Patterns; Dairy Products; Dump Digging; Dumpster Diving; Farms; Food Consumption; Garbology; Organic Waste; Pollution, Air; Pollution, Land; Pollution, Water; Water Consumption; Worms.

Further Readings

Allison, A. "Japanese Mothers and Obentos: The Lunch-Box as Ideological State Apparatus." *American Anthropologist*, v.64/4 (1991).

Kantor, L., K. Lipton, A. Manchester, and V. Oliveira. "Estimating and Addressing America's Food Losses." *Economic Research Service, United States Department of Agriculture Food Review*, v.2/11 (1997). http://www.ers.usda.gov/Publications/FoodReview/Jan1997/jan97a.pdf. (Accessed November 2010).

Levenstein, H. *Paradox of Plenty: A Social History of Eating in Modern America*. Berkley: University of California Press, 2001.

Rathje, W. and C. Murphy. *Rubbish! The Archaeology of Garbage*. Tucson: University of Arizona Press, 2001.

France

Quand c'est usé, je jette, je jette et je rachète (when it's worn, I throw it, I throw it and buy another) was the chorus of a popular song ("quand c'est usé, je le jette") of a popular singer, Jacques Dutronc, broadcast in 1969. It was an ironic charge against the consumer society and its abuses. Like in other industrialized countries at the same time, France was in the 1960s in the paradoxical situation of advanced development and significant economic growth, with one of the highest standard of living in the world. However, the Western counterculture movement was contesting the bourgeois ideals of progress and consumption, pointing at the forthcoming disastrous effects on nature. In the early 21st century, the last years of the so-called Glorious Thirty (years of economic growth), this division between the logic of economic and industrial development and the promotion of more ecological values remains strong. France's 65 million people continue to enjoy a high standard of living. But what has changed in the meantime is the haunting resonance of ecological motives in French culture and society.

Ecological Debate and Programs

In the French contemporary ideological context, the matters (and problems) of environment, pollution, overconsumption, and waste management are embedded in two broader issues: sustainability and ecology. Political ecology and the consciousness of the human threat against nature began to gain visibility in the early 1970s (after the United Nations' first Conference on Human Environment in Stockholm, 1971). But it had, at the time, limited impact on society, despite continuous official status of environment and ecology in the French state administration after 1971.

However, the two oil crises (1973 and 1979) suddenly highlighted France's dependence on energy and fragile economy and left room for the expression of alternative scenarios of development. Pollution, the greenhouse effect, biological diversity, and impacts of overconsumption largely infused French society and compelled economic and political sectors to adjust to these new cultural models in the 1980s. France was engaged in the first international Earth summit in Rio de Janeiro in 1992, and from the 1990s onward, national institutions promoted "green" labels to motivate ecologically oriented behaviors (in food, industry, consumption, automotive, energy, construction, and waste).

Issues of waste and consumption are rather new in France but crucial debates on the relevance of the "ecological crisis" theory and the economic and social costs of converting the country to other economic models have occurred. Since the late 1990s, the political force of green movements became more significant—although still having a modest demographic surface in the political landscape. The social pressure on the French government to allow more ecological policies brought about a series of symposiums and conferences. Held in October 2007, the "Grenelle de l'environnement" (relating to the "Accords de Grenelle" that took place in the 1968 political crisis) was supposed to define new parameters for agriculture, transportation, energy, health, and biodiversity in order to adjust government policies to the worldwide ecological crisis. It was quickly followed by the first policy (also labeled "Grenelle 1") in August 2009, but this did not succeed in convincing ecologist movements and parties. The setting up of ecological programs in the political agenda of France gave rise to bitter controversies, and the French government was accused of political duplicity. While France was supposed to champion, in the international arena, the defense of standards of development more respectful of the ecology, it reversed and eventually did not apply these same standards to the domestic economic agenda. Intellectuals and scholars are thus divided between the supporters of radical sustainability, based upon a significant reduction of industrial production (degrowth, or *décroissance* theory), and the promoters of an "ecofriendly" economic growth. Far from closed, the debate is still topical in the early 21st century, but at least all decision makers agree on the need for transformation.

The environmental policies of France are torn between domestic issues in growth, sustainability, ecology, and the normative standards of the European Union (EU), which has urged the countries of the continent to align with dramatic reduction of waste and energy consumption. As Claude Allegre—a much-debated geochemist and former minister of education who refutes the global warming theory and degrowth—put it in a 2007 opus, "we have to make a clear distinction in between worldwide scale problems, and the ways France can contribute to their resolution, and those which are typically French, whose solution depends essentially on us." The solutions are threefold: organization (management of waste), innovation (the search for alternative modes of consumption), and communication (broadcasting messages for the prevention and the collection of waste).

Funding

The expenses for protection of the environment have dramatically increased, from 30 billion euros in 2000 to 44 billion euros (according to the National Institute of Statistics and Economic Studies) in 2010. The most important sources of pollution and waste are, in order of importance, agriculture (43 percent), industry (41 percent), and households (3.5 percent). State administration (public sector) and industry (private sector) have therefore devoted considerable effort to the improvement of chains of waste prevention, production, collection, and recycling, and allowed important funding for research and development (3.5 billion euros in 2008). In the mid-2000s, the national expense for the management of waste amounted to 11.5 billion euros (14 billion euros in 2008), of which 63 percent was funded by households and collectives. Taxes for the collection of waste make up to 83 percent of the total amount. Public expense for waste management has increased up to 5 percent per year. While the major sources of waste and pollution are located in systems of agrarian and industrial production, public campaigns and media coverage of information on pollution, waste, and other ecological topics, as well as systems of collection and recycling, mainly target households and consumers.

Waste Sorting

If Eugène Poubelle was the first to introduce the dustbin—and waste sorting—in France in 1884, (his name has since then been associated with the garbage can), the very first attempts to establish a nationwide system of waste sorting occurred in the mid-1970s, following the first oil crisis. Substantial progress did not come before the 1990s and the 1992 legal dispositions for the creation of recycling channels for every type of waste. The aim was to reduce landfills and incineration of waste, and these witnessed a slow but significant decline.

The recycling of packaging (5 percent of waste) ranges among the fastest expanding (up to 57 percent of the total amount in 2007), but the household consumption of manufactured goods has nevertheless increased (almost twofold) since the 1990s (up to 22 billion euros a year in 2010), and waste generated by individuals increased consequently (7 tons per year, including 425 kilograms of domestic waste).

Innovation

As for research, the foundation of the Agency for the Environment and the Management of Energy (ADEME) in 1990 offered a chance for decision makers and enterprises to design their ecological policies on reliable information. The missions of the ADEME encompass prevention, management, and research for ecological purposes. This is one of the main nonacademic sponsors for research on pollution, alternative sources of energy, and waste in France and is one of the main sources of information. The ADEME's publications provide regular reports and databases, for instance, on the evolution of green behaviors in France and innovation in new energy.

Attitudes

The impact of campaigns in the broadcast media had qualitatively observed impacts on individual attitudes toward environmental crisis, consumption, and waste. But inquiries conducted or funded by the Ministry of Ecology, Sustainable Development and Energy or by the ADEME unveil ambivalent responses.

They demonstrate that only half of the French population (52 percent) is explicitly inclined to introduce dramatic change in ordinary life and habits as a response to the environmental crisis and in search of new models of behaviors toward consumption and waste. The appeal of new technologies for low-consumption energy is high (more than 80 percent) However, the sense of individual responsibility in the ecological crisis and responses is decreasing (from 87 percent in 2005 to 83 percent in 2009) and the sense of guilt for a lack of active participation in ecological behaviors remains low (33 percent), and most people (73 percent) are reluctant to be sermonized. They nevertheless claim to be attentive to sorting their waste (92 percent), and half of them do it in accordance with the norms of sorting (45 percent) in their area of residence or by bringing sizable or hazardous waste to civic amenity sites. The same also have the feeling of participating in a widespread movement (84 percent) but recognize they throw small waste in public spaces (10 percent), whereas they recycle their unused medics (59 percent) in the proper way (bringing them back to the chemist's store).

Overall, France has taken the issue of waste seriously in an attempt to arrange a nationwide system of waste management, but the lack of information or the misunderstanding of the directives often fetters the act of recycling. When available, local information on waste arrangements is preferred by households (74 percent of respondents in a 2010 poll).

Little attention has otherwise been paid to the existence and development of unofficial channels for recycling, based on free gifts, or alternative modes of the "motion" of waste and its reuse in specific circuits of redistribution. However, a few renowned nonprofit organizations (like Les compagnons d'Emmaüs, or Les Restos du coeur), and many more socially discreet ones, epitomize a widespread movement of civil society in the management of the distribution of surplus, the management of waste under the "umbrella" idea of reducing economic inequalities, and the reduction of consumption, in a voluntary, often community-based social economy in the direction of the deprived strata of society. Studies demonstrate that, for individuals and households, the very nature of waste determines its destination and confirm that furniture (33 percent of respondents of a 2010 poll) and especially textiles (67 percent) live a second life in these circuits (an often-discussed point, since the figures are highly controversial).

Conclusion

These few examples illustrate the way France has captured and tackled the issue of waste by means of active politics and the mobilization of economic and social (state and nongovernmental) actors in a debate and movement toward a green economy. Moreover, the infusion in French cultural, economic, and industrial sectors of the values and ethics of environmentalism and its corresponding practices, while flourishing, renders a complex mosaic of sometimes conflicting policies and behaviors.

Lionel Obadia
Université Lyon

See Also: Economics of Waste Collection and Disposal, International; European Union; Politics of Waste; Recycling Behaviors.

Further Readings

Allegre, Claude. *Ma Vérité sur la Planète*. Paris: Plon, 2007.

Barles, S. *L'Invention des Déchets Urbains, France, 1790–1970*. Seyssel, France: Champ Vallon, 2005.

Bertolini, G. *Le Marché des Ordures. Économie et Gestion des Déchets Ménagers*. Paris: L'Harmattan, 2003.

European Commission. "Waste Generated and Treated. Data 1995–2003." http://epp.eurostat.ec.europa.eu/

cache/ITY_OFFPUB/KS-69-05-755/EN/KS-69-05-755-EN.PDF (Accessed July 2010).
Pierre, M., ed. *Les Déchets Ménagers, Entre Privé et Public*. Paris: L'Harmattan, 2002.
"Waste Figures for France: 2009 Edition." ADEME Publications. http://www2.ademe.fr/servlet/getDoc?cid=96&m=3&id=69417&p1=00&p2=05&ref=17597 (Accessed July 2010).

Freeganism

Combining the words *free* and *veganism*, *freeganism* suggests an alternative economy, diet, and worldview. Freeganism can be best described as a loosely bound movement of individuals who protest consumer society and the market economy by living a life that produces no demand for goods. Born out of the antiglobalization and environmentalist movements of the 1960s, freegans take the expression "waste not, want not" to an extreme. Freegans feed upon economies of excess and luxury that produce loads of waste. This social movement is redefining what is meant by the term *waste*. What is considered waste is relative in the world of freeganism—someone's garbage can be another person's next meal or home furnishings.

Freegans

Freeganism has been described as a reaction to "industrial eating." This can mean a rejection of process and heavily processed foods. However, this is not always the case; many freegans will eat just about anything that they find that is still comestible, whether it is Twinkies or carrots. Some freegans are devout vegetarians or even vegans. What the individuals of this diverse group have in common is a desire to break from the capitalist cycle—they are opting out of consumer society. At the same time, freegans are often skeptical of charity and government programs. They have also spoken critically about the foods made available through government aid. Ultimately, freegans realize that the majority of a person's life is spent working for money to buy consumer goods, something that often takes them very far from food production. In their own way, freegans are connecting with their food source through foraging. Anthropologist Joan Gross has noted that this group models its subsistence strategy on preagricultural hunter-gather societies. At the same time, there are many aspects of the freegan movement that are distinctly contemporary.

Methods

Many freegans are outspoken political activists, while some individuals use freegan tactics as a survival strategy to overcome social and economic exclusion and others substitute their diets and consumer habits with some forms of freeganism. Freegan activities include dumpster diving and other forms of foraging. Some people also include guerrilla gardening (planting crops on unclaimed land) and growing food in community gardens as part of their freegan activities. These activities, unlike stealing, underline resourcefulness and a degree of self-sufficiency. Most importantly, freegan subsistence tactics stand in direct opposition to consumerism and mainstream distribution channels.

Organization

There are freegan gatherings such as the Really, Really Free Markets, where individuals can give away their excess foraged goods rather than disposing of them. Many freegans may choose to forage alone, but in cities in the United States and the United Kingdom, a sense of community grows among freegans.

One of the more organized expressions of freeganism is a group called Food Not Bombs (FNB). Founded in Boston in 1980, this group gathers vegan food through donations and from dumpsters and distributes meals to the homeless and poor each week. There are hundreds of FNB groups operating throughout the United States and Canada. FNB has been seen by authorities as a symbol of anarchism for its nonconformist food distribution activities.

Groups such as FNB and freegans more generally raise the question of what waste is and who it belongs to. Why is dumpster diving and getting things for free so offensive and even appalling for much of mainstream society? Milton Friedman is famous for saying "There is no free lunch," and people believed him. Perhaps this is why freegans rub mainstream society the wrong way—there is a free lunch, if one is willing to go out and look for it. That said, there is great debate over who waste belongs to and whether it is a crime to skim off the leftovers of an overconsuming, wasteful society. Freegans bring into question and challenge this wastefulness, whether through anarchic political statements or peaceful community gatherings. North Americans and Europeans may begin to question if it is ethical or even legal to produce so much waste or to dispose of the excess when people are going hungry each day.

Rachel Black
Boston University

See Also: Consumerism; Crime and Garbage; Definition of Waste; Dumpster Diving; Food Waste Behavior.

Further Readings

Food Not Bombs. http://www.foodnotbombs.net (Accessed June 2010).

Gross, Joan. "'Freegans' and Foragers Form New Foodways." http://oregonstate.edu/dept/humanities/Newsletter/Spring%2008/Joan%20Gross.htm (Accessed June 2010).

Shantz, Jeff. "One Person's Garbage ... Another Person's Treasure: Dumpster Diving, Freeganism and Anarchy." *Verb*, v.3/1 (2005). http://verb.lib.lehigh.edu/index.php/verb/article/viewArticle/19/18 (Accessed June 2010).

Fresh Kills Landfill

Staten Island's Fresh Kills Landfill, which first began receiving New York City's trash in 1948 and officially closed in 2001, is the largest landfill in the world as of 2010 and, arguably, the largest human-made structure as well. Due to the awesome size of this mound and the equally daunting scope of the waste disposal problem confronting the world's most affluent industrial societies, Fresh Kills, in popular literature and environmental discourse alike, has become an imposing material reminder of the consumption and waste driving consumer culture and economy. It is a towering monument to the everyday practices contributing to the maintenance of a materially prosperous and environmentally careless U.S. society.

Plans and Operation

The symbolic potential of Fresh Kills was recognized and discussed from its very inception, albeit in a less critical vein than is commonly taken up by 21st-century critics of the site. From its origins, Fresh Kills was meant to serve as a model facility that showcased advanced strategies of municipal waste disposal. The landfill's charter subscribed to a modern code for "sanitary" landfills that called for the placement of a clean daily cover (such as dirt) over the working face at the end of each day, the construction of a 24-inch-thick final cover, and the prevention of waste material from entering open water.

The landfill occupies 2,200 acres and, at the peak of its operation in the late 1980s, received around 29,000 tons of waste every day. The landfill closed on March 22, 2001, but was temporarily reopened later that year following the terrorist attacks of September 11. Approximately 2 million tons of rubble from Ground Zero, which included the remains of more than 3,000 people who died in the attack, were brought to Fresh Kills for sorting, identification, and disposal.

Despite the high hopes of certain New York officials that appropriate management of the site would make it a positive example of waste disposal, a good deal of resentment has attended the operation of the landfill over the years. Residents of Staten Island have grudgingly borne the brunt of New York's steadily mounting waste output since the 1950s, especially after it was discovered that the landfill, initially unlined, was failing to live up to its promise of keeping toxic leachate out of the water (alterations in the landfill's design have since been incorporated to address the problem).

Controversy

Residents of Manhattan Island's lower-income neighborhoods have also had sufficient reason to complain. Garbage from Manhattan and its neighboring boroughs travels to the landfill via the Harlem transfer station where the traffic from nearly 100 heavy trucks converges on a daily basis. The diesel exhaust generated by this traffic has created concern about air quality in the area, with many environmental justice advocates arguing that the increased risks of Harlem residents acquiring diseases like childhood asthma,

Vehicles destroyed in the September 11, 2001, terrorist attacks in New York City are heaped at Fresh Kills Landfill, which was officially closed in March 2001 but was temporarily reopened following the attacks. Approximately 2 million tons of rubble from Ground Zero, which included the remains of more than 3,000 corpses, were brought to Fresh Kills for sorting, identification, and disposal. A proposed park on the site would be three times the size of Central Park and would provide amenities such as walking trails, a memorial, and a wildlife refuge.

bronchitis, cancer, and emphysema have become too great to ignore. Though the landfill has closed, the Harlem and Staten Island transfer stations continue to process most of the city's solid waste, which is (as of 2010) exported by rail to distant states rather than buried.

The resentment certain New York residents have felt toward this landfill is ironic in light of the optimistic rhetoric espoused by the site's original architects. In the early years of the landfill's operation, New York City parks commissioner Robert Moses proposed that Fresh Kills Landfill offered a singular opportunity, not only for waste disposal but also for community planning. A substantial portion of lower Manhattan, as well as several other once-marshy areas in the metropolitan area, had already been constructed on top of landfill. The Fresh Kills Landfill therefore promised to create even more habitable real estate on previously water-logged stretches of Staten Island. Moses's development project for the island envisioned the creation of an integrated community of residential, recreational, and industrial sites all built on the solid foundation of New York's discards.

Moses's plan, however, was based on the assumption that the landfill would only be in operation for a short period of time. The landfill was supposed to close once the salt and interstitial marshes had been filled, a process expected to take anywhere from three to 12 years; its operation, however, continued unabated for the next 50 years. The land created from dumped garbage not only filled in the marsh but it also eventually came to tower above it in four mounds that range from 90 to 225 feet high.

Closing

Following the site's official closing in 2001, at least a part of Moses's vision—that of building a large park—has been revived. Thanks to its relative isolation and lack of development, Fresh Kills offers an unexpected refuge for wildlife in the midst of the nation's largest urban center—birds and bird watchers, in particular, have flocked to the site. Plans for an official Freshkills Park were announced in 2003 (the subtle name change attempts to distance the site from the negative connotations attending its original use). The proposed park will be three times the size of Central Park and, when completed, will include walking, biking, and hiking trails as well as wildlife refuges, a 9/11 memorial, and other recreational amenities. The conversion from landfill to public green space is expected to take 30 years to complete, a period of time that coincides with the number of years that the city of New York is responsible for monitoring changes to the site's topography that stem from the slow decomposition of organic material contained within.

Apart from its continuing utility as a transfer station, the site also captures and sells a portion of the methane it produces (a by-product of decomposition) to utilities that use the gas to generate electricity. The landfill gas recovery operation claims to process enough methane to power more than 20,000 homes on Staten Island. Though to some the site is still an eyesore and a public health concern, considering the concerted efforts of its curators today, the site—as an energy producer and recreation area—may yet transcend its negative association and live up to the optimistic expectations of its founders.

Kevin Trumpeter
University of South Carolina

See Also: Archaeology of Modern Landfills; Garbage Project; Landfills, Modern; New York; September 11 Attacks (Aftermath).

Further Readings

DeLillo, Don. *Underworld*. New York: Scribner, 1997.

Miller, Benjamin. *Fat of the Land: The Garbage of New York—the Last 200 Years*. New York: Basic Books, 2000.

Rathje, William and Cullen Murphy. *Rubbish!: The Archaeology of Garbage*. New York: HarperCollins, 1992.

Royte, Elizabeth. *Garbage Land: On the Secret Trail of Trash*. New York: Little Brown, 2005.

Fuel

The word *fuel* describes a material employed to produce heat, power, or light. Wood, coal, oil, natural gas, uranium, and biomass are currently the most used conventional fuels. Because the processing of these materials generally involves continuous combustion, the word has lent itself to the idea of revitalizing emotions. The word *fuel* can represent an inner force that motivates humans, a force that boosts activities and emotions.

A key difference between "material fuel" and "fuel of will" stems from the fact that the latter is supposed to be strong enough to sharpen the everyday life of ordinary humans so that it is renewed time and again. In contrast, the most important materials employed in the 21st century to produce heat, power, or light are exhaustible, nonrenewable, and have pollution as a side effect: they cannot be used without paying the corresponding costs.

Finite Fossil Fuels

Coal, oil, and natural gas are called fossil fuels because they are formed when decomposed plants, animal matter, and marine organisms are combined with heat and pressure beneath the surface of the Earth. A process of chemical change sustained over millions of years transforms the fossil remains into carbon and compounds of carbon and hydrogen, which release a large amount of heat under combustion. They therefore constitute the fixed stock of fuels on Earth, a stock that, once expended to produce energy, cannot be restored. To keep track of the world's fuel reserves, an array of specialized institutions that operate in the private sector and in international forums publish annual assessments of the proven reserves, probable (recoverable) reserves, and unknown or undiscovered resources. "The World Energy Council Survey," the "International Energy Agency Information Report, "The Oil & Gas Journal Report," and the "BP Statistical Review" were the most quoted sources of information as of 2010.

In spite of its technical dimension, the estimation of the stock of fossil fuels was hotly debated in the past. The possibility that the world might be running out of oil provoked great concern among both the authorities and the public at large. Two moments stand out in the trend for geological pessimism: the first quarter of the 20th century in the United States and the 1960s and early 1970s across Europe, Japan, and North America. Fears of an oil shortage played an important role in triggering public awareness of the need to conserve natural resources. Ultimately, the conservationist currents, which ripened within the wealthy, educated segments of the population, evolved toward more politically oriented ecological movements. To face Earth's challenges, they called for energy-saving policies on the demand side and tighter regulation of private business on the supply side. However, the gloomy predictions of a looming age of scarcity were dashed by unfolding events: first, untapped oil reserves were discovered in the United States, which turned anxiety about a shortage into glut management (1926–29); and second, the price hike triggered by the Middle East Arab oil embargo and production cuts (1973–74), whereby long-term ecological pessimism was replaced by a critical, immediate energy crisis.

Climate Change

Although these political questions have not vanished, they have been displaced by the more urgent, overarching, and globalized problem of climate change. In the late 20th century, fuel economy and fossil fuel economy, in particular, became closely linked to the fate of the planet and this led to a redefinition of previous agendas. Burning coal, oil, or natural gas to produce energy releases large emissions of carbon dioxide (CO_2) into the atmosphere and is a major source of the greenhouse gases that contribute to global warming. Multilateral settlements on reducing CO_2 emissions thus became a part of the agenda of international diplomacy, even though the concrete results achieved at the 1997 Kyoto and 2009 Copenhagen summits fell short of the expectations of leading European nations and the scientific community.

Fuel Consumption

Apart from the United States, the phenomenon of "addiction" to fossil fuels goes back as far as the 1950s. This golden era of reconstruction and recovery from war spurred a phase of industrialization in developing countries, favoring further technological innovation in Europe and Japan and increasing investment in heavy production within the Soviet Union, followed by some stimulus to light industry. Moreover, economic growth became progressively grounded on trade liberalization, financial stability, inflation control, and full employment. Together, these forces prompted an increase in the global demand for energy, which henceforth could be provided from the Middle East at very low production costs. This was accompanied by a downward trend in the price of coal and natural gas and by incentives to substitute coal with liquid fossil fuels. Overall, the effects of low energy prices on economic growth revealed a bidirectional causality, with each factor reinforcing the other.

Up to the first oil crisis (1973–74), world consumption of energy grew at an annual rate of 5.2 percent. On a per capita basis, this evolution translated into a twofold increase in the average per capita world fuel consumption, from 5 barrels of oil equivalent (BOE) per year in 1950 to 10.5 BOE in 1973. Because of the disturbing macroeconomic effects of the sudden rise in fuel prices, this was followed by a deceleration in primary world energy consumption to 1.7 percent per year, and it continued around 1.6 percent per year even after readjustments that drove fuel prices lower in the second half of the 1980s. At the turn of the 20th century, world per capita consumption had reached 11 BOE per person per year, which means that energy growth had almost paralleled population growth. Later, another industrialization boom generating from the late-starter countries (notably Russia, India, and China) increased the pace of growth. By 2004, China was the largest coal producer in the world, well ahead of the United States, and the acceleration of its energy demand placed increasing strains on the multilateral agenda for CO_2 reduction.

Changing Lifestyles

The urgency of changes in patterns of fuel consumption went far beyond the agenda of governments and international institutions to become a core political issue and a matter of citizenship and individual behavior. How big is one's ecological footprint? That is: what is the fuel consumption and related CO_2 emissions that result from one's lifestyle? Once in the realm of individual-collective options, a difference in lifestyles matters, because personal vehicles and residential energy are responsible for about 40 percent of energy end-use in Organisation for Economic Co-operation and Development (OECD) countries.

The U.S. way of life has been in the spotlight because of its high-energy regime based on sprawling cities with outlying areas; the dominance of single-family households, large cars, and exurban living; sizable air-conditioning and high-quality energy services; affluent consumption; ever-rising consumer expectations; and extensive packaging of goods. The result is that the United States exceeds all boundaries in energy demand, even after the necessary adjustments for relative levels of development. On a per capita basis, in 2009, U.S. consumption was nearly twice that of the French and the English, three times that of the Polish and Iranians, four times that of the Argentineans and Romanians, and around 20 times that of Pakistanis and Indians. In addition, the benefits of a high-energy regime have proven highly addictive. Moral calls for the U.S. public to practice restraint in order to save energy, such as Jimmy Carter's speech delivered by television in 1979 when the oil crisis was at its peak, have produced little or no effect.

Energy-Efficient Technology and the Rebound Effect

Technological improvements combined with incentives for clever behavior provide another path for cutting fuel demand. The "energy efficiency" concept has been put forward to describe the delivery of larger amounts of useful work, heat, or light from decreasing quantities of fuel.

If this goal is pursued systematically and successfully through every stage of energy conversion, global fuel reductions are likely to occur without a loss of consumer welfare. The energy efficiency solution has nonetheless been rebutted by some sectors of the scientific community. They have argued that the effect of improving the efficiency of a productive factor, like energy, is to lower its implicit price and hence make its use more affordable, which in turn leads to either greater consumption or the purchase of other goods and services that also require energy.

More specifically, if the price of gasoline per mile steadily declines for some technical reason, the driver is likely to increase mileage, anticipate the purchase of a new car, or buy goods with the money saved. Although increased energy efficiency at the microeconomic level fosters a reduction of energy use at this level, it does not necessarily bring about a reduction at the national or macroeconomic level. In the worst-case scenario, microeconomic savings are completely offset by macroeconomic rebounds.

The view that economically justified, energy-efficiency improvements will increase rather than reduce energy consumption was first proposed by British economist William Stanley Jevons in 1865. The effect he singled out is dubbed the "rebound effect" and its foundations are known as Jevons's paradox. An economy-wide rebound effect of 100 percent means that 100 percent of the potential energy savings are "taken back" through the above-mentioned mechanisms.

Much of the debate on the rebound effect is encumbered with difficult technical questions, including discrepancy in basic assessment criteria, noncomparability of empirical studies, and different methodologies to assess the macro implications of microtechnological events. However, two conclusions stand out from the debate. First, the global rebound effect is not high enough to completely outweigh the importance of energy efficiency as a means of reducing carbon emissions.

Second, the rebound effect is likely to be greater in the early stages of development and diffusion of the technology, under conditions of increasing returns, in the presence of network externalities, and when the technology is used in a wide array of economic applications. In other words, political action and international economic regulation do not alleviate the need for individual action and community endeavors. Reducing CO_2 emissions and the demand for fuel is as much a concern for the individual as it is for nations.

Nuno Luis Madureira
University of Lisbon

See Also: Carbon Dioxide; Clean Air Act; Gasoline; Power Plants; Resource Conservation and Recovery Act; Sustainable Development.

Further Readings

Jevons, W. *The Coal Question: An Inquiry Concerning the Progress of the Nation, and the Probable Exhaustion of Our Coal-Mines.* 2nd ed. London: Macmillan, 1866.

Nye, D. E. *Consuming Power. A Social History of American Energies.* Cambridge, MA: MIT Press, 1998.

Salvador, A. *Energy: A Historical Perspective and 21st Century Forecast.* Tulsa, OK: American Association of Petroleum Geologists, 2005.

Sorrel, S. "Jevons' Paradox Revisited: The Evidence for Backfire From Improved Energy Efficiency." *Energy Policy*, v.37 (2009).

Funerals/Corpses

Human corpses are treated as garbage only in exceptional circumstances. Ritualized disposal of human cadavers is standard fare around the globe and has been so since the beginning of humanity. Why is it that corpses have rarely been viewed as waste? The answer to this question deals with notions of dignity and being human.

Furthermore, the ambiguity of feelings toward the corpse—both love and fear—qualify it for special treatment. In addition, it is generally believed that departure from life is followed by an assumed, continued existence in one way or another. The condition of the corpse, according to Robert Hertz, stands for the fate of the soul. The awe inspired by corpses relates to the human body as a root metaphor. When alive, it is the only intrinsic symbol, referring to itself. On the occasion of death, however, a rite of passage is required to mark its change of status, referring to something else. It implies a shift from the category of the living to the category of the dead. Funerals mark this transition.

The journey or initiation of the spirit of the dead takes some time and may (if one accepts Hertz's theory) correspond with the decomposition of the corpse or the process of emotional detachment by survivors. Nothing was more intimately related to the once-living person than their body and nothing resembles that person more; therefore, magical properties are attributed to the corpse. The veneration of relics provides a good example. Human remains as relics are perceived of as contact points with another realm of existence. The dead, so it is believed, can still exert their influence on the living. Consequently, corpses should not be tampered with. The idea of the returning dead, revenants, illustrates this point. To curb the danger of mixing up the living and the dead, corpses are ritually disposed of.

"Improper" Disposal

The denial of a proper funeral counted as a severe additional punishment of executed criminals in the European past, for it meant a denial of the transition to the afterworld. These criminals were buried at crossroads, a ritual gesture to prevent the soul from moving elsewhere. Occasionally, their corpses were left hanging at the gallows to decompose. Early anatomists also preyed on these corpses for the purposes of dissection. Survivors frequently fought over the executed corpse with the authorities in order to give the deceased relative their last respects. This was a matter of dignity, of course, but given the close relationship, it might also have been motivated by the fear of a haunting ghost. Ritual measures most probably were also taken with ancient corpses discarded in peat swamps, later discovered by archaeologists, in northwestern Europe. The bog bodies, some strangled with a noose, seem to have been ritually killed. Their disposal in an uninhabitable area indicates their social exclusion, but the places concerned did not serve as garbage dumps. The intentional mutilation of enemy corpses left on battlefields, such as the noses being cut off—a literal and symbolic "loss of face"—is another form of humiliation and dehumanization. Makeshift mass graves resulting from war atrocities come close to the treatment of corpses as garbage. Unintentional occurrences, such as floods, earthquakes, and other disasters with mass loss of life, might at times give the impression of the corpses floating or lying around being merely waste—even more so when it takes considerable time to recover them. There are also historical examples of epidemics in which the death toll was too great for the survivors to organize proper funerals. Recently, Sherpas and others decided to clean the garbage left behind by mountaineers on Mount Everest. The corpses of some 200 climbers who had died and remained there at great height were finally recovered. The exposure of the dead mountaineers to the elements had been the unavoidable result of their tragic deaths in places from which they could not easily be removed.

An intentional exposure of corpses to the elements occurs at a so-called body farm. Most famous is the Forensic Anthropology Center of the University of Tennessee in Knoxville, Tennessee, founded by William Bass. Volunteers donate their bodies to the center, and these are placed in the open air at the "body farm," allowing scientists to establish the nature and various rates of decomposition. No matter the increased visibility of corpses in popular television series, such as the *CSI* series, sensitivity toward real-life confrontations with human remains seems to have increased. In the Netherlands, for instance, a man who near his wife's tomb stumbled upon an open communal grave containing bones, skulls, and the remains of coffins and clothing founded a Vigilance Committee. This committee drew attention to the presumptuous way in which human remains were deposited in communal graves in Dutch graveyards, and a change of policy took place. Furthermore, publicly accessible collective monuments have been raised for reburied human remains in several graveyards across the Netherlands.

In the same vein, media reports on discoveries of human remains (such as fetuses, organs, and other body parts) from hospital waste that ends up at garbage dumps tend to provoke scandals. Such exceptional incidents that were reported on indicate increased sensitivity. In Basel, Switzerland, cremated human remains from hospital waste are buried in a special field at Hörnli cemetery. For the lack of alternative procedures, the city allows a certain measure of pollution to be caused by the burning of body parts preserved in formaldehyde, especially those of body donors to medical science. At the field, an Anatomical Monument has been created to commemorate the body donors. Besides two sculptures in the back, it has a flat memorial stone in front. Freely translated, the inscription reads "The dead who helped you, the living," and below these words it says, "Anatomical Institute of the University of Basel." The commemorative monument is part of a general trend that also applies to previously neglected remains of fetuses and stillborns. Such monuments on graveyards provide the survivors with a place of remembrance. At the same time, they entail a redefinition of the status of the remembered ones through a recognition of their humanity.

Regulations and "Proper" Disposal

Regulations concerning bodily disposal vary from country to country. Environmental concerns increasingly influence these regulations with regard to materials used and grave gifts allowed. The aim is to prevent pollution of the environment with unwanted and damaging substances. Preparation of bodies for burial may include toxic embalming fluids such as formaldehyde. Cremation may release toxins in modern bodies, such as heavy metals found in dental fillings and pacemakers. The desire for a more "natural" treatment of the dead, in addition, is exemplified by the recent popularity of woodland burials in the United Kingdom. The scattering of ashes in forests and nature reserves of people wanting to be "close to nature" in death as well as commemorative objects brought to the respective places by survivors, however, have been a nuisance in the eyes of conservationists.

In India, environmentalists have been alarmed by barely cremated bodies (due to the scarcity and costs of firewood) floating downstream in the holy Ganges River. In the ethnographic record, there are various examples of the disposal of human cadavers, such as on charnel grounds or higher platforms, by leaving them exposed to the elements.

In Tibet, so-called sky burials are practiced by Buddhists where the dead are left on mountaintops, exposed to the elements, animals, and birds. Both burial and cremation are not feasible in the high altitudes. Exposure of the dead by the Parsi of India, following their Zoroastrian tradition, occurs on top of Towers of Silence to the sun and birds of prey. As the number of birds has rapidly decreased due to a poison used in agriculture, the custom has come under pressure.

Funerals vary according to the nature of death as well as the social status, age, and gender of the deceased. Religious ideas affect the means of bodily disposal considered appropriate. Belief in bodily resurrection, for example, mostly coincides with earth burial. In general, notwithstanding mortuary variation, human cadavers are rarely seen as waste. The move of graveyards to the borders and outskirts of the community in 19th-century Europe did, however, have to do with beliefs among sanitary reformers in miasmic or polluting air causing disease that later proved to be wrong. Contemporary concerns relate to increasingly popular cremation, particularly its energy use and production of carbon dioxide and other pollutants, so that in the future it may be replaced by more environmentally favorable techniques. Lynn Åkesson speaks of a moral dimension of waste, and probably the riddance of human corpses most strongly expresses this dimension of "wasting" as a cultural process. Ritualized disposal of human cadavers, rather than doing away with them like garbage, is accepting by most people.

Eric Venbrux
Radboud University Nijmegen

See Also: Culture, Values, and Garbage; Human Waste.

Further Readings

Åkesson, Lynn. "Wasting." *Ethnologia Europaea*, v.35/1–2 (2005).

Hertz, Robert. *Right and the Death Hand*. Glencoe, IL: Free Press, 1960.

Janssen, Jacques and Theo Verheggen. "The Double Centre of Gravity in Durkheim's Symbol Theory: Bringing the Symbolism of the Body Back In." *Sociological Theory*, v.15/3 (1997).

Penfold-Mounce, Ruth. "Consuming Criminal Corpses: Fascination With the Dead Criminal Body." *Mortality*, v.15/3 (2010).

Stroud, Ellen. "Dead Bodies in Harlem: Environmental History and the Geography of Death." In *The Nature of Cities: Culture, Landscape and Urban Space*, Andrew Isenberg, ed. Rochester, NY: University of Rochester Press, 2006.

Furniture

The term *furniture* usually refers to movable objects that support the body (for example, a bed, chair, or sofa) or provide storage for smaller objects like books, clothes, or kitchen tools. Furniture is everywhere—in homes, offices, schools, shops, and restaurants. Furniture is so embedded in everyday life that people often forget how important it is. In fact, the presence or absence of furniture determines the function of a room in a house. For example, bedrooms contain beds and dining rooms contain dining tables. Working spaces are also defined by furniture; offices consist of desks, chairs, and bookshelves. Furniture has both functional and symbolic roles. For instance, the larger chair and desk of a teacher in a classroom symbolize a hierarchical position between the teacher and students. Although some authors affirm that there is a homogenization of furniture in the world due to the growing success of global furniture brands, many differences can still be noticed between different cultures and countries. For example, dining tables have different heights in Western countries compared to Asian countries, such as Japan, where traditional tables are relatively low and people do not use chairs but instead sit on the floor when they eat. Although furniture is part of everyday life, certain examples of furniture are very valuable and are considered to be pieces of art for display in museums and art galleries. Valuable furniture can be contemporary, produced by famous designers or artists, or antique, handcrafted masterpieces from past eras and cultures. Given that furniture is so important in daily life, it is important to understand its life cycle—how it is produced, consumed, and disposed.

Production

Data related to the production and export of furniture are varied, but production trends can still be analyzed. Since the production of furniture became global, Asian countries such as China, Taiwan, Malaysia, and Indonesia have become the dominant producers and exporters. Raw materials and cheap labor forces make these countries very competitive, attracting foreign investment. During the 1990s, a number of U.S. and European companies started moving their production to these countries and to China in particular. According to the China

Building Decorator Association, in 2006, Chinese furniture exports reached $17.4 billion, ranking it first in the world; exports for 2010 are expected to reach $48 billion. China competes fiercely with both Western countries such as Canada, Germany, and Italy, and Asian countries such as Thailand, Taiwan, and Malaysia. During the first decade of the 21st century, China overtook Canada as the largest exporter to the United States, providing more than half of the furniture consumed by Americans. In 2000, China became the largest supplier of furniture to Japan, overtaking Thailand. Besides the U.S. and Japanese markets, China's export targets are Europe, Hong Kong, and the Middle East. According to the Statistical Office of the European Community (Eurostat), in 2005, China supplied almost half of the furniture consumed in European Union countries. Although China leads the export of furniture, countries such as Indonesia, Malaysia, and Thailand are aggressive competitors. In fact, in these countries, the cost of labor is cheaper than in China, causing Chinese companies to move their production to these countries.

Consumption and Consumers

Richer countries are bigger consumers of furniture. Among them, the United States is the largest consumer, accounting for one-quarter of the world's furniture consumption. According to the American Furniture Manufacturers Association, U.S. consumption of furniture increased significantly before the 2008 economic downturn and recession. In 2007, China exported 43.5 percent of its production to the United States. With the 2008 recession, U.S. demand for furniture decreased, but (as of 2010) the United States remains the largest consumer. Some consider this to be a temporary scenario, however. Because of stringent environmental regulations in the United States and the downturn of the U.S. housing market, some large Chinese companies are looking to other markets. Similarly, in the European market, consumption of furniture has been hit by the 2008 economic downturn. Because of this shift in the U.S. and European markets, Chinese companies are looking at domestic as well as Middle Eastern and African markets. The Chinese domestic market is growing rapidly. In fact, the booming economy and the consequent higher disposable incomes and improved living standards caused an impressive growth in the internal demand for furniture, which consumes most of the industry's output. Given that Asian markets are growing, Sweden, Italy, Germany, Spain, and the United States are trying to increase their furniture exports to these countries and to China in particular.

According to various marketing studies in the United States and Europe, women ages 25–45 are most interested in buying furniture. The most popular reason for buying furniture is the purchase of a house. Because of the 2008 economic downturn and consequent recession, fewer people bought houses in the United States and Europe, meaning fewer people bought furniture. Compared to other areas of household expenditure, buying new furniture is often a low spending priority. In other words, people preferred going on holiday or buying a new television to buying a sofa, which seems to be purchased only for a new home or to replace a damaged one. According to some sociological studies, many people no longer regard furniture as a long-term investment, and thus fewer people buy antique or designer furniture. People regard furniture as a short-term investment, opting for low prices and good-quality brands. This explains the increasing success of globally famous companies or do-it-yourself chains highlighting the low prices and durability of their products.

Disposal

The disposal of furniture is particularly relevant in rich countries where people change furniture more often than in poor countries. In the 21st century, reusing and recycling furniture are more popular practices than in the past, in both rich and developing countries. The secondhand-furniture sector size in the United States is $100 billion per year. In Ghana, the secondhand sector employs over 15,000 people. In the United Kingdom in 2006, over 3 million items were donated to charity shops, two-thirds of which were household furniture. In the 21st century, people reuse their furniture more than in the past (through charity shops, friends and family, Websites, garage sales, and for-profit retails), but they also seem to change it more often. Although furniture styles change less frequently than clothing and shoes, its relatively low prices (compared to past years) encourage people to change furniture. A variety of furniture disposal methods exist, and people decide the best method based on what they are disposing of. For instance, an antique chair is likely to be either donated to family members or sold for a profit through retail or through a Website; a low-value chair can be donated to friends or charity shops; and a broken chair is generally thrown in the garbage. Because people are more aware of the value of their furniture, they donate fewer objects to charity shops than in the past and they sell their objects through garage sales or Websites. This means that fewer people are helping charitable organizations, but it testifies to a cultural shift in looking at disposed objects as sources of income as well as valuable purchases. This also shows that, although people increase their consumption of furniture, they are becoming more aware of environmental and ethical concerns about the management of waste in their household.

Benedetta Cappellini
Royal Holloway University

See Also: Appliances, Kitchen; Consumerism; Home Appliances; Materialist Values.

Further Readings

Curran A. and I. D. Williams. "The Role of Furniture and Appliance Re-Use Organizations in England and Wales." *Resources Conservation and Recycling*, v.54 (2010).

Edwards, C. *Turning Houses Into Homes: A History of the Retailing and Consumption of Domestic Furnishings*. Burlington, VT: Ashgate Publishing, 2005.

Kaplinsky R., M. Morris, and J. Readman. "The Globalization of Product Markets and Immiserizing Growth: Lessons From the South African Furniture Industry." *World Development*, v.30 (2002).

Fusion

Most nuclear power, including all current commercial nuclear reactors, uses nuclear fission—the absorption of a neutron by an atomic nucleus, resulting in its splitting into two or more smaller nuclei. Fusion power, which has been pursued experimentally since the early atomic age in the 1950s, generates power instead through nuclear fusion in which two atomic nuclei fuse together to form a single, heavier nucleus, releasing energy in the process. As with fission reactors, most fusion power plant designs use nuclear fusion to generate heat, which operates a steam turbine, which in turn drives generators to produce electricity. Fusion power has successfully been generated in small amounts. However, a fusion power plant design had not yet been devised and implemented as of 2010 that was efficient and commercially viable. This is in contrast to the brief period from the inception of fission power to its first commercial applications, which took less than a decade.

Fusion Designs

Early fusion power research revolved principally around the idea of the "pinch," the compression of plasma by magnetic forces. When an electric current is run through plasma, a magnetic field is generated, which creates an inward-directed force that causes the plasma to collapse and become more dense. Denser plasmas then create denser magnetic fields, increasing the inward force further in a chain reaction. Carefully controlled, this can generate the right conditions for nuclear fusion. However, even running the current through the plasma in the first place is no simple task; usually, it is induced by an external magnet. Pinch devices were used experimentally in the 1940s and 1950s, including the imaginatively named "Perhapsatron series" of fusion power devices developed at the Los Alamos National Laboratory in 1952. However, the Perhapsatron devices, like other pinch devices, became unstable at critical points, and beginning in the mid-1950s, the suggestion was made that pinch devices might be inherently unstable. Most work on them ended by the 1960s, but the flurry of experimentation and exchange of ideas laid the groundwork for fusion research.

Various other designs have been attempted since. Some fusion designs attempt to reach a level of high temperature and density for a short period, like the pinch designs or the inertial confinement devices, which initiate fusion through heating and compressing a fuel pellet. One such inertial confinement fusion device was designed by Philo T. Farnsworth in 1968, 40 years after his invention of the electronic television. Inertial confinement remains one of the two most popular areas of fusion research; the other is magnetic confinement, which attempts to confine hot fusion fuel in the form of plasma and maintain a steady state. Occasionally, there is also work on devices that approach fusion from neither of these angles, but instead focus on producing low quantities of fusion at a low cost. While physically possible, these approaches have not proven fruitful or to have many useful applications. The term *low cost* is relative to the cost of the larger fusion power plants, not to other existing energy options.

Safety and Containment

The safety of nuclear fusion was not fully understood as of 2010 because of the lack of full-sized fusion power plants and the ability to study their operations. However, because of the differences between nuclear fusion and nuclear fission, the catastrophic accident that fission reactors are capable of (such as experienced in Chernobyl in 1986) is not a possibility with a fusion reactor. Fusion reactors require precise temperature, pressure, and other parameters in order for energy to be produced. Damage to the reactor would upset those parameters and cease the production of energy. Fission reactors, in contrast, produce fission products that will continue to generate heat even after the reactor is shut down, which can result in a meltdown of fuel rods as heat accumulates. The dangers of a fusion reactor would involve radiation exposure of staff and the release of radiation to the immediate vicinity but not the calamitous disasters of a Chernobyl-like incident. Further, unlike the fission reaction, the fusion reaction cannot "run away" and produce excess heat and waste products—fusion only works in a narrow window of parameters and exceeding them will simply result in the fusion process ceasing. No failsafe mechanisms such as the many implemented in fission reactors are necessary; further, the amount of fuel used is very small.

Magnetic containment fusion reactors could theoretically explode in the event of a failure of the structure that holds in place the coils where the magnetic fields are generated. A containment building would be sufficient to confine the damage of such an explosion, which would be similar to the explosions of MRI machines. It is also possible that the liquid lithium suggested as a coolant for fusion reactors could add to the damage caused by a fire, as it is highly flammable.

Radioactivity

The by-products of fusion reactors are helium (which is harmless) and tritium, an isotope of hydrogen. The amount of tritium released by a single fusion reactor's operations is thought to be negligible in its effect, but it is not clear what the cumulative effect would be of an energy economy based significantly on the operations of fusion reactors. Tritium is both volatile and biologically active, but it has a short half-life of 12 years. Also, unlike many other radioactive contaminants, it does not bioaccumulate.

Operating a fusion reactor causes the structural materials of the reactor to become radioactive from the flux of high-energy neutrons. The half-life of the radioisotopes produced is shorter in general than those associated with fission reactors, and the most dangerous part of the reactor—the reactor core—will be reduced to low-level waste after 100 years, and at 300 years will be as radioactive as coal ash. In comparison, fission reactor waste remains radioactive for thousands of years. However, for the first 50 years, the reactor core is much more radioactive than any fission waste and would need to be stored very carefully. Helping this process, though, is the fact that fusion reactors can be made from a broader range of materials than fission reactors, including many like carbon fiber and vanadium, which becomes less radioactive than the stainless steel used in fission designs. It seems fair to estimate that overall, fusion reactors are less dangerous and produce less dangerous waste than fission reactors.

Production

The greatest benefit of fusion power is that it provides significantly more power per weight of fuel than any current method of generating power. Further, while fission reactors use radioactive materials like uranium and plutonium, fusion reactors rely principally on deuterium—an isotope of hydrogen common in seawater. Working large-scale fusion reactors could provide the world's energy needs for hundreds of thousands—perhaps even millions—of years. Further, there is no reason to think that fusion power suffers from the diseconomy of scale that many alternative power sources suffer from (wind and solar, for instance, become increasingly expensive per unit of electricity generated, once their use reaches the point that all the optimal locations have been used). Using seawater to fuel fusion reactors would also make it easy to use desalination plants to produce fresh water, an anticipated problem for future generations. The Fusion Torch concept introduced in 1968 by Atomic Energy Commission program managers Bernard Eastlund and William Gough also suggests ancillary benefits for fusion power. The Fusion Torch is the use of the plasma of a fusion reactor to break down waste materials and convert them into useful elements, such as hydrogen fuel.

Cold Fusion

For a time, especially in the 1980s and 1990s, cold fusion received the lion's share of attention, both public and scientific. Cold fusion is nuclear fusion that takes place at temperatures close to room temperature, rather than in the high-temperature, high-pressure conditions within stars and known fusion reactions. The idea received a good deal of publicity in 1989 when Martin Fleischmann and Stanley Pons conducted an experiment using heavy water and a palladium cathode that they believed generated energy—through fusion, they hypothesized—that could not be accounted for through chemical reactions. Both the public and the scientific community were receptive to the idea of such a breakthrough because of the recent similar breakthrough in the discovery of high-temperature superconductivity. However, attempts to replicate the Fleischmann-Pons experiment were unsuccessful or problematic, and a special panel convened by the Department of Energy concluded that no convincing evidence could be established for cold fusion or for a useful energy source based on the experiment. Interest had been stimulated, though, and cold fusion supporters within the scientific community continue to pursue their experiments and present their papers, including Nobel laureate Julian Schwinger, who resigned from the American Physical Society when his paper on cold fusion was rejected from its scientific journal. Twenty years later, cold fusion continues to be a matter of study, though considered a dead end by the bulk of the scientific community. Researchers sometimes use the terms *low-energy nuclear reactions* or *condensed matter nuclear science* to refer to their work to avoid what has become a disused label.

Bill Kte'pi
Independent Scholar

See Also: Atomic Energy Commission; Environmentalism; Hanford Nuclear Reservation; Hazardous Materials Transportation Act; High-Level Waste Disposal; Nuclear Reactors; Uranium.

Further Readings

Atzeni, Stefano and Jurgen Meyer-Ter-Vehn. *The Physics of Inertial Fusion*. Oxford: Clarendon Press, 2004.

Braams, C. M. and P. E. Stott. *Nuclear Fusion*. New York: Taylor & Francis, 2002.

Harms, A. A., K. F. Schoepf, G. H. Miley, and D. R. Kingdon. *Principles of Fusion Energy*. Singapore: World Scientific Publishing, 2000.

Kragh, Helge. *Quantum Generations: A History of Physics in the Twentieth Century*. Princeton, NJ: Princeton University Press, 2002.

Mahaffey, James. *Atomic Awakening*. New York: Pegasus, 2010.

Garbage! The Revolution Starts at Home

Garbage! The Revolution Starts at Home is a 2009 documentary film directed by Andrew Nisker. The title is partially taken from a conversation between Nisker and Bob Hunter, the founder of Greenpeace, who said "The revolution starts at home." Nisker's film chronicles the consumption and waste patterns of Toronto residents Michele and Glen McDonald and their three children. During a three-month period, Nisker asked the family to store all of their garbage in their garage. A goal of the film is to raise awareness in the viewer about the garbage they produce and to see the impact they have on the environment. In addition to chronicling the garbage and consumption practices of the family, Nisker aimed to connect these practices to other issues, each a segment of the film. These included organics, garbage, packaging, recycling, transportation, water, electricity, and the "human dump."

Film and Pollution Issues

In the first segment of the film, the patterns of the McDonald family are analyzed through their consumption and resultant organic waste, including leftover food, pet waste, and diapers. The practices of an organic recycling center are profiled, including efforts to sort and remove plastic garbage bags from the materials so that they can be recycled. According to the film, nearly 1 trillion plastic bags are used worldwide for an average of only five minutes, and in the United States, 100 billion plastic shopping bags are used annually. Next, the film focuses on the issue of garbage. Specifically, there are over 3,000 landfills in North America, and one of the major issues faced in Toronto is the lack of space for storing garbage. Toronto, like many other major cities, ships its garbage elsewhere—specifically to the state of Michigan. Huron Township is profiled. The multiple dumps in the township have created massive increases in traffic and have severely impacted the quality of resident's lives. Dump dust, which often contains asbestos, is one of the many environmental issues associated with such landfills. The film then turns to the issues of packaging and recycling. In the first case, Nisker points out that families like the McDonalds are overly reliant on convenient and overpackaged foods, and that many of the packages of such products are not recyclable. In addition, families often overconsume during the holidays, and this has a major environmental impact. In terms of the second area, the film emphasizes that while recycling has a positive effect, many items, such as toothpaste containers, are made of composite materials and thus cannot be recycled. Some plastic laminates cannot be directly recycled back into the same product and are instead downcycled into other, often less-useful, materials.

The film then focuses on issues related to transportation, including the McDonalds' two SUVs, their CO_2 production, and the problem of road runoff, which is petroleum and water. Water is the next emphasis, with attention to the pollution caused by the family doing the dishes, flushing the toilet, and doing laundry. Household cleaning products (including bleach, phosphates, and others) are profiled for their negative impact on marine life. The case for moving away from bottled water is also made. The impacts include a depletion of indigenous aquifers in other parts of the world and the amount of plastic that is used to produce the bottles. The film then focuses on the issues of electricity (profiling the environmental impacts of West Virginia coal mines) and the "human dump" (focusing on dangerous chemicals, such as brominated flame retardant, that are impacting the human body).

At the start of the project, the family estimated that they would produce 42 to 65 bags of trash during the three months. In actuality, they produced 83 bags of trash (including landfill garbage and recycled garbage) and 320 pounds of wet organic waste. By requiring the McDonald family to focus intently on every piece of garbage that they throw out, as well as on the consumption practices that produce such garbage, the viewer gains an awareness of how much of consumption is invisible to people. As Nisker explains, "when the garbage man picks [garbage] up, it disappears like the rest of my pollution—out of sight and out of mind."

Scott Lukas
Lake Tahoe Community College

See Also: *Garbage Dreams*; Household Consumption Patterns; Shopping Bags.

Further Readings

Nisker, A. *Garbage! The Revolution Starts at Home.* Documentary film. Toronto: Take Action Films, 2009.

Nisker, A. *Garbage! The Revolution Starts at Home.* http://www.garbagerevolution.com (Accessed June 2011).

Garbage, Minimalism, and Religion

Discussions of minimalism immediately call to mind the religious renunciant—one who eschews worldly possessions, relations, and passions to abide by "high-living" principles. Renowned renunciants include St. Francis of Assisi, Buddha, Mother Theresa, and Mahatma Gandhi. However, the renunciant is only one of the most recognizable figures rejecting worldly things for religious or spiritual reasons. There are, in fact, a variety of ways to examine the complex relationship between religion and the stuff called "garbage" or "waste," including household trash, human excrement, industrial waste, airborne pollutants, and other undesired, obsolescent, putrid, or worn things. Well-established connections exist between religious stewardship and "saving the Earth," recycling, and consumer waste, as well as the spiritual qualities of conservation. But waste is much more than simply discarded stuff: it is a dynamic cultural category. Perhaps the more interesting discussion arises when one looks closely at the symbolic, moral, and religious values attached to waste forms, such as impurity; their relationship to advanced capitalism; and the everyday practices of handling, expelling, or avoiding waste forms across religious traditions.

What motivates practices of minimalism, restraint, or asceticism among religious practitioners? Much of the answer to this question requires an appreciation of the individual body as a site of rich symbolism and cultural instruction within religious traditions. As temple, gift, or vessel, scholars have shown that the body sometimes operates as a model and microcosm of the cosmic universe and a physical foundation for the individual's relationship to a divine force. It is widely recognized that behavioral restrictions in a variety of religions attempt to impose order on processes of decay and death. Bodily impulses, sensations, and experiences, ranging from eating to sexual activity, therefore become important places for demonstrating self-control and mastery over desire, hunger, disintegration, suffering, and, ultimately, human mortality. Concepts of waste derive meaning through associations with decay, death, and degeneration. It is important to note at the outset that approaching religion only as a static body of tenets, rules, or symbols yields limited insight. While practitioners strive to live according to such authoritative models, religious engagement is filled with multiple points of ambiguity and cultural variability. This should not dissuade anyone from proceeding; rather, it simply shifts the focus to human behavior and makes religion about much more than a series of rites, creeds, or traditions.

One way to explore the relationship of minimalism, garbage, and religion, therefore, is by investigating how laypeople navigate religious forms of taboo, prohibition, and moral censure concerning waste forms. Anthropologist Mary Douglas has shown that dirt and other waste matter derive their power through not simply being waste or having a kind of negative value. Rather, as "matter out of place," things deemed dirty, spoiled, or noxious carry polluting effects for the person, transmitted through bodily substance or through touching. By avoiding impure or defiled forms, practitioners shore up purity and associate themselves with a path of holiness or sacredness. Hindu-inflected caste hierarchy in India historically associated the handling of garbage and waste forms with impurity and relegated this work to the "untouchables," or *dalit*. Though formally banned by the Indian state in 1950, the caste structure reinforced status differences in those who had the greatest ability to spiritually elevate themselves and avoid forms of contagion.

Food Taboos and Regimens

Food regimens, dietary rules, and their "abominations" can also been seen as a way religious practitioners reduce hazardous associations with polluted, impure forms. Scholars of religion point out that these practices simultaneously fashion religious identity, group allegiance, and social distinctiveness. For example, pious Muslims abstain from alcohol and pork according to halal; observant Jews abide by kashrut, or keep kosher; and strict Mormons prohibit the intake of tobacco, alcohol, caffeine, and other unwholesome foods. These dietary practices are not only motivated by a concern with waste, impurity, and pollution but they also demonstrate self-restraint and piety. Food taboos form simply one area of behavioral restrictions concerning purity and impurity, which extend to religious bans on using profanity ("trash talk") and abstinence from premarital sex and lustful thoughts (for example, getting one's "mind out of the gutter").

Sometimes, eating practices also serve a poignant critique of social inequities, symbolically linking the individual body to the social body with each bite. In the late 1960s and early 1970s, Elijah Muhammad of the Nation of Islam banned among followers such foods as mustard greens, corn bread, refined sugars and flours, and scavenger fish. Instead, African American Muslims were encouraged to eat fresh vegetables, brown rice, prime cuts of beef and lamb twice per week, navy beans, butter, and milk. Building on a vibrant debate on southern cuisine among African Americans, anthropologists Carolyn Rouse and Janet Hoskins describe how Muhammad connected eating directly to the adversity and struggle of slavery in which many slaves wrested access to whites' discarded cuts of meat. Muhammed argued that southern cooking was a tool of racism, sapping the vitality of African Americans. He delineated a classification scheme for healthy, dangerous, and sacred foods in his regular written column, "How to Eat to Live." By politicizing eating and radically altering consumed foods, Muhammad worked to cleanse his followers of the pollution and profanity of enslavement and racism.

Fasting, hunger strikes, frugality, and specific diets, such as vegetarianism, veganism, and locavorism (eating locally produced foods), have also used religious justifications to call attention to food and resource waste, labor conditions, and inequalities of the world food system. The American Jewish World Service encourages the purchase of kosher-certified, Fair Trade coffee and chocolate for celebrations and events, linking "ethical consumption" to the Torah's vision of just society. Beginning in 1998, People for the Ethical Treatment of Animals (PETA) debuted the controversial slogan "Jesus Is a Vegetarian" to persuade Christians to adopt a new dietary model and spread awareness of the conditions of industrial meat slaughter and resource use. Fasting has long been a religious practice plumbing the depths of human discipline. Scholar Carolyn Bynum tells famously of how pious women refused the carnal pleasures of food in late medieval Christianity through harrowing periods of voluntary starvation. Contrasted with these severe fasts, comprising daily watered wine or a bite of apple, was the largesse of a heavenly banquet, which starkly revealed the deep symbolism and metaphor of food as divinity in Christianity.

Rituals and Waste

Still other religious communities, movements, and practices overtly stress certain behaviors to align their faith with pressing social concerns involving waste. Like rich, tempting foods that can lead to gluttony, excess money or wealth can also induce polluting effects for the person (for example, greed, sin, and wastefulness) if withheld and left unshared. According to Islamic law, Muslims should participate in almsgiving, or *zakat*, ranging from 2.5 to 10 percent of their annual income, which is redistributed to "the needy." Social reformers identify religion's influence on personal conduct as the foundation for tremendous social change in a number of areas, from ethical consumption to political environmentalism. Among Hindus, the Ganges River is known to be a sacred waterway; pilgrims bathe in the water's purifying substance and some place deceased loved ones on floating pyres to ensure their safe passage to heaven. Yet, much has been written about the polluting effects of rapid industrialization on the Ganges's water quality. Anthropologist Kelly Alley describes conflicts between officials designing waste management programs for such pollution (*pradushen*) and the more widely used ritual notion of pollution (*gandagi*) among river workers and pilgrims, who argue for the Ganges's essential purity. She suggests that the contentious public debate in Benaras misses the significance of varying cultural knowledge of waste drawn upon by officials, boatmen, merchants, and pilgrims.

Minimalism and Donations

Often times what one might call "minimalist" practices are embedded in an intricate moral economy, where the practitioner receives seemingly immaterial returns, such as blessing, purity, prosperity, or nobility, for their discipline and piety. While encouraging the scrupulous use of material things, some religions simultaneously promise great material wealth for pious behavior or imagine the afterlife as a place of staggering material abundance. A crop of U.S. televangelists earned media scorn in the mid-1980s as seductive swindlers, banking on their followers' willingness to make immediate financial sacrifices for promises of great riches. Scholars point out that the oscillating orientations (for example, sacrifice and prosperity, minimalism and excess) widely attributed to the prosperity gospel in Pentecostalism need not be contradictory for the practitioner. Donations form part of a sacred economy of giving, where self-sacrifice pays dividends in frequently unpredictable ways. Some suggest increasingly well-documented attempts to control unpredictable market forces through recourse to spirits, witches, or divine beings reveal pervasive contradictions of capitalism in contemporary life.

Religious practitioners, however, engage in a continual process of reinterpreting and emphasizing some principles and behaviors over others according to their cultural context. German sociologist Max Weber famously observed the historical relationship between capitalism and the spiritual "calling" to work in Calvinism. But even in Protestant Christianities, the unflinching embrace of capitalism has not been a foregone conclusion. Early 20th-century Christian charity organizations in the West, such as the Salvation Army, Goodwill, and St. Vincent de Paul, salvaged discards like scrap textiles and secondhand clothes for the benefit of the "less fortunate" in their home countries and a variety of non-Western locales. Saving and reusing was not simply an economic boon but also a moral practice confirming thrift and worth. Charity donations offered middle classes the privilege of feeling good about getting rid of worn or old things. Such programs, however, tacitly acknowledged long-emerging inequities of Western capitalism and material abundance.

Simple Living

Some religious movements practice a collective version of renunciation by spatially separating themselves and forming what scholars call "intentional communities." Nineteenth-century communitarians in the United States, such as Shakers, Millerites, and Oneida Perfectionists, espoused principles that restricted individual wealth and material possessions at a time when the factory-industrial complex was being established and a growing cash economy quickened consumerism. For example, Shakers living across some 18 settlements took a vow of celibacy and practiced communal landholding and profit-sharing in their agrarian business ventures, including seed selling. Old Order Amish continue to observe the *ordnung*, the communal body of rules that limit individual status goods, especially automobiles and electric-powered radios and telephones. Amish theology contests modern individualism and places value on *gelassenheit*, the clustered ideals of submission, obedience, and self-sacrifice. Scholars like David Shi maintain that, while they perhaps appear to distance themselves from "mainstream" U.S. society, communitarians actually exemplify the widely found themes of "simple living," self-restraint, and frugality long circulating in U.S. political and social life.

Social scientists point out that the desire for a simplified existence, which Old Order Amish exhibit and perform for tourists, is a particularly modern condition. Unrelenting material accumulation (prompted by planned obsolescence) and the marketplace's impersonal nature bring into relief the absences produced by "conspicuous consumption." Getting back to basics—to a more simple life unencumbered by material abundance and filled with the "real" values of friendship, sharing, and community—has been an increasingly pervasive theme in some quarters of Western societies.

Anxieties encircle not only people's relationship with their cast-off things but also the moral implications of the "throw-away society," including fleeting relationships, a weak social fabric, and hyper-individualism. The skeptical argue that nostalgia for a simpler life assuages middle-class concerns over abundance in advanced capitalism. Still others imply that simple living rings hollow without the guiding hand of organized religion. It is important to recognize, however, that even broad cultural messages against "affluenza's" hazards contain distinct spiritual qualities, origins, and underpinnings.

Simple living, or the voluntary simplicity movement, is embraced by a wide range of practitioners in the United States and western Europe, from seekers of heightened spirituality to concerned conservationists observing industrial capitalism's environmental costs. Specific practices can include buying and wasting less, growing one's own food, making hand-produced objects, limiting use of technologies like computers, eating a vegan or vegetarian diet, and adopting other ecofriendly activities, like bicycling. Voluntary simplicity draws loosely from a diverse array of Eastern and Western religious influences, such as the teachings of Buddha and of John the Baptist. Ironically, marketing in the United States and elsewhere, ranging from the magazine *Real Simple* to the banal Calgon Bath commercial (and its tagline, "Take Me Away!"), capitalizes on simple living's popular goals. Such advertising frequently appeals to a middle-class (feminized) desire for near-transcendent experiences of simplicity, austerity, and tranquility begotten from one's release from material possessions, time pressures, and responsibilities.

Conclusion

The renunciant frequently operates as a moral exemplar for the layperson, symbolizing discipline, completeness, and piety. The renunciant's life is especially intriguing and desirable, yet elusive in an era of staggering consumption (e.g., the invitation to "shop till you drop"). What the image of the religious renunciant obscures, however, are the small yet momentous struggles of restraint in the religious actor's everyday life. By examining the moral dilemmas of waste, wastefulness, and pollution, one can better appreciate the position of the religious practitioner assessing taboos, prohibitions, and other behavioral restrictions as well as their hard-earned, luxuriant rewards. Ethnographers maintain that religious practices do not simply oppose capitalist accumulation, or "materialism," wherein the spiritual is contrasted with the material, religion to economy, and spirit to flesh, but often coexist in an ambivalent relationship with them.

Garbage is an ongoing human process of classifying, sorting, and reordering. Middle-class Americans spend much time and money extricating themselves from waste's immediacy, visibility, and physicality: scheduling weekly municipal trash collection, moving garbage swiftly to an opaque bag; installing quick-flush toilets, and building bigger and bigger out-of-sight dumps. But lingering moral problems remain. Practitioners use their religious engagements to question, mitigate, and make sense of capitalism's advance and the oft-neglected aspect of consumption: waste.

Through the robust and varied research concerning garbage, religion, and minimalism, anthropologists propose that the much-less grand, more-ordinary lives of practitioners striving for purity, holiness, sacredness, or oneness hold perhaps the greatest intrigue and warrant our undivided attention.

Britt Halvorson
Colby College

See Also: Consumerism; Culture, Values, and Garbage; Food Waste Behavior.

Further Readings

Alley, Kelly. "Ganga and Gandagi: Interpretations of Pollution and Waste in Benaras." *Ethnology*, v.33/2 (1994).

Bynum, Carolyn Walker. "Fast, Feast, and Flesh: The Religious Significance of Food to Medieval Women." *Representations*, v.11 (1985).

Douglas, Mary. *Purity and Danger: An Analysis of Concepts of Pollution and Taboo*. London: Routledge and Kegan Paul, 1966.

Hawkins, Gay. *The Ethics of Waste: How We Relate to Rubbish*. Lanham, MD: Rowman & Littlefield, 2006.

Miller, Vincent J. *Consuming Religion: Christian Faith and Practice in a Consumer Culture*. New York: Continuum, 2003.

Rouse, Carolyn and Janet Hoskins. "Purity, Soul Food, and Islam: Explorations at the Intersection of Consumption and Resistance." *Cultural Anthropology*, v.19/2 (2004).

Shi, David E. *The Simple Life: Plain Living and High Thinking in American Culture*. New York: Oxford University Press, 2001.

Strasser, Susan. *Waste and Want: A Social History of Trash*. New York: Metropolitan Books, 1999.

Garbage Art

Garbage art (alternatively known as trash art or recycled art) is art created from materials including post-consumer and other waste, collected debris, or objects previously used for other purposes. It can be viewed as a special form of recycling, and the growth of this genre is a reflection of the increasing importance of environmentalism in all segments of society, including the arts. Creating art from garbage involves transforming the meaning of objects by placing them in new, aestheticized contexts. This practice is not new; tribal peoples have adapted bits of trash from industrialized societies into their traditional arts since coming into contact with products of the developed world. Reworking worn-out or otherwise-useless materials into new objects with aesthetic or functional value also has a long history in folk art traditions around the world. What is relatively new is the widespread popularity of garbage art in industrialized consumer societies in both fine art and community art contexts. Burgeoning interest in the new genre of garbage art is countercultural in several ways; work in the genre expresses critical perspectives on the materialism, commodification, environmental destructiveness, and individualism characteristic of modern consumer culture. Despite its critical political associations, garbage art has enjoyed rising popularity with the broad acceptance of environmentalist ideas that were once considered "radical" into mainstream society.

Garbage art may be seen as a subset of the broader category known as environmental art. The latter also includes works created from natural materials or designed to interact with environmental forces, rather than only those using waste-stream media. Sculpture that produces musical tones when the wind blows through it, or work that decoratively enhances natural objects, are examples of environmental art. Even more broadly, environmental art includes pieces designed for display or performance in natural settings, work inspired by nature, and other conceptual art about environmental relations.

Creation and Materials

Creating art from trash involves "consuming" garbage in the sense that artists appropriate and rear-

Garbage art (or trash or recycled art) has been in practice for centuries but has remained on the fringes of the art world. It is often seen as a subgenre of what is known as environmental art. Dumpster diving is a popular method of gathering materials.

range the materials in personal ways, transform their meanings, utilize them to their own ends, and represent them in new ways. It involves taking unwanted materials out of their "waste" context and recontextualizing them as "art." The previous uses and effects of such objects may be symbolically employed to evoke meaningful associations and to communicate messages about consumer society or environmental damage. This remaking of the meaning of objects is a defining characteristic of the garbage art genre and it is often what makes it so enjoyable and accessible to the art-consuming public. The playful and surprising effects of recombining evocative and often familiar objects in ways that transform their meanings provoke a high level of engagement in viewers that does not depend on conventional art appreciation knowledge.

Broadly speaking, garbage art includes phenomena from the margins of industrialized society to its very center. The adoption of cast-off items of industrial origin for their ornamental value by people in nonindustrialized, tribal societies around the world can be considered instances of garbage art; for example, where pop-tops from beer cans, wire paper clips, or keys from sardine tins have been incorporated for their exotic novelty value into traditional jewelry forms. In postcolonial societies that are more thoroughly integrated into global systems of consumer political economy, items of industrial waste such as aluminum cans, telephone wire, and candy wrappers are used to produce artisanal creations such as toy cars, woven baskets, and handbags for both domestic consumption and the international ethnic art market. Such items are often marketed in the industrialized world as Fair Trade commodities produced in workers' collectives, and they fetch high prices for this reason as well as their "ethno-chic" associations. The reuse of materials otherwise destined for the trash heap has also had a place in folk art produced by nonprofessional artists in developed nations. Traditional quilting, for example, arose as a frugal means of recycling the material from worn-out clothing to create warm bed coverings. Other examples are rugs woven from strips of waste cloth and sculpted figures (sometimes used to advertise auto-repair shops) made by welding together scrap car parts.

Brief History and Counterculture

The use of trash as a fine art medium dates back at least to the work of early-20th-century artists such as Fortunato Depero and Kurt Schwitters. Use of found materials, including garbage, has been associated with assemblage art since the 1950s and has been practiced by other well-known artists, including graphic artist Christian Boltanski, sculptor Louise Bourgeois, and photographer Andres Serrano. Art made from garbage has since become much more common in fine arts venues such as museums, galleries, and high-profile installations, including H. A. Schuldt's famous "Trash People," which has traveled around the world since 1996.

As an offshoot of the broader environmentalist movement, the rise of garbage art as a distinct genre since the latter decades of the 20th century has mirrored the course of environmentalism from being a marginal—and in its more radical forms a subversive—movement to gaining mainstream popular and institutional acceptance. Over the course of this period, artists working in waste media have developed the genre as a self-conscious, countercultural expression. The garbage art movement has been a countercultural force in at least three ways. First, it has served as a vehicle for critical commentary on the consumer culture and unsustainable lifestyles that characterize industrialized societies and the globalized political economy. This is the case, for example, of works that either explicitly reference the negative effects of pollution and global warming on natural ecosystems, or are linked to programmatic efforts to inspire resource conservation and recycling in the public. Second, this genre manifests a critique of capitalist culture in terms of its construction of value. In other words, garbage art can deconstruct capitalist culture as a system that depends on maintaining the appearance of monetarized value as a natural, relatively stable basis for conceptualizing worth. The artist can do so by demonstrating the mutable nature of value in taking supposedly worthless objects and revalorizing them by transforming them into art with aesthetic value, which may itself be sold as a commodity with monetary value. This sometimes-playful sleight of hand challenges the art-consuming public to view value and its construction (literally and figuratively) in new and critical ways.

A third way in which the practice of creating garbage art has been deliberately countercultural is the prominent place held by collective and community-based production, as opposed to the dominant "fine art" model of the individual professional artist. Many garbage art projects, such as "trash-to-treasure" art-making parties at community centers or ad hoc sculpture assembly in association with urban neighborhood cleanup events, are organized as inclusive productions, inviting the public to participate and involving people who do not self-identify as "artists." This model of alternative art making was nurtured in the 1960s counterculture, and it remains a strong theme in the garbage art movement linked to grassroots community and political organizing. An early example of collective trash art production is the unsanctioned, yet highly visible, Emeryville mud flats sculpture garden on the eastern shore of San Francisco Bay. Here, individuals and groups have been creating sculptures from driftwood, trash flotsam, and the toppled remains of earlier sculptures since the 1960s. Many 21st-century garbage art projects are collective undertakings, with participants recruited from all over the world to contribute sections of a large additive piece. The result demonstrates community solidarity behind the artwork's critical message and de-emphasizes the contribution of individual artists. An example is the "Hyperbolic Crochet Coral Reef" exhibit by Christine and Margaret Wertheim in Santa Monica, California, in which coral reef representations made from trash by artists in many countries were brought together to dramatize the destruction of these ecosystems by pollution and climate change. Similarly, an exhibition called "World Reclamation Art Project" (WRAP) in Syracuse, New York, organized and assembled by Jennifer Marsh, entailed covering an abandoned gas station with a patchwork of panels incorporating trash and messages critical of petroleum dependence. Panels were created by a worldwide network of amateurs, hobbyists, and professional artists organized by Marsh called the International Fiber Collective. Projects organized according to this collective model are characteristic of the interface between art and social movements, and they run counter to the dominant culture of the fine art world in which professional training is institutionalized and individual artistic achievement is celebrated.

Becoming Mainstream

The messages and practices of garbage art challenge conventional ideas about art and artists. It has thrived mainly on the fringes of the fine arts world. However, just as many environmentalist ideas once considered radical have been embraced by mainstream society, garbage art has gained acceptance by both the public and the fine arts community. In the 21st century, community-based "trash-to-treasure" events have become common, such as garbage art-making parties, markets, and exhibits, as well as trash fashion wearable art contests.

Government and nonprofit foundation support for garbage artists has increased worldwide, along with other opportunities for professional development, such artist residencies dedicated to artists using waste-stream media. Garbage art is becoming increasingly institutionalized and accepted for its positive educational and transformative effects, as well as its aesthetic quality.

Caroline Tauxe
Le Moyne College

See Also: Consumerism; Environmentalism; Post-Consumer Waste; Recycling.

Further Readings

Sorensen, A. "Trash Culture: Your Garbage May Be a Work of Art." Metroactive.com. http://www.metroactive.com/papers/sfmetro/09.96/garbage-96-9.html (Accessed June 2011).

Vergine, L. *When Trash Becomes Art: Trash Rubbish Mongo*. Milan, Italy: Skira, 2007.

Whiteley, G. *Junk: Art and the Politics of Trash*. New York: I. B. Tauris, 2010.

Garbage Dreams

Garbage Dreams is a 2009 documentary film directed by Mai Iskander. The film focuses on the *zabbaleen*, an Egyptian group of Coptic Christians who live on the outskirts of Cairo. The 60,000 zabbaleen reside in Mokattam, a "garbage village" that once served as the primary waste relocation spot for most of Cairo. Cairo's 18 million residents were highly dependent on their waste management work, especially since the city lacked a sanitation system. The zabbaleen model, which is over 100 years old, is considered by many to be the most efficient recycling system in the world.

Summary and Waste Management Methods

The film, shot by Iskander over four years, follows a biographical style and focuses on the lives of three teenage boys who inhabit Mokattam. Osama is a 16-year-old who is portrayed as moving from one job to the next. Nabil is an 18-year-old who dreams of owning his own apartment and getting married. Adham is a 17-year-old whose work involves shearing off the tops of soda cans and who strives to modernize the Zabbaleen's waste management practices. The film opens with an illustration of the Zabbaleen waste management system. The Zabbaleen take great pride in their traditions, one stating that they "turn garbage into raw materials." They collect every type of waste from Cairo residents, including paper, plastic, metal, and other forms of waste. Using plastic granulators, cloth grinders, and paper and cardboard compactors, the Zabbaleen are able to successfully recycle 80 percent of the waste they receive. The materials that they generate are sold to countries including China, Belgium, and France. While they serve an integral role in Cairo's economy, they are often shunned. In addition, the nature of their waste work results in cases of hepatitis and the need for tetanus vaccinations.

The viewer is introduced to Laila, a passionate organizer of the Zabbaleen, who warns the residents of Mokattam that their recycling work is coming under attack. In 2005, Cairo sold $50 million in annual contracts to three private waste management companies from Italy and Spain. City managers were concerned that the practices of the Zabbaleen were too traditional and argued that a new, more modern system was needed. These modern waste companies lack the efficiency of the Zabbaleen, with only 20 percent of the collected waste being recycled and the remainder being incinerated or placed in landfills. A poignant segment of the film involves a school trip to one of these private landfills. One of the teenage boys is asked what he would do if he were given a landfill. "I'd dig it out. It's all a gift from God to be recycled and reused," he replies. The boys then inspect the landfill site and are shocked to discover many useful items in the piles.

The film shifts to a study abroad trip. Nabil and Adham are sent to Wales to study contemporary advancements in recycling. Upon arrival, they discover that Wales recycles only 28 percent of its waste. They then take part in a curbside recycling program and tour a recycling center. One of the boys grabs some of the very fine bits on the center's conveyor belt, exclaiming that these, too, no matter how small, can be recycled. Their tour guide explains that these bits will be incinerated, much to the disappointment of the boys. They then understand that while Wales has new recycling technology, it lacks the "precision" that characterizes their work in Mokattam. Upon their return home, the boys, particularly Adham, are inspired to use their experiences in Wales to solve the crisis posed by the corporate waste management firms. Adham explains to Laila that one potential solution is to conduct source separation, a practice in which residents presort their waste prior to pickup. Laila and others go door-to-door and ask residents if they would support such a practice of essentially putting out one bag of food waste and a second with everything else. Most are supportive of this idea, but the Zabbaleen discover that the modern waste firms are mixing these bags anyway.

The film also profiles the Recycling School, an effort to educate young people in geography, computers, business skills, and recycling knowledge. The films concludes with uncertainty about the future of the Zabbaleen recycling practices.

Scott Lukas
Lake Tahoe Community College

See Also: Cairo, Egypt; Dump Digging; Dumpster-Diving; Recyclable Products; Recycling Behaviors.

Further Readings

Iskander, M. *Garbage Dreams*. Documentary film. Iskander Films, 2009.

PBS. "Garbage Dreams." http://www.pbs.org/independentlens/garbage-dreams (Accessed July 2010).

Smith, S. "Cairo's Devoted Refuse Collectors." *BBC News* (June 2, 2005). http://news.bbc.co.uk/2/hi/middle_east/4602185.stm (Accessed July 2010).

Garbage in Modern Thought

When addressed in a conceptual sense, garbage allows a deeper understanding of the human condition. Garbage has always been an integral part of society. As William Rathje and A. J. Weberman write, garbage is a form of revelation. Through material means, it allows for the understanding of the essence of human nature. Some of the specific conceptual issues associated with garbage and modern thought are garbage as a central/secondary concept, socially significant, a form of misunderstanding, categorization and separation, value/no value, as ephemeral, and excess and loss.

One issue has been the question of whether garbage occupies a central philosophical location in human consciousness, or whether it exists at the periphery, beyond the realm of comprehension. Philosophers and intellectuals have expressed the need to focus on the centrality of garbage, but for everyday individuals, the understanding of garbage is often as something "out of sight, out of mind." Rathje and Cullen Murphy express that garbage, unequivocally, represents the sign of human presence and thus illustrates the centrality of garbage to human thought. For many people, the thought that comes to mind when holding a piece of trash and preparing to dispose of it is simply a lack of thought. Modern humans, as part of their penchant for consumption and unsustainable living, often think very little about the waste that they produce. As Rathje's research in garbology has illustrated, people have unrealistic understandings of why and when they waste, and their ability to see garbage as central to their lives is lost as a result. Like many aspects of capitalist living, the person throwing away a piece of trash does not connect the various levels of production, consumption, and post-consumption involved in the trash. It becomes a secondary matter—an afterthought.

Central Concept

Martin O'Brien, among many thinkers, argues that the understanding of garbage should be a central concept, especially since garbage typically correlates with social change, social roles, and institutions. Thus, beyond the level of individuals and their relationship to garbage, there is an interest in understanding the central role that garbage plays in all of society's roles, institutions, and forms of change. Major cities depend on whether the trash gets picked up on time or not. Politicians can lose their offices if they do not effectively deal with refuse issues in their locality. Much like the minds of individuals who view garbage in a secondary light, societies (as amalgamations of individuals) proceed on the basis of viewing trash as a secondary condition. Many societies approach the idea of garbage, much like actual garbage, as an afterthought. Philosopher

Georges Bataille wrote of the ways in which early societies, like the Mayans, used sacrifice as a means of expelling the accursed share. Garbage is excess—it is a part of society that society no longer desires.

Social Roles

Dominique Laporte suggests that a major step in the transformation of the ways in which people dispose of garbage was how garbage disposal was domesticated, privatized, and essentially more intimately connected to the family as a unit. Garbage, understood in reference to social spheres, is a reflection of the relations among people. Entire social systems—of refuse collection and recycling—are dedicated to the social problems associated with garbage and its disposal or reuse. Politicians, activities, social agencies, and new social movements organize collective efforts around the issues associated with waste. Families, whether they approach garbage through effective measures or not, focus on garbage as aspects of their daily lives.

Reflection of the Individual

Garbage is also a reflection of more specific and intimate details of individuals. Weberman infamously used techniques of what he deemed *garbology* to uncover what he saw as the essential nature of people. He once said, perhaps indirectly referencing Jean Brillat-Savarin's quote about food, "You are what you throw away." Weberman saw his pilfering through the trash of Bob Dylan, Jacqueline Kennedy Onassis, and others as a means of evening the score with the powerful. Through clandestine methods, truths could be uncovered, and secrets could be discovered. Perhaps a document discovered in the trash of a politician could expose a hypocrisy, or a lack of documents could illustrate how an artist-activist was politically inauthentic.

Misunderstanding

Garbage, as Rathje has argued, expresses a disconnect. Even though people are connected to their waste because they produced or consumed what is eventually thrown out, reused, or recycled, they somewhat ironically are often unable to accurately understand the significance of what they waste and what is actually wasted. When asked about their own garbage, people give inaccurate answers; when asked about their perceptions about what is present in landfills, they overestimate certain products (like diapers) and underestimate others (like construction debris). Garbage, then, is misunderstanding. It is believing one thing when another is actually true. It illustrates how contemporary societies often value lifestyles that are governed by "invisibility," or a way of living that promotes looking only at the immediacy of the things around an individual. A person is never encouraged to investigate the natural and human resources used to produce things or to focus on the harms that a product will pose to the environment after its use. One of the dangers of thinking of garbage in such incomplete and incorrect ways is being unable to critically reflect on, and alter, dangerous consumer lifestyles.

Categorization and Value

Garbage is categorization, according to Susan Strasser. Societies devise entire conceptual systems that are geared to delineate between objects that have value and objects that are without value. The latter set of items is referred to as garbage, trash, and rubbish. Just as the world of valuable objects is subject to intense forms of categorization (sometimes called "folk taxonomies"), the world of garbage is subject to the same forms of ordering. In recycling programs and in places of refuse disposal, items of trash are categorized depending on their potential value, possible environmental harm, or time of decay. Consumers have become accustomed to the categories that are often applied to garbage. Many cities require people to dispose of their garbage in an orderly fashion—perhaps separating wet household waste from dry—and recycling programs ask individuals to divide their recyclable items into sets (such as plastic, glass, aluminum, and paper) and smaller subsets (such as PET or 01, PE-HD or 02, and PVC or 03). Garbage is an illustration of how humans use mental categories to order the material world.

According to John Scanlon, garbage is indicative of a separation of the world—the desirable from the unwanted. Michael Thompson uses the riddle of the rich and poor person's approach to snot (one keeps his in a handkerchief, the other disposes of it with a tissue) to underscore the curious ways in which garbage is connected to the issue of value. While garbage is universal—all societies, extinct and extant, have produced or produce garbage—the conditions under which garbage is understood are culturally determined. Many non-Western societies attach a much greater value to items after they are discarded. In the United States and many other nations, garbage often results not because something no longer has utilitarian value but because the item in question is defined as something of no value. Thus, garbage is not only an objective condition of material culture, but also a subjective one of mentalist culture. People define what is trash and what is valuable.

Popular Culture

In popular culture, the terms *garbage*, *trash*, and *rubbish* have an especially potent semiotic context. In popular writing (such as novels), in television, films, music, and other forms of mass expression, the term *trash* is used to signify work that is of especially low value. A novelist who writes work that is considered too popular and is without literary merit may be referred to by the terms *garbage* or *trash*. Individuals who read "trashy" novels may do so because they desire escape and, thus, their time expended may be considered "wasted." In rock and roll music criticism, the term *trash* is also used to indicate an artist who either had potential but did not effectively express it or sells out to commercial interests. The Scottish/American rock group Garbage, in part, took their band name because of a comment that their music sounded like garbage.

In popular linguistic idioms in the United Kingdom and the United States, the terms *rubbish* and *trash* are also commonly used. In the United Kingdom, rubbish is the term used to describe forms of argument that are nonsensical or without value, while in the United States, "trash talking" refers to a form of competitive discourse (sometimes found in sports) that reflects an excessive boastfulness or hyperbole. The pejorative term *white trash* originated in contexts of U.S. slavery in which black slaves referred to white servants using the term. In contemporary social contexts, the offensive term refers to European American (white) people who are considered to be of low socioeconomic class and who lack education and sophistication.

Garbage is commonly associated with the idea of the ephemeral—of something that lasts only a short amount of time. In modern societies, people often throw their household items away without regard for the consequences, and thus they promote a form of living that emphasizes the ephemeral. Poets, including Derek Mahon, use the subject of garbage in their work to draw attention to the condition of modern life, including transiency, waste, and alienation. Author Don DeLillo also uses the idea of garbage in his book *Underworld*. One of the characters in the novel describes the state of garbage and the world, saying "We make stupendous amounts of garbage, then we react to it, not only technologically but in our hearts and minds. We let it shape us. We let it control our thinking." There are cases of famous artists, including Marcel Duchamp and Damien Hirst, who found their art objects—sometimes called "ready-mades" for the fact that they appear to be everyday, household objects—being thrown out in the trash. It turned out that people did not understand that they were not trash and were, in fact, art.

Garbage also indicates issues of excess and loss. Georges Bataille's "The Solar Anus" uses the image of the sun to reflect on the inherent tendency of all things being geared toward excess. While the sun can be viewed as a source of energy, another interpretation is that the sun, as an anus and through excretion, is involved in massive expenditures of energy. Garbage is often viewed as a form of society's excess—as the unwanted things that are thrown out without regard. Even foreign policy, such as a country's invasion of another, could be viewed as garbage, or the state's excess of political and military energy that is released outside its territories. Garbage can also be thought of as a form of loss. According to John Scanlon, once a product losses its ability to be a useful product, and once it is defined as garbage, it loses its ability to stand as something related to something else. Once something is identified as garbage, unless it is reclaimed by someone prior to disposal, it loses its material and conceptual value. In the world of computer science, the term *garbage* also refers to situations of loss in which data or objects in memory go unused in computer operations.

Scott Lukas
Lake Tahoe Community College

See Also: Culture, Values, and Garbage; Hoarding and Hoarders; Sociology of Waste.

Further Readings

Bataille, G. *The Accursed Share, Vol. 1: Consumption.* New York: Zone Books, 1991.

Cerny, C. and S. Seriff. *Recycled Re-Seen: Art From the Global Scrap Heap.* New York: Harry N. Abrams, 1996.

DeLillo, D. *Underworld.* New York: Scribner Paperback Fiction, 1997.

Hamelman, S. "But Is It Garbage? The Theme of Trash in Rock and Roll Criticism." *Popular Music and Society*, v.26/2 (2003).

Haughton, H. "The Bright Garbage on the Incoming Wave: Rubbish in the Poetry of Derek Mahon." *Textual Practice*, v.16/2 (2002).

Laporte, D. *History of Shit.* Cambridge, MA: MIT Press, 2000.

O'Brien, M. "Rubbish-Power: Towards a Sociology of the Rubbish Society." *Consuming Cultures: Power and Resistance*, J. Hearn and S. Roseneil, eds. Houndmills, UK: Macmillan, 1999.

Rathje, W. and C. Murphy. *Rubbish! The Archaeology of Garbage.* New York: HarperCollins, 1992.

Scanlon, J. *On Garbage.* London: Reaktion, 2005.

Shanks, M., D. Platt, and W. Rathje. "The Perfume of Garbage: Modernity and the Archaeological." *Modernism/Modernity*, v.11/1 (2004).

Strasser, S. *Waste and Want: A Social History of Trash.* New York: Metropolitan Books, 1999.

Thompson, M. *Rubbish Theory: The Creation and Destruction of Value.* Oxford: Oxford University Press, 1979.

Weberman, A. J. *My Life in Garbology.* New York: Stonehill, 1980.

Garbage Project

The Garbage Project was founded in 1973 by William Rathje and a group of students at the University of Arizona. Since its inception, it has produced numerous studies supported by a variety of government agencies, nonprofit institutions, and private companies. Over the years, the Garbage Project has examined samples of fresh sorts—garbage fresh off the garbage truck and landfill sorts across the country—by meticulously sorting, coding, and cataloging their contents. Garbage, while a persistent consequence of human activity, is to a large extent invisible, disappearing in trash cans and landfills, and it has been subjected to numerous myths and speculation. The Garbage Project not only unearthed and shed light on garbage itself but also on the human mind and various behaviors.

The Garbage Project was inspired by an anthropology class at the University of Arizona in 1971 designed to teach principles of archaeological methodology. Students undertook independent projects to show links between various kinds of artifacts and various kinds of behavior, which included a comparison of garbage samples from different households, an approach that seemed to hold great promise. By 1973, the Garbage Project entered an arrangement with the city of Tucson, whereby randomly selected household pickups from designated census tracts were delivered to a site for analysis. The garbage was sorted into 150 specific coded categories. For each item, the information recorded included the date collected, the census tract from which it came, any information available from packaging (such as original weight or volume, its cost, or brand), and weight. In 1987, the Garbage Project began excavations of landfills.

Fresh Sorts

Several interesting insights in what and how people consume have been provided by the Garbage Project. Health trends have been reflected in the discards, with a certain irony: attempts to restrict consumption of certain food contents are often counterbalanced by extra consumption of the same food contents in hidden form.

At the beginning of 1983, a sudden widespread increase in the percentage of fat trimmed off and discarded was noticed. This correlated with the publication at the end of 1982 of the National Academy of Science's report and subsequent media reports identifying fat in the diet—particularly fat from red meat—as a significant cancer risk factor. These findings indicated that people made efforts to reduce their fat intake from fresh, red meat.

However, as people bought less fresh meat, their consumption of processed red meat, which tends to

The Garbage Project study found that foods eaten on a daily basis are much less likely to be wasted. For example, the relatively few ingredients in Mexican cuisine are easily reused, producing less waste. Ingredients for very specific uses are wasted at higher rates.

have an even higher fat content in hidden forms, increased.

A common way to study consumer behavior has been via self-report measures. However, an inaccuracy of self-report measures has been suspected, as people's memories tend to be unreliable and biased. A comparison between the reported behavior and sorted garbage revealed these discrepancies. Alcohol consumption tends to go underreported. Reported food consumption tends to be skewed in ways that have been labeled the Good Provider Syndrome and the Lean Cuisine Syndrome. The Good Provider Syndrome is reflected in the tendency of homemakers to almost uniformly report that their family consumes more than sorters can find evidence of in the garbage. Also, homemakers tend to underreport the amount of prepared food. The Lean Cuisine Syndrome is reflected in the tendency of people to both consistently underreport the consumption of items high in sugars and fats and overreport the healthier foods.

Through the examination of garbage, the traditional model of ethnic assimilation—the assumption that immigrants to a new culture will exhibit a cultural style lying somewhere between the normatively prescribed behavior prevalent in the culture of origin and those prevalent in the culture of residence—was challenged. A study conducted by Michael Reilly and Melanie Wallendorf, focused on the daily household consumption of Mexican Americans in Tucson, Arizona, and a comparison with the households of Anglo Americans in Tucson and Mexicans in Mexico City, revealed some surprising findings. With regard to beef consumption and sugar-based soda, the average Anglo American household consumed considerably more than the average Mexican household. However, the average Mexican American household consumed considerably more of those items than the Anglo-American households; this was also the case regarding coffee, convenience foods, high-sugar cereals, and white bread. At the time of the study, Anglo-Americans became more health conscious about their foods, while it appeared that Mexican Americans had overassimilated to their prior conceptions of the U.S. lifestyle encountered before migrating and influenced by inferences drawn from the mass media. This also has been referred to as the Hollywood Hypothesis.

Principles of Waste

The examination of food wastes, or, more specifically, the amount of edible or once edible food that has been thrown away, revealed a counterintuitive tendency to waste more of what is in short supply than what is plentiful. One example comes from examining beef waste, an ideal subject for investigation, because meat packaging is labeled with detailed information such as the type of cut, weight, price, and date. Beef waste was determined by not counting fat or bone. During a widely publicized beef shortage in the spring of 1973, a comparison of the amount of beef waste during the shortage and after the shortage ended showed that there were much higher amounts of beef waste during the shortage. It has been hypothesized that people responded to the media coverage by buying all the beef they could find, including cuts they did not know how to cook. Coupled with not knowing how to store large amounts of beef, these behaviors resulted in greater waste.

Garbage Project studies have found that U.S. families waste between 10 and 15 percent of the food they buy and that various factors account for food waste. One of these factors is called the First Principle of Food Waste: the more repetitive the diet—the more is eaten of the same things day after day—the less food is wasted. For example, waste of standard breads—the most common sandwich bread continually used for many meals—is virtually nonexistent. Specialty breads, such as hot dog buns, are wasted at high rates because they are used for very specific kinds of meals, then are stored away, and eventually go to waste. The Mexican American census tracts have generally less food wastes than the Anglo-American census tracts. While Mexican American cuisine offers a diverse array of dishes, the ingredients are relatively few, so the same ingredients are easily integrated into a variety of meals.

The First Principle of Food Waste also applies to other items, such as hazardous waste. Products used on a regular basis, such as cleaners, are less likely to be wasted than products used in special renovation jobs. The Garbage Project discovered that about 1 percent of household garbage by weight consists of hazardous waste and that the type of hazardous waste is related to socioeconomic status. The hazardous waste for low-income households tends to be automobile care items, such as motor oil; for middle-income households, it is home improvement items, such as paint; and for high-income households, it is garden items, such as pesticides, herbicides, and fertilizers. Among the hazardous wastes in landfills coming from households, in order of amount are household cleaners and pesticides, followed by paint, then motor oil.

Efforts to reduce hazardous waste in household refuse via special collection days have also led to unexpected consequences. A 1986 study that sorted and compared the garbage a month before and two months after the first well-publicized "Toxic Away!" in Marin County, California, revealed that the garbage contained more than twice as much hazardous waste by weight after the special collection day. This phenomenon has been confirmed by studies in Tucson and Phoenix, Arizona. Homeowners, who were made newly aware of the hazardous waste in their homes but missed the collection day, rid themselves of their hazardous wastes in the conventional way.

Landfills

The excavation of landfills began in 1987 with the goals of seeing if the garbage fresh off the truck could be cross-validated by the data from municipal landfills and to seeing what happens to garbage after it has been interred. It began at a time when an adequate knowledge about landfills and their contents did not exist, and when news of a mounting garbage crisis entered into the national consciousness.

The Garbage Project found that the rate of natural biodegradation in landfills was much slower than anticipated, even though paper contributed 40 percent of the waste and fast food containers, disposable diapers, and Styrofoam amounted to less than 3 percent.

Concerns about accelerating rates of garbage generation and landfills filling up and running out of space for new ones raised questions of what fills up the landfills. This had been speculated with calculations based on national production figures and assumptions about rates of discards, but it had never really been examined by actually digging into landfills and recording with details its contents.

The Garbage Project sampled landfills from varying climates, levels of rainfall, varying soils and geomorphology, and varying regional lifestyles across the United States. Despite different environmental contexts, the contents of the landfills examined appeared relatively uniform. Variation in terms of weight percentage of different categories of the refuse samples seemed negligible. Since landfills close when they are full, consideration of volume, rather than weight of the contents, becomes more important. Since most garbage tends to expand once it is extracted from deep inside a landfill, the garbage was subjected to compaction of 0.9 pounds per square inch in order to reflect the volume of the garbage when squashed and under pressure inside a landfill.

The actual makeup of landfills has been shown to be different from popular imagination. Common perceptions of what fills up the landfill have been fast food packaging, expanded polystyrene foam (commonly referred to as Styrofoam), and disposable diapers. Fast food packaging of any kind, including bags used to deliver food, consisted of less than 0.5 percent by weight, and no more than one-third of 1 percent of the total volume of a landfill's content. Expanded polystyrene foam amounted to no more than 1 percent of the volume of the samples. Disposable diapers amounted to no more than 1 percent per weight and an average of no more than 1.4 percent of the contents by volume. As for the possibility of effects on public health, the amount of pathogens in landfills is so enormous that the diapers seem relatively insignificant. Landfills receive sludge from sewage treatment plants and normal household garbage includes residues of personal hygiene, as well as medical wastes of every kind. It also has been documented that bacteria and viruses tend to expire in landfills.

In the 1980s, the volume of all plastics, after excavation and compaction to replicate conditions inside a landfill, was less than 16 percent. Though larger amounts of physical objects are being compared to earlier decades and the number of plastic objects has increased, the actual proportion it takes up in a landfill has not changed.

Due to "light-weighing," the process of making objects to retain the necessary functional characteristics with less material, more plastic takes up less space. For example, a PET soda bottle from 1981 in comparison to one from 1989 is considerably thicker and stiffer.

The major contents in landfills have been found to be paper, at well over 40 percent, with newspapers alone taking up about 12 percent. Construction debris takes up about 12 percent. Contrary to popular belief, not much decomposing occurs in landfills. Even supposedly biodegradable materials do not biodegrade much in landfills. Newspapers have been preserved legibly, so they are commonly used in dating garbage deposits. Organics that have been dated back two decades have remained very much intact.

While food and yard waste are most vulnerable to biodegradation, they only account for between 10 and 20 percent of the organic material in landfills and 5 to 10 percent of total landfill contents. However, even after two decades in a landfill, about one-third to one-half of these organics have been found in recognizable condition.

Tomoaki D. Imamichi
City University of New York

See Also: Archaeology of Garbage; Arizona Waste Characterization Study; Garbology; Landfills, Modern.

Further Readings

Rathje, W. L. and S. Dobyns. "Handbook of Potential Distortions in Respondent Diet Reports. The NFCS Report/Refuse Study." *Final Report to the Consumer Nutrition Division*, S. Dobyns and W. L. Rathje, eds. Washington, DC: U.S. Department of Agriculture, 1987.

Rathje, W. L. and C. Murphy. *Rubbish! The Archeology of Garbage.* New York: HarperCollins, 1991.

Reilly, M. D. and M. Wallendorf. "A Longitudinal Study of Mexican-American Assimilation." *Advances in Consumer Research*, v.11 (1984).

Stanford University. "Garbology Online: The Garbage Project." http://traumwerk.stanford.edu:3455/36/48 (Accessed July 2010).

Garblogging

The term *garblogging* refers to the growing body of work by environmentally conscious bloggers (writers of online Weblogs) that addresses the political, environmental, personal, or social impact of waste, trash, garbage, and refuse. Garblogging is practiced by a broad spectrum of both professional online journalists and environmentally concerned bloggers. Although there are a few blogs that focus exclusively on trash or garbage, there are many more blogs that include a significant number of posts about the production and environmental impact of waste and garbage. Environmental bloggers—as distinct from professional environmental journalists who contribute to newspapers like the *New York Times* or online periodicals such as Slate.com, Salon.com, or the *Huffington Post*, all of which have dedicated sections to environmental matters—produce a range of blogs. These blogs include noimpactman.com and greenasathistle .com, which are journal-like accounts about how a single individual experiments with decreasing consumption and thus wasteful practices; wastedfood .com, which is written by a researcher who evaluates the impact of wasted of food in households, restaurants, and stores; and 365daysoftrash.blogspot .com, which began as an experiment in trash reduction and has become a wide-ranging exploration of the practices of waste disposal and the global trade in garbage.

Garbloggers

Garbloggers, like all environmental bloggers, are comprised of a range of people; stay-at-home parents, organic farmers, activists, scholars, historians, and artists all keep environmental blogs that include posts on garbage. Many of these blogs have hybrid content—bloggers repost conventional environmental journalism and photographs, as well as their own observations and photos.

Although the blogs differ in focus (some bloggers are interested in simplifying their lives for personal reasons, some are interested in critiquing the wasteful economy of planned obsolescence, many are interested in how one repurposes items to keep them out to the waste stream), almost all share an emphasis on linking the problem of garbage and waste to excessive consumption in the material world. They are also especially interested in discussing the impact of household choices on global issues, linking the personal form of the blog genre to the global effects of garbage.

Environmental Role

Garblogging's role in the environmental movement as well as in environmental coverage in the media is still evolving. Media theorists and historians date the advent of blogging sometime in the mid-1990s, and according to techorati.com (an online site devoted to tracking the impact of user-generated blogs on the Web), since then, the number of blogs that are added daily has increased dramatically. As of 2010, there are well over 2 million blogs on the Internet, many of which are geared toward specialized audiences. Although critics like Andrew Keen have argued that blogging has created a culture in which poorly researched information is disseminated to and by amateurs who do not have the tools to evaluate its content, other critics like Scott Rosenberg and Clay Shirkey have argued that the Internet is self-correcting and that flawed information will be addressed and amended by users as it is disseminated more widely.

Rosenberg and Shirkey also praise the way that the Internet has spurred activism by creating virtual communities of like-minded people who share information. Certainly, blogs have formal features that allow bloggers to reach out to like-minded people; even if bloggers choose not to use these features, all blog templates allow bloggers to create a list or "blogroll" of other blogs and all allow comments on posts. More conventional online periodicals devoted to environmental issues (for example, treehugger.com and grist.com, two of the most widely read) are blends of blogs and periodicals, and they often rely on reader questions, reader-submitted posts, and "citizen journalist" responses, all of which owe as much to a blogging culture as they do to older forms of environmental media such as *Mother Earth News*.

Although garblogs might be seen to enlarge the environmental community and contribute to the creation of a lively online do-it-yourself culture, they are not generally subject to editorial control, the postings can be sporadic, and the blogs themselves are easily abandoned. They are, however, an excellent resource for those readers who are seeking accounts of how virtual green communities exchange information about environmental issues that have an impact on daily life.

Stephanie Foote
University of Illinois

See Also: Culture, Values, and Garbage; Garbage in Modern Thought; Sociology of Waste.

Further Readings

Keen, Andrew. *The Cult of the Amateur: How Blogs, MySpace, YouTube, and the Rest of Today's User-Generated Media Are Destroying Our Economy, Our Culture, and Our Values*. New York: Crown Publishers, 2008.

Rosenberg, Scott. *Say Everything: How Blogging Became What It's Becoming and Why It Matters*. New York: Crown Publishers, 2009.

Shirky, Clay. *Here Comes Everybody: The Power of Organizing Without Organizations*. New York: Penguin Press, 2008.

Garbology

The field of garbology involves the study of refuse and waste. It enables researchers to document information on the nature and changing patterns of modern refuse, hence assisting in the study of contemporary human society or culture. According to the *Oxford English Dictionary*, the term was first used by waste collectors in the 1960s. A. J. Weberman popularized the term in describing his study of Bob Dylan's garbage in 1970. It was pioneered as an academic discipline by William Rathje at the University of Arizona in 1973. The term is used interchangeably for the "science of waste management," and refuse workers are referred to as "garbologists."

In addition to helping municipalities understand the dynamics of the waste products generated in their communities, the best way to manage them and whether or not they have any salable value, industries and major firms are also avid followers of this research. It enables them to comprehend whether or not the packaging and other discards associated with their products are indeed harmful to the environment. The field of garbology often intersects with archaeology, since fossilized or otherwise time-modified trash may be the only remnant of ancient populations.

Garbology has also been used as an investigative tool of law enforcement, corporate espionage, and other types of investigations. This involves not just a physical sorting of papers but also files from the "trash" e-mails in computers. Journalists often use such surreptitious methods to investigate stories. Special intelligence services have also used garbology to combat crime—illegal in many countries, unless used by the government's intelligence units.

Garbology in Contemporary Times

In 1987, William Rathje initiated the Garbage Project at the University of Arizona. The goal was to determine what was below the landfills and how much of it was biodegradable. The project demonstrated that there were major disparities between what was actually in the landfills and what Americans perceived was in them. Most people mistakenly believed that the landfills were filled with fast food containers, disposable diapers, and Styrofoam, although these amounted to less than 3 percent of landfill volume. Plastic was 20–24 percent of waste, while paper contributed 40 percent. The project also surveyed different regions of the country to better comprehend what types of garbage survive under different climates and found very little difference between sites, since the garbage is compacted.

The Garbage Project found other misconceptions about landfills. Specifically, the rate of natural biodegradation was found to be much slower than anticipated. It was also found that plastic bottles that were crushed at the top were more easily inflatable than those that were at the bottom. It was also predicted that 50 percent of the landfills that were open would close within five years. Finally, in an effort to avoid large landfills, states often ship their trash to other states.

Garbology is used to assess waste and ascertain new and innovative ideas for waste management. For example, scientists are studying the best way to dispose of a floating mass of trash in the Pacific Ocean. They are also studying the impact of trash on marine life and the process of changing waste into energy. There is also the potential for using methane in landfills to generate small amounts of electricity.

Abhijit Roy
University of Scranton

See Also: Archaeology of Garbage; Archaeology of Modern Landfills; Construction and Demolition Waste;

Culture, Values and Garbage; Dating of Garbage Deposition; First Principle of Waste; Garbage Project; Paper and Landfills.

Further Readings

Blumberg, Louis and Robert Gottilieb. *War on Waste: Can America Win Its Battle With Garbage?* Washington DC: Island Press, 1989.

Carey, Susan. "A Neat Stunt: How Alaska Airlines Beat Back Challenges From Bigger Rivals—It Trimmed Costs Carefully by Selling Aging Planes, Dropping Weak Routes: Some Lessons in 'Garbology.'" *Wall Street Journal* (May 19, 1997).

Cote, Joseph A., James McCullough, and Michael Reilly. "Effects of Unexpected Situations on Behavior-Intentions Differences: A Garbology Analysis." *Journal of Consumer Research*, v.12/2 (1985).

Hirsch, Rod. "Talking Trash in Mercer County: Garbology 101." *Mercer Business*, v.85/1 (January 2009).

Rathje, William and Cullen Murphy. *Rubbish!: The Archaeology of Garbage*. New York: HarperCollins, 1992.

Richards, Jennifer Smith. "Valuable Lessons Hidden in Trash: Freshmen Use Math, Science Skills to Analyze Garbage." *Knight Ridder Tribune Business News* (October 27, 2006).

Young, Mitchell, ed. *Garbage and Recycling*. Detroit, MI: Greenhaven Press, 2007.

Garden Tools and Appliances

Given both the productive potential of domestic gardens and the scope they offer for a diverse range of activities that rely little on consumption, it is ironic that they have become increasingly unproductive and dependent for their maintenance on the consumption of an ever-larger range and scale of commodities. Many of these are garden tools and appliances and, despite their transformative and productive promise and potential, are destined to become underused, unused, or identified for disposal.

Definitions

Garden tools are implements roughly equivalent to unpowered hand tools and are used to materially work a garden for functional or cosmetic purposes. Examples of garden tools are spades, hoes, saws, and manual lawn mowers. Appliances include both powered versions of tools as well as those powered devices that are employed in the garden but are not used to materially work it. For example, both gasoline-driven lawnmowers and outside lighting systems are garden appliances. There is some overlap between these categories.

History

One marked characteristic of garden tools is the historical longevity of their form, despite much historical variation in the purposes, forms, and representation of gardens. For example, one 12th-century treatise on the subject advocated the acquisition of an essential garden tool kit, which would include knives, a shovel, a billhook, and a wheelbarrow. The recognizable antecedents of many such forms are traceable back into human prehistory and across cultures. The mattock, for example, is the tool of choice across much of the world and is used to chop, clear, dig, furrow, and weed. Appliances, on the other hand, are—with exceptions—historically recent inventions. The ability to power traditional hand tools has only come about via the development of portable power: batteries, electric cables, small gasoline engines, and gas canisters. The widespread consumption of those devices, which are not used to work the garden but have become clearly identifiable as garden appliances and accessories, is an even more recent innovation, one that has gathered pace since the 1980s. These include leaf blowers and powered hedge cutters, as well as water features, outdoor lighting, patio heaters, and even outdoor air conditioning.

The historical immutability of the material form of many garden tools is, to a large extent, inevitable. They endure because the tasks and labor associated with working the garden both for productive and cosmetic ends endures. Consequently, the proliferation of garden tools into multiple lines of increasingly differentiated forms from competing manufacturers did not gather pace until the 19th century. This was a result of the technological possibilities of industrialization, the productivity of reorganized labor, and the economic imperatives of capitalist political economy. Despite this expansion, there were relatively few genuine technical innovations. The reworking of existing technologies and incremental technical change was the basis of most differentiation. The appearance in the mid-1800s of the manual lawn mower was a notable and, for many, very welcome exception. Throughout the 20th century, the proliferation and differentiation of recognizable tools into product lines has intensified both quantitatively and qualitatively. This includes the addition of power (such as the gasoline-driven lawnmower and later the electric lawnmower) and the development of alternative tools to achieve the same ends; for example, the widespread availability of the hover lawn mower since the 1960s whereby a cylinder of blades rotating around a horizontal axis was replaced by a flat, circular blade revolving around a vertical axis.

Waste

The problem for those who wish to profit from selling large numbers of garden tools that will be "wasted" through redundancy, underuse, or disposal is that of durability—not the durability of their largely enduring form but that of durability in use. There is a widespread expectation that garden tools should last. After all, many of them are made as a result of heavy industrial processes such as forging. Moreover, they tend to contain few, if any, moving or otherwise vulnerable parts or mechanisms. The common materials used in their construction (steel, iron, and hardwoods) also suggest durability. Despite this expectation of the material qualities associated with garden tools, they have been prey to many processes in terms of design, manufacture, materials, and parts that work against this.

First, there are several forms of identifiable, built-in obsolescence. Built-in technological obsolescence is discernible in relation to cheap or value ranges of garden tools; their relatively low price is a direct consequence of decisions to use less-durable, poorer-quality materials, inferior design, and quite often weak points, especially in the joining of components or materials. The old joke, "I've had this broom for 20 years. It's had five new heads and four new handles but I have had it 20 years" is actually a testament to the lack of durability of the tool in question rather than a testament to it. Moreover, garden tools and appliances have become increasingly styled, often with function being compromised by superficial stylistic elements. Even where functional efficacy is not compromised, highly styled garden tools are deemed by various promotional industries to have worn out aesthetically and to be in need of replacement by functionally equivalent but new and superficially differing alternatives.

The wasting of functionally intact tools via stylistic obsolescence or fashion is at odds with the idea of durability. Second, the overelaboration of traditionally simple garden tools is apparent; for example, the invitation to replace the use of twine in attaching plant stems to canes with molded plastic, spring-loaded clips.

Injunctions to replace simple hand tools with powered alternatives are also commonplace, for example, the electric hedge trimmer for hand shears, or the leaf blower for the rake. The assumed and unquestioned universal benefits of technology have played a key role in the promotion of such alternatives. Third, "new" garden tools and appliances proliferate. One garden catalogue boasts several dispensers for garden twine (both complex and functionally vulnerable), specialized tools for pruning different shrubs, and a wooden ruler for seed planting, which claims some specific garden relevance. Across all of these categories of tools and appliances there has been a general proliferation of choice—30 different kinds of bird feeders, 10 different systems of waste composting.

Given the amount and range of such less-than-durable garden paraphernalia, combined with the durability of many other garden tools, garden tools and appliances add to the amount of waste generated in consumer societies. Many garden tools and appliances are wasted through redundancy—they lay unused in sheds, cellars, outhouses, and the like.

Others are wasted through underuse; for example, the mass ownership of lawn mowers makes very little social sense, as an individual mower is likely to be used just once per week for a few months of the year. Shared ownership, lending, or hiring would be much less wasteful in such circumstances. Increasing numbers of unwanted garden tools and appliances find their way into the circuits of waste disposition: giving away, economic recycling through rummage sales and Internet auctions, donations to charities, and disposal.

One may also argue that repeat consumption of garden tools and appliances, worth, for example, 300 million pounds annually in the United Kingdom, is wasteful, both in itself and in terms of the other consumption necessary to undertake this, such as transport costs. The result often tends to be a spiral of wasteful consumption; for example, the ownership of many duplicated or functionally very similar tools often requires further consumption of garden sheds for storage.

One development has been the increasing deployment of garden appliances that are intended solely to enhance the garden environment. The widespread rhetoric since the 1960s of the garden as an outside room underlies much of this wasteful consumption and includes the installation of elaborate outside lighting and entertainment systems, ever more-power-hungry barbecues, water features and pumps, hot tubs, patio heaters, and even outdoor air conditioning.

Finally, there is a notable, and ironic, development in the function of the garden itself. From being productive spaces and spaces of contemplation, entertainment, and leisure, the garden has increasingly become a space of storage and disposal—a resting place for the wasted products of consumption. Not only garden tools and appliances but also other unwanted or ambiguous objects are found here. Often initially seen as transitional spaces, where objects can be placed while their long-term usefulness is assessed and the options for disposal and disposition considered, many gardens have become recognizable marginal spaces of legitimized, domestic fly-tipping.

Neil Maycroft
University of Lincoln

See Also: Consumerism; Fly-Tipping; Overconsumption.

Further Readings

Lucas, Gavin. "Disposability and Dispossession in the Twentieth Century." *Journal of Material Culture*, v.7/1 (2002).

Sanecki, Kay N. *Old Garden Tools*. Buckinghampshire, UK: Shire Publications, 1987.

Uglow, Jenny. *A Little History of British Gardening*. London: Chatto & Windus, 2005.

Gasoline

The term *gasoline* is a generic name for a liquid mixture derived from petroleum. It is primarily used as a fuel in internal combustion engines. Before its first use in engines, the market for gasoline was almost nonexistent. In the 1800s, it was considered mostly waste or used as paint solvent, a treatment against lice and their eggs, and a cleaning fluid to remove grease stains from clothing. The first U.S. refineries processed crude oil primarily to recover the kerosene, and some dumped gasoline as waste directly into rivers or fields in the nearby area.

Health Hazards

As a hydrocarbon, gasoline is considered a hazardous substance and is regulated in the United States by the Occupational Safety and Health Administration. The material safety data sheet for unleaded gasoline shows more than 15 hazardous chemicals occurring in various amounts, including benzene (up to 5 percent by volume), toluene (up to 35 percent by volume), naphthalene (up to 1 percent by volume), trimethylbenzene (up to 7 percent by volume), methyl tert-butyl ether (up to 18 percent by volume, in some states), and about 10 others. Many of the nonaliphatic hydrocarbons naturally present in gasoline (especially aromatic ones like benzene), as well as many antiknock additives, are considered carcinogenic. Because of this, large-scale or constant leaks of gasoline pose a threat to the public's health and the environment in the event the gasoline reaches a public supply of drinking water. According to the Environmental Protection Agency (EPA), one gallon of gasoline can pollute 750,000 gallons of water.

Refining

In addition to the extraction of oil and its transportation to refineries, the first step to obtain gasoline is the refining process. This process releases a number of substances with extensive air-polluting potential into the atmosphere. A considerable odor normally accompanies the presence of a refinery. Aside from air pollution impacts, there are also wastewater concerns; risks of spills, fires, and explosions; and health effects due to industrial noise. A persistent problem linked to hydrocarbons is corrosion of the transportation lines and storage devices. Corrosion occurs in various forms in the refining process, such as pitting corrosion from water droplets, fractures in metals due to exposure to hydrogen, and stress corrosion cracking from sulfide attack. To prevent corrosion, carbon steel is normally used for upward of 80 percent of refinery components, which is beneficial due to its low cost.

In the United States, there is strong pressure to prevent the development of new refineries, and no major refinery has been built in the country since Marathon's Garyville, Louisiana, facility in 1976. Since the 1980s, over 100 refineries have closed due to obsolescence or merger activity within the industry. Around the world, environmental and safety concerns have led to the construction of oil refineries some distance away from major urban areas. Nevertheless, the pace of urbanization makes this practice unsustainable. The possibility of disasters due to hydrocarbon spills in the vicinity of densely populated areas increases every year.

Spills and Cleanup

Most spills involving gasoline occur at the distribution phase. On June 10, 1999, a gasoline pipeline in Bellingham, Washington, owned by the Olympic Pipe Line Company ruptured, discharging approximately 236,000 gallons of gasoline into Hanna Creek. It then leaked into Whatcom Creek, a 3.5-mile-long coastal stream that runs through a city park, residential neighborhoods, and urban industrial areas before emptying into Bellingham Bay. As the gasoline was carried down the creek, the fumes were ignited, killing three people and affecting a variety of natural resources along the creek's path.

When a spill involving gasoline occurs, there are remediation-performance monitoring methods that can be applied to the contaminated site. Methods include phytoremediation, bioremediation, in situ chemical oxidation systems, or mechanical cleanup of the polluted materials. Phytoremediation is the use of vegetation to remediate contamination by the uptake (transpiration) of contaminated water by living plants. Foliage can be used to contain, remove, or degrade contaminants.

Bioremediation processes use microorganisms, fungi, green plants, or their enzymes to return the polluted environment to its original condition. It can be employed in areas that are inaccessible without excavation. It is especially used in hydrocarbon spills (mostly oil) or where chlorinated solvents may contaminate groundwater. Bioremediation attacks specific soil contaminants, such as degradation of chlorinated hydrocarbons by bacteria. An example of a more general approach is the cleanup of oil spills by the addition of nitrate and sulfate fertilizers to facilitate the decomposition of crude oil by endogenous or exogenous bacteria. Chemical analysis is required to determine when the levels of contaminants and their breakdown products have been reduced to below regulatory limits.

Bioremediation is typically much less expensive than excavation followed by disposal elsewhere, incineration, or other ex situ treatment strategies, reducing the need for "pump and treat," a common practice at sites where hydrocarbons have contaminated clean groundwater. Pump and treat involves pumping out contaminated groundwater with the use of a submersible or vacuum pump and allowing the extracted groundwater to be purified by slowly proceeding through a series of vessels that contain materials designed to absorb the contaminants from the groundwater. For gasoline-contaminated sites, this material is usually activated carbon in granular form. Chemical reagents such as flocculants followed by sand filters may also be used to decrease the contamination of groundwater. Air stripping is a method that can be effective for volatile pollutants such as benzene, toluene, ethylbenzene, and xylenes (BTEX compounds) normally found in gasoline spills.

In situ chemical oxidation (ISCO) is the injection into the subsurface of liquid or gas that causes oxidation and can result in the direct destruction of gasoline contamination. The chemical reaction produced, called oxidation, is a chemical reaction characterized by the loss of one or more electrons from an atom or molecule. When an atom or molecule combines with oxygen, it tends to give up electrons to the oxygen in forming a chemical bond. In contrast to other remedial technologies, contaminant

A puddle of waste gasoline, which must be managed as a hazardous waste because it is ignitable and toxic. Waste gasoline is gasoline that has been mixed with water or other products or is too old to be used. The U.S. Environmental Protection Agency (EPA) enforces specific rules for disposal, which include a limited time of storage, hiring a licensed transport hauler, and a detailed shipping manifest to be submitted to environmental authorities. The EPA warns that one gallon of waste gasoline can pollute 750,000 gallons of water.

reduction can be seen in short time frames (such as weeks or months). Although many of the chemical oxidants have been used in wastewater treatment for years, only recently have they been used to treat hydrocarbon-contaminated groundwater and soil in-situ. This process can also result in the indirect decrease of petroleum contamination by increasing the dissolved oxygen content in groundwater, which enhances biodegradation.

The oxidant (such as hydrogen peroxide) reacts with the contaminant causing decomposition of the contaminant and the production of relatively-innocuous substances such as carbon dioxide and water. Carbon in the form of organic carbon and manufactured hydrocarbons are common substances readily oxidized (reductants). For ISCO to effectively reduce contaminant concentrations, contact between the oxidant and the contaminant must be direct.

In case of a relatively contained and small spill, mechanical removal of contaminated soil is carried out by excavating visible contamination, screening the excavation site for hot spots with a photo ionization device, and then taking confirmation samples to show that the contaminated soil has been removed. Samples for gasoline spills should be analyzed for pollutants.

Waste Gasoline

An important source of pollution is waste gasoline. This is gasoline that has been mixed with water or other products or is too old to be used. In those cases, waste fuel has to be managed as a hazardous waste because it is both ignitable and toxic. Hazardous wastes must be managed on-site and disposed of by following specific EPA rules. Those rules include a secure custody chain "from cradle to grave," a limited time of storage depending upon the amount of waste, hiring a licensed hauler to transport the waste gasoline off the site, and a detailed shipping manifest to be submitted to environmental authorities.

Whenever burned in engines, gasoline poses a number of risks for health. Lead pollution from automobile exhaust is expelled into the air and is easily inhaled. Tetraethyl lead (TEL) was widely used as an antiknock agent and to increase the fuel's octane rating until the 1970s. Concerns about health effects eventually led to the ban on TEL in automobile gasoline in many countries.

Lead is a toxic metal that accumulates and has subtle and insidious neurotoxic effects, such as low IQ and antisocial behavior, even at low exposure levels. It has particularly harmful effects on children. The EPA issued standards in 1973 that called for a gradual phase-down of lead to reduce the health risks from lead emissions from gasoline, culminating in the Clean Air Act Amendments of 1990 and EPA regulations banning lead in motor vehicle gasoline after 1995.

Beginning January 1, 1996, the U.S. Clean Air Act prohibited the sale of leaded fuel for use in on-road vehicles. Possession and use of leaded gasoline in a regular on-road vehicle carries a maximum $10,000 fine. However, fuel containing lead is authorized for off-road uses, including aircraft, racing cars, farm equipment, and marine engines.

Benzene, an aromatic hydrocarbon used as a gasoline additive to increase the octane rating and reduce knocking, is carcinogenic. Gasoline contained high proportions of benzene before the 1950s, when tetraethyl lead replaced it as the most widely used antiknock additive.

With the prohibition of leaded gasoline in the 1970s, benzene returned as a gasoline additive in some nations. In the United States, concern over its negative health effects and the possibility of benzene entering the groundwater have led to strict regulation of gasoline's benzene content, with limits typically around 1 percent. The EPA issued new regulations to lower the benzene content in gasoline to 0.62 percent by 2011.

Toluene, another carcinogenic, aromatic hydrocarbon can be used as an octane booster in gasoline fuels. Inhalation of toluene may irritate the upper respiratory tract. Overexposure can be associated with fatigue, confusion, headache, dizziness, and drowsiness. Peculiar skin sensations (such as pins and needles) or numbness may occur. Very high concentrations may cause unconsciousness and death.

Studies suggest that the amount of gasoline that evaporates from an automobile's fuel system when the vehicle is being used on a hot day may be 5–25 times greater than the amount of unburned gasoline that is permitted to escape in exhaust. The "running losses" may amount to 2–10 grams per mile, compared to 0.41 grams out the tailpipe. Ten grams of gasoline is about half a liquid ounce. Beginning in 1989, the EPA required gasoline to meet volatility standards to decrease evaporative emissions of gasoline in the summer months when ozone levels are typically at their highest.

According to the EPA, regulations require that each manufacturer or importer of gasoline, diesel fuel, or a fuel additive have its product registered prior to its introduction into commerce. In some cases, the EPA requires testing of these fuels and fuel additives for possible health effects. The EPA also requires that gasoline contain a certified detergent in order to reduce emissions.

In the early 1990s, the EPA began monitoring the winter oxygenated fuels program implemented by the states to help control emissions of carbon monoxide during the winter months and established the reformulated gasoline (RFG) program to reduce emissions of smog-forming and toxic pollutants. The EPA promulgated new regulations, setting standards for gasoline performance levels and for low-sulfur gasoline to reduce harmful air pollution and help ensure the effectiveness of advanced emission control technologies in vehicles.

A source of pollution is the fuel filter used in the fuel line that screens out dirt and rust particles from the fuel. These are normally made into cartridges containing a filter paper. Trace impurities in the gasoline and other incomplete combustion products contribute to the emission of volatile organic compounds (VOCs). In 1995, nine metropolitan areas in the United States were required to switch to reformulated gasolines (RFGs) for VOC controls. RFG requirements resulted in limits on benzene, sulfur, and aromatics to achieve reductions in VOCs, nitrogen oxides, and toxic emissions. In addition, oxygen content of at least 2 percent by weight was required to improve cold-start emissions from older cars.

Mauricio Leandro
Graduate Center, City University of New York

See Also: Automobiles; Car Washing; Emissions; Engine Oil; Fuel; Tires.

Further Readings

Colorado Department of Labor and Employment, Division of Oil and Public Safety. "Petroleum Hydrocarbon Remediation by In-Situ Chemical Oxidation at Colorado Sites." June 2007. http://oil.cdle.state.co.us/oil/Technical/In%20Situ.pdf (Accessed January 2011).

Environmental Protection Agency. "Gasoline Fuels." January 2011. http://www.epa.gov/otaq/gasoline.htm (Accessed January 2011).

Georgia

Established in 1732, last of the original 13 colonies, Georgia was also one of the original Confederate states and the last state to be restored to the Union. The ninth most populous state, in 2007–08, Georgia had 14 of the nation's 100 fastest-growing counties. The capital, Atlanta, is the largest city and metropolitan area, and it is a major center in both the state and the southeastern United States as a hub of communications, industry, transport, tourism, and government. The agricultural and industrial economy of Georgia is diverse; many corporations (including Coca-Cola, Home Depot, and UPS) have their headquarters in Georgia, including 15 Fortune 500 companies and 26 Fortune 100 companies.

Georgia has one of the nation's highest electricity generation and consumption rates, the industrial sector being the chief energy consumer. While industry is a heavy consumer of electricity due to energy-intensive wood and paper product manufacturing, Georgia is also one of the nation's top producers of power from wood and wood waste.

The 16th Nationwide Survey of MSW Management in the United States found that, in 2005, Georgia had an estimated 11,549,889 tons of municipal solid waste (MSW) generation, placing it 13th in a survey of the 50 states and the capital district. Based on a population of 9,342,080, an estimated 1.24 tons of MSW were generated per person per year (ranking 26th). The state landfilled 7,195,075 tons (ranking 12th) in the state's 56 landfills, a figure that includes 1,738,964 imported tons. The state sent 81,535 tons (ranking 23rd) to its waste-to-energy (WTE) facility and was the only state to not report its recycled MSW tonnage.

In 2006, Georgia had 429,202,431 cubic yards of landfill remaining and was increasing its capacity; it was ranked seventh out of 44 respondent states for number of landfills. Whole tires, used oil, and lead-acid batteries were reported as banned from Georgia landfills, with yard waste banned from landfills built to Subtitle D specifications. Georgia is ranked seventh in the United States for number of landfills.

History of Waste Disposal

Georgian archaeology has engaged in detail with waste disposal on historical sites, even at the farmstead level. It is known that farmyards were swept frequently, especially on African American farms, and that this contributed to sheet midden forming around the yard's rear edges. Trash disposal is one of the key archaeological features of farm life, and Georgian archaeologists were able to apply the four patterns of disposal defined at Finch Farm, South Carolina, to Georgian farmsteads. These four patterns are (1) the Brunswick Pattern, refuse accumulating around the structures' rear doors, (2) accumulation of rear yard sheet midden, (3) the Piedmont Pattern, throwing trash down gulleys and ravines found near the farmsteads, and (4) the widespread practice of open burning.

The four patterns are thought to have followed each other to some extent. Sheet midden (late 18th and 19th centuries) shows a more hygienic approach than the Brunswick Pattern (18th century). The Piedmont Pattern (19th century) is seen as a response to erosion and the increasing use of disposable glass, which had to be discarded away from trampling hooves and feet. Open burning is thought to be contemporary with the Piedmont Pattern and reflects the growing use of paper packaging and changing attitudes toward household waste disposal.

As urban centers developed (in Georgia, every incorporated town is a city, regardless of size), informal and formal dumps were adopted. As the cities grew, they developed municipal garbage collection and disposal, some in the late 19th century and all by the mid-20th century. Two dumps in the Atlanta area dating from this period have been archaeologically investigated: Edgewood and Maddox Park.

Edgewood Dump, on the outskirts of Atlanta, is a ravine infilled with rubbish in the space of a few years up to ca. 1911, when a house was built on the same spot. Partial excavation produced an assemblage of 522 whole bottles that were used to study the time lag in bottle use (time elapsed between manufacture and disposal). Paradoxically, the greatest lag was in bottles produced for fresh beverages, probably because the bottles were recycled after their original contents had been used. Beer also had a long lag for which there was no apparent explanation. Long time lags in medicine and wine bottles were expected, as there was no necessity for prompt consumption.

As a whole assemblage, the Edgewood Dump bottles had a longer time lag deposition than three comparable dump sites, possibly because of the lower socioeconomic status of the Edgewood community. Analysis of ceramics and faunal remains from the site were also carried out.

Maddox Park was Atlanta's "Sanitary Dumping Ground" from 1884 to 1910, and the report is considered an outstanding study in the development of municipal garbage collection in Atlanta as well as in garbage collection and disposal technology. The site was eligible for National Register of Historic Places nomination, but no further work was required in this case, which involved a proposed rail line.

Hercules 009

The Hercules 009 Landfill is one of the best-known Superfund sites and is an example of one of the scheme's successes. The federal Comprehensive Environmental Response, Compensation, and Liability Act (CERCLA, or Superfund) allows the government to police and clean up toxic waste sites. As of 2010, Georgia had 15 Superfund sites, which are placed on the National Priority List (NPL); four sites have previously been cleaned up and delisted.

Hercules Incorporated chemical manufacturing site, near Brunswick, Glynn County, is a 16.5-acre facility and was granted a permit in 1975 to dispose of contaminated waste from manufacture of the agricultural pesticide toxaphene. Seven acres at the northern end of the site were turned into six 100–200-foot-wide by 400-foot-long cells, which have become known as the Hercules 009 Landfill. These cells were reportedly lined with a soil-bentonite clay mixture. Produced by Hercules since 1948, toxaphene was one of the most heavily used insecticides in the United States until 1982, when the Environmental Protection Agency (EPA) canceled its registration; by 1990, it was banned completely. The Hercules 009 Landfill was used between 1975 and 1980 to dispose of toxaphene-contaminated sludge, empty drums, glassware, rubble, and trash.

In 1980, state investigation showed that toxaphene in soil and water samples from drainage ditches around the site had reached unacceptable levels (15,000 ppm) and the landfill permit was revoked. The landfill was closed by 1983. In 1984, the EPA added Hercules 009 to the Superfund NPL. The threat of groundwater from the landfill flowing into private drinking-water wells was countered by connecting some properties to the Brunswick municipal supply in 1991. Toxaphene was detected in front yards of nearby residences in 1992; these homes were evacuated while the contaminated soil was excavated and replaced. A record of decision (ROD) for the site cleanup operation was put in place in 1993. The remedial work was designed and implemented by Hercules Incorporated.

The landfill was treated by remediation contractors using in situ methods. Contaminated soil from outside the landfill was also treated onsite and used to cap the landfill. The remediation included solidification/stabilization (S/S) treatment using Portland cement as a binding agent mixed into the contaminated waste to immobilize and physically or chemically alter hazardous constituents. The EPA upholds S/S treatment as the Best Demonstrated Available Technology (BDAT) for a range of hazardous wastes listed in the Resource Conservation and Recovery Act (RCRA). Breakthrough mixing techniques have been developed and applied at full-scale U.S. remediation projects, and these have made a large contribution to the adaptability of S/S technology, which can be used in situ or ex situ.

At Hercules 009, 15 percent Portland cement by weight was added to the landfill contents using an excavator with attachments to mix dry cement into the in situ–contaminated deposits. Up to six 25-by-25-feet subcells could be treated at once with this method. Horizontal rotary mixers and deep-soil mixing augers can also be used for this purpose. The treated depth penetrated beyond the sludge zone in the bottom of the landfill and in most subcells into the regional groundwater table; 67,394 cubic meters (88,148 cubic yards) were reported as being treated. Post-treatment testing showed that a compressive strength exceeding the 50 psi specification was reached within three to five days, and Toxicity Characteristic Leaching Procedure (TCLP) testing showed no toxaphene present in the leachate. The site was regarded and revegetated following remediation, which could be considered "green remediation," as the improved construction properties of the treated matrix allowed reuse, and in situ treatment avoided 5,000 dump truck round-trips.

Jon Welsh
AAG Archaeology

See Also: Beverages; Comprehensive Environmental Response, Compensation, and Liability Act (CERCLA/Superfund); Dating of Garbage Deposition; Pesticides.

Further Readings

Arsova, Ljupka. Rob van Haaren, Nora Goldstein, Scott M. Kaufman, and Nickolas J. Themelis. "The State of Garbage in America: 16th Nationwide Survey of MSW Management in the U.S." *BioCycle*, v.49/12 (2008).

Dickens, Roy S. *Archaeology of Urban America: The Search for Pattern and Process*. New York: Academic Press, 1982.

Olvey, Whitney. *Where Did They Dump It? Refuse Disposal at the Ford Plantation Sites*. Paper presented at the Society for Historical Archaeology's 35th Annual Conference. Mobile, Alabama, 2002.

Reed, Mary Beth, J. W. Joseph, and Thomas R. Wheaton. *Archaeological and Historical Survey of the Maddox Park Site (9Fu114): Atlanta's Sanitary Dumping Ground" 1884–1910*. Report submitted to the Metropolitan Atlanta Rapid Transit Authority. West Chester, PA: John Milner Associates, 1988.

Germ Theory of Disease

The germ theory of disease evolved in the latter half of the 19th century. This fundamentally new understanding of infectious diseases became accepted with vital results for the health of humankind. The theory is anchored in the independent research of Louis Pasteur (1861) and Robert Koch (1862–63), although bacteria were discovered in 1675.

Louis Pasteur found microbes to be behind the fermentation of sugar into alcohol and the souring of milk and developed a heat treatment (pasteurization) that killed microorganisms in milk, which then no longer transmitted tuberculosis or typhoid. Pasteur also developed new vaccines against such infections as anthrax and rabies. His work on fermentation from the 1860s became familiar to Joseph Lister, who introduced antiseptic techniques in hospitals with the crucial positive result of decreasing the spread of infection and resultant mortality.

Another great success in bacteriology and human disease in the late 19th century was achieved by Robert Koch thanks to his implacable belief that germs cause diseases and to his almost superhuman tenacity. In two consecutive years (1882–83), he identified the bacteria that cause tuberculosis and cholera, respectively. Koch and his students opened the door for the golden age of microbiology. Four fundamental postulates about microorganisms were published in 1890; they are partially true from the perspective of later knowledge.

There were many famous scientists and social activists who did not accept germ theory, initially, or at all. Rudolf Virchow (1821–1902) denied its validity for most of his life. Florence Nightingale ridiculed the idea of germs until her death in 1910. However, in 1886, the theory entered the leading American pediatric textbook by Job Lewis Smith (1827–97) in which Koch's research on the tubercle bacillus was named the most brilliant discovery of the last decade.

Toward the end of the 19th century, the growing world of microbes included discoveries by

Louis Pasteur found that microorganisms in wine were responsible for spoilage and developed a unique heating technique to prevent the process. He applied his findings to raw milk, which stopped the transmission of tuberculosis and typhoid.

Dmitri Ivanowski (1892), and Martinus Beijerinck (1898), who revealed tiny infectious agents ("filtrable viruses") that were too small to be seen with the conventional microscope and could pass through bacteria-stopping filters. In 1928, Sir Alexander Fleming discovered penicillin.

Both the miasma and germ theories of disease causation stimulated an expansion of the policy of increasing personal, surgical, and public hygiene in Europe and the United States in the late 19th to early 20th centuries. Germ theory's preventive health innovation included protective inoculations.

Reassessment

Koch's postulates were reproduced decade after decade as a traditional belief that living microorganisms cause infection and contagious disease. However, in 1972, Stanley Prusiner identified a protein as an infectious agent. It was labeled a *prion*, a term derived from PRotein Infection ONly, for which he received the Nobel Prize. Recognition of transmissible disease mediated by a misfolded version of a normal cellular protein that is not associated with a microbe or nucleic acid is revolutionary. It modifies the germ theory of disease since it adds a new cause of disease.

Important Dates

- 1546—Girolamo Fracastoro (1478–1553) proposed that *seminaria contagiosa* could infect people three ways: by direct contact, through contaminated food or clothing, or through the air.
- 1675—Antonie van Leeuwenhoek (1632–1723), the father of microbiology, saw bacteria under his microscope. Van Leeuwenhoek did not author any books, although he did write many letters.
- 1846—Ignaz Semmelweis (1818–65) concluded some unknown "cadaverous material" caused childbed fever.
- 1861—"Experiments and New Views on the Nature of Fermentations," *Comptes Rendus*, v.52, in French) by Louis Pasteur (1822–95).
- 1882–83—Robert Koch (1843–1910) identified the bacteria that cause tuberculosis and cholera.
- 1867—"Antiseptic Principle of the Practice of Surgery," *British Medical Journal*, v.2299) by Joseph Lister (1827–1912).
- 1868—*Studies on Tuberculosis* (Paris, in French) by Jean-Antoine Villemin (1827–92).
- 1886—*A Treatise of the Disease of Infancy and Childhood*, 6th ed. (1827–97).
- 1928—Sir Alexander Fleming (1881–1955) discovered penicillin.

Lolita Petrova Nikolova
International Institute of Anthropology

See Also: Bubonic Plague; Hospitals; Medical Waste; Miasma Theory of Disease; Microorganisms; Public Health.

Further Readings

Barnard, F. A. P. "The Germ Theory of Disease and Its Relations to Hygiene." *Public Health Papers and Reports*, v.1 (1873).

Baxter, A. G. "Louis Pasteur's Beer of Revenge." *Nature Reviews Immunology*, v.1/3 (2001).

Buchen, L. "Microbiology: The New Germ Theory." *Nature*, v.468/7323 (2010).

Cooter, R. "Of War and Epidemics: Unnatural Couplings, Problematic Conceptions." *Social History of Medicine*, v.16/2 (2003).
Dubos, R. J. *Pasteur and Modern Science*. Washington, DC: ASM Press, 1998.
Fry, D. E. "Prions: Reassessment of the Germ Theory of Disease." *Journal of the American College of Surgeons*, v.211/4 (2010).
Herbst, J. *Germ Theory*. Minneapolis, MN: Twenty-First Century Books, 2008.
Kim, Sh. K. "An Antiseptic Religion: Discovering a Hybridity on the Flux of Hygiene and Christianity." *Journal of Religion & Health*, v.47/2 (2008).
Patterson, A. "Germs and Jim Crow: The Impact of Microbiology on Public Health Policies in Progressive Era American South." *Journal of the History of Biology*, v.42 (2009).
Prusiner, S. B. "Prion Diseases." *Scientific American*, v.272/1 (1995).
Rothstein, W. G. *Public Health and the Risk Factor*. Rochester, NY: University of Rochester Press, 2003.
Thagard, P. "Concept of Disease: Structure and Change." *Communication and Cognition*, v.29 (1996).
Tomes, N. J. "American Attitudes Toward the Germ Theory of Disease: Phyllis A. Richmond Revisited." *Journal of the History of Medicine*, v.52 (1997).
Waller, J. *The Discovery of the Germ: Twenty Years That Transformed the Way we Think About Disease*. Cambridge: Icon, 2002.

Germany

Germany is one of the largest industrial powers on Earth; by 2010 it was the most populous member state of the European Union, with over 80 million people. As with other industrialized societies, Germany experienced a first "modern waste crisis" at the end of the 19th century in its cities. Because of exploding population figures and new consumption patterns in urban areas, the wasting practices of town dwellers began to substantially change over time. Toward the end of the 19th century and in a further effort of urban sanitation, larger cities such as Berlin, Munich, Frankfurt, and Hamburg installed municipal waste collection services and, with it, the compulsory domestic trash bin. The municipal disposal services, for which Prussian cities had imposed a tax since 1893, gradually put an end to the traditional backyard fosses in which leftovers were ditched. It also eliminated informal evacuation by local traders or farmers, who had retrieved the urban waste as manure.

Early 20th Century

During the first half of the 20th century, waste remained an urban hygienic problem. The waste's content consisted mainly of ashes, sweepings, and leftovers, with heavy seasonal and regional variations because of particular eating and heating patterns. Local regulations, combined with a transnational knowledge transfer among European cities, dominated waste treatment, while collecting and disposal technologies differed from city to city. Albeit such disparities, the simple dumping of waste in the urban periphery was omnipresent. Waste was used for landscaping, to create hills, gain land, or for soil amelioration. A special form of this was "waste flushing" (*Müllspülung*), practiced in the 1930s in the outskirts of Berlin. Evacuated by barks, the waste was applied on surrounding wetlands with the help of water and pumps.

In 1896, Hamburg, which had just undergone a cholera epidemic, was the first German city to implement (British) incineration technology. In general, German cities were eager to introduce the hermetic collection system, which meant that bins were equipped with lids and fitted the notches of the collection vehicles so that nearly no dust (identified as the main peril to hygiene) would escape.

In the pre–World War I era, it was assumed that an urban dweller produced about one liter of waste (or half a kilogram) per day. The waste was collected two or three times a week. Some rare cities such as Potsdam and Charlottenburg, experimented with a separate collection, as it was known in U.S. cities: edible waste and scrap (such as leather, paper, textiles, or rags) were evacuated separately in order to feed swine or to sell the scrap to scavengers. In Munich, the urbanites' waste traversed a separation facility, where scrap materials were sorted out manually. Scrap materials were rarely dumped but instead were reused or recuperated by ragmen since they had an economic value in the scavengers' trade.

World Wars

World War I, the crisis of the 1920s, Nazi politics, and World War II reinforced such cultures of thrift and turned the recovery of secondary materials into an issue of patriotism and national politics. Beginning in1914, some cities recovered leftovers for hog feeding, often supported by women's organizations. In 1916, any city with more than 40,000 people was officially obliged to separately collect edible waste, which would compensate for the deficit in foodstuffs. In addition, because of the shrinking calorific value of waste, incineration plants were shut down.

The national socialist regime saw the systematic collection of scrap as a way to meet requirements of economic self-sufficiency and its policy of rearmament. From the mid-1930s onward, the regime thus restructured the system of scrap collection in an overly bureaucratic way.

Moreover, a large number of scrap chandlers were forced out of the business because of their Jewish background. The new policy allocated certain districts to each scrap collector and stipulated a collection from private households on certain days. Dwellers were increasingly disciplined, as material became scarcer during the later years of the war.

For a number of years, the national socialist party had already propagandized the collection of scrap as a national duty and it had carried out collections through its organizations, such as the Hitler Youth. After 1939, however, the police were legally allowed to inspect the obligatory collection of specifically leftover foodstuffs and fine contempt. With regard to recycled material, the system was not entirely successful. Specifically, for the collected scrap paper, buyers could often not be found. Only scrap metal and rags always found customers.

Cold War Era

The communist German Democratic Republic (GDR) in many regards enforced a very similar policy as the national socialist regime. The collection of scrap material was seen as a national duty because of material scarcity. The mass organizations conducted public collections of scrap, and people were disciplined through a semipublic collection system within the apartment building community. The system also differed in many ways. The separate collection of different waste fractions for recycling was not a legally imposed obligation for households.

Instead, the system worked with economic incentives, such as exchange of scrap for new products or money. Most importantly, collection and recycling were gradually socialized until the early 1970s. In practice, this meant that both collection and recycling of scrap from industry and households were organized by governmental institutions according to the five-year plan. In a sense, the production of waste thus needed to be planned in advance, not unlike it is done in the 21st century by microeconomic materials management that has integrated waste as both a cost and a potential material resource. The statistics of the plan, which need to be treated with care, indicate that until the late 1970s, the GDR collected a higher rate of (household) scrap per capita than West Germany. This early implementation of a large-scale recycling system is one of the reasons why, after the fall of the Berlin Wall, many people regretted its fast disappearance.

The GDR thus had a somewhat exemplary role with regard to recycling, which West Germany, however, never copied and only studied briefly in the 1990s. In contrast, the GDR was particularly known in the West for its slack but nevertheless proclaimed environmental policy since the early 1970s. This included the import of industrial and household waste from West Germany, particularly from West Berlin, as well as from the rest of western Europe. The operation of the landfill Schönberg/Ihlenberg became known with this respect after the *Wende*, for imported toxic wastes had been disposed of for years without any kind of sealing.

In West Germany, the recuperation of waste materials still had its place in the mid-1950s, and even composting experienced some governmental funding. Dumping, however, dominated waste disposal, since both financial support and any public awareness on disposal methods were lacking. In the 1960s, about one-fifth of municipal waste was incinerated, only a few percent was composted, and the rest was dumped in unregulated landfills. At this time, West Germany had fully turned into a mass consumer society, and its consumers produced new waste fractions such as junk cars, bulky refuse, plastics, and packaging (the one-way glass bottle, soon followed by the plastic bottle, were introduced in the 1960s).

More packaging, an increased paper consumption, and the deprivation of private hearths (which once had absorbed a substantial allotment of household waste) led to an alarming rise in municipal waste volumes and to the notion of an impending "waste avalanche" (*Müllawine*). The subsequent need for higher disposal capacities, a scarcity of tips, and the first legislation on water protection resulted in a growing awareness of the waste problem that soon would be reframed as an environmental problem.

In the 1960s, incineration had a resurgence despite technical and gas emission problems due to plastic waste. In the following decade, landfills were steadily regulated. In 1972, West Germany issued its first Waste Management Act, followed three years later by a comprehensive program to restructure the waste business. Since then, there has been a growing effort in West Germany's national politics toward recycling, influenced by citizens' initiatives that called for a separate collection of glass and paper.

In the 1980s, private disposal contractors increasingly entered the waste market via the field of salvage materials. In 1986, a law on the avoidance and removal of wastes came into force as an amendment to the 1972 waste law. It tried to institutionalize a new hierarchy in the consumption-wasting cycle, namely, to avoid waste in first place, recycle it in second place, and dispose of it only in third place.

1990s and Beyond

In 1991, the packaging ordinance was issued, which has been continuously revised. It aimed at using less packaging and at recycling. In 1996, the German law *Kreislauf- und Abfallwirtschaftsgesetz* was implemented. This law can be considered a milestone for Germany's garbage policy because it merged partly conflicting laws. In practice, the 1996 Waste Act furthered the systematic dealing with waste produced on a massive scale in consumer society, mainly by regulating both its recycling and disposal.

Furthermore, since officials and companies had only slackly considered prior laws, a range of measurements for controlling the disposal system were implemented. Since 1996, the law was at various points adapted to and supplemented by legislation of the European Union. In Germany, major adaptations concerned the regulation of waste disposal in landfills, the implementation of the polluter pays principle with respect to electronic scrap, the export/import of waste and the regulation of waste disposal from mining.

When compared to other European states, Germany's waste system was not more advanced or more hygienic or "sustainable" than, for example, the British system before the 1970s. Only through legislation since then and the citizens' rising environmental awareness did (West) Germany achieve high recycling numbers. In this respect, the GDR can be considered a forerunner that implemented a system of recycling on a national scale. In the 21st century, Germany, together with countries such as Belgium, Sweden, and Austria, leads in respect to the waste amounts that are either recycled or composted.

Heike Weber
Technische Universität Berlin
Jakob Calice
Leeds Metropolitan University

See Also: Culture, Values, and Garbage; Recycling; Social Sensibility.

Further Readings

Calice, Jakob: "Garbage Recycling Rhetoric in the GDR: An Environmental Historic Perspective." *Journal of Emergence*, v.3/2 (2005). http://textfeld.ac.at/text/713 (Accessed June 2010).

Fishbein, Bette K. *Germany, Garbage, and the Green Dot: Challenging the Throwaway Society*. New York: Inform, 1996.

Lekan, Thomas and Thomas Zeller, eds. *Germany's Nature: Cultural Landscapes and Environmental History*. New Brunswick, NJ: Rutgers University Press, 2005.

Mauch, Christof, ed. *Nature in German History*. New York: Berghan Books, 2004.

Uekōetter, F. *The Age of Smoke. Environmental Policy in Germany and the United States, 1880–1970*. Pittsburgh, PA: University of Pittsburgh Press, 2009.

Gillette, King C.

Widely praised or blamed as the entrepreneur who inaugurated throwaway products, as a utopian socialist reformer, King Camp Gillette (1855–1932) spent his life excoriating capitalism's inequalities and the inefficiencies of competition. Published a decade before the patent for disposable razor blades, Gillette's 1894 book *The Human Drift* dreamed up a vast hydropowered megalopolis that would consolidate U.S. urban populations and industrial production. His experiences running the Gillette Safety Razor Company informed later monographs advocating the establishment of a single, publicly owned corporation that would produce and distribute the entirety of humanity's material needs, putting an end to war and social inequality. Gillette's business practices and socialist advocacy were both marked by a materialist pragmatism riveted on exploiting possibilities of efficiency and waste inherent in commodity chains and industrial systems.

Gillette was born in Fond du Lac, Wisconsin, and raised in Chicago, Illinois. His father worked as a patent agent after the fire of 1871 destroyed his hardware store, and young Gillette became a traveling salesman. His mother published a celebrated cookbook that perhaps inspired Gillette's lifelong preoccupation with writing. Despite financial success from his razor empire, Gillette was nearly bankrupt at the time of his death in the midst of the Great Depression.

Disposable Products

In 1895, Gillette envisioned his namesake invention while shaving with a straightedge razor. It had grown too dull to be stropped and needed honing at a barbershop. He picked up the concept of repeat-purchase disposable products in the 1890s as a salesman of William Painter' s patented Crown Cork bottle-capping system. Working with MIT graduate William Emery Nickerson, Gillette spent a decade developing thin, stamped steel blades before patenting his safety razor kit in November 1904. The product embodied ideals of genteel masculinity, cleanliness, convenience, and independence (from barbershops) while getting consumers used to planned obsolescence (manufactured objects designed to wear out and throw away). Although the blade opened the doors for all manner of disposable products, it was not the first. In 1810, Nicolas Appert innovated canning methods for the Napoleonic army's long marches that inspired Peter Durand's tin can the same year; following the success of Painter's 1892 bottle caps, Johnson & Johnson's menstrual pads were first marketed in 1896.

Now celebrated as a mythical entrepreneur, Gillette's "freebie marketing" or "razor and blades" business model of inexpensively mass-produced disposable parts earned him a fortune. With global distribution networks and production facilities in France, Germany, and England, Gillette's face, integrated into packaging design, became internationally known. A World War I government contract for 3.5 million razors and 32 million blades issued as "khaki sets" introduced a generation of men to Gillette's product. Little piles of rusty blades can still be found beneath houses built between the world wars that included small slots in bathroom walls for the safe disposal of spent razors. However, the reuse of spent razors through steel recycling does not figure into the invention's history.

Efficiency

While the success of his invention depended on waste and disposability, as a utopian socialist reformer, Gillette was obsessed with efficiency. *The Human Drift* advocated the centralization of U.S. cities and industrial production in a single, master-planned urban core called Metropolis to be located between Lake Erie and Lake Ontario and powered by Niagara Falls. Centralization would eliminate waste involved in continent-wide distribution networks while streamlining the production of life's necessities: food, clothing, and shelter. Gillette's Metropolis consisted of a honeycomb pattern of circular 25-story apartment complexes covered by glass domes. The gridded layout and prevalence of greenhouses anticipated both Ebenezer Howard's Garden Cities (1898) and Le Corbusier's Ville Radieuse (1924). Subterranean layers provided for sewerage, transportation, and food distribution. Metropolis would be the hub of a "United Company" (later called "the People's Corporation") that would outcompete capitalism through superior efficiency. Gillette conceived of the incorporated city as an enormous machine responsible for producing and distributing not only life's necessities but also individuals.

Scott Webel
University of Texas, Austin

See Also: Consumerism; Disposable Diapers; Disposable Plates and Plastic Implements; Steel.

Further Readings

Adams, Russell B., Jr. *King C. Gillette: The Man and His Wonderful Shaving Device*. Boston: Little, Brown, 1978.
Gillette, King C. *The Human Drift*. Delmar, NY: Scholars' Facsimiles & Reprints, 1976.

Gluttony

The word *gluttony* has especially Eurocentric origins; the Latin *gula*, meaning "throat" or "eating to excess"; *gluttire*, meaning "to swallow" or "gulp down"; and the 12th-century Middle English glotonie or glutonie, which refer most directly to the practice of eating or drinking in excess and in such a manner as to lose control of one's mental and physical faculties, or in fact to do great harm to the body. To eat or drink simply for the pleasure of the experience, or to withhold food from those who are in dire need of it, is also considered gluttonous. The term is inextricably related to social and political relations, religion, and spiritual practices—especially those of Christianity. Views on gluttony also reflect ever-changing ideas about discipline, the body, and the ethics of pleasure.

Ancient Greece and Rome

In the ancient Western world, Greeks and Romans embraced polytheistic mythologies and pantheons of gods that embodied complex matrices of social relations, rituals, and stories of origin. The Roman Empire was especially invested in ritual events that fostered a sense of duty and loyalty to the state and the emperor, who was in many ways a human extension of the deities. Religious festivals honoring the gods were often extravagant, public events that provided an occasion for the norms of society to be temporarily suspended, and participants were encouraged to indulge in the excesses of food, drink, dance, and all manner of conviviality. These traditional belief systems and pagan practices stood in sharp contrast to the emergent tenets of Christianity.

Early Christians

Romans generally combined elements of other belief systems—most notably that of the Greeks—into their religious practice. However, early Christian theologians openly condemned and directly refuted the festive celebrations integral to Roman religious expression. Monks, communities of men devoted to the ascetic practices of various Christian religious orders, vowed to live somewhat apart from the secular world and its trappings and relinquished any claim to worldly goods. In their earliest recorded writings, Christian monastic leaders advocated more temperate, contemplative acts of devotion, modeled after the example of the martyred Jesus Christ, whom they recognized as the only son of a singular and omnipotent God. The vices of gluttony permeated the writings of Christian monks as early as the 4th century.

During the 3rd century, the Roman government aggressively persecuted Christian extremists on the grounds that their teachings challenged the authority of the state and disrupted society. Christian worship rituals such as communion, which called for the symbolic consumption of the body and blood of Christ, were largely misunderstood and viewed with great skepticism. The average Roman also may have felt that Christian forms of devotion angered the ancient Gods and had the potential to threaten the balance of nature. Their fears sporadically materialized in the form of threats and violence against Christian practitioners; as a result, significant numbers of Christians fled Rome, seeking safety in the far reaches of the empire. Those who settled in Egypt near the Nile Delta subsisted on little sleep or food, solitary work, prayer, and meditation. Their primary objective was to follow a strict interpretation of Christ's teachings without seeking public recognition or validation of their efforts.

Writing mostly to direct their fellow brethren on the path of spiritual enlightenment through personal acts of sacrifice and prayer, monastic theologians issued lengthy treatises on ascetic practices to counter the indulgences of the flesh. A number of scholars trace formative uses of the term *gluttony* to Evagrious of Pontus, known as one of the "Desert Fathers" or "Desert Monks" of Egypt. At the time of Evagrious's writings, Egypt was an occupied territory of the Roman Empire and its main supplier of grain. The scribe dedicated the first of eight chapters in his brief treatise: "The Eight Spirits of Wickedness," to gluttony, which he characterizes as an "evil thought" that must be contemplated and combated. In his later treatise: "On the Vices Opposed to the Virtues," addressed to the monk Eulogios, Evagrious encourages his brother to combat the evil of gluttony with its antithetical virtue, abstinence.

John of Cassius further expanded the Desert Fathers' writings on gluttony in the voluminous work known as *De institutis coenobiorum* (The Institutions). In books 5–12, John issues rules on morality to help those in monastic life navigate the ethical terrain of the eight principal faults, which he lists as gluttony, fornication, avarice, anger, dejection, weariness of heart, vainglory, and pride. In Book V: Of The Spirit of Gluttony, John counsels his brethren to reference Egyptian customs and traditions as models of discipline and self-control as they prepare to do battle with the "pleasures of the palate."

He suggests that gluttonous desires can be overcome through a methodically enacted combination of fasting, participating in vigils, reading, and mentally disciplining oneself to think of food as a bodily necessity and not a pleasurable concession. He offers practical and spiritual guidance, noting that a one-size-fits-all approach does not adequately accommodate physical differences, such as health and age, which can impact individuals' capacities for restraint. Likewise, he advises those seeking to guard themselves against gluttonous impulses to be mindful of not only how much food they consume but also to guard against eating luxurious or indulgent meals, as both in his opinion have the ability to "dull the keenness of the mind."

Deadly Sin

During the 6th century, Pope Gregory the Great included gluttony among the Seven Deadly Sins, which he ranks in order of increasing severity and identifies as the crux of all other sins: pride, envy, gluttony, lust, anger, greed, and sloth. While food and drink in and of themselves are neither good nor bad, the act of indulging in them is, in Gregory's estimation, the sinful element of the vice. Citing biblical examples of resistance to gluttony (Adam resists the devil's temptation to eat of the tree of knowledge; Christ does not yield to the devil's temptation when he is fasting in the wilderness), Gregory closely aligns the sin of gluttony with the sin of lust, noting Adam's eventual act of eating the forbidden fruit after being tempted by his wife, Eve. In his opinion, the sinful nature of gluttony is its predilection toward excess in all forms.

For Gregory, gluttony poses a threat to one's spiritual balance in a number of ways; no good can come of wanting any food—especially if it is extravagant, expensive, elaborately prepared, served in an unnecessarily generous proportion, sought after too much, or eaten out of wanton desire, rather than bodily need.

Later theologians proposed the Seven Heavenly Virtues as foils to counter these vices; in this formulation, temperance—the notion of restraint and delayed gratification—serves as a counterpoint to gluttony. When Gregory sent the monk Augustine to lead a Roman mission to Kent to convert the pagan Anglo-Saxons to Christianity, Augustine, who would later become the first archibishop of Canterbury, continued to consult the pope on matters of temperance as he advised the newly converted in spiritual matters.

The Seven Deadly Sins and Heavenly Virtues figure prominently in medieval philosophy, theology, art, literature, and everyday religious life. Thomas Aquinas equates the inordinate and irrational desire for food with spiritual depravity and takes great pains to distinguish between the innate biological or "natural" need to eat for sustenance and the desire to eat for pleasure, condemning the latter. Throughout the Middle Ages, religious canon laws designated certain days for ritual fasting—reverent occasions of personal sacrifice during which all Roman Catholics were to abstain from sex and meat ingestion. These "lean days" initially included all Fridays, the 40 days of Lent, the days before feast days (known as vigils), and other religious observances, which equated to more than half of the days of the year.

Gluttony

From the 12th through the 14th centuries, the notion of enjoying one's food too much was considered especially gluttonous. As a result, poorer people steeled themselves against the pleasures of the plate, whereas affluent folk unapologetically indulged in considerably more lavish meals. During this era, the glutton became an archetype of sorts, alternately personified by a pig or a corpulent man stuffing himself to the point of delirium and incontinence.

The Pardoner introduces readers to all manner of gluttonous consumption in the prologue to Chaucer's *Canterbury Tales*; a three-headed monster terrorizes the gluttons relegated to the muddy grounds that comprise the third circle of hell in Dante's *Inferno*. Conversely, the chivalrous knight and the virtuous woman emerged as models of temperance in literary and visual arts. Never greedy or drunk, the knight embodied all that is fair and just. In a similar fashion, the image of a woman pouring water from a pitcher became a popular illustration of the virtue of temperance.

In the 21st century, gluttony continues to yield insights into beliefs and attitudes about health, body image, and other cultural markers. Unlike the medieval glutton, a modern-day obese person may simply eat too much of the wrong foods and lack exercise.

Renaissance

A number of scholars have compellingly argued that attitudes toward gluttony began to undergo a significant transformation as a result of the shift in thinking inspired by the humanist movement during the Italian Renaissance, which began in Florence and emanated throughout Europe. Humanist thinkers took a renewed interest in the pagan classics and rituals of ancient Greek and Rome as models for how to live an ethical and fulfilling life. Renaissance art became more lifelike and engaged pleasurable themes and aesthetic practices; food was considered among these.

As the Renaissance spread, so did revolutionary changes to cuisines throughout Europe. When the adolescent Italian princess Catherine de Médicis came to Paris as the wife of King Henry II, she brought with her an entourage of Florentine cooks, expertly trained in the fine arts of Renaissance cooking. A number of French soups, sauces, and other culinary delicacies are attributed to the Italians. Catherine also brought with her to the French table a manner of refined aesthetic sensibilities, which served to enhance the dining experience. Under her hospitality, "ladies," who except for special occasions previously had dined in private, were admitted to the royal court on a regular basis. Décor and table settings were meticulously fashioned and selected by artists, and beverages were served in fine Venetian crystal stemware. Meals were served and eaten from artfully glazed dishes especially crafted for the occasion. During the medieval era, such an elaborate manner of dining would have been considered gluttonous.

Modern Conceptions

In a similar fashion, the technological innovations of the Industrial Revolution and 18th-century rationalism are credited for refocusing attention from the heavenly realm to that of the more immediate materialism of the lived experience. Throughout history, class differences, not religious edicts, often determined who ate what.

Landless agricultural workers in the Old World often raised crops for export or for the upper classes; poor laborers usually had limited control of and access to provisions for personal consumption. Poor Europeans often relied upon the benevolence of some of their more affluent countrymen to put food on their tables.

However, it was not until 1962 that progressive bishops acting on behalf of their local constituencies earned the right to modify rules of fast and abstention for Catholics in their dioceses. A combination of Vatican II Council reforms and the 1966 Apostolic Constitution Paenitemini of Pope Paul VI on fast and abstinence further relaxed the strict Roman Catholic fasting and abstinence requirements according to local social and political conditions.

In the 21st century, gluttony continues to yield fruitful insights into beliefs and attitudes about health, body image, and other cultural markers of identity, such as race, class status, gender, and nationality. Those who are obese do not necessarily fit the image of the medieval glutton; they may not necessarily be obsessed with food but may simply eat too much of the wrong kinds of foods and not get enough exercise.

Likewise, those with compulsive eating disorders may obsess about food not because their souls are bereft but because of deep-seated emotional and psychological turmoil. In some cultures, the corpulent body is both a sign of wealth and a marker of social obligation to those in need. Fascist dictators, drug lords, and corporate "fat cats" perhaps embody the image of the modern glutton—those who have more than what they need have acquired it at the expense of those who do not have enough, and appear to be wasteful in their consumption habits.

Not unlike ancient engagements with the concept, gluttony remains a source of fodder for philosophical, religious, and artistic musings.

Lori Barcliff Baptista
University of Chicago

See Also: Consumption Patterns; Culture, Values, and Garbage; Food Consumption; Garbage, Minimalism, and Religion.

Further Readings

Bickerton, E. *The Desert and the City*. Stonville, Australia: Eloquent Books, 2009.

Cassian, J. "The Twelve Books of John Cassian on the Institutes of the Coenobia and the Remedis for the Eight Principal Faults." http://www.osb.org/lectio/cassian/inst/inst5.html (Accessed July 2010).

Fisher, M. F. K. *The Art of Eating*. New York: Vintage, 1976.

Prose, F. *Gluttony: The Seven Deadly Sins*. Oxford: Oxford University Press, 2006.

Tillby, A. *The Seven Deadly Sins: Their Origin in the Spiritual Teaching of Evagrious the Hermit*. London: SPCK Publishing, 2009.

Goodwill Industries

Goodwill Industries is a network of 180 affiliated organizations in the United States, Canada, and other nations, organized under Goodwill Industries International, Inc. Its mission is to provide employment opportunities to those who have disabilities, have a lack of education or job experience, or face other challenges. Throughout its history, Goodwill has defined its social role as far greater than the trade in used goods. However, as one of the largest and oldest nonprofit reuse organizations in the country, it stands at the vanguard of this method of waste reclamation. Nearly two-thirds of Goodwill's revenue comes from the collection and sale of donated clothing, household items, and furniture in its 2,400 thrift stores. In 2009, it took in close to one billion pounds of material that might have gone to disposal, using proceeds to serve close to 2 million individuals with employment and training, and spending 83 percent of its $3.7 billion revenue directly on programs.

Beginnings and Goals

Goodwill was founded in 1902 in Boston, Massachusetts, by the Methodist clergyman Edgar Helms. Helms had come to Boston in the late

1890s to minister to Polish, Italian, and German immigrants, many of them converts from Catholicism. He settled at Boston's South End Memorial Chapel, which had been established in 1859 by Henry Morgan, a renegade Methodist missionary. In 1900, Helms began to go door-to-door to collect unwanted clothing and household goods from wealthy Bostonians. According to official histories, parishioners were too proud to accept the items as handouts, preferring instead to refurbish and sell them. This formed the core of the Goodwill mission: to aid the poor via the redemptive quality of work, offering a hand up, not a handout. This model of welfare provision had strong roots in Protestant Christian doctrine, which emphasized vocation as central to the individual's relation to God. It contrasted with Catholic models of charity premised on giving alms. In this regard, Goodwill, along with other Protestant charities originating in this period, brought religion, welfare, commerce, and recycling of materials together in a working institutional model.

Spread and Secularization

In the following decades, hundreds of affiliated Goodwills were established across the United States under the loose administration of the Goodwill national office. After World War I, as disabled servicemen returned home, the organization began outreach to the physically handicapped. Goodwill worked closely with the National Recovery Association throughout the Great Depression and supported federal initiatives during World War II to gather small quantities of scrap to support the nation's war economy. Goodwill began a long process of secularization in the 1940s, both as a result of factionalism within Methodism and in response to federal funding requirements. It prospered in the 1950s and 1960s as booms in material production channeled increasingly short-lived goods toward donation.

Recycling

Goodwill's relationship to recycling since the 1970s has been complex. Prior to 1970, Goodwills had at times collected both reuseable and recyclable commodities, competing with the for-profit rag and paper scrap industries. With the emergence of urban recycling as an activity whose goals included saving the Earth, along with job creation and urban order, Goodwill began to redefine some—though not most—of its mission to include environmental considerations. Throughout the 1970s, Goodwills in some states established recycling drop-off centers for glass, metal, and plastic containers. These programs were small-scale and short-lived as markets for recycled commodities plummeted through the 1980s. By the latter part of that decade, municipal curbside recycling programs were beginning to proliferate, as were container deposit laws in some states, making Goodwill's involvement in recycling collections redundant.

Reuse

Goodwill pioneered institutionally organized reuse as an alternative to disposal. In the early 20th century, woodworking and sewing had been core skills for vocational program participants and were an important step in revalorizing discards. Some Goodwill stores continued to train clients in furniture and equipment repair through the early 1990s. Donation guidelines in the 21st century explicitly require all items to be clean and in complete working order. The demise of clothing and product repair as a trade, in tandem with the growth of irreparable furniture and other household goods, was in keeping with overall trends of low-quality mass production and planned obsolescence. Such trends rendered the repair aspect of Goodwill work obsolete. At the same time, Goodwill began taking in increasing quantities of used goods each year and began to stand as a model for community-based social enterprises focused on reuse.

The organization has embraced the information technology revolution in both virtual and practical ways. In 1999, it launched an online auction site, with revenue growing from $63,000 that year to over $10 million in 2009. In 2004, a Goodwill in Austin, Texas, established "Reconnect" in partnership with Dell Computers to collect used residential electronics. By 2010, some 1,900 shop locations had joined the partnership, with an estimated 96 million pounds of e-waste recycled since the program's founding. Goodwill's online impact calculator estimates diversion of close to 1 billion pounds of usable goods from landfills in 2010 alone. On

April 22, 2010, the 40th anniversary of Earth Day, Goodwill president and CEO Jim Gibbons noted the organization's long history as an environmental pioneer and its role in diverting materials from disposal, continuing the organization's secular role uniting social services, retail commerce, and materials recovery in response to evolving societal needs and economic conditions.

Samantha MacBride
New York University

See Also: Consumption Patterns; History of Consumption and Waste, U.S., 1900–1950; History of Consumption and Waste, U.S., 1950–Present; Salvation Army.

Further Readings

Goodwill Industries International. "About the Donate Movement." http://donate.goodwill.org (Accessed June 2010).
Goodwill Industries International. "Our Mission." http://www.goodwill.org/about-us/our-mission (Accessed June 2010).
Lewis, J. F. *Goodwill: For the Love of People.* Washington, DC: Goodwill Industries of America, 1977.
Strasser, S. *Waste and Want: A Social History of Trash.* New York: Metropolitan Books, 1999.
Weber, M. *The Protestant Ethic and the Spirit of Capitalism.* New York: Penguin Books, 2002.

Greece

From its roots as a classical civilization, modern Greece has developed into one of the more developed economies in the Mediterranean. Garbage in Greece has been a central subject of debate between various officials, members of parliament, European Union (EU) representatives, citizens, scientists, and environmentalists. Various difficulties in the treatment and disposal system—in combination with the rising consumption of mass commodities, especially after the 1960s—the development of the Greek industry, the increase of the general population, and the pollution of the Mediterranean basin are the main concerns of the agents involved. While the debate has mainly been focused on garbage as consequence of the development of the country and only technological and organizational solutions have been proposed, there is a need for an understanding of garbage as a cultural, political, and economic phenomenon. A major problem has been the divergent views between environmentalists, communities, and state officials. The Greek informal garbage economy, for example, has been largely neglected as a source of collection, management, and recycling of waste.

In the 21st century, Greece enjoyed an average annual growth of almost 4 percent before a debt crisis in 2010 led to austerity measures and social unrest. Over 10 million people live in Greece, about one-third in the Athens urban area. As a member of the European Union, Greece has benefited from the influx of European cohesion funds and subsidies. A significant part of EU cohesion funds and public investment, as well as major public works for the 2004 Olympic Games, were dedicated to Greece's infrastructure network. Environmental projects have been initiated, including various efforts to invest in alternative energy, as both the government and businesses have to take certain measures to meet the country's obligations to the EU. In particular, recycling, solid and hazardous waste treatment, and alternative energy sources in Greece fall short of EU averages.

Hazardous Waste

A new legislative framework for hazardous waste treatment was adopted in 2005. As such, a set of specific, systematic measures for dealing with these waste streams still remained to be implemented as of 2010. In Greece, there are very few hazardous waste treatment and final disposal sites, which are not sufficient for dealing with existing demand. According to the dominant political views, there is a strong need for new technologies and transnational collaborations.

As is the case with hazardous, industrial, or hospital waste management, there is also organizational demand and significant space for improvement in the field of animal by-products waste treatment and management. Finally, as the agro-food sector is one of the major industrial sectors in the country, organizational systems are needed for the treatment of food-related waste streams for the production of usable energy.

Waste Management

Greece is part of the developed European Union, but the amount of garbage produced is relatively low in relation to other European countries, especially because the Greek industrial sector is one of the smallest in Europe. Moreover, the average per capita income of Greeks is much lower in comparison to other European countries, and average consumption of mass commodities is lower as a result.

A major factor in the increase of garbage production has been population. The population in Greece, according to the 2001 census, is 11,275,312, and each inhabitant generates 411.5 kg of municipal waste per year (1.12 kg per day). Despite the fact that the birthrate is low, urbanization and immigration have contributed to an increase in the population density, mostly in the urban centers of Athens and Thessalonica. Almost half of the country's population dwells in these two cities.

Waste management in Greece has been upgraded since the 1990s, mostly in urban areas and in some large parts of rural areas. More recently, significant improvements have been accomplished in terms of facility development, collection, and recycling. However, the management of municipal solid waste (MSW) in Greece needs to be further improved in order to achieve the quantitative targets posed in the European Union Directives, with the landfill directive especially requiring a restructuring of many components of the waste management system, as Greece relies on landfilling for over 90 percent of its waste.

Various regional waste management plans foresee the construction of mechanical biological treatment plants, but many of the proposed projects had not entered the actual planning phase as of 2010. The possibility of revising these waste management plants to include other options, such as thermal treatment or source separation, has been taken under consideration. In the early 21st century, there are no facilities to process source-separated organic waste. As far as composting is concerned, no source separation schemes are in place for the organic fraction of MSW, therefore, there are no composting facilities producing quality compost. The very low charges for disposal do not act as an incentive for the implementation of other options, such as recycling and composting. The Greek Ministry for the Environment, Physical Planning and Public Works reports that the annual generation of solid waste in Greece is exceeding 4 million tons per year. The overall production of MSW in Greece is estimated around 4,600,000 tons per year. New sanitary landfills and treatment plants, as well as composting facilities and waste transfer stations, had been constructed as of 2010, but there are many more to be completed. For example, there is significant potential in the creation of landfill gas plants. Incineration is not broadly accepted and is illegal. Greece is one of the few countries in Europe in which incineration plants are not yet operating. At the same time, complete recycling is not applied because of infrastructure and logistics insufficiencies.

The most serious solid waste pollution problem is in the greater area of Athens, with a population close to four million inhabitants. In 2003, the Greek government decided to implement a development plan that includes the building of new sanitary, legal waste burial sites and waste processing centers in Attica. Similar projects are planned for Thessalonica, the major cities on the island of Crete, and many other cities throughout Greece. In addition, the Ministry of Environment has put forward legislation regarding recycling packaging materials and more specifically glass, plastic, metal, and paper and promoting new waste management methods, such as converting decomposing refuse into biogas to produce electricity. This legislation has supported two major projects in the area of Attica: the construction of a new recycling facility and the development of a biogas station. Establishing more recycling facilities and landfills is among the main priorities of the state.

Pollution and Environmental Protection

A problematic view of environmentalists and conservationists as well as policy makers is related to an understanding of nature as separate from society. An example of such policies comes from the conservationists and state organizations that tried to initiate a project in Lake Prespa in northern Greece in order to protect a rare species of pelican. The project initiated the removal of a few small settlements of fishermen in the area, including their waste disposal sites, so no human intervention would take place in the "wild" protected area. The results of this policy were devastating when the settlements were moved, as the garbage disposed by the fishermen fed the birds. Garbage is not necessarily viewed as polluting by all social groups but rather can be a source of income. Greek Roma travel regularly around the countryside in order to collect garbage that has not been removed by municipal authorities. Old cars, household devices, iron, steel, and plastic are collected on a regular basis and resold to various companies to be recycled. Another divergent view of garbage has been established between local communities and state officials. Local communities have opposed large environmental projects, including the wind turbine industrial plan in parts of Greece, mainly because there is a significant concern over the scale and maintenance of such projects. As a result, European Union and environmentalist views of preserving the environment and reducing pollution contradict the views of various Greek communities that realize these efforts as production of pollution and garbage in their own communities. Past failed projects and poor or nonexistent maintenance of various environmental efforts that were caught up in state politics legitimize such views.

Therefore, many environmentalists and state officials consider the environment and garbage as separate from society. The role of Greek Roma, for example, or other social groups who participate actively in the informal garbage economy is widely neglected. As a consequence, environmentalists and policy makers have been proposing technological solutions in order to deal with the treatment and disposal of garbage. However, such plans have resulted in conflicting views with local societies that are either obliged to live with the disposal of garbage close to their communities or face the failure of environmental policies.

Finally, a view of garbage as a social, cultural, and political process might be able to bring more insights into its production, consumption, and disposal. A focus solely on the organizational and technological aspects of waste treatment has not been as productive in Greece; on the contrary, it has resulted in various social conflicts, divergent views, and misconceptions.

Tryfon Bampilis
University of Leiden

See Also: Environmentalism; European Union; Politics of Waste; Power Plants; Recycling; Waste Treatment Plants.

Further Readings

Karapostolis, V. *Consumption Behaviour in Greek Society 1960–1975*. Athens: EKKE, 1983.

Klok, W. and K. Blumenthal. "Environment and Energy: Generation and Treatment of Waste." *EUROSTAT* (2009).

Theodossopoulos, D. *Troubles With Turtles: Cultural Understandings of the Environment on a Greek Island*. New York: Berghahn Books, 2003.

Greenpeace

One of the most visible environmental activist organizations in the world, Greenpeace has engaged in several campaigns against nuclear arms, toxic waste, water pollution, and global climate change. It has advocated zero waste practices in developing nations across the world. Its methods are nonviolent, but confrontations with several nations (among them Japan and France) and industrial interests have aroused both controversy and acclaim over the organization's history.

Greenpeace was formed in 1971 when some members of the Quaker faith in Vancouver, Canada, decided that they wanted to protest the underground testing of nuclear bombs taking place on the tiny island of Amchitka off the west coast of Alaska. One of the tenets of the Quaker faith is to "bear witness," or to observe situations that one feels are morally wrong. The small group of activists, concerned about the consequences of nuclear tests, sailed off in a boat named the *Phyllis Cormack*, but they did not make it to the Amchitka site before the nuclear test. Far from being a failed endeavor, however, the attempt to witness the nuclear test garnered so much media, public, and political attention that the activists realized that the attention to the issue of nuclear testing was, in itself, a success.

At a meeting soon after the *Phyllis Cormack*'s sailing, someone left the room saying, "peace" and someone else replied, "make it a green peace," and the international environmental organization Greenpeace was born. In the 21st century, Greenpeace continues to maintain as one of its core values "bearing witness to environmental destruction in a peaceful, nonviolent manner." It has grown around the world, according to the Greenpeace international Website: there are Greenpeace offices in 41 countries, almost three million members, and a long list of victories. For example, the attention that the first Greenpeace sailing garnered, along with the subsequent work, was successful in pressuring the United States to abandon its Amchitka nuclear testing site in 1972.

Exposing Issues of Waste

Greenpeace's real strength has been in exposing consumption and waste-related issues that would otherwise be hidden from view. For example, Greenpeace's campaign against the killing of whales (a practice that often takes place in remote ocean waters) is perhaps its best-known campaign and remains a central focus for the organization. As whaling ships have aimed their harpoons at the whales, Greenpeace activists have positioned their small inflatable boats between the whales and the harpoons. In 1982, Greenpeace was instrumental in pressuring the International Whaling Commission (IWC) to adopt a whaling moratorium, but Greenpeace continues to fight to stop whaling, as one of the most prolific whaling countries, Japan, continues to both kill whales and lobby to have the IWC's moratorium rescinded.

Another well-known Greenpeace campaign has addressed a devastating practice that has taken place throughout modern history: the use of the world's oceans as a waste dump. This practice has included the dumping of radioactive nuclear waste into the remote—and seemingly endless—expanses of the world's oceans. Greenpeace has drawn attention to such dumping by having activists place their inflatable boats in the path of the barrels of waste that sailors try to dump overboard. As these actions have drawn attention, the media, members of the public, and politicians have begun to pay attention. In 1983, the parties to the London Dumping Convention called for a moratorium on radioactive waste dumping at sea. According to Greenpeace, this marked the first year since the end of World War II that no radioactive waste was dumped at sea. Subsequently, in 1993, the London Dumping Convention created a permanent worldwide ban on the dumping at sea of radioactive and industrial waste.

One of the most notorious chapters in Greenpeace's history also relates to the campaign against ocean dumping of toxic and radioactive wastes. During the 1980s, the organization used its *Rainbow Warrior* ship in the Pacific Ocean to assist in the evacuation of 300 Rongelap Atoll residents from their community that had been contaminated by U.S. nuclear tests during the Cold War. Subsequently, the vessel was to help lead protests against French nuclear testing in 1985, but it was sunk in Auckland Harbor, killing Dutch photographer Fernando Pereira. An investigation by New Zealand police revealed the boat had been bombed on orders of the French government, ultimately resulting in both financial compensation and unprecedented international publicity for Greenpeace.

These Greenpeace campaigns against nuclear testing, whaling, and ocean dumping represent a small selection of the consumption and waste-related work that Greenpeace performs as part of its six core campaigns of climate change, forests, oceans, agriculture, toxic chemicals, and ending nuclear power. Greenpeace continues to use the tactics of nonviolent action to draw attention to the issues, but its work also includes political lobbying, public outreach, and scientific testing.

Greenpeace's development of campaigns within and across national boundaries represents an important historical precedent for the work of environmental nongovernmental organizations worldwide. Its campaigns to establish zero waste programs in nations ranging from Argentina to Lebanon are important shapers of 21st-century waste policies.

Jennifer Good
Brock University

See Also: Consumerism; Ocean Disposal; Overconsumption; Radioactive Waste Disposal; Zero Waste.

Further Readings

Dale, S. *McLuhan's Children: The Greenpeace Message and the Media*. Toronto: Between the Lines, 1996.

Good, J. "Shop 'Til We Drop?': Television, Materialism and Attitudes About the Natural Environment." *Mass Communication and Society*, v.10/3 (2007).
Kasser, T. *The High Price of Materialism*. Cambridge, MA: MIT Press, 2002.
Millennium Ecosystem Assessment. *Living Beyond Our Means: Natural Assets and Human Well-Being*. Washington, DC: World Resources Institute, 2005.

Grocery Stores

For many people, grocery stores serve as the primary—if not the sole—outlet by which they acquire food. Grocery stores create the illusion that food is easily and readily available. Grocery stores sell a variety of household goods, including both fresh and packaged foods. Climate-controlled grocery stores complete with refrigeration and freezer units are often disconnected from the conditions under which food was grown, harvested, processed, packaged, and shipped. In some instances, the quality of foods is measured not in terms of taste or freshness, but rather by the extent that they can endure being shipped hundreds or thousands of miles.

Small grocers evolved from food peddlers in many cities in the early 20th century. In several U.S. cities, stores developing within local immigrant populations were commonplace by the 1930s, and small chains were evident as early as World War I. As the United States suburbanized after World War II, large grocery stores became staples in shopping malls, allowing consumers to drive and park, taking care of their shopping needs once a week, rather than purchasing small amounts of food daily. This development substantially increased the convenience and flexibility of housewives to provide food for their families, as both the volume and variety of goods in supermarkets grew in the postwar era. Large grocery stores have been described by some commentators as contributing to the undermining of local food systems, particularly small grocers who are unable to compete with large grocers. Larger grocers have the advantage of being able to buy large quantities of food and hence receive large-volume discount prices. The savings are passed on to consumers, who usually choose the larger store in greater frequency to the smaller store. This decrease in price also means that consumers are able to spend less of their income on meeting their caloric needs, hence freeing up more income for either necessities or other consumables.

Food Deserts and Locations

As an alternative to this narrative of large grocers dominating the market and pushing smaller stores out of business, some commentators have touted large grocery stores as the solution to food deserts. Although there are tens of thousands of grocery stores in the United States alone, many people and communities may find themselves without ready access because grocery stores—particularly larger grocery stores—are not evenly distributed in relation to the population. Food deserts are those areas where residents have limited to no access to fresh produce. These food deserts are mapped onto landscapes so as to make claims about relative food accessibility, and hence food security, in both rural and urban regions. By focusing solely on the presence or absence of large grocery stores, these studies discount the importance of local and regional sources of food, which can also provide high-quality foods in terms of nutrient-to-calorie ratio.

Local and regional sources of food include farmers markets, community-supported agriculture operations, box schemes, community gardens, neighborhood gardens, school gardens, household gardens, farm stands, local butchers, and smaller grocers. Informal networks of sharing and exchange are often not considered in analysis of food deserts. These studies of food deserts have contributed to a corpus of knowledge to argue for the development of new, large grocery stores as the solution to food deserts and inequities in the food system in terms of nutrition and food availability. This has led to a legislative push to enact economic incentives to increase the number of large grocery stores being built. As an example, Maryland passed a law to allow grocery stores to be built tax-free in given locations.

The location of grocery stores is of particular importance to those who do not have access to personal transportation and instead have to rely upon public transportation infrastructure or social networks for transportation needs. People without reliable access to transportation must rely on whatever food stores are available in their neighborhoods; for those in poorer neighborhoods, these are often convenience stores, which stock little—if any—fresh foods. In some cases, the relative absence of larger grocery stores is due to historical socioeconomic shifts in neighborhoods and the construction of large grocery stores in suburban areas has also led many smaller grocers in urban settings to go out of business.

Nutritional Quality and Price

It is not simply the mere absence or presence of grocery stores that determines nutritional outcomes and options for a population. The foods found on their shelves, as well as the price, also determine the food options available to consumers. The presence or absence of a greater variety and selection in terms of fresh, nonprocessed foods in comparison to what Anthony Winson refers to as "pseudo-foods," or heavily processed foods, also impacts the overall availability of high-quality foods in terms of nutrition, taste, and preference. As has been shown by Winson, the quality of grocery stores, in terms of their selection of low-nutrient, high-calorie foods compared with nutrient-dense foods, varies even in the same city. As Winson has demonstrated in Canada, it is generally the poorer regions of a given area that have a greater concentration of pseudo-foods in their grocery stores.

The goods present in grocery stores rely upon corporate infrastructure, including shipping and handling of fresh and processed foods over long distances. In some instances, this transportation, particularly to more remote areas, can compromise the overall quality of fresh foods. In addition, higher food miles (the average distance that food is transported) increase the price of foods for more remote areas.

As an example, foods in urban Alaska are generally 15–30 percent higher in price than in the continental United States, and food prices in rural Alaska can be 200–300 percent higher than they are in the continental United States. Since the presence of food is reliant upon transportation infrastructure and fossil fuel consumption, the price of foods fluctuates with the price of oil.

Effects on Health and Culture

The arrival of grocery stores in a region can dramatically impact the health of people. Often, the arrival of grocery stores marks a key transitional stage for rural communities. The shift from a subsistence-based economy to one revolving around the purchase and consumption of foods from grocery stores can, over time, contribute to marked changes in the health of differently situated populations. In some instances, the utilization of grocery stores has led to a movement away from traditional subsistence activities, which require extensive labor to acquire food resources. The utilization of grocery stores as a part of integration into the cash economy meant that some populations spent less time in food procurement and, with this, a decrease in caloric expenditure. This decreased involvement in subsistence activities has led to shifting cultural understandings over generations, which contributes to a loss of previously shared, traditional, ecological knowledge concerning subsistence resources. At the same time, grocery stores have, in some instances, been able to serve local communities by selling traditional foods. Regulation as well as centralization of food distribution has kept many grocery stores from realizing their full potential to sell such foods. As centralized units of food distribution, grocery stores, while adapting to meet the regional demands for foods, often do not carry locally grown, lesser-known foods. However, as consumer demand calls for specialty products and labeling such as third-party certifications for Fair Trade or organic, grocery stores not only carry these items but also dedicate and arrange specific spaces or sections for these foods.

Marketing

Grocery stores and the products in them are not just passive spaces and objects of mere consumption but are rather spaces and objects that engage consumers. Grocery store managers and food industry executives help fashion consumer perception through the deployment of advertising, specials, and product availability, and these products are also themselves fashioned by consumer perceptions and preferences. This has not been lost on grocery store managers, marketing firms, and corporate executives who attempt to engage the consumer in a variety of ways. Perhaps the most prevalent of these mechanisms and the least intrusive is the utilization of rewards or loyalty cards, which have been adopted by every major grocery store and provide valuable information to grocery store managers and chains. These individual records of purchases are utilized to perform sophisticated analyses of consumer shopping preferences and patterns. This is done in hope of adjusting marketing strategies and store selection in order to increase profits. At the same time, there is a positive feedback loop wherein grocery stores attain a greater share of the market and enter into the process of assessing and predicting consumer preferences and shopping patterns.

Hence, grocery stores are not passive spaces where consumers make selections based on what is available, but rather they are active in selecting what products the grocery stores sell. Grocery store managers and executives deliberately try to create a pleasant shopping experience in higher-end retail settings with lighting, music, free samples, and in-store dining options all designed with the intent of extending the length of time shoppers will spend in the store. In some cases, specialty grocery stores, such as natural food stores, go well beyond this by offering cooking classes, holistic health speakers, or in-store mini-massage therapy and bodywork sessions.

There is not, however, infinite choice as the range of potential products is rather narrow, with many products being very similar, with only slight variations of one another. In addition, grocery store chains buy large quantities of foods, hence products that may be highly desirable to a small percentage of shoppers might not necessarily be available. At the same time, grocery stores have developed their own brands, often undercutting the price of national brands for oft-consumed food products.

Some commentators have claimed that large grocery stores undermine local food systems, especially small grocers who are unable to compete. One of the biggest threats to the growth of grocery stores comes from the rise of supercenters and warehouse stores, where bulk purchasing allows retailers to lower prices on a variety of products traditionally purchased at grocery store outlets. The savings are passed on to consumers, who tend to choose the larger store more often than the smaller store.

Prices

The economies of scale drive down the price food processors pay to producers and ultimately lower the profit-margin on the farm. At the same time, this has resulted in a decrease in the real price of foods for consumers, although the prices are typically higher in smaller grocery stores. The "price wars" between grocery stores has also contributed to decreased food prices paid by consumers. While analysis of these general trends is possible, in-depth individual consumer preferences are generally regarded as proprietary business information and as such are usually only available to company insiders and not to outside researchers.

Uniformity and Consolidation

The corporations that own grocery stores and other aspects of the food system are increasingly able to exert greater control over the food supply of those who do not produce their own foods. This is particularly true in both Europe and the United States, where a handful of corporations are responsible for a greater percentage of total food sales each year. In order to produce a uniform product, stores and processors dictate the types of plants and animals to be raised and the conditions under which they are grown. This emphasis on uniformity leads to increased vulnerabilities within agricultural systems, which can lead to increased on-farm consumption of chemicals and other farm inputs with their resulting waste issues.

A food system that rewards uniformity, scale, consolidation, and ready availability of globally produced goods calls into question the future of grocery stores. One of the largest threats to traditional grocery stores comes from the rise of supercenters and warehouse stores, whose prices are markedly lower than those of traditional grocery stores for dairy and a variety of other products. Grocery stores led a shift in the way food has been distributed but they are not permanent fixtures. Increasing consolidation by retailers has led to an increase in the variety of goods sold by each store; this, coupled with Internet-based sales of food, has the potential to dramatically alter the distribution of food in the future. In addition, the alternative food movement, including alternative distribution mechanisms, is challenging grocery stores, albeit to a limited extent economically. The more serious challenge to the current paradigm of food distribution via grocery stores is that these alternative food movements may present an ideological shift in the future wherein consumers continue to demand not only safe and nutritious foods but also locally and community-grown foods.

Large grocery operations such as those found in Walmart stores can pressure suppliers into affecting the cost and quality of their food. This has often led to a "race to the bottom" decline in the quality of foods on shelves. In 2010, Walmart announced a sustainability initiative that included favoring small, local farms and reducing the environmental impact of farming. Time will tell if this initiative reverses or slows environmental damage associated with large grocery stores, but at the very least, it represents heightened awareness of unsustainable food production and distribution practices.

David Fazzino
University of Alaska

See Also: Carbon Dioxide; Certified Products (Fair Trade or Organic); Dairy Products; Farms; Food Consumption; Organic Waste; Slow Food; Sustainable Development.

Further Readings

Deutsch, Tracey. *Building a Housewife's Paradise: Gender, Politics and American Grocery Stores in the Twentieth Century*. Chapel Hill: University of North Carolina Press, 2010.

Fazzino, D. and P. Loring. "From Crisis to Cumulative Effects: Food Security Challenges in Rural Alaska." *NAPA Bulletin*, v.32 (2009).

Hawkes, C. "Dietary Implications of Supermarket Development: A Global Perspective." *Development Policy Review*, v.26 (2008).

Winson, A. "Bringing Political Economy Into the Debate on the Obesity Epidemic." *Agriculture and Human Values*, v.21 (2004).

Hanford Nuclear Reservation

The Hanford Nuclear Reservation (Hanford), the largest nuclear waste site in the Western Hemisphere, is located in the state of Washington. Maintained by the U.S. Department of Energy (DOE) since having been taken over as part of the Manhattan Project, by the time production stopped in the 1980s, Hanford had made most of the plutonium produced in the United States.

Hanford reactors produced plutonium for the United States' defense program for more than 40 years (1944–87). In addition to the liquid and solid waste generated from the production of plutonium, the facilities and structures associated with Hanford's defense mission must also be deactivated, decommissioned, decontaminated, and demolished.

History

The area along the Columbia River has been home to several, typically nomadic Native American tribes, including the Nez Perce, Umatilla, Wanapum, and Yakama. In the mid-1800s, pioneers and settlers began to arrive, eventually forming the town of White Bluffs, home to approximately 900 people by the 1940s, and the small community of Hanford. In 1943, under eminent domain, residents of White Bluffs and Hanford were given 30 days and a small payment to evacuate for the purposes of "important war work." The war department needed a remote area with access to cold water and electricity (to be provided by the recently completed Grand Coulee hydroelectric dam) to develop atomic weapons.

After the residents of White Bluffs and Hanford had been evacuated, the war department began recruiting workers to build the nuclear reactors and processing facilities required to produce plutonium for atomic weapons. Ultimately, a 51,000-person workforce was formed, creating the fourth-largest city in Washington at the time, with few of the workers having knowledge of what was being built or what the completed facilities would do.

Three areas were formed: the "100 Area" for transforming uranium into plutonium, the "200 Area" for plutonium processing and waste storage, and the "300 Area" for manufacturing work and experiments. Workers began building the first three of what would ultimately be nine plutonium production nuclear reactors at Hanford, the first of their kind in the world. Within 13 months, work was completed on the B Reactor, the world's first nuclear reactor, as well as the T Plant, the world's first facility to extract plutonium from irradiated fuel rods. In

August 1945, plutonium from the B Reactor and the T Plant was used in the Fat Man bomb detonated over Nagasaki, Japan.

In 1959, construction began on the ninth Hanford reactor, the N Reactor, with the dual purpose of producing plutonium for atomic weapons and steam for generating electricity. N was the only dual-purpose reactor in the United States. For two years, from 1963 to 1965, all nine reactors were producing plutonium for the U.S. defense program. By the mid-1960s, some of the older reactors began to be shut down, with B, C, D, DR, F, and H Reactors being deactivated by 1970. The K-East Reactor ceased production in 1970, followed by the K-West Reactor in 1971. The N Reactor continued producing both plutonium and electricity until 1987, when the Department of Energy placed N in a standby status and it has not been reactivated.

Cleanup Initiatives

That which is considered solid waste is broad, ranging from broken reactor equipment to contaminated clothing. Solid wastes were buried in the ground in pits or trenches, with or without containment (steel drums or wooden boxes), and, depending on when the waste was buried, records about what was buried and where are highly variable.

In addition to millions of tons of solid waste, hundreds of billions of gallons of liquid waste generated during the plutonium production days were intentionally disposed of by pouring them directly onto the ground or into trenches or holding ponds. Unintentional spills of liquids also took place. As with the solid wastes, records of spills, including what, where, and volume, are highly variable.

A majority of the solid wastes, contaminated soil, and building debris will be taken to the Environmental Restoration Disposal Facility (ERDF) located on the Hanford Site and regulated by the Environmental Protection Agency. This includes some of the more hazardous chemical or radioactive solid wastes stored in the Canister Storage Building at Hanford. Solid transuranic (TRU) waste—debris that is contaminated with plutonium or other materials that may remain radioactive for hundreds of thousands of years—is packaged and shipped to the Waste Isolation Pilot Plant in New Mexico. Much of Hanford's liquid waste will ultimately be vitrified (transformed into a stable glass product) at a facility being constructed at Hanford.

Liquid waste that had been poured onto the ground or held in ponds or trenches has caused soil and groundwater contamination. Spreading at variable speeds (depending on chemical composition and rock and soil type), plumes require a variety of containment strategies. Remediation techniques include barriers and biostimulation, a new technology in which organic materials are pumped into the ground to be eaten by soil microorganisms, thereby altering the chemistry of the groundwater and rendering the contaminants harmless to the environment.

Conclusion

Government officials recognize that they still have a weak grasp of how much plutonium is contaminating the environment as a new analysis indicates that the amount of plutonium buried at Hanford is nearly three times what the federal government previously reported. This suggests the need for a more challenging cleanup initiative, perhaps requiring technologies that were not created as of 2010. While the DOE has been weighing to what degree remediation should occur, given the relative cost and extent of devastation, the preferred option is 99 percent remediation.

Leslie Elrod
University of Cincinnati

See Also: Environmental Protection Agency (EPA); Nuclear Reactors; Radioactive Waste Disposal; Radioactive Waste Generation; Uranium.

Further Readings

Department of Energy. "Hanford." http://www.hanford.gov (Accessed September 2010).

Hanford Watch. http://www.hanfordwatch.org (Accessed September 2010).

Wald, Matthew L. "Analysis Triples U.S. Plutonium Waste Figures." *New York Times* (July 27, 2010). http://www.nytimes.com/2010/07/11/science/earth/11plutonium.html?_r=3&pagewanted=1 (Accessed September 2010).

Hawaii

The U.S. state of Hawaii is a volcanic archipelago comprised of hundreds of islands spread over some 1,500 miles in the Pacific Ocean and it is the only state comprised entirely of islands. Along this great expanse, only the southeasternmost islands, generally termed the main islands, of Niihau, Kauai, Oahu, Molokai, Lanai, Kahoolawe, Maui, and Hawaii are settled or typically considered in discussions about Hawaii.

This set of main islands lies over 2,000 miles from the nearest continent, making the state the remotest part of the world that humans have settled. In fact, the city of Honolulu on the island of Oahu is considered the remotest urban area in the world. Home to roughly 1.3 million people (most of whom live in Honolulu on the island of Oahu), plus millions of tourists who visit the islands every year, the state faces challenges in terms of both consumption and waste because of its isolation.

Production History

Historically, Hawaii had an economy based on agricultural export; however, as pineapple and sugar became cheaper to produce elsewhere in the world, agriculture in the islands declined markedly. As a result, the islands import approximately 85 percent of their food in the 21st century, making the islands dependent upon the global market to supply their food. Likewise, the lack of nearly all natural resources in the islands means that nearly all materials needed for construction, transportation, and personal consumption are produced elsewhere and imported. Because of this dependence on imported goods, Hawaii's food security is quite vulnerable to shipping delays or natural disasters that affect the infrastructure, such as tsunamis and hurricanes.

Waste Incineration and Export

The large volume of goods imported to Hawaii produces a significant waste stream that poses serious challenges to the state. Specific aspects of waste collection in the state differ by island and county but share a number of similarities across the state. For instance, nearly all household and commercial waste is deposited in either landfills or incinerated for energy. As many of the landfills have reached capacity or will reach capacity in the next several years, there has been increased pressure to expand the landfills. However, given the strong public opposition to expanding the landfills, due to both the fragile nature of the island ecosystems and the proximity of the landfills to areas of rural poverty, local governments have been pursuing alternatives.

One alternative is expanding the garbage-to-energy plant on the island of Oahu, which incinerates trash to create steam for a turbine. As of 2010, the garbage-to-energy plant (H-Power) is capable of producing 46 MW of power through consuming up to 2,160 tons of garbage per day. This volume of garbage reduces the volume of municipal solid waste that goes to the landfill by 90 percent. Because of this decrease in solid waste, the garbage-to-energy plant is seeking to add a third boiler, thereby increasing annual capacity from 600,000 to 725,000 tons of garbage incinerated per year. The islands of Maui and Hawaii are exploring garbage-to-energy plants as their landfills are reaching capacity.

Hawaii's economy used to depend on agricultural export, primarily pineapples and sugar. However, global price declines for these commodities have considerably suppressed agricultural production on the islands, which now import 85 percent of their food.

A second alternative is to export waste out of Hawaii to the continental United States. Similar to other municipalities across the United States, Hawaii has explored the potential to export a portion of its municipal waste to a landfill in Columbia Gorge, Washington. In fact, trash on the island of Oahu was collected and baled for shipment beginning October 2009. However, after failing to receive the necessary permits needed to transport and dump the waste, the program was suspended in September 2010, with no waste having left the island. All of the baled waste is now being processed for either the garbage-to-energy plant or the landfill.

Recycling

Aside from addressing the waste removal process, Hawaii has been exploring other avenues to reduce the waste stream. For example, island-wide recycling programs have started on several islands, including Oahu. Recycling programs use single stream collection, accepting newspaper, aluminum, glass, several forms of plastic, and cardboard. While steel and tin are noticeably absent from the recycling collection system, these items are removed further down the waste stream on islands such as Oahu, where the trash is sorted prior to incineration.

A second form of recycling that was implemented in 2005 was a bottle deposit for aluminum, bi-metal, glass, and plastic (PET and HDPE only) containers up to 68 ounces, with milk and several alcoholic beverage containers exempted. The $0.06 deposit is attached to beverages at time of purchase, with a $0.05 return rate (the extra $0.01 is a nonrefundable container fee paid to the Deposit Beverage Container fund to help pay redemption centers' handling fees). According to current estimates, more than 75 percent of beverage containers covered by the deposit are returned.

A third form of recycling that greatly reduces input to the waste stream is green waste collection. Green waste refers to plant material that can be mulched or composted. Given the tropical climate and rapid plant growth, removing green waste greatly reduces inputs to landfills.

Other Legislation

Besides recycling programs, new laws have recently been codified that will remove important components from ever entering the waste stream. In particular, Maui has enacted a law, similar to Washington, D.C., that bans the use of plastic bags by most types of stores. Likewise, there is increased interest in eliminating plastic bags from the other islands due to their nonbiodegradable nature and marked impact on the marine ecosystem (for example, entanglements with animals) surrounding the islands.

Conclusion

The insular nature of Hawaii, its dependence on imported goods, and its unique ecosystems ultimately yield few options for addressing waste removal and disposal. While garbage-to-energy offers a way to reduce landfill inputs and lessen the state's dependence upon imported energy, it alone is not a solution. The plant still produces toxic material, some air pollution, and continues to add waste to the near-capacity landfill. Likewise, recycling programs have reduced inputs to the waste stream but have not been as fully implemented as could be possible. Thus, other options likely will need to be explored in the 21st century. Such options could include additional deposit laws on other consumer items, expansion of the existing bottle law, banning certain products that cannot be disposed of properly, and duties or fees on nonrenewable packaging.

Christopher A. Lepczyk
University of Hawaii at Manoa

See Also: Beverages; Composting; Food Consumption; Incinerators; Landfills, Modern; Recycling; United States.

Further Readings

Eckelman, M. J. and M. R. Chertow. "Using Material Flow Analysis to Illuminate Long-Term Waste Management Solutions in Oahu, Hawaii." *Journal of Industrial Ecology*, v.13/5 (2009).

Kasturi P. and T. A. Loudat. "Socio-Economic and Environmental Impacts of Landfill Sites on Oahu, Hawaii." *Journal of Environmental Systems*, v.31/4 (2004).

State of Hawaii. Department of Health. "Hawaii Beverage Container Deposit Program." http://hawaii.gov/health/environmental/waste/sw/hi5/index.html (Accessed October 2010).

U.S. Census Bureau. "2006–2008 American Community Survey." http://factfinder.census.gov/servlet/DatasetMainPageServlet?_program=ACS (Accessed October 2010).

Hazardous Materials Transportation Act

The Hazardous Materials Transportation Act (HMTA) was signed into law in 1975 in order to better regulate the safe transport of dangerous materials and to harmonize conflicting state and local laws regarding the transportation of hazardous materials (also known as hazmats). The law represents the continued attempt by the Department of Transportation and other regulating bodies to respect the freedom of interstate commerce while guaranteeing public safety. The HMTA is routinely at the center of high-level legal battles and has been criticized both as too nebulous to allow proper enforcement and too stringent for municipalities and businesses to properly follow.

History, Goals, and Guidelines

The HMTA was born out of public concern arising from the large number of highway accidents involving hazardous cargo in the 1960s and from raised awareness of the role of toxic substances in daily life inspired by growing environmental movements. Early environmental battles were often waged by those concerned with the location of facilities using potentially dangerous chemicals, but it soon became clear that these substances also posed a serious danger while in transit. Particularly given the more dispersed nature of commerce and industry and the expansion of the national highway system during the 1950s and 1960s, cross-state regulation became crucial to preventing accidents. The HMTA necessitated enhanced safety precautions as well as special permits for the transport of hazmat materials. The law was an attempt to tighten safety, prevent illegal disposal, and standardize varying state and local laws by applying minimum safety thresholds that could be expanded upon locally.

The HMTA designates specific materials such as oil, nuclear, and chemical substances as "hazardous" and provides guidelines for transport. There are specific requirements for the packaging, handling, and delivery of the hazardous materials, as well as training requirements for those licensed to transport such materials. The act also specifies worker safety, such as protective suits and respiratory devices that must be used by those handling hazardous materials. These rules apply to all forms of shipment and are also general guidelines that may be augmented by local policies, for instance, by regulating times of delivery, adding special precautions in populous areas, or limiting the use of specific roads and highways.

Controversy

From the first years of the HMTA, legal battles often involved local regulations that were challenged on the basis of infringing interstate commerce because they sought limitations on transport through their jurisdictions. Often, the issue at hand was the use of urban roads that were particularly vulnerable to an accident because of high density and narrow maneuvering space. Some localities received protection by the courts because of their high-risk potential, but others were denied because they prescribed unnecessary diversions based on local interests. In general, these efforts have required regions to compete against one another in order to avoid potentially dangerous uses of highways and other transportation centers.

The HMTA has been criticized as insensitive to local realities given the broadness of the Department of Transportation's administration of the law, thus excluding those with local knowledge from decision making. Some also find fault with the approach of setting only minimum requirements, because stricter laws are only enacted from lower levels of government where there can often be a lack of political will to encumber business and pass new legislation that goes beyond federal law. Accordingly, there is a disincentive for cities or states to make the law stricter because it depicts regulators as deliberately raising the cost of doing business. However, local politicians are also sensitive to the needs of their constituents, and in areas with increased environmental activism or high risk potential, considerably stricter local ordinances have been successfully passed.

Amendment

The HMTA was amended in 1990 in order to clarify discrepancies in the previous law and to enact stricter penalties for noncompliance. Key areas of public interest were industrial pollution, the management of oil spills, and the transportation of radioactive waste. Incidents such as the *Exxon Valdez* spill in 1989 highlighted the need to enact more specific legislation for special activities such as coastal oil extraction and transport. Although the 1990 amendment specifically addressed the transportation of oil, it also set liability limits, which were vaguely defined and contingent upon the financial resources of the owner. This loophole became an issue for the compensation of Gulf Coast residents as well as the costs of cleaning the Deepwater Horizon spill in 2010. Finally, the revised HMTA is still administered by a number of federal agencies, such as the Department of Transportation, the Department of Energy, the Nuclear Regulatory Commission, and the Occupational Safety and Health Administration, leading to bureaucratic conflict and inconsistencies in enforcement. Some consider this fragmentation as a potential threat to the integrity of the act, because responsibility is split even at the highest levels and there is the possibility that proper regulation may be encumbered by lack of communication or coordination.

Since September 11, 2001, HMTA has again been the subject of debate because of renewed interest in hazmat transportation as the potential target of terrorism. Municipalities and states began to include the protection of key transportation sites in new emergency plans that prepared for both accidental chemical disasters and the intentional targeting of nuclear or chemical facilities meant to disrupt transportation or cause injuries. Although no substantial changes were made to the law, the risk of transporting potentially dangerous materials through highly populated areas continues to be a subject of safety debates in the fields of disaster planning, terrorism prevention, and the building and maintenance of transportation hubs.

Max Holleran
New York University

See Also: Environmental Protection Agency (EPA); Radioactive Waste Disposal; Toxic Wastes.

Further Readings

Environmental Protection Agency. "Hazardous Materials Transportation Act." http://www.epa.gov/oem/content/lawsregs/hmtaover.htm (Accessed July 2010).

Moses, Leon N., et al., eds. *Transportation of Hazardous Materials: Issues in Law, Social Science, and Engineering*. Boston: Kluwer Academic Press, 1993.

Wallace, Todd. "Preemption of Local Laws by the Hazardous Materials Transportation Act." *University of Chicago Law Review*, v.53/2 (1986).

High-Level Waste Disposal

Nuclear energy is a paradox. It is both a high-tech industry and extraction based, with the strengths and weaknesses of both. It is one of the most debated, analyzed, and policy-driven issues in the 21st century, but it is also one of the most vaguely defined. It is considered by some to be the cleanest energy source in the world and by others to be the dirtiest. It is both a weapon that can potentially destroy mankind and a source of potentially perpetual industry.

For the first part of its history, the threat of nuclear war and, later, nonproliferation overshadowed nuclear waste disposal. Since some radioactive wastes have half-lives of one million years or longer, most waste was placed in above-ground storage in the hope that a disposal solution would be found within the next 100 years. Some of the first high-level nuclear waste was 60 years old as of 2010, and many nuclear power plants, like many older manufacturing facilities around the world, are reaching the end of their operating lives. Some countries are dealing with leftover radioactive waste from decommissioned cold war–era nuclear missiles. While nuclear technology has advanced, nuclear waste disposal still remains a waiting game of keeping it as far from humans as possible until it is not toxic. When many countries built nuclear reactors, they expected that their nuclear fuel would be reprocessed and stored in a foreign country. As the world becomes increasingly crowded, it becomes more difficult to find places to handle and store potentially harmful waste. In the 21st century, several new ways of disposing of high-level nuclear waste are being developed, and many countries have begun to reprocess or recycle nuclear waste. The biggest challenge is not only technical but also political and societal because of competing beliefs, goals, and impacts. Nuclear waste and the subsequent issues of storage are as diverse as the policy debates on the issue. To fully comprehend the implications of the latter, one must first have a basic understanding of the former. Nuclear waste can be delineated into three categories: extraction waste, low-level waste, and high-level waste.

Extraction Waste or Milling Waste

Extraction waste, or milling waste, is a by-product of uranium mining. Waste of this sort comes from the extraction of the uranium ore from the Earth and the subsequent process of concentrating it into yellowcake uranium. National or regional regulatory agencies like the U.S. Nuclear Regulatory Commission (NRC), France's Agence nationale pour la gestion des dechets radioactif, or the International Commission on Radiation Protection often regulate both the processes of extraction (to what degree depends on the process used) and waste management. Uranium mining can be broken down into two categories. The first, conventional mining, is the process of removing uranium from the Earth via underground shafts or open pits. The second category is in situ recovery, also known as solution mining, which chemically alters the uranium by pumping a solution, usually made up of water mixed with hydrogen peroxide or oxygen, called lixiviant, through a series of wells. This process causes the ore to dissolve into the solution, which is then pumped to a series of recovery wells at the surface.

Like most types of mineral and resource mining, extracting uranium produces four types of waste: waste rock, tailings from the ore processing, industrial waste, and wastewater. Throughout the mining process, the most common of these four categories of waste is waste rock and tailings. Waste rock is simply defined as rock with no commercial value that must be extracted from the mine in order to get access to the uranium ore. Above-ground mines, commonly known as strip mines, produce significantly higher amounts of waste rock than the below-ground mines. The second most common waste product from the mining process is tailings. Where waste rock stems from the mining process itself, tailings are a product of the process of separating metals from ores. For most of the waste products associated with the mining process, the goal is to dispose of them in a manner that would require no upkeep.

High-Level Nuclear Waste

High-level nuclear waste is defined as the extremely radioactive by-products of either the fission process found in nuclear power plants, weapons, or fuel reprocessing. High-level waste makes up over 95 percent of the total radioactivity produced in the process of nuclear electricity generation. A typical nuclear power plant produces about 27 tons of nuclear waste every year. Some examples of high-level nuclear waste include spent nuclear fuel, the highly radioactive solid or liquid materials that are products of nuclear reprocessing, and any other materials that have become highly radioactive through their proximity to the reactor that need to be stored in isolation. In most countries, spent nuclear fuel (SNF) is the most common type of high-level nuclear waste. SNF can be defined as used fuel from a nuclear reactor whose fission process (the process in which a uranium atom is split, releasing a small number of neutrons that then collide with other uranium atoms, thus causing a chain reaction) has slowed to the point where it is no longer efficiently producing electricity. Moreover, uranium is not the only radioactive element that must be dealt with. The fission process releases other radioactive elements, including strontium 90 and cesium 137, which also require storage. This release of elements intensifies the radioactivity of the fuel rods (metal rods with ceramic pellets containing uranium inside).

Governments around the world use similar waste management and disposal approaches, which are often dependent on the age of legacy nuclear power plants, the amount of nuclear energy they generate, and the proximity of neighboring countries that use nuclear energy. France, for example, produces 75 percent of its electricity from nuclear reactors, the highest percentage of any country. France has had one of the most advanced high-level management approaches, since the country began using nuclear power in the early 1970s in response to the first oil crisis. It reprocesses and recycles spent reactor fuel and even recycles fuel from other countries, which is returned to the country of its origin. While a few countries, such as France and Japan, provide reprocessing for other countries, many more provide underground disposal sites. Although it has no nuclear sites itself, the Association for Regional and International Underground Storage is located in Australia and provides storage for Belgium, Bulgaria, Hungary, Japan, and Switzerland. The Central Organisation for Radioactive Waste (COVRA) in the European Union (EU) is working on a European-wide waste disposal system with single disposal sites that can be used by several EU countries. Russia operates as an unofficial international repository system, with South Africa, Argentina, and China considering taking on the same international role. Other countries, such as Finland, prohibit the import or export of radioactive waste.

High-level waste that is not reprocessed is stored in two very different ways by nuclear power plants. The first and most common way is in water-cooled pools. These pools are large basins in which the spent fuel rods are placed beneath roughly 20 feet of water. At that depth, the water is able to act as an insulator, thus protecting the surrounding areas. The typical water-cooled storage facility can hold roughly five times the amount of fuel that the core can. The second type of high-level waste storage facility is commonly called dry cask storage. It was originally developed out of necessity. During the 1970s and 1980s, the general consensus among experts was that cooling pools could not keep up with the amount of waste being produced. As an alternative, dry cask storage typically stores spent fuel in steel cylinders that are either welded or bolted closed and are then surrounded by additional concrete, steel, or other material to act as a radiation barrier. The challenge is to find storage materials that retain their integrity for several thousand years. Ceramic is currently the most advanced method, while nanostructured materials are the future. Because spent fuel, and other high-level waste, comes out of the reactor at extremely high heat, it needs to be placed in a storage pool for between one and three years for cooling purposes before it has the option of being placed in a dry storage facility. Dry cask storage also affords the user some fairly significant benefits. For instance, many of the cylinder designs can be used for both storage and transportation. Moreover, this style of storage facility can be—and has been—designed to allow for either vertical or horizontal storage, thus maximizing any given space.

Several countries, most notably the United States, Canada, China, Germany, and Sweden, have begun researching and siting underground facilities or long-term geological depositories. In the United States, disposal of high-level nuclear waste is done at nuclear power plants because there was no permanent disposal facility within the United States as of 2010. This means that every power plant that is operating within the United States has its own storage facilities. However, the 1982 Nuclear Waste Policy Act states that all high-level nuclear waste should be disposed of deep underground in a geologic repository. The Nuclear Regulatory Commission (NRC) has stated that it will only grant the proper permits and licenses to operate such a facility if it can be proven that (1) the installation can be safely constructed and (2) it can be operated without any significant risks. As of 2010, the only site proposed was Yucca Mountain in Nevada, which would hold roughly 77,000 tons of high-level waste. It would fall under the jurisdiction of the NRC, Department of Energy (DOE), and Environmental Protection Agency (EPA). Preliminary work began on the site in 1991.

In 2002, the DOE, after completing about 10 years' worth of tests, determined that Yucca Mountain would be a suitable location for a storage facility. President George W. Bush and Congress accepted the DOE's findings and allowed it to submit a license application for construction, which it did in June 2008. Later in the same year, the NRC issued notice of a hearing on the Yucca Mountain site, which was to be adjudicated by the Atomic Safety and Licensing Board Panel. In total, about 319 contentions were filed against the proposed facility by 12 different groups.

Low-Level Nuclear Waste

While many people are aware of the potential impacts of high-level nuclear waste, low-level nuclear waste suffers from the same challenges as high-level nuclear waste. Where high-level nuclear waste is defined as the waste material produced from the fission process, low-level waste consists of items that become contaminated with radioactive material or have become radioactive through their proximity to radiation. Examples of low-level waste include anything from shoe and clothing covers to medical tubes and filters. All generators of low-level nuclear waste must possess a specific or general license that allows them to have and use radioactive materials. Moreover, low-level waste comes from a variety of places, including medicinal use; decommissioning, research, and development activities; and industry.

Low-level waste can be broken down into three categories that range in severity from background radiation levels that can be found in nature to highly radioactive: A, B, and C. Dividing waste into the three classifications is a complicated process that consists of measuring the concentrations of a number of radionuclides. If the concentration of a specific radionuclide is below 0.1 times the specified concentration in a cubic meter of waste, it is considered to be low-level waste class A. If the concentration exceeds 0.1 but does not exceed the specified concentration for a cubic meter, it is considered to be class C. Class B lies somewhere in between. If the concentration exceeds the set limit, it is then bumped into the high-level waste category. Some other aspects taken into consideration when classifying low-level waste include the total of all nuclides with less than a five-year half-life and whether the waste in question has a combination of radionuclides. Class A waste makes up roughly 96 percent of the total volume of low-level waste.

Storage for low-level waste varies based on the level of radioactivity found in the item. Items with concentration levels that are too high to allow them to be thrown away in the trash are shipped via Department of Transportation–approved containers to commercially operated disposal facilities. These facilities are licensed through the NRC or, based on the Atomic Energy Act of 1954 mandate, an Agreement State, as the act allows the NRC to delegate some of its regulatory authority to the state. An Agreement State is one that has petitioned the NRC to have the ability to regulate nuclear by-product, source material, and certain quantities of special materials as well as to issue licenses to those using or storing radioactive material. To become an Agreement State, the state in question must first prove to the NRC that it has a regulatory program that is comparable to that of the NRC and, second, the governor of the state must officially sign on to the program through a formal agreement with the commission's chairman. In all, 37 states have committed to being Agreement States, the first, Kentucky, signed on March 26, 1962.

The most common way to store high-level nuclear waste is in water-cooled pools such as this one, which cover the rods with about 20 feet of water. The water insulates the waste and protects the surrounding area.

Barbara MacLennan
Mylinda McDaniel
West Virginia University

See Also: Hanford Nuclear Reservation; Hazardous Materials Transportation Act; Medical Waste; Nuclear Reactors; Uranium.

Further Readings

Johnson, Genevieve Fuji. *Deliberative Democracy for the Future: The Case of Nuclear Waste Management in Canada*. Toronto: University of Toronto Press, 2008.

Melfort, Warren S., ed. *Nuclear Waste Disposal: Current Issues and Proposals*. Hauppauge, NY: Nova Science Publishers, 2003.

Parker, Frank L. "A New Paradigm for High Level Radioactive Waste Disposal." Disposal

Subcommittee of the Blue Ribbon Commission, July 7, 2010.
U.S. Department of Energy Office of Civilian Radioactive Waste Management. "Oklo, Natural Nuclear Reactors." Yucca Mountain Project, DOE/YMP-0010. November 2004. http://www.ocrwm.doe.gov/factsheets/doeymp0010.shtml (Accessed September 2009).

History of Consumption and Waste, Ancient World

The first evidence of object production by humans begins around 2.5 million years ago, when the earliest humans collected stones for making crude chopping tools. Stone tools are made by striking a blow that fragments the stone, leaving a sharp edge on the main piece of stone and resulting in the discard of the associated chips that are the by-product of the flaking process. The manufacture of other early objects made of bone, wood, and other perishable materials would similarly have resulted in waste materials as an integral part of the production process.

Food Waste

The process of preparing and eating food also produces waste. Almost all foods have some inedible part, such as shells, peels, seeds, husks, and bones, which are discarded as part of preparation, serving, and consumption. The earliest humans would have continually been picking through their food items and discarding the unwanted portions, discarding them in patterned ways that can be recovered through archaeological excavation. The rate of discard of food items increased after the development of pottery and metal vessels, which also broke and were thrown away as part of food-related waste.

The process of discard diversified when populations grew and as residential sites became larger. Starting 10,000–12,000 years ago, human group sizes began to increase from 50–100-person mobile foraging groups to sedentary village societies with hundreds or thousands of inhabitants. Often, this transition to fixed-place living coincided with the development of agriculture, which required a whole new range of production and processing tools such as farming equipment, grinding stones, storage bins, and cooking pots. The process of settled village life also resulted in regular patterns of discard behavior as people determined the locations of appropriate refuse disposal.

Early Technologies

The technologies of the sedentary period comprised many composite tools such as plows, harnesses, and other agricultural equipment in which elements could be replaced to continually upgrade and maintain the utility of the object. This resulted in fragments of "partible waste" that represented the discarded or replaced elements of the composite object. Other types of objects, such as pottery, were usually discarded rather than repaired. The notion of selective repair resulted in the increase in categories of "waste" objects as individuals had to decide how and whether to fix items or simply discard them altogether.

New technologies such as metallurgy and pottery production produced large quantities of waste in addition to producing changes in the landscape through the extensive use of trees as fuel. In the Indian subcontinent as well as in many parts of western Africa, the production of iron resulted in large and distinct piles of slag as part of the human-made landscape. Waste from production of these new technologies and the discard of surplus objects piled up around settlements, resulting in a landscape in which waste was a constant component of everyday life.

Waste Management

The process of biological decay also resulted in the innovation of new techniques to proactively address the production of waste. Fermentation can be regarded as a process in which people deliberately managed natural decomposition processes to produce cheese, wine, beer, and other edible products. Other processes such as smoking, wrapping, and sealing off air are meant to forestall decay in fresh foods. The process of making value-added foods through additional work can be seen in analogous developments in the 21st century, such as the creation of croutons from stale bread or the invention of "baby" carrots through the shaving down of otherwise odd-shaped vegetables.

Archaeological sites have large amounts of detritus that show deliberate accumulation, even when the inhabitants had many options for removing waste from habitations. In the stone-tool period, sites regularly are found with thousands or even hundreds of thousands of waste flakes littering the living areas. Sites of the early food-producing period also regularly show the accumulation of large amounts of broken objects and implements. In the sedentary period, starting with the first villages and up until nearly the present day in urban contexts, archaeologically recovered food remains such as bones, pottery, eggshells, and inedible plant parts often are found in close proximity to dwellings.

Archaeological sites near seacoasts or rivers often are marked by dumps of food-related debris, including large amounts of shells of edible mollusks. These "shell mounds" constitute some of the earliest settlement types of every inhabited continent and in some cases represent hundreds of years of accumulations that afterward served as elevated areas for housing and other activities. One shell mound, at Indian Knoll, Kentucky, was so large that when excavated, it was found to contain over one thousand human burials.

Recycling

The recycling of waste objects for the manufacture of new goods also can be seen in ancient contexts. In the process of manufacturing that resulted in flakes, peelings, and trimmings, people evaluated the potential for those waste materials to be turned into useful objects. The development of the first composite tools, consisting of stone flakes hafted into a handle, may have been one of the first innovations that made use of otherwise discarded materials to make a tool that could be renewed through the insertion of fresh blades.

The term *waste* can be generative of new commodities when objects are recycled. The contents of archaeologically recovered metal hoards worldwide show that smiths deliberately set aside worn and broken objects for melting down to produce new ones. Ancient pottery sometimes also deliberately incorporated crushed broken pottery as a means of strengthening the clay of new vessels. In the 21st century, the use of undisguised waste materials as art can be seen in the use of "found" objects for collages and even for large architectural creations (such as the Watts Towers in Los Angeles, California).

Conspicuous Waste and Deliberate Discard

The ancient attitude toward waste included not only the discard of by-products and worn-out elements but also the deliberate discard of usable goods as a sign of social power and wealth. Conspicuous waste was an important aspect of feasting and other celebrations even in the earliest village societies starting 10,000–12,000 years ago. Ancient burial sites often are accompanied by deposits of smashed cups and bowls, which are usually interpreted as signs of a funeral feast in which objects that might otherwise be reused are instead consigned as commemorative deposit. More rarely, elite burials include multiple bodies representing the sacrifice of otherwise able-bodied individuals whose premature death might otherwise be considered "wasteful."

In some historically documented cultures, serving dishes were meant to be utilized only once, resulting in the potential for significant quantities of discard. While discarded items might have been regarded as ritually "polluted," an ethos of deliberate discard was part of many ancient festivities. Other practices that might be considered "wasteful" of productive capacity include warfare (which often disproportionately results in the deaths of young men whose strengths would also otherwise be needed for farming, fishing, and provisioning their households and communities) and sequestering into religious institutions (which often disproportionately results in females removed from the reproductive pool).

Thus, the notion of waste is not merely about trash or detritus but also encompasses humans' cognitive abilities to assess the relative value of different states of material or physical being. The assignment of the philosophical or economic category of "waste" is a variable one, with some occasions (such as daily meals) including a focus on parsimony, while other occasions (such as celebrations) marked as "successful" by the quantity of excess food and drink that is afterward divided among the participants or simply thrown away.

Urban Waste

The creation and disposal of waste accelerated by the time of the appearance of the first cities, starting around 6,000 years ago. Ancient cities are some of the best places to understand trash behavior, as there are more people in cities and each individual has more types of objects to choose from, acquire, and use. Often times, this waste was piled up in residential areas and became part of the living landscape. Research at ancient Mayan cities shows consistent patterns of trash disposal around households, with some areas devoted to ordinary trash and other zones with chemical signatures of special-purpose waste, including signatures of heavy metals from the production of pigments. Monte Testaccio, in Rome, is a hill almost entirely made of discarded pottery vessels from the early centuries C.E.

In many cities, rubble from dismantled buildings was reused for the construction of platforms and building foundations. This practice was significantly different from the modern tendency to remove the debris of demolished buildings. The habit of knocking down older structures and using the elevated rubble as a foundation for new buildings resulted in the vertical accumulation of archaeological deposits, as seen most distinctly in the elevated cities known as "tells" in Mesopotamia.

In ancient cities, people recaptured solid and liquid waste as a marker of urban cleanliness. Archaeological investigations of the Harappan Bronze Age culture (ca. 2500 B.C.E.) of the western Indian subcontinent show that there was a collection of waste water from individual houses that led to larger-scale street drains. At the site of Taxila (ca. 6th–2nd centuries B.C.E.) in Pakistan, the excavator noted the presence of refuse bins for waste materials such as bones and broken pottery.

In other ancient cities, accumulated human waste was collected for agricultural and industrial purposes. The archaeologist Tony Wilkinson has suggested that the low-density artifact scatters seen around urban sites in Mesopotamia were the result of using domestic waste to enrich hinterland fields. In ancient Rome, the emperor Vespasian had urine collected, taxed by the state, and sold as a cleansing agent.

Around the world at the household level, the urban response to waste has included the use of wood trimmings and other dry organic waste for household fires (a precursor to more modern forms of waste energy capture such as modern cogeneration plants, which capture remnant heat energy that would otherwise be unused). Urban waste and its treatment thus represent a significant potential for both the individual and the collective approach to material culture.

Monica L. Smith
University of California, Los Angeles

See Also: Archaeology of Garbage; Food Waste Behavior; Funerals/Corpses; Recycling in History; Sewers.

Further Readings

Hayden, Brian. "Funerals as Feasts: Why Are They so Important?" *Cambridge Archaeological Journal*, v.19/1 (2009).

Hutson, Scott R. and Travis W. Stanton. "Cultural Logic and Practical Reason: The Structure of Discard in Maya Households." *Cambridge Archaeological Journal*, v.17/2 (2007).

Ingold, Tim. *The Appropriation of Nature: Essays on Human Ecology and Social Relations*. Iowa City: University of Iowa Press, 1987.

Scanlan, John. *On Garbage*. London: Reaktion, 2005.

Stearns, Peter N. *Consumerism in World History: The Global Transformation of Desire*. London: Routledge, 2006

Wilkinson, T. J. "The Definition of Ancient Manured Zones by Means of Extensive Sherd-Sampling Techniques." *Journal of Field Archaeology*, v.9/3 (1982).

History of Consumption and Waste, Medieval World

Over the medieval period (ca. 500–1500), the European population rose dramatically, people began to move into urban areas, and towns and regions were well connected via trade routes. This situation facilitated booming consumption and the concomitant problems of waste disposal.

Two types of sources are available for medieval consumption and waste disposal: written and archaeological. The documentary evidence is wide-ranging, including manorial records of harvests and goods consumed, town and guild regulations, wills and property transfers, and cookbooks. The material remains of consumption—the leftover material from craft production, broken pots in pits, and worn-out clothing—are typical objects of archaeological investigations. In fact, waste is the main source for much understanding of daily life, since objects like spoons, cups, buckets, shovels, clothing, and shoes were often in service until they were no longer usable and were then thrown away. Typically, only high-end artistic—and thus expensive—objects and those used in funerary rites survive as non-waste.

The objects of consumption fall into several categories: building materials, everyday craft goods, luxury goods, and food. There were differences between rural and urban consumption, with urban residents tending to depend on specialized craftsmen and merchants to supply the majority of their goods, whereas rural people produced more—but not all—of their own goods.

Building Materials

Building techniques varied widely over the medieval period, with everything from turf houses to grand stone cathedrals. Materials used in construction included stone, marble, brick, wood, and window glass. Building projects in much of Europe intensified after 1000 C.E. as more and more people began moving into urban areas. Old building materials could be reused as foundation or fill for new buildings and streets or to create new land in harbor areas.

Everyday Craft Goods

Everyday craft goods included a variety of items needed for the average household, including pottery, iron cauldrons, leather shoes, fabric clothing, bone combs, and wooden furniture. In general, craftsmen tried to maximize the amount of raw material from a given source; for example, a slaughtered cow generated meat for the butcher, a hide to work into leather, tallow (fat) for candle and soap making, and bones and horns for combs and utensils. However, the production of household goods for sale also generated large quantities of waste, which is often the main evidence for the location of particular craft shops.

Luxury Goods

The higher classes, including secular rulers and churchmen, consumed luxury goods, including gold and silver objects, paintings, sculptures, and books, throughout the medieval period. Many of the medieval objects on display in 21st-century museums such as the Louvre and the British Museum fall into this category of luxury object. The consumption of private books increased dramatically in the later half of the Middle Ages as university education expanded (increasing the demand for textbooks) and private religious devotional books came into use.

Food

Food consumption, while variable over space, time, and social station, was not as sparse in the medieval world as might be assumed. Documents show that beef, pork, fish, bread, produce, and spices were all sold in town markets across Europe, and town governments often attempted to control the quality and prices of the food sold. The consumption quantities were not insignificant: *Le Ménagier de Paris*, a book written at the end of the 14th century, records that 3,080 sheep, 514 beef cattle, 306 calves, and 600 pigs were slaughtered in Paris butcher shops every week to feed the 200,000 residents, and the royal and noble households consumed more on top of that. Much of the food was grown locally, including produce from urban and extra-urban gardens and livestock raised on nearby fields, but not everything was short-traveled. Even in the Middle Ages, extensive trade networks existed for goods like spices and fish, which came from limited geographical areas. Spices were imported via exchange networks from southeast Asia and Africa. Many locations needed to import fish in order to meet the requirement of fasting from meat during the Christian Lenten holidays.

Guilds

In the second half of the medieval period, craftsmen organized into guilds by the type of material they worked with—there were guilds for stonemasons, carpenters, dyers, and bakers. Guilds maintained monopolies on craft production in a particular town, ensuring that outsiders did not enter the market. By setting standards and monitoring the production of their members, guilds controlled the quality of goods, prices, and membership in the trade.

Waste

All of this consumption, particularly in the growing urban areas after the 12th century, led to waste disposal challenges. Waste sources included craft by-products, household rubbish, food scraps, livestock dung, and human excrement. Waste disposal regulations appeared by the 13th century and became commonplace by the end of the 15th century.

Craft By-Products

Archaeological investigations often discover rubbish pits co-located with craft workshops. Considerable amounts of waste are often associated with craft processes. For example, in the shoemaking area of medieval Bergen, Norway, archaeologists have found approximately 27,000 individual pieces of leather scraps weighing 200 kg from the 12th century. The pottery industry also generated thousands of kilos of waste materials and broken pots at sites such as the pottery kilns from the 13th and 14th centuries outside the city walls of Brugge, Belgium.

Urban Waste

Town and central governments throughout Europe issued ordinances forbidding the disposal of waste in urban rivers or vacant property within the town walls. The earliest regulations appear in the mid-1200s in several places: guild regulations from Berwick, Scotland (ca. 1249), contain an ordinance against filth in the town; Bassano, Italy, had sanitation regulations by 1259; and Verona, Italy, had similar regulations by 1276. Records show that local authorities handed out citations and fines to those who violated waste disposal regulations. In spite of the illegality of urban waste disposal in unacceptable places like streets and rivers, there were numerous complaints about waste strewn in the towns. Butchers were particularly notorious for throwing offal into urban rivers and received special treatment in laws in England, France, and Italy. Commoners and elites alike regularly complained about the stench of the urban rivers in metropolitan cities such as London and Paris.

Waste Collection and Disposal

Barrels and waste bins appear in towns beginning in the 1200s in northern Europe. Residents placed barrels in the ground or at the corner of the house or in the courtyard. It would appear that householders took the barrel contents for final disposal off-site and the archaeological material that remains is from the final time the barrel was filled. By the 15th century, the most common acceptable waste disposal locations were community rubbish pits outside the town walls. For example, documents from Coventry, England, include mention of five acceptable waste pit locations outside the gates by 1427. In Stockholm, a 1482 proclamation required individuals to take all waste to a specifically marked area in the hills outside town.

To aid in waste collection, some towns hired street cleaners and weekly waste disposal carts by the end of the 1400s. Carters went through town once or twice a week and collected waste that had been piled up by residents the night before on the street in front of their doors. The local governments collected taxes from residents to pay for these services. The carter services often focused on urban livestock dung, which was either sold or given to farmers for use as fertilizer on their fields.

Latrines

Latrine pits were commonly used for human waste collection in European towns. Privately owned pits were often located within the residential area on back sections of plots or between houses. When possible, they were built to overhang ditches or rivers that would wash the excrement away. Latrine houses often had more than one seat and were shared among residents. Some urban governments maintained common latrines—London had at least 13 common latrines by the early 1400s. Many latrine pits were emptied and the contents disposed of off-site. Human excrement may have been mixed in with animal waste for disposal in some locations, although the documentary evidence is often unclear, with no distinction between human and animal waste. Cleaning these pits was a dangerous job, as medieval coroner's reports in London reported several deaths caused by falling into one.

Dolly Jørgensen
Norwegian University of Science & Technology

See Also: Archaeology of Garbage; History of Consumption and Waste, Renaissance; Human Waste.

Further Readings

Dyer, Christopher. *Making a Living in the Middle Ages.* New Haven, CT: Yale University Press, 2002.

Hoffmann, Richard. "Frontier Foods for Late Medieval Consumers: Culture, Economy, Ecology." *Environment and History*, v.7 (2001).

Jørgensen, Dolly. "'All Good Rule of the Citee': Sanitation and Civic Government in England, 1400–1600." *Journal of Urban History*, v.36/3 (2010).

Jørgensen, Dolly. "Cooperative Sanitation: Managing Streets and Gutters in Late Medieval England and Scandinavia." *Technology and Culture*, v.49/3 (2008).

Sabine, Ernst L. "Latrines and Cesspools of Mediaeval London." *Speculum*, v.9/3 (1934).

Zupko, Ronald E. and Robert A. Laures. *Straws in the Wind: Medieval Urban Environmental Law: The Case of Northern Italy.* Boulder, CO: Westview Press, 1996.

History of Consumption and Waste, Renaissance

From the vantage point of the 21st century, the Renaissance is a period that is remembered for art, literature, music, science, and exotic luxuries, which can be seen as hallmarks of the expansion of the human mind and spirit. The Renaissance, literally a "rebirth," signaled a stage in European history when man sought to understand not only the world but also the human condition, in new and vividly creative ways. By the middle of the 16th century, Leonardo da Vinci had created fanciful machines and intricate anatomical drawings, Martin Luther had challenged the authority of the Roman Catholic Church, and Christopher Columbus had proven that the world is round. International commerce expanded on an unprecedented scale, and the wealth of monarchs, clergy, and noblemen fueled widespread decadence and competitive patronage of the arts. Universities were established to foster education in literature, science, philosophy, and history. With advances in translation and printing, books became accessible for many.

But to consider just the riches of the Renaissance is to adopt a myopic view. Europe during the Renaissance was also a very restless and volatile part of the world. Political upheaval forced the redefinition of once-feudal states into centralized territorial states, commercial battles ensued over trade routes, and corruption and debt cast shadows across society. The church often exerted influence on the state in domestic and international arenas. The discovery of distant lands resulted in the exploitation of indigenous peoples and their resources, ultimately leading to colonization and slavery. Death was an ever-present reality that took the guise of war, famine, or disease, affecting both rich and poor with equal consequence. In Renaissance Europe, misery served as a counterweight to opulence. To understand this dichotomy and the socioeconomic forces behind it, one must consider the long-term effects of climatic changes originating in the previous period. For the majority of Europeans, patterns of consumption and waste during the Renaissance hinged directly on subtle, but lasting, shifts in the weather that had far-reaching consequences on land, at sea, and even in the religious imagery of the afterlife.

For a span of four centuries, from about 800 to 1200, Europeans enjoyed a time of mild and stable weather known as the Medieval Warm Period. These years witnessed great trends of exploration and activity: the Norse established settlements in distant lands of the North Atlantic, majestic cathedrals were erected in France and England, and resources were amassed to support the Crusades. Both poor and rich found sustenance in the annual harvest, and during the Medieval Warm Period, regular, predictable seasons meant that harvests were plentiful enough to feed people and livestock. Some were able to supplement their diet with fish or game, and hearty farm animals ensured that there would be plenty of manure to fertilize the next cycle of crops. This also meant that wool and leather were widely available commodities. Human and animal populations rose dramatically during the Medieval Warm Period, and villages sprang up on formerly open tracts of undeveloped land. Despite localized conflicts and the struggles of vassal relationships, life was generally good. Christianity provided a social order that was felt across Europe, and people sought to demonstrate their devotion and gratitude to God in their daily lives.

In the late 1200s, temperatures shifted and grew increasingly colder. Around 1300, a phenomenon

Country life in France from the 1517 book Manners, Custom and Dress During the Middle Ages and During the Renaissance Period. *Contrary to the popular notion of the Renaissance as an idyllic time of discovery, the period was also marred by strife and disease.*

known as the Little Ice Age began and held Europe firmly in its grip until the 19th century. Conditions during the 1300s were the most severe. Cold rains that lasted well into the summer months created flooding that limited crop yield and prevented harvested hay from drying. Inevitable rot and diminished returns eventually brought on illness and hunger. The wettest years sparked a widespread famine lasting from 1315 to 1321. Those people and animals that survived were so nutritionally deprived and physically weakened that bubonic plague easily wiped out whole towns and villages, culminating in the Black Death of 1348. Glaciers began to advance in the Alpine regions of Europe during the late 16th century, but the coldest period of the Little Ice Age did not occur until the late 17th century. Drier weather during the intervening three centuries, from roughly 1400 to 1700, allowed the international commerce and intellectual advancement of the Renaissance to flourish. The abundant stores of the medieval period were never matched, however, so winters were hard on livestock, which meant there would not be enough manure to use as fertilizer. Famine and disease remained ever-present threats throughout the Renaissance, particularly among the poor.

Food preparation and cooking during the Renaissance continued the traditions of the medieval period, but as trade increased, so too did the European appetite for sugar, spices, and foods from foreign lands. Cakes, candies, and other sweets were in high demand. Among the foods from the New World that gained a foothold during the Renaissance was the potato, which became such an essential staple in Europe that it devastated Ireland during the notorious potato famine of the 19th century. Though a variety of vegetables were enjoyed by the elite, the poor were usually limited to the few root vegetables and herbs that grew locally. Meat was scarce. By far, the most widely available crops were grains, and when they were viable, they formed the foundation of the diet for both humans and livestock. Some scholars argue that an overreliance on grain and a general lack of nutrition triggered a mental agitation that fostered the religious fanaticism and political upheavals that framed the period.

Christianity was severely tested during the Renaissance. The Ottomans seized Constantinople in 1453 and, with it, the Orthodox Church. In the early 1500s, the Catholic Church was rocked by the Reformation. In its most extreme, Protestantism gave rise to Puritanism. As the New World became accessible, the Mediterranean began to lose primacy as the center of the known universe, and Protestantism flourished in distant lands. Those deemed witches, and what were interpreted as the pagan ways of witchcraft, became easy scapegoats for otherwise inexplicable problems such as crop failures or natural disasters. Evangelism and apocalyptic fervor swept across northern Europe, fueled by mass printing of the Bible.

Health and Hygiene

Not surprisingly, "disease" was considered synonymous with "filth," and both were deemed impure and sinful. By the 16th century, the church had had a long history of caring for the sick, and being physically clean and healthy was linked to being spiritually pure. This concept likely generated popular support for advances in medicine and insight into the spread of disease, which then gave rise to public sanitation practices. Hospitals kept contagious people isolated and clothing worn by the sick was prohibited from being resold or otherwise reused. Still, epidemics raged through military ranks and other tightly housed communities.

Personal hygiene was desired and attempted, but usually too expensive for the average person to sustain on a regular basis. The components of soap, such as olive oil and plant-derived fragrances, made the final product so costly that soap was usually reserved for washing linens and clothing, rather than the body. Keeping one's body clean was achieved by changing clothes and bed linens as often as could be afforded. In Germany, people could enjoy subsidized public baths, but elsewhere, heated water and other components were too difficult or expensive to be maintained.

No matter how earnest the attempt to keep clean, bugs and vermin were ever-present facets of personal space. Attempts were made to dispel offending creatures and odors, but filth was everywhere. In an effort to stem the spread of disease, some cities enforced legislation requiring that streets be cleaned, sewers drained, garbage removed, and inspections made of food and imported goods. The rich had private toilets designed for comfort and proper drainage, and hired workers to keep them clean. They also hired women to launder their clothes and linens. Such workers sometimes doubled as cooks, which gave rise to a separate variety of hygiene problems. The social dilemmas of Renaissance sanitation and hygiene, complete with associated raunchy puns, are vividly detailed in contemporaneous art and literature, such as Dante's *Divine Comedy*.

Conclusion

The topic of consumption and waste in the Renaissance would be left incomplete without touching on some ideas that are harder to quantify. What were the effects of deforestation to provide the wood used to construct ships, granaries, and musical instruments? How expensive was it to use egg albumin to bind pigments or to create gold leaf for illuminated manuscript pages? Were European Christians more receptive to Muslims in commercial arenas than they had been during the Crusades? Can the consumption of ideas and cultural exchange be measured? How much does imagination drive consumption, or ignorance fuel waste? Such questions only scratch the surface of a period that served to set the stage for the political revolutions, industrial innovations, and ground-breaking inventions that ushered in the modern era.

Nancy G. DeBono
Independent Scholar

See Also: Bubonic Plague; Culture, Values, and Garbage; History of Consumption and Waste, Medieval World; History of Consumption and Waste, World, 1500s; History of Consumption and Waste, World, 1600s; Human Waste.

Further Readings

Biow, Douglas. *The Culture of Cleanliness in Renaissance Italy*. Ithaca, NY: Cornell University Press, 2006.

Burckhardt, Jacob. *The Civilization of the Renaissance in Italy*. New York: Harper & Row, 1958.

Cunningham, Andrew and Ole Peter Grell. *The Four Horsemen of the Apocalypse: Religion, War, Famine and Death in Reformation Europe*. Cambridge: Cambridge University Press, 2000.

Fagan, Brian. *The Little Ice Age: How Climate Made History 1300–1850*. New York: Basic Books, 2000.

Ferguson, Wallace K. *Europe in Transition 1330–1520*. Boston: Houghton Mifflin, 1962.

Hale, J. R. *Renaissance Europe 1480–1520*. 2nd ed. New York: Blackwell, 2000.

Jardine, Lisa. *Worldly Goods: A New History of the Renaissance*. New York: W. W. Norton, 1996.

Maiorino, Giancarlo. *At the Margins of the Renaissance: Lazarillo de Tormes and the Picaresque Art of Survival*. University Park: Pennsylvania State University Press, 2003.

Pinkard, Susan. *A Revolution in Taste: The Rise of French Cuisine 1650–1800*. New York: Cambridge University Press, 2009.

Quataert, Donald, ed. *Consumption Studies and the History of the Ottoman Empire, 1550–1922: An Introduction*. SUNY Series in the Social History of the Middle East. Albany: State University of New York Press, 2000.

Thorndike, Lynn. "Sanitation, Baths, and Street Cleaning in the Middle Ages and Renaissance." *Speculum*, v.3/2 (April 1928).

History of Consumption and Waste, U.S., Colonial Period

The intellectual roots of consumer culture during the U.S. colonial period can be traced back to 17th-century western Europe and the antimercantilist idea that domestic markets were sufficient to sustain national economies. By the mid-18th century, as capitalist ideology and early industrialization spread through England and the rest of Europe, it also enhanced the culture of consumption. Specifically, the 17th- and 18th-century English culture had a dominant influence on consumption patterns of colonial America. Colonists considered English-made goods as indicators of status and respectability in society.

The abundance of land in the colonies prompted a focus on agriculture. By around 1715, the colonies had also achieved extensive community stability—family formations had reached levels that allowed for self-sustaining growth. During the first half of the 18th century, the colonies enjoyed ample harvests and, increasingly, a better quality of life. Regional patterns of agriculture and agricultural exports evolved. The south specialized in products such as tobacco, rice, and indigo. Mid-Atlantic farmers grew wheat, which in raw and processed form comprised the second-leading export in the mid-18th century.

Because of rocky soil, horticulture was constrained in New England, and it had to rely more on exporting fish and whale products along with foodstuffs, livestock, and rum. There were differential patterns of reinvestments; the southerners focused on reinvesting their profits in the land without diversifying their holdings, while the northerners invested in various enterprises, hence causing economic differentiation.

Social Stratification

With the accumulation of wealth, social class structures became more distinct, the greatest stratification occurring in coastal areas linked to transatlantic markets. Perhaps fewer than 20 percent of colonists lived at or close to the subsistence level, but the majority of them lived in artisan or yeoman families that earned a decent livelihood. With the rise of disposable income and more efficient ocean transportation, more households bought British household goods. More than half of properties were owned by the upper-class whites who also owned the lucrative plantation and counting houses. They aspired to have the status and power of the British upper classes and did so by buying the urban luxuries imported from that country.

Growth and Consumption

The population of the American colonies grew at a very fast pace after the mid-18th century, and along with it grew the per capita consumption of British imports. More than half the colonists were younger than the age of 16 and were primarily responsible for the exploding demand. The market for import goods rose 120 percent between 1750 and 1773. Even in rural regions, consumers began relying on them. For example, a study found women in rural North Carolina willing to pay a very steep price for Irish soap. The colonial consumer was faced with an unprecedented variety of choices in the marketplace.

For example, in 1720, merchants in New York City rarely advertised more than 15 different imported items per month. By the 1770s, the list had expanded to over 9,000 different manufactured goods. With the expansion of the number of items, the descriptions of the categories became more vivid. Between the 1740s and the 1760s, the New York merchants went from simply advertising for paper to listing 17 varieties by color, function, and quality. Similarly, customers went from expecting a generic variety of satin in the 1730s to being offered a dozen different varieties by the 1760s. No carpets were available in the 1750s, but by the 1760s, brand names such as Axminster, Milton, Persian, Scotch, Turkey, Weston, and Wilton were carried by stores. Since the 1750s, gloves were sold as "orange," "purple," "flowered," "white," "rough," "chambois," "buff," "Maid's Black Silk," "Maid's Lamb Gloves," "Men's Dog Skin Gloves," and so on.

Consumption and Taxes

For most of the colonial period, individual citizens rarely paid any federal taxes, since federal revenues were mostly derived from excise taxes, tariffs, and customs duties. Before the U.S. Revolutionary War, the colonial government had only a limited necessity for revenue, while each of the colonies had their individual revenue demands, which were met with different types of taxes. For example, the southern colonies principally taxed imports and exports, the middle colonies from time to time imposed a property tax and a "head" or poll tax levied on each adult male, while the New England colonies raised revenue primarily through general real estate taxes, excise taxes, and those based on occupation.

England's need to generate revenues to pay for its wars against France necessitated the passage of the Stamp Act in 1765, which was the first tax imposed directly on the American colonies, followed by Parliament's imposing a tax on tea. The colonists were asked to pay these taxes without representation in the English Parliament, leading to the rallying cry of the American Revolution and establishing a persistent skepticism regarding taxation as a part of U.S. culture.

Hygiene and Waste Management

The American colonies lacked organized public works for street cleaning, refuse collection, water treatment, and human waste removal prior to the mid-18th century. Recurring epidemics mandated initiatives to improve public health and the environment. Early European visitors to the American colonies in the 17th and 18th centuries frequently complained about the absence of reliable supplies of soap and water, the prevalence of mud and manure, flies and insects, and disgusting tobacco stains (from both spitting and chewing). More than 80 percent of the colonists lived in hygienically primitive situations on small farms or in country villages. Even in larger cities like Philadelphia and Boston, epidemics such as cholera and typhoid necessitated efficient water and sewer systems and stimulated massive public works construction, yet changes in personal and domestic cleanliness practices came slowly.

Some of the earliest citywide sanitary systems in colonial America were built in the early part of the 19th century. Few communities boasted of sophisticated systems, and much of the responsibility lay in the hands of the individual. The casting of rubbish and garbage on the streets was commonly practiced during most of the 18th century. Animals such as pigs, rats, and raccoons ate most of the garbage, while natural systems, such as the sun and wind, eventually eliminated the trash. Since the colonies had plenty of land and natural resources at their disposal, early waste management practices did not severely damage the environment. Urban area dwellers also had a higher tolerance for filth.

Diseases during the colonial period were mainly attributable to garbage because it contaminated water supplies and served as a breeding place for flies, rats, and other disease-carrying vermin. Diseases such as dysentery, bubonic plague, cholera, and typhoid fever increased because of appalling sanitary conditions. On hot summer days, cities smelled of odors from open-air markets, rotting trash, dead animals, animal waste, and leaking privy vaults in the basements of apartment buildings in the city slums. Even in rich neighborhoods, piles of garbage and the sickening odors of decaying refuse were evident. Dust and soot were very common. Sewer systems typically ran into the nearest rivers and oceans, contributing to the stench of daily life. Town thoroughfares were filled with manure from horses and animals scavenging through garbage thrown on the streets. Rainstorms turned streets into rivers of slime.

According to the American Public Works Association (APWA), Benjamin Franklin was responsible for designing the first colonial waste management system and municipal cleaning program in Philadelphia in 1757. The Corporation of Georgetown (now in Washington, D.C.) passed one of the first ordinances in 1795 outlawing dumping in the streets and strong refuse on private property. President John Adams hired a private refuse carter when he moved into the newly constructed White House in 1800. Refuse from other federal buildings was burned on the grounds of the buildings. The smell of burning garbage sickened President Thomas Jefferson while riding in his carriage, which led him to institute a plan to have the solid waste collected from the federal buildings and burned elsewhere. By the early 1800s, private waste collectors were responsible for cleaning the streets of Washington,

D.C., periodically and keeping the solid waste at tolerable levels, at least according to the standards of the time.

Abhijit Roy
University of Scranton

See Also: Consumption Patterns; Culture, Values, and Garbage; Household Consumption Patterns; Recycling.

Further Readings

Breen, T. H. *The Marketplace of Revolution: How Consumer Politics Shaped American Independence.* New York: Oxford University Press, 2004.

Firth, A. "Moral Supervision and Autonomous Social Order: Wages and Consumption in 18th-Century Economic Thought." *History of the Human Sciences*, v.15/1 (February 2002).

Glickman, Lawrence B. *Consumer Society in American History*. Ithaca, NY: Cornell University Press, 1999.

Hoy, Suellen. *Chasing Dirt: American Pursuit of Cleanliness*. New York: Oxford University Press, 1996.

Leach, William. *Land of Desire: Merchants, Power, and the Rise of a New American Culture*. New York: Vintage Books, 1993.

Louis, Garrick E. "A Historical Context of Municipal Solid Waste Management in the United States." *Waste Management Research*, v.22/4 (August 2004).

Melosi, Martin V. *The Sanitary City: Urban Infrastructure in America From Colonial Times to Present*. Baltimore, MD: Johns Hopkins University Press, 2000.

Perkins E. "The Consumer Frontier: Household Consumption in Early Kentucky." *Journal of American History*, v.78/2 (September 1991).

History of Consumption and Waste, U.S., 1800–1850

By 1800, Euro-Americans were already accustomed to the variety of consumer choices, moral ambivalence, and standardization of consumer behavior that accompanied the consumption of ready-made commodities. However, this period saw significant social, economic, and technological changes that made it unique from the patterns of commodity consumption and waste preceding it. Transportation projects, urbanization, and the redistribution of wealth had some of the most profound effects on consumption and waste. New transportation networks enabled commodities to reach more consumers.

While both waste and consumption increased slightly, because of urbanization, waste became much more concentrated in cities. While the miasma theory of disease linked filth and garbage with the epidemics raging through cities, concepts of urban ecosystems or the political structures that could tackle citywide material problems had not yet developed. The 1900s would usher in both an environmental consciousness and, slightly later, a commodity culture. Finally, the stratification of social classes due to the redistribution of wealth resulted in very different consumption patterns and waste management practices for different groups of people.

Transportation Networks

Massive canal and railroad projects between 1820 and 1850 revolutionized markets and market relations as goods circulated more widely and became cheaper. Not only could commodities now reach rural farmsteads and villages in larger quantities more regularly but also the savings in costs-per-ton-per-mile of transporting goods was passed onto the consumer in a "price revolution," making consumer goods more affordable. However, while there was a fully commercialized network of production in urban areas in 1830, mass production and mass consumption would not occur until the 1900s. Instead, the circulation of goods (and waste) remained largely local or regional, akin to cottage industries, even though U.S. import merchants were introducing increasing numbers of foreign-made goods.

Urbanization and Wealth Redistribution

Simultaneously, the saturation of settled rural land along the eastern seaboard resulted in an urbanizing trend that was to characterize the century. In the first federal census in 1790, 5.1 percent of the recorded population lived in 24 cities. By 1840, this number had risen to 10.8 percent in 131 cities, though only New York City's population exceeded 250,000. Thus, consumption patterns, access to commodities, and experiences of waste were different for rural and urban dwellers. Cities were generally sites of filthy streets, polluted water, and epidemic diseases, but rural areas could bury, burn, and otherwise dilute their trash with fewer immediate repercussions.

Finally, wealth was radically redistributed during this time. By 1850, the top 10 percent of the population owned 90 percent of the wealth, and the most privileged 1 percent owned 50 percent. Most of this wealth came from importing, exporting, and finance. At the local level, this gap in wealth and resources led to more economic distance between social classes, which differentiated both consumption and waste patterns.

Barter, Thrift, and Reuse

The major form of commodity distribution in the first half of the 19th century, especially in rural areas, was the peddler. The peddler did not only bring ready-made commodities to households but also enabled a two-way barter economy between domestic settings and industry. Industrial production relied heavily on scavenging for raw materials, and rural consumers—particularly the women who managed their household economies—had little cash. Rags, bones, ashes, and fat could be traded for tinware, cloth, tools, and kitchenware. Thus, people obtained manufactured goods by saving waste, though "waste" might more appropriately be thought of as scraps or useful materials within this context. The concept of "garbage" as an undifferentiated mass of undesirable cast-off materials did not yet exist.

Not all scraps were traded for ready-made goods, however. Many things were still made at home, both rurally and in cities. Historian Susan Strasser has written about how women's domestic manuals taught women and servants how to reuse, mend, and repurpose everyday objects and leftovers. Just as the case of saving scraps for barter allowed people to participate in commercial consumption, so did thrift and reuse. Domestic frugality and what Strasser calls the "stewardship of objects" was meant to prolong the life and use of new commodities. It included remaking clothes or display objects to participate in changing fashions, for example. In other words, thrift was not necessarily an ideal or end in itself but instead was one means to achieve greater comfort and consumption.

Barges await coal loads along the Delaware and Hudson Canal. Massive canal projects in the mid-1800s revolutionized markets as products became cheaper to transport and were more widely distributed. Consumers ultimately benefited from the cost savings.

Charity

The prolonged life of objects allowed them to enter into informal and formal economies beyond the peddler barter system. For example, before the sewing machine was invented in 1846, little ready-made clothing was affordable; all women knew how to sew, regardless of their social class or standing. Clothes had long lives, despite fashion trends, and were constantly darned, mended, remade, and passed down. While a significant quantity of clothing was sold to a large secondhand clothing market, upper-class women gave clothes to servants, slaves, or those "less fortunate" than themselves. This charity was considered part of their domestic duties. Goodwill was established in the United States in 1845; the Salvation Army began two decades later. However, the concept of charity through passed-down clothing was common long before it was institutionalized in the 21st century.

A second domestic duty that tied charitable domestic economies and scraps was food waste. Iceboxes remained an expensive luxury after their invention in 1827, until technological advances in ice cutting allowed affordable, urban ice carts to become a regular sight in city streets by 1850. While many food scraps were saved for future meals, "pan-toting" (the practice of domestic servants receiving food scraps) was also common. It was sometimes considered either stealing or charity, and other times it was arranged by contract. In rural areas, food scraps provided excellent animal feed, and slop buckets were a permanent fixture on front steps. In cities, food scraps also went to livestock of sorts. Once scraps were thrown out the window onto the streets, roaming pigs, goats, and dogs performed the Herculean task of garbage disposal. In addition to animal scavengers, many more-organized structures existed to collect food waste and slops for transport and sale to regional farms.

Urban Waste

However, pigs, goats, and more-organized collections of waste could not keep up with the trash in city streets. Throughout the 19th century, locals and travelers recounted the overwhelming stench of "filth" that made most urban roads impassable. Seafaring travelers could smell New York City before they could see it. Cities were sites of epidemic disease, polluted wells, and overflowing privies. By mid-century, public health boards had formed in many cities. They reiterated laws that had long been in the books, passed new ones prohibiting dumping in streets and waterways, and outlawed roaming scavenging animals, all with little effect. Municipalities that set aside money for sporadically "purging" the streets of waste or for hiring cartmen to pick up unlawful waste also struggled to keep streets clean and clear. One main issue was that there were no clear lines of responsibility or enforcement for refuse collection and disposal. A second issue was government corruption and graft, where money, inspection positions, and private collection contracts usually went to political allies, rather than effective street cleaning enterprises. Ad hoc systems of waste removal developed, using both private contracts (mainly for businesses and wealthy neighborhoods) and public agents, which rarely offered comprehensive community-wide programs. Waste collection was thus organized according to class and social standing.

Consumption

Consumption patterns were also heterogeneous across and within class divisions. Not only was there no "mass" consumption but also there was not a steady increase in average per capita consumption from 1800 to 1850, if such a thing could be said to exist. The upper class solidified as 90 percent of U.S. wealth was redistributed to 10 percent of the population. Within the working class, earnings rose between the 1820s and 1840s but fell in the 1850s. Both skilled and unskilled workers did not have enough resources to support their families with basic comforts like regular food and heat. This resulted in reducing consumption of meat and other expensive foodstuffs, fuel, and new, ready-made commodities. When the working class did purchase ready-made goods, their use and symbolism was appropriative, rather than strictly emulative of the upper classes. The working class did not have access to the bourgeoisie "dream worlds" of leisure entertainment and shopping centers where the symbolism of ready-mades would have been articulated. Mass production, mass consumption, and mass communication would radically change this in the following century.

Max Liboiron
New York University

See Also: History of Consumption and Waste, U.S., Colonial Period; History of Consumption and Waste, U.S., 1850–1900; Miasma Theory of Disease; Street Scavenging and Trash Picking.

Further Readings

Agnew, Jean-Christophe. "Coming Up for Air: Consumer Culture in Historical Perspective." In *Consumption and the World of Goods*, John Brewer and Roy Porter, eds. New York: Routledge, 1993.

Laurie, Bruce. *Working People of Philadelphia, 1800–1850*. Philadelphia: Temple University Press, 1980.

Strasser, Susan. *Waste and Want: A Social History of Trash*. New York: Metropolitan Books, 1999.

History of Consumption and Waste, U.S., 1850–1900

The second half of the 19th century was the era of railroad transport in the United States. Railroads interconnected the states from coast to coast (after 1869) and sped economic development and urbanization. Railroads were the only efficient and economical means of moving goods and people across the country. The railroads reduced the use of steamboats in the midwestern and southern states served by the Mississippi River system (including the Mississippi, Ohio, and Missouri rivers).

At the time, there were several types of cities: (1) cities that showed enormous increases in population and commerce as a result of industrialization and that also experienced a huge migration wave from Europe, (2) small, old cities, and (3) newly founded cities. In large cities, horse wagons coexisted and competed with interurban (trolley or "street running") rail lines. In San Francisco, California, the very first successful cable-operated street railway was the Clay Street Hill Railroad, which opened on August 2, 1873.

Industrialization, Consumption, and Waste

Railroads helped many industries, including agriculture. Farmers had a new way to send wheat, grain, and other products to ports. From there, ships could carry the goods around the world. Trains had special container cars with ice to keep meat, milk, and other goods cold for long distances on the way to market. People could get fresh fruits and vegetables throughout the year. Locally grown crops could be sold nationally. By 1880, there were more than 160,000 miles of railroad in the United States. Rail traffic spurred the growth of industrial cities in the midwest, especially Chicago, which became a center for trading grain, lumber, and meat in the late 19th century.

Environmental consequences for these industries were concentrated. Chicago became known as the "hog butcher to the world" after 1865, slaughtering thousands of animals each day for distribution across the nation. Wastes from the slaughterhouses dumped into the adjoining fork of the Chicago River created the infamous Bubbly Creek, where methane from organic wastes bubbled constantly, and a thick skin on the water allowed both chickens and humans to stand on the surface.

Demographics and Statistics

In 1850, the U.S. population was 21,191,786. Farm population consisted of 11,680,000 farmers—about 64 percent of the labor force. In just 30 years, the population increased to 50,155,783, while the farmer demographic decreased to 49 percent.

Industrialization influenced the growth of the value of products of certain manufacturing industries differently:

- The highest growth was seen in iron and steel production, from ca. $200 million in 1870 to ca. $800 million dollars in 1900. Slaughtering and meat packing grew from less than $25 million in 1850 to almost $800 million in 1900. Flour and grist mills grew from ca. $150 million in 1850 to ca. $550 million in 1900. Similar increases were seen in the clothing and textile, cars (steam railroad), and cotton and wool industries.
- The United States was the world's leading producer and consumer of zinc in 1850–1900 and afterward. U.S. mines and smelters produced 15 and 20 percent, respectively, of world output, and U.S. zinc consumption accounted for about one-fourth of the world total—63 million tons (1850–1990). Dissipative uses and landfill disposal have accounted for about 73 percent of the potential zinc losses to the environment, followed by mining and smelting (22 percent), and manufacturing (5 percent).
- The production of leather increased to ca. $200 million in 1900. The production of paper and wood pulp, silk, agricultural implements, clay products, hosiery, and knit goods increased from $50 million to ca. $100 million dollars.
- The statistics show a considerable increase in the consumption of water in the United States toward the end of the 19th century, from 42 gallons per capita in 1850 to 101 gallons per capita in 1994 (Boston). In Detroit, the consumption of water almost doubled from 120 gallons per capita in 1882 to 209 gallons per capita in 1889, but it fell to 150 gallons per capita by 1891, in particular, because of elimination of waste at the pumps, among other reasons.
- In the 1860s, the California Volunteers under Col. Patrick E. Connor discovered copper in the mountains of the Salt Lake Valley. The Walker brothers, Salt Lake City merchants, hauled the first wagonloads of copper from Bingham Canyon in 1868. Throughout the 1870s and

1880s, copper was essentially a by-product of the lead-silver ores mined in Bingham and elsewhere.

Innovations

During the second half of the 19th century, many essential innovations became embedded in people's everyday consumption, including the first refrigerator, electricity available for households, and the telephone. In 1859, during the Ginseng Rush, for one week only, about 12,000 pounds of ginseng were exported through Faribault, Minnesota. This was also the period when the first commercial chewing gum, Coca-Cola, breakfast cereals, and self-rising flour were introduced. Other inventions from this period included condensed milk, dry milk, and milking machines; potato chips; margarine; and the oil hydrogenation and methanol industries that grew out of the research by Paul Sabatier on catalytic organic synthesis. In addition, this period spawned the first vegetable oil used in the United States (cottonseed oil), the first large vineyards planted (Sonoma Valley, California), the first tin can with a key opener, pascal celery, and the first Nestle chocolate bar, among numerous other innovations.

From the perspective of food, the second half of the 19th century was a period of diversity in food styles. Queen Victoria ruled England, but her way of life was imitated all over the world, including in the United States. Immigrants—Europeans and Chinese—introduced new traditions and ingredients to U.S. dinner tables. In addition, many traditional foods and recipes were brought from west Africa through slave foods and preparations.

Sustainability, Health, and Sanitation

During this period, people generally avoided wasting things. According to narratives, even middle-class people traded rags to peddlers in exchange for teakettles or buttons. Most meats, fruits, and vegetables were bought fresh from markets or were raised and processed by the household. Canning was the common method for preserving food.

In 1852, the first public lavatory opened for business in London. Five years later, H. N. Wadsworth received the first U.S. toothbrush patent. Mass production of toothbrushes in the United States began in 1885. In 1972, Silas Noble and James P. Cooley of Massachusetts patented a toothpick-making machine.

Waste and Disposal

From 1850 to 1900, the main waste strategy was to burn garbage or deposit it in rivers and oceans. Typical problems with garbage were smoke, odors, rodents, flies, litter, and ground and surface water pollution. One of the results of burning garbage was air pollution. This practice was one of the components of increasing sulfur emission—from 311 to 9,345 ($GgSO_2$). The increase of garbage, in the course of industrialization, created a new problem: how to find the best management strategies. This was resolved in the early 20th century through the systems of landfills.

One of the newly founded cities at that time was Alpine (former Upper Dry Creek, Lone City, Mountainville) in Utah. Just several years after its founding, the garbage situation was out of control. On June 25, 1862, a $1.00 per month assessment for garbage pickup was to be payable with the water bill.

In 1885, the first U.S. garbage incinerator was built on Governor's Island in New York. The first rubbish-sorting plant for recycling was built in 1898 in New York. Since people were commonly frugal, household wastes were much less in volume compared to the 21st century.

As late as the U.S. Civil War, pigs, goats, and stray dogs were free to roam the streets as "biological vacuum cleaners." The theory that filth could contribute to human illness began gaining popularity and gradually made its way from England to the United States. As a result, local governments began setting standards for the protection of human health. The nation's first public health code was enacted in New York City in 1866.

By the late 1800s, the germ theory of disease, and its correlation to sanitary conditions, was reaching its peak largely because of three epidemics in the 1870s. A cholera epidemic in the Mississippi Valley in 1873 killed approximately 3,000 people, while New Orleans and Memphis were both struck with yellow fever epidemics. Then, in 1878, the south was struck with the worst yellow fever epidemic in the country's history. Death from disease in the Old West was typically due to infections from typhoid, yellow fever, diphtheria, malaria, measles, tuberculosis, cholera, and dysentery. Death from typhus and typhoid were particularly common because of overcrowding and poor sanitary conditions in many early camps and towns. In large part because of these epidemics, the federal government finally began to realize it should play a role in ensuring sanitation and created a National Board of Health in 1879.

Nevertheless, the streets of New York City were several feet deep in horse manure and garbage by 1882. The city reeked, and any garbage that was collected was dumped into rivers, lakes, and oceans. Private businesses specializing in waste hauling and trading grew substantially during this period. The rag trade, in force since the early 19th century, remained a major industry, although the advent of wood pulp in paper manufacture had begun to affect markets for cotton and linen rags by the end of the century. Trade in old iron, fueled in large part due to the expansion of the railroads, grew in the last third of the century, with scrap peddlers, yards, and brokerages growing in most northern industrial cities. While work in much of the waste trades was subsistence based, industrial demand for secondary materials (including copper and precious metals) allowed businesses to grow. First-generation immigrants from southern and eastern Europe founded many of these businesses; they would grow the waste trades as the United States entered the 20th century.

The period between 1850 and 1900 was a bridge to the contemporary world of diversity in consumption and of overconsumption, technology, public health concerns, and the highly developed technology of waste and garbage. It was an innovative period in the field of consumption and a period when contemporary waste and garbage technology began as one of the most essential tendencies in developing sustainable social sensitivity.

Lolita Petrova Nikolova
International Institute of Anthropology

See Also: History of Consumption and Waste, U.S., 1800–1850; History of Consumption and Waste, U.S., 1900–1950; Industrial Revolution; Industrial Waste; Meat; Population Growth.

Further Readings

Barbalace, R. C. "The History of Waste." http://environmentalchemistry.com/yogi/environmental/wastehistory.html (Accessed July 2010).

Cronon, William. *Nature's Metropolis: Chicago and the Great West.* New York: W. W. Norton, 1991.

Dunson, Ch. L. "Waste and Wealth: A 200-Year History of Solid Waste in America." *Waste Age* (December 1, 1999). http://wasteage.com/mag/waste_waste_wealth/index.html (Accessed July 2010).

Grover, K. *Dining in America, 1850–1900.* Amherst: University of Massachusetts Press, 1987.

Melosi, Martin V. *The Sanitary City: Urban Infrastructure in America From Colonial Times to Present.* Baltimore, MD: Johns Hopkins University Press, 2000.

Tarr, Joel A. *The Search for the Ultimate Sink: Urban Pollution in Historical Perspective.* Akron, OH: University of Akron Press, 1996.

Zimring, Carl A. *Cash for Your Trash: Scrap Recycling in America.* New Brunswick, NJ: Rutgers University Press, 2005.

History of Consumption and Waste, U.S., 1900–1950

From 1900 to 1950 in the United States, consumer expenditures increased sixfold, while the amount thrown away rose steadily. However, it is inaccurate to conclude that Americans simply became more wasteful. This period witnessed a dramatic transformation in practices concerning consumption and waste resulting from numerous intersecting changes and their unforeseen consequences. In the burgeoning mass market, large-scale production generated new kinds of waste, such as packaging and disposable items.

Estimating garbage increases, however, is problematic because until the 1930s individuals were largely responsible for managing their household waste. At the same time, mass production created greater efficiencies in homes, factories, and farms. Innovations for household use allowed consumers to enjoy more free time. New technologies enabled producers to produce in greater numbers and variety. Invention also rendered some goods obsolete.

Broad access to myriad goods held the social benefit of diminishing class lines. While the perception of waste as improper to use or something without value did not change, what was deemed wasteful remained a matter of debate. Both consumption and production practices came under fire. Critics argued that manufacturers profited by squandering materials and labor on luxury goods or low-quality items that would need to be replaced, instead of producing long-lasting basic products. Under this strategy, manufacturers relied on limited supplies rather than maximized output in order to increase prices, which many perceived as exploitative and inefficient. Others expressed concern that consumers bought things that they did not need or carelessly discarded perfectly good items for more fashionable versions. Many even viewed the Great Depression as punishment for the profligate spending of the 1920s. Whether mass consumption improved U.S. quality of life or was outweighed by its negative impact on the environment and social relations remains debatable. Regardless, from 1900 to 1950, consumption of mass-produced commodities emerged as the U.S. way of life, reuse changed from an ordinary practice to a symbol of poverty, and consumers ascended as a political group as vital to the nation's wealth and identity as industrial and agricultural producers.

Impact of Mass Production

Several factors rendered 19th-century household habits of repurposing and saving rags, bones, and scrap metals for sale to peddlers less practical and unnecessary. Nationwide distribution via the railroad system and standardized manufacturing enabled consumers to purchase products that previously had been homemade, too expensive, or inaccessible. At the same time, industries recycled many by-products of mass production, selling them in huge quantities to one another. For instance, soap manufacturers purchased grease from meat processors, and paper mills bought textile scraps. This effectively eliminated some outlets for household discards, as peddlers could not compete in terms of quantity with corporate salvage at a time when consumers no longer needed to spend time making things such as soap. Likewise, the steady increase in manufacturing jobs resulted in a population shift to urban centers; by 1920, more than half of the U.S. population lived in cities. These citizens lacked storage space as well as time for remaking odds and ends.

While the feasibility of reuse decreased and thus changed leftovers from resources to garbage, rapid advances in design and assembly-line production increased manufacturers' capacities and product variety. In the digital age, it is easy to imagine the excitement an ever-growing parade of new products from radios to refrigerators stimulated and how quickly a new technology might prompt consumers to toss outmoded items. The rise in consumption of mass-produced goods reached new heights by the end of the 1920s. The largest increases in expenditures were on items that promised more leisure time or recreation. The purchase of processed foods, household appliances, and ready-made clothing freed women from hours of domestic labor each week. By 1929, one in five Americans owned a car, cutting travel times from farms to towns and enabling tourism. Additionally, products manufactured for one-time use, like toilet paper and napkins, made disposability, not durability, a desired product feature. Mass-produced products became associated with convenience and a better quality of life; more stuff was bought and then discarded for more fashionable or technologically improved models.

Packaging and Marketing

Packaging represented an integral part of changing consumption practices and was a new form of waste. Manufacturers used packaging to market their products as superior to the competitors' in terms of quality, style, and experience. High subscription rates to magazines, littered with advertisements, created product branding. Advertising associated products with values ranging from wholesomeness to rugged masculinity; using certain products offered a way for consumers to express their identities. Although packaging also improved product cleanliness and minimized damage, it proved fundamentally worthless once stripped from a product.

Charity and Salvage

Charitable institutions established at the turn of the century, such as the Salvation Army and Goodwill, allowed consumers to unload unwanted items and justify buying new ones. In the process, they fostered the association of secondhand with poverty.

A classic Model T Ford from the early 1900s. By 1929, one in five Americans owned a car, cutting travel times and boosting tourism. The largest increases in expenditures during that time were on items promoting leisure time or recreation. By the 1930s, the Great Depression derailed consumerism and dramatically altered Americans' perception of production. During this period, ample supply stood in contrast to a lack of purchasing power. To reignite consumption, New Deal policies attempted to restore the balance of production and demand.

The experiences of the Great Depression and two world wars demonstrated castoffs' potential value, simultaneously reinforcing reuse as a method of making do in times of crisis. Salvage campaigns conducted during the wars connected reuse with patriotism. However, war represented a disruption to ordinary life.

Great Depression and the New Deal

The Great Depression not only disrupted growing consumerism but also radically changed how Americans thought about production, consequently politicizing consumers. Americans believed that producers formed the basis of the nation's wealth by literally creating material abundance, whether by cultivating land or making automobiles. Even Progressive-era legislation designed to protect consumers from adulteration and price hikes limited the government's regulatory power because free market competition was assumed to be the best regulatory measure. By the time President Franklin D. Roosevelt took office in 1933, however, the "paradox of want amid plenty" belied faith in laissez-faire capitalism and the value of ever-expanding production. The paradox described the situation of ample supply and high demand without purchasing power. It seemed most visible in agricultural commodities rotting in fields because the unemployed could not afford to buy them and farmers could not afford to harvest them. New Deal policies attempted to end underconsumption by restoring purchasing power and bringing production into balance with demand.

The National Industrial Recovery Act (NIRA) and the Agricultural Adjustment Act (AAA) curtailed production, focused on increasing consumers' purchasing power by mandating a minimum wage and maximum work hours and integrating consumer councils that lobbied for fair prices. Policy makers considered consumption vital to economic recovery and so began to recognize consumers as an interest group distinct from industry, labor, and agriculture.

Wartime production of the 1940s helped to end the Depression, but it also caused consumers to save rather than spend. Rationing limited what could be legally purchased, and industry shifts to military production created a scarcity of consumer goods. After more than 15 years of making do and going without due to economic crisis and war, consumers with huge savings accounts and high employment were ready to spend. The ability to do so had become so integral to how Americans understood their capitalist democracy that during the cold war, mass consumption was touted as evidence of the United States' superiority to communist nations.

Waste Management

Municipalities' responsibility for waste handling grew alongside the rise in mass consumption; together, they ultimately changed how the country dealt with and thought about waste. At the turn of the 20th century, rural households and city dwellers were still largely responsible for waste removal. This often entailed garbage being thrown out windows onto streets, burned, composted, or sold. Some cities at this time began organized trash collection to control stench and disease, as well as to improve aesthetics. Collection practices initially mirrored household methods of sorting and salvaging materials; private waste collection firms and city agencies required households to sort trash so as not to waste a revenue source. By the time that publicly funded waste collection became widespread in the 1930s, however, reuse as part of waste management had largely given way to engineering cheaper and pleasanter disposal. Municipalities used various methods including burying, reduction, incineration, and river, lake, and ocean dumping until the 1950s. Military technologies developed during World War II improved landfills; inexpensive, sanitary, and the least offensive to the senses, landfills became the dominant form of trash disposal. As opposed to unsightly and malodorous methods, landfills could mask garbage as green spaces. Though the amount of garbage increased, this kind of waste became less visible.

Ann Folino White
Michigan State University

See Also: Consumerism; Economics of Waste Collection and Disposal, U.S.; History of Consumption and Waste, U.S., 1850–1900; History of Consumption and Waste, U.S., 1950–Present; Household Consumption Patterns; Overconsumption; Underconsumption.

Further Readings

Cross, Gary. *An All-Consuming Century: Why Commercialism Won in Modern America*. New York: Columbia University Press, 2000.

Donohue, Kathleen G. *Freedom From Want: American Liberalism and the Idea of the Consumer*. Baltimore, MD: Johns Hopkins University Press, 2003.

Melosi, Martin V. *The Sanitary City: Urban Infrastructure in America From Colonial Times to Present*. Baltimore, MD: Johns Hopkins University Press, 2000.

Rogers, Heather. *Gone Tomorrow: The Hidden Life of Garbage*. New York: New Press, 2005.

Strasser, Susan. *Waste and Want: A Social History of Trash*. New York: Metropolitan Books Henry Holt, 1999.

Strasser, Susan, Charles McGovern, and Matthias Judt, eds. *Getting and Spending: European and American Consumer Societies in the Twentieth Century*. Cambridge: Cambridge University Press, 1998.

Tarr, Joel A. *The Search for the Ultimate Sink: Urban Pollution in Historical Perspective*. Akron, OH: University of Akron Press, 1996.

History of Consumption and Waste, U.S., 1950–Present

In 1950, Americans had begun to emerge from the austere Great Depression and rationing of World War II to enter a new consumptive era. If wartime rationing had meant a moratorium on domestic housing construction, automobile manufacture, and the distribution of luxury goods, the postwar era saw accelerated rates of consumption and waste beyond any the United States had seen before the market crash in 1929.

Changes in policy and corporate structure aided what historian Lizabeth Cohen calls the "Consumers' Republic." During World War II, federal military contracts allowed a few corporations such as General Motors, Boeing, and U.S. Steel to grow far larger than they had before the war. In the 1950s, these behemoths employed hundreds of thousands of white-collar and blue-collar workers who were at once producers and consumers. Without serious economic and financial competition, the United States enjoyed not only unparalleled productivity but also global dominance, at least for a time. Buoyed by consumer spending and unprecedented "peacetime" military budgets (inflated by the cold war with the Soviet Union, in which American materialism provided ideological ammunition for both sides), the nation's material production expanded so explosively that Harvard scholar John Kenneth Galbraith's *The Affluent Society* in 1958 gave a name to a phenomenon captivating writers in the United States for years.

Affluent Society

Never in the history of humankind had so many people in any nation enjoyed so much prosperity, even though the impoverished continued to number in the millions. According to William Leuchtenburg, the United States had only 6 percent of the planet's people at mid-century, yet it was producing and consuming over 30 percent of the world's good and services. The phenomenal domestic production growth that occurred after the war depended on international political and economic networks of which it formed a constituent part, but it also depended on a series of domestic compromises and repositionings on the part of the major actors. A complex of dynamic industries arose, including automotive, steel, petrochemical, and construction. Some of the propulsive engines of growth, based on technologies that had matured in the interwar years, had been prompted after the war by both new extremes of rationalization and links to academic research and development.

From 1946 to 1958, writes Leuchtenburg, corporations put an average of $10 billion a year into new plants and machinery. Consequently, large corporate power was further deployed to assure steady growth of managerial expertise in both production and marketing and the mobilization of economies of scale through product standardization. The state, for its part, assumed varying social and political obligations. Given that mass production required heavy investment in fixed capital, it also required relatively stable demand conditions to be profitable. The state strove to curb domestic business cycles through a mix of fiscal and monetary policies directed toward public investment in transportation, suburbanization, and public utilities; in the establishment of social wage agreements, social security benefits, and federal guarantees of mortgages earmarked for a mass of ex-GIs—all pursuits that were vital to the perpetuation of the flows of mass production and mass consumption. In the meantime, the civilian labor force expanded, as Leuchtenburg put it, from "54 million in 1945 to 78 million in 1970."

Suburban Nation

Increasingly, these people lived in new housing built in suburban subdivisions away from aging city centers. Federal mortgage policies developed to stimulate the housing market during the New Deal allowed banks to offer millions of Americans 25- and 30-year mortgages, providing the dream of home ownership with a modest down payment. Between 1950 and 1970, the exodus to the urban periphery transformed the United States from an urban to a suburban nation; historian William Leutchenburg sums up the demographic shift by noting the following:

> . . . by 1950 there were 37 million suburbanites; two decades later, the number had nearly doubled. As many moved to the suburbs each year as had come to the United States in the peak year of transatlantic migration.

These new houses were built on what had been farms, fields, forests, meadows, and prairies. Wildlife habitats from Long Island, New York, to Malibu, California, were demolished to make way for new human communities with new demands on local resources. Historian Adam Rome summed up the transformation of the landscape:

> In 1950 the U.S. Census concluded only 5.9 percent of the nation's land as urban or suburban, whereas in 1960 the figure was 8.7 percent, and in 1970 it reached 10.9 percent. Throughout the 1950s the nation's cities and suburbs took a million more acres—a territory larger than Rhode Island.

The subdivisions had effects on the land, water and air: aside from planting millions of houses on uninhabited land, the new homes required new sanitary landfills to dump their garbage. Too far from existing sewer systems in central cities, the new houses required septic tanks to handle human wastes. Leaking tanks could contaminate groundwater. The new modern conveniences such as air conditioners, dishwashers, and television sets required more and more energy to power, which had effects on local air quality around power plants.

Air pollution was more commonly associated with the millions of automobiles the new suburbanites required to commute to jobs in the central cities. Tailpipe emissions were blamed for visible smog, ambient lead (before the removal of lead as a fuel additive in 1972), and (by the end of the century) greenhouse gases associated with global warming. The automobile became integral to American life as the population sprawled across the continent; aside from the journey to work, Americans (inspired by marketing campaigns to "see the USA in your Chevrolet") used their automobiles for vacations, to see movies at drive-ins, to shop at the giant suburban malls developers hastily constructed in the 1950s, and to engage in most recreational and cultural activities that were beyond a walk in postwar land-use planning. By 1961, essayist John A. Kouwenhoven noted the centrality of the automobile to American life when he titled his collection *The Beer Can by the Highway*. In it, he observed:

> Two aspects of American civilization strike almost everyone: the abundance it enjoys, and the waste it permits (if it does not enjoy that too). As a *London Times* reviewer observed, in a discussion of three important books on American history, the occupation and development of this country has been a wasteful process, and the American god was not, is not thrifty. Nor is he, or was he tidy.

Not all Americans participated equally in the new, highly consumptive suburban way of life; federal mortgage policy and longstanding social discrimination (including attitudes of leading housing developers such as William Levitt) barred African Americans from using mortgage insurance to purchase housing in the new subdivisions. Unequal access to housing, education, restaurants, transit, and other public accommodations inspired consumer-based protests such as the 1955 Montgomery Bus Boycott, in which the African American population in Montgomery, Alabama, successfully forced the Montgomery Bus Line's segregated seating policies to change by withholding their business. Consumer actions, including the use of sit-in protests to desegregate lunch counters, became important tools in the emerging civil rights movement.

Those Americans who did enjoy all the modern conveniences of suburban life grew disenchanted with aspects of their consumption. If new suburban homes were marketed as escapes from polluted central cities, the environmental destruction of the subdivision aroused anger in many of its residents. Worries about environmental health, catalyzed by Aldo Leopold's 1949 book *A Sand County Almanac* and Rachel Carson's 1962 book *Silent Spring*, led suburbanites to join existing preservation groups such as the Sierra Club and create new environmental protection groups such as the Natural Resources Defense Council.

Worries about consumer safety increased, especially after consumer advocate Ralph Nader published a scathing exposé of Chevrolet's Corvair automobile in 1965 titled *Unsafe at Any Speed*. Interest in impartial evaluations of products by *Consumer Reports* magazine increased, and Nader founded state-based public interest research groups (PIRGs) and the national organization Common Cause to hold manufacturers accountable for the safety of their products.

Concern about garbage disposal, littering, and waste in general inspired many municipalities to develop recycling programs, appropriating a long-used efficiency measure of manufacturers for the purposes of environmental protection. Pressure on elected officials (especially in the northeast and on the Pacific Coast) led to bipartisan federal efforts to protect land, air, water, wilderness, and wildlife in the 1950s, 1960s, and 1970s under several acts signed into law by presidents Eisenhower, Kennedy, Johnson, Nixon, Ford, and Carter. Republican support for environmental protection effectively ceased with the presidency of Ronald Reagan, ending a period of national innovation in environmental protection laws.

Wilderness Recreation

Before the 1950s, few Americans ventured into the pristine backcountry. By the 1960s, however, a burgeoning population driving automobiles across the nation's interstate highways stampeded the wild woodlands. Wilderness managers, who faced ascending popular interest in outward-bound pursuits by the 1970s (visits to rural areas during the 1960s and 1970s, claims Roderick Nash, grew 12 percent annually), presumed that uncontrolled wilderness recreation was just as much a threat to the backcountry as economic development.

Wilderness management, some scholars reason in a mingled spirit of Thoreau, Leopold, Muir, and Brower, is a contradiction in terms. By Old English etymology and tradition, the word *wilderness* means *wildeorness*, a space of wild beasts. By contrast, the word *management*, derived from the Latin *manus*, meaning "to handle." With this interpretation, Nash quips that the "National Wilderness Preservation system might be regarded as a kind of a zoo for land." Marvin Henberg scoffs, moreover, that the four federal agencies responsible for administering wilderness lands have been forced into a wilderness management that smacks of "a paradox if ever there was one."

Even before Nash and Henberg wrote, others considered the shrunken, despoiled landscape. "Take a last look," William Whyte wrote in a 1959 issue of *Life* magazine, which entertained the woes of suburban sprawl.

> Some summer morning drive past the golf club on the edge of town, turn off onto a back road and go for a short trip through the open countryside. Look well at the meadows, the wooded draws, the stands of pine, the creeks and streams, and fix them in your memory. . . . this is about the last chance you will have.

Only a few years later, the writer Wallace Stegner lamented:

> Something will have gone out of us as a people if we ever let the remaining wilderness be destroyed; . . . if we pollute the last clear air and dirty the last clean streams and push our paved roads through the last of the silence, so that never again will Americans be free in their own country from the noise, the exhausts, the stinks of . . . automotive waste.

The remarks by Nash, Henberg, Whyte, and particularly Stegner show an acute sense of concern about the contradiction between the natural environment and too much built environment. But it is Stegner's concern that touches most closely the paradoxical emergence of progress and chaos embodied in the stupendous surges of productive consumption and productions of waste. Productive consumption and waste were accelerating, while the life span of products was decreasing. The years from 1950 to 1970 can be called the age of throwaway items and an epoch of plastic and paper packaging. The years after 1980, with their proliferating productions of e-waste, add to the cascading mess. The best available quantitative measurement of waste production perhaps comes from the U.S. Office of Technology Assessment of 1989, which, basing its figures of national municipal waste in the United States on the EPA/Franklin model, pronounces that in 1979, 136 million tons of waste was produced (though while this model in 1979 had estimated 136 million tons, its 1986 version estimated that the waste for 1977 was 122 million); that in 1986, there arose 158 million tons; and that, by 2000, waste generation was to reach 198 million tons.

The EPA/Franklin model captures well enough the production of waste from the late 1970s through the 1980s, but its prediction for the turn of the century was not on the mark. According to David Pellow, writing in 2007, "The volume of municipal waste generated in the United States between 1998 and 2001 grew by 6.6 million tons, or 20 percent, to a total of 409 million tons per year," a striking figure, since it "has only 5 percent of the world population but generates 19 percent of its waste."

Waste Disposal

While the United States had come of age as a supremely progressive technological country, its major disposal solutions became problematic. Sanitary landfills emerged between the great wars. They multiplied after the war as supposedly the most effective solution to the twin problem of the putrid open dump and the one-time highly touted but belching incinerator, even though Louis Blumberg and Robert Gottlieb (quoting the EPA) note that as many as 14,000 communities were likely still using open pits during the 1970s.

Martin Melosi reports, the landfill in the 1970s became the "central symbol of the garbage crisis" because availability of landfill space plummeted, especially throughout the northeastern states. Availability can be interpreted as scarcity of land, for, as Melosi shows, the National League of Cities and the U.S. Conference of Mayors issued a report in 1973 that claimed in part "[w]ith most of our cities running out of current disposal capacity in from one to five years, American urban areas face a disposal crisis."

Although trash mounted rapidly after 1950, a sole focus on space availability confounds the complex social and historical predicament of waste disposal, As reported in William Rathje and Cullen Murphy's *Rubbish!,* in 1889, the chief health officer of Washington, D.C., claimed "appropriate places for garbage are becoming scarcer year by year." Available space mattered, but so did the contentious sociopolitical struggles buzzing around causes such as NIMBY (Not in My Backyard), property values, political opportunism, costs, and privatization of waste disposal.

It was not until the 1980s—when the Environmental Protection Agency, through its Resource and Recovery Act, began to protect social and environmental health by requiring landfill operators to either close their landfills or restructure them by instituting leachate- and methane-collection systems—that a major waste disposal squeeze surfaced nationwide.

Facing more stringent enforcement, operators of many of the nation's sanitary landfills failed to embrace the mandatory structural improvements. Hence, many landfills ceased operation, so that, as Elizabeth Royte declares, the nearly 8,000 sanitary landfills that functioned nationwide in 1988 plummeted to 2,314 in 1999 and to 1,777 in 2002. Over the course of these years, when state authorities could find precious few solutions beyond instituting stringent requirements to the waste problem, many communities and regions found waste disposal fixes either in an expanding web of intrastate waste exchanges, as Edward Repa aptly shows, or in a thriving global trade of waste, as a headline in a 1983 issue of *New Scientist* suggests: "U.S. Steps Up Trade in Toxic Waste."

Recycling

As a solution to waste disposal, the nation's experience of recycling, as important and useful as it remains, has been no less disenchanting than its experience with the sanitary landfill. Recycling was practiced during the 1960s as a grassroots approach to resource reduction, but it gained mainstream popularity during the 1980s amid a recurrent debate between sanitary landfill and incinerator proponents. Melosi affirms that, before 1980, fewer than 140 communities boasted door-to-door recycling collection service, and that, by 1995, more than 9,000 communities established curbside collection of recyclables.

Recycling developed strong social and political appeal as aggressive programs sprouted throughout states like Connecticut, New York, New Jersey, Pennsylvania, Rhode Island, Florida, and Oregon. Even amid the ascent of national recycling pursuits, it was wondered whether creative incentives could be formulated to prompt broader household compliance; whether markets could be developed to take burgeoning volumes of recyclables, especially since recycling (like incineration) almost always results in production of waste that must be landfilled; and whether merit existed in John Tierney's *New York Times Magazine* article of 1996 "Recycling is Garbage," which argues that recycling costs more in labor than is saved, that shortages in natural resources are so mythical that recycling is pointless, and that the nation embraces it "as an act of moral redemption."

Morals have not always reigned supreme in the recycling business, particularly considering the e-waste fiasco during the first decade of the 21st century. Joseph Ladou and Sandra Lovegrove declare that there is an escalating trade in obsolete, discarded computers collected in the United States, where as much as "80 percent of the e-waste collected for recycling . . . is not recycled domestically, but instead exported to developing countries." Catherine Komp, writing in 2006, charges that there is "a monstrous e-waste trade in which municipalities and businesses blindly give their electronics to 'recycling companies,' which, in turn, sell the toxic trash to 'developing countries' at a profit." Why is this so? Simply, the United States refused to sign the 1989 Basel Convention treaty, which was designed to reduce the transfer of hazardous waste to less developed countries, and was ratified by 172 countries.

The United States has also refused to sign the Basel Convention's 1995 amendment, which strictly prohibits the exportation of hazardous waste from developed countries to developing countries—an amendment implemented by 32 of the 39 developing countries to which it applies. Not surprisingly, then, the Basel Action Network claims that "informed [recycling] industry insiders have indicated that what comes through their doors will be exported to Asia, and 90 percent of that is destined for China . . . and as recycling rates are expected to increase 18 percent per year, we can also expect the amount going for export to increase as well."

Conclusion

Environmental consciousness did not reduce American consumption at the end of the 20th century so much as transform it. The advent of municipal recycling programs, historian Susan Strasser argues, did not inspire Americans to change their consumption of goods so much as to separate the disposal of goods, which they put into the municipal waste stream with "the belief and expectation that the material would be reused."

Expanded access to credit, in the forms of credit cards and loans for real estate, education, and goods, increased consumer spending. Where consumers spent their money also changed. As suburban malls developed, the model of the downtown department store gave way to the big box store with a giant surface area and huge parking lots. Walmart, an Arkansas-based chain, became the model for the big box retailer, becoming the largest corporation in the United States in 2002, with an inventory that recalled the variety of the old county store on a far larger scale. Using the slogan "always the low-priced leader," Walmart developed supply-chain and inventory practices that reduced labor costs and compelled its suppliers around the world to engage in what critics called a "race to the bottom," reducing the quality of life of workers so consumers would have the cheapest possible goods. Walmart was not alone using this model, but it became the largest retailer practicing it.

Supermarkets also grew and adapted to changing consumer values. Concern about food quality fostered the development of organic co-operative stores in the 1970s. In 1980, an Austin, Texas, store named Whole Foods was founded and developed into a national chain of supermarkets selling organic foods. A far cry from the small local co-op, the Whole Foods store resembled the giant supermarkets found in 1960s suburbs, only with foods labeled organic. The foods evolved from products of small family farms to nationally distributed brands such as Cascadian Farm, Annie's Naturals, and Santa Cruz Organic. Even Walmart expanded its organic food offerings by 2010. The corporation with the most sophisticated inventory system in the world understood that foods labeled organic were highly valued by American consumers, and reacted accordingly.

The mass marketing of organic food joined the ever expanding world of goods and services Americans bought in the 60 years after World War II ended. Suburban sprawl was fueled by the use of automobiles consuming so much gasoline that the United States' imports of oil grew substantially from the 1960s onward. The geopolitical dimensions of this consumption were made clear when a 1973 Organization of the Petroleum Exporting Countries (OPEC) embargo led to long lines at gas stations, and when criticism of two wars the United States waged with Iraq involved allegations that the United States wanted to secure supplies of oil from the Middle East.

Houses evolved in the 1980s and 1990s into larger structures that earned the name *McMansions*. These houses contained more amenities (including home computers and computerized appliances) that consumed ever-greater amounts of energy. The rise of the Internet brought with it the rise of online shopping, made easier by the expansion of credit to make purchases. In the early 21st century, American consumers could even sell their unwanted possessions online via eBay.

All this consumption outpaced income after 1965 and levels of consumer debt rose. Sociologist Juliet Schor observed in her book *The Overspent American* that the number of goods and services Americans deemed necessities rose, and that even upper-middle-class Americans had concerns about being able to afford what they "needed" to consume, increasing loans for homes, education, automobiles, and balances on credit cards. One aspect of consumer debt hit a crisis when expanded mortgage lending led to a wave of home foreclosures in 2007, the collapse of several banks in 2008 and 2009, and years of depressed real-estate values in the United States.

In 2010, the U.S. Congress passed the Dodd-Frank Wall Street Reform and Consumer Protection Act, creating the U.S. Consumer Financial Protection Bureau (CFPB) organized by consumer advocate Elizabeth Warren in 2011. Time will tell what (if any) effect this law has on American consumers, but its passage reflected widespread concerns about American consumption in the early 21st century.

Blaise Farina
Rensselaer Polytechnic Institute
Carl A. Zimring
Roosevelt University

See Also: Barges; Culture, Values, and Garbage; Economics of Waste Collection and Disposal, U.S.; Environmental Protection Agency (EPA); Landfills, Modern; Post-Consumer Waste; Recycling; Recycling Behaviors; United States.

Further Readings

Cohen, L. *A Consumers' Republic: The Politics of Mass Consumption in Postwar America*. New York: Knopf, 2003.

Cross, G. *An All-Consuming Century: Why Commercialism Won in Modern America*. New York: Columbia University Press, 2000.

Gottlieb, Robert and Louis Blumberg. *War on Waste: Can America Win Its Battle With Garbage?* Washington, DC: Island Press,1989.

Komp, Catherine. "Barely Regulated, E-waste Piles Up in U.S., Abroad." *New Standard News* (May 11, 2006).

Kouwenhoven, J. A. *The Beer Can by the Highway: Essays on What's American About America*. Baltimore, MD: Johns Hopkins University Press 1988.

Ladou, J. and S. Lovegrove. "Export of Electronics Equipment Waste." http:/www.ban.org/Library/Features/080101_export_of_electronics_waste.pdf (Accessed June 2011).

Leuchtenburg, W. A. *Troubled Feast: American Society Since 1945*. Boston: Little, Brown, 1983.

Nash, R. *Wilderness & the American Mind*. New Haven: Yale University Press, 1982.

Pellow, David N. *Resisting Global Toxics: Transnational Movements for Environmental Justice*. Cambridge, MA: MIT Press, 2007.

Rathje, W. and M. Cullen. *Rubbish! The Archaeology of Garbage*. Tucson: University of Arizona Press, 2001.

Repa, Edward W. "U.S. Solid Waste Industry: How Big Is It?" *Waste Age* (Dec. 1, 2001).

Rogers, H. *Gone Tomorrow: The Hidden Life of Garbage*. New York: New Press, 2005.

Rome, A. *The Bulldozer in the Countryside: Suburban Sprawl and the Rise of American Environmentalism*. New York: Cambridge University Press, 2001.

Royte, E. *Garbage Land*. Boston: Little, Brown, 2005.

Schor, J. *The Overspent American: Why We Want What We Don't Need*. New York: Harper Perennial, 1999.

Stegner, W. "The Wilderness Idea." In *Wilderness: America's Living Heritage*, David Brower, ed. San Francisco: Sierra Club Books, 1961.

Strasser, S. *Waste and Want: A Social History of Trash*. New York: Metropolitan Books, 1999.

Tierney, J. "Recycling is Garbage." *New York Times Magazine* (June 30, 1996).

History of Consumption and Waste, World, 1500s

The 16th century marked a substantial increase in contact and trade across the Earth's civilizations. In particular, the expansion of European economies into the Americas developed the Columbian Exchange (named after Christopher Columbus's voyage to the New World), with significant consequences for resource extraction, nutrition, disease vectors, and redistribution of wealth and political power around the globe.

Expansion of the Islamic world in the 15th century, including taking control in 1453 of the vital trading city of Constantinople (subsequently renamed Istanbul) allowed Muslim traders to control most of the land and shipping routes between Europe, north Africa, and India, developing substantial amounts of wealth in Istanbul. One effect this had on European economies was a redirection of efforts to find western sea routes to India and China. As a consequence, several European nations established contact with the Americas, and in the 16th century, they extracted a variety of resources and foods from the New World. Foods, including maize, potatoes, and tomatoes, began their spread through Eurasia during this period. In return, infectious diseases from Eurasia had devastating consequences on indigenous American peoples, in many regions killing off more than half the existing population during this period. American foods had rather more beneficial effects on the rest of the world's population.

Consumption

In the 1500s, in the Americas, Europe, and other countries, consumption was dependent on necessities. Around the world, people were generally farmers or herders, raising more plants and animals each passing year. For example, consumption was based on eggplant, parsnips, turnips, and other vegetables in Europe. Peasants and nobles had different consumption habits because of different sources and opportunities. Peasants could mostly consume porridges made from grains with vegetables like onions, cabbage, and turnips. Bread was the basic food, and water sources were generally polluted because of not having a sewer system. Nobles, on the other hand, could also eat meat from pigs, chicken, or fish. Therefore, in the 1500s, waste was mainly based on food scraps, coal ash, and a small proportion of simple manufactured products, such as paper and glass.

Garbage

The garbage crisis occurred when people determined to move the garbage into dumps, which meant disposal everywhere. Larger pieces of waste were thrown into the streets and were eaten by semidomesticated animals; the rest was simply left or burned. Dumping, slopping, and scavenging were the methods of dealing with wastes in the Americas and Europe until the 1800s. A slopping and scavenging system remained the same for most developing countries.

In the 1500s, people dumped garbage into valleys or on the outskirts of cities. The trash was burned slowly, covered with dirt, or left in the open. This approach created health and environmental problems by attracting rats, dogs, cats, insects, and scavengers. Garbage was also scattered by wind, creating litter everywhere in towns.

The most preferred option of disposal was burying garbage, because it was the cheapest method. In Australia and China, for instance, food leftovers were fed to pigs, chickens, and dogs, and the rest was buried in the soil or left for composting. The people of Australia tried to protect the environment. For example, they used tools produced from natural materials and repaired or adapted them for other uses. They threw away the tools whenever there was no use left.

Reuse

People also used leather and feathers as recovered materials and green wastes as fertilizers. Timber was usually reused in shipbuilding and construction. Precious metals such as gold have always been melted down and reworked several times. Later, scrap metal and paper were also included in the recovery activities. For example, in the 1500s, for cementation of copper, Spanish copper mines used scrap iron. In England, Elizabeth I gifted special rights for the people who collected rags to make paper. Households kept rubbish indoors until it was removed by the pickers in Britain, which also prohibited keeping garbage in public waterways and ditches, because of increasing numbers of people throwing wastes out of windows and doors. For instance, garbage was also piled outside the Paris gates so high that it nearly restricted city defense in the 1500s. In Germany, the government used wagons to carry products into the cities and waste outside the cities. In Philadelphia, the Rittenhouse Mill produced paper from recycled waste papers and rags. In Germany, people started to place their papers in wrappers. In the 1500s, people preferred to throw away their waste in random ways and governments tried to prevent those behaviors by law, with little effect.

Energy

In the 1500s, varied energy sources had different impacts on the environment. For example, people used animal-fat candles for light, peat and charcoal for cooking and heat, wind for sailing, and water wheels for grain crushing. They also manufactured a small amount of goods in their houses. Charcoal is one of the most environmentally damaging fuels in the world and is a significant source of indoor pollution. In the 1500s, it was produced in traditional kilns and mounds that gave off dangerous smoke.

Disease

Trash has played a significant role in world history. Diseases such as cholera, bubonic plague, and typhoid fever changed the populations of continents in the Middle Ages. The diseases were caused by contaminated water sources and spread by rats. For example, the plague killed 30–60 percent of Europe's population—almost 400 million people. The disease was caused by garbage on unpaved streets, which were attractive places for rats. Fleas on the rats then spread the disease to humans. The plague created religious, economic, and social problems throughout the world, and Europe needed 150 years to recover.

Hasret Balcioglu
Cyprus International University

See Also: Bubonic Plague; Germ Theory of Disease; Miasma Theory of Disease; Open Dump; Recycling.

Further Readings

Brewer, John and Roy Porter, eds. *Consumption and the World of Goods*. New York: Routledge, 1993.

Glover, M., ed. *SWID Handbook*. Australia: Solid Waste Infrastructure Development Group of the Grocery Manufacturers of Australia, 1995.

Haynes, Douglas E., et al., eds. *Towards a History of Consumption in South Asia*. Delhi: Oxford University Press, 2009.

McCracken, Grant. "The History of Consumption: A Literature Review and Consumer Guide." *Journal of Consumer Policy*, v.10 (1987).

History of Consumption and Waste, World, 1600s

Major events in the 17th century that affected production and consumption included the Little Ice Age, the plague, and European colonization of Africa, Asia, and the Americas. Crop failures, famines, and the plague reduced the European population and increased the value of labor. Production, trade, and consumption grew with the incomes of the rising numbers of wealthy merchants and middle-class craftsmen and women, whose families produced an increasing diversity of consumer goods, with more venues for purchasing them, as guilds lost control.

Women's domestic production, including textiles, butter, cheese, eggs, chickens, beer, and cider were important parts of economies. International trade was dominated by men, although some Dutch women were renowned in the North American fur trade. The Little Ice Age increased the demand for warm woolen clothing, fur coats and hats, and led to the innovation of fireplaces, starting in elite homes in the1450s.

Colonies

Feudalism began to break down as landlords enclosed common lands for sheep grazing, forcing tenant farmers and dairymaids into the towns or to colonies, despite resistance. Many poor people paid for their passage to colonies by indenturing themselves to work for several years for wealthier colonists. Governments shipped their destitute and criminals to colonies. Inexpensive European consumer goods, such as glass beads, copper beads, rings, pendants, kettles, iron knives, hoes, red cloth, and old guns, were exchanged with Native Americans for valuable beaver furs that were used in Europe to make expensive top hats. Native American women who were traders or "country wives" of European traders consumed trade goods for status display for themselves and their families. Most of the trade in consumer goods within the American colonies was conducted through barter because European money, or specie, was rare.

Production and consumption increased as the inexpensive or free resources of colonies enriched many European colonists, trading companies, royalty, and aristocracy. A triangular trade developed: African slaves were shipped to the Americas, where they and enslaved Native Americans inexpensively produced new crops such as tobacco and potatoes, as well as cotton, rice, and sugarcane in the Caribbean. The sugar was manufactured into molasses and rum to sell in New England and Europe. The great popularity of tobacco, and the development of the European clay pipe industry, is apparent at archaeological sites, where quantities of clay pipes suddenly appear in layers from the later 17th century. Colonial timber was in demand in Europe, where the use of charcoal for small-scale industrial production of goods such as glass, copper, and iron had so depleted the supply of wood that in the 1550s King Henry VIII reserved the remaining wood to build ships for the English navy. New England households were assigned a quota of required textile production by women and children.

Pottery

In the 1600s, food was stored and prepared in woodenware, redware, or stoneware pottery vessels. Food was consumed using predominantly woodenware and horn cups in the lower to middle classes, pewter in the upper-middle class, and silver in the elite class. Chinese porcelain increasingly replaced silver and pewter over the century. Shared bowls, plates, and mugs were common, especially in the lower classes and in the colonies. Cutlery included knives and spoons. Forks were introduced to Europe by a Byzantine princess who had married the doge of Venice in 1075. Forks became popular in Italy with the elite in the 1300s, spread to the merchant classes by 1600, then to Western Europe by the mid-17th century, and its colonies in the early 18th century.

Seventeenth-century pottery produced and consumed in Europe and its colonies consisted of earthenware and stoneware. Traditional lead-glazed redware pottery was manufactured for everything from milk pans for dairies to plates and mugs of slipware made in England or Germany. Seventeenth-century slipware decoration often involved finer lines of white slip than in the previous century. Since the 16th century, lighter pink earthenware covered with a white opaque tin glaze and decorated in multicolored designs on large status-display plates was called faience in France, galleyware and blue-dash chargers in England, and majolica in Spain and Italy. In sgraffitoware, patterns of lines were scratched through the tin glaze to reveal a dark redware body. Germany, the Netherlands, and England produced brown and gray stoneware mugs, tankards, pitchers, jugs, and food-storage vessels. Fine small brown stoneware teapots appeared shortly after tea was imported from China ca. 1650. In the 17th and 18th centuries, the Dutch imitated Chinese porcelain with blue hand-painting on white tin-glazed earthenware, called Delft for the Dutch town that first manufactured it. The number of factories in Delft increased from 14 in 1650 to 30 in 1670, as the trade became international. Starting in the late 17th century, potters such as the Elers brothers became famous for their small, thin, fine red stoneware teapots with applied vines and crabstock handles. Pottery consumption increased markedly after 1650 along with the increasing importation of Chinese porcelain and increasing production of Delftware and specialized stoneware, such as Westerwald and Bellarmine jugs.

Trading Companies

European merchants grew wealthy and supplied an increasing diversity of consumer goods from trade forts and colonies from Africa to the Far East, starting with Portugal in the 1400s. In the 17th century, the English and Dutch East India Companies ended Portuguese monopolies in the major trades for African gold, ivory, coffee, and slaves;

Indian cotton, ivory, tea, cinnamon, indigo dye, saltpeter for gunpowder, and opium; Indonesian pepper, nutmeg, and other spices; and Chinese and Japanese porcelain, silk, and tea. The Dutch East India Company imported the most Chinese porcelain, rapidly increasing from 50,000 pieces per year in 1605 to 100,000 pieces by 1610 to 200,000 pieces by 1620. When the Chinese porcelain trade was interrupted by unrest during 1644–64, Japanese Imari-style porcelain was imported to Europe. The king of Portugal outlawed the trade in Chinese slaves in 1624. Until it was outlawed in the 19th century, the African slave trade provided free labor both in Europe and in the American colonies, enriching Europeans.

Wealth Disparity

Increasing wealth among Europeans led to increasing demand for consumer goods, starting in Dutch towns and then spreading to surrounding farms. In the first half of the 17th century, wealthy urban Dutch merchants began to display Chinese porcelain, silk, ivory, and new styles of furniture. After 1650, they started to build brick townhouses with separate drawing rooms, dining rooms, and bedrooms. A study of probate inventories in and around a small industrial and market town near Amsterdam found that half of the town's households had Chinese porcelain in the 1650s, and half also had Delft by the 1680s. In contrast, by the 1690s, only 20 percent of the surrounding farms had porcelain, and 75 percent had Delft. Towns contained more wealthy households than the countryside.

The new architecture and consumer goods spread throughout Europe and its colonies in wealthy and then middle-class households. A study of 3,000 British median-probate inventories from 1670 to 1730 found that during this period the percentage of inventories at least doubled, listing a clock, pictures, looking glasses, curtains, tea and coffee utensils, saucepans, and earthenware dishes. Other studies found an increase in interior decoration and new furniture. British inventories in County Kent listed five new types of furniture: court and press cupboards, chests of drawers, cabinets, and new styles of chairs and tables. In the early 17th century, just 30 percent of households owned any of these items, compared to 80 percent of households by 1750. In the same period, the median number of pieces of furniture in wealthy and merchant households doubled, from 12 to 24. Working-class and many rural middle-class houses retained the medieval vertical plank, wattle and daub, or half-timber house styles, with few if any separate internal spaces or specialized rooms and rarely any of the new furniture.

Tea Consumption

The English East India Company defeated Portugal to dominate the trade with India. Tea first arrived in Englishmen's coffeehouses in the 1650s. Catherine of Braganza, wife of Charles II, introduced tea to the women of the English royal court in 1662, where it became very popular as the most high-status beverage. Tea became associated with women in the court and aristocracy, and then spread to wives in the merchant, artisan, and laboring classes, becoming the most popular beverage in England by the end of the century. By the 1690s, tea was already considered an "ancient custom" among ladies. Women controlled all aspects of the tea ceremony, from buying tea and selecting teaware to inviting the guests and serving the tea. Documentary research has revealed that through their tea ceremonies, women displayed their family's status and controlled the reputations of neighbors.

Within patriarchy, 17th-century wives in Europe and its colonies held a few rights as independent adults, such as the ability to own land, control their inherited property, and operate businesses, ranging from goldsmiths, blacksmiths, printers, and sailmakers to weavers, brewers, vintners, bakers, cooks, fishmongers, peddlers, and storekeepers. Women produced and sold a wide range of consumer goods, most often in partnership with husbands or as widows. Wigs also became increasingly fashionable from the second half of the 16th century through the 17th century. Sumptuary laws were abandoned in England and Holland by the early 17th century, while in France and Spain, greater restriction of fashionable dress to elites lasted until the late 18th century. Cotton-printed calicoes from India democratized fashion in the 18th century because the cloth could be afforded and was popular from the elite to the servant classes. As the Little Ice Age abated in the 18th century, clothing shifted from heavier to lighter wools, mixed linsey-woolsey, linen, and cotton. Consumption of clothing increased due to the shortened fashion cycle of clothes as the elite instituted new styles to overcome the trickling down of older styles into the middle classes.

Human and Animal Waste

Women and girls were usually responsible for house cleaning and discard, whether by sweeping trash and pet wastes out the doors, or throwing garbage, or human waste from chamberpots, out of windows. In towns and cities, pedestrians walking beneath windows could be hit with waste from windows as well as splashed with mud and filth by carriages. Despite late-17th-century laws requiring fencing of domestic animals, pigs, cattle, sheep, and goats roamed freely, depositing manure in fields and village and town streets alike. This situation led to the chivalrous custom of men walking on the outside edge of sidewalks and sheltering women under the second-story overhangs of colonial garrison houses. Dogs and cats defecated freely in private and public buildings and there was no municipal garbage collection. Pigs and goats roamed through neighborhoods, eating garbage and gardens. Farmers' homes and yards were filled with surface scatters of garbage and trash.

Since the 1400s, widespread plagues led to the European belief that bathing caused disease. The normative practice involved washing only hands. Starting around 1550, changing into a clean linen undershirt for men, or a chemise for women, was considered safer, more reliable, and scientifically superior to washing with water because white linen was believed to attract and absorb sweat.

In the 17th century, a rage for clean white linen developed among the royalty, the aristocracy, and then the rising middle classes. At least one clean linen shirt or chemise was required daily per person, developing a growing market for white linen. Medieval paintings of bathhouse scenes were replaced by paintings of clean linen undershirts spread in bleaching fields, in neat stacks in linen chests, and peaking out from under overshirt necklines, cuffs, slashed sleeves, and doublets. The normative practice of monthly washing by slaves or servants meant that probate inventories and wills listed more linen shirts and chemises than any other article of clothing. In contrast, a piece of flannel was commonly pinned up over the lower bodies and legs of infants from shortly after birth until the flannel envelope fell off.

Further, some people tightly bound infants against a board to straighten their backs and lessen crying by slowing their metabolisms. Starting in the late 16th century, etiquette books urged people not to relieve themselves in public (suggesting that this had been a common practice), but to instead use privies. In urban areas, people were most likely to maintain privies by hiring scavengers to clean them out at night, while in rural areas, full privies were simply covered with dirt and a new privy hole was dug. People were accustomed to the smells of unwashed bodies and waste that surrounded them.

Suzanne M. Spencer-Wood
Oakland University

See also: Bubonic Plague; History of Consumption and Waste, Renaissance; Human Waste.

Further Readings

Anderson, Bonnie S. and Judith P. Zinsser. *A History of Their Own: Women in Europe From Prehistory to the Present, Vol. 2*. New York: Harper & Row, 1988.

Ashenburg, Katherine. *The Dirt on Clean: An Unsanitized History*. Toronto: Vintage Canada, 2008

Burton, William. *A History and Description of English Earthenware and Stoneware (to the Beginning of the 19th Century)*. London: Cassell, 1904.

Collins, Gail. *America's Women: 400 Years of Dolls, Drudges, Helpmates, and Heroines*. New York: Harper Perennial, 2003.

Corson, Richard. *Fashions in Hair: The First Five Thousand Years*. London: Peter Owen, 1965

De Vries, Jan. *The Industrious Revolution: Consumer Behavior and the Household Economy, 1650 to Present*. Cambridge: Cambridge University Press, 2008.

Hildyard, Robin J. C. *European Ceramics*. Philadelphia: University of Pennsylvania Press, 1999.

Ukers, William H. *All About Tea*. Eastford, CT: Martino, 2007.

Wiesner, Merry E. *Women and Gender in Early Modern Europe*. 2nd ed. Cambridge: University of Cambridge Press, 2000.

History of Consumption and Waste, World, 1700s

The 18th century brought an increase in the world of goods as trade expanded and industrialization in Europe emerged to increase that continent's economic and political power. The resulting economic activity had important and enduring consequences for consumption patterns across the Earth.

While economies in the Islamic world, India, and China remained vibrant, a significant development during the period was continued increase in the strength of European wealth and power that had emerged with the dawn of the Columbian Exchange (between the New World and Old World) at the end of the 15th century. In the second half of the 18th century, major increases in consumption and discard resulted from the Industrial Revolution, in which factories and mechanization greatly increased production and decreased costs of consumer goods, including textiles, ceramics, buttons, buckles, boxes, and toys. Water-powered machinery was increasingly replaced by steam-powered machinery, spreading from British factories after 1775. The Industrial Revolution was led by the shift from charcoal to coal and coke, permitting large-scale production of iron and steel used for the construction of steam engines, water pumps, and bridges. Industrial waste was discarded across landscapes. New kinds of status-display goods were introduced to the elite in European colonies, such as individual place-settings of white ceramic tableware, tea caddies, forks, brass candlesticks, clocks, and scientific instruments such as thermometers, barometers, globes, compasses, and telescopes. However, consumption and discard of these status-display items was rare except for ceramics; whiteware ceramics are the most commonly excavated artifacts and were discarded by all classes.

Ceramics

New kinds of ceramics invented in the 18th century were discarded in trash pits and surface scatters that are excavated at historic sites. While traditional craft production of lead-glazed redwares and salt-glazed gray and brown stonewares continued, the Enlightenment led to scientific experiments, especially by English potters, resulting in an increasing variety of whiteware ceramics. Tin-glazed, hand-painted Delftware was produced until 1790 to imitate expensive, Chinese porcelain. Expensive white, salt-glazed stoneware was produced until the 1770s. Thomas Wheildon invented and produced creamware with a mottled glaze as well as rococo vegetable and fruit-shaped tureen lids and teaware from 1740 to 1770. In 1750, Josiah Wedgwood invented a further-refined creamware with a slightly green glaze that he marketed to the elite as Queensware after Queen Charlotte admired it in 1765. Subsequently, Wedgwood marked down the price of plain creamware to sell it first to the middle class and then to the working class. In 1780, Wedgwood invented pearlware by further refining the creamware body and its glaze by adding cobalt, which made pearlware appear whiter, although it was actually greenish-blue. Undecorated, edged, or hand-painted creamware and pearlware predominate at sites in England and its colonies in the second half of the 18th century.

Consumer Culture and Choices

Consumer choices expanded and diversified as factories decreased the cost of producing many goods, resulting in lower prices. The amount of discard increased in the 18th century as fashions changed rapidly with inventions of new consumer goods, from clothing styles to ceramics. Consumer choices about what to buy and whether things were carefully curated or carelessly used and frequently broken and discarded were conditioned by many interrelated factors, such as the availability of goods, wealth, prices, taste, symbolic meanings of objects, desire to display socioeconomic status, and intersection with gender ideology, religious ideology, race, or ethnicity. Socioeconomic status was important, but it was not the only factor in consumer choices. For instance, increasing ownership of scientific instruments by the elite not only displayed their ability to afford these instruments but also symbolized their interest in the development of science in the Enlightenment. In another example, some religious groups, such as Puritans, Quakers, and Shakers, were against status display. Consumer choices could also be conditioned by the complex intersections of multiple identities, such as gender, class, race, ethnicity, or religion.

The availability of consumer goods and their falling prices also led to changes in consumer choices by people in different classes. Communal eating with hands from a shared bowl was common among the working classes until at least the late 18th century. But by 1730, the elite added the innovations of individual plates, glassware, and forks to their previous use of spoons and knives for eating. Over the century, the elite kept up with rapidly changing fashions in dress and the latest styles of porcelain, glass, and silver tableware, while the middle class demonstrated respectability with pewter and ceramics imitating porcelain, and the working classes, impoverished by low wages in laissez-faire capitalism, predominantly used secondhand or old styles that were out of fashion.

Feminist historical archaeologists have argued for the importance of analyzing gender power dynamics in the past, including who controlled household consumer choices and discard of material culture. The person who controlled the selection of a household's artifacts also controlled how the family displayed its status to friends and business acquaintances. In 18th-century Europe and its colonies, patriarchal men ruled families. Wives and children had the status of minors, dependents, and chattel and had very few civil rights. While husbands had the legal right to control their families, feminists still found evidence of the social agency of wives as producers and distributors of consumer goods. Wives were often their husbands' business partners and traders of their own domestic products in Europe and its colonies. Women's production of butter, cheese, textiles, beer, and cider were important items of exchange both within the American colonies and in international markets. Many women operated stores selling a variety of goods, from imported exotic items such as coffee and chocolate to ordinary food items, nails, medicines, textiles, or crockery. Some women produced consumer goods that they sold in their communities by working in occupations such as blacksmith, tanner, printer, seamstress, and occasionally tanner or slaughterhouse operator. It was also common for wives to operate their family businesses when husbands were absent or dead. In contrast to the dominant gender ideology that women and their domestic sphere did not participate in men's sinful public-sphere capitalism, women often produced and distributed consumer goods for sale in men's public sphere.

Women also often made family consumer choices of status-display items, including ceramics, since at least the 18th century. Documentary evidence indicates that although men often bought ceramics and other consumer goods because they legally owned all the earnings of family members, fiancées and wives often selected the ceramics acquired by men. Women also often bought ceramics, especially teaware. In the mid-18th century, a letter in the *London Gazette* discussed how all the elite ladies went to Warburtons to buy the most popular queensware ceramic tableware. While elite wives usually bought the teaware that they used in serving tea, their husbands often bought the large sets of matched dinnerware. In the Americas, elite men such as George Washington and Thomas Jefferson bought large sets of ceramic tableware, but it is possible that their wives influenced the selection of ceramics. Throughout the 18th century, women bought household status goods and performed status rituals such as serving tea, as well as carving and allotting dinner meat to guests.

Females continued to perform house cleaning and discard of trash, garbage, and human and animal wastes. Excavations of four 18th-century American house sites have dated ceramic dumps to 12 changes in female heads of household, who each discarded the whiteware of the previous female head of household and bought new whiteware. These house sites provided excavated evidence of several wives, single women, and widows who controlled their household acquisition of whiteware ceramics for status display at dinners and teas. Since discard did not occur until there was a change of the woman in charge of the housework, this evidence indicates that women, rather than men, made the decisions to discard old ceramics and select new ones.

Waste Management

As early as 1700, city and town ordinances were passed prohibiting people from discarding waste in streets. After 1750, Europeans increasingly dumped trash and garbage in square backyard pits, often seven-feet-deep, probably in reaction to widespread cholera epidemics. These orderly pits may have developed from the Enlightenment belief in masculine rational-scientific conquest and ordering of irrational mother nature. This belief also underlay the construction of elite Georgian houses, with half basements raised above leveled land, shaping female nature to men's cultural constructions. At less wealthy house and farm sites, surface scatter continued, and there was still no garbage collection. However, gardens and fields began to be fenced to prevent animals from eating crops.

Bathing was still considered unhealthy and washing was usually limited to hands and face, with feet becoming recommended by French etiquette books ca. 1750. Etiquette books did not recommend a complete bath until 1820. Some scholars believe that 20–30 years could pass between bathing and hair washing. Wigs declined in fashion after the death of King Louis XIV in 1715 and became unfashionable after the French revolution. Daily wearing of a clean white undershirt or chemise was still considered the epitome of cleanliness. During the 18th century, the production and consumption of linen continued to grow and most gentlemen owned a minimum of 12 undershirts; only one set of clothing might be owned by a frugal academic. Clothing was washed at most once a month, and often not at all on the American frontier. Human waste of infants often accumulated in diapers that were not washed but might be placed by the fire to dry and then put back on the baby. Doctors began to decry this practice in the 19th century. Adults relieved themselves in the bushes more often than in privies, which landowners did not clean often enough. Although three water closets were invented in the 1770s, few people installed them in their homes. Waste disposal and cleanliness were only slightly improved from the 17th century.

Suzanne M. Spencer-Wood
Oakland University

See Also: History of Consumption and Waste, U.S., Colonial Period; History of Consumption and Waste, World, 1500s; History of Consumption and Waste, World, 1800s; Human Waste; Residential Urban Refuse.

Further Readings

Agnew, Aileen B. "Women and Property in Early 19th Century Portsmouth, New Hampshire." *Historical Archaeology*, v.29/1 (1995).

Ashenburg, Katherine. *The Dirt on Clean: An Unsanitized History*. Toronto: Vintage Canada, 2008.

Collins, Gail. *America's Women: 400 Years of Dolls, Drudges, Helpmates, and Heroines*. New York: Harper Perennial, 2003.

Cumbler, John T. *Northeast and Midwest United States: An Environmental History*. Oxford: ABC-CLIO, 2005.

Goodwin, Lorinda B. R. *An Archaeology of Manners: The Polite World of the Merchant Elite of Colonial Massachusetts*. Boston: Kluwer Academic/Plenum Publishers, 1999.

Leone, Mark P. "The Georgian Order as the Order of Merchant Capitalism in Annapolis, Maryland." In *The Recovery of Meaning: Historical Archaeology in the Eastern United States*, Mark P. Leone and Parker B. Potter, eds. Washington, DC: Smithsonian Institution Press, 1988.

Martin, Ann Smart. "Frontier Boys and Country Cousins: The Context for Choice in Eighteenth-Century Consumerism." In *Historical Archaeology and the Study of American Culture*, Lu Ann De Cunzo and Bernard L. Herman, eds. Knoxville: University of Tennessee Press, 1996.

Spencer-Wood, Suzanne M. "Introduction." In *Consumer Choice in Historical Archaeology*, S. M. Spencer-Wood, ed. New York: Plenum, 1987.

History of Consumption and Waste, World, 1800s

The 19th century was a critical period in the development of modern modes of consumption and waste disposal. Europe had begun intensive industrialization in the 18th century; by 1900, this had extended to much of the world, transforming economies from Japan to the United States. Transportation, including the expansion of rail and transition from wooden to iron ships, expanded the distribution of foods and durable goods worldwide. Factory production of textiles, machinery, and increasingly specialized housewares transformed the consumption habits of millions of people, not to mention their workplaces and residences. With rapid industrialization came exponential population growth centered in urban areas. This led to serious strain being placed on existing sanitary infrastructure. Coupled with a growing awareness by reformers of the role of environmental factors in the spread of disease, a sanitary movement emerged, which championed the enforcement of strict hygiene and the planning of citywide waste disposal networks. In the first half of the century, these developments were focused in Europe and particularly in its largest city, London; by the end of the century, they had spread worldwide, particularly to the emerging metropolises in North and South America.

Sanitation

Throughout the 19th century, London was the largest city in the world, with over twice the population of its nearest rival, Paris. Even before the rapid deterioration of London's sanitary state in the first half of the 19th century, the city's watercourses were never particularly salubrious, as testified by Ben Jonson's mock epic poem describing a nauseating journey along the river Fleet in the early 17th century. But with the population of London increasing almost threefold from 1800 to 1850, the city's sanitation, or the lack of it, became a dominating concern. This exponential growth of the city's population, largely concentrated in already densely crowded areas of the city, led to serious strain being placed on a once effective and sustainable system of natural drainage. London's many rivers—tributaries of the Thames, such as the Fleet, Westbourne, and Tyburn—had, up until the beginning of the 19th century, provided a ready means of draining rainwater within the built-up area.

The gradual expansion of this built-up area led to these rivers being systematically built over, putting an ever-increasing strain on this existing system, which had been in place for centuries. The usual method of disposing household sewage—even up until the mid-19th century—was to empty it into pits, known as cesspools, located close to dwellings, with most households having access to one. Workers, known as nightmen, removed the sewage from cesspools at night and were able to dispose of it at a profit to farmers, whose fields were close to the city limits. The city's population also benefited from developments in the supply of piped water in the early 19th century. The substitution of cast iron for wood in the manufacture of water mains meant that the new London water companies were able to deliver a more regular supply at higher pressure. Coupled with the invention and subsequent popularity of the water closet, the increased availability of water resulted in much greater volumes of water and sewage being discharged into both London's rivers and its existing sewers.

By the 1840s, densely populated areas of London became the focus for sustained investigation by would-be sanitary reformers, particularly insalubrious areas of the city, such as the Kensington Potteries and parts of the parish of St. Giles in Westminster. These reformers, such as Edwin Chadwick (1800–90), began to investigate the detrimental effects of poor sanitation and to use the evidence they collected—often in the form of lurid descriptions—to argue the case for reform. Chadwick's interest in sanitation arose partly as a result of his work as secretary to the Poor Law Commission, established in 1834 under the Poor Law Amendment Act.

The new commission attempted to reform London's administrative boundaries by setting up a centralized, government-funded body to administer poor relief to its less fortunate citizens. The commission overrode London's existing system of governing welfare, which was legally and administratively a loose conglomerate of some 300 individual parishes and wards. From his involvement in the commission until the publication of his "Report on the Sanitary Conditions of the Labouring Population of Great Britain" in 1842, Chadwick used his influence to focus the attention of the commission on the relationship between poor sanitation and disease, arguing that those who required poor relief were often victims of the unsanitary environments in which they lived.

In 1848, a new centralized governing body, the Metropolitan Commission of Sewers, headed by Chadwick, was set up in late 1847 to replace the heterogeneous group of governing bodies that had existed for centuries, the London Sewer Commissions. The Metropolitan Commission of Sewers was formed with an explicit goal: to plan and construct a new citywide sewer system for London. Even though such a system would not be built for another decade, 1848 marked the genesis of the main drainage system in that it saw the emergence of a new way of seeing the city's sanitation. The period from 1848 to 1868 represents the years when London's new system of sewers was first planned and then constructed in its first and most important phase. Designed by the engineer Sir Joseph Bazalgette (1819–91), the main drainage system consists of five distinct, large-scale sewers that cross London either side of the Thames from west to east, converging in outfall sewers running roughly parallel to the river. These are known as "intercepting" sewers in that they intercept waste from existing street sewers and divert it to outfalls on the Thames outside the city limits. The sewage was originally discharged into the river at high tide to prevent it from flowing back into the city area. Included in the main drainage system are four principal pumping stations, situated at points in the system where the sewage needed to be lifted from low-lying areas up to the level of the river.

Throughout the 19th century, in both Britain and Europe, experts debated how best to dispose of urban sewage. Many were guided by one overarching principle: the hope that the agricultural use of urban sewage could finance much urban improvement. Using human wastes as agricultural manure reflected a growing interest in Britain in the 1840s regarding the possibility of turning waste into profit and formed the subject of many parliamentary committees and debates in the second half of the 19th century. Ideas on sewage utilization, or recycling, were derived from the German chemist Justus von Liebig (1803–73), who was one of the leading promoters of sewage recycling as an essential requirement for the long-term sustainability of agricultural productivity. On the one hand, sewage recycling made economic sense; as Liebig argued, it would release a previously untapped mine of gold and simultaneously sustain rapid population growth in urban areas and the agricultural productivity necessary to sustain that growth. And yet, sewage recycling was also a solution to an important theological dilemma: was human waste really part of God's bountiful creation? In the mid-19th century, natural theology made commonplace the notion that the character of God was evident in the laws of nature. Waste and decay were seen as a perversion of natural cycles, which occurred only when organic wastes were not quickly returned to the soil. The almost obsessive interest in sewage recycling in this period was one way of resolving the theological dilemma raised by the idea of human waste. However, despite the power of these arguments, sewage recycling was an expensive and inefficient process and—although tried in limited areas of London and Paris—was eventually replaced by the wholesale flushing of waste into urban rivers. From the 1880s onward, the development of increasingly sophisticated methods of treating sewage meant that the question of sewage disposal lost its previous sense of urgency; human waste was accepted as just that—waste—and the emphasis shifted to how best to limit its polluting nature.

London's main drainage system was the first citywide sewer system and perhaps the most influential example in terms of its impact on other developing urban centers. Bazalgette's work as a consultant demonstrates this; from the 1850s onwards, he advised on sewerage schemes for both provincial towns and cities in Britain, including Bristol (1863), Oxford (1865), Belfast (1866), and Northampton (1871), as well as the foreign capitals of Budapest (1869–74) in Hungary and Port Louis (1869) in Mauritius. By the early 20th century, Bazalgette's main drainage system had become the model, whether directly or otherwise, for any metropolis aspiring to be "modern" in terms of the disposal of its wastes and was widely adopted in the rapidly expanding cities of Europe and the Americas.

Paul Dobraszczyk
University of Reading

See Also: Population Growth; Public Health; Sewage; Sewers.

Further Readings

Allen, M. *Cleansing the City: Sanitary Geographies in Victorian London*. Athens: Ohio University Press, 2008.

Dobraszczyk, P. *Into the Belly of the Beast: Exploring London's Victorian Sewers*. Reading, UK: Spire Books, 2009.

Goodman, D. C. and C. Chant, eds. *The European Cities and Technology Reader: Industrial to Post-Industrial City*. New York: Routledge, 1999.

Hamlin, C. *Public Health and Social Justice in the Age of Chadwick: Britain 1850–1854*. Cambridge: Cambridge University Press, 1998.
Melosi, M. V. *The Sanitary City: Urban Infrastructure in America From Colonial Times to the Present.* Baltimore, MD: Johns Hopkins University Press, 2000.
Reid, D. *Paris Sewers and Sewermen: Realities and Representation.* Cambridge, MA: Harvard University Press, 1991.

History of Consumption and Waste, World, 1900s

The 20th century was the most devastating period of world history in terms of consumption and waste. It began with the declining age of European empires and ended with the increased pace of globalization and worldwide commodity consumption. During this period, large political, social, and technological changes resulted in new forms of consumption and commoditization. Waste emerged as a major problem inherited from the Industrial Revolution and was intensified to the extent that, at the end of the century, a social and political ideology known as environmentalism or "green" politics emerged in many nations. New forms of waste appeared as a result of various technological and scientific innovations. The 20th century, in the words of historian John McNeill, was unusual for the intensity of environmental change and the centrality of human effort in provoking it.

Mass Commodity Culture

The intensification of a mass commodity culture and the increasing waste as a result of industrialization, consumption, and urbanization bewildered various social and economic thinkers at the beginning of the 20th century. The establishment of a world capitalistic system, the innovation and circulation of large numbers of new commodities, the growing sizes of the cities and their populations, and new technological discoveries became influential processes of social and economic change.

The intensification of capitalist culture, which would become focused on mass production and consumption, had been predicted by Karl Marx. Marx argued persuasively that human labor had become alienating and commodities had come to replace human relationships. At the end of the 19th century and the beginning of the 20th century, another economist and sociologist argued that waste was the result of a leisure class that had developed through business and had as its main preoccupation conspicuous consumption. Thorstein Veblen developed a theory of a leisure class, which was based on a lavish acquisition of goods for the purpose of displaying wealth, social status, and social differentiation. Such attitudes were generally viewed as negative consequences of mass consumption in the 20th century.

Two World Wars

The most devastating events of the 20th century in terms of consumption and waste were the two world wars and the arms race, which brought large social, political, and technological changes. This first period of the century could be described as the age of the catastrophic wars. During this period, the capitalist and liberal Western societies lost faith in their economic and constitutional values, a bourgeois culture was replaced by an emerging middle class, material and moral progress was interrupted by acts of violence and intolerance, and the Eurocentric colonial empires whose soldiers had conquered large parts of the world began to break apart into nation states and independence movements. The first four decades of the century brought new forms of waste that would have been unimaginable in the 19th century and resulted in a huge consumption of continuously developing weapons that would be used even in the most remote islands of the Pacific.

In World War I, tanks, grenades, machine guns, chemical weapons, and aircraft were used for the first time in a war and completely changed the strategies of the empires involved. The use of chemical weapons such as tear gas and mustard gas, along with lethal agents that included phosgene and chlorine, became a regular part of warfare. Despite the fact that only a minority of soldiers were killed due to chemical warfare, the pollution of the environment during this period was large scale. In France, for example, at the end of the war, 40,390 square miles were cordoned off because of unexploded materiel. Shells of chemical weapons are still found regularly by farmers or building operations in areas where the war took place. The use of chemical agents in warfare was internationally condemned with the Third Geneva Convention signed in 1925. However, more than a decade later, in World War II, poison gas played an important role in the Holocaust.

As a result of the chemical innovations of World War II, new forms of plastic material were developed. Polystyrene was among the first and most widely used plastic materials in everyday life. Polyamide, also known as nylon, was developed a few years later and became widely used in various new consumer goods. Stockings, for example, became widely popular and consumer demand resulted in the nylon riots of 1945. In the United States, 4 million pairs nylon stockings a day were sold in 1939, but when the country entered the war, production companies interrupted their operations and produced only war materiel. As a result, a stocking panic emerged as the few nylon stockings that could be found in the market were produced in the prewar era.

The wide use of polystyrene in food packaging, clothing, furniture, and everyday consumer goods had an inconceivable environmental impact. A major cause is the fact that discarded polystyrene does not biodegrade for hundreds of years, it is resistant to photolysis, and the majority of these products cannot be recycled.

Warfare was further advanced during World War II, with major innovations in aircraft, tanks, machine guns, rockets, and nuclear weapons. The use of the atomic bomb in Hiroshima and Nagasaki at the end of the war signaled a new era of nuclear warfare that had devastating effects on humans and the environment. The discovery of nuclear fission during World War II resulted not only in an arms race between the Soviet Union and the United States but also in commercial energy use by various countries such as France, Japan, and Germany. Since then, the extended use of nuclear energy and the waste that such nuclear power plants produce has been the most threatening and irreversible process of environmental destruction. Large-scale nuclear accidents in the 20th century include the Chernobyl disaster in 1986, which released large quantities of radioactive contamination in the Soviet Union and Europe.

During the first period of catastrophic wars (1900–45), a large proportion of the world's population lost faith in liberal capitalism and adopted socialist and communist ideas and political systems. In Russia, the revolution of 1917 resulted in the creation of the Soviet Union, the world's largest state, covering an area over 13,918,715 square miles, and which became head of a later socialist network that included China and amounted to one-third of the human population. Within this constitutional socialist state, commoditization and consumption took different trajectories from liberal democracies, as Western commodities were scarce, limited, and even prohibited.

The unease with liberal capitalism grew greater with the world economic crisis of 1929 that brought instability and insecurity not only to the weakest but also to the strongest capitalist countries of the time. Even the economic system in the United States that was not ruined by World War I almost collapsed. It was within this context that fascism and its sphere of influence advanced in most parts of the world and resulted in World War II.

Prosperous Age

The second period of the 20th century, which can be identified as the prosperous age of world history, are the two decades following World War II (1947–73). This era was extraordinary in terms of social, economic, and cultural transformation. It included the period of the cold war, the decolonization process, and the division of the world into spheres of socialistic and capitalistic influence. Within this context, production, consumption, and waste emerged as fundamental categories in an effort to assess and understand the impact of those transformations. This period was probably the most influential ever recorded in terms of consumption and waste, as industrial capitalism expanded the Fordist system of production worldwide and had devastating effects for the environment. The standardization of production, the establishment of the assembly line, and the increased salaries of workers resulted in the invention of an enormous number of mass commodities that could be bought and consumed by the workforce.

The main philosophy of the period was the combination of mass production with mass consump-

The most devastating events of the 20th century in terms of consumption and waste were the two world wars and the arms race. During World War I, new military hardware and chemical weapons completely changed wartime strategies. Environmental pollution from chemicals was widespread, and large tracts of land in Europe were unusable because of unexploded material. The discovery of nuclear fission during World War II resulted in the atomic bomb and ushered in a new era of nuclear warfare and commercial nuclear energy use.

tion, sustained economic growth, and material progress. These ideas became widespread in liberal capitalistic economies globally and gave rise to a golden age of capitalism. Mass commodities like cars, the refrigerator, the television, and an enormous number of household devices appeared in the market. A new cultural industry emerged, and consumption became a synonym for capitalistic freedom. Moreover, the first supermarkets were established shortly after World War II and became highly popular.

The emergence of the supermarket in contemporary society signaled a type of consumption revolution, as consumers were able to purchase a wide selection of products from near their homes. More innovations and new products appeared in the market, such as frozen and microwavable food, so that styles of dining and eating rapidly changed. Despite some positive effects of the supermarket culture, the increased pace of production and consumption within this context resulted in large amounts of waste deriving from food packaging and household chemical substances and products.

During this period, various innovations in communication and transportation changed society as well. The telegraph was replaced by the telephone and opened up new possibilities for communication from one part of the world to another. Television and cinema became the means of a new process of consumption and reception, and the culture industry appeared as a major force of social and economic transformation.

In transportation, mass-produced cars became an everyday part of the social life of humans, and the innovation of the jet engine in commercial airliners made long transatlantic flights easier. In fact, by the 1960s, all large civilian aircraft were jet powered, which resulted in compressing large distances around the globe and a changing conceptualization of space and time. Despite these major changes in transportation, the jet engine would become a major pollution problem in the decades to follow. Aircraft emissions have been polluting the air and resulting in large amounts of carbon dioxide in the atmosphere. Similarly, the pollution of automobiles has been a major problem in the 20th century, attributed to the internal combustion engine and the large amounts of gasoline used.

The golden period of liberal capitalism collapsed with a new economic crisis that began in 1973 as a result of the increased price of oil, and this affected most sectors of the economy. Most international economies viewed the crisis as a temporary lapse in their development and governments tried to apply short-term solutions. However, this last period of the 20th century (1973–99) was one of long-term structural problems and irreversible waste encounters. The ideas of the free economy, mass consumption, political emancipation, and waste management seemed to be failing. Most of the liberal countries were soon faced with mass unemployment, high inflation, poor housing, and increasing pollution problems.

Globalization

Within this context, a new term emerged: *globalization*. In terms of economic capital, globalization resulted in a specialization of the world economy and, more particularly, in movement that occurs across national and political boundaries. This internationalization was nothing new, but the speed of the circulation of capital, commodities, and markets was surprising and unique. Capital, labor, and consumer markets shaped the world economy. Furthermore, the domination of financial capital shaped the main directions and trends globally and led to increasing inequalities across countries. Globalization increased the circulation of people, capital, images, and concepts around the globe.

Globalization was characterized by an intensification of social and economic life that was further expanded in an effort to overcome the problems of the Fordist regime of production. Within this context, the shift to a post-Fordist era of flexible accumulation was characterized by the appearance of new sectors in the economy, increasing and new consumption patterns, new sectors of production, new markets, and an intensification of information networks that aimed to speed up the processes of consumption and production. In addition, the flexibility of production, consumption, and labor became central in the establishment of a new form of capitalism. The speeding up of social and economic life resulted in the expansion of large multinational corporations and a well-known process: transnational capitalism.

It has been observed that this form of capitalism brought large-scale social inequalities around the globe and created fears of a global consumer culture that would homogenize the world. Consumption was intertwined with globalization primarily because globalization was thought of as "McDonaldization." This entailed a world connected by trade and information technologies, a global village that consumed similar images and shaped similar identities, and a process of "time-space compression" with a major goal of speeding up the production and consumption of transnational capitalism on a global basis. Commodities such as Coca-Cola and whiskey could be found almost anywhere in the world; music was becoming increasingly globalized; global movements followed similar styles; and issues of global meaning, such as the environment and the greenhouse effect, circulated around the world.

One of the most influential changes within the context of consumption and society during this last period of the 20th century was the electronics revolution that included the establishment of the personal computer in the 1980s and, during the last decade of the century, the World Wide Web. The personal computer in the form of desktops or laptops became an everyday device in almost every household that could afford it, introducing new possibilities of working and communicating, such as word processing, e-mailing, hypertext, and video conferencing. Moreover, the Internet became a global source of information and communication during the 1990s and a large socializing arena through social networking sites. As a result, new possibilities of social interaction emerged through the Internet, which became a political, economic, cultural, and social network that would influence almost every aspect of the social life of humans.

By the end of last period of the 20th century, two major consequences of the social and economic transformation were apparent. The first change was in relation to population growth. The world's population had exploded in size since the beginning of that century and almost tripled, despite the fact that the 20th century was the bloodiest, with 187 million victims as a result of various conflicts. The second consequence of the mass production and consumption processes was ecological. Nature became a central topic of discussion during and after the 1970s, with governments, international organizations, scientific communities, and ecological movements trying to identify the problems of accelerating growth and production of waste.

Within this context, waste became a central argument in relation to environmental catastrophe, even if the majority of pollution came from the rich and developed countries of the Western world or Asia. A large part of the environmental movement and scientific community argued for sustainable development, which meant a balance between the resources that a society consumed and the waste produced and recycled. In that way, the effects on the environment would be controlled and minimized. However, it was not clear how such a plan could materialize in relation to population levels, production rates, technology, and consumption. What was the right proportion of population growth and death rate or consumption and waste rate? It became apparent that the solutions to environmental problems could not be found in scientific formulas or big innovations but were intertwined with political, economic, and social interests. Despite that fact, the realization that society and nature are not two entirely different spheres and cannot be kept apart had a limited range.

By the end of the 20th century, waste and consumption were radically different concepts in comparison to earlier centuries. Europe was no more the center of mass commodities, industrialization, and global domination. On the contrary, the European empires had been dissolved into small nation-states without colonies, and other countries such as the United States and China had emerged as influential powers and waste producers. Globalization had changed the types of commodities that were consumed by expanding and encompassing the entire world and bringing, even to the most remote islands, mass-produced products, the Internet, and satellite television. Consequently, even the smallest communities had to deal with waste treatment. Therefore—despite the fact that in the 20th century many people became literate, innovations in medicine extended life expectancy, natural science and technology made the biggest steps ever in human history, including nuclear fission and the exploration of the moon—humanity went through the new millennium with a deep feeling of uneasiness about the future and what it could bring. Modern consumption and waste with all their consequences, were a significant part of this uneasiness.

Tryfon Bampilis
University of Leiden

See Also: Capitalism; Commodification; Consumerism; Environmentalism; Human Waste; Industrial Revolution; Nuclear Reactors; Politics of Waste; Sociology of Waste.

Further Readings

Harvey, D. *The Condition of Postmodernity*. Oxford: Blackwell, 1989.

Hobsbawm, E. *Age of Extremes. The Short Twentieth Century: 1914–1991*. London: Abacus, 1995.

Latour, B. *We Have Never Been Modern*. Cambridge, MA: Harvard University Press, 1993.

McNeill, J. R. *Something New Under the Sun: An Environmental History of the Twentieth Century World*. New York: W. W. Norton, 2000.

Miller, D. *Material Culture and Mass Consumption*. Oxford: Basil Blackwell, 1987.

Wallerstein, E., J. Arrighi, and T. Hopkins. *Antisystemic Movements*. London: Verso, 1989.

Hoarding and Hoarders

The term *hoarding* describes the excessive acquisition of relatively worthless things to the extent that it compromises the living space or daily activities of the affected person. This individual is referred to as a "hoarder." Hoarding is considered a symptom of obsessive-compulsive disorder (OCD), but many people who hoard show no other OCD traits. In the 21st century, researchers are working on understanding hoarding as a distinct and separate mental health problem, rather than one aspect of OCD.

Hoarding is a condition that can strike anyone of any age, gender, or economic level. It is difficult to determine how common the disorder is because researchers have only recently begun to study it, and hoarders often hide their condition. Many hoarders do not recognize that they have a problem, which can make treatment difficult. There are various levels of hoarding behavior, ranging from the relatively harmless "pack rat" to intense acquisition that causes life-threatening conditions.

Signs and Symptoms

There are a number of risk factors or common characteristics found in individuals exhibiting hoarding behavior. Hoarding often starts in early adolescence and worsens with age. It is more common in individuals who have a family member who also hoards. A stressful life event such as a death, divorce, natural disaster, or fire can trigger hoarding in a person already harboring the tendency. While hoarding often leads to social isolation, lonely or withdrawn people may hoard to find comfort and control in having their possessions surrounding them. Hoarders are frequently perfectionists, wanting to make exactly the right decision on what to do with their possessions. Hoarders have increasingly cluttered living spaces because they cannot bear to throw things away but cannot stop acquiring more. A fear of losing information or knowledge that may be needed in the future results in massive piles of newspapers, magazines, and even junk mail. The fear of throwing out something that may be useful later causes anxiety. Clothing, broken appliances, bits of string, foil, gift wrap, and almost any item has value and represents endless possibilities for reuse. The inability to organize and worry about making the wrong decision compound until the hoarder is incapable of making any decision at all.

Quality of Life

Hoarding eventually impacts the quality of life of both hoarders and their families. Unlike most homes where a varying amount of untidiness is normal, the excessive clutter in the homes of hoarders makes the use of rooms for their intended purposes an unattainable goal. Doors and hallways may be impassible. The kitchen may not be functional because the counters, stove, and table are covered with clutter. The bathtub may be filled with stuff, making bathing impossible. When there is no more room inside, the clutter often overflows to the garage, yard, and vehicles.

Hoarding at this level becomes dangerous. There is a greater risk of fire when so much debris is stacked in the home, and, in the event of a fire, it is more difficult for people to get out and for firefighters to get in. There is the danger of falling and getting hurt, especially for elderly hoarders. Large stacks of newspapers can topple over, trapping people underneath. Respiratory problems and other health risks are factors if the clutter includes rotting food, mold, or other organic debris. Health issues are exacerbated when animals are hoarded in addition to (or instead of) objects. The health and safety of the animals is compromised to the same extent as the owner's. Overcrowding among pets often leads to malnutrition, neglect, and disease.

Treatments

There is no cure for hoarding, and treatment can be challenging. Many people who hoard do not see their activity as something that has a negative impact on their lives. Some acknowledge they have a problem but do not see a need to stop. If they receive comfort from their possessions or animals, why should anyone interfere? The pressure to change this behavior is usually initiated by family, friends, or neighbors attempting to deal with the mess.

There are two types of treatment for hoarding: psychotherapy and medication. Cognitive behavioral therapy, a specialized form of psychotherapy, involves a series of strategies to help the hoarder control the urge to acquire and save possessions. The therapist will help the hoarder explore the reasons they feel the need to hoard, teach organizing and categorizing skills, foster decision-making skills, teach relaxation exercises, and help the individual learn to let go of a few items at a time. A therapist will go to the patient's home and help them apply what is learned in therapy sessions. While cognitive behavior therapy has some success helping hoarders declutter and feel better about themselves, few people are actually cured. Even after a full course of therapy, which can be intense and time consuming, periodic visits and ongoing treatment may be necessary to keep the individual from slipping back into old behaviors.

Medications used in the treatment of obsessive-compulsive disorder are often prescribed to people who hoard. Selective serotonin reuptake inhibitors (SSRIs), a type of antidepressant, are most commonly prescribed. Drugs may not cure hoarding, but they can help an individual manage a hoarding problem by alleviating other conditions that worsen the behavior. Many hoarders have additional conditions or health problems, such as depression or attention deficit disorder, that can interfere with the ability to focus and to stay on task while sorting and organizing possessions.

What drives the need to hoard? Researchers do not yet understand what causes this condition, making prevention impossible. However, hoarding is a mental condition that displays a known series of tendencies or habits. Getting treatment early, as soon as symptoms begin to appear, may help prevent hoarding from becoming so severe that it affects quality of life.

Jill M. Church
D'Youville College

See Also: Consumption Patterns; Culture, Values, and Garbage; Household Consumption Patterns; Magazines and Newspapers; Personal Products.

Further Readings

Frost, Randy O. and Tamara L. Hartl. "A Cognitive-Behavioral Model of Compulsive Hoarding." *Behaviour Research and Therapy*, v.34/4 (1996).

Frost, Randy O. and Gail Steketee. *Stuff: Compulsive Hoarding and the Meaning of Things*. Boston: Houghton Mifflin Harcourt, 2010.

Neziroglu, Fugen, Jerome Bubrick, and Jose A. Yaryura-Tobias. *Overcoming Compulsive Hoarding: Why You Save and How You Can Stop*. Oakland, CA: New Harbinger, 2004.

Tolin, David F., Randy O. Frost, and Gail Steketee. *Buried in Treasures: Help for Compulsive Acquiring, Saving and Hoarding*. New York: Oxford University Press, 2007.

Tompkins, Michael A. and Tamara L. Hartl. *Digging Out: Helping Your Loved One Manage Clutter, Hoarding and Compulsive Acquiring*. Oakland, CA: New Harbinger, 2009.

Home Appliances

Most households in 21st-century Western societies employ an array of labor- and time-saving devices to accomplish the basic tasks associated with day-to-day living practices. These objects, designed and acquired with a particular instrumental (as opposed to strictly aesthetic) purpose in mind, are referred to as home appliances. Often, they are designated in terms of their primary role (e.g., clothes dryers, washing machines, or cd-players), although this is not always the case (for example, the term *tea kettle*). Home appliances play an important role in the lives of consumers and present a number of possibilities in terms of their production, use, and disposal.

While each appliance is a distinct object it its own right, it is also embedded in the broader utilities framework of the household. The most obvious cases of these are the provision of electricity and water and the means by which these appliances are able to connect to these utilities. Without an appropriate degree of integration with these services, the ability of appliances to fulfill their intended purpose is severely hampered, if not entirely diminished. Thus, a washing machine is only recognizable as such due to its ability to wash clothes with the aid of plumbed water, electrical current and the appropriate connections, and some type of cleaning agent.

Development

Many contemporary home appliances are the result of relatively recent technological innovations and are often electrical or computerized improvements on manually operated appliances that performed a similar function in the past. Examples include the electric shaver replacing the razor blade, the air-conditioning unit replacing the ceiling and standing fans, and the CD player replacing cassette tape player. In similar cases, newer appliances may be acquired to supplement, rather than entirely replace, older devices. For example, electric clothes dryers have not entirely replaced outside clotheslines (or drying racks), but nonetheless are designed to perform a similar task.

Energy Consumption

Home appliances have come under the scrutiny of various regulators and independent groups concerned with energy security and environmental sustainability. It has been estimated that these appliances comprise approximately 17 percent of household energy consumption. Further, 10 percent of this electricity is consumed by appliances that are not even in use in what is known as standby power. Appliances are frequently assessed by independent programs, such as Energy Star, which partner with various public and private organizations in order to offer energy and water efficiency ratings for a range of home appliances. Such rating systems are designed to both inform consumers about the energy requirements for a given appliance on the market, as well as offer a means to compare the efficiency of different models of a particular appliance type.

Effects on Perceptions and Behavior

Social scientists have attempted to understand the role that appliances play in the lives of their users beyond the immediate instrumental tasks they are intended to perform. For instance, there has been an increasing prevalence of air conditioning units inside the home, as well as in other in spaces such as the workplace, car, and shopping mall. Sociologist Elizabeth Shove has noted that this widespread use of air-conditioners has contributed to changing expectations about what is considered normal ambient temperature.

This includes a reluctance to tolerate temperature fluctuations commonly found in climate uncontrolled environments and the almost universal acceptance of a mean comfortable temperature of around 22 degrees Celsius (71.6 degrees Fahrenheit). This development is subsequently implicated in changing patterns of dress and the decline of practices previously used to insulate people from the heat of the day, such as the siesta. As such, home appliances are not only labor-saving devices but are also increasingly associated with more abstract notions of domestic life, such as comfort and convenience.

Manufacture and Design

Technology has clearly played a critical role in the story of appliances in the home. The manufacturers' ability to economically manipulate and integrate steel with plastic is a rather basic, but essential, development in the mass production of home appliances. Globalization and the liberalization of trade agreements opened access to cheap labor markets, thereby making many appliances affordable for large proportions of consumers throughout the world.

In addition to the application of various design features for ergonomics and usability, many home appliances are increasingly becoming computerized. Users are often able to program appliances with more specificity than past iterations of the same appliance; for example, the length and temperature of a washing machine cycle, or controlling the emission of steam from a clothes iron. However, the complex circuitry that comprises many computerized home appliances is difficult to comprehend for those without specialized knowledge and equipment. Consequently, these items are more difficult to repair and maintain for the average consumer. Repair of home appliances has been found to be financially unviable in many cases, since the cost of repair labor outweighs the cost of simply replacing the broken device. This suggests that many newer appliances, while offering a range of new performance features, are also more prone to disposal due to their structural complexity.

Reuse, Recycling, and Discard

The replacement of appliances with updated models is one of the driving forces in the creation of waste from home appliances. While appliances, by definition, are primarily utility-serving devices, the sale and marketing of sleek and stylish home appliances as a key part of the modern home suggests that they have an aesthetic component to them. As such, they are often discarded on this basis, having lost their original allure through general wear and tear or by being seen as dated through possessing particular stylistic appearances and features (or deficiencies).

Charity groups often encourage consumers to donate old and unused appliances. This stands as an alternative to merely disposing of appliances from the household, as the charity groups attempt to sell unwanted goods for a reduced price, thus preventing them from going directly into the waste stream. However, the willingness of charities to accept these goods has occasionally led to consumers' dumping unusable appliances and other objects into their hands as a means of disposal rather than donation. Consequently, the capacity of these groups to absorb unwanted but still usable goods is compromised as the waste removal costs are merely transferred from the consumer to the charity organization.

Increasingly high commodity prices have led to discarded appliances being salvaged for their resale value as raw materials. Copper recovered from power cords, gold recovered from computer circuits, or steel from appliance bodies can all be viable sources of income for those willing to acquire the discarded objects and separate their constituent parts. These practices are undertaken at a variety of scales, both automated and by hand.

Home appliances are ubiquitous for those engaged in the domestic practices of day-to-day life. An appliance can embody many ideals, including simplicity, style, durability, or usability. The production, use, and disposal of these goods are critical for issues of consumption and waste, as well as modern life generally.

Andrew Glover
University of Technology, Sydney

See Also: Appliances, Kitchen; Computers and Printers, Personal Waste; Household Consumption Patterns; Personal Products; Sociology of Waste; Ventilation and Air-Conditioning.

Further Readings

Cowan, Ruth Schwartz. *More Work for Mother: The Ironies of Household Technology From the Open Hearth to the Microwave*. New York: Basic Books, 1983.

Molotch, H. *Where Stuff Comes From*. New York: Routledge, 2005.

Shove, E. *Comfort, Cleanliness and Convenience*. Oxford: Berg, 2003.

Home Shopping

The rise of home shopping—the ability of individuals to purchase goods and services from their homes, generally over the Internet—has not only changed patterns of consumer spending but has also changed the environmental impact of consumption. In large part, this is because home shopping—itself an element of a larger trend in retailing called e-commerce—necessitates a reevaluation of how carbon miles are calculated in the cycle of purchasing goods and services. Carbon miles—the cost of transporting goods and services—is only one element of the calculations involved in totaling the carbon footprint of a given artifact over its life cycle. For consumers, it is often the most visible of the many factors that contribute to the total amount of energy embodied in an object over its life and expended in its creation, distribution, and consumption. The total carbon footprint of home shopping, or e-commerce more broadly, is difficult to determine, but in assessing its impact in terms of how much environmental waste it produces, it is generally most important to look at the distribution and transportation of an object or service. The production of a given service or object is not affected by the mode of retail and distribution; home shopping's innovation lies in changing patterns of distribution.

Home shopping is a relatively recent phenomenon, and although in the United States it bears similarities to the late-19th- and early-20th-century phenomenon of catalogue shopping or purchasing goods from door-to-door salesmen who represent a larger company or organization, the catalogue mail-order services were mainly invented to serve rural populations who had little access to urban retailers. Home shopping over the Internet, on the other hand, is as likely to serve urban populations as it is to serve rural consumers. It has contributed to the remarkable globalization of consumption—goods are available to anyone who can access the Internet and afford to purchase and have items shipped to them. It has also paradoxically reaffirmed local networks of service and consumption that many critics of globalization have argued are among the most significant casualties of a rootless, placeless global economy.

E-Commerce

But if home shopping can be criticized as a symptom of globalization—goods, for example, that might be made in China can be imported to the United States, warehoused, and shipped on demand to customers, and then redistributed to other countries in the form of waste—it has also presented environmentalists with a conundrum. Does the centralized storage and distribution of e-commerce and home shopping reduce overall waste by saving on transportation costs and consolidating the energy involved in warehousing products, or are those savings lost when increased packaging or the hidden costs associated with e-waste are considered?

It is therefore worth making distinctions between the kinds of shopping available over the Internet, for each has a different local and environmental impact in terms of energy expended or wasted, as well as in their implications for post-consumer waste. One can broadly categorize the goods available to consumers in their homes as those offered by big corporations that distribute and market, but do not fabricate, a range of objects from different manufacturers. Perhaps the most famous example is Amazon.com, but in this category one can include those sites that retail objects from a range of producers, such as shoes, jewelry, housewares, and linens. The most famous of these is Overstock.com. Such retailers, large and small, are noteworthy because they often do not have their own retail stores; rather, they store goods in warehouses (or, in the case of etsy.com, in homes or workshops), and ship on demand. For such retailers, the costs of storage and shipping and packaging represent the waste produced by online shopping. The second sort of online retailer with which many patrons are familiar are those retailers that have retail outlets in addition to online sales. Such retailers include major box stores like Target or Walmart, which are consolidated in local geographic areas but will ship goods to local stores on demand for pickup by consumers. This list of corporate online retailers can include those corporations that sell or rent digital media directly to consumers in place of actual artifacts, for example, digital books that are downloaded directly to an e-reader or digital music and films downloaded directly to an mp3 player or smartphone.

While the growth of home shopping over the Internet has contributed to the globalization of consumption, it has also reaffirmed local networks of service and consumption such as shipping. Warehousing may also be more energy efficient.

Swap and Barter

But corporatized e-commerce does not represent the whole of home shopping; there is a great deal of home shopping that takes place through brokers and is dedicated to used goods. The most famous site for this sort of retailing is eBay, which is among the most successful Websites in the world. eBay was founded in 1995 and, according to its own Website, is committed to sustainability because it offers a venue that "extends the useful life of millions of products, keeping items out of landfills and reducing the need for new manufacturing." Similarly, an important venue for home shopping is organized around barter sites such as Craigslist.org (founded in San Francisco in 1995, with over 50 million users and 700 local sites in 70 countries, according to its Website) and freecyle.com, a local nonprofit Website of objects for sale, trade, or for the taking. Craigslist and Freecycle are locally operated, with various geographic areas sponsoring their own sites (though eBay now owns 25 percent of Craigslist.org). Freecycle, like eBay, is self-consciously committed to "reuse and keeping good stuff out of landfills."

Of the various home shopping venues, the local swap and barter sites appear to be the most environmentally friendly in terms of their overall carbon footprint, especially because they operate locally, traffic in used items, and tend to reduce packaging and shipping costs by requiring that interested buyers pick up their purchases from the seller directly. However, it is still difficult to come by any real data about the success of e-commerce and home shopping or about its environmental impact. There are a handful of companies that attempt to track online commerce, including Forrester Research, which produces an annual report called "The State of Retailing Online," or *Plunkett's E-Commerce and Internet Industry Annual Almanac*. Forrester Research, for example, is mainly focused on marketing strategies and is geared toward increasing retailer profitability, rather than measuring the environmental impact of e-commerce and home shopping.

Although there are industry standards for the regulation of Internet commerce, it is not easy to find information about the environmental impact of home shopping, even when a corporation trumpets its concern with its environmental impact. In part, this is because not all home shopping can be included in any calculations. While it is possible to find information about how much music is downloaded on an mp3 player and the impact of digital music on the music industry's conventional ways of distributing its product in stores, it is not easy to determine, for example, how the savings in making, shipping, packaging, and retailing a CD compares to the overall environmental impact of producing an mp3 player or of disposing of it.

Similarly, it is easier to find information on business-to-business e-commerce—how online sales affect a major retailer's expenditures for transportation and storage—than it is to determine the environmental waste or benefits involved in peer-to-peer or business-to-consumer home shopping. Despite the difficulties of finding and coordinating such data, scholars tentatively argue that shopping at home through the Internet can offer substantial savings in terms of how much waste is generated by consumers in terms of packaging and transportation and are currently devising models that can satisfactorily account for the environmental costs and benefits of e-commerce and home shopping.

Stephanie Foote
University of Illinois at Urbana-Champaign

See Also: Household Consumption Patterns; Packaging and Product Containers; Shopping.

Further Readings

Abukhader, Sajed. "Eco-Efficiency in the Era of Electronic Commerce: Should 'Eco-Effectiveness' Approach Be Adopted?" *Journal of Cleaner Production*, v.16 (2008).

Cohen, Nevin. "Greening the Internet: Ten Ways E-Commerce Could Affect the Environment." *Environmental Quality Management* (Autumn 1999).

Cullinane, Sharon. "From Bricks to Clicks: The Impact of Online Retailing on Transportation and the Environment." *Transport Reviews*, v.29/6 (November 2009).

Edwards, Julia and Alan McKinnon. "The Greening of Retail Logistics." In *Logistics and Retail Management: Emerging Issues and New Challenges in the Retail Supply Chain,* 3rd ed, John Fernie and Leigh Sparks, eds. Philadelphia: Kogan Page, 2009.

Fichter, Klaus. "E-Commerce: Sorting Out the Environmental Consequences." *Journal of Industrial Ecology*, v.6/2 (2003).

Velasquez, Marcelo, Abdul-Rahim Ahmad, Michael Bliemel, and Mohammad H. Imam. "Online vs. Offline Movie Rental: A Comparative Study of Carbon Footprints." *Global Business and Management Research: An International Journal*, v.2/1 (2010).

Hospitals

A hospital is a facility where healthcare services are available. A hospital can be either a private or a public facility. It can produce two types of waste. Some of it is regular waste (which can be disposed of through regular garbage disposal), while the other type of waste is clinical waste. Some facilities have recycling programs for all types of garbage, such as paper, glass, and specialized materials such as electronic waste and medical waste. Medical waste necessitates specific disposal procedures to ensure that there is no contamination of staff, other patients, or the general population. Disposal methods include incineration and in-house treatment before regular disposal. Increasingly, hospitals are dedicating resources to recycling waste as a way to reduce garbage and costs. Nonetheless, disposal remains a challenge, especially in developing countries where limited funds require facilities to cut corners, endangering the local population.

Background

A hospital is a facility where healthcare services are dispensed to patients by dedicated staff using specialized equipment. Presence of specialized staff and equipment means that it acts as a hub for healthcare in a community, and it can have numerous patients and employees frequenting it at regular intervals. Having a high number of people in a relatively small space makes it a fertile place for cross-contamination. Hence, hygiene and waste management are key components of maintaining a healthy facility.

Hospitals can be funded either by the public sector or through private funding (foundations or by insurance companies, for example). As such, each type of facility might have a different approach to patient care and waste management. Different types of hospitals include general hospitals, specialized hospitals (specializing in specific illnesses), and academic hospitals, which are related to institutions of learning and act as practice environments for future healthcare personnel.

Types of Waste Generated by Hospitals

Hospitals generate two types of waste: regular (general) waste and clinical waste. Regular waste is akin to waste generated by most standardized businesses, such as food or paper waste, for example. Clinical waste is waste that can be harmful to human beings and has to be disposed differently than regular waste. It is also sometimes called regulated medical waste (RMW). Clinical waste is regulated to ensure that hospitals dispose it in a safe manner, and hospitals are usually required to maintain logs and inventory of all clinical waste they produce.

Clinical waste can include pathological waste such as tissues, body parts, materials from patients with communicable diseases, soiled linens, human blood, blood products, and needles. The Environmental Protection Agency (EPA) defines medical waste as "any solid waste that is generated in the diagnosis, treatment, or immunization of human beings or animals, in research pertaining thereto, or in the production or testing of biologicals." This definition includes blood-soaked bandages, discarded surgical gloves, surgical instruments, and needles, as well as removed body organs (for example, tonsils, appendices, or limbs). On average, between 10 and 15 percent of waste generated by a hospital is believed to be RMW waste.

Disposal of Hospital Waste: Regular, Clinical, and Recycled Waste

Since there are two types of waste generated in hospitals, facilities are responsible for managing them distinctly. Some hospitals go further, implementing recycling programs to incite personnel to recycle materials such as paper, glass, cardboard, and plastics. Some hospitals have more elaborate recycling programs dedicated to specialized waste, such as managing electronic waste (such as outdated computers or diagnostic machines) or medical waste.

Regular waste does not require any specific waste management treatment. As long as the waste does not qualify as clinical waste, it can be disposed of through regular waste channels. Since clinical waste is a regulated material, hospital staffs are usually well advised of which waste can be thrown away in regular waste and which waste needs special attention.

Most hospitals separate clinical waste from regular waste at the source. The most common system is the implementation of specially labeled containers throughout the facility. Many hospitals color-code waste containers to increase compliance by staff. Most hospitals prefer to separate the waste on-site, rather than collect it in a mixed container with regular waste and then separate it. This is done to reduce the risk of contamination to staff.

It is strictly forbidden to mix clinical waste with regular waste because there are many different populations who are at risk of contamination. First, there is a direct risk of contaminating the hospital staff or the people collecting the waste. These individuals are in direct contact with the waste, and it only takes a moment's inattention to accidentally touch a stray needle, leading to a potentially dangerous infection. There is also a secondary risk of contamination to the surrounding area where the waste is disposed of (such as the general population or the land) if no formal steps are taken to properly treat the waste. Therefore, hospitals found to have disposed of clinical waste through regular waste channels are subject to heavy fines.

Disposal methods used by hospitals include waste incineration or its treatment before being sent to a landfill. RMW treatment guidelines can vary by jurisdiction and can sometimes include both national and state-level guidelines. Nonetheless, incineration of medical waste is increasingly controversial because of negative environment impacts as well as secondary infections that can result from improperly maintained facilities (such as organic dust toxic syndrome, an illness contracted following the inhalation of dust).

New technologies are continually being developed to reduce the environmental impact of medical incineration. Less frequent alternatives include autoclaving (sterilization by high-pressure, saturated steam), microwave sterilization, and chemical and irradiation processes (sterilization through radiation).

Hospitals are also seeking to better manage the total waste they generate through the reduction of waste volume generated. The purchase of green products and reusable products is favored by some institutions, but costs can increase by 25–30 percent for the purchase of a green or reusable product. While hospitals face increasing budget constraints, there have been limited opportunities to use green-friendly products in hospitals.

Recycling programs for regular waste are also becoming more common. There are basic recycling programs that target everyday waste, such as paper, cardboard, and plastic. These are more popular since they are less costly and less complicated to implement. Some facilities put into practice much more specialized recycling programs, which target specific hospital waste such as electronic waste (medical devices and computer components) as well as programs for recycling medical waste. Most hospitals implementing recycling programs do it to save on disposal costs and reduce municipal waste, yet incur some costs in terms of staff training and extra steps in terms of waste handling.

On-Site Waste Disposal

Some studies estimate that 25–30 percent of hospital facilities dispose of waste on-site directly, rather than disposing of it off-site. These methods can include incineration and autoclave technologies. The reasons for having on-site waste management can vary, but most believe that processing on-site is more economical and more reliable. Some other facilities also believe that managing on-site allows better control over the disposal of the waste. Since there is an inherent responsibility in the disposal of medical waste, some hospitals prefer to take a cradle-to-grave approach to RMW waste management.

Secure Disposal of Medical Records

Hospitals also deal with another sensitive waste product—expired patient records. Many jurisdictions impose important constraints on hospitals regarding how they can dispose of hospital records, since these often contain sensitive information about patient heath and treatment. Hospitals often implement specific programs to manage the disposal of patient files. Often, hospitals outsource this responsibility to a third party, which shreds sensitive documents on-site (enabling the hospital to monitor the proper disposal of the documents). Other hospitals shred sensitive documents on-site themselves, while a minority of facilities outsource to companies that pick up the sensitive documents and shred them off-site. This is somewhat rare because the facilities are unable to directly monitor the destruction of patient records, a legal responsibility in some jurisdictions.

Challenges in Developing Countries

While waste management in hospitals might be sensitive issues in developed countries, it is even more complicated for hospitals in developing economies.

While central authorities have some defined rules for waste management, even going as far as supplying some of the needed infrastructure (color-coded containers, for example), staff compliance is uncertain and medical waste is sometimes mishandled. This creates a risky environment for the surrounding population. There can be poor or no segregation of different types of hospital waste, with general and infectious wastes often being mixed in the same primary container at the source. Training and advertising campaigns to educate workers continue to meet resistance. Disposal is also elementary, with smaller facilities practicing on-site burial and dumping of medical waste in municipal waste facilities. Onsite incinerators are also present in some facilities, but rather rare, and some lack the necessary controls to measure toxic emissions. Some hospitals dispose of waste with their generic waste, and it is not unheard of for liquid medical waste to be disposed in the local sewer system, potentially contaminating the local water supply. In addition, hospitals often lack the resources and expertise to properly track the waste once it leaves the site, endangering local populations because the waste can be incorrectly disposed of. Since the legislative framework is sometimes vague, and the enforcement of legislation is lacking, proper disposal is a long-term challenge in emerging economies.

Jean-Francois Denault
Independent Scholar

See Also: Incinerator Construction Trends; Medical Waste; Toxic Wastes.

Further Readings

Bernstein, Donald, et al. "Hospital Waste Management: An Informational Assessment." *Academy of Health Care Management Journal*, v.5/2 (2009).

Harhay, Michael, et al. "Health Care Waste Management: A Neglected and Growing Public Health Problem Worldwide." *Tropical Medicine & International Health*, v.14/11 (November 2009).

Healthcare Waste Management (HCWM). "Safe Management of Wastes From Health-Care Activities." World Health Organization. http://www.healthcarewaste.org/en/documents.html?id=1 (Accessed September 2009).

Household Consumption Patterns

At the core of household consumption patterns is the interrelation between scale and composition of the household, income, savings, expenses, and enculturation patterns.

There are many psychological factors that influence the household consumption sensibility—from the willingness to achieve and/or reproduce a specific social status, to demonstrating material culture in a competition between neighbors who have to have more and better. A series of oppositions (e.g., wealth–poverty, urban–rural, small family–big family,) represent extremes of household structural characteristics and of the character of consumption. Additionally, the specific historical and regional processes of education, enculturation, and socialization influence household decision makers as to what to consume. The transforming of American culture from a culture of consuming toward a culture of sustainability and caring redirects the evolution of household consumption patterns. However, the gradual cultural reproduction of endless wants has resulted in the emergence of forms of compulsive buying that may relate to an impulse-control disorder that was documented as long ago as the early 20th century.

The household is an elementary social unit and includes common expenses and individual expenses. The typology of households includes single member, childless households (including married and nonmarried partners), two generations (parents or grandparents and children), three generations (parents, grandparents, and children) and others.

Housing, nutrition, and clothing belong to the general common consumption pattern. Heating, cooking, lighting, cooling, and water supply have also been described as primary household consumptions (operations). Telephone, laundry, and dry cleaning services and supplies, babysitting, maid services, holiday decorations, and others belong to secondary household consumptions. They vary by household, although the main confounding factor is the number of children. Health, mobility, and recreation are among individual specific consumption patterns.

Households affect both the structure of consumption and the environment through their day-to-day decisions on what to buy and how they use goods and services, where to live and work, what kind of dwelling to have, how to manage their waste, and where to go on vacation.

Consumption and Globalization of Economy

Globalization of economy is one of the important factors that drive household consumption, along with growing incomes, technological innovations (such as the Internet and mobile phones), decreases in the number of household members, an aging population, and habits and cultures. In the European Union's 27 member states, between 1990 and 2007, consumption expenditure increased by 35 percent. Growth has been more rapid in the west Balkan countries and Turkey, rising by 130 percent and 54 percent, respectively, between 1990 and 2007. Households spend between two and six times more than the public sector. Increasing knowledge of health issues and the opportunity for free sharing of information via the Internet have begun creating a household consumption pattern of healthy nutrition that contrasts to late-20th-century overconsumption and an increasing tendency toward obesity. At the same time, in many countries of the world, smoking cigarettes is a common household problem. Results from a survey in China indicate that spending on tobacco affects other categories of consumption such as education and health, household economic productivity (e.g., farming equipment and seeds), and financial security (e.g., saving and insurance). Smoking is often found in combination with alcohol consumption and exacerbates the impact of addictive substances on household common consumption patterns. The same conclusion can be proposed for many other parts of the world—eastern Europe, for instance. On the other hand, tobacco smoking has become less popular in the United States.

One of the main roles of households that is shaping and reshaping consumption patterns globally is tourism. International tourism has grown at an annual average rate of 7 percent since the 1950s and is projected to continue to grow at 4.3 percent per year to 2020. In 2020, long-haul travel will be as frequent as nearly 70 percent of all tourism travel in 1995. Car and air travel dominate, while rail and maritime travel account for a comparatively smaller number of tourism miles. Patterns of vacation travel are changing as many households make shorter, more frequent tourism trips, although the long (over two-week) holiday is still the norm in many countries, especially in Europe.

Rural and Urban Seasonality

A preliminary study of household consumption patterns in Orissa, India, infers that in both rural and urban areas, cereals, edible oil, vegetables, spices, and fuel and light are considered necessities. Pulses and beverages are necessities in urban areas, while eggs, fish and meat, sugar, education, and medical care are treated as luxuries in both rural and urban areas. Micronutrient deficiencies reported in Nigeria are rife, in spite of the reasonable consumption of vegetables during the peak season of production. Among the reasons are the seasonal use of vegetable production and a culture that may limit the adequate consumption of leafy vegetables, even when they are in abundance.

Seasonal fluctuations in food consumption are also a serious problem in rural Mozambique, where community isolation is high, and market integration, use of improved inputs, and access to off-farm income are low. An analysis of the total expenditure elasticity of food groups reveals how precarious food security is in rural households in the poorest quintile, even in regard to the most basic staples of maize and cassava.

Households and Sustainable Consumption

During the consumer revolution (late 16th–early 19th centuries), households developed patterns of mass consumption of luxury and exotic goods. The evolution of these patterns resulted in overconsumption tendencies in the late 20th century. From critical analysis of the mass culture and mass consumption were born the models of sustainable culture that became key issues in the early 21st century. Sustainable consumption refers to goods and services, the use of which respond to basic needs and bring a better quality of life. It minimizes the use of natural resources, toxic materials, and emissions of waste and pollutants over the life cycle, guaranteeing successful social reproduction without jeopardizing the needs of future generations.

Households play a primary role in the embedding of sustainable consumption worldwide, since they are made up of decision makers about what lifestyle to have and what to consume. Among the economic tools are deposit-refund schemes and taxes on disposable products and packaging. Regulatory tools include ecolabeling, while social tools like environmental education and information on green purchasing give support to voluntary initiatives.

Development of sustainable consumption may help to resolve the problem of poverty. Surveys infer that there are means to get rid of poverty and to improve the quality of life worldwide. Since the middle of the 20th century, the global population has consumed more goods and services than the combined total of all previous generations. However, the richest one-fifth of the world accounts for 86 percent of consumption, while the poorest one-fifth accounts for about 1 percent of consumption.

One of the ways to reduce poverty is to develop household ecotourism in low-income countries and to use the services of small household hotels and houserooms. Such services, for instance, have been developing as a consumption pattern in China and in the Balkans.

Lifestyles

Lifestyles link social structure and status to attitudes and behavior by showing how people live and interpret their lives. Depending on the criteria and goal of research, there might be only one (e.g., American style of life) or numerous lifestyle choices, depending on the consumption behavior and types of preferences. Food consumption is one of the main criteria for defining different lifestyles. It does not always depend on income and social status. The same level of calorie consumption may characterize very different income levels, although consumption structure differs widely even at the same high-income level.

Household Consumption Patterns and Waste

In developed countries, growing demand for energy and water services is tied to larger homes. Municipal waste is projected to grow by 43 percent from 1995 to 2020, to reach approximately 700 million tons per year in some of the economically leading countries in the world (Organisation for Economic Co-operation and Development [OECD] members), including the United States, Canada, and Australia. The highest waste-generators among the OECD countries are the United States, Ireland, Iceland, Norway, and Australia.

During the last few decades, recycling rates have increased in an attempt to slow the rate of growth of waste destined for final disposal. Reductions have not been seen in the total volume of waste generated, although there is diversification of the waste stream.

Economists' theories of consumption (e.g., the permanent income hypothesis of Milton Friedman, the life cycle theory of Franco Modigliani-Richard Brumberg-Albert Ando) focus on the economic backgrounds that almost completely depend on household income and decision making. In turn, psychological and anthropological approaches help to understand the individual or group motivation of consumers in depth. Complex research shows that household consumption patterns have dynamic characteristics and their evolution resembles a spiral with economic, social, and cultural variable parameters.

Lolita Petrvoa Nikolova
International Institute of Anthropology

See Also: Consumption Patterns; Economics of Consumption, International; Economics of Consumption, U.S.; Food Waste Behavior; Overconsumption; Recycling Behaviors; Underconsumption.

Further Readings

Cohen, L. *A Consumers' Republic. The Politics of Mass Consumption in Postwar America.* New York: Alfred A. Knopf, 2003.

European Environment Agency. "About Household Consumption." http://www.eea.europa.eu/themes/households/about-household-consumption (Accessed June 2011).

Gram-Hanssen, K. "Standby Consumption in Households Analyzed With a Practice Theory Approach." *Journal of Industrial Ecology,* v.14/1 (2009).

Hart, A. D., C. U. Azubuike, I. S. Barimalaa, and S. C. Achinewhu. "Vegetable Consumption Patterns of Households in Selected Areas of the Old Rivers of Nigeria." *African Journal of Food, Agriculture and Nutritional Development,* v.5/1 (2005).

Moll, H. C., K. J. Noorman, R. Kok, R. Engström, H. Throne-Holst, and Ch. Clark. "Pursuing More Sustainable Consumption by Analyzing Household Metabolism in European Countries and Cities." *Journal of Industrial Ecology*, v.9/1–2 (2005).

Nikolova, L. "Global Society and Poverty Reduction." National World Culture/Examiner. http://www.examiner.com/world-culture-in-national/global-society-and-poverty-reduction-tourism#ixzz1EknIb5Aa (Accessed February 2011).

Organisation for Economic Co-operation and Development (OECD) Publications. *Towards Sustainable Household Consumption? Trends and Policies in OECD Countries*. Washington, DC: OECD Publications, 2002.

Tudor, T., G. M. Robinson, M. Riley, S. Guilbert, and S. W. Barr. "Challenges Facing the Sustainable Consumption and Waste Management Agendas: Perspectives on UK Households." *Local Environment*, v.16/1 (2011).

Wang, H., J. L. Sindelar, and S. H. Busch. "The Impact of Tobacco Expenditure on Household Consumption Patterns in Rural China." *Social Science and Medicine*, v.62/6 (2006).

Household Hazardous Waste

Household hazardous waste consists of a variety of commercial products used in the home, such as cleaning supplies, pesticides, and pool chemicals, that are dangerous to human health or the environment and end up in the waste stream. The disposal of these products by homeowners is not regulated by most state and local governments, as long as the products are meant to be used in the home by residents. Conversely, a waste brought home from a worksite would not be considered household hazardous waste and would be regulated differently.

Development and Spread

Since the early 1900s, a wider variety of consumer goods have been made available to the general public. Many of these contain chemical components that pose a hazard to human health or the environment. These materials are marketed similarly to other consumer goods: they are inexpensive, relatively available, and ubiquitous. Some of these products have taken the place of inexpensive alternatives that were traditionally made in the home. For example, simple cleaning solutions can be made much cheaper and safer than many of their commercial counterparts. Although these commercial products are inexpensive and available in many locations, they can be very dangerous to the environment or to human health—even in small doses. A quick inventory of products in most homes will produce a list of dangerous chemicals that can do great harm to adults, children, air, water, and soil. Many of them can be used for bomb making.

Waste Management

The fact that household hazardous wastes are dispersed in hundreds of thousands of homes in small quantities makes them difficult to manage. Waste collection options are limited, and the wastes can become part of the general waste stream in a community unless there are programs designed to collect these special chemicals. Most of the collection programs focus on particular pickup dates, which can pose an inconvenience, thereby limiting participation in household hazardous waste collection programs. The end result is that a large volume of household hazardous waste winds up in the regular trash.

Particular problems occur when household hazardous waste is stored for long periods of time within homes, garages, or storage areas. Many have experienced finding a can or bottle of a household product with a missing or difficult-to-read label. When this occurs, the waste is expensive and difficult to dispose of properly. In addition, household hazardous waste stored for long periods of time may chemically transform into more dangerous materials.

Types of Household Hazardous Waste

Household hazardous waste may be corrosive, toxic, ignitable, or reactive. Labels classify consumer products using these terms if they have the potential to be harmful to human health. Examples of corrosive materials include bleach, battery acid, or oven cleaning products. Examples of toxic materials include a number of products that can cause harm or fatality if ingested or absorbed, such as antifreeze, pesticides, and some pharmaceutical products. Ignitable waste is comprised of materials

Most homes have an informal stock of dangerous chemicals that are potentially harmful to humans, animals, and the environment. Many of them can even be used for bomb making. Waste collection options for these products are limited and usually focus on certain pickup dates, which is inconvenient and discourages participation, especially when bottles and cans lose the labeling that would provide hazard and disposal information. As a result, a large volume of household hazardous waste winds up in the regular trash.

that will readily catch fire, such as gasoline, varnishes, and some paints. Finally, reactive ingredients can give off toxic gases or explode and consist of lye and pool and spa chemicals.

Along with the classification of the product's type of harm (corrosive, toxic, ignitable, or reactive), products are also classified by how dangerous they are. There are three levels of danger: caution (least dangerous), warning, and danger (most dangerous). Products listed with a caution warning can cause burning of the skin or dangerous fumes. The term *warning* is used for similar, but more health-threatening products and for products that can catch fire. The appellation *danger* is used for products having the potential to cause highly significant health problems upon exposure or ingestion, for explosive products, and for materials that can cause blindness.

These products can be found in kitchens, garages, basements, barns, sheds, bathrooms, and other settings in many households. In addition to the products listed, many electronics are also considered household hazardous waste since they contain a number of trace elements and also contain a variety of plastic products. Additionally, ammunition, aerosol cans, fuels, batteries, and automobile waste—such as used motor oil—are considered household hazardous wastes. Some unusual mercury-containing products, such as thermometers and fluorescent lighting, are also hazardous waste. Each year thousands of individuals are harmed by exposure to household hazardous waste.

Health and Environmental Problems

A variety of health and environmental problems can occur with household hazardous waste. Due to improper handling or storage, residents of a home may incur health problems upon exposure to the material. For example, improper storage of a pesticide may cause a container to rupture, thereby exposing members of a household to potentially dangerous conditions for long periods of time due to the slow release of the material. In contrast, poor handling or use of a material may cause sudden exposure, resulting in bodily harm. An example of this is the improper use or handling of oven cleaner, leading to bodily burns. When materials are stored or released outside the home—in a yard, carport, or other area exposed to the elements—they can be released into the near-home environment and can cause localized environmental damage; pet and wild animal poisoning; air, soil, and water pollution; and other unforeseen problems.

Waste workers, such as garbage collectors and workers in dumps, may be exposed to the waste that is unloaded within the general waste stream. In addition, landfills and waste-to-energy facilities may be damaged by the hazardous material. Deposition of waste in landfills or the burning of waste in incinerators may release the hazardous waste into the ground or atmosphere. Due to the high variability of these wastes, it is difficult to predict health or environmental consequences from their release, particularly since they can be mixed with other household hazardous materials, resulting in unforeseen consequences.

Regulation of Household Hazardous Waste

Household hazardous waste is difficult to regulate. The best approach to managing household hazardous waste is to not purchase or use such products. Since the 20th century, the public has become more interested in purchasing products that are safer and less damaging to the environment. The growth of the green cleaning-product industry, for example, demonstrates the viability of the commercialization of replacements for commonly used products that are considered household hazardous waste. It is relatively easy to find or make green cleaning products, pesticides, and other products. The Internet is a particularly useful tool for consumers seeking to limit their exposure to harmful chemicals or who are interested in having a gentler footprint on the planet. Nevertheless, tons of household hazardous waste are disposed of every year. One way residents can minimize the volume released is to find others who may be able to use the products. For example, if one no longer needs a pesticide, friends or relatives may be able to use the product. Some communities have established household hazardous waste swaps where individuals can bring in their unneeded products and swap them for a product that is needed. In addition, some charitable organizations, such as Habitat for Humanity, will take unused products, such as paint, for their use or, in some instances, resale.

Another approach to managing hazardous waste at the community level is to have a collection day affording residents the opportunity to bring unwanted hazardous materials to a central collection facility where the waste can be organized and disposed of in appropriate ways. The benefit of this approach is that it enables all residents to get rid of waste during regular intervals; furthermore, the management of the waste can be handled in bulk. Sometimes, communities organize days for the collection of specific materials, such as waste motor oil or batteries. In contrast to this, other communities have routine curbside collection of hazardous waste. The danger with this approach is that public workers must handle the material and transport it to a collection facility, and, because collection usually occurs more frequently than drop-off programs, the waste is often stored at a central collection facility. Some communities have specially designed mobile units that will gather household hazardous waste by appointment or that will collect within communities during scheduled collection days. Special wastes, such as batteries, electronics, used motor oils, and fluorescent lights, are oftentimes collected by private businesses for ease of recycling.

Regardless of management options, most communities have some type of household hazardous waste program in place. One key aspect of these programs is the education of the public about the environmental and health problems associated with the waste and the ways that individuals can manage the waste. Many cities have educational Websites with information about disposal options available to residents, as well as information about alternative products. In addition, it is recommended by the Environmental Protection Agency that those involved with hazardous waste collection programs use licensed hazardous waste haulers to transport the waste after it reaches a collection facility. Individuals may transport their own household waste in vehicles. However, once at a collection site, the volume of material often requires special handling and transport to an appropriate disposal site.

Robert Brinkmann
University of South Florida

See Also: Cleaning Products; Household Consumption Patterns; Personal Products; Post-Consumer Waste; Public Health; Toxic Wastes.

Further Readings

Environmental Protection Agency. "Household Hazardous Waste." http://www.epa.gov/osw/conserve/materials/hhw.htm (Accessed July 2010).

Slack, R. J., et al. "Household Hazardous Waste in Municipal Landfills: Contaminants in Leachate." *Science of the Total Environment*, v.337/1–3 (2005).

Human Waste

The term *human waste* refers broadly not only to the by-products of human physiological processes, most commonly to urine and feces, but also to sweat, phlegm, and flatus, among other bodily excretions. Urine and feces are commonly regarded as being dirty across cultures, while the mixing of human waste with food is almost always regarded as taboo. More than any other development in the history of excretory experiences, the 19th-century sanitation movement and the developing discourses of hygiene and public health transformed people's experience and relationship to human waste. Flush toilets and sewers in urban centers organized human waste disposal into a centralized system under public management, allowing human waste to be carried away by water, making excreta invisible in public spaces.

This method of human waste removal, now prevalent in the 21st century, carries tremendous environmental consequences. Flush toilets are highly water intensive, and the release of untreated sewage pollutes local water sources. Sewers and flush toilets have replaced previously common methods of handling human waste. Night soil collection, where excreta in urban centers was systematically collected and processed into fertilizer for agriculture, was widely practiced throughout east Asia and selectively in Europe and North America. In the 21st century, in an effort to provide adequate sanitation, developing countries are exploring low-cost, closed-looped sanitation systems that use human waste as a resource.

Social Stigma

Human waste has been regarded cross-culturally as a prototypical offensive substance. Various social theories have been offered for the prevalence of aversion to human waste across cultures. Some theories categorize the disgust associated with waste as an evolutionary response to guard against disease, while others hold that the aversion to human waste reflects disgust at the "animalness" of the act and product of defecation. Freudian theories suggest that the disgust toward human waste originates from the emotional trauma of the toilet training experience. Various cleansing rituals exist to separate human waste from daily life and especially from coming into contact with food. Across many cultures, social stigma is often attached to groups that handle human waste, most notably night soil collectors throughout east Asia and untouchables in south Asia.

Early History of Human Waste Disposal

Archaeological remains and archival documents of ancient cities illustrate that organized systems of human waste removal have existed for millennia. Dating back to 2500 B.C.E., excavation from the Indus basin and Mesopotamia reveal highly developed brick structures where waste from each house was directed into drains. Significant social variations in defecation practices also exist throughout history. Defecation has been both a solitary and a social activity. While most Western-style bathrooms in the 21st century contain stalls to provide privacy, the design of Roman bathrooms displays a more social setting for defecation. Excavations of public latrines in the city of Ostia indicate that toilet seats were lined up closely next to one another, forming a square with no barrier between users.

While human waste has typically been regarded to some extent as offensive, historically, both urine and feces have also been considered a useful resource. In large agrarian societies such as China and India, human waste was often collected in towns and cities to be transformed into fertilizers for the countryside through what was known as night soil collection. Human feces, collected with or without urine, was placed outside buildings to be picked up and transported by night soil collectors to the countryside, where it was processed and used as fertilizer. It is difficult to tell how far back the use of human manure extends in history, particularly because the Chinese word for "manure" (one of the most prominent regions practicing night soil collection), *fen*, does not distinguish between human and animal sources. Agricultural manuals in China dating from the 12th century offer a variety of techniques for the conversion of waste to fertilizer, demonstrating the awareness that human waste needed to be processed and purified before it could be applied to agricultural crops. The composting and conversion of human excreta into fertilizer to sustain agricultural production was practiced widely throughout east Asia, in France, and selectively in North America in the mid-19th century. Urine, often regarded as having disinfectant properties, was also applied widely as a pesticide, cleaning agent, and sterilizer. It was also not uncommon in 15th-century France to use urine for the cleansing of draperies and clothes.

Centralized Sewer Systems

The sanitation revolution in 19th-century Britain changed the experience of defecation and cast human waste treatment as an environmental and social risk to be managed by a centralized system of disposal. Medieval European cities were awash with filth, as it was common for human waste to be collected in chamber pots and dumped into streets. However, with the growth of cities in the 19th century because of industrialization, urban filth was associated with disease and epidemics. The miasma theory of disease, popular during the Medieval era up to the mid-19th century, held that poisonous vapors, or "bad air," was the cause of various epidemics of the 19th century, most notably cholera. Dr. John Snow's work helped in part to dispel the miasma theory of disease by demonstrating that the 1854 London cholera epidemic was spread through drinking water. The germ theory of disease (which held that disease was transferred through microbes) and the passing of Edwin Chadwick's Public Health Act of 1848 marked the beginning of the sanitation movement, which aimed to provide clean water and sewage treatment to Europe's most crowded cities. Sewers, as a means of transporting filth away from cities, also became an issue of public concern as public health bureaus became established for the regulation of sanitation projects.

Aside from the more obvious benefits of waterborne sewage systems for sanitation, the widespread adoptions of flush toilets and sewers were also linked to the changing sensibility of modern elites who sought to be released from the filth of medieval cities. Sewers and Victorian water closets facilitated the domestication of waste, where waste was cast as a morally offending substance that needed to be removed from the public sphere. The creation of hidden sewer networks below the surface of the city rendered morally offending substances invisible and transformed the sensual experience of the city, especially in relation to smell and sight.

Flush toilets became the primary mode of waste treatment in the United States and Europe. However, they require a vast amount of water, and the discharge of untreated waste from sewage systems has caused pollution and contamination, especially to local water sources. According to the United Nations, in 2008, more than two billion people still lack access to improved sanitation. Inadequate sanitation infrastructure and untreated wastewater can also transmit diseases such as cholera, and unprocessed excreta often leads to intestinal infections and diarrhea, one of the biggest killers of children younger than 5 years old. Moreover, a lack of sanitation facilities often has an uneven social impact on women, as open defecation poses shame and difficulty during menstruation.

Ecological Sanitation

International sanitation movements are experimenting with ecological sanitation as a cost-effective and environmental alternative method of human waste disposal. Ecological sanitation borrows the idea of a closed loop waste treatment system exemplified by traditional night soil collection to treat waste and wastewater as a source of nutrients for agriculture. Contemporary ecological sanitation projects are wide ranging and include methods such as dry waste treatment, greywater reuse, and human excreta reuse in aquaculture (using human waste as a source of fertilizer for raising fish as well as aquatic crops such as water spinach). Dry waste treatment refers to both dehydration and composting treatments.

Dehydration treatments require the separation of urine and feces and the addition of lime, ash, or earth to feces in order to produce a soil conditioner. The collected urine can be diluted and used as a source of nitrogen fertilizer for plants. Urine and feces can also be composted, adding worms and other materials to the human waste mixture to help decomposition. Attempts to introduce new ecological sanitation practices often require overcoming developed social taboos against the use of human waste for fertilization. The success of ecological sanitation projects requires attentiveness to the social and cultural experience of human waste, and proper usage depends upon sustained research into the health impacts of sanitation projects.

Amy Zhang
Yale University

See Also: Composting; Germ Theory of Disease; Miasma Theory of Disease; Public Health; Sewage; Sewage Collection System; Sewers; Sustainable Waste Management.

Further Readings

Duncker, L. C., G. N. Matsebe, and N. Moilwa. "The Social/Cultural Acceptability of Using Human Excreta (Feces and Urine) for Food Production in Rural Settlement in South Africa." *WRC Report*, v.310 (2007).

Inglis, D. *A Sociological History of Excretory Experience: Defecatory Manners and Toiletry Technologies*. Lewiston, NY: Edwin Mellen Press, 2001.

King, F. H. *Farmers of Forty Centuries or Permanent Agriculture in China, Korea and Japan*. Madison, WI: Mrs. F. H. King, 1911.

Laporte, D. *The History of Shit*. Trans. by N. Benabid. Cambridge, MA: MIT Press, 2000.

Idaho

Admitted to the United States in 1890, Idaho is the 43rd state and 14th-largest state by area. Boise, the capital, is the largest city and metropolitan area. Landlocked and mostly mountainous, Idaho's northern limit forms part of the Canadian border. With 60 percent of the state's land held by the National Forest Service or Bureau of Land Management, Idaho leads the United States for forest service land as percentage of total area. The second-largest unit of the National Wilderness Preservation System can also be found in Idaho—the Frank Church River of No Return Wilderness. Agriculture features predominantly in the state's economy; leading industries include processed food, lumber and wood, machinery, chemical and paper products, electronics manufacturing, mining, and tourism. Idaho has seen two significant anthropological studies of consumption and waste disposal carried out in recent years: the Campus Trash Project and the archaeological investigation of the Minidoka Relocation Center Dump.

The 16th Nationwide Survey of MSW Management in the United States found that in 2004, Idaho generated an estimated 1,238,394 tons of municipal solid waste (MSW), placing it 44th in a survey of the 50 states and capital district. Of this, 99,590 tons were recycled, placing Idaho 45th in the ranking of recycled MSW tonnage. Based on the 2004 population of 1,463,878, this is an estimated 0.85 tons of MSW generated per person per year, the lowest per capita MSW generation in the United States. Idaho landfilled 1,138,804 tons (40th ranking) and recycled 99,590 tons (45th ranking). Only whole tires and lead-acid batteries were reported as being banned from Idaho landfills.

Campus Trash Project

At the University of Idaho in Moscow, the Campus Trash Project yielded valuable information on the human and environmental factors contributing to patterns of litter accumulation and distribution. The project provided an evaluation of the effectiveness of university waste management policies and also served as a valid training exercise for archaeology undergraduates. The use of archaeological methodologies highlighted the mundane, human, and geomorphic factors that created the pattern of littering on campus.

Research carried out as part of the project showed that university campuses produce an equal or greater amount of trash than the urban centers around them. It also highlighted the problem that, while universities have created permanent positions and initiatives to reduce campus waste problems, they often overlook everyday human and environmental processes by focusing on macroeconomics. Trash signatures at the University of Idaho were found to be similar to those of a similar study at the University of Louisville. Approximately 19 percent of all campus waste was recycled, compared to 11 percent of recyclable waste in the rest of the city (Moscow), while only 8 percent of the state's overall waste was recycled. The state, city, and campus were all below the national recycling average of 25 percent.

The project began in 2008 as an assignment for archaeology students. Four trash-ridden zones on the campus were identified, and a group was sent to each with a map of their zone and handheld GPS units, which they used to plot relevant features, such as bins and parking bumps. Having mapped the features in the zones, a surface collection of litter was carried out and every litter artifact was bagged and located with spatial coordinates. An ethnographic element included observing and interviewing the students and staff who used the zone. The groups were given instructions on how to write up their findings structured in a way similar to an archaeological report.

Findings of the Campus Trash Project showed that prominently placed "butt pipes" for cigarette waste disposal were ignored by students smoking directly next to them. The problem was identified as being a case of the cigarette waste disposal pipes blending into their surroundings to such an extent they went unnoticed, and students lacked awareness that cigarette butts constituted litter.

The football tailgating parking lot returned 388 litter artifacts, of which 76 percent (298 artifacts) were alcohol related. The waste management procedures in place (handing out plastic bags to all vehicles entering) failed. Dividing the tailgating lot into quadrants for observation and recording during tailgating events showed that alcohol intoxication combined with poor dumpster location and coloring were the major factors contributing to litter buildup. Proposed solutions included allocated parking bays to allow the issuance of fines and more distinctive, easily-located dumpsters.

The factors accounting for accumulation of litter in a creek (Paradise Creek) and swale land were pinpointed by participant observation and the mapping of trash scatters. It had been suggested that students littering on entry and exit of the nearby recreation center created the accumulating garbage in this ecologically sensitive area. However, the actual source was three dumpsters at a neighboring student accommodation block, which were regularly left open. The conclusions reached by the Campus Trash Project included the need for more research on campus trash. University campuses were vaunted as an ideal testing ground for new waste management policies, allowing them to be observed in practice. The unpredictability of human behavior was highlighted, as was the need to test changes on the macroscale in a microscale environment.

Minidoka Dump

Near Twin Falls, Jerome County, on Bureau of Land Management–administered public land is the site of the Minidoka Relocation Center Dump. Minidoka Relocation Center was built in 1942 to intern 10,000 Japanese Americans during World War II. One of the largest and most densely populated settlements in Idaho during the war, the relocation reserve covered 33,000 acres, and its peak population reached 9,400. The U.S. Supreme Court declared internment illegal in 1944. The last internees, made destitute by their relocation, were evicted from Minidoka in 1945. In 1947–49, suitable areas of the reserve were cleared and given to World War II veterans as family farms. The administrative area, a small part of the original central area, survives in the 21st century as the Minidoka Internment National Monument, one mile south of the dump.

The original dump, used by the Relocation Center for daily rubbish between 1942 and 1945, was around 2.5 acres. When the center was abandoned, the area used for dumping increased to over 26 acres as demolition debris, construction debris, and redundant institutional furnishings were dumped. The speed of demolition and vast amount of material remaining even after reusable materials had been removed turned the original dump into a much larger wasteland. This wasteland could not be farmed, and it was seen as an ideal dumping ground by nearby residents. Until 1982, there was regular trash disposal at the Minidoka Dump, with the last disposal made in 1988.

In August 2004, the National Park Service mapped and recorded 229 trash features and over 260 distinct surface deposits consisting of piles and scatters of debris and coal residue. Features ranged in size from 1 m^2 to 4,300 m^2; the largest features and the only buried features were associated with the relocation center. Disturbance of the deposits was noted in the form of bottle digging, trampling, and vehicle damage. The debris includes piles of bricks, rocks, and concrete. Often, one type of building material is dominant, and many of the piles are linear, up to 225 feet long. The relocation-era (1942–45) rubbish was systematically deposited into a prepared pit and other nearby concentrations, and these are the only features that have stratified deposits. The dump was half a mile north of the residential area. The main feature is a partly filled 370-by-100-foot trash pit from the camp; there is also potentially an earlier capped dump from the camp.

The most useful artifacts for dating were found to be (1) glass manufacturer date codes, (2) drink can types, (3) vehicle license plates, and (4) brand names and trademarks. The recent date and narrowly separated phases of the dump meant that artifacts that are usually regarded as useful date indicators in U.S. historical archaeology were of little help. Ceramic backstamps and glass manufacturer's marks remained unchanged for decades in the 20th century, and milk can dimensions were inconsistent.

It was found that many features had artifact types from more than one decade, and these were counted in each decade represented, even where they were likely to be single-dumping events. While this method of tabulation inflates the number of features per decade, it was thought more desirable than arbitrarily assigning a decade or ignoring the feature altogether. Eighty-four features were considered undatable. It was possible to date some of the relocation center dump features to within a single year.

Significant findings in the post–relocation era garbage include increasing automobile-related artifacts between the 1950s and 1960s and a notable preponderance of alcoholic drink containers associated with the World War II veteran farmers. The presence of artifacts thought to be from the relocation camp in post-relocation refuse is also surprising, as usable goods from the centers were routinely destroyed to avoid flooding local markets. The largest numbers of features containing toys dated to the 1950s, thought to coincide with the postwar Baby Boom. It is possible that just one or two households deposited all of the post–relocation era trash. Recommendations for further work included land acquisition and fencing, additional excavation of relocation-era features, oral history collection, more recording of post–relocation center features, and hazardous materials (hazmat) evaluation.

Jon Welsh
AAG Archaeology

See Also: Archaeological Techniques, Modern Day; Archaeology of Garbage; Construction and Demolition Waste; Dating of Garbage Deposition; Dump Digging; Race and Garbage.

Further Readings

Arsova, Ljupka, et al. "The State of Garbage in America: 16th Nationwide Survey of MSW Management in the U.S." *BioCycle*, v.49/12 (2008).

Burton, Jeffery F. *The Fate of Things: Archeological Investigations at the Minidoka Relocation Center Dump*. Tucson, AZ: Publications in Anthropology, Western Archeological and Conservation Centre, National Park Service, U.S. Department of the Interior, 2005.

Camp, Stacey Lynn. "Teaching With Trash: Archaeological Insights on University Waste Management." *World Archaeology*, v.42/3 (2010).

Nagawiecki, T. *University of Idaho Waste Characterization*. Moscow: University of Idaho Press, 2009.

Illinois

A hub of transportation, industry, and agriculture, Illinois is a diverse state. Featuring the third-largest city in the United States, much of the population is concentrated in the Chicago metropolitan area along the Lake Michigan coast in the northeastern corner of the state. The entire western border of the state follows the path of the Mississippi River, and the Ohio River joins the Mississippi at the southernmost tip of the state. Almost 13 million residents lived in Illinois in 2010, 5,194,675 of them in Cook County. From a consumption and waste standpoint, downstate Illinois differs considerably from the Chicago area. Most of the land area in Illinois is south of Chicago and is often just given the general designation of "downstate." Central and southern Illinois are largely agricultural areas interrupted by a few larger cities. In addition to agriculture, southern Illinois has a considerable amount of coal mining as well as some oil fields. Both of these extraction activities have led to significant waste and pollution issues downstate. Northern Illinois and Chicago have also been long-known for their steel mills. The vast majority of the iron ore is actually mined out-of-state and transported in by railway.

History

Illinois became a state in 1818. Due to its proximity to midwest railways as well as shipping routes in the Great Lakes, the Chicago area quickly became the most densely populated region of Illinois. Given its size, consumption and waste has played a primary role in a surprisingly large number of Chicago's major historical events. As early as 1849, the city had official "scavengers" (the historical term for *garbage men*). One of the seminal events in early Chicago history was the Great Fire of 1871, which destroyed nearly four square miles in the heart of the city. Disposing of the tremendous amount of waste from the fire was the first step in the rebuilding process. Debris from the fire, and later, general city garbage was pushed from the shoreline into Lake Michigan. In the 21st century, most of what is Grant Park on Chicago's Lakefront—including Soldier Field—is built on top of old Chicago fire debris and garbage. One of Chicago's long-standing core industries is the slaughterhouse and meatpacking business. Following the Great Fire, these previously scattered businesses were consolidated into the 100-acre Union Stock Yards, southwest of the central business district.

By 1890, 12 million head of cattle were slaughtered annually, which represented a staggering 150 million pounds of animal waste. In the early days of the slaughterhouses, the majority of this animal waste was simply disposed of into the Chicago River. In fact, one fork of the Chicago River, which received much of the animal waste, was nicknamed "Bubbly Creek" because of the bubbles that would rise to the surface from decaying animal matter. Fieldwork in the Chicago slaughterhouses helped Upton Sinclair write *The Jungle* in 1906. His novel helped to expose atrocious working and unsanitary conditions in the slaughterhouses and eventually led to the federal Meat Inspection Act. A profit-motivated push for greater efficiency in the slaughterhouse process eventually led to the repurposing of slaughterhouse waste into categories like fertilizer, lard, leather, soap, and tallow. This greatly reduced the amount of physical waste, but instead created large amounts of air pollution and hazardous working conditions in many of the rendering plants.

Environmental justice is a relatively recent reform movement, but Jane Addams of Chicago's Hull House fame was tackling environmental inequality issues in the early 1900s. Corrupt politicians and lax scavengers had allowed small mountains of garbage to accrue in the Hull House's 19th Ward. Connecting the trash problem to high levels of sickness and disease in her ward, Addams began a trash crusade by reporting thousands of trash ordinance violations to the city's Health Department. These tireless efforts eventually resulted in the mayor appointing Addams as an official city trash inspector.

Reversing the flow of the Chicago River is another major event in the history of Chicago that has its roots in waste management. In the late 19th century, Chicago had the ability to pull more than 100 million gallons of water per day from Lake Michigan. The water came from offshore intake pipes. For all of Chicago's early history, the Chicago River was used as the primary city sewer. As the city grew and wastes became more concentrated, the need to protect the city's drinking water became pressing. The solution was to create the Chicago Sanitary and Shipping Canal. This engineering feat involved excavating a 28-mile-long canal that would send Chicago's wastes downstream away from Lake Michigan and create a convenient shipping lane to the Illinois River. The opening of the canal in 1900 effectively reversed the natural flow of the Chicago River and caused decades of strife with cities downriver that now had to deal with Chicago's wastes.

As Chicago continued to grow, so did its waste generation. By mid-century, waste collection had become a lucrative business, and organized crime tended to follow profitable industries. Dutch immigrants had a long history of managing most of the city's waste hauling contracts. Their market share grew even more when, in 1959, the Dutch Mafia consolidated its waste hauling businesses under the name Chicago and Suburban Refuse Disposal Association. Not to be left out of the graft, the Italian Mafia was quick to follow suit under the leadership of Willie Daddano. Daddano, nicknamed "Willie Potatoes," formed the West Suburban Scavenger Service in 1960 and used mob contacts, threats, and intimidation to take over waste hauling contracts. Corruption and graft in the garbage business did not end with the mob's involvement. From 1992 to 1996, the Federal Bureau of Investigation conducted an investigation called Operation Silver Shovel. This investigation uncovered a corrupt system of bribes and money laundering in relation to the illegal dumping of construction debris. Operation Silver Shovel eventually led to the conviction of 18 people, including many Chicago aldermen and inspectors. From the 1960s through the 1980s, incineration became the favored method for garbage disposal. When it went into service in 1971, the Northwest Incinerator was the largest in the world and handled 20 percent of Chicago's garbage. As the environmental impacts of incineration were more fully understood, contemporary Chicago again had to contend with its waste problems.

Illinois in the 21st Century

In 2008, 45 active landfills in Illinois accepted more than 50 million gate cubic yards of municipal waste. Chicago has gained the dubious distinction of having more landfills per square mile than any other city in the world. By one estimate, every Chicagoan generates one ton of waste per year. In 2010, Illinois also had 481 Superfund sites as designated by the Environmental Protection Agency. The majority of these can be found in and around Chicago—a testament to the city's past struggles with waste disposal. Within the city, there is an unequal distribution of these landfills and Superfund sites. For example, a 1983 study of the Southeast Side of Chicago revealed cancer rates double those of the rest of the city. Not by coincidence, this area also has over 25 square miles of landfill, in addition to other environmental problems. The residents of this area are predominantly poorer African Americans and immigrants. The environmental justice movement has called this unequal exposure of a particular social group to pollution, toxins, and other hazards "environmental racism." More than a century ago, Jane Addams noted that not all of Chicago's wards were treated equally when it came to trash removal. Her reform ethic has gained hold in 21st-century Chicago with early environmental justice groups such as People for Community Recovery and the Chicago Resource Center.

The Chicago Resource Center (CRC) was founded in 1975 by then University of Chicago philosophy student Ken Dunn. The CRC attempts to merge social and environmental causes. The center began with a program that paid the homeless to collect recyclables. In the 21st century, the CRC uses a multifaceted approach to help achieve social and environmental justice that includes turning vacant lots into community gardens, curbside recycling, managing a creative reuse warehouse, composting, and environmental education.

After many years of dragging its feet on a citywide recycling system, the Blue Bag Recycling Program was initiated in 1995 by Mayor Richard Daley. The concept was to have residents purchase blue plastic bags that could be filled with recyclables and tossed out with the rest of the trash. The blue bags would be collected along with the regular trash and were then to be sorted out at "recovery facilities." From the beginning, the system was plagued with problems. Many bags never made it to the sorting facilities, and those that did were often broken and unusable. The system was eventually somewhat improved by the replacement Blue Cart system. In 2010, however, a study commissioned by Chicago's Department of Environment found that only 8 percent of the waste from the 600,000 homes with city garbage service was recycled (in part because Blue Carts were distributed to fewer than half of households), and few of Chicago's many highrise buildings enjoyed recycling services. The city announced a new pilot program in 2011 to modestly expand the Blue Cart system, but at the beginning of 2012, most Chicagoans lacked recycling pickup services.

Christopher Sweet
Illinois Wesleyan University

See Also: Crime and Garbage; Environmental Justice; Race and Garbage; Waste Management, Inc.

Further Readings

Biles, Roger. *Illinois: A History of the Land and Its People*. DeKalb: Northern Illinois University Press, 2005.

Grossman, James, ed., et al. *The Encyclopedia of Chicago*. Chicago: University of Chicago Press, 2004.

Pellow, David Naguib. *Garbage Wars: The Struggle for Environmental Justice in Chicago*. Cambridge, MA: MIT Press, 2002.

Incinerator Construction Trends

An incinerator is an industrial unit used to treat waste by combusting it at high temperatures. In the 1980s and 1990s, incinerator construction slowed down because of concerns over air pollution. The use of incinerators was limited to existing structures and for the treatment and disposal of toxic and medical waste. In the 21st century, there has been renewed interest in new facilities, driven by the growing interest in using waste as an energy source, the emergence of such technologies, and new regulations regarding landfills. Nonetheless, critics continue to argue against the building of new incinerators, asserting that their usage does not give consumers any incentive to recycle, reuse, or reduce consumption, and that they are more polluting than coal power plants.

An incinerator is an industrial unit used to treat waste by combusting it at high temperatures. The largest incinerators handle and dispose of municipal waste, while smaller incinerators are used for specialized materials (such as toxic and medical waste). There has been an informal moratorium in many geographical locations on the building of new incinerators because of the potential harm to the environment. Nonetheless, incinerators continue to be built and used for hazardous and clinical waste because high temperatures are necessary to destroy pathogens and toxic contaminants. With the emergence of environmentally friendly incinerators, several municipal and regional governments are studying the possibility of resuming construction, since they are an efficient method to dispose of waste in geographies where there is a high density of population or without suitable space for landfills.

Construction Considerations and Trends

Multiple considerations are taken into account when building an incinerator. One of the decisive factors is the local waste; the current and future quantity of waste generated is the prime consideration, but other factors, such as the composition of the waste, are also considered. The local political and public environment must also be carefully measured; legislation on emission controls and public perception both play an important role in choosing a locality for establishing a new facility. Other considerations include available infrastructures, locally available materials, and expertise.

Construction of new incinerators slowed (or, in the case of the United States, completely halted) from 1995 to 2006. The main reason for this moratorium was environmental concerns because of the pollution allegedly produced by these structures. Since 2006, there has been a renewed interest in the construction of such incinerators. This interest has been led by the growing use of waste as an energy source, the emergence of new technologies, and new regulations surrounding landfills.

One of the main reasons for renewed interest is that waste incineration has been granted qualification for renewable energy (RE) production tax credits in the United States, and it has obtained equivalent certification in Europe. Since many solid waste components contain hydrocarbons, their incineration generates steam and heat, which can be harnessed to generate electricity. Furthermore, the "renewable" nature of municipal waste has enabled it to qualify for renewable energy status. Hence, its designation as an RE has led to project expansions as well as feasibility evaluations for new plants. In addition, new projects devote extra attention to energy production in their proposal to evaluators and investors and take advantage of new waste-to-energy technologies in their design.

New technologies have also been developed to reduce greenhouse gas emissions and improve air pollution control. For example, a study compared the air quality in three communities with incinerators and three with no such structures. The research-

An incineration plant in Italy. From 1995 to 2006, construction of new incinerators slowed—or was completely halted, as was the case in the United States. This moratorium was due to environmental concerns of the pollution allegedly produced by these structures. However, renewed interest in building incinerators has been spurred by the prospect of using waste as an energy source. In the United States, incineration has been granted renewable energy production tax credits, and European builders can receive equivalent certification.

ers did not detect any differences in concentrations of particulate matter among the communities. These technologies, as well as the continued efforts of recycling ashes into construction materials, have defused some of the concerns put forth by environmental critics. Finally, there is less and less space available for landfilling. Land is fundamentally a finite resource. Incinerators are an ideal technology for locations with dense population because they have limited space available to bury garbage. This has convinced authorities to once again look at incinerators as a viable waste management technique.

Critics

Even using improved technologies and considering the benefits obtained from reusing waste (such as energy generation or ash recycling), critics argue that incinerator use should be far more restricted. It is believed that the proliferation of incinerators discourages consumers to reduce, reuse, and recycle because they have little incentive to do so in the first place. Cheap waste disposal is believed to inhibit consumer incentive in green programs, and the emergence of a green side to incinerators only reinforces those negative behaviors. Its designation as a renewable energy has also been criticized, as opponents allege that on a pound-for-pound basis, incinerating waste generates more pollution than a coal plant. Finally, critics remain unconvinced that the pollution and greenhouse gases emitted by an incinerator can ever be contained, managed, or mitigated.

Jean-Francois Denault
Independent Scholar

See Also: Hospitals; Incinerator Waste; Incinerators; Landfills, Modern.

Further Readings

Shy, Carl, et al. "Do Waste Incinerators Induce Adverse Respiratory Effects? An Air Quality and Epidemiological Study of Six Communities." *Environmental Health Perspectives*, v.103/7–8 (1995).

Stehlík, Peter. "Contribution to Advances in Waste-to-Energy Technologies." *Journal of Cleaner Production*, v.11/10 (July 2009).

Incinerator Waste

Incinerator waste is a broad term for residues that result primarily from controlled incineration activities in large-scale facilities. Incineration is considered one of the most effective waste treatment operations because of high reduction and decontamination rates of physical and chemical composition of wastes. But not all substances are destroyed by the industrial furnaces, cement kilns, or other thermal installations used for incineration. Some noncombustible matters persist and are classified as residual waste streams, dealt with by sterilization, transport and landfilling actions, or recovery activities like downcycling or energy generation.

Statistics on the percentage of waste not destroyed by incineration are usually disparate, with figures anywhere between 5 and 50 percent, given distinct applied technologies, original matters, and other factors. Sizable amounts of solid components, gaseous effluents, liquid substances, and air particles remain, however, as by-products of incineration. Among these, fly and bottom ash are regularly identified as main residues, but matters like grate siftings, frequently amalgamated with fly and bottom ash, as well as slag with vitrified metals, or cleansing waters and sludges, should also be counted as wastes.

Fly and Bottom Ash

Fly and bottom ash are the major residual traces of incineration in key scientific debates and literature on thermal treatment systems. The first corresponds to small-size, light particles that emerge as flue gas from pollution control processes. It contains pollutants that change in number and concentration, given desired or required emission standards, combustion technology, and physical properties of primary residues. The second matches the unburned organic or inorganic materials at the outlet of burning chambers. Bottom ash is typically made of heavy and solid elements that settle by gravity, such as ceramic-like matters, and often has less pollutants and heavy metals than fly ash.

Much of the debate over incineration waste is focused on toxicity as in analogous arguments on other environmental "bads." Thus, matters like fly and bottom ash are often evaluated and managed with a focus on hazardous elements that are largely toxic in small amounts, such as dioxins and furans or specific chlorides and halides. When not regulated or controlled, these find their way into living environments and produce extended collective and individual disorders on health and ecological fronts. Moreover, toxicity problems also connect these wastes with geographical and social inequalities, not only from the location of incineration facilities, but also considering factors such as occupational diseases exposure in waste treatment workers, who often belong to minority populations.

Emissions Treatments

Legal and technical regulations prevent the occurrence or amplification of disturbances related to waste hazardous properties. They are in place from the development of international or national emission standards and legislation on leachability of combustion wastes and technical control devices that decrease ash emissions or reduce perilous materials in remaining bottom ashes, like the newest pollution filters made by living systems or simple scrubbers for acid neutralization. Nonetheless, after combustion, there are also considerable precautions regarding transport and deposition of residues. Measures are designed to avoid fugitive emissions and assure low toxicity in residues to be landfilled, for example, closed systems to manage fine particles and later groundwater tests.

At the conclusion of incineration, residues like fly and bottom ash may be landfilled or treated as suitable resources for matter recoveries, depending on the assessment of optimum scenarios and the foreseen environmental impacts of each option. Among waste hierarchies, recent models often choose burned wastes as reusable substances, reducing pressure on landfills and promoting sustainability in incineration. When reuse is impossible by lawful and environmental constraints or the lack of commercial markets, residues end up in landfills. But incinerator wastes have gained a wide acceptance as building materials, like bottom ash, or even nutrient sources, as in fly ash. The first is retrieved, for example, as landfill cover and secondary aggregate in asphalt pavings or bulk fill.

Incineration as a source of waste recuperations is, in fact, ordinary if seen from a thermodynamic viewpoint. These wastes are seldom detached from recoveries, as most 21st-century incineration facilities function with waste-to-energy schemes. Combustion systems may reuse incinerator wastes from the beginning by converting burning processes into electricity or heating production, through entropic steam or waste masses as fuels. Prospects regarding incinerator waste management are even related to these procedures as routes for higher environmental and economical ratios. Nonetheless, other future treatment possibilities are also at play, designing not only better legislation and technical control of combustion but also improved thermal technologies that augment burning efficiency and extended producer responsibility platforms, considering both incineration increase and decrease scenarios.

Alexandre Pólvora
University Paris 1 Panthéon-Sorbonne

See Also: Incinerator Construction Trends; Incinerators; Scrubbers; Sustainable Waste Management; Toxic Wastes; Waste Treatment Plants.

Further Readings

Committee on Health Effects of Waste Incineration. *Waste Incineration and Public Health*. Washington, DC: National Academies Press, 1998.

Roberts, Stephen M., Christopher M. Teaf, and Judy A. Bean, eds. *Hazardous Waste Incineration: Evaluating the Human Health and Environmental Risks*. Boca Raton, FL: CRC Press, 1999.

Incinerators

As applied in solid waste management, an incinerator is a facility designed for the efficient, controlled combustion of wastes at a high temperature. Incinerating waste reduces the volume of waste over a short period of time. Incineration can destroy harmful chemicals and pathogens and can be used to produce electricity and heat. Modern incinerators are designed to completely combust waste products and minimize and treat emitted air and solid pollutants. Many types of wastes can be burned in an incinerator, including municipal solid waste (MSW), hazardous waste, and refuse-derived fuel (RDF), which is pellets made from the high-energy fraction of waste.

Brief History of Incineration

Incinerators have evolved greatly since the early 20th century, when open burning of wastes (either in a chimney or outdoors) was the norm for the first half of the 1900s. Incinerators in the 21st century are highly centralized facilities designed to completely combust wastes, produce energy, and minimize and treat the resulting air emissions.

The first waste incinerator was constructed in England in 1874, and the first incinerator in the United States was built in 1885. In the early 20th century, in-house incinerators were very common in the United States, resulting in a remarkably high ash fraction in U.S. garbage (43 percent in 1939). The fast growth of incineration in the United States was halted by a growing environmental movement, which led to both increased legislation and a powerful grassroots movement that fought to keep incinerators from being sited in their communities because of concern about the emissions they produced. Both the Resource Conservation and Recovery Act of 1976 and the Clean Air Act of 1990 set strict standards with which incinerators must comply. Modern incinerators are equipped with various air pollution control devices that treat and minimize the harmful emissions from burning waste. Though popular resistance to waste incineration is strong in the United States, incineration is accepted in other parts of the world; the waste management technology is used widely in Europe and Japan, which combusted 75 and 90 percent of its MSW in 2000, respectively.

Incineration Process

Incineration is the oxidation of materials at a high temperature. During the burning of wastes, moisture evaporates from the fuel and organic compounds are ignited in the presence of oxygen. The incineration process is designed to attain complete combustion of wastes; this means that all carbon in the waste is converted to carbon dioxide (CO_2), all the hydrogen to water (H_2O), and all the sulfur to sulfur dioxide (SO_2). By-products include ash, air emissions, heat, and energy.

For an efficient combustion process, incinerated wastes should have a low moisture content (less than 50 percent) and should have a relatively high heating value (greater than 5 MJ/kg). If moisture contents are higher and heating values are lower, the wastes will require additional fuel to sustain combustion. Wastes that contain inorganic salts, high sulfur or chlorine contents, or radioactive materials must be treated in specially designed facilities. Generally, the two wastes used to produce electricity are MSW, which is unsorted waste, and RDF, which is comprised of a subset of MSW that has a higher average energy content.

A simplified representation of the complete combustion of waste can be expressed as

$$C_nH_mO_pN_qS_r + x(O_2+3.76N_2) \rightarrow nCO_2 + aO_2 + (m/2)\,H_2O + cNO + dN_2 + rSO_2.$$

As shown on the left side of the equation, waste is represented as some combination of carbon, hydrogen, oxygen, nitrogen, and sulfur, and it is burned in the presence of air, which is largely composed of oxygen and nitrogen. When completely burned, the process emits carbon dioxide, oxygen, water, nitrogen oxides, nitrogen gas, and sulfur dioxide.

To achieve complete combustion, the incinerator must provide sufficient oxygen, high temperature, adequate retention time, and turbulence. The appropriate temperature in the incinerator is maintained by balancing the feed rate of waste and the aeration rate. To assure complete combustion, excess air is provided (usually 100 percent more than strictly needed for complete combustion); the percentage of extra air provided in the incinerator is represented by x in the above equation and is a major design parameter for the facility. The target temperature for incineration is usually between 750 degrees Celsius and 1,000 degrees Celsius.

One of the goals of burning waste is to recover energy. Incinerating waste produces hot water and steam, and the steam can be used to generate electricity. Heat from the process can also be used for district heating.

Modern Incinerator Types

For unprocessed waste, there are two main types of incinerators: mass burn and modular. Both of these incinerators take waste as an input, have a combustion chamber where it is ignited, a boiler that captures the heat from the flue gases from which electricity is generated, and a number of air pollution control technologies. Mass burn facilities usually use a moving grate system to move the trash along the incinerator; as the waste passes through the combustion chamber, air is blown on the waste to achieve complete combustion. Modular incinerators are usually smaller than mass burn facilities and are comprised of two combustion chambers: the first with low oxygen levels (to prevent formation of NO_x), and the second with excess air to achieve complete combustion.

For preprocessed waste, such as RDF, incinerators usually take the form of rotary kilns or fluid bed incinerators, though RDF can also be burned along with fossil fuels in conventional power plants. In a rotary kiln incinerator, the main combustion chamber is a rotating horizontal cylinder; the waste enters at one end of the cylinder and is converted to ash by the other end. In a fluid bed incinerator, the waste is burned in a turbulent bed of hot, inert materials (for example, sand or limestone). The waste is suspended by upward flow of high-speed air.

Incomplete Combustion: Pyrolysis and Gasification

Pyrolysis and gasification are two other thermal processes used to convert waste to energy. Where conventional incineration uses excess air to completely combust fuels, pyrolysis and gasification burn fuel in an oxygen-deficient environment. Both are endothermic processes, meaning that heat must be provided to the process to sustain it.

Gasification occurs in a hot (hotter than 650 degrees Celsius) and "air-lean" environment; there is not enough oxygen to allow complete combustion of the fuel. The process results in two products: syngas (a combination of CO, CH_4, and H_2) and a solid (unburned waste and char, a carbon-rich solid). The syngas can then be burned as a fuel, and the resulting char can be used as a fuel or as a soil amendment.

Pyrolysis is the oxidation of waste in the absence of oxygen. The process has been used widely, ranging from Amazonian indigenous people who use *terra preta* (char) as a soil amendment to 21st-century commercial processes that use pyrolysis to produce charcoal, methanol, and coke. The overall process can be expressed as

$$C_aH_bO_c + heat \rightarrow H_2 + CO_2 + CO + CH_4 + C_2H_6 + CH_2O + tar + char.$$

Both tar (a carbon-rich liquid) and char (a carbon-rich solid) can be used as fuel. Higher pyrolysis temperatures (hotter than 760 degrees C) favor the production of the gases (H_2, CO_2, CO, CH_4), and lower temperatures (450–730 degrees C) favor the production of the solid (char) and liquid products (tar).

Air Emissions and Management

All combustion processes result in the production of gases and particulates, which require control strategies and technologies to meet air quality standards. Prior to advances in air pollution control technologies, incinerators were a major health hazard; this history has resulted in sustained resistance to the siting of incinerators near residential areas.

The emissions from waste incineration depend on a number of factors, including the type of waste burned, type of incinerator, and conditions under which waste is combusted (esp., temperature and the amount of excess air provided). Incineration of waste produces the same basic by-products as the combustion of any hydrocarbon: carbon dioxide (CO_2), water (H_2O), and particulate matter (PM). The sulfur in waste gets converted to SO_2, whose emission is implicated in the formation of acid rain. In the presence of high temperatures and oxygen, the nitrogen in waste gets converted to NO_x, which plays a role in the production of ozone (O_3). Heavy metals in waste, such as mercury (Hg), lead (Pb), cadmium (Cd), and arsenic (As), also volatilize and condense onto fly ash particles; these metals are harmful to human and ecological health. Incineration of chlorine-containing fuel (such as plastics) can result in the emission of dioxins and furans (polychlorinated-dibenzofurans, and polychlorinated-dibenzodioxins), which are chlorinated hydrocarbons that are persistent, toxic, and bioaccumulating.

To minimize the emission of harmful pollutants formed during combustion, a number of air pollution controls have been developed and are standard in 21st-century incineration facilities. These include cyclones, electrostatic precipitators, and fabric filters, which all act to remove particulate matter from the flue gas. Sulfur dioxide and other acid gases are removed by scrubbers, which use alkaline mists to neutralize the flue gas. Selective catalytic reduction (SCR) and selective noncatalytic reduction (SNCR) use ammonia (NH_3) to convert the emitted NO to N_2. The emission of NO_x can also be minimized by reducing the temperature in the incinerator or reducing the amount of oxygen available in the chamber. Finally, the injection of activated carbon into flue gases removes dioxins, furans, and heavy metals by binding to the harmful chemicals.

Ash Production and Management

Combustion of wastes produces ash, as well as some waste products from air pollution control technologies. Two types of ash result from incineration: bottom ash, which is the residue left over from burned waste, and fly ash, the ash that is removed from the flue gas. While bottom ash makes up 90 percent of the total ash produced, the fly ash contains most of the toxicity from incinerator waste.

Ash can be disposed of or it can be reused. Disposal normally occurs in a specialized landfill (called an ashfill) because codisposal with MSW can produce toxic leachate when the acids produced by decomposing MSW lower the pH of the leachate, which increases the solubility of toxic metals. For ash to be reused, it must first be treated to reduce the amount of leachable metals and salts as well as to improve its chemical and physical stability. Ferrous metals, making up about 15 percent of ash, can be recovered using magnets. To make metals insoluble, fly ash can be mixed with lime and Portland cement. Once stabilized, ash can be used as either aggregate for road construction, part of asphalt mixtures, or landfill cover. In Europe, ash is commonly used as a construction material.

Sintana E. Vergara
University of California, Berkeley

See Also: Incinerator Construction Trends; Incinerator Waste; Incinerators in Japan; Toxic Wastes.

Further Readings

Kreith, Frank, Calvin R. Brunner, and Floyd Hasselriis. "Waste-to-Energy Combustion." In *Handbook of*

Solid Waste Management, 2nd ed., G. Tchobanoglous and F. Kreith. New York: McGraw-Hill, 2002.
Louis, Garrick E. "A Historical Context of Municipal Solid Waste Management in the United States." *Waste Management Research*, v.22/4 (August 2004).
Rhyner, Charles R, Leander J. Schwartz, Robert B. Wenger, and Mary G. Kohrell. *Waste Management and Resource Recovery*. Boca Raton, FL: Lewis Publishers, 1995.
Rootes, Christopher and Liam Leonard. "Environmental Movements and Campaigns Against Waste Infrastructure in the United States." *Environmental Politics*, v.18/6 (2009).

Incinerators in Japan

Disinfection, stabilization, and volume reduction: these have been the core agendas of 21st-century solid waste management practice in Japan. It is not surprising that incineration has been adopted to achieve these goals. Japan thermally treats 38 million tons, or 78 percent, of its solid waste every year. Japan had more than 1,500 incinerators as of 2010. This represents two-thirds of all the incinerators in the world, located in an area as large as the U.S. state of California, of which more than 70 percent of the area is mountainous and not suitable for human settlement. Incineration is considered a thermal recycling process and will continue to be a core technology even under the initiative of transforming Japan to a sound material-cycle society.

Brief History

Adaptation of incineration technology began in the late 1800s, as Japan lifted restrictions on international commerce. At that time, cholera and other commutable diseases were still major threats to public health. The outbreaks of plague from 1885 to 1887 also pushed the Japanese government to enact a series of laws enhancing sanitary practices, including the Waste Cleaning Act of 1900. The Waste Cleaning Act was the basis for modern solid waste management in Japan and appointed local municipalities responsible for managing refuse. The law also stated that incineration was the preferred option for treating solid waste and thus set the path for Japan to become the most incinerator-laden country of the world.

The first incinerator in Japan was built in 1887, in Tsuruga, a port city thriving on commerce with China at that time. Other major cities followed with their own incinerators. At first, incinerators were batch incinerators—essentially, a large-scale coal stove. These incinerators required daily removal of ash and caused intense odor and air pollution problems. The first incinerator with a draft furnace was built in Osaka in 1916. A series of research projects on various topics, including thermal energy recovery, air scrubbing, and pyrolysis furnaces were conducted. The outcomes of this research were only partly successful but greatly contributed to the development of incineration technologies in Japan. Incorporating the findings of this research, the Fukagawa Refuse Treatment Plant was built in 1929 in Tokyo, capable of treating more than 700 tons of waste per day.

Waste management practices ceased as a result of World War II in the early 1940s but resumed soon after the end of war in 1945. In order to accommodate the growing urban population and the increase in per capita waste generation due to rapid economic growth, more efficient waste incinerators became a necessity. Technology development again occurred in Osaka, and the first stoker-type incinerator was introduced in the 1960s, followed by the fluidized bed incinerator in the 1980s.

Japanese Incinerators in the 21st Century

By 2008, there were 1,567 incinerators and 91 gasification plants treating general waste in Japan. Half of these facilities operate in continuous combustion, treating 86 percent of general waste, while the remainder of the incinerators found in Japan operate on a much smaller scale. The breakdown of these facilities is 965 stoker type, 241 fluidized beds, 158 fixed beds, and 213 other type of incinerators. Incinerators are also used for industrial wastes, such as medical and food processing wastes, as a means of disinfection and stabilization. Incineration is also the major technology of treatment for human waste: 54 percent of all human waste is incinerated. Energy recovery from incineration facilities is also encouraged in the form of thermal recycling, and 980 facilities have adopted systems for electricity generation or waste heat utilization. From these sources, 1,868 MW of electricity are generated.

Health Hazards

Incineration, however, also imposes health concerns. Flue gases contain standard air pollutants, such as particulate matter and nitrogen oxides, as well as organic micropollutants such as dioxin. Some scholars claimed that as of the late 1990s, Japan was more polluted by dioxin than any other country in the world and called for mitigation. The emission of dioxin from the incinerators is regulated under the revised Water Management and Public Cleansing Law of 1997 and the Law Concerning the Special Measurement Against Dioxins. In accordance with these standards, all facilities constructed before this time were mandated to install bag filters. Further, 66 percent of them are equipped with catalytic reactors; existing facilities and by 2010 were also required to meet the new regulation combustion temperature of 800 degrees Celsius and cooling temperature of 200 degrees Celsius. In a study of 107 municipal solid waste incinerators and gasification melting facilities from 2001 to 2003, it was found that none of the facilities violated the legal discharge limit of flue gas, regardless of the size of plant.

Small batch incinerators and open burning of waste have been banned in accordance of the ordinance set by the Ministry of Public Health. Some claim that dioxin pollution is overly addressed and was used to protect the interests of plant makers. Others point out that in accordance with these regulations, local municipalities are encouraged to form a regional association and share larger and more sophisticated incinerators. This created public opposition against both the high construction cost for new incinerators and economic and environmental burdens of waste transportation. There have also been some movements toward using more holistic approaches in order to consider a wider range of alternatives for intermediate waste treatment practices.

Hiroko Yoshida
University of Wisconsin, Madison

See also: Incinerator Waste; Incinerators; Japan; Osaka, Japan; Pollution, Air; Tokyo, Japan.

Further Readings

Imura, Hidefumi and Miranda Schreurs, eds. *Environmental Policy in Japan*. Cheltenham, UK: Edward Elgar, 2005.

Inoue, Kenichiro, et al. "Report: Atmospheric Pollutants Discharged From Municipal Solid Waste Incineration and Gasificaiton-Melting Facilities in Japan." *Waste Management and Research*, v.27 (2009).

Ministry of Environment. Government of Japan. "State of Discharge and Treatment of Municipal Solid Waste in FY 2008." http://www.env.go.jp/en/headline/headline.php?serial=1333 (Accessed July 2010).

Mizoguchi, Shigeru. *Gomi no Hyakunenshi*. Tokyo: Gakugei Shorin, 1984.

India

The Republic of India is the seventh-largest country in the world and the second most populous country, with 1.2 billion people. It has one of the fastest-growing economies and is undergoing major reform and development. Therefore, how India charts its development in the 21st century will have a significant impact on the future of a large proportion of the world's consumption, along with environmental, financial, and social implications. Development and consumption in India have been linked to many positive changes, such as rising incomes and living standards; but new issues have arisen, such as environmental degradation, social inequality, and increasing amounts of domestic and industrial waste.

Consumption in India: Past and Present

Until the end of the 20th century, consumption in India was at a low level and based on meeting basic needs. During the precolonial period, which lasted up to the 18th century, almost all of the population resided in rural villages. Subsistence agriculture was the mainstay of the economy, alongside networks of commerce and manufacturing. Hand-based industries were common, such as crafts, food processing, textiles, or toolmaking. There was also trade between cities and exports of agricultural products and textiles to overseas markets in Europe, the Middle East, and southeast Asia.

Consumption patterns changed when India was colonized by the United Kingdom in the early- to mid-19th century. The traditional, agrarian way of life was gradually replaced by national systems of economic, legal, and social organization. Consumption was transformed by development and urbanization, including the establishment of industries, markets, and towns. These changes brought new ways of living. The colonial period coincided with major changes in the world economy, particularly industrialization, production, and trade. However, the impact of British rule on India's economy and consumption is a controversial topic. Some argue that policies implemented by the British Raj were exploitative and led to the demise of domestic industries, causing agricultural production to be insufficient for feeding the population. When India gained independence from British rule in 1947 and became a republic with a new constitution in 1950, it was a deeply impoverished country—one of the poorest in the world. Much of the population was still surviving on subsistence levels of consumption.

From the 1950s to the 1980s, India was a socialist state, generally isolated from the world economy. It focused on self-sufficiency and encouraged consumption of Indian-made products, rather than imported goods. Economic and social development was limited by the government's adherence to socialist policies, state ownership of many sectors, and extensive regulation.

Manufacturing accounts for about half of India's gross domestic product. The country is well known for textile manufacturing, and areas such as Ludhiana and Tirupur have gained recognition as leading sources of hosiery, knitted garments, and sportswear.

The tide began to turn in the 1990s, when economic reform and liberalization opened new markets, as India moved toward a market-oriented economy. Changes included privatization of certain public-sector industries, new policies on international trade and investment, and tax reforms. Consequently, in the intervening years, India has transformed into one of the fastest-growing economies in the world, with an average annual gross domestic product (GDP) growth rate of 5.8 percent since the 1990s.

India's growing economy has fueled increasing consumption. Household spending power is greater as a result of rising wages. Similar to historical patterns in other countries, India's consumers are beginning to spend proportionally less on basic necessities. Income growth, as a result of strong economic growth, will likely continue to increase consumption in the 21st century and will most likely turn India into one of the world's largest consumer markets. Increasing consumption, in turn, creates more business and employment opportunities, further fueling the economy and consumption. Domestic consumption has played a key role in India's growth, in contrast to other Asian countries such as China, which are more dependent on exports.

However, although incomes have increased, household saving rates hinder consumption in the early 21st century. The first priority of many families is not to buy new goods but instead to save money. The Indian tradition of frugality with money, coupled with the lack of a social security safety net, encourages saving for children's education, healthcare, and old age. It is also common practice to invest savings in physical assets, such as cattle, houses, or land.

Another key driver of consumption is India's large population, which has increased rapidly since the 1960s as a result of advances in agriculture and healthcare. Population growth has been particularly rapid in urban areas, supplemented by rural–urban migration.

Cities are major centers of consumption. The largest cities in India are Delhi, Kolkata, and Mumbai—

each which a population of over 10 million—which dominate India economically. Many leading companies are concentrated in urban areas but absent in rural areas because of market inaccessibility, limited infrastructure, and low incomes. However, rural households, due to their majority share of the population, are collectively India's largest consumer group. More than 70 percent of the population still lives in rural villages of varying sizes and depends, to a large extent, on subsistence agriculture.

Consumption differs across economic and social groups. Many people in India, such as unskilled and low-skilled workers, are poor and typically spend a large proportion of their income on basic necessities. Higher up the income scale are college graduates, professionals, mid-level government officials, and traders, who have spending power above the subsistence threshold. The highest group of earners is the emerging middle and upper class, with preferences for expensive and luxury goods, but this group constitutes a small proportion of the overall population. As incomes continue to grow, the structure of consumer society will change significantly.

There is a wide variation in consumption across India's 28 states and seven union territories, which differ in per capita income, development level, poverty, and infrastructure. Economic growth rates are higher in states such as Delhi, Gujarat, and Haryana and much lower in states such as Chhattisgarh, Jharkhand, Madhya Pradesh, Orissa, and Uttar Pradesh.

Main Areas of Consumption

Increasing purchasing power of consumers has transformed India's retail market into a pillar of the economy. There are more than 12 million retail outlets operating across the country in the form of either licensed retailers (such as large retail businesses and hypermarkets) or traditional, small-scale retailers, such as local convenience stores, *kirana* shops, open markets, and street vendors. Although absent at the end of the 20th century, the shopping mall has become a feature of consumer life in cities, with New Delhi and Mumbai leading the way in terms of glamour and glitz. The retail industry is currently in a state of flux, undergoing enormous growth. There is also investment by multinational companies, as the world's largest retailers by sales, such as Costco, Tesco, and Walmart, are planning to open stores across the country. Indian retailers are also increasing their brand presence overseas. The largest categories of spending are alcohol and beverages, food, tobacco, transportation, and housing. Other categories, such as education, healthcare, personal products, services, and recreation, are likely to be popular in the 21st century.

Growth in consumption has led to higher demand in the service sector. The best-known example of India's service sector is information technology, which has grown in response to availability of highly skilled, low-cost, English-speaking workers. Bangalore is regarded as the country's information technology capital, and other important centers include Chennai, Hyderabad, Kolkata, Jaipur, and Mumbai. There is a strong demand for banking and investment services. Since liberalization of the economy, banking reforms opened up the market to private and foreign companies. Mumbai is regarded as the commercial and financial capital of India.

The healthcare system is one of the largest systems in the country in terms of revenue and employment and consists of both a government-funded sector, ranging from primary to tertiary services, and a private sector that focuses mainly on primary care. Private healthcare accounts for the majority (more than 80 percent in 2010) of Indian hospitals and hospital beds, and standards of care and quality are significantly higher than in the public sector.

India's services-led economic growth model is distinct from other Asian countries such as China, Indonesia, and Malaysia, which have followed industry-based growth. However, manufacturing is important in India, accounting for about half of GDP, although one-third are engaged in simple household industries. Major industries include chemicals, food processing, mining, petroleum, steel, telecommunications, and textiles. India is well known for textile manufacturing, and areas such as Ludhiana and Tirupur have gained recognition as leading sources of hosiery, knitted garments, and sportswear. However, the global financial crisis at the beginning of the 21st century slowed economic growth, particularly in the manufacturing sector.

Development, increasing population, and urbanization have led to increasing resource consumption. Agriculture is the largest sector of the economy, and allied sectors such as fishing, forestry, and logging are also important. India's farm output ranks second worldwide. Major agricultural products produced and consumed in the country include cotton, fruit, milk, oilseed, potatoes, rice, sugarcane, tea, tobacco, and wheat. India has the world's largest cattle population, and animal husbandry plays an important role in the rural economy, providing eggs, hides, meat, and milk. India is the world's largest consumer of silk, the majority of which is produced in Karnataka, particularly Bangalore and Mysore.

Agricultural yields have increased since the 1950s as a result of emphasis on agriculture in national development plans, the green revolution, improvements in agricultural practices and technology, and projects promoting linkages between farmers and consumers. However, low productivity was still a problem in many parts of the country by 2010 because of factors such as agricultural subsidies, which hamper productivity; overregulation; economically unsustainable practices; poor irrigation; and inadequate market infrastructure.

India was one of the world's largest energy consumers as of 2010. Coal accounts for more than half of total energy consumption, followed by oil (31 percent), natural gas (8 percent), and hydroelectric power (6 percent). Nuclear power provides a very small proportion of total energy consumption, but it is expected to increase in the 21st century. The country has also invested in renewable energy sources. However, India lacks sufficient domestic energy resources and is dependent on imports. For example, in 2009, India imported 2.56 million barrels of oil per day, making it one of the largest buyers of crude oil.

Tourism is relatively undeveloped. As well as domestic tourism, India is a popular destination for international travelers from places such as Bangladesh, Europe, the Middle East, the United States, and Pakistan. Another type of tourism, medical tourism, is experiencing high annual growth rates, and the government has initiated tourism incentives such as marketing campaigns, improved airport and transport infrastructure, and tax incentives.

Society, Environment, and Waste

Rising living standards have transformed people's everyday lives. Poverty has significantly declined. There have been improvements in educational standards, food security, health, life expectancy, literacy rates, living standards, and overall quality of life. Millions of people have access to goods and services that were unavailable in the 20th century.

However, despite India's impressive economic growth and excitement about the potential of its consumer market, there is some distance before its income and consumption levels reach world standards. India is still very much a developing country, with the largest concentration of the world's poor. Pressing problems include meeting basic needs, high illiteracy, and malnutrition. Millions of Indians do not have access to modern toilets, regular electricity, and running water, especially in rural areas. In cities, high-rise, modern buildings are next to slums and other shabby areas that are home to urban poor, such as migrant laborers seeking job prospects. Moreover, since independence, India continues to face challenges such as naxalism (extremist communist groups) and regional separatist insurgencies.

There are also numerous factors that curb consumption and growth, including poor infrastructure; insufficient energy, as electricity supply outstrips demand; regional variation between cities, states, and territories; inequalities in income and wealth distribution; and corruption. To maintain long-term growth, there needs to be agricultural and rural development; energy security; opening up of the manufacturing sector; modernization of industries, such as retail; reforms in the financial and public sectors; and incentives to attract international trade and investment.

Consumer complaints are common and include fake products, poor services, poor product safety, and substandard goods. The most common counterfeit and pirated goods are CDs, DVDs, cigarettes, and clothing. To deal with these issues, the Consumer Protection Act of 1986 and the Department of Consumer Affairs were created to ensure the rights of consumers. It should be noted, however, that not everyone in India is positive about its new consumerism. Some worry that traditions are eroding, the poor are being abandoned, and Indian companies are being taken over by multinational companies.

India's consumption has placed a heavy burden on the environment. Issues include drought, floods, deforestation, desertification, habitat destruction, and soil erosion. Rising demand for energy has led to air pollution, climate change, and water pollution. Air pollution from industry emissions and vehicles is a problem in cities. Rising demand for natural resources, such as forest-based products, is causing deforestation and encroachment onto forest land. These problems extend to other countries, as India's huge consumption has led to imports of natural materials from elsewhere. For example, India is a large importer of palm oil, mostly from Indonesia and Malaysia, causing environmental and social problems there.

Waste and sanitation are problems. Most urban household waste is buried in open landfills and only a small proportion is composted or incinerated, although composting is widespread in rural areas. Although some cities operate good sanitary landfills, most are poorly operated and in many cases are simple, open dumps. Many areas do not have adequate waste treatment facilities and technical expertise. Large amounts of household and industrial waste are untreated, causing pollution and threatening groundwater quality. Many cities dump untreated sewage and even partially cremated bodies into rivers, even though the water is used downstream for bathing, drinking, and washing. For example, the Ganges River, a national symbol of India, is polluted with sewage and industrial waste, even though it is used by more than 400 million residents and an estimated 2 million pilgrim bathers.

Household waste is dominated by organic matter, although the proportion of waste glass, paper, and plastic is rising as the country adopts modern ways of living. Ash is another major component in household waste because many homes use coal for cooking and heating. Recycling of household waste is underdeveloped. There are few recycling companies, laws, regulations, and policies. However, a great deal of recycling is done by waste collectors who sort through rubbish in streets, households, and even landfill sites to collect recyclable products to sell to traders.

Gareth Davey
Hong Kong Shue Yan University

See Also: Delhi, India; Developing Countries; Kolkata, India; Mumbai, India; Slums; Street Scavenging and Trash Picking.

Further Readings

Jaffrelot, Christophe and Peter van der Veer. *Patterns of Middle Class Consumption in India and China*. New Delhi: Sage, 2008.

Reddy, Krishna. *Indian History*. New Delhi, India: Tata McGraw Hill, 2003.

Wilhite, Harold. *Consumption and the Transformation of Everyday Life: A View From South India*. New York: Palgrave Macmillan, 2008.

Wolpert, Stanley. *Encyclopedia of India*. Detroit, MI: Charles Scribner's Sons, 2005.

Indiana

A midwestern state in the Great Lakes region, Indiana is the smallest state in the continental United States west of the Appalachian Mountains. However, the capital and largest city Indianapolis is the second-largest state capital, and there are several metropolitan areas with populations greater than 100,000. Indianapolis has a diverse economy and is part of the Corn Belt (an intensively agricultural region) and the Rust Belt (a manufacturing region, recovering since the 1970s) of the United States.

Statistics and Rankings

The 16th Nationwide Survey of MSW Management in the United States found that in 2004 Indiana had an estimated 13,570,231-ton municipal solid waste (MSW) generation, placing it eighth in a survey of the 50 states and the capital district. Based on the 2004 population of 6,302,646, an estimated 2.15 tons of municipal solid waste (MSW) were generated annually per person, the highest in the United States. Indiana landfilled 8,469,912 tons (ranking 9th) in the state's 35 landfills. It imported 2,165,429 tons of MSW, and export tonnage was unreported. In 2006, Indiana was increasing its 245,570,987-cubic-yard landfill capacity, and it was ranked joint 16th out of 44 respondent states for number of landfills. Only whole tires and lead-acid batteries were reported as being banned from Indiana landfills, with a partial ban on yard waste (no leaves or coarse wood debris longer than three feet). Indiana's average landfill fees per ton were $29.57, where the cheapest and most expensive average landfill fees in the United States were $15 and $96, respectively. Indiana has one waste-to-energy (WTE) facility, which processed 569,263 tons of MSW (14th out of 32 respondents), and 4,531,056 tons of MSW were recycled, placing Indiana eighth in the ranking of recycled MSW tonnage.

Archaeology of the Homeless

Professor Larry Zimmerman at Indiana University-Purdue (IUPUI) has used archaeology to study the homeless subculture in Indianapolis. Working with Jessica Welch, a former student and ex-homeless person, Zimmerman used previous experience from investigating a homeless campsite in Minnesota. He stated, "homeless people, often invisible to those around them, have, use and dispose of material culture as they move across the landscape. But because people are homeless, many Americans think they lack material culture." Discarded food remains, containers, personal items, and bedding at former campsites, as well as caches of material left for future use, are often regarded as refuse by society at large. Abandoned cars, discarded mattresses, and personal belongings stored in garbage bags can appear to be trash when homeless people are actively engaged in recycling them.

The Indianapolis study is completely unlike past anthropological studies of homelessness, which have mainly been ethnographic and taken place in controlled settings like shelters and hostels. Avoiding interacting—and therefore interfering—in the lives of vulnerable people, Zimmerman and Welch located camps and shelters, which they photographed, and inventoried the items left and discarded there. They did not open caches of goods sealed in trash bags and hidden in discreet places, as these had been hidden by homeless people for future use. The refuse of the homeless showed that aid given to the homeless often constitutes society's preconceptions of what it thinks homeless people need, rather than what they actually require. For example, Zimmerman and Welch found numerous hotel-sized bottles of shampoo, conditioner, and toothpaste, but only the toothpaste had been used by the homeless, as dental care is a greater priority to hair care and access to water is limited. Large numbers of used food cans were found that had been opened by hitting them with rocks or heating them until they exploded—ownership of a can opener was rare.

Landfills

As Indiana generates more MSW per person than any other state, disposal is a major issue. Two of the most contentious cases in recent years have been the landfills at Randolph Farms and Mallard Lake.

Randolph Farms Landfill in Modoc, Randolph County, is owned by the Balkema family of Michigan. It serves six Indiana counties and two Ohio counties, and 2,000 cubic yards of waste are buried daily in 4-by-6-foot lifts. In 2004, the Balkemas began petitioning for a third time to expand the landfill from 120 acres to 340 acres to allow a further 50 years of operation. This expansion was controversial to residents of the largely rural county because only 6 percent of the landfill's contents came from their county, the majority derived from Miami County, Ohio. There was concern that the expansion would allow the landfill to reach a height of 50 meters, which would make it one of the state's largest landfills and the highest point of elevation in Indiana, visible from over a mile away. The landfill is on a limestone hill, directly over highly permeable sand and gravel aquifers near the drainage basins of the White River and Whitewater River, which tests have confirmed are at high risk of water pollution. Testing in the 1990s showed low levels of volatile organic compounds from landfill gas in groundwater monitoring wells. There is a legend that Modoc, Indiana, was named by Henry Conley, one of its first settlers, after he picked up a cigar box thrown as refuse from a passing train. The cigar box contained the name Modoc, which Conley suggested would be a good name for the new town.

The Mallard Lake Landfill has been the subject of a 30-year legal battle since the Reed family applied to Madison County to rezone their farm in the 1970s. The Killbuck Concerned Citizens Association formed in 1979 with the aim of stopping the landfill, which would be sited opposite an elementary school. Repeatedly, the Reed family (JM Corp.) has had environmental law judges rule in their favor in proceedings with the Indiana Department of Environmental Management (IDEM) and the opponents of the landfill, only for the court's decision to be appealed.

IDEM issued an operating permit in 1988, which the landfill's opponents appealed, resulting in a 10-year litigation, which ended with a 1998 order for the IDEM to process a final operating permit. A dispute over the depth of the landfill's lining prevented this, and, in 2003, JM had to renew its landfill permit. This permit was applied for, but the IDEM requested additional information as landfill design regulations had changed in the interim period, and a request for additional time to supply this information was granted. However, when a potential buyer of the landfill, Consolidated Waste Industries, Inc. (CWI) began trying to get a permit, requests for further time to supply the additional information were denied by the IDEM. Negotiations broke down and CWI had to take the IDEM to court to appeal the decision, to which the opponents of the landfill filed petitions to intervene. In 2004, a court order forced the IDEM to reinstate the permit renewal application. A state panel was set for two days of hearings on the permit's renewal in late 2010.

The proposed facility at Mallard Lake is a 13-acre, 44-foot-deep landfill that will receive another six feet of composite liner final cover. The landfill is projected to have a uselife of around four years. The liner system consists of three feet of compacted soil and a 60-millimeter high density polyethylene (HDPE) geomembrane, backed up by a leak detection system. A series of perforated pipes under the landfill collect leachate, which is then pumped out and taken to a treatment facility.

Industrial and Farming Pollution

Lake County, in the northwestern corner of the state, is heavily industrialized with significant water, air, and land pollution problems caused by more than a century of steel production, chemical production, coal-burning power plants, and illegal waste dumping. While industrial capacity at U.S. Steel's massive Gary works declined in the 1970s, existing wastes continue to affect the health and land values of the community. The population of Gary became heavily African American between 1950 and 1980 as white middle-class residents moved to cleaner environments, and the community suffers from high rates of asthma and infant mortality.

Another controversial waste disposal issue that has created citizen concern in Indiana is factory farming waste from concentrated animal feeding operations (CAFOs) and confined feeding operations (CFOs). Initially, IDEM only recognized the CFO category (more than 600 hogs or sheep, 300 cattle, or 30,000 fowl fed in pens, sheds, or buildings). The more recent CAFO category threshold is defined as 700 mature dairy cows, 1,000 veal calves, 1,000 nondairy cattle, 2,500 swine over 55 lbs, or 10,000 swine under 55 lbs. CFO is an Indiana categorization, CAFO is a federal category, and both categories are regulated by the state of Indiana.

The biggest problem with factory farms is the amount of manure produced when around 20,000 pigs are confined at a single facility. Sows can spend two years in a sow barn without leaving the building, during which time their excreta is channeled from beneath the floor into huge outdoor storage lagoons to be used as fertilizer. When animals are moved out of the buildings after such long periods, both have to be power washed, creating even more liquid waste. One of these manure lagoons in Randolph County covers 7.2 acres and holds around 20 million gallons of manure. The tile drainage systems used to control surface water in Indiana make it virtually impossible to keep manure pollutants out of the water system.

The IDEM 1998 State of the Environment Report noted that 82 percent of Indiana's rivers, lakes, and streams were unfit for swimming due to high levels of E. coli bacteria. The CAFO pig-farming explosion is attributed to Governor Mitch Daniels's goal of doubling the state's pork production and a moratorium on new CAFOs in North Carolina. Indiana citizens have been vocal in the press about the IDEM's failure to regulate the pollution from CAFOs/CFOs and other issues with the factory farm industry that has spread across the state.

Jon Welsh
AAG Archaeology

See Also: Archaeology of Garbage; Farms; Landfills, Modern.

Further Readings

Arsova, Ljupka, Rob van Haaren, Nora Goldstein, Scott M. Kaufman, and Nickolas J. Themelis. "The State of Garbage in America: 16th Nationwide

Survey of MSW Management in the U.S." *BioCycle*, v.49/12 (2008).
Hurley, Andrew. *Environmental Inequalities: Class, Race, and Industrial Pollution in Gary, Indiana, 1945–1980*. Chapel Hill: University of North Carolina Press, 1995.
Zimmerman, Larry J., Courtney Singleton, and Jessica Welch. "Activism and Creating a Translational Archaeology of Homelessness." *World Archaeology*, v.42/3 (2010).

Indonesia

Many Asian countries have large, growing populations, rapidly rising consumption levels, and massive increases in waste production. Indonesia has the fourth-largest population in the world and the largest in southeast Asia. Per capita production of solid waste (SW) is not high by world standards, but it has increased exponentially since the 1970s. It also has one of the less-developed infrastructures for waste management (WM) in the region. Consequently, problems of waste disposal are massive and growing. Government, local communities, civil-society groups, international aid agencies, and industries are all involved in developing solutions. Waste-to-energy and composting are emerging as promising directions but ultimately reduction in waste production will be essential.

Brief History of Consumption and Waste

The basis of the Indonesian economy was, historically, small-scale agriculture—mostly subsistence production for family consumption. Nonfood items were also largely produced from local natural materials, including bamboo, timber, and banana leaves. Anything useful was reused, while surplus, unused, or abandoned materials were simply left wherever they fell. Waste management consisted of periodic sweeping of organic material into piles out of the way to be eaten by chickens, dogs, and pigs, or to simply decompose. Quantities sufficient to cause inconvenience or ritual pollution were burned. Rubbish in the sense that it is known in industrial economies did not exist. Neither did the notion of "waste management."

Industrial processing began during the colonial period, largely of agricultural products for export. European-manufactured goods were also introduced in small quantities. In the latter part of the 20th century, local industrial production increased, and more manufactured goods were imported. Bicycles were replaced by motorcycles and then cars. Radios, then televisions, and computers became commonplace. Prepackaged foods and drinks replaced ones wrapped in banana leaves or served in glasses. With urbanization came dependence on consumer goods. Economic growth created new prosperity and a middle class with tastes and appetites for international levels and styles of consumption. The mass media fed these appetites, and spread to all levels of society and parts of the country. All this has led to new kinds and ever-growing quantities of waste. Traditional ideas and practices provided little precedent for dealing with the changing reality.

Waste Statistics

Quantification of waste is never easy and statistics on Indonesia are notoriously unreliable, but the following figures give some indication of the scale of the issues and patterns of growth:

- National population: 220–240 million
- Proportion served by WM Authority (2006): 56 percent
- Per capita production of SW (1989): 0.4–0.76 kg/day; (2006): 1.12 kg/day
- National production of SW (daily): 20,000–186,366 tons
- National SW production (annual) 22 million tons (2007), 38.5 million tons (2006), 106 million tons (2010)
- Increase in SW production between 1971 and 2000: tenfold
- Projected national annual production of domestic SW (2020): 53.7 million tons
- Proportion of household waste (2006): 43 percent
- Proportion of waste collected and managed: 40–69 percent
- Proportion of waste recycled: <2 percent
- Proportion incinerated: 35.49 percent
- Proportion into landfills: 7.54 percent

- Per capita daily generation of urban waste: (2001) 0.8 kg, (Jakarta, 2000) 0.65 kg, (Jakarta 2000) 1–2 kg/day
- Daily production of urban waste: 55,000 tons
- Proportion of urban waste collected (Jakarta): <66 percent
- Increase or urban domestic waste (annual): 2–4 percent
- Increase of urban waste (daily, Jakarta): (1985) < 20,000 cu m;(1991) 23,708 cu m; (2001) > 25,600 cu m
- Amount of waste arriving at final disposal sites: 13.6 million tons
- Proportion of organic material: (1989) 87 percent, (2006) 62 percent
- Proportion of plastic: (1989) 3 percent, (2006) 14 percent
- Proportion of toxic and hazardous materials: <10 percent
- Number of people employed by WM authorities: 73,500
- Number of scavengers at official landfill sites: (2006) 14,538
- Number of scavengers in Jakarta: (1995) 10,000–40,000

These figures support several generalizations: per capita production is not high; total production is large; both are growing fast; the waste is relatively high in organic matter but this is decreasing; the most rapid increase is in plastic waste; and, finally, waste management policy, practice, and capacity lag far behind waste production.

Waste Management

New understandings, policies, and practices of waste management have simply failed to keep pace with the growth in quantity and complexity of waste. In most rural areas, waste is simply dumped on vacant land or burned. In most urban areas, there is some form of collection by local authorities and transport of at least some of it to a landfill or incinerator site. The rest is simply dumped on vacant land, or in rivers, or sold for filling building sites. Because of the high moisture content of the waste, incineration tends to be inefficient and highly polluting.

Most landfills are primitive: tip and cover with soil. Because of the high organic content, high temperatures, and high humidity, it decomposes rapidly and mostly anaerobically, producing large amounts of methane—a foul-smelling, explosive, and powerful greenhouse gas. Because of poor design, construction, and monitoring, many landfills leak toxic leachate into surrounding farmland, streams, and water tables. Landfills also support large populations of rodents and flies.

In Indonesian urban areas, some of the trash is collected by local authorities and taken to incinerators or landfills, where high humidity and temperatures generate large volumes of methane. The rest is dumped on vacant lots or rivers, or sold as construction filler.

Recycling takes place at all levels of the system. An informal economy of "scavengers" or "waste-pickers" make their living collecting recycleable material (mostly glass, metals, paper, and plastics) out of the waste stream and selling it to commercial recycling plants. By the time waste arrives at a landfill, it has been well picked over, but most larger landfills, especially in large overcrowded cities, support armies of scavengers, many of them children. Working conditions are dangerous and unhealthy. Nevertheless, this informal sector forms an integral part of the waste management system and provides livelihoods for thousands of people.

Government control of waste management is weak. Environmental legislation has existed since the 1980s, and since the late 1990s, a series of environmental laws have come into force. However, enforcement has been haphazard and is often undermined by confusion between multiple levels of government and departments, serious underfunding, and corruption. In 2005, a landslide at a landfill in the city of Bandung killed 140 people and forced the government to pay attention to waste management. In

2008, a new Solid Waste Management Act came into force, and with it the beginning of a new approach to waste management involving new parties and a more complex and diverse WM system.

While the informal sector and very-small-scale private enterprise have long been involved in waste, the formal private sector has only been involved since the mid-1990s, but private sector participation is still low (less than 10 percent by 2010). As government gets to grips with modern WM practices, it turns increasingly to the private sector, often internationally, for technical expertise and capital. Likewise, international development and aid funding has played a part in improving systems. For example, loans and technical expertise from Japan were used in 2002 to upgrade the existing collection system and landfill in Jakarta. Such projects, while useful, do not address the fundamental unsustainability of landfilling. Other projects do this either by developing composting systems or waste-to energy (WTE) approaches. The logic of WTE in Indonesia lies in the surplus of waste and a growing shortage of energy, both created by increasing levels of consumption. The conversion of waste into fuel (via either methane capture or direct burning) for electrical generation has the potential to mitigate both problems. There are at least three large-scale internationally designed and funded WTE projects in Indonesia, but none were fully operational by 2010.

At the other end of the spectrum are many small-scale, community-based projects for improving local WM. Most involve combinations of improved collection, recycling, composting, and in some cases methane digestion for energy. Many of these have been initiated and supported by international aid, mostly via nongovernmental organizations (NGOs), usually in partnership with local NGOs. Some have proven unsustainable because of lack of follow-up, but others have been integrated into or imitated by government initiatives.

While the problems and the need for new solutions are apparent, there are many differences between proposed solutions. Proponents of large-scale, hi-tech industrial approaches argue that the problem calls for large-scale approaches and that only economies of scale enable the complex and expensive technologies necessary. Proponents of small-scale, lower-tech, community-based approaches begin with the needs of local households and communities and are skeptical of the ability of top-down approaches to recognize, let alone address, these issues.

Prospects for the 21st Century

Given global patterns at the beginning of the 21st century, the trend of increasing consumption and waste seems unlikely to change for some time. One alarming aspect of the global economy of waste is the export of (especially toxic) wastes, from more- to less-developed countries. Indonesia is already an importer, but the government is moving slowly to control it. On the positive side, global awareness of climate change has added impetus to the transfer of waste management ideas and technologies from north to south. At the same time, the growing carbon economy offers potential for funding of better solutions via the Clean Development Mechanism and other initiatives. While these are not beyond criticism, they have helped fund innovative waste management projects in Indonesia.

The most optimistic scenario is that Indonesia will learn from the mistakes of most of the rest of the world and turn around its growing levels of consumption and waste before they become unmanageable.

Graeme MacRae
Massey University

See Also: Composting; Consumption Patterns; Developing Countries; Incinerators; Landfills, Modern.

Further Readings

Chaerul, M., M. Tanaka, and A. V. Shekdar. "Municipal Solid Waste Management in Indonesia: Status and the Strategic Actions." *Journal of the Faculty of Environmental Science and Technology*, v.12/1 (2007).

Medeiana, C. and T. Gamse. "Development of Waste Management Practices in Indonesia." *European Journal of Scientific Research*, v.40/2 (2010).

Republic of Indonesia. State Ministry of the Environment. "Indonesian Domestic Solid Waste Statistics." (2008).

Sicular, D. T. *Scavengers, Recyclers and Solutions for Solid Waste in Indonesia*. Berkeley: University of California, 1992.

Industrial Revolution

The Industrial Revolution was a period that spanned the 18th and 19th centuries with deep and profound social consequences and large transformations in population, economy, culture, production, consumption, technology, agriculture, mining, and transportation. These changes began in the 18th century in the United Kingdom and gradually spread to the rest of Europe, the United States, and the rest of the world. A major trend within this period was the transformation of rural and agrarian societies to predominantly urban ones, based on manufacturing and industry.

Brief History of Innovations

During this period, an industrial capitalistic consumption culture emerged and various new commodities became available to large parts of the population. Another major consequence of the Industrial Revolution was the industrial and urban refuse problem that had devastating consequences for the majority of the population. For the first time, humans were faced with unprecedented problems of pollution, which required effective and immediate policies. The problems at first were seen merely as an inconvenient annoyance and only later as a serious health risk related to the public and the environmental crisis in general. Consequently, while the Industrial Revolution has been thought of as the major force of social and economic development, during this period excessive production and consumption as well as garbage and pollution resulted in an environmental and refuse problem. Moreover, a consumer revolution took place, and new conceptualizations of hygiene, cleanliness, and public sanitation policy emerged as major social categories. These trends resulted in the application of science to the problems of pollution and waste.

The roots of industrial capitalism in Europe originate in colonialism, the period when exploitation took the form of many new commodities that were widely circulated and traded all over the globe. In the Caribbean, for example, the mass production of sugar was structured in terms of factory culture—with different types of skilled workers divided by age and gender, tight supervision, scheduling, time consciousness, and disciplinary mechanisms—long before such industrial organization was known in Europe.

The Industrial Revolution began with small technological inventions of machinery in the sectors of textiles and metallurgy and effected large production changes in various other sectors. Steam engines, water wheels, and other powered machinery were used for the first time and gradually were spread throughout the colonial empires. Such technologies were improved and gradually incorporated into other spheres of society, such as steamships and trains in the transportation sector. Moreover, during this period the internal combustion engine and electric power generators were invented and widely used. The chemical industry also developed, with new inventions such as sulfuric acid, sodium carbonate, and cement.

The development of production resulted in a new network of transportation that was needed to connect urban trade centers, where most factories were situated, with mines and the areas that provided raw materials. Within a few decades, in the United Kingdom alone, a national network of canals was constructed, along with a more elaborate and efficient plan of roads, and, finally, a large railway system. Similarly, in the United States, steamships were widely used, the railways expanded rapidly, and transatlantic commercial lines were established. Transportation was further developed with the invention of the bicycle and automobile.

Consumption

The increased pace of production, in combination with the low cost of new mass commodities, resulted in increased consumption. For the first time, products were available in outstanding quantities, at relatively low prices, and available to anyone. A new cultural industry also emerged as steam power was applied to the printing press. Books became cheaper, newspapers were founded and published in large numbers, and magazines emerged. Mass consumption was represented in what became known as the epitome of the Industrial Revolution, the Great Exhibition of the Works of Industry of all nations held first in Hyde Park, London, in 1851. The Great Exhibition suggested the significance of industry and tried to universalize the idea that progress and development in society were tied to industrial achievements. Colonial ideas were also prominent as various exotic subjects and objects were on display. Gradually, other exhibitions followed the pattern and later on tried to include not only machinery but also large architectural structures like the Eiffel Tower in the Exposition Universelle in 1889.

Moreover, such exhibitions served as representations of commodities that celebrated a consumption-oriented modernity. The commodities on display were not necessarily useful, but they constituted a new form of visual consumption. As Walter Benjamin noted, these industrial exhibitions were actually commodity worlds that people entered to experience this new phantasmagoria, and they erected the universe of commodities. In a perfect capitalist manner, the sphere of production was isolated from the sphere of consumption, and commodities exhibited their deep fetishistic character. Commodities appeared more powerful, illuminating, and dream-like. These Industrial Exhibitions became a capitalistic spectacle and a celebration of an emerging consumer culture.

The Industrial Revolution coincided with a kind of consumer revolution in which an increased amount of standardized consumption goods became available to different social strata. While various commodities had already been popularized in the preindustrial world, such as tobacco, coffee, sugar, and tea, during the Industrial Revolution large amounts would be massively produced, consumed, and circulated. For example while the average Briton consumed 20 pounds of sugar in 1815, by 1890, this amount increased to 90 pounds. Moreover, conspicuous consumption became intense among the middle class, and various new commodities would be used in order to assign status.

The consumer revolution became possible with a doubling of wages in Western countries, especially after the middle of the 19th century, stimulating a demand for various new commodities. While until the beginning of the Industrial Revolution most luxury and exotic goods would be available solely to the elite and restricted to others; gradually, as production increased and the cost was lowered, they would potentially become available to all different social strata. That potential did not bring immediately a better quality of life for the lower classes. In reality, there were vast income differences between people, especially in class-oriented societies. Poverty and starvation would increase in the new urban environments, and the quality of life of laborers would take many years to improve.

Urbanization

One of the most noteworthy effects of the Industrial Revolution was the demographic change. From 1815 to 1914, the population of Europe enlarged threefold, from 250 million up to 750 million. That increase took place while the death rates decreased tremendously as a result of new medication and vaccines, changes in diet and hygiene, and increased sanitation in cities. That demographic shift, which was first encountered in England, affected the size of the city and resulted in serious problems of overcrowding. The most determinant factor in the expanding size of industrial cities was the fact that factories were concentrated around them. The continuous need of capitalists for labor and increasing production brought large numbers of workers into the urban environment.

First, only the Netherlands was more urbanized than England, but gradually that trend changed. While before the 19th century, Great Britain was an agrarian society, the transformation to an urban, industrialized society brought almost half of the population to cities by 1851. During the reign of Queen Victoria, the population of Great Britain increased twofold and the majority of the population lived in cities.

Sanitation and Pollution

The increase of city dwellers resulted in serious sanitation and garbage problems, which were encountered for the first time. The increase of the population density and the concentration of factories around cities produced living and working conditions of incredible hardship, deprivation, and despair, especially among the poor. The factory, for example, introduced new ways of surveillance and discipline comparable to the "panopticon," also a product of the Industrial Revolution. Laborers were exploited in an inhumane manner by working 12–16 hours per day, and children as young as 6 years old worked as miners and factory workers. Pollution and hygiene were not emphasized, as there was no running water or toilets within households. Large, poor neighborhoods had to share outdoor pumps and public toilets, and cities became overcrowded to the extent that several people or families shared a room. Cellars and underground storages became mainstream accommodation, and household garbage was thrown into the streets. Moreover, factories and household chimneys produced a large amount of smoke, which blocked daylight and covered the streets. In these terrible conditions, epidemics of cholera and typhoid, as well as respiratory and intestinal disease, became regular phenomena.

The concentration of factories in and around cities intensified the environmental problems. Iron and textile mills as well as chemical factories were constructed next to waterways in order to transport their commodities efficiently and to provide the large quantities of water needed in steam engines and chemical solutions. These factories disposed their waste in waterways, a habit that resulted in serious water pollution. Manufacturers were also responsible for land pollution by dumping garbage, ash, iron, and other rubbish onto land.

Waste Management, Public Health, and Hygiene

At the beginning, each municipality tried to independently resolve the refuse and hygiene problems that industrial society faced. An increasing number of small-scale municipal services were established, such as fire and police protection, water supply, waste collection and disposal, and public hygiene works. However, the recognition of a serious environmental crisis came at the end of the Industrial Revolution. An emphasis was placed on clean water and a sewage system, while the attempts to limit air and land pollution were not taken seriously. A consequence of the appraisal of municipal responsibility was the initiation of public health agencies and departments.

Furthermore, the establishment of modern public health science and the foundation of the Sanitary Commission in England in 1869 provided the foundation of public health legislation and policy. Sanitation emerged as a new concept to combat various communicable diseases, and scientific methodology was used to understand waste and public health issues. The first observations related communicable disease with filthy environments and waste, without a clear understanding of the factors of such diseases. Moreover, scientists tried to develop a link between mortality rate and degree of wealth. The turning point of the refuse problem was the application of science to the issue of pollution. Public health social policy, for example, became an undisputed antidote to the increasing refuse problem and the raging diseases, even if hygienists were not totally aware of what was truly "pollution." Similar programs in Europe and the United States acknowledged the need for public policy and brought relief to the overcrowded factories and the heavily industrialized urban environment.

During the Industrial Revolution, a new mentality in relation to hygiene emerged, predating the discovery of microbes. City dwellings were thought to be of poor quality, and cities were imagined as death chambers with ignorant and contagious poor workers. Within this context, cities should be transformed into clean spaces with large parks and outdoor areas, gardens, and good sewage systems.

During the 19th century, laborers were inhumanely exploited by being made to work from 12 to 16 hours a day. Pollution and hygiene were not emphasized, and garbage was thrown into the streets. In these conditions, disease epidemics became common.

Hygiene became, therefore, connected to a healthy way of life in opposition to the wealthy, industrial urban environment that brought disease and degeneration. The movement of hygienists influenced to a large extent the understandings of cleanliness and pollution that were disseminated through media, public debate, scientific methodology, and large exhibitions.

In the hygiene exhibition of London in 1884, for example, crowds could admire how a hygienic and healthy style of life could be adopted. Refrigerated meat, pasteurized milk, orthopedic shoes, healthy clothes, filters to purify water, flush toilets, and even plans of airy and heated hygienic houses with drainage and convenient-to-clean furniture were on display.

Bacteriological science emerged with general microbiology, and new inventions followed in this area, such as pasteurization. The microbe personified pollution, dirt, and all social and cultural problems that had emerged out of the environmental crisis. As such, the hygienic achievements in the years to follow would signal a clear separation between scientific and social understanding of industrial pollution.

Since then, the need to manage pollution and the environmental crisis as a result of the Industrial Revolution and the continuous industrialization of various other areas of the world has been widely recognized by the international community. While industrialization brought various technological and economic achievements, its consequences were unprecedented. By the end of the Industrial Revolution, people suspected that the level of industrialization of each country was proportional to the level of the refuse problem. But even then, society focused on the scientific aspects of problems and the new conceptualizations of hygiene, pollution, and microbes. The responsibility of each society and the social, economic, cultural, and political causes of the industrial environmental crisis would take many years to realize.

Tryfon Bampilis
University of Leiden

See Also: Capitalism; Commodification; Consumerism; Garbage in Modern Thought; Industrial Waste.

Further Readings

Hobsbawm, E. *The Age of Revolution: 1789–1848*. New York: Vintage Books, 1996.
Latour, B. *The Pasteurization of France*. Cambridge, MA: Harvard University Press, 1993.
Melosi, M. *Garbage in the Cities: Refuse, Reform and the Environment*. Pittsburgh, PA: University of Pittsburgh Press, 2005.

Industrial Waste

Industrial wastes are unintended by-products of industrial production and may contain hazardous substances. While the disposal of hazardous industrial wastes is well regulated in developed countries, many other aspects of industrial wastes remain underaddressed by environmental policy. Data on generation rates, material composition, and disposal methods are uncertain or lacking in most cases. Problems of contamination from industrial waste releases, as well as their export from developed to developing nations, have been somewhat ameliorated through national and multilateral policies since the 1970s. These practices, however, still pose risks to populations worldwide, disproportionately affecting poor and nonwhite people. Policies to promote reuse and recycling of industrial wastes are underdeveloped in contrast to those for municipal solid wastes (MSW), about which a great deal of data are regularly reported. The disparity in attention between MSW and industrial wastes does not correspond to quantity; industrial waste are generated in far greater tonnages than MSW. Promise for better management of industrial wastes lies in continued activism to counter unsafe industrial waste practices, along with research and development of new and efficient forms of industrial waste minimization using the model of industrial ecology.

Definition of Industrial Wastes

In 21st-century societies all over the world, the extraction or harvesting of materials from nature, and their transformation into finished products, takes place primarily through processes of industrial production. Industrial production differs from other forms of production carried out either by hand or with personalized tools in that it entails relatively large-scale, routine, and automated activities. The unintended material by-products of industrial production are industrial wastes.

In their broadest definition, industrial wastes encompass any by-products that are emitted directly into the atmosphere or released into waterways as well as those temporarily or permanently deposited on or in the ground. Typically, however, airborne emissions are referred to as air pollution, and waterborne releases as wastewater or water pollution. Both are considered separately in policy and discourse from solid or semi-liquid industrial wastes. The latter group of materials is more commonly what is meant by terms such as *industrial wastes* or *industrial solid wastes*.

Industrial wastes may arise from mining, petroleum extraction and refining, agriculture, energy production, construction and demolition, transportation, or manufacturing activities. In some cases, industrial wastes, through statutory definition, exclude wastes from agriculture and extractive industries; for example, such wastes are not classified as industrial under the U.S. Resource Conservation and Recovery Act. Depending on the agency and jurisdiction, additional categories may be considered as falling under the umbrella definition of industrial wastes. Overall, however, industrial wastes can always be defined by what they exclude: municipal solid wastes and nuclear wastes. No country's definition of industrial wastes includes MSW, which represents instead the discards of human settlements, including households, public institutions, and nonindustrial commercial enterprises such as offices, retail, and food service. Nuclear waste is also excluded from all definitions of the term.

In many industrialized nations, a subset of industrial wastes possessing empirically demonstrated danger to health and safety is referred to as hazardous industrial wastes, hazardous wastes, or toxic wastes. These wastes are considered separately from industrial wastes not deemed hazardous and are subject to state regulations concerning transport and disposal that do not apply to other industrial wastes. The designation of some industrial wastes as hazardous may be made on the basis of rough characteristics such as ignitability or corrosiveness, may reflect intended product use (e.g., as explosives or solvents), or may reflect chemicals constituents that have been demonstrated to pose risks to health and safety.

Ongoing research in fields of public health and toxicology periodically brings to light new categories of hazardous compounds that were previously considered safe. Such research may respond to patterns of illness discovered or reported near sites where industrial wastes containing such compounds are stored or disposed. The existence of research does not guarantee, however, the classification of additional industrial wastes as hazardous without a process of administrative rule change. Such processes are highly contested, with different perspectives from industry, medicine, and citizen groups arguing different scientific points of view regarding actual or perceived risk. For this reason, even in countries in which hazardous industrial wastes are regulated to protect health and safety, industrial wastes that are not designated as hazardous may still pose a risk.

Quantity and Composition

One of the many ways in which industrial wastes contrast with MSW is in the level of public knowledge about their quantity and composition. While MSW tonnages and composition are routinely tracked and reported in detail by most developed nations—and many developing nations as well—tracking of industrial waste statistics is only required in European Union (EU) nations. Even then, data is self-reported by industrial trade groups and often reflect only a subset of the true total of industrial waste generation. One of the most far-reaching estimates of industrial waste quantity around the globe is the 2006 World Waste Survey, which cautions readers about the incomplete and inconsistent nature of industrial waste reporting. It estimates that industrial wastes not explicitly classified as hazardous are being generated at a rate between 1.3 and 2.8 billion metric tons annually, noting that such estimates reflect only selected nations (the EU, United States, Canada, Japan, South Korea, Australia, Mexico, Brazil, Thailand, Taiwan, and China) and exclude potentially large quantities in other nations, particularly nations of the former Soviet Union and India. The 2006 World Waste Survey estimates that hazardous industrial wastes total roughly 200 million metric tons annually for a different set of countries with available statistics (the EU, United States, Canada, Japan, South Korea, Thailand, China, Mexico, India, and South Africa). In comparison, world quantities of MSW are estimated at roughly 1.2 billion metric tons for all continents and are considered more reliable, though far from perfect.

In the United States, one study conducted in 1987 quantified nonhazardous industrial wastes at 7.6 billion tons in that year. While this tonnage is still reported by the Environmental Protection Agency (EPA) in the early 21st century as the only official statistic on nonhazardous industrial waste generation in the United States, its datedness makes it of little value. The 7.6-billion-ton estimate is not reproduced in statistics compiled by the Organisation for Economic Co-operation and Development (OECD), which form the basis of the 2006 World Waste Survey reporting. The OECD's total national waste statistics for other North American nations, including Canada and Mexico, also omit industrial waste reporting. By comparison, nearly all European and Asian OECD member states provide data for waste generated by manufacturing, construction and demolition, and other industrial sectors. National reporting on hazardous industrial wastes is generally of better quality. In the United States, the EPA reports state-by-state statistics on generation, with tonnages totaling 47 million in 2007. Other OECD members have provided data for at least one year between 2000 and 2008, enabling the calculation of an annual generation rate of roughly 111 million tons for all members.

Data on the composition of industrial wastes, hazardous or otherwise, is even less developed, with few comprehensive statistics about the materials that make up reported tonnages. The Intergovernmental Panel on Climate Change (IPCC) is concerned with estimating degradable organic carbon (DOC) in industrial wastes because it is one of the main sources of methane emissions from land disposal. Its assessment of industrial wastes arising from selected countries in Asia, Europe, and Oceania estimated DOC content of up to 43 percent of all nonhazardous industrial wastes, with food, textile, wood, paper, and rubber industries generating the most organics-rich by-products.

Another way of thinking about quantities of industrial wastes is to use materials flow analysis, a methodology that tracks the complete flow of raw materials through the processes of extraction, production, use, disposal, and recovery within a nation, factoring in exports and imports. Such approaches, particularly those carried out by the World Resources Institute (WRI), use trade data in conjunction with information on extraction and manufacture to account for the flow of materials by weight through each nation's economy. A study completed by the WRI in 2000 estimated that manufacturing activities in Austria, Germany, Japan, the Netherlands, and the United States resulted in the release of some 2.4 billion tons of waste in 1996. However, this methodology suggested that the majority of releases were to the air in the form of direct emissions and incineration, with relatively less going to land disposal.

Regardless of methodology or region, industrial waste quantity far outweighs municipal solid waste. The production of a finished good requires a long series of steps starting with extraction, continuing through refining and fabrication, with transportation and energy inputs at each step along the way. Every consumer commodity is a point at the summit of a pyramid of materials flows that end in its provision on the market, and much of that pyramid consists of industrial wastes.

Disposal Methods

Industrial materials that are not recovered for reintroduction into manufacture may be disposed of in a variety of ways. Hazardous waste disposal is highly regulated in all OECD nations, with strong provisions requiring environmentally sound management controls from the point of industrial generation to disposal. Such provisions usually include special permitting requirements for hazardous waste disposal facilities that specify infrastructure protections against releases to the environment. They may also include prohibitions against the mixing of different types of hazardous wastes—or their dilution—prior to disposal, and requirements for separation, stabilization, or destruction of hazardous waste constituents by mechanical, thermal, or chemical methods. Common forms of hazardous waste disposal include combustion or incineration, particularly for wastes containing organic chemicals. Hazardous wastes may also be deposited in landfills, waste piles, land application units, or surface impoundments (diked areas holding semisolid wastes such as sludges). Sludges and liquid hazardous wastes may be injected underground into wells or storage tanks, or they may be contained above ground indefinitely in tanks or barrels.

Disposal of nonhazardous industrial wastes, which are far greater in quantity than those classified as hazardous, is much less regulated. Dry wastes, if not falling under the hazardous definition, may simply be dumped or stockpiled on the industrial property. In some cases, oils, plastics, or wood wastes may be incinerated, with or without production of energy. When trucked off-site for disposal, industrial wastes come under enhanced regulatory control. Landfill and incinerator regulations for nonhazardous industrial wastes are generally similar to those for MSW in any country. There is relatively little comprehensive data on a national level about disposal methods for nonhazardous industrial wastes.

Alternatives to Disposal

The term *recycling* refers both to a process of materials recovery and a practice aimed at fostering ecological and social benefits, including resource conservation, pollution mitigation, and job creation. Most of the recovery of industrial waste materials is recycling only in the process sense—it takes place when and where it makes financial sense for the firm. Industrial wastes of a heterogeneous, semiliquid, or hazardous nature are the least likely to be recovered, while by-products that can be easily reintroduced into the same industrial process that gave rise to them are the most likely. Recycling methods in industrial settings will therefore first and foremost take place within the industrial plant; for example, scraps of waste metal, plastic, or other materials will simply be added to incoming material inputs for a second round through the production process. In other cases, industrial by-products that are clean and homogenous may be sold on the open scrap market; these will compete favorably against those coming from municipal sources, which tend to be mixed, contaminated with residue, and of lower quality. Examples of such commodities include scrap paper, metal, plastic, wood, and textiles—all of which have a long history of trade in most nations.

The exchange of by-products among industries, especially those located in proximity to one another, is a field of active research, policy development, and focus within environmentalism. The pre-eminent model of industrial by-product exchange arose between the 1960s and 1980s in Kalundborg, Denmark. In the 21st century, a constellation of symbiotic firms there includes a coal-fired power plant that provides surplus heat to local residences and a nearby fish farm and sells steam to a large pharmaceutical plant. Sulfur dioxide from the power plant's scrubber yields gypsum, which is then sold to a local manufacturer of drywall. The fish farm converts its waste sludge to fertilizer used by local farms. Finally, ash from the power plant is used for local road building and cement production.

Governmental and nongovernmental organizations pursuing sustainable waste management see great promise in fostering intentional industrial symbioses through the development of eco-industrial parks. Research has found, however, that such projects are less likely to succeed if intentionally planned than if they arise spontaneously, adapting to particular economic and material niches that vary tremendously from place to place. In the United States, Canada, and the United Kingdom, a related, though not geographically organized, area of environmental programming seeks to develop industrial materials exchange networks, often in an online format, to encourage trading of by-products among industries at a regional or national scale. Free-market exchange, often entailing global export and import of scrap materials, also takes place through online networks not concerned with environmental outcomes.

Governments regulating hazardous wastes encourage, but do not require, generating industries to apply the waste reduction hierarchy, which promotes prevention and reuse first, recycling and composting second, energy recovery third, with disposal as a last resort. Hazardous waste recycling is much more practiced in EU member nations, which average a 27 percent rate, than in the United States or Canada, which each have rates lower than 5 percent. Most such recycling involves the

The 1989 Basel Convention was developed to combat a practice that developed during the 1970s and 1980s: firms in developed countries, coming under increased hazardous waste disposal regulations, would disburden themselves by exporting discarded electronics abroad, in particular to poor developing nations in Asia, Africa, and the Pacific. Organizations such as the Basel Action Network closely monitor these transnational shipments and have made progress in limiting unsafe recycling, incineration, and disposal practices in importing nations.

separation and extraction of chemicals and metals from hazardous compounds for reuse in industrial processes. While the EU and North American hazardous waste regulations impose similar sets of restrictions, North American industrial trade groups more often call such regulations a deterrent to hazardous waste recycling, and they sometimes use this argument to lobby to reclassify or exempt certain wastes from designation as hazardous. An alternative to recycling is the prevention of hazardous waste generation in the first place. Efforts to promote this practice are organized under governmental or nongovernmental programs bearing the name "pollution prevention" in North America and "cleaner production" in Europe. Such initiatives, pursued on a nonregulatory basis, feature an array of collaborative partnerships between governments and firms to minimize waste in all forms, through product and process redesign.

Global Trade in Industrial Waste

The Basel Convention on the Control of Transboundary Movements of Hazardous Wastes and Their Disposal of 1989 is a multilateral treaty governing the export and import of hazardous industrial wastes. It was developed in response to tendencies that arose in the 1970s and 1980s for firms in developed countries, coming under increased hazardous waste disposal regulations, to disburden themselves through proliferating channels of export abroad, in particular to poor developing nations in Asia, sub-Saharan Africa, the Caribbean, and the Pacific. The convention was implemented in 1992, and its original terms allowed the exportation of wastes only if exporting countries lacked proper disposal infrastructure and importing countries did have such capacity, or if the wastes in question could be construed as raw materials to be used in production in the importing nation.

It also required written consent from governments in importing nations. In 1994, parties to the convention strengthened its reach by voluntarily agreeing to ban all waste exports from OECD nations to non-OECD nations. In 1991, other multilateral treaties were created, including the Bamako Convention that bans most waste export to Africa, meant to protect poor nations in the Caribbean and Pacific.

These conventions have ameliorated but not eliminated the transfer of wastes from the developed to the developing world. There is a brisk illegal trade in hazardous waste exports that exploits unstable regimes in resource-poor nations. The Basel Convention exempts industrial scrap, allowing the trade in waste paper, metal, plastic, and other materials if not mixed with refuse or hazardous materials. In many cases, especially in the case of paper and metal scrap, such exports supply input materials that are in high demand in rapidly industrializing countries. In other cases, particularly with electronic waste and post-consumer plastics, such exports may contain hazardous materials and often require extensive sorting and disposal of unmarketable residue in the receiving country.

Problems with scrap exports have been documented in Europe, particularly in Germany for plastics and in those countries as well as the United States and Canada for electronics. In both cases, the export of mixed materials containing hazardous constituents as "industrial scrap" arose as an unintended consequence of extended producer responsibility policies. Such policies are meant to transfer some or all of the burdens of waste collection and disposal from taxpayers and consumers to producers of packaging and electronics. These policies require industries to literally or virtually (through financing of third-party collections) take back finished commodities at end of life, introducing a new form of industrial waste to a sector that had no precedent in handling its own spent products. As a result, industries and trade associations representing them have sought export as a method to move along product returns. Organizations such as the Basel Action Network are actively monitoring the transnational shipment of electronics and have made strong inroads into curtailing unsafe recycling, incineration, and disposal practices in importing nations such as India, Ghana, and China. The fate of plastics is less studied, although also undergoing scrutiny.

Social Movements and Industrial Waste

Several high-profile cases of hazardous waste contamination took place in the 1970s and 1980s. In the late 1970s, residents of the small town of Love Canal, New York, organized to protest health problems arising from the improper land disposal of thousands of tons of hazardous waste by a chemical company. In 1982, the EPA confirmed that recycled waste oil sprayed on roads as a dust suppressant in Times Beach, Missouri, contained dioxin from a chemical manufacturing plant. In 1984, a malfunction at a Union Carbide pesticide plant in Bhopal, India, resulted in a leak of methyl isocyanate gas, resulting in the exposure of over 500,000 people to severe risk and the immediate death of thousands, making it the worst industrial accident in history. While Love Canal and Times Beach have been remediated, the Bhopal site has not. By 2010, hundreds of tons of hazardous wastes were still stored in tanks at the closed facility.

The Love Canal case is credited with spawning the grassroots antitoxics movement in the United States, which broke from prior traditions of conservation-focused environmentalism to highlight health risks to people from industrial activities, particularly hazardous waste disposal. The antitoxics movement, with its focus on the rights of "regular people" to identify patterns of illness, brought attention to the fact that working-class residents were both burdened by and excluded from planning around the siting of industrial facilities nationwide. In 1982, residents of Warren County, North Carolina, whose minority population was largely low-income, organized to oppose the state's siting of a hazardous waste landfill with inadequate groundwater protections in Shocco Township. Protesting residents were joined by national civil rights and environmental groups in opposing the facility. This struggle formed the basis for the emergence of the environmental justice movement. Its concerns include the displacement of the dangers of industrial waste from rich to poor nations as well as displacement of risk from more to less powerful people within nations, drawing attention to power divisions based on race as well as class.

Within the United States, the antitoxics and environmental justice movements have led much if not all of the activism that casts industrial wastes as an environmental problem. These movements are, to various degrees, active in other nations along with the green parties in Europe. Other branches of environmentalist movements, particularly those concerned about resource conservation and jobs creation through recycling, have paid relatively less attention to industrial wastes in comparison to more visible materials in municipal discards. The zero waste movement, however, supports many of the voluntary approaches to industrial waste minimization featured in pollution prevention policies, seeking to foster collaborations and interchange among firms that prevent waste and maximize by-product exchange for reuse.

Industrial Ecology

A framework for considering how to reduce industrial wastes systemically, for environmental as well as economic reasons, is being developed and promoted by planners, architects, and engineers working in the field of industrial ecology. Industrial ecology is an approach to the design of industrial production that takes as its model the study of natural systems and the cycles of materials in those systems. Its goal is to make industrial processes less harmful to the environment as they become more efficient. The reduction of industrial wastes generated through and as a result of production is one of its goals. Major proponents of industrial ecology see design of processes and products as the primary mechanism to achieve this reduction. They promote the use of all industrial by-products, either within the manufacturing firm itself or as inputs to nearby, linked industries. They seek to reevaluate processes within the factory to eliminate waste and make efficient use of energy and resources. They promote research into new materials that are less toxic and easier to recover, and they also advocate the design of finished products that can be taken back and repaired or used for parts.

By 2010, industrial ecology had not been integrated into comprehensive national industrial policy in any nation, although its influence was felt in more traditional forms of environmental policy making that emphasized data gathering and technical assistance to industries to promote efficient and sustainable practice. According to some scholars, industrial ecology's emphasis on loop closing in all aspects of economy has led to increasing interest among industry and governments in what is called "waste valorization" in Europe, Asia, and South America and "beneficial use or reuse" in North America. Both terms refer to the recovery of non-hazardous industrial wastes for use as inputs to production or fuel sources. In the 21st century, state regulatory focus on industrial waste disposal lags advances in industrial ecology. By 2010, research on various aspects of industrial ecology was vibrant, and there was considerable pilot testing of innovative ways to recover large, homogenous quantities of industrial wastes, including sludges, metal slags, wood wastes, foundry sands, and combustion ash—none of which has been traditionally addressed in the scrap trade. As concerns grow over the connection between industrial production and climate change, the field of industrial ecology offers promise to address industrial wastes through efficient, data-driven, systems-based public policy.

Samantha MacBride
New York University

See Also: Africa, Sub-Saharan; Construction and Demolition Waste; Developing Countries; Environmental Justice; Environmental Protection Agency (EPA); European Union; Hazardous Materials Transportation Act; Industrial Revolution; Love Canal; Mineral Waste; Pre-Consumer Waste; Producer Responsibility; Race and Garbage; Radioactive Waste Disposal; Radioactive Waste Generation; Resource Conservation and Recovery Act; Scrubbers; Sustainable Waste Management; Toxic Wastes; United States; Zero Waste.

Further Readings

Ayres, R. U. and L. Ayres. *Industrial Ecology: Towards Closing the Materials Cycle*. Brookfield, VT: E. Elgar, 1996.

Brown, P. and E. Mikkelsen. *No Safe Place: Toxic Waste, Leukemia, and Community Action*. Berkeley: University of California Press, 1990.

Dernbach, J. "Industrial Waste: Saving the Worst for Last?" *Environmental Law Reporter*, v.20 (July 1990).

Eggleston, H. S., et al. *2006 IPCC Guidelines for National Greenhouse Gas Inventories*. Kanagawa, Japan: IGES, 2006.

Environmental Protection Agency. *Draft Final Report: Screening Survey of Industrial Subtitle D Establishments*. Rockville, MD: Westat, 1987.

Lacoste, E. and P. Chalmin. *From Waste To Resource: 2006 World Waste Survey*. Paris: Economica, 2007.

Mathews, E., et al. *The Weight of Nations: Materials Outflows From Industrial Economies.* Washington, DC: World Resources Institute, 2000.
Nshihou, A. and R. Lifset. "Waste-Valorization, Loop-Closing, and Industrial Ecology." *Journal of Industrial Ecology*, v.14/2 (2010).
Pellow, D. *Resisting Global Toxics.* Cambridge, MA: MIT Press, 2007.
Tarr, Joel A. *The Search for the Ultimate Sink: Urban Pollution in Historical Perspective.* Akron, OH: University of Akron Press, 1996.

Insulation

Insulation is a nonconductor material or substance that prevents the passage of heat, electricity, or sound from, for example, a room. Societies and governments have increasingly recognized the importance of insulation as a significant means of reducing household and building energy consumption levels. This is becoming ever more important in view of the need to stabilize climate change through the responsible and efficient use of natural resources. Carbon emissions can be significantly reduced with the installation of building insulation, which reduces the demand for energy to heat or cool an interior space. This results in less demand for electricity, which reduces emissions associated with production, and it has the added bonus of saving the householder money.

Energy Loss

In the United States, the heating and cooling of homes and buildings accounts for 50–70 percent of the energy used in the average home. In the United Kingdom, the energy consumption level, for heating in particular, is even higher at 80 percent. The Kyoto Protocol and subsequent commitment by member states of the European Union (EU) to reduce greenhouse gas emissions by 20 percent of 1990 levels by 2020 has led to a variety of measures seeking to slash emissions. For example, through the emissions trading scheme (also known as cap and trade) as well as the introduction in recent years of smaller-scale but nonetheless effective national policy drivers, householders have incentives to make their homes more energy efficient. Small grants, subsidies, and tax breaks are provided for those carrying out certain energy-saving improvements on their properties. However, the longevity of such schemes and the eligibility criteria for approval is likely to become less certain in view of the economic downturn. The importance of insulation can be illustrated by the substantial contribution (25 percent) that energy lost in homes and buildings adds to the total carbon emission of the United Kingdom. In the United States, the energy loss for residential properties alone is responsible for 18 percent of total emissions. Taken together with commercial buildings, the level of emissions rises to 38 percent.

Insulation and Heat Transfer

Insulation works by slowing the movement of heat and absorbing heat and inhibits or resists the displacement of heat from a hot space to a cooler one. Likewise, insulation inhibits cold or warm air penetrating from the outside to the inside of a building. This thermal resistance to heat movement is referred to as the "R-value." The best nonconductors for insulation purposes are those with the highest R-values per inch of thickness. Movement of heat can occur through radiation, conduction, or convection. Radiation is the transfer of heat through electromagnetic light waves, such as radiant heat from a wood fire. Conduction is the transfer of heat as it moves through an object, and convection is the transfer of heat in air or water by upward and outward movement.

Installation

Walls, roofs, floors, windows, and doors can all be insulated to help prevent the escape of either hot or cool air from the building to the outdoors. However, walls and roofs are the biggest heat loss culprits. Roof and wall insulation is used within the fabric of the building and is laid between the outer cladding and an internal lining profile. The easiest time to upgrade the thermal resistance of a building through the use of insulation is during construction. However, insulation can be added or upgraded retrospectively, although the choices of insulation may be narrower in scope.

The best type of insulation depends upon the local climate and whether one needs to keep heat in, out, or both. A building's orientation, design, and materials are also a factor. Materials used as insulation for the fabric of buildings may be grouped as organic or inorganic insulants. Inorganic insulants are made from naturally occurring materials formed from fiber, powder, or cellular structures that have a high void content, such as glass fiber, rock wool, cellular glass beads, vermiculite, calcium silicate and magnesia, or compressed cork. They are generally incombustible, highly heat resistant, and are rot and vermin proof. Generally, these insulants are in the form of loose fibers; mats and rolls of felted fibers; and semirigid or rigid boards, batts, or slabs of compressed fibers. Inorganic insulants are based on hydrocarbon polymers in the form of thermosetting or thermoplastic resins to form structures with high void content, such as polystyrene, polyurethane, isocyanurate, and phenolic. The problem with this type of insulants is that they tend to be combustible and have a low melting point. They are mostly used in the form of expanded polystyrene (EPS), either as beads or boards.

In terms of insulating walls, insulation material can be placed in one of three positions in a masonry cavity wall: on the outer face, on the inner face, or filling the cavity. Applying insulation to the outer face of a wall using specialist fixings and a weather-resistant finish can be done as part of a refurbishment. Insulation can be built into the building in the cavity as the walls are raised. This will be held in position not only by the sandwiching of the two leaves of the cavity wall but also by the wall ties. Materials that can be used for cavity walls and the inner faces of walls include glass fiber, mineral wool, and foamed plastics (such as polystyrene beadboard or loose beads, extruded polystyrene, urea formaldehyde foams, phenol formaldehyde foams, and polyurethane foams). The most effective way of insulating an existing cavity wall is to fill the cavity with some insulating material that can be blown into the cavity through small holes drilled into the outer leaf of the wall. Glass fiber and granulated rock wool of EPS beads are used for injection of insulation for existing cavity walls.

Sustainability and Environmental Effects

It is important to consider the overall environmental impact of insulation material to ensure that the material used contributes positively toward the achievement of reducing emissions and mitigating climate change. While the function of insulation material is to help reduce the movement of hot or cold air to a different place, thereby reducing energy used in heating or cooling a specific space, energy is used during the manufacture and transportation of insulation. The use of energy during manufacturing operations and transportation is referred to as the "embodied energy." Similarly, at the point at which a building is deconstructed, account should be taken regarding the potential of the insulation material to be reused or recycled. The materials that represent the greatest benefit to environmental sustainability are those that have the ability to remain within the materials economy beyond the life of the building.

Fiber glass and rock wool use a high percentage of recycled material, which helps to reduce the use of natural resources. In fact, in the first decade of the 21st century, manufacturers of fiber glass and rock wool insulation have diverted more than 20 billion pounds of glass and blast furnace ash from the U.S. solid waste stream. Both fiber glass and rock wool are highly durable and, if used in blatt, slab, roll, or sheet form, can be easily removed and reused elsewhere. On the other hand, polystyrene, polyurethane and polyisocyanurate are produced by blowing gases into the material. This operation causes the release of ozone-depleting substances into the atmosphere. While they have the best insulation value of any commonly available insulation material, they have a high embodied energy.

New insulation materials, offering more efficient insulation with lower embodied energy, are constantly coming onto the market. For instance, insulation that uses recycled paper (Warmcel 100) and blended sheep wool (Thermafleece), which has a low economic value since it is unsuitable for many applications but makes efficient insulation, is recyclable and reusable.

Hazel Nash
Cardiff University

See Also: Construction and Demolition Waste; Floor and Wall Coverings; Household Consumption Patterns; Styrofoam; Ventilation and Air-Conditioning.

Further Readings

Emrath, P. and H. F. Liu. "Residential Greenhouse Gas Emissions." *Special Studies* (April 2007). http://www.homebuyertaxcreditusa.net/fileUpload_details.aspx?contentTypeID=3&contentID=75563&subContentID=105106 (Accessed November 2010).

North American Insulation Manufacturers Association (NAIMA). http://www.naima.org/main.html (Accessed November 2010).

Pew Center on Global Climate Change. *Buildings and Emissions: Making the Connection.* Arlington, VA: Pew Center on Global Climate Change, 2009.

Iowa

Located in the Corn Belt of the midwest, Iowa is often known as the American Heartland. The capital, Des Moines, is the largest city and largest metropolitan area. The state owes its name to the Ioway people, one of the many Native American tribes that occupied the state when European exploration began in the late 17th century. At this point, the Iowan Native Americans were virtually all settled farmers living in a society with complex economic and political systems. After the Louisiana Purchase (in which the United States bought the territory from France), settlers created an agricultural economy, which endured until the 1980s Farm Crisis. Since the mid-1980s, Iowa has reemerged as a diverse, mixed economy, which now includes advanced manufacturing, biotechnology, and green energy production. Agriculture remains a large part of the Iowa economy in the 21st century, but has a less direct role.

Agriculture

Most of the state is used for agriculture—60 percent is crop-covered—and 30 percent is grassland, consisting mainly of pasture and hay. Iowa has had to use legislation to protect its agricultural interests from the adverse effects of waste disposal, such as open burning. This legislation has had instant impact on the way people in Iowa dispose of their garbage. In 1953, a state law was passed that required garbage fed to pigs to be cooked at 212 degrees F for 30 minutes in order to reduce swine disease pathogens in the feed. Prior to the law going into effect in June 1953, there were over 400 garbage feeders in Iowa, mostly small-scale operators running their operations as a side business. A month later, there were only around 60, and 70 percent of these were using homemade garbage cookers. Of the 64 percent of Iowan garbage cookers that were direct fired, 66 percent were homemade. Non-farmers operating these usually set up on or near garbage dumps and typically use wood or waste tires for fuel.

Ethanol production uses around one-third of the Iowan corn crop, and Iowa is the biggest producer of ethanol in the United States. In the 21st century, doubts have been expressed in the scientific community about the amount of logistics required to harvest, transport, and store the crop biomass needed for making cellulosic ethanol. This problem, however, is avoided by using municipal solid waste (MSW) and paper production waste as the feedstock. In 2010, Fiberight, in Blairstown, began production at the first commercial cellulosic ethanol plant in the United States to use enzymatic conversion technology and industrial and municipal MSW as a feedstock. The biorefinery was converted from a former first-generation corn ethanol plant in a $24 million investment projected to reach a commercial production capacity of 6 million gallons of renewable biofuel in 2011. Fiberight believes its core extraction and processing technology can potentially derive 9 billion gallons of renewable biofuel from 103 million tons of nonrecyclable MSW generated each year in the United States. On reaching full production, the Blairstown plant should be able to process over 350 tons of waste per day, producing biofuel at less than $1.65 a gallon. This use of MSW is believed to offer significant advantages over other waste-to-energy (WTE) methods, as low temperature and closed-loop systems exploit the MSW for a higher value while avoiding the emissions created by other methods.

Statistics and Rankings

The 16th Nationwide Survey of MSW Management in the United States found that in 2004, Iowa had an estimated 4,341,454-tons of MSW generation, placing it 29th in a survey of the 50 states and the capital district. Based on the 2004 population of

2,972,566, an estimated 1.46 tons of MSW were generated per person per year (ranking 14th). Iowa landfilled 2,822,563 tons in the state's 59 landfills, making Iowa the 29th-largest landfill-using state in the United States. It exported 92,156 tons of MSW. Landfill tipping fees across Iowa averaged $39, where the cheapest and most expensive average landfill fees in the United States were $15 and $96, respectively. By 2010, Iowa had 96,753,114 tons of landfill capacity remaining and was increasing its landfill capacity; it was ranked joint sixth out of 44 respondent states for number of landfills. Yard waste, whole tires, used oil, lead-acid batteries, and white goods were reported as being banned from Iowa landfills. Iowa's single WTE facility (the Arnold O. Chantland in Story County) burned 54,496 tons of MSW. The state recycled 1,464,395 tons of MSW, placing Iowa 23rd in the ranking of recycled MSW tonnage.

Middens in Iowan Prehistory

Native Americans in Iowa first began to move away from hunter-gatherer subsistence in the Late Archaic Era (5,000–2,800 years ago), when the first domesticated plants and large enduring settlements appear. During the subsequent Woodland period, divided into Early (800–200 B.C.E.), Middle (200 B.C.E.–400 C.E.), and Late (400–1250 C.E.), many Iowan Native Americans abandoned hunting and gathering and made greater use of agriculture. With sedentary living came substantial deposits of midden, which became an important part of the archaeological record of these cultures.

At the Gast Spring site, midden features predominate; a substantial Middle Archaic midden was associated with a feature interpreted as a sunken floor with an oval pit in the center. The floor surface and the pit dug into it were also repositories for midden deposits, containing the remains of fish, turtles, birds, mammals, seeds, gastropods, fire-cracked rock, and lithics. The abundant charcoal in the midden deposits provided a radiocarbon date of 5200–4500 B.C.E., suggesting that if the Gast Spring site was a house, it would be the earliest-known house in Iowa.

Two of the three known ring midden settlement sites belonging to the Weaver culture of the Woodland period were found in Iowa: Gast Farm and Oak Village. These sites are defined by a circular concentration of midden and pit features around an open central area, and they occur throughout a large part of the midwest, from southern Wisconsin and eastern Iowa to Kentucky and Tennessee. At Gast Farm, the diameter of the ring midden was 100 meters and the depth of the midden deposits was up to 0.5 meters in places and thought to represent the location of dwellings, while the central area was used for communal activities such as feasting.

The Late Prehistoric (900–1600 C.E.) Mill Creek culture of northwest Iowa is characterized by its large midden mounds—rich deposits that have attracted archaeological investigation since the early 20th century. Mill Creek sites have been described as miniature *tels*, borrowing from Near Eastern archaeology the term for a large mound created by successive occupation on a site over a period of centuries. The sites have been elevated up to 3 meters above the original ground level by the accumulation of debris and crumbled mud-walls.

Arnold O. Chantland Resource Recovery Plant

The Arnold O. Chantland Resource Recovery Plant (RRP) in Ames, Story County, was the United States' first full-scale municipal WTE facility. It opened in 1975, 90 years after the first incinerator in the United States. Initial planning began in 1971. Money was raised by general obligation bonds against the background of peak oil, the 1973 energy crisis, and growing concern over loss of farmland to landfilling. The project was a gamble because there were few operating facilities and massive upfront expenditure was required. In the 21st century, however, the RRP is cited as the model recycling system. Continuous development and innovation has allowed the Arnold O. Chantland RRP to exceed federal and state targets for recycling.

The plant recovers all recyclable materials from garbage and sells recovered ferrous metal (5 percent of waste received) to a scrap dealer. The RRP calculates that they recover enough ferrous metal annually to make 1,200 car bodies. Garbage that cannot be reused is shredded and divided into burnable and nonburnable categories. Around 65 percent of waste received is burnable. Nonburnable material is sent to a landfill, and the shredding process significantly reduces its volume. Burnable material becomes refuse derived fuel (RDF), which is piped to the Ames city power plant where it is used as a renewable supplemental fuel. By burning the RDF with coal, fossil fuels are conserved and sulfur dioxide emissions are decreased. The RDF makes up 10 percent of the fuel used at the Ames city power plant. The facility has thus far saved 80 acres of farmland from becoming landfill and provides enough electricity for 4,600 homes per year. The processing capability of the Arnold O. Chantland RRP is such that there is no landfill in Story County; what MSW there is remaining to be landfilled is taken to neighboring Boone County by arrangement. An environmental landfill for construction and demolition waste is operated by Story County.

Garbage is taken to the plant by private citizens and waste disposal contractors and is deposited onto a tipping floor, which can hold 450 tons of garbage. The separation process begins on a conveyor into the primary shredder, where steel hammers reduce the garbage to pieces around 1 foot in diameter. These pieces are then passed under an electromagnet to remove ferrous metals before being deposited onto a series of two disc screens to separate out smaller particles. A second shredder reduces the garbage to pieces around 2 inches in diameter, which are then sorted on conveyors by air knife into burnable RDF and heavier nonburnable rejects. Conveyors then take the RDF to the twin screwfeeders, which load the material into the pneumatic tube that sends it to the power plant.

The RRP also provides glass, car battery, and waste oil recycling (with only a minimal charge for waste oil over five gallons), as well as disposal of hazardous household materials (free) and appliances ($20 each). The glass recycling program was initiated in 2006 to divert glass from the RDF and landfill streams. Glass ending up in the RDF melts in the power plant furnace and solidifies into slag in the boiler tubes, necessitating expensive repairs, while glass being sent to landfill increases transportation and dumping costs, uses landfill space, and is a waste of resources. By 2009, the glass recycling program had diverted over 100 tons of glass from the landfill. The recovered glass is crushed to sand-grain size and used for landscaping, industry, and construction. Financial assistance for the Yard Waste Site also comes from the RRP, allowing five free days of seasonal disposal a year: one day in spring for leaves and brush; and four in fall for leaves and grass.

Jon Welsh
AAG Archaeology

See Also: Economics of Waste Collection and Disposal, U.S.; Farms; Incinerators; Recycling.

Further Readings

Adams, S. Keith and John C. Even Jr. "Seasonal Flow Stream Analysis in a U.S. Post-Consumer Resource Recovery Plant: A 1977–1990 Comparison." *Resources, Conservation and Recycling*, v.17/1 (1996).
Alex, Lynn M. *Iowa's Archaeological Past*. Iowa City: University of Iowa Press, 2000.
Arsova, Ljupka, et al. "The State of Garbage in America: 16th Nationwide Survey of MSW Management in the U.S." *BioCycle*, v.49/12 (2008).
Norton, Glenn A., et al. "Chemical Characterization of Ash Produced During Combustion of Refuse-Derived Fuel With Coal." *Environmental Science and Technology*, v.22/11 (1988).

Iran

One of the world's oldest civilizations, Iran is a Middle Eastern Islamic republic on the border of Eurasia and is an important power in the world economy because of its large oil and natural gas reserves. Oil was nationalized in Iran in 1953, and the industry is entirely state owned and operated by the National Iranian Oil Company. The national economy is a mixture of state-owned industries (mainly oil), central planning, rural agriculture, and small private endeavors, including companies in the biotechnology and nanotechnology industries. In the Middle East, Iran is a leading manufacturer of vehicles, construction materials, appliances, pharmaceuticals, and petrochemicals. Oil contributes significantly to the economy, though not as greatly as in even more oil-dependent countries like Saudi Arabia. Oil revenues constitute a little less than half of the government's budget, compared to about one-third from taxes and fees. Unemployment, inflation, and budget deficits are ongoing problems, the latter ascribed to state subsidies for food and gasoline. Energy subsidies alone cost the government close to $100 billion annually, and a 2010 economic reform plan calls for gradual reductions of subsidies and replacement with more focused social assistance, as the country attempts to move toward a more free market by 2016 while still maintaining a concern for social justice.

Oil and Natural Gas Production and Consumption

Iran has the third-largest oil reserves (10 percent of the world's proven oil reserves) and the second-largest natural gas reserves (15 percent of the world's reserves) in the world, and the country is the fourth-largest oil exporter. Output has been constrained by the country's inefficient technology; the 21st-century average output is only two-thirds the peak of six million barrels per day, which was reached in 1974, and the drilling of exploratory wells has slowed considerably. Like much of the Middle East, a dilemma in 21st-century Iran has been the combination of plentiful petroleum resources, yet vast inefficiencies and steadily rising demand in domestic energy usage. Iran's per capita energy consumption is 15 times that of Japan and 10 times that of the European Union. Even relative to the rest of the Middle East, the country's energy intensity (units of energy used per unit of the gross domestic product) is two and a half times that of the regional average. In gasoline consumption, Iran ranks behind only the United States in per-vehicle consumption. The subsidies that have kept prices low have encouraged wasteful consumption, and an illegal sub-industry has developed to smuggle cheap, subsidized gasoline from Iran into Iraq, Pakistan, Afghanistan, and Turkey, where local prices are high enough to make smuggling profitable.

Oil and natural gas combined account for about half of the country's energy consumption. Coal power plants continue to be an important contributor and have recently been streamlined for greater production. To help meet its energy demands, the country has been building new hydroelectric power plants since 2004, along with its first wind-powered and geothermal plants. The first solar thermal plant came online in 2009. There have also been tentative experiments with generating electricity from waste and sewage as well as assessing Iran's tidal power potential, which is believed to be considerable on the Persian Gulf coastline.

Despite Iran's vast amounts of oil, oil imports actually rose in the first decade of the 21st century, increasing from 30,000 barrels per day in 2000 to 180,000 barrels per day at the start of 2005, and originating mostly in India. The issue is the country's refining capacity. Imports shrank in 2010, and the Iranian government claimed in September 2010 that imports had stopped entirely—but because only months before, imports had supplied 30 percent of the country's domestic consumption, that announcement was met with skepticism.

At least some of the reduction in domestic oil consumption can be ascribed to the increased use of natural gas; reserves were largely untapped until 2005. The government has subsidized natural gas prices to encourage natural gas consumption in the interest of reducing oil imports. By 2010, Iran had the third-largest natural gas consumption in the world, after the United States and Russia, and the highest growth rate.

Electricity consumption has risen steadily since the 1980s and is heavily reliant on oil- and coal-burning energy, with a small but growing number of hydroelectric plants and the recent additions of alternative sources. Because of technological problems and inefficiencies, about 18.5 percent of electricity produced in Iran is wasted before it is consumed, lost on the way from generator to consumer.

Nuclear Power

Iran's nuclear program was launched in the 1950s when the country was an important ally of the United States and the West. Because of the withdrawal of Western support after the 1979 Iranian Revolution, which deposed the shah the Westerners had helped put in power, elements of the nuclear program were temporarily disbanded. Nuclear research in Iran stagnated for a long time; gestures toward working again on nuclear power were frequently met by suspicion from the West that the country was interested in developing nuclear arms. As a signatory to the 1968 Nuclear Non-Proliferation Treaty, Iran's nuclear program is subject to verification and inspection by the International Atomic Energy Agency (IAEA), which it has not only accepted but also on occasion invited

The Iran Bushehr Nuclear Power Plant (scale model depicted above) was fueled and officially launched on August 21, 2010, and hailed as a major coup by the Iranian government, igniting concern from many Western governments. Because of the controversial nature of Iran's leadership and its provocative political statements, it is difficult to accurately interpret the legitimacy of Iran's stated ambitions to use nuclear technology only as a source of energy and medical radioisotopes. Currently, the Bushehr plant is staffed by Russian specialists.

in order to dispel these nuclear weapon rumors. Both Iran and the IAEA have denied reports of Iran's blocking IAEA inspectors.

In the 21st century, reporting its activity to the IAEA as required, Iran has constructed two nuclear-material enrichment facilities, which are necessary to produce fuel for nuclear power plants, since it considers foreign sources of such fuel unreliable due to politics. Iran has produced enriched uranium contrary to a series of United Nations Security Council resolutions forbidding it from doing so, resulting in sanctions being imposed on the country, culminating in a complete arms embargo on Iran starting in the summer of 2010. The purpose of the sanctions is to limit Iran's ability to turn nuclear technology into nuclear weapons.

Iran has explicitly declared itself a nuclear state, and it intends to produce enriched uranium both for a medical reactor (to produce medical radioisotopes) and for a nuclear power plant. Iranian president Mahmoud Ahmadinejad went so far as to declare, "If we wanted to manufacture a bomb, we would announce it," and implied that the country had the ability to do so. Given the president's controversial nature and previous provocations of the West, it is difficult to know how much credence to give either half of the statement.

The Bushehr Nuclear Power Plant was fueled and officially launched on August 21, 2010, and hailed as a major coup by the Iranian government, prompting tentative retroactive support from many Western governments that stressed that their concerns were over illicit nuclear arms, not legal nuclear power. The plant is operated by Russian specialists, fueled for the time being with Russian fuel, and the spent fuel will be sent back to Russia for storage. The plant was expected to meet 2 percent of Iran's national electricity demand upon connection to the grid in early 2011.

Waste

Solid waste in Iran is primarily landfilled. Iran's capital city Tehran had a population of over 8 million people and generated 2,626,519 tons of solid waste in 2005. Most of this went to three municipally operated landfills in the city, although some was scavenged through informal channels. In the western Kurdistan province, unsegregated collection and open dumping of municipal solid wastes is common, and material reclamation goes through scavengers, rather than formal mechanisms.

Bill Kte'pi
Independent Scholar

See Also: Africa, North; Gasoline; Middle East; Nuclear Reactors; Saudi Arabia.

Further Readings

Ali Abduli, M. and T. Nasrabadi. "Municipal Solid Waste Management in Kurdistan Province, Iran." *Journal of Environmental Health*, v.69/7 (2007).

Axworthy, Michael. *A History of Iran*. New York: Basic Books, 2010.

Damghani, A. M., et al. "Municipal Solid Waste Management in Tehran." *Waste Management*, v.28/5 (2008).

Polk, William R. *Understanding Iran*. New York: Palgrave Macmillan, 2011.

Iron

Iron is one of the building blocks of industrial society. An element found beneath the Earth's surface, the extraction of this material scarred the land and killed many people who harvested it. Once above ground, the pliability and strength of ferrous metal allowed it to structure modern life, from skyscrapers to automobiles. The properties of iron have allowed its users to remelt and refashion it over and over, making it one of the most recycled materials in history.

Iron (Fe) ore is abundant across the planet, the fourth most common element in the Earth's crust. Artifacts containing iron (such as spear tips and beads) exist from human societies as early as the Paleolithic Age more than 30,000 years ago. Intensive mining and smelting (in which iron oxide is separated from the ore to form metallic iron) of iron to fashion tools and weapons developed throughout civilizations in China, India, and the Mediterranean between 1200 and 500 B.C.E. Biblical verses refer to the remelting and reshaping of old iron implements for agriculture and warfare, converting plowshares into swords (Joel 3:10) and swords into plowshares (Isaiah 2:4 and Micah 4:3). Iron and the alloys made with iron were vital to the rise of the Islamic world (including the development of the legendarily strong and flexible swords made of Damascus steel) and the guns and agricultural implements leading to the rise of nations in Europe.

English and colonial blacksmiths used old iron for horseshoes and other goods in the early modern period, getting the metals mostly from local sources. The American colonies produced iron from ore as early as 1645; by 1700, American iron accounted for almost 2 percent of world production and about 10 percent of British production. Due to restrictions from the mother country, the colonists could not establish new ironworks to produce finished products, limiting most of the colonial blacksmith's work to mending existing goods.

Iron was crucial to the fighting of the U.S. Civil War; the U.S. system of rifle manufacture (a precursor to modern mass production) depended upon the malleability and durability of iron. As the United States industrialized in the decades after the war, iron—and its harder, lighter alloy, steel—allowed cities to rise vertically, supplying the skeletons for skyscrapers. Iron formed the trains and the rails they rode on from coast to coast. Iron revolutionized naval and cargo ships. Demand for this ferrous metal spurred the digging of more mines, the opening of more foundries to turn iron into steel, and more scrapyards to reclaim disused metals.

Changing production techniques led to increased recycling of iron and steel. The Bessemer process, used widely in the late 19th century, used a limited amount of scrap iron, up to about 10 percent of the charge in order to regulate temperature. Open-hearth furnaces operated at higher temperatures than Bessemer converters, and they burned off phosphorus and impurities found in scrap that Bessemer converters could not remove, allowing for use of more scrap iron. The change was important because steel producers coveted ways to reuse old iron and steel, rather than mining virgin ore. Mining incurred great costs in capital and labor to extract ore from under the Earth's surface, and the hazards involved led to frequent injuries, deaths, and labor disputes. Environmental effects of mining range from the creation of sinkholes to the release of toxic chemicals into the groundwater, and mining remains one of the most dangerous occupations in industrial society.

Iron and steel producers with access to ore had great advantages over their competitors. The interests that controlled the Iron Range in northeastern Minnesota effectively controlled the virgin ore market in the United States, and Carnegie Steel monopolized the vast ores of high-phosphorus iron in Minnesota in the 1890s. This advantage catalyzed the transformation of the firm into the gigantic U.S. Steel, which then built the world's largest steel mill in northwestern Indiana. The Gary Works built several open-hearth furnaces between 1908 and 1920, and many of U.S. Steel's smaller competitors switched over from Bessemer furnaces to open-hearth furnaces in order to survive. The transition to open-hearth steel combined with expanding markets for steel to trigger new demands for scrap iron and steel.

Scrap and Recycling

With production techniques that could use more scrap, steelmakers could rely upon the discarded ferrous metals found in industrialized society. Dumps, independent junkyards, and demolition sites all became sources of scrap metal; instead of digging into the Earth thousands of miles away, a mill could find its materials in scrapped ships, railroads, and automobiles on the Earth's surface. Railroads acted as suppliers and customers of scrap iron on the open market. Old tracks and engines were made of heavy iron and became valued commodities for railroads. Steel mills collected in-house scrap (also known as "prompt industrial scrap" or "home scrap") in order to eliminate waste and use the cheapest-available materials for production.

Scrap collection and sorting is hard, dangerous work. Scrap metal does not come in neat, symmetrical sizes and shapes; it comes in whatever form it was in when it was discarded. In order to sell these materials, dealers have to convert them into portable shapes and sizes of material free from impurities. Ship hulls may be encased in rust that has to be scraped or cut off; potentially valuable cables have to be separated from worthless refuse. Processing scrap involves cutting it with shears or acetylene torches, either of which could easily maim a man. Sharp or jagged edges of scrap metal can cut flesh; if rusty, they can cause tetanus. Like mining, scrap processing was—and is—a dangerous occupation.

Thousands of workers engaged in this work because it was profitable. By 1917, *Scientific American* reported that the annual business in scrap iron increased from $100 million to $1 billion. The Great Depression stunted demand for iron and steel, but the United States consumed more ferrous metal than ever during World War II. Mines and scrapyards were mobilized for the war effort, and citizens' groups conducted scrap drives to maximize the amount of metal put into military production.

Scrap drives ended with the war, but U.S. consumption of iron and steel did not abate in 1945. Military use of the metal continued as the United States mobilized for a cold war that expanded the peacetime military and its purchase of weaponry and vehicles. The domestic economy grew substantially between 1945 and 1965, and with it came greater metal consumption. Iron and steel continued to shape new buildings, as well as the millions of new automobiles sold every year. It shaped consumables ranging from cans of food to air conditioners, as the material remained integral in modern conveniences.

Innovations

Further changes in steelmaking, including the advent of the electric arc furnace and continuous casting, reshaped the industry in the late 20th century as giant steelworks gave way to smaller minimills in both the United States and abroad. Another change in technology worked to counter iron's influence. The rise of plastics after World War II produced new, flexible materials that often matched or surpassed the malleability of ferrous metal but weighed less. A person drinking soda would be as likely to sip from a plastic bottle as from a steel can at the end of the 20th century. Even automobiles began to contain growing amounts of plastics in their bodies and interiors. While iron remained a widely used material in daily life in the United States, the age of plastic meant that petroleum's influence extended well beyond fuel to power vehicles once made primarily of iron and steel. Despite these changes, iron remains a crucial building block of industrial society, and its properties make it one of the most recycled materials on Earth.

Carl A. Zimring
Roosevelt University

See Also: History of Consumption and Waste, U.S., 1850–1900; History of Consumption and Waste, U.S., 1900–1950; History of Consumption and Waste, World, 1800s; Industrial Revolution; Junkyard; Recycling.

Further Readings

Misa, Thomas J. *A Nation of Steel: The Making of Modern America, 1865–1925*. Baltimore, MD: Johns Hopkins University Press, 1995.

Rogers, Robert P. *An Economic History of the American Steel Industry*. New York: Routledge, 2009.

Seely, Bruce, ed. *The Encyclopedia of American Business History and Biography: Iron and Steel in the Twentieth Century*. New York: Facts on File, 1994.

Wertime, Theodore A. and James.D. Muhly, eds. *The Coming of the Age of Iron*. New Haven, CT: Yale University Press, 1981.

Zimring, Carl A. *Cash for Your Trash: Scrap Recycling in America*. New Brunswick, NJ: Rutgers University Press, 2005.

Italy

Bordered by the Mediterranean and the Alps, Italy is a 116,346-square-mile peninsula with limited space and natural resources. In addition, it is one of the European countries with the highest population density (517.4 people per square mile). The national territory also includes several islands, the largest being Sicily and Sardinia. This Mediterranean country has the fourth-largest economy in Europe by 2010 and experienced rapid industrial growth after World War II. The natural environment and rapid social and economic change have had an impact on Italian attitudes toward waste and environmental management. Not unified into one country until 1861, in many ways Italy remains regionally divided in both cultural and economic perspectives. The largest division is the north from the south. The north has developed a strong industrial economy and modern infrastructure, while the south remains largely agrarian with pockets of high poverty. When discussing Italian attitudes and approaches to consumption and waste, these divisions must be taken into consideration.

World War II Era

Since the 1940s, Italy has come nearly full circle when it comes to dealing with waste. Prior to and immediately following World War II, the average Italian experienced a life of great scarcity. Almost nothing, from water to manure, was wasted. Everything was reused until it could be used no more. For example, Italian pig slaughtering and processing used every part of the animal, from the snout to the tail. Italian culinary culture developed to make efficient use of the pig, from blood sausage to be consumed immediately after slaughter to long-aging hams that would be consumed the following year. Meat was scarce and often only a small piece of fat would be used to flavor a dish of pulses. Popular Italian cuisine, particularly in central and southern Italy, was referred to as *cucina povera* (meaning "poor or frugal cuisine"). Scarcity during times of war also led Italians to forage for wild foods such as mushrooms, chestnuts, and tubers. Scarcity was most felt when it came to subsistence, but the same frugality carried through the rest of Italian consumption patterns.

Italy manufactured very few consumer goods until the second half of the 19th century. As an agricultural economy, most farms were organized in the south, mainly as large shareholder farms, and in central Italy, a sharecropping system called *mezzadria* existed. This meant that the majority of Italians produced most of their own food and consumed very few consumer products. It was Benito Mussolini's dream to make Italian agriculture self-sufficient.

Postwar Economic Growth

Production and consumption changed drastically after World War II as Italy began to industrialize rapidly in the north. Helped along by funds from the U.S. Marshall Plan, Italy rapidly went from being one of the least to one of the most developed countries in Europe. There was a great wave of migration from south to north to the growing factories in cities like Turin, Genova, and Milan. The period from the mid-1950s to the 1970s is often called the "economic miracle." In particular, car manufacturing took off as one of the main industries, drastically changing Italy's economy.

The Turin-based company Fiat came up with a design for a small economic vehicle, the 500. The Fiat 500 was produced from 1957 to 1975 and was a car that even working-class Italians could afford. Previously, cars had been strictly luxury items to which only the privileged few had access. With the growth of the automobile market, Italians could now travel longer distances. They were no longer limited to the range of a bicycle or the limited network of railways lines. Cars like the 500 offered Italians a new autonomy and outlook on their country. Owning a car was also a major status symbol and public display of upward mobility. By the 1960s, Italy was moving from a culture of poverty and scarcity to a culture of consumption, thanks to the rise of manufacturing and heavy industry.

Much Italian consumption was focused on the domestic sphere. Italy manufactured domestic appliances such as washing machines, clothing, shoes, food products, and other consumer goods. Along with industrialization came urbanization, and this also contributed to Italy's growing consumption. Once people moved off the land, they were no longer able to produce their own food or fuel. During the second half of the 20th century, there was also a great desire to modernize and move away from "backward" rural ways. Shopping in supermarkets, rather than tending a garden or going to an open-air market, was one of the ways in which "modern" families could display their growing prosperity. Women began to enter the workforce in larger numbers by the end of the century, increasing the average family budget. Consequently, women had less time to devote to household chores, such as cooking and cleaning, but innovations such as vacuum cleaners, washing machines, and prepared foods helped lift some of the domestic burden from women's lives. Women's entry into the workforce had serious repercussions for social and gender roles in the family: women were no longer economically dependent on men, which caused tension as the power dynamics shifted in households. Many women stopped preparing elaborate dishes except during holidays because there was not enough time in the workday. Italian cuisine was affected by the introduction of mass-produced foods and the homogenizing experience of the supermarket. In the 21st century, groups such as Slow Food are attempting to protect Italian culinary heritage by encouraging Italians to turn back to traditional forms of agricultural production and a more meaningful form of consumption.

Waste

With the growth in urbanization and the rapid expansion of consumer culture, Italians began to produce more garbage and use more fuel. The largest polluting factor in Italy in the 21st century is heating fuel, and Italy is struggling to meet its goals for the Kyoto Protocol because of the expense and challenges of moving to greener fuel sources. Italy has banned nuclear power production and must buy electricity from neighboring countries, such as France.

From a waste perspective, many Italian municipalities have adopted effective recycling programs and sustainability programs that feature the composting of organic waste. Waste management programs vary widely from town to town. There is once again a north-south divide in popular attitudes about waste and municipal management of this issue. In 2008, the city of Naples and the surrounding areas experienced a terrible garbage crisis. The media revealed that the local Mafia, the Camora, controlled garbage disposal in all of the area and had been dumping the city's waste in illegal, unauthorized sites. Landfill sites were full, and there were no modern incinerators in the area. The Mafia opposed the building of more incinerators because it did not want to lose control of its waste disposal enterprise. The local government also feared that the Mafia would take control of newly built incinerators and would start disposing of toxic waste from northern Italy, a practice that they had historically engaged in.

The crisis culminated when garbage stopped being collected and began to pile up in city streets, and citizens began burning the refuse piled up in the streets. The national government and armed forces had to step in to restore order, and many municipalities in the north offered their incinerators and trucked waste north from Naples. By 2010, the issues affecting Naples and other municipalities in southern Italy had not been concretely addressed or resolved.

Italy's rapid industrialization and urbanization have made it into one of the largest consumer cultures and economies in Europe, and the waste that this type of economy produces must be dealt with. On January 1, 2011, the Italian Environmental Ministry began enforcing a ban on polythene shopping bags in an attempt to reduce plastic waste. Italians, particularly in the north, are beginning to come full circle in the early 21st century in their search for more sustainable consumption. Italy struggles to protect its natural resources, which are essential to some of its major industries (fishing, agriculture, and tourism). Without strong national or local government control, environmental catastrophes cannot be averted and abuses often go unchecked. With the serious disparity that exists between north and south, Italy has serious challenges to confront in the 21st century as it tries to manage its waste while maintaining economic growth.

Rachel Black
Boston University

See Also: Capitalism; Consumerism; Culture, Values and Garbage; Economics of Consumption, International; Food Waste Behavior; Fuel; Grocery Stores; Home Appliances; Household Consumption Patterns; Incinerator Construction Trends; Incinerator Waste; Politics of Waste; Recycling; Slow Food.

Further Readings

"Garbage Crisis in Naples Provokes Protest" MSNBC. (January 14, 2008). http://www.msnbc.msn.com/id/22505149 (Accessed July 2010).

"Italian Garbage Crisis Leads to Murder." NPR (June 5, 2008). http://www.npr.org/templates/story/story.php?storyId=91180857 (Accessed July 2010).

Melosi, Martin V. *Garbage in the Cities: Refuse Reform and the Environment*. Pittsburgh, PA: University of Pittsburgh Press, 2004.

Scrivan, Paolo. "Signs of Americanization in Italian Domestic Life: Italy's Post-War Conversion to Consumerism." *Journal of Contemporary History*, v.40/2 (2005).

United Nations Department for Policy Coordination and Sustainable Development. Division for Sustainable Development. "Italy: Country Profile." 1997. http://www.un.org/esa/earthsummit/italy-cp.htm (Accessed July 2010).

Japan

This island nation of some 127 million people (the 10th-largest population in the world as of 2010) was a rising power at the beginning of the 20th century. It turned into a major power in World War II; from defeat, it rose to prominence as a global economic player. As such, it was considered by some to be the exotic "Japan, Inc.," which was difficult for Westerners to understand (but had to be emulated and engaged with). It experienced economic decline beginning in the 1990s from which it had yet to recover as of 2012. Japan is a highly industrialized, urbanized, and consumptive nation with land scarcity, and a detailed discussion of waste management practices in a separate entry focuses on incineration rather than landfill. Environmental constraints, including severe urban air pollution in the mid-20th century and continued exposure to extreme seismic activity, shape life in Japan.

Culture, Production, and Consumption

Located at the eastern fringes of the Eurasian landmass, part of the "Ring of Fire" that circles the Pacific Ocean, Japan consists of four major and thousands of smaller islands, with an interior dominated by volcanic mountain ranges. Japan's development has been strongly influenced by its natural environment and location. Traditional culture even sees the world as animated. Especially in Japan's native religion, Shinto, every object is seen to be imbued with spirits. These do not reside only in nature but are also to be found in man-made objects.

Nature is prized most highly when it is cultivated, and human behavior is seen as the most highly cultivated when mastering learning to such an extent that it becomes the seemingly natural way of doing things. This blurring of the natural and artificial is often used to explain why there is a deep appreciation for nature and its beauty and power in Japan. However, a desire to intermingle these two aspects of the world—in ways that seem mutually exclusive in Western cultural terms—creates an attitude very open to novelty, not least in technological innovation. Such an inclusive view of human activities as a part of nature also influences the consumption of natural resources, for example, when it is argued that whaling should be allowed as a normal part of human activities, especially as whales' supposed overconsumption of other fish needs human intervention (to create harmony in nature).

Both traditional and modern culture have been shaped by the scarcity of resources that accompanies Japan's geography. The islands that make up

Japan offer forested, mountainous interiors and proximity to the resources of the Pacific Ocean surrounding them but few industrial resources and comparatively little land suited for agricultural cultivation and human habitation. These geographic factors help explain why Japan's population is especially concentrated in rather few areas. The human domination of these areas is also particularly great. Green areas in these conurbations are often times left only in the environs of Shinto temples because their association with these religious buildings gives them spiritual meaning and thus protection. The Greater Tokyo Area, also known as the "Tokyo-Yokohama urban corridor," is particularly noteworthy. It contains over 35 million people—28 percent of the country's total population. This makes it the world's largest metropolitan area, and it is increasingly merging with other coastal cities.

The rural population, in contrast, declined to only 21 percent of total population in 2000, and the agricultural sector, although strongly protected by government measures, only makes up 4.4 percent of total employment (according to International Labour Organization statistics from 2002). This is not to say that Japan's natural environment had been outside human colonization of nature; there have been ecological crises due to deforestation and remediated by active reforestation efforts.

Japan's geography is also one of the reasons (besides the large population and the highly developed economy) why Japan was the fifth-largest importer in the world and the third-largest importer of agricultural products by 2010. Agricultural imports had a value of over $40 billion in 2009 and constituted some 60 percent of the calories consumed in the country. The trade balance also shows this effect, as agricultural products made up 12.3 percent of Japan's imports in 2009, but only 1.4 percent of exports (according to the World Trade Organization trade profile of Japan).

Japan, in spite of its economic problems, achieved a 2010 estimated nominal gross domestic product (GDP) of over $5 trillion (or over $4 trillion by purchasing power parity), making it the world's second- or third-largest economy after the United States and China. Thus, it is also one of the world's largest consumers of raw materials because of its economic activity. It is also a major exporter of manufactured products. For example, in 2009, fuels and mining products constituted 34 percent of Japan's imports but only 4.4 percent of exports, and manufactured goods made up 51.8 percent of imports and 87.5 percent of exports.

Ecological Impact

Japan's economy grew rapidly after World War II, and intensive industrialization produced acute air, land, and water pollution problems relating to heavy metals, sulfur dioxide, and nitrogen dioxide. Public complaints in the late 1960s led the government to develop several environmental protection laws, with varying levels of success in abating pollution concerns. Comparing between major economies, Japan is still something of an outlier among industrialized countries. Its ecological footprint (the balance between national productive area and in-country ecosystem's capacity to absorb pollution, on the one hand, and resource use and concomitant emissions and pollution, on the other hand) is above the worldwide average of 1.8 to 2 global hectares of productive land available per person, standing at 4.73 global hectares per person in 2007. However, considering the relationship between ecological footprint and economic productivity, as measured by the Human Development Index (HDI) for Japan (which is at 0.884 points for 2010, putting Japan in 11th place), the country is highly effective at converting consumption of resources into value. The United States, in comparison, has an HDI of 0.902, a rank of fourth, but with an ecological footprint that is much higher at about 8 global hectares per person.

Japan ranked third among the world's nation in electricity production in 2008, trailing only the United States and China. Japan has relied upon nuclear energy to power industry and households since the 1960s, having 53 active reactor units in 2009, again ranking third worldwide behind the United States and France. The nation's vulnerability to earthquakes and tsunamis has shaped architecture and safety standards throughout Japanese society, including some of the most stringent building codes on Earth. Existing safety measures, however, have not prevented accidents at nuclear plants. To date, the most severe of these was the catastrophic meltdown at the Fukushima Daiichi power plant following the March 11, 2011, earthquake and tsunami. That disaster spread radioactive cesium hundreds of kilometers into the Pacific Ocean and forced tens of thousands of nearby residents to abandon their homes. The continuing environmental damage from the meltdown brought into question the nation's reliance on nuclear power, with public opposition soaring, and Prime Minister Naoto Kan stating "We will aim at realizing a society which can exist without nuclear power." It remains unclear whether Japan can replace nuclear energy for its considerable energy demands.

At least part of Japan's declining impact on the world can be explained by the stagnant economy, but there is also a distinct focus on greener technologies and more efficient production. Japan's lack of resources is credited with raising awareness of the need for a shift to less-consumptive practices (which are, therefore, less dependent on outside inputs). It is also a factor in the Japanese approach to wastes, which sees them very much as potential resources and aims to ultimately achieve a "zero waste society."

Japan, as a part of east Asia, also belongs to the region heavily influenced by Chinese culture. Its contemporary society intermingles aspects of traditional culture and global consumer culture in ways that are at the forefront of modernity, and it has increasingly exerted an influence on other societies. At the same time, Japan, in its economic development from postwar rise to feared economic dominance during the 1980s and on to stagnation, serves as an example of how economic and social dynamics interact, sometimes shift quickly, and may also change in other countries. This is all the more so as Japan's population development toward declining total numbers and rising numbers of seniors shows a path that other developed countries, and the rest of the world, will follow if world population to reaches a high point in the latter half of the 21st century and declines afterward.

Consumption patterns, both industrial and individual, have shifted and will shift in parallel with these economic and social developments. First, total fertility rates have been close to or below replacement level since 1957, meaning that Japan is now at the forefront of the global "silvering of society" (rapid aging) and showing an inverted population pyramid in which dependency ratios (of children and mainly retirees to economically active population) are increasingly high. This is particularly pronounced as the Japanese have the highest average life expectancy in the world. As a result, consumption declines along with population, increasing age, and perceived need to save for one's support in old age. In Japan, this challenge—and reduced productivity and consumption—is further exacerbated by the economic stagnation that has been near-continuous since the boom of the 1980s ended. One effect of it has been that the current young generation has only known a stagnating Japan and worrisome personal prospects (along with deflation) and has shifted its consumption habits accordingly, away from luxury brands to a concern with quality and, particularly, price. On the other hand, there is a concurrent trend toward urban living in single (or at least, smaller) households, which typically drives up consumption and production of waste per person.

Japan (as opposed to China) has the advantage of having become a rich, developed economy before becoming a society in which old people increasingly dominate. It also has cultural traditions valuing quality and aesthetics. In an aging world having to move toward less consumptive ways of life, these traits put it in the forefront of development—whether for better or worse will, of course, remain to be seen.

Gerald Zhang-Schmidt
Independent Scholar

See Also: China; East Asia (Excluding China); Economics of Consumption, International; Incinerators in Japan; High-Level Waste Disposal; Osaka, Japan; Sustainable Development; Tokyo, Japan.

Further Readings

Biello, D. "Fukushima Meltdown Mitigation Aims to Prevent Radioactive Flood." *Scientific American* (June 24, 2011). http://www.scientificamerican.com/article.cfm?id=fukushima-meltdown-radioactive-flood (Accessed July 2011).

Japan for Sustainability. http://www.japanfs.org (Accessed November 2010).

Karan, P. P. *Japan in the 21st Century: Environment, Economy, and Society.* Lexington: University of Kentucky Press, 2005.

Reader, Ian and George J. Tanabe Jr. *Practically Religious: Worldly Benefits and the Common*

Religion of Japan. Honolulu: University of Hawaii Press, 1998.
World Trade Organization Statistics Database. "Trade Profile Japan." http://stat.wto.org/CountryProfile/WSDBCountryPFView.aspx?Language=E&Country=JP (Accessed June 2010).

Junk Mail

Junk mail, sometimes referred to as direct mail or advertising mail, is unsolicited mail that is sent to people through the postal system. Junk mail may include letters, catalogs, and flyers from companies; credit card applications from banks; CDs and other forms of merchandising materials; and correspondence from politicians, candidates, and other organizations. In the United States, junk mail has been a facet of the postal system since the 19th century, but its popularity rose dramatically in the late 20th century, in part due to bulk mail permits.

Junk mail accounts for 100 billion pieces of U.S. mail per year, or approximately 848 pieces per household, weighing 41 pounds per year. U.S. junk mail accounts for 30 percent of all the mail in the world. The U.S. Postal Service estimates that advertising mail went from 35 billion pieces of mail sent by companies in 1980 to 64 billion pieces in 1990, to 90 billion pieces in 2000. Junk mail is received less than half as much by families making $35,000 or less per year than families making $100,000 or more. The rate of receiving junk mail also increases as the number of adults in the household increases. An additional factor is education.

Waste and Disposal

An estimated 44 percent of all junk mail goes to landfills unopened, and the estimated response rate to junk mail is less than 2–3 percent. Credit card companies report that a response rate of 0.25 percent is acceptable. Because of the acceptability of such a low response rate, environmentalists point out the immensity of the impact of junk mail. Some studies report upward of 100 million trees to produce all of the junk mail in the United States in a given year. Compounding this, the greenhouse gas effect of all of this junk mail is over 51 million metric tons of greenhouse gases. According to the Environmental Protection Agency, less than 50 percent of the junk mail received is actually recycled.

Direct Mail

A new phenomenon is the emergence of direct mail. Through the use of databases, direct mail is customized and individualized and sent to specific consumers. Direct mail may involve products beyond the paper that is commonly used in junk mail. Plastic bags, catalogs, mailings that resemble paperback books, and even small boxes are examples. The potential environmental impacts of direct mail are large considering the volume of mail that is sent to consumers and the types of products used for the mailings. Nineteen billion consumer catalogs, including those printed on glossy paper, are sent on average to U.S. consumers. The environmental effect of catalogs printed on materials that are less easy to recycle is substantial.

Reduction Efforts

Consumers have called on politicians and advocacy groups to curb the expansion of junk mail. The city of San Francisco called on the state of California to create a statewide Do Not Mail Registry; other states, including Florida and New York, have brought forward similar legislation. In the United States, consumers are often unable to opt out of mass junk mailings because of the postal service's insistence that it is providing a service for companies that pay to send their messages to consumers. Thus, the economic motivations of capitalism can be attributed to the challenges faced by consumers and advocacy groups that wish to curtail the junk mail phenomenon.

A small number of consumers use techniques of resistance and transgression, including sending back reply cards with false information, as a response to the problem of junk mail. On the Internet, a number of free and pay-for-service organizations offer opt-out options for consumers who wish to limit the amount of junk mail that they receive. A number of these organizations have raised public awareness about the environmental impacts of junk mail, to which other organizations, including those representing direct mail companies, argue that they have minimized their environmental impacts by printing many of their mailings on recycled paper.

Internet Spam

Junk mail is also a growing phenomenon on the Internet. Although this form of mail is sent through the electronic means of the Internet, environmental impacts do exist. A study by McAfee Avert Labs concluded that 62 trillion spam messages are sent each year, with each e-mail associated with 0.3 grams of carbon dioxide released as greenhouse gas. The combined spam emissions of all e-mail users around the world accounts for 17 million metric tons of carbon dioxide per year, or 0.2 percent of total greenhouse gas emissions. The total energy used to transmit, process, and filter spam is 33 terawatt hours, or the amount of electricity used in 2.4 million homes. Much like junk mail sent through the postal system, those who receive spam messages report negative effects of annoyance, decreased work productivity, and in some cases, crime and identity theft.

Popular Culture

Junk mail has also impacted public popular culture. The television comedy show *Saturday Night Live* also once featured a parody of direct mail advertisers. In 2007, artists Barbara Hashimoto and Nancy Spiller produced the artistic piece *Reverse Trash Streams: The Junk Mail Project*. The piece was shown at the LA Contemporary gallery. The artists produced the piece in response to concerns related to junk mail, including the impact on global warming, deforestation, and injuries to postal carriers. The piece featured hundreds of pounds of shredded junk mail that the artists arranged in a heap at the rear of a gallery.

Scott Lukas
Lake Tahoe Community College

See Also: Consumerism; Material Culture Today; Paper and Landfills; Paper Products.

Further Readings

Burke, R. "Barbara Hashimoto Gets All Trashy. Reverse Trash Streams: The Junk Mail Project." *Staticmultimedia.com* (November 5, 2007).

ForestEthics. "Junk Mail's Impact on Global Warming." http://forestethics.org (Accessed July 2010).

Henkin, D. *The Postal Age: The Emergence of Modern Communications in Nineteenth-Century America*. Chicago: University of Chicago Press, 2006.

McAfee. "The Carbon Footprint of Email Spam Report." Santa Clara, California, 2009. http://newsroom.mcafee.com/images/10039/carbonfootprint2009.pdf (Accessed July 2010).

Reynolds, G. and C. Alferoff. "Junk Mail and Consumer Freedom; Resistance, Transgression and Reward in the Panoptic Gaze." In *Consuming Cultures: Power and Resistance,* J. Hearn and S. Roseneil, eds. Basingstoke, UK: MacMillan, 1999.

U.S. Postal Service. "The Household Diary Study: Mail Use and Attitudes in FY 2008." Austin, TX: NuStats, 2008.

Junkyard

The junkyard is an iconic symbol of waste in industrial life, and it is one of the more complex spaces devoted to post-consumer materials. Instead of simply being an end-stage sink for materials like a landfill, it offers the potential for returning materials to production through recycling. At any one time, a yard may host materials at the end of their lives or near the beginning. Messy and complex, even the term *junkyard* is contested, belying the battles for identity and respect that these businesses have fought for decades.

History of Junkyards

Yards with scrap materials such as copper, iron, or silver have existed as long as markets for those metals. In the American colonies, Paul Revere kept a yard of scrap metal for his work as a metalsmith. The number of such yards, and the name *junkyard* increased with the Industrial Revolution. Shipbuilders and railroads kept salvage operations on their own sites, and independent yards containing rope, rags, metals, and any other materials desired by industrial producers became common by the end of the 19th century, containing the detritus of farm machinery, household tools and appliances, scrapped vehicles, and retired industrial machinery. A consumption ethic based on style rather than functionality spurred disposal of durable goods in the late 19th and early 20th centuries, increasing the volume and variety of objects found in junkyards.

In the United States, thousands of junkyards—most found in industrialized areas of growing cities—welcomed millions of mass-produced metal goods that could be resold as secondhand appliances or processed as scrap. Many yard owners started as junk peddlers. In the United States, a common path to ownership during the period of growth between the Civil War and World War I was to emigrate from Europe, become a peddler, and with the money raised, rent or purchase land to create a junkyard. Demand for secondary materials fueled the growth of junkyards, especially for materials containing iron and steel. The number of yards listed in city directories in Detroit, New York, Philadelphia, and Boston almost doubled between 1880 and 1921; in Chicago, the total listed in the annual *Lakeside Directories* increased from 140 in 1890 to 471 in 1917.

The trade in junk became an organized industry during this period, as yard owners created trade publications such as the *Waste Trade Journal* and professional associations such as the National Association of Waste Material Dealers (NAWMD) and the Institute for Scrap Iron and Steel (ISIS) in the first three decades of the 20th century. These institutions reflected the pervasiveness of yards in modern life and their role in both housing scrapped materials and reintroducing them to industrial production. In the 1920s, the automobile graveyard became a new, specialized junkyard, where customers could purchase obsolete automobiles for scrap or purchase individual parts off junked automobiles in order to repair other automobiles.

Technology

Junkyard technology changed over time. In the early 20th century, workers sorted and processed materials by hand or with small shears or acetylene torches. By the late 1940s, common technology included conveyors, dust-collecting systems, forklifts, hand trucks, rag-cutting machinery, rag shredders, shears of various sizes, torches to reduce items into manageable sizes, balers to turn scrap into symmetrical cubes, cranes and magnets to move heavy materials, and a variety of other tools to process materials. With the mass disposal of the automobile in the 1950s, several yards adopted specialized automobile shredders that disassembled a car within minutes, allowing yard owners to harvest No. 1–grade ferrous scrap. The torches and shears used in yards were too slow to profitably separate steel from the rest of an automobile, but by 1960, machinery designed to quickly harvest scrap metal from automobiles evolved to hammer and shred automobiles, using magnets to separate the ferrous scrap from other materials. The automobile shredder became a staple of the junkyard by the late 1960s, as its ability to hammer down an automobile and quickly separate light iron and steel from the many other materials found in automobiles gave operators the ability to process a massive source of consumer-generated scrap. Larger yards adopted other devices such as balers and cranes to process and transport materials. Smaller yards focused on securing goods. Among the tools to provide security were barbed wire, electrical fences, and intimidating junkyard dogs.

Social Conceptions

Despite the role junkyards play in returning disused materials to industrial production, cultural and social perceptions of junkyards have been consistently negative, associating them with disease, dirt, dangerous dogs, and blight. Progressive reformers at the turn of the century saw junkyards as blights upon urban neighborhoods; advocates such as Jane Addams and Jacob Riis warned of the dangers of junkyards to children's moral and physical health. Police worried about the difficulty of finding stolen goods in yards. Due to these concerns, several cities established zoning regulations after World War I that barred junkyards from residential areas. In fact, yards often moved to areas where police did not enforce zoning regulations. By 1956, geographer Gerald Gutenschwager found that most of Chicago's scrap iron and steelyards were located in majority African American residential neighborhoods, revealing environmental inequalities based upon race decades before the term *environmental justice* was first used.

Regulation

State and federal governments began to take an interest in regulating junkyards in the late 1950s.

As consumers in the late 19th and early 20th century shifted their focus from the functionality of household goods to the style and degree of modernity, the volume and variety of discarded durable goods found in junkyards increased. The junk trade became an organized industry, complete with trade publications and professional associations. Over time, technological advances in sorting and processing increased the usability and value of scrap. As such, the industry sought to redefine its image and promoted titles such as "salvage yard."

The Highway Beautification Act of 1965, however, was the first federal law regulating the scrap industry on aesthetic grounds. It sought, according to President Lyndon B. Johnson, to eliminate or screen unsightly, beauty-destroying junkyards and auto graveyards along U.S. highways. During the battle over the legislation, representatives of the scrap industry complained about the name *junkyard*, insisting that the materials they contained were valuable secondary materials and not worthless junk. Industry leaders promoted alternate names like *salvage yard* and *scrapyard*, with limited success. After the act passed, the Department of Transportation estimated that over 3,300 illegal junkyards were removed or screened from view by 1979.

Federal regulation of junkyards in the late 20th century focused more on environmental hazards than aesthetic blight. Under the Comprehensive Environmental Response, Compensation, and Liability Act (CERCLA), better known as Superfund, yard owners could be held liable for any hazards caused by the materials they contained. This was true of hazards both brought to the yards, such as rusting cans of petroleum, or hazards produced by the yards. Yards that shred the bodies of junked automobiles, computers, refrigerators, or other complex machinery produce residue that might contain a variety of carcinogenic, flammable, corrosive, or otherwise hazardous materials. Junkyards are at once areas containing both valuable recyclables and hazardous wastes, revealing the dangers of production techniques that do not include designs to promote full recycling of products.

Despite regulatory attempts to limit its presence, the junkyard remains a common and iconic presence on the landscape as yards containing automobiles, appliances, industrial machinery, and other scrapped artifacts remain common in industrialized society. They will remain as long as society designs and purchases disposable commodities.

Carl A. Zimring
Roosevelt University

See Also: Automobiles; History of Consumption and Waste, U.S., 1900–1950; History of Consumption and Waste, U.S., 1950–Present; Iron; Recycling.

Further Readings

Gutenschwager, Gerald A. "The Scrap Iron and Steel Industry in Metropolitan Chicago." Ph.D. diss. University of Chicago. Chicago: University of Chicago Press, 1957.

Strasser, Susan. *Waste and Want: A Social History of Trash*. New York: Metropolitan Books, 1999.

Zimring, Carl A. *Cash for Your Trash: Scrap Recycling in America*. New Brunswick, NJ: Rutgers University Press, 2005.

Kansas

Named after the Kansas River and, ultimately, the Kansa Native American tribe, the U.S. state of Kansas is located in the midwestern region. Until the first European settlement in the 1830s, the state's inhabitants were a variety of Native American tribes, settled agriculturists in the east, and seminomadic hunter-gatherers in the west. The state emerged from a chaotic period of political wars over the slavery issue in 1861 to join the Union and grew rapidly after the Civil War as immigrants turned the prairie into farmland. The state remains one of the most agriculturally productive in the 21st century.

Statistics and Rankings

The 16th Nationwide Survey of MSW Management in the United States found that, in 2006, Kansas had an estimated 4,089,591 tons of municipal solid waste (MSW) generation, placing it 30th in a survey of the 50 states and the capital district. Based on the 2006 population of 2,755,817, an estimated 1.48 tons of MSW were generated per person per year (ranking 12th). Kansas landfilled 3,271,773 tons (ranking 24th) in the state's 52 landfills. The state exported 140,939 tons of MSW, and the import tonnage was 770,650 tons. Kansas was ranked eighth out of 44 respondent states for number of landfills, but it had no plans to increase its landfill capacity and no waste-to-energy (WTE) facilities. Only whole tires were reported as being banned from Kansas landfills. It recycled 817,818 tons of MSW, placing Kansas 26th in the ranking of recycled MSW tonnage.

The historic disposal methods of pig-feeding and open burning ceased between the 1950s and 1970s as U.S. life became more urban and trash became an increasing problem in Kansas as well as elsewhere. The first statewide solid waste regulations were not passed until 1970 in Kansas, although the same is true for many other states. The nationwide garbage crisis was also felt in Kansas with a 1969 *US News & World Report* article describing 770 paper cups, 730 cigarette cartons, 590 beer cans, 100 whiskey bottles, and 90 beer cartons being found dumped on one mile of a two-lane Kansas highway.

Quindaro

Situated on the south bank of the Missouri River in Wyandotte County in what is now Kansas City, Quindaro was a short-lived (1856–62) community established after the Kansas-Nebraska Act to create a free-state port of entry into Kansas. Originally

Wyandot Indian land, the town's founders were former members of the tribe who chose to remain in the area and become U.S. citizens when the tribe disbanded. The town was named after one founder's wife, Nancy Quindaro Brown Guthrie, whose husband was a European American adopted into the tribe on marriage.

Close to the Missouri River, the town was an active part of the Underground Railroad, helping slaves escape to freedom. A boomtown with a population that peaked at 600, the town's fortunes declined when a nationwide economic depression hit and efforts to attract a railway line to the town failed. Much of the male population left to fight in the U.S. Civil War, and those who remained continued farming while the town drifted into abandonment. Freed African American slaves settled in the area after the war and buildings of Freedman University (later Western University) occupied the bluff over the site of the old town. These buildings would also become ruined within a century.

In the late 1980s, a projected landfill on the site was countered by the Kansas Antiquities Commission Act. As Kansas City owned part of the site and was a permitting authority, an archaeological evaluation had to be conducted. Two years of archaeological work uncovered a cistern, three wells, and the remains of 22 buildings. The level of archaeological remains raised a public outcry, which halted the landfill development, but this had the effect of removing the funding source for the site archive's storage and analysis. An agreement was eventually reached, which transferred the archive to the Kansas Historical Society.

Bill Compton and Kansas Pyrolysis

During the late 1960s and early 1970s, the U.S. government funded various resource recovery technologies in the hope that the crises of garbage accumulation and oil shortage could be alleviated by WTE facilities. Pyrolysis (which cooks trash, rather than incinerating it, to release oil, carbon, and gas, was seen as the most promising method of disposal, but it was also the most expensive, technologically complex, and heavily funded method. Even with federal funding, the hope of creating a pyrolysis plant remained beyond the financial and intellectual capabilities of most communities. Many of the resource recovery programs established by the 1970 Resource Recovery Act and the 1976 Resource Recovery and Conservation Act failed, and federal research was dropped when the energy crisis threat waned in the early 1980s. Loss of government interest made it seem less likely that technology could combat the garbage accumulation problem.

Bill Compton, a former World War II fighter pilot from Wichita who used the GI Bill to obtain chemistry and physics degrees, designed, built, and operated his own small pyrolysis plant in his yard from 1992 to 1998. Running the plant for 350 hours converted 700 pounds of trash into usable oil, carbon, and gas. Compton and his supporters constantly petitioned the government to consider building full-scale pyrolysis plants, without success. Compton's supporters place the blame for their initiative's continual knockbacks on bureaucracy, politics, apathy toward resource recovery, and bad publicity from the Baltimore Pyrolysis Plant affair. Compton's supporters continue their campaign into the 21st century after Bill Compton retired due to advanced age and health problems.

Project Salt Vault

Another controversial waste management program vaunted for Kansas was a plan formulated in the early 1970s to use an abandoned salt mine in Lyons as a disposal site for high-level radioactive waste from the nuclear weapon program and commercial nuclear power plants. The Atomic Energy Commission (AEC) reacted swiftly while under pressure to establish the Lyons site's suitability and, in doing so, provoked a backlash from Kansas's scientists and politicians. The AEC sought to find a way to contain the most hazardous form of nuclear waste, the highly radioactive liquids that were by-products of reprocessing uranium fuel to recover plutonium. It was suggested that salt formations were the best geological siting for radioactive liquid waste, being dry, impervious, and having a plasticity that sealed itself around fractures or anything deposited in it. Salt formations were also abundant, found in areas of low earthquake risk, and cheap to mine. The depth, size, and thickness of the central Kansas salt deposits were thought ideal.

The AEC began preliminary experiments in a disused section of the Carey Salt Company mine in Hutchinson. Between 1959 and 1961, scientists injected nonradioactive liquids into cavities bored into the mine floor to simulate the heat produced by nuclear liquid waste. The results of these tests were inconclusive but deemed encouraging. A second battery of tests was announced in 1963, which were to be conducted between 1965 and 1968 at another Carey salt mine in Lyons. Called Project Salt Vault, this test used solid radioactive waste as the test substance. In 1970, the AEC was ready to move forward with construction, but ambivalence toward the project became a sensitive political issue in the upcoming gubernatorial election. As the debate simmered, relations soured, and political support for the project was lost as Kansas' officials tired of the political blundering by the AEC. The incident generated bad press nationwide and made finding the solution to the nuclear waste problem an even more arduous problem for the AEC. A major stumbling block had been the AEC's eagerness to press onward while overlooking serious concerns of the Kansas Geological Survey and American Salt Corporation about the risks posed by earlier workings nearby.

Jon Welsh
AAG Archaeology

See Also: Atomic Energy Commission; High-Level Waste Disposal; Radioactive Waste Disposal.

Further Readings

Arsova, Ljupka, Rob van Haaren, Nora Goldstein, Scott M. Kaufman, and Nickolas J. Themelis. "The State of Garbage in America: 16th Nationwide Survey of MSW Management in the U.S." *BioCycle*, v.49/12 (2008).

Gumm, Angie. "Bill Compton Builds Wichita a Pyrolysis Plant." *Kansas History: A Journal of the Central Plains*, v.31 (2008).

Rome, Adam. *The Bulldozer in the Countryside: Suburban Sprawl and the Rise of American Environmentalism*. New York: Cambridge University Press, 2001.

Schmits, Larry J. "Quindaro: Kansas Territorial Free-State Port on the Missouri River." *Missouri Archaeologist*, v.49 (1988).

Walker, J. Samuel. "The Controversy Over the Lyons Radioactive Waste Repository, 1970–1972." *Kansas History: A Journal of the Central Plains*, v.27 (Winter 2006–07).

Karachi, Pakistan

As one of the world's largest and rapidly growing cities with a population exceeding 18 million in 2007, Karachi has high levels of consumption and waste. Given the financial constraints typical for a developing country as well as limited systematization and control, the waste management system is inefficient, with at least half of the waste remaining uncollected and untreated. As is the case with most of the urban waste generated in Pakistan, a large part of the garbage is sorted and recycled by street scavengers and trash pickers. This often amounts to improper treatment of waste, which increases the city's pollution levels. Landfills and open dumping remain the primary means of disposing all kinds of solid waste, exacerbating the issue. Water is not only contaminated by the leachate from landfills but also by the open discharge of industrial waste in drains, canals, rivers, and the sea. For a rapidly and somewhat haphazardly growing city like Karachi, where shantytowns appear intermittently, there is the additional problem of increasing pollution encroaching on residential areas and further endangering the health of inhabitants.

Known since the times of Alexander the Great, Karachi is named, according to the most popular legend, after Mai Kolachi, one of the earliest settlers in the fishing village, who established a matriarchal regime over the area. In the 21st century, the city sprawls over 2,192 square miles and contains Pakistan's two main ports. Serving as Pakistan's capital from the country's independence till 1960 and currently the provincial capital of Sindh, Karachi is the most developed region in the country, attracting immigrants from rural areas and abroad—particularly Afghanistan since the Taliban occupation of Kabul in the mid-1990s.

Consumption

In spite of the difficult economic conditions and the low standard of living, with the majority of the population situated below the poverty line, the transition toward a consumer culture is evident in the densely populated urban region of Karachi. Indicating the country's lax import policy, extensive varieties of foreign goods are easily available in stores throughout the city, ranging from cosmetics and clothes to food items. Franchises of many multinational fast food chains crop up frequently and, as a result, Karachi had seven McDonald's restaurants as of 2010 and an even larger number of Pizza Hut branches. The kinds and varieties of products marketed have affected eating and living habits and, ultimately, the kind of waste generated. Subsequently, a general shift from organic to inorganic substances is discernible in garbage composition, combined with an overall rise in packaging material, particularly plastics. In 2001, around 40 percent of the waste was biodegradable, roughly 10 percent was paper, and the amounts of metal and glass as well as textiles and leather were even less.

Waste Management

In accordance with its size, Karachi produces more than 10 times the waste generated in the Punjabi cities of Islamabad and Multan combined, and this amount has an annual growth rate of 2.4 percent. Varying between 6,000 and 7,000 tons daily per person in 2001, the city's waste production is almost double that of New Delhi and is among the highest in south Asia. From this, according to the country's environmental department, less than half succeeds in reaching a landfill site, even though more than 1 billion Pakistani rupees are spent annually by the city on solid waste management.

Karachi's waste collection is divided among the municipal councils of its five districts for the south, east, west, and center of the city, as well as Malir in the north. From these, the districts of southern and central Karachi, which encompass Karachi's most commercial and affluent regions, generate the largest amount of waste. The district municipal councils are responsible for transporting the waste to one of the two landfill sites located within the northern limits of the city. These sites are operated by the Solid Waste Management Department of the Karachi Metropolitan Corporation (KMC) and have the collective capacity of around 2,000 tons of waste per day for two decades, consequently lacking the capacity for absorbing all of the city's waste.

As Pakistan's main port city, Karachi's waste management also involves tasks specific to harbors and beaches, such as collecting and disposing of seashells on the beach. The Karachi Port Trust and the Port Qasim authorities are involved in the solid waste management of their respective ports. Similarly, industrial areas, including Pakistan Steel Mills, Sindh Industrial Trading Estate, and the Export Processing Zone, have almost autonomous control over the management of their waste. The newest port, Port Qasim, generates approximately six tons of waste per day, which is the same as the amount produced by Karachi's Jinnah International Airport. The pre-independence Ports Act from 1908 still governs the discharge of waste at the ports. While oil is frequently discharged from the ports into the sea, a considerable proportion of industrial waste from the Export Processing Zone as well as other industrial areas on the coast, like Korangi and Landhi, is still dumped in the sea. Since Landhi contains the Bhains or Landhi Cattle Colony, one of the five cattle colonies in Karachi and the largest buffalo cattle colony in the world, the majority of the waste emptied into the Arabian Sea is animal dung. In 2007, a waste-to-energy plant was opened in Landhi, funded by New Zealand Aid, with the original initiative taken by the International Union for the Conservation of Nature (IUCN). Converting the waste into biogas provides up to 30 megawatts of energy and simultaneously produces 14,000 tons of organic fertilizer.

Medical Waste

The city has more than 200 hospitals, which is the highest concentration of hospitals in the country; in contrast, the entire province of Punjab contains approximately 250 hospitals. In addition, Karachi houses about 1,400 healthcare units (HCUs), which include establishments like laboratories or small clinics. These generate more than 3,000 kilograms of waste per day, of which at least one-sixth is hazardous. Owing to limited concern and awareness, as well as the absence of strict rule enforcement by the Sindh Environment Protection Agency, less than one-tenth of the HCUs send their waste to the incineration plants set up, like the landfill sites, by the Karachi City district government. The incineration plants were only installed in 1996, and both of

Karachi, Pakistan, is one of the world's largest and rapidly growing cities, with a population over 18 million in 2007. As such, it has high levels of consumption and waste. In the shantytowns that appear haphazardly within the city (left), street scavengers and trash pickers (right) sort through trash, leading to improper treatment of waste and an increase in the city's pollution levels. Solid waste is primarily disposed of via landfills and open dumping, from which tainted water finds its way into drains, canals, rivers, and the sea.

them are located in Mewa Shah in the west. Thus, most of the hospital waste is mixed with municipal waste. Given that dumping is usually open and frequented by trash pickers and stray animals, which can also include cattle, the health risks of such disposal practices are extensive.

Organic Waste

Two of the city's largest markets, namely, the Vegetable Market and the Empress Market, generate 70 and 100 tons of waste per day, respectively, a considerable proportion of which is biodegradable. The possibility of composting such waste has triggered several initiatives. On the other hand, Karachi's high temperatures also reduce the amount of moisture in green waste that is necessary for composting. Given the scarcity of water in the city, the reduced moisture can also pose problems. Karachi's sole composting plant, the Farooq Compost Fertilizer plant, was set up during the 1980s in the northern part of the city. Because of its inefficiency and the high costs procured, the plant was eventually shut down. In 2010, one of the prominent nongovernmental organizations (NGOs) in Karachi that aimed to collect and reuse organic waste was Gul Bahao. Also functioning as a research institute, Gul Bahao buys organic waste and other items that cannot be sold to *kabaris* (the local waste dealers). This waste has been transformed into a variety of items, including fertilizer, animal feed, possible fuel for power plants, and even low-cost houses (as part of the Chandi Ghar, or Silver House, project), which have been made out of plastic waste. In addition, the Pakistan Environment Welfare and Recycling Program (PEWARP) concentrates on collecting organic waste from the vegetable market and converts the vegetal solid waste into pesticide and liquid fertilizer.

The waste from high-income housing areas contains a large amount of food remains, which is rarely the case with low-income areas. Garbage from affluent residential areas also contains a larger proportion of garden waste, like dry leaves.

Organic waste from well-off areas can amount to almost 22 percent of total waste, whereas in poorer households it can be less than 9 percent.

Household Waste

Due to the lack of awareness and means, household waste is rarely sorted. Only one-tenth of domestic garbage—usually paper, plastic, and metal—is separated in households and sold for recycling to neighborhood dealers, who then sell it to recycling plants for processing. Since many of these plants are located in the neighboring province of Punjab in the north, much of Karachi's sorted waste is sent there. A lot of plastic material, for instance, is processed in Gujranwala, and it is usually converted into lower-quality plastic products. Only one-quarter of the recycling industries are located within the city's industrial zone, the majority of which make cardboard out of wastepaper. However, such factories are ill equipped and have the drawback of causing heavy pollution.

Waste from houses is usually brought to the nearest municipal garbage containers, which are known as *katchra kundi.* Since these are almost always open, they attract trash pickers and scavengers like crows or stray dogs. While litter around the container is common, and usually the result of careless throwing, some containers overflow since they are not emptied regularly. Both trucks as well as bull or donkey carts are used for collecting the waste, none of which are covered, causing some of the garbage to fall out during the transportation process.

Scavenging

Trash pickers, who frequently live near the waste disposal sites or along the garbage transportation route, scour the garbage for metal, recyclable plastic, and glass, which is then sold for further recycling. To separate metals from other components of the waste, the trash pickers often burn the garbage, thereby polluting the air. While the total number of trash pickers in Karachi has not been calculated, approximately 100,000 are Afghans. Furthermore, a large number of children are also involved in trash picking, and it is estimated that in Karachi alone, more than 20,000 children work as trash pickers. Street sweepers are also involved in waste separation but, in contrast to the trash pickers on the landfills, they usually separate and sell paper and cardboard collected from the streets and garbage containers.

Although proposals have been made to incorporate trash pickers into the waste management system—most prominently by the Karachi-based NGO Shehri—these were yet to be implemented as of 2010. However, the success of private initiatives shows that with more widespread motivation and support, the waste management issue can be resolved in such a way as to contribute to ameliorating some of Pakistan's economic and energy issues.

Maaheen Ahmed
Jacobs University

See Also: Composting; Developing Countries; Medical Waste; Open Dump; Organic Waste; Pakistan; Street Scavenging and Trash Picking.

Further Readings

Akhtar, W., I. Ali, S. Jilani, and S. S. H. Zaidi. "The State of Pollution Levels of Karachi Harbour and Adjoining Coastal Water." *Water, Air, & Soil Pollution*, v.94 (1997).

Hakeem, R., A. H. Shaikh, and M. Ziaee. "Socio-Economic Differences in Frequency of Food Consumption and Dietary Trends in Urban Areas of Karachi, Pakistan." *Pakistan Journal of Biological Sciences*, v.7 (2004).

Hasan, A. *Understanding Karachi: Planning and Reform for the Future.* Karachi, Pakistan: City Press, 2002.

Visvanathan, C. and U. Glawe. "Domestic Solid Waste Management in South Asian Countries—A Comparative Analysis." http://www.faculty.ait.ac.th/visu/Prof%20Visu%27s%20CV/Conference/25/3R-MSWM.%20Visu.pdf (Accessed August 2010).

Kenilworth Dump

From the time it opened in 1942, Kenilworth Dump (KD) served as the waste management center for residential and commercial waste generated within the Washington, D.C., metropolitan area. In 1968, following an incident that resulted in the death of a young boy, KD was effectively shut down, and in the early 1970s, it became a recreational area and continues as such into the 21st century. When considering the social process of consumption and waste, KD is an important discussion point because its entire history epitomizes the complexities involved in the seemingly simple practice of throwing something away.

For 26 years, the land along the eastern shoreline of the Anacostia River in northern Washington, D.C., was used for the disposal and burning of municipal waste. At the time of its construction, no physical barriers had been included to protect the river from the waste products that would migrate from KD and subsequently contaminate both the soil and water. Furthermore, by the time the Resource Conservation and Recovery Act (RCRA) was signed into law in 1976, the landfill was officially closed; KD operated without a permit for its entire existence.

Operations and Problems

The primary feature that negatively affected KD was its operation as an open burning dump site. An open dump is an uncovered space that serves as a repository for refuse, with all waste material exposed. As an open-burning site, KD created two main problems: sanitation and air pollution. Without capping a layer of soil atop the rubbish piles or using a lining system to act as a barrier between the refuse and ground—both of which are consistent with sanitary landfills—KD became a breeding ground for vermin and other pests. Additionally, the uncontrolled incineration of waste material released vast amounts of contaminated ash into the air. By 1967, roughly 80 percent of all the refuse produced in the Washington metropolitan area was being burned, while the remaining 20 percent was buried in landfills. Of approximately 1.5 million tons of refuse produced each year, an estimated 1.2 million tons were burned; and of this amount, roughly 160,000 tons were burned in open dumps, most of it at KD.

Residents had been complaining about KD for several years when the U.S. Surgeon General convened a conference on solid waste management in 1967 to address the growing problems at KD. One important conclusion drawn from the conference was that processes for dealing with safe and sanitary solid waste disposal were mainly political and economic in nature, rather than technology based. In other words, the inability to develop technology that could safely dispose of waste in a permanent manner was not the primary reason waste management practices were fraught with issues and challenges. Rather, it had to do with the lack of—and thus, essential need for—jurisdictional cooperation. For example, by 1967, Washington, D.C., disposal systems were overwhelmed by increasing amounts of solid waste—a result of both the growing population and the single-use disposability mentality that gained a stronghold directly following World War II—so simply shutting down KD without a replacement site would have been extremely difficult. Herein lies the political problem: for KD, jurisdictional cooperation was predicated on the understanding that regardless of whether one experiences the smoke from KD directly or not, everyone in the Washington community was involved because everyone breathed the same air, drank the same water, and, most importantly, everyone contributed to the accumulation of waste, most of which was burned at KD. However, deciding which existing site would take on the waste, how the materials would be transported (what routes it would be allowed to follow and through which communities it could travel), and other costs created a challenging atmosphere in which to make a decision.

Closure

Consensus held that KD needed to be closed, but translating that into reality was not so simple. To help solve the problem, the short-term solution of reincarnating KD from an open-burning dump site into a sanitary landfill was suggested and approved. However, because of various delays in the process, open burning continued past the set deadline of January 1, 1968. Then, in February of that same year, a young boy was playing in KD (KD was often referred to as Landfill Mall because it was commonplace for local residents to search through the piles of waste in search of items they could find useful) when a change in the wind pattern abruptly altered the course of a fire that had been burning waste material that afternoon. Unable to escape the flames, the boy was killed. As a result of this incident, Mayor Walter E. Washington agreed to shut down KD. By 1970, the new sanitary landfill was nearly filled and was closed and capped with clay and topsoil. At this time, the landfill, which was owned by the U.S. government and administered by the National Park Service (NPS), comprised about 145 acres and contained roughly 4 million tons of raw refuse, incinerator ash, and other burned residue.

Continuing Contamination

In the 21st century, the KD landfill site is known as the Kenilworth-Parkside Recreation Center and features a gymnasium, swimming pool, soccer and baseball fields, and basketball courts. However, despite the goal of turning KD from something negative into something positive, the land space continues to suffer environmental damage. By May 1980, KD had successfully been turned into a park. However, in 1997, NPS permitted two companies, Driggs Corporation and Barrett Tucker Corporation, to deposit on the Kenilworth park space excavation and construction debris from their work sites. In 1998, when it had been determined that neither company had an official permit for such activities, the Environmental Health Division of the district's Department of Health issued a Notice of Violation to the NPS. A year later, additional concerns arose that the companies had dumped asbestos, polychlorinated biphenyls (PCBs), and other hazardous waste in violation of their contract. The Environmental Protection Agency, under the Comprehensive Environmental Response, Compensation, and Liability Act (CERCLA, Superfund), subsequently performed inspection tests, but preliminary sampling found no presence of PCBs. Despite the Superfund findings, however, issues related to remediation of contaminants in the underlying dump and questions of soil quality throughout the park continue to exist into the 21st century and have slowed the development of proposed recreational field projects.

By serving as evidential proof that open burning systems are both unproductive and unsafe, KD made an important contribution to waste management practices in the United States by catalyzing a movement away from utilizing open burning facilities. On the surface, waste seems a self-evident truth: when an item is no longer useful, it becomes waste and it is thrown away. While this is a simple process in theory, KD demonstrates the complex nature of simply throwing something away.

Rebecca Estrada
University of Manchester

See Also: Comprehensive Environmental Response, Compensation, and Liability Act (CERCLA/Superfund); District of Columbia; Incinerator Waste; Landfills, Modern; Solid Waste Disposal Act.

Further Readings

Blumberg L. and R. Gottlieb. *War on Waste: Can America Win Its Battle With Garbage?* Washington, DC: Island Press, 1989.
Rogers, H. *Gone Tomorrow: The Hidden Life of Garbage*. New York: New Press, 2005.
Strasser, S. *Waste And Want: A Social History of Trash*. New York: Metropolitan Books, 1999.

Kentucky

Bordered to the north by the Ohio River, Kentucky is a state with natural amenities ranging from the Mississippi River to the west and the Appalachian Mountains running through most of the eastern half of the state. In 2010, 4,339,367 residents lived in Kentucky, with the bulk of the population clustering in the metropolitan areas on the shores of the Ohio River. Louisville is the largest city, and Kentucky is also home to Cincinnati, Ohio's southern suburbs. Kentucky's consumption and waste issues vary widely depending upon the region of the state. With such diversity, be it in Louisville or Lexington, at the University of Kentucky, or in the southeastern Appalachian coalfields, the issues facing this state vary largely depending upon location. While cities like Lexington and Louisville are making headway in recycling programs, the rural areas with industry continue to have major problems because of industrial waste. Consumption and waste problems in Kentucky are closely related to the coal mining industry. With large-scale coal mining in the southeastern part of the state, Kentucky has easy access to abundant local coal. This has led to low coal prices, which in turn has resulted in cheap electricity for Kentucky residents from coal-fired power plants.

Kentucky is one of the most coal-dependent states in the United States. Kentuckians spend the same percentage of their income on electricity as residents of any other U.S. state, even though they are paying a lower rate. This means that Kentucky residents use more electricity per person than residents from other states. The cheap access encourages irresponsible energy use. Other than coal-fired power plants, the remaining electricity generation within the state is mostly provided by petroleum-fired and hydroelectric power plants. Kentucky is one of just a handful of states remaining that uses no nuclear power. With their resource-laden mountains, Kentuckians use a lot of energy with little cost and care, leading them to confront significant consumption and waste issues.

Energy From Coal

After Wyoming and West Virginia, Kentucky is the third-largest producer of coal in the United States. Kentucky is also sixth in the nation in coal power generation, with 56 operating coal-fired power plants in 21 stations across the state. In total, the state produces 16,510 megawatts (MW) of energy from coal-fired power plants.

Energy produced at coal-fired power plants is dirty business. Contributing significantly to air and water pollution, both locally and globally, Kentucky's production and consumption of coal has had destructive effects on both the environment and the health of local residents.

Coal-fired power plants are recognized as some of the heaviest-polluting forms of energy production, causing major pollution of air quality. When coal is burned for energy, a harmful pollutant gas, sulfur dioxide, is released into the atmosphere. Coal naturally contains sulfur, but when the coal is burned, the sulfur combines with oxygen to form harmful emissions. These emissions far exceed safe or natural levels of production and contribute to significant health and environmental problems in Kentucky, as well as to the growing crisis concerning global warming.

Coal in eastern Kentucky is widely considered to contain less sulfur than coal from the western part of the state. This means that the coal from eastern Appalachia is increasingly sought after, as regulations begin to further restrict the amount of pollutants that can be legally released when coal is burned. While less environmentally damaging when burned, all burning of coal emits harmful pollutants and contributes to local and global environmental problems.

Once coal is burned to make energy, large amounts of waste are left behind in the filters at coal-fired power plants. Increasing environmental standards, particularly the Clean Air Act, have regulated power plants to catch coal ash before it is released into the air during the coal burning and energy-producing process. This is beneficial to the public, as fewer pollutants are released into the communal air. However, this waste is now caught in filters at the power plants and needs to be disposed of. Known as coal ash, this waste contains extremely large amounts of dangerous toxins. In Kentucky, coal ash is generally stored near power plants or recycled for use in highway building. Some power plants store their coal ash in liquid form in man-made lagoons.

Surface and Underground Mining

Water contamination is one of the biggest waste problems associated with both strip coal mining and underground coal mining in Kentucky. While mountains naturally contain high levels of minerals, these minerals, like selenium, for example, only become harmful when the mountain's rock is detonated in order to mine coal. When the rock explodes, the contaminants and minerals are disturbed and exposed. As a result, runoff from the mining location can carry this toxin to the local water source. While there are regulations in place to try to prevent this sort of water contamination, regulations are not always followed, and dangerous contamination happens. Water contamination, in the 21st century, continues to adversely threaten the health of residents who use water affected by mining.

As coal is processed—from hard rock in a mountain to burnable fuel—more toxic waste is accrued. Some minerals are removed to increase the coal's power-burning potential. The removed mineral then becomes mining waste and is mixed with water. The resulting mix, known as coal slurry, is a cement-like paste. Slurry is primarily stored in man-made sites near mining sites. Also referred to as coal waste or coal sludge, slurry is filled with dangerous toxins, including mercury, lead, arsenic, copper, and chromium. Coal companies generally dispose of this waste by constructing dams from the solid mining refuse to store the liquid waste. Slurry impoundments are generally located in the valleys near coal processing plants and are tenuous at best. Life threatening disasters have occurred in Kentucky because of problems with coal slurry impoundments.

On October 11, 2000, 306 million gallons of liquid slurry came pouring out of an impoundment at the abandoned Massey Energy mine site in Martin County, Kentucky. The slurry polluted hundreds of miles of the Big Sandy and Ohio rivers. Over 27,000 residents were unable to use their water, and all aquatic life in the nearby waters was killed. Heavy metals were found in the water supply, and, according to the Environmental Protection Agency, the spill was 30 times larger than the *Exxon Valdez* oil spill. As the Martin County disaster makes clear, Kentucky's relationship with the coal mining industry and their waste has led to some devastating environmental problems.

Mining waste continues to affect the region long after an active mine site has closed down. The coal-producing regions of the United States are not only forced to deal with mining waste while the mine is active but also continue to face the long-term reality of living near toxic slurry impoundments and the destroyed mountains. There may not always be a disaster or waste spill to mark the issue, but the threats of water contamination and the long-term risks of large amounts of toxic chemicals are a concern for all Kentuckians. A form of surface mining, referred to often as mountaintop removal mining, is a popular and cheap form of coal mining in Kentucky in the 21st century.

The cheapest and fastest way to get at seams of coal, mountaintop removal mining literally removes the tops of mountains in order to access the coal. These large-scale mining operations create significant amounts of waste. As the mountaintops are removed, the blown-off rock is dumped into the valleys and headwaters of streams nearby. Known as valley fill by regulators, the former mountaintops become waste material. Now laden with harmful minerals that have become exposed during the explosion process, this rock is dumped in valleys next to the mountain where mining occurred. When blasting, the technique changes the chemical composition and the physical structure of the ore or mineral. This waste then resides at the heads of streams and disrupts the local water flow and source.

Consumption and Waste Management

Kentucky's per capita consumption of residential electricity is among the highest in the United States. In Kentucky, the cheap price of coal makes coal-fired power plants the most popular form of energy for electricity. The U.S. Energy Information Association (ETA) claims that coal accounts for more than 90 percent of the electricity produced in Kentucky. This makes Kentucky one of the most coal-dependent states in the United States. According to the ETA, the remaining electricity generation within Kentucky is mostly provided by petroleum-fired and hydroelectric power plants. Kentucky's per capita energy consumption is ranked seventh in the United States, making the 462 million BTUs of per capita energy consumption a large part of the country's dependency on fossil fuels.

Waste management in Kentucky is regulated by the Waste Management Department within the state's Department of Energy. The Waste Management Department is the largest department in the state. All counties across the state claim to offer universal collection of residential solid waste. Counties are given the freedom to contract with private waste management companies, allowing county governments to have leverage in the costs associated with pickup.

Individual energy consumption rates in Kentucky are high, and Kentuckians also produce a lot of garbage. This is in many ways linked to the mining industry. Kentucky's extremely cheap garbage disposal costs are kept low because of the 35 years of already-permitted landfill space awarded to the Waste Management Department. The availability of landfill space on former strip mines continues to keep prices low. Permitting for landfills at former strip mining sites is generally much less difficult than on previously undamaged land. The land has already been disturbed and deemed "worthless," and officials are unlikely to challenge permits.

Recycling efforts in Kentucky continue to grow. Increasing from 29.5 percent to 34.6 percent in 2008 for common household items, the government sees recycling as an important area for improvement. Many Kentucky counties have been receiving help from state grants to invest in the capital-intensive machinery required for recycling centers. Kentucky was above average in per capita household recycling rates as of 2010. The state is continuing to facilitate the further development of recycling infrastructure into the 21st century. Kentuckians must recognize that if the per ton cost of garbage continues to remain as low as it was in 2010, the recycling effort will remain ineffective.

Kentucky is a state confronting many of the contemporary evils associated with an economy and population heavily dependent on fossil fuels. As a state with diverse regions with very diverse needs, it is important to think about the way Kentucky can move forward in more sustainable ways. Because of low electricity rates in the state, Kentucky has encouraged energy-intensive practices by producers, consumers, industries, and residents. In order to reduce consumption of energy, preserve the undamaged landscape, and begin to restore the already environmentally devastated areas, the state government needs to take leadership. Regulations are changing, and Kentucky will continue to be at the center of this conversation because of its vast coal resources and irresponsible consumption and waste practices.

Stephanie Joy Friede
Columbia University

See Also: Clean Air Act; Clean Water Act; Coal Ash; Mineral Waste; Pollution, Water.

Further Readings

Kentucky Department for Environmental Protection. "Division of Waste Management." http://waste.ky.gov (Accessed March 2011).

U.S. Government Accountability Office. "Surface Coal Mining: Characteristics of Mining in Mountainous Areas of Kentucky and West Virginia." Report to Congressional Requesters, December 2009.

Kolkata, India

Kolkata (Calcutta) is the capital of the Indian state of West Bengal. Kolkata was once the wealthiest and most opulent city in India, the center of the British East India Company, and the second-largest city in the British Empire after London. In the 21st century, with some of the worst air pollution in the world, Kolkata epitomizes the popular image of the third world city, ravaged by industrialization, overpopulation, large-scale poverty, agricultural stagnation, political instability, and weak environmental regulations. At the same time, the city is a bustling hub of revolutionary politics, literature, arts, and film. Kolkata has been an industrial center since its foundation, from jute production in the mid-1800s to 21st-century information technology, electronics, and petrochemicals.

History

Kolkata is situated on the east bank of the Hooghly River and serves as an important global seaport. Located on a vast wetland, the small Bengali village of Kalikata was occupied sequentially by French, Portuguese, and British traders. In 1690, Kolkata became a trade settlement for the British East India Company. By 1698, the East India Company bought Kalikata and two adjoining villages from a local landlord and began developing Calcutta. In 1772, Calcutta became the capital of British India, and the Crown relocated all administrative offices to the new city. In 1912, after several nationalist uprisings, the British colonial government moved its capital to Delhi, which is still the capital of postcolonial India. Calcutta was the center of revolutionary politics during the movement for Indian independence, and it continues to have active leftist and trade union movements. The city is also home to the longest-running, freely elected communist government in the world. In 2001, the city was renamed "Kolkata" to break with colonial heritage and better reflect Bengali pronunciation.

Once the center of British opulence and oppression, Kolkata still provides a sensory overload for visitors. Kolkata is often described in the pejorative: a corrupt, dying, and decaying city. Thousands of people live in the streets, fixing their tarpaulin ceilings to the iron gates of colonial-era villas. Popular literary representations of the city, including those of Claude Levi-Strauss, Dominique Lapierre, and Günter Grass, describe Kolkata as an urban disaster, filled with street-dwelling, emaciated bodies teetering at the edge of survival. Kolkatans attribute the popularization of this negative stereotype not only to Western authors but also to the work of Mother Teresa and her Missions of Charity, whose Nobel Peace Prize brought global attention to the destitution of Kolkata. Despite the protests of many of Kolkata's residents, contemporary press accounts of life in Kolkata draw Western audiences' attention to the poverty, chaos, and disease of the city.

In the mid-1800s, the areas surrounding Kolkata were described as jungle with a few intermittent villages. In 1855, the first Indian jute mill was founded in Kolkata, and by the early 1900s, Kolkata jute production surpassed European production. Since jute was used to make gunny sacks, burlap bags, and other packing materials, jute production in Kolkata expanded along with other British export enterprises such as opium and tea. The jute boom continued through World War I, after which additional industries such as tanneries, glassworks, ceramic factories, chemical plants, and textile mills began to dot Kolkata's periphery.

A hand-pulled rickshaw carries a customer through the streets in Dalhousie Square South, Kolkata. Many migrants have sought employment in Kolkata. While some have found industrial jobs, others struggle as rickshaw pullers, ragpickers, or street sweepers.

These factories were constructed to the north and south of the city, particularly along the Hooghly River. West Bengal's rural coalfields fueled Kolkata's growing industries.

After India's Partition in 1947, refugees from East Pakistan (now Bangladesh) fled to Kolkata and sought work in the factories of the city. According to official censuses, Kolkata's population grew from 4.4 million in 1961 to over 15 million by 2010; however, since many migrants from within West Bengal maintain ties to their ancestral rural villages, these population estimates neither adequately account for these migrant populations nor properly represent migrants who live in small informal settlements, know as *bustees*. Migrants from other economically depressed Indian states, particularly Bihar and Orissa, come to Kolkata in search of employment and are also not counted in official censuses.

Many argue that Kolkata's industrial areas have changed little since the British-led industrial boom of the late 19th century. Bustees in the industrial areas that surround the city have little ventilation, open drains, few toilets, and minimal municipal facilities. Continued political strife and the decline of traditional smallholder agriculture in rural West Bengal have driven people to the city to find food and work. While some find employment in the industrial sector, others work as rickshaw pullers, street scavengers, or informal "sweepers." Responding to the protests of wealthier Kolkatans who resent the influx of migrants, city officials have organized the demolition of bustees in or near the city center. As a result, many bustee residents often become pavement dwellers.

Waste Disposal

There is little formalized solid waste management in Kolkata. Waste is carted to open pits, or "dust bins," which also double as public toilets. For pavement dwellers and bustee residents, street scavenging in and around these bins has become a common way to make a living. Street scavenging also helps the municipality, which permits the practice because of the amount of waste scavengers remove. The same officials that drive bustee residents into the streets rely on them to help contain municipal waste.

Kolkata's only sewer system lies in the city center. The bustees surrounding the city are not attached to the sewer system. During the monsoon, the city center floods when drains and sewers overflow, forcing Kolkatans, rich and poor, to wade through waist-high contaminated water. Kolkata is surrounded by marshland, which acts as the city's primary sewage treatment facility. In the 21st century, both government and nongovernment agencies have begun to develop and support wastewater-fed aquaculture within fisheries in the wetlands. If successful, feeding wastewater to fish could simultaneously solve the sewage problem and provide local people with seafood, which is a staple in Bengali cuisine.

Pollution and Health Problems

Many Western accounts of Kolkata focus on the air and noise pollution caused by the fleets of yellow Ambassador taxis (remnants of colonial-era transportation) or the speeding municipal buses. Air pollution caused by cars, buses, and auto-rickshaws is significant; however, local Kolkata development workers and environmentalists argue that while the danger posed by transportation is great, the municipality is ignoring the pollution caused by the city's numerous coal-fired power plants and factories. Coal mining is a major industry in India and West Bengal, and there is minimal national or local government initiative to place environmental controls on these factories and power plants.

A study by a prominent Indian research organization, the Chittaranjan National Cancer Institute (CNCI), highlighted the alarming incidence of lung cancer in the city and tied this directly to air pollution. The rates of suspended and respiratory particulate matter (SPM and RPM) in Kolkata are high, even compared with other Indian cities, often exceeding safe levels (140 for SPM and 60 for RPM) by two to three times. Street dwellers, hawkers, and scavengers are the worst sufferers, as they spend most of their days in congested intersections where the air pollution is most concentrated. The CNCI study claims that close to 80 percent of street dwellers have deteriorated lung function or damaged lungs.

Kolkata's lung cancer incidence is far higher than that of Delhi, the country's capital and largest city. In Kolkata, the rate is at least 18.4 cases per 100,000 people, compared to Delhi's 13.34 cases per 100,000, according to the CNCI study. In the 21st century, Indian cities have felt strong pressure from government agencies and the High Court to reduce air pollution; however, unlike Delhi, Kolkata had not been able to significantly reduce its air pollution bys 2010. In May 2005, the Indian government ordered all vehicles in Kolkata manufactured before 1990 to convert to greener fuels such as liquid pressurized gas (LPG). Had the government followed through on this directive, nearly 80 percent of the city's buses, many of which are city owned, and nearly 50 percent of taxis and auto-rickshaws, would have been taken off the streets. Arguing that the ban would cripple the city's transportation system, the Kolkata High Court quashed the central government's mandate.

It was not until August 2009 that the municipality began to crack down on polluting vehicles. City managers and activists are using initiatives like the curbs on vehicle emissions to improve infrastructure and sewage and join other Indian cities such as Delhi and Mumbai in rectifying decades of unregulated consumption, waste, and urban expansion.

Sarah Besky
University of Wisconsin, Madison

See Also: Automobiles; Delhi, India; Developing Countries; Emissions; India; Mumbai, India; Noise; Power Plants; Public Health; Slums.

Further Readings

Hutnyk, John. *The Rumor of Calcutta: Tourism, Charity, and the Poverty of Representation*. London: Zed Books, 1996.

Mukhopadhyay, Bhaskar. "Crossing Howrah Bridge: Calcutta, Filth, and Dwellings: Forms, Fragments, Phantasms." *Theory, Culture & Society*, v.23 (2006).

Roy, Ananya. *City Requiem, Calcutta: Gender and the Politics of Poverty*. Minneapolis: University of Minnesota Press, 2003.

Thomas, Fredric. *Calcutta Poor: Elegies on a City Above Pretense*. New York: M. E. Sharpe, 1996.

Laidlaw, Inc.

Laidlaw, Inc., was a multifaceted company providing tour and transit passenger services, healthcare transportation, and waste transportation in the United States and Canada. Traditionally a waste management company, Laidlaw, Inc., divested itself of waste collection in the 1990s and served this sector through a 35 percent ownership of Safety-Kleen Corporation (formerly Laidlaw Environmental Services). Safety-Kleen was involved in collection and recovery services, treatment and disposal services, and landfill services. In 2007, First Group plc., the United Kingdom's largest surface transportation company, acquired Laidlaw International, which had become one of North America's largest providers of school and intercity bus transport services and a leading supplier of public transit services with companies including Laidlaw Education Services, Greyhound Lines, Greyhound Canada, and Laidlaw Transit.

Laidlaw Transit was named after Robert Laidlaw, who started a small trucking company in Ontario, Canada, in 1927. In 1959, Michael DeGroote purchased Laidlaw Transit for $300,000, keeping the established and respected Laidlaw name and turning it into a multiservice transportation firm. In 1969, Laidlaw, Inc., went public, and by the time it was sold in 2003, it was Canada's number one growth stock for 17 straight years. Laidlaw was North America's largest solid waste disposal and hauling, chemical and hazardous waste management, school bus, and public transit transportation industry. The owner, Michael DeGroote, a self-made millionaire twice over, has been involved with all of the major waste haulers, including Republic Waste Industries, Allied Waste, and Browning-Ferris Industries. It is impossible to discuss the emergence and challenges of modern solid waste management without discussing Laidlaw, Inc., and DeGroote.

1970s and 1980s

In the 1970s, increased environmental awareness and subsequent environmental regulation coupled with a declining economy to create the rise of the modern solid waste industry. The era of the trash crisis was initiated as stricter landfill regulations and governmental oversight precipitated the closing of many municipal dumps, opening the market for competition among waste transportation companies. In the United States, stricter regulation, including the 1976 Resource Conservation and Recovery Act (RCRA), fostered a situation whereby municipalities could no longer afford to upgrade or build new facilities to meet the more stringent standards. Concern over a lack of landfill space further opened the market for private corporations to turn waste into an economic commodity through interstate commerce and transportation.

Laidlaw, Inc., was at the forefront of waste transportation because DeGroote saw trash disposal as a basic need, making it a "recession-proof" industry. Laidlaw became one of the first vertically integrated solid waste companies by handling waste hauling, landfills, hazardous waste, and other methods of disposal. Laidlaw grew through acquisition of smaller companies, drawing criticism for aggressive takeovers, instead of using partnerships. By 1978, Laidlaw, Inc., had become a major player in the U.S. solid waste industry and the U.S. school bus and public transportation industry. By 1988, when DeGroote sold his controlling stake in Laidlaw to Canadian Pacific, it had become a world leader in several industries. DeGroote went on to become the head of Republic Waste Industries, a competitor to Laidlaw, Inc.

1990s

In 1990, Laidlaw, Inc., was listed on the New York Stock Exchange, where it continued to grow by acquiring school bus, public transit, security, emergency, and solid waste contracts, including Mayflower; National Bus Service; Greyhound Lines' U.S. operations; Safety-Kleen; and Attwoods, a UK waste management company. Like the small municipalities and haulers before them, Laidlaw, Inc., and other solid waste corporations were impacted by increasingly tougher environmental laws and oversight. However, Laidlaw, Inc., had the added burden of acquiring too much debt along with companies and, in 1999, started to take corrective action, including getting out of the solid waste industry.

2000s

In 2000, Laidlaw was the key player in two precedent-setting legal actions. In these lawsuits, Laidlaw has been said to represent the problems of the solid waste industry at the time. In *Friends of the Earth, Inc. et al. v. Laidlaw Environmental Services, Inc.*, the U.S. Supreme Court held that residents in the area of South Carolina's North Tyger River could sue Laidlaw Environmental based on the fact that the residents alleged they would have used the river for recreational purposes but could not because of pollution. Despite the fact that Laidlaw Environmental claimed the case was moot because they had closed the factory responsible for the pollution, the court agreed with Congress that civil penalties in Clean Water Act cases do more than promote immediate compliance by limiting the defendant's economic incentive to delay its attainment of permit limits—they also deter future violations.

In 2002, Laidlaw defended securities fraud class action litigation that alleged the company and the other defendants disseminated false and misleading financial statements and press releases concerning the company's financial condition. In 2003, Laidlaw settled a class action lawsuit that alleged that Laidlaw's corporate officers had misrepresented the financial condition of the company. Soon after, Laidlaw declared bankruptcy and became Laidlaw International.

Barbara MacLennan
West Virginia University
Michael P. Ferber
The King's University College

See Also: Browning-Ferris Industries; Canada; Clean Water Act; Mega-Haulers.

Further Readings

Crooks, Harold. *Giants of Garbage: The Rise of the Global Waste Industry and the Politics of Pollution Control.* Toronto: James Lorimer & Co., 1993.
Friends of the Earth, Inc., v. Laidlaw Environmental Services, Inc., 528 U.S. 167 (2000).

Landfills, Modern

Next to incineration, the disposal of solid waste by landfilling in quarries or via artificial hills constitutes the oldest form of disposing of waste. The Romans created the Monte Testaccio in Rome, a 35-meter-high hill created by the shards of containers that were used to merchandise goods such as wine, oil, and grain. Albeit grown in dimension, modern landfills will similarly last as memorials of the 20th-century consumption culture. The Fresh Kills Landfill of Staten Island, New York, which from 1948 to 2001 absorbed New York City wastes and was then briefly reopened to handle the debris of the World Trade Center after the September 11, 2001, attacks, was said to be the world's largest landfill and can be considered to be the largest human structure ever built.

Modern landfills, conceived as controlled, hygienic, and applying modern engineering principles, accompanied the rise of modern mass-consumer society, but the wild, crude, or uncontrolled dump also proliferated. At first placed in the urban periphery, landfills served to absorb the growing waste amounts produced by urbanites. Although landfills for a long time were considered to reclaim land, their placing also was an issue of environmental injustice, when poor areas were chosen to absorb other people's trash. When ruralists turned into consumers in the postwar boom years, landfills also encroached into rural areas, in particular, in their crude form. Seen over the course of the 20th century, the changing discourse on and the changing practices of landfilling reflect major turning points in the history of city–periphery relations, of public health and hygiene, and in the history of environment and consumption.

Beginnings of Modern Landfills

Disposing of leftovers and street cleanings on dumping sites for a long time resembled a natural process. As long as wastes consisted of organics and minerals, dumped in small piles, they would turn into a sort of "waste humus," which local farmers would then use as dung or soil conditioner. The distinctions between dumping and producing this so-called waste soil were vague, as long as both practices were devoid of specific methods and the fermentation process was yet unexplored. Moreover, in many regions, ancient landfills were exploited to retrieve the fermented soil and fresh waste was used to reclaim land, for example, by filling it into previously idle wetlands or by creating artificial hills for sports and recreation. The engineering method of modern landfilling was rooted in such practices, even though the 19th-century sanitary movement had hinted at the hazards of water and soil contamination by waste piles.

In the 1920s and 1930s, the "sanitary landfill" and "controlled tipping" were developed in the United Kingdom and United States, and France developed the *dépôt contrôlé*. These novel landfilling concepts were supposed to modernize tipping, which became considered as unhygienic or wild. The main alterations between the traditional and the modern landfill initially consisted of (1) the use of mechanical traction for transport, tipping, and planning; (2) the method of dumping in compacted layers; and (3) the regular—at best daily—covering of the wastes with soil, ashes, or dirt in order to reduce vermin, smell, fires, and wind-blown litter. In addition, some forms of waste compaction were practiced, and scavengers were increasingly banned from the landfills' territory.

While up to the 1960s, the trenching method was often undercut, the regular covering with dirt finally constituted the chief distinction between control, hygiene, and engineering principles versus disorder. Because modern landfills were assumed to feature hygienic conditions (which, however, never existed and were approached only by the adoption of insecticides, among them DDT), they were sometimes sited next to residential zones. The main advantages ultimately lay in cost-effectiveness, low investment costs, and the possibility to run a landfill with just a few workers and an excavator, or, later on, a bulldozer. Besides, modern landfills were used to gain land, for example, in Robert Moses's 1930s sanitation, reclamation, recreation plan for New York.

European Landfills

Controlled tipping was pioneered in 1915 by the city of Bradford, England. In United Kingdom (UK) cities, incineration was largely in place, but the new method promised to provide a cheaper means of disposal. During the following three decades, it would turn into the dominant method of UK municipal waste disposal. State regulations described its modes of operation, ruling, among other things, that the waste layers should measure less than 1.8 meters (about six feet), the covering was to be applied at least every 24 hours, and the covering should measure about 0.25 meters (0.8 feet). Each waste layer should rest for some time to let it ferment and sink down before the next waste layer was to be spread on top of it. Later on, inclined layers were applied. French engineers and hygienists visited the Bradford landfill just before siting the first French *dépôt contrôlé* in Liancourt-Saint-Pierre in 1935. It received waste from the capital of Paris, roughly 37 miles away, for which additional and less-distant landfills were soon installed. On French controlled tips, the new waste layer should be applied only after the temperature of the old waste layer had declined to that of the surrounding soil, thus indicating that fermentation had reached its peak. Other continental European countries—or, more accurately, their cities—did not implement modern landfilling until the 1960s, or even later. European engineers studied the site of Bradford and, to a lesser degree, that of Liancourt, but not the U.S. model. Some European waste experts assumed the European model to be superior, since it was said to guarantee an aerobic fermentation process. The latter was a centerpiece in interwar UK and postwar European landfill concepts, for it would destroy pathogen germs and render putrescible matter into harmless, even valuable soil.

U.S. Landfills

In the United States, the first sanitary landfill opened about three miles away from Fresno, California, in 1937. The landfilling site was closed in 1987 and subsequently designated a National Historic Landmark. The concept soon was also used in New York and San Francisco. During World War II, the method was practiced and improved by the U.S. Army. Members of the military's Corps of Engineers, as well as the U.S. Public Health Service, spread this knowledge throughout the country. After then, and similar to the United Kingdom, sanitary engineers praised the sanitary landfill as the universal solution of the waste problem. In the 1950s, the first manuals and guidelines were formulated by organizations such as the U.S. Public Health Service, the Sanitary Engineering Division of the American Society of Civil Engineers, or the American Public Works Association. These would describe the method as a safeguard for public health, since it was said to avoid all the nuisances produced by wild tipping. In retrospect, the U.S. sanitary landfill method and its appraisal were similar to the UK or French cases. What distinguished them was mainly the waste. Since U.S. waste contained more packaging and paper, landfilling applied more compression, which became an issue in Europe only after the 1960s.

Alternatives

Even if French engineers were among the pioneers of modern landfilling, in the postwar decades, incineration and composting were the two primary engineering methods of waste disposal until the early 1960s. The same is true for the West German waste experts' discourse. By then, because of an exponential increase in waste volumes and changing waste contents, there was a wide understanding that modern consumer societies were in need of modern landfills, since incineration and composting also left behind leftovers such as slags and nondegradable rests. Novel plastic wastes yielded problems in incineration, but they were seen as inert—and, thus, as innoxious as glass powder or stones—for landfills. Detached from the waste experts, the daily practice of municipalities was predominantly mere crude dumping.

Modern Landfill Challenges

Since the 1960s, waste experts and politicians have advocated modern landfilling, mainly as a tool to limit wild tipping and to site new landfilling space, which had become a problem due to a shortage of space and citizen resistance. European cities had experienced a shortage of close-by dumping space for some time already, and, in the United States, siting became problematic during the 1970s. In the long run, a decreased number of large, centralized landfills substituted for the many small, wild landfills, which resulted in longer transport distances, or, in the case of the United States, in interstate waste transport. Modern landfilling was further regulated, while applied treatment technologies had hardly been advanced. The regulations—in West Germany by means of sanitary guidelines, in France by means of a 1963 Circulaire, and in the United States in connection with the 1965 Solid Waste Disposal Act—resulted from previous legislation on water protection. Accordingly, a controlled landfill mainly meant to choose a site under geological and hydrological considerations so as to exclude potential contamination of water systems. Spontaneous fires, smoke, dust, vermin and flies, and wind-blown drifts of waste were normal occurrences on controlled landfills. Nevertheless, landfilling still was considered an issue in urban planning, greening, and recreation.

A further challenge of modern landfilling lay in compacting and homogenizing the solid waste. Household waste was less dense than in the 1930s. The latter had consisted mainly of ashes and humid organic leftovers, while paper usually had been burned in domestic fireplaces and scrap had been given to ragmen. Together with the arrival of the so-called throwaway society and its packaging, such as the one-way glass—and later plastic—bottle, waste volumes were rising at a faster rate than waste weights. In addition, municipalities began to entomb bulk waste, car tires, and industrial wastes. Industrial waste was turning into a major municipal concern, and waste became classified according to potential hazard levels, with special landfills for toxic wastes.

To compact and homogenize the heterogeneous waste masses into the desired layers of the average sanitary landfill, mills and successively specialized compactor engines were introduced in landfilling. Despite the substantial change of waste contents and despite known accidents caused by methane emissions stemming from anaerobic decay, the intractable paradigm of an aerobic fermentation that would prevail in the guts of modern landfills still persisted.

New Technologies

In contrast to its naming, until the 1970s, modern landfilling was lagging behind the scientific and technological progresses realized in the production, distribution, and consumption of those goods it would absorb. Both crude and modern landfills had never been closely studied over time, and no scientific scrutiny had been conducted on their inner processes or long-term effects.

Modern landfilling can be interpreted as a genuine experiment that was conducted under real conditions and that carried unknown results. Only since the 1970s were science, technology, and professionalism applied to landfilling. The beginning ecological era came along with a rise in environmental awareness. Besides, next to municipalities, the state became a main actor in landfilling and funded research and data collection or issued waste abatements. In the United States, for instance, the 1976 Resource Conservation and Recovery Act imposed criteria for groundwater protection and landfill gas migration control. As a consequence, in many countries, the 1970s constituted a crucial era of retooling modern landfilling. Recent explorations of ancient landfills had indicated slow degradation processes and the importance of anaerobic fermentation. Decade-old paper was conserved, while plastics showed signs of disruption. Hence, waste experts became aware that modern landfills represented long-term storage, rather than transient reactors—a fact that anthropologist William Rathje explored from the late 1980s onwards in his archaeological surveys (or *garbology*) of modern landfills.

Landfill engineering now included mineral as well as plastic liners, diverse containment structures, leachate collection, and drainage systems. Eventually, the concept of the perfectly isolated landfill was developed that would physically encapsulate the waste and thus inhibit any contact with the circumjacent water, air, or soil. As a reaction to leakage water occurrences, multiple barriers were used to hermetically seal the bottoms of landfills; to seal the top, among others, plastic foams were tested.

Retooled in such ways, the modern landfill was said to resemble the most environmentally friendly disposal method, since no interaction between waste and its environment would occur. This proved to be an illusion, and sanitary engineers soon realized that landfills needed decade-long aftercare procedures due to ongoing interior reactions and the need to extract methane. Many modern landfills that operated at the end of the 20th century had been installed before these new paradigms gained influence, and they thus lacked a coherent avoidance of pollution by leachate. Next to aftercare plans for current landfills, the remediation of old landfill sites has turned into a further field of waste management.

Due to its simplicity and cost-effectiveness, landfilling—both wild and controlled—remains the most common way to treat waste in the early 21st century. Meanwhile, its adversaries have added its methane emissions to the long list of potential environmental hazards, while its adherents keep alluding to the need for sinks to hold unavoidable leftovers. Or they envision—in line with the previous exploitation of waste earth from old dumps—landfills as mines for future exploitation, for the recovery of gas, or of materials such as plastics and metals. In the beginning of the 21st century, the United States landfilled more than half of its municipal waste. In developing countries, most waste is landfilled, often with interposed scavengers who sort out recyclable materials. In European countries, where European Union (EU) politics put recycling on top of the five-stage waste hierarchy (reduce, reuse, recycle, recovery, and disposal), the percentage of landfilled waste differs greatly, amounting to more than 70 percent in Greece, the UK, or Portugal and to less than 20 percent in the Netherlands, Denmark, Belgium, and Sweden.

The EU hurried to find a coherent waste treatment regulation, and, in 1999, the European Landfill Directive was issued. This directive asks the member states to apply strict requirements for future landfilling, such as aftercare plans or the pretreatment of waste before its tipping to destroy any organic content. High landfilling taxes are supposed to provide compliance incentives for the favored waste policies. Europe's largest and oldest open landfill in Entressen, in the French Crau plains, measuring about 80 hectares of dumping surface and emerging from an agrarian reuse of waste as humus that Marseille had begun there in 1912, only closed in 2010.

Heike Weber
Technische Universität Berlin

See Also: Composting; Garbology; Packaging and Product Containers; Paper and Landfills; Trash to Cash.

Further Readings

Colten, Craig E. "Chicago's Waste Lands: Refuse Disposal and Urban Growth, 1840–1990." *Journal of Historical Geography,* v.20 (1994).

Cooper, Timothy. "Burying the 'Refuse Revolution': The Rise of Controlled Tipping in Britain, 1920–1960." *Environment and Planning* A, v.42/5 (2010).

Groß, Matthias, et al. *Realexperimente: Ökologische Gestaltungsprozesse in der Wissensgesellschaft.* [Real Experiments: Ecological Design Processes in the Knowledge Society]. Bielefeld, Germany: Hanser Verlag, 2005.

Melosi, Martin V. "The Fresno Sanitary Landfill in an American Cultural Context." *Public Historian* v.24/3 (2002).

Melosi, Martin V. *Garbage in the Cities: Refuse, Reform, and the Environment,* rev. ed. Pittsburgh: University of Pittsburgh Press, 2005.

Rathje, William L. and Cullen Murphy. *Rubbish! The Archaeology of Garbage.* Tucson: University of Arizona Press, 2001.

Rogers, Heather. *Gone Tomorrow: The Hidden Life of Garbage.* New York: New Press, 2006.

Lighting

The introduction of artificial light has allowed humans to extend their ability to perform tasks. This basis became externalized in urban building patterns that gave rise to the street light as an organized system of lighting.

Ancient Lighting

Lamps have existed from ancient times. An example is the Inuit oil lamps, admired for their wide range of heat and light, a by-product of the relationship between hunter and prey that provided critical fuel for heat and light for people living beyond the Arctic Circle year-round.

As early as 5,000 years ago, architecture became formalized for a significant section of cities, and the desire and need for rooms to have lighting after nightfall led to a market for early candle factories. This also extended to fine oils for lamps. Resources for these lighting products are examples of early biofuel use and came from cattle and other animals as tallow, a waxy fat, and from certain plants from which burnable oils were pressed. Waxes were learned about over time and incorporated as the major component by many candle makers, along with scents and specialty wicks from certain fibers.

Candles can be made in sizes that burn in tens of days, or are gone in minutes, and, while ancient in technology, are still preferred for the atmosphere they deliver as they illuminate for an intimate meal or other gathering.

Industrial Age

Candles were not very practical for the steam-powered Industrial Age. Cities have long been susceptible to fires that spread catastrophically. The Industrial Age brought new fuels, including gases, and also brought many refinements. These improvements came in the form of woven wicks, reflectors, glass globes and lamps with control of both oil and air, and, most important, with lenses, a practical way to control beams of light. Light physics is tied to lighting and illumination, and lenses became a way to amplify the effectiveness of the lumens by defining the focal length and shape for a desired purpose.

Early Industrial Age light physics produced focusing and diffusing lenses and mirrors, then the Fresnel lens design. This lens is so efficient in the parallelization of light from an oil lamp or candle that it quickly became a standard in lighthouses to extend their feeble flame into the night. Good use can be made with this design today in solar collectors, concentrating the light and producing 1,315 degrees Celsius/2,400 degrees Fahrenheit temperatures on a clear day from a thin plastic lens of 2 square meters, with output about 750 watts at the focal point, 2.5 cm in diameter.

With early direct current (DC) came the solution lab battery, then a damp mixture called a "dry-cell," a self-contained battery in a casing common as a 6-volt size, so reliable and efficient that even today they are used globally. These early batteries, and others that ended up powering the electric hand torch in sizes from AAA to D, supplied an eager, growing industrial society where electric lighting allowed processes to continue through the night. But these needed a bulb.

Thomas Alva Edison is a name still known for electricity, and his development of the incandescent light bulb was the major event in modern lighting. His lightbulbs soon became widespread for a variety of uses, from room lights to street lights, replacing the gas lamps of the era, and then made into small hand torches.

Soon the "electrical" world became politicized, and in the battle for networked power, Nikola Tesla's alternating current (AC) became the standard on transmission lines. Today, nearly all generation systems have been built for AC. By losing less power over distance, AC allows generators to be miles away from where the current is needed, but is not very energy efficient when the end use is to heat something.

Over time, lighting has taken many forms for many uses: from lighting for photography and movies to industrial lighting for machine automation and control. Lights are used to add energy to a scene, and to fulfill illumination requirements whether to set a mood or to see a can on a conveyor.

Heat Pollution

From the point of view of sustainability, the present use of power for lighting is the main concern as it relates to waste, in this case defined as "heat pollution," "heat not used in a process and released to the environment," or "entropy." Entropy comes from not using the heat as it moves from the source to the end use, so any heat not used is lost forever from the system. Incandescent bulbs release a lot of energy as heat for the amount of illumination that is returned.

What this means, in terms of heat pollution or waste heat, is significant on a large scale, because lighting uses a massive amount of energy. To reduce both heat emissions and the amount of electricity needed for lighting, it becomes important to look at using fluorescent tubes, modified to be usable replacements for standard bulb sockets and uses. Their trace mercury is the drawback, but their lumens per watt efficiency, along with a long life expectancy versus incandescents, are benefits.

Efficiency

Lighting a home using electricity suffers losses in terms of wasted heat at every transformation, generator, or inverter. There are massive losses in original heat output to create that much electricity per hour from a process 50 percent efficient at best. Converting power to lumens is costly, and while an offshoot of solid-state products has been the development of a practical light-emitting diode (LED), their costs are a concern.

These devices turn amperes into lumens at much higher rates than ever thought possible and part of their importance is that they light a home using photovoltaic-battery systems very well and so are very appropriate to use in a macroeconomic view. The LED, if it can be made for significantly less per unit, is ready to eliminate the need for an antiquated and expensive system developed a century ago when resource management and energy efficiency were not yet conceptualized.

The greatest advancement in lighting came with solid-state materials that led to the LED, which greatly reduces the watt-per-lumen ratio. This is an ongoing field, and as organic LEDs (OLEDs) have become feasible, any of these can be integrated into products for assorted lighting needs. The speed of development in off-the-grid illumination products for housing is advancing quickly within the LED industry. Many nations have initiatives for innovation and progress in energy-conserving products, and using renewable processes and methods is becoming an economic advantage over the ever-increasing costs of using nonrenewable resources or thermodynamically wasteful processes in the marketplace.

Lighting in the future may well be based on quantum physics and solid-state developments in materials and manufacturing methods to deliver even more lumens-per-watt-per-unit-cost, so that photovoltaics (PVs) will more easily supply home lighting. New LED lightings that run on rechargeable batteries are pragmatic and cost-effective for homes never wired for electricity, as is the case in most of the world's housing. For existing homes, luminaires—LEDs arrayed to act as standard lightbulbs—are being worked on in practical sizes, and reducing costs is the priority.

Today, industrial designers and architects are faced with the choice of energy status quo in their work or the move to far less costly ways of providing light. This can be from physical sun ports or openings in the roof to light hallways, the placement of windows and doors, anything that can supply energy-neutral modern living. Because of this and from the push for hybrid vehicles, new lithium batteries will end up in many household items. These fast-charging batteries are replacements, and, with

Luminaires, LEDs arranged to function like standard lightbulbs, are being developed for practical use. LED lights requiring only rechargeable batteries to run are pragmatic and cost-effective for homes that are not wired for electricity, which is common in most of the world. Fast-charging lithium batteries are projected to become more commonly used in household items, and when used with LED technology will enhance the usability and cost effectiveness of both products.

LEDs, become even more effective for the people depending on them. Product lines that are helping this transition are replacements for common lightbulbs with LED models, a critical issue in sustainable product design and where high costs prevent the technology from filtering down to where it can have a statistically significant impact on thermal emissions reductions globally.

The percentage of power used only for lighting would be hard to estimate, from oil lamp and candle to industrial assembly lines, yet all of it is thermally expensive to produce. As a recent U.S. National Aeronautics and Space Administration (NASA) Earth Observatory image at night of the Nile Delta artfully captured, some lighting is so intense that it can be easily seen from space, a huge expenditure of energy pointing to why lighting is a primary energy conservation issue. Nearly half of all energy use can be associated with lighting and illumination.

Conclusion

An electricity grid sprawls over thousands of kilometers, involves capital expense and ongoing costs of all that infrastructure, and has effects on sedimentation and runoff. In contrast, an onsite system in the home collects, stores, and uses solar power for much of its electricity, along with microgeneration for high-current needs. An important issue with respect to lighting is the true return per lumen for an original watt of power used to create it, and the impact to the environment in the delivery of those lumens to where they are being used.

Like Telsa's AC versus Edison's DC, from a cost standpoint, the LED-battery system is far better, and the realistic ratio in cost-per-watt-per-lumen is 100 times more for watts over-the-wire than a simple PV-battery system, once refined. While lighting is not the primary use of power worldwide, eliminating the need for over-the-wire power for residential use will mean a great gain in thermodynamic efficiency to conserve heat.

Tom Mallard
Independent Scholar

See Also: Economics of Consumption, U.S.; Household Hazardous Waste; Industrial Revolution; Sustainable Development.

Further Readings

Hunt, Charles. *A History of the Introduction of Gas Lighting*. Charleston, SC: Nabu Press, 2010.

Lenk, Ron and Carol Lenk. *Practical Lighting Design With LEDs*. Piscataway, NJ: IEEE Press, 2011.

Whitehead, Randall. *Residential Lighting: A Practical Guide to Beautiful and Sustainable Design*. Hoboken, NJ: Wiley, 2008.

Linen and Bedding

Linen and bedding make up a large component of the textile industry. Household textiles, or soft furnishings, are fabrics used in the home. The term *linen* is also used to classify bath towels, dish towels, table linens, sheets, pillowcases, mattresses, blankets, comforters, and other bedding. These materials are not only used in private residences but are also found in vast quantities in hotels, hospitals, office buildings, restaurants, nursing homes, and countless additional commercial establishments. The textile industry is one of the largest industries in the world.

The volume of textiles produced to support 21st-century society has a large impact on the environment. The processes and materials used to create and eventually destroy these fabrics require a great deal of resources, energy, and space.

Environmental Impact

The production of textiles releases several forms of pollution into the environment. Unconverted raw materials, by-products of the manufacturing process, and shipping of the finished products release solid, liquid, and gaseous pollution into the water and air. Textile manufacturing consumes water, fuel, and chemicals in a complex sequence of processes that vary depending on the type of fabric being produced. The main environmental problem during the production phase involves contaminated water, but air pollution and excessive noise or odor also impact the health and safety of workers.

Textiles undergo a series of processing steps, and different products create different waste streams. The variation in type of fiber used to create the product, the dyes and chemicals used, and the technology of the facility producing the material all impact the amount and type of waste generated. In the case of linen and bedding, cotton and polyester are most commonly used, but bedding incorporates many different fabrics.

Natural fibers have the greatest environmental impact during the growth and harvesting of the raw materials. Synthetic fibers have the greatest environmental impact during processing, when large quantities of chemicals and energy are being used. There are many stages in the life cycle of both natural and synthetic fibers.

Growing the crops to create household textiles is resource intensive. Cotton plants are the most pesticide-intensive crop in the world, but any plant used to create textiles uses agricultural land that could be used for food. Fertilizers, herbicides, and pesticides applied to crops of cotton, jute, bamboo, and flax (for linen) contaminate soil and groundwater. Animals and insects are also exploited to harvest wool, fur, leather, and silk.

Harvesting the raw materials uses additional chemicals in the application of defoliants to the plants and the exploitation of cheap labor or use of fuel-powered machinery to gather the materials. Production cleaning of the raw fibers uses detergents, soaps, or bleaches and creates waste by-products that are shipped to a landfill, and the machinery to perform these tasks uses fuel or electricity. Noise and dust are frequently health issues with employees in the plant.

Additional processing to create fabric from the fiber results in toxic fumes and by-products from dyes and finishing chemicals. Heavy metals used to fix dye to fabric or solvents to seal waterproofing to material also contaminate wastewater, which often flows into local rivers. Virtually all polycotton (polyester/cotton blends), especially bed linen, and any garments labeled "wrinkle resistant," "easy care," or "permanent press" are treated with toxic formaldehyde to give them that quality. Much of this industry has been shipped overseas, where labor and resources are less expensive and environmental health and safety regulations are often lax.

Synthetic fibers consume finite resources such as petroleum, coal, and oil in their creation. The by-products eliminated during the production process are often toxic and require extensive treatment before being released into the waste stream. Air pollution is especially prominent when these substances are in use. After the textiles are created, there is additional waste and pollution generated as the linens, garments, and other household products are fashioned. Packaging and shipping the finished products by land or sea uses additional resources and fuel.

Sustainable Product Development

Steps are being taken by the textile industry to minimize harm to the environment. There is much written in the media in the early 21st century about the "greening" of the industry. Ecofashion is a movement to take the health of the environment and consumers into account. Fabrics are made using organic raw materials. Chemicals and bleaches are not applied when dying the material. The industry is taking steps to reduce wastewater. The careful selection of fabric and dye combinations can maximize dye effectiveness. Separating wastewater streams allows selected by-products to be reused. Using biodegradable enzymes in place of chemical agents, where possible, improves wastewater quality and reduces treatment costs. Design and production are being reconfigured to reduce energy use. Simple changes, like insulating boilers and pipes or using cold-temperature dyeing and finishing, improves efficiency and reduces cost.

Recycling or Reusing

There are many obvious benefits to recycling or reusing textiles. It reduces the need for landfill space. Textiles discarded in the 21st century are more problematic because of the development of synthetic fabrics, which do not decompose. Reusing textile products reduces the pressure on finite raw materials and reduces pollution and energy needed to process them. Medical facilities and the hospitality industry are heavy users of textile products. Many hospitals, hotels, motels, and restaurants have implemented recycling programs for linens.

Textile recycling encompasses reusing or reprocessing used clothing, linens, fabrics, manufacturing scraps, and other textiles. The Environmental Protection Agency (EPA) estimates that 12.4 million tons of textiles were generated in 2008, approximately 5 percent of total municipal solid waste generation. There are a growing number of textile recycling companies that divert these products from the municipal waste stream. These operations recover good-quality clothing and linens that can be resold in secondhand outlets. More than half of the clothing recovered for resale is exported to foreign countries. Damaged textiles can be converted to rags and industrial wiping cloths. Used textiles can also be broken down into fiber that is refashioned into new products. Fiber reclamation mills sort incoming materials by type and color, then shred and blend the fibers into new yarn for weaving or knitting. The fibers can also be compressed into filling material for mattresses, insulation, roofing felts, and furniture padding. Synthetic fabrics can be cut into small pieces that are granulated into pellets that can be melted and spun into new polyester fabrics.

Regardless of their final destination, used textiles have a relatively stable and high price. They remain a valuable commodity, even after heavy use. Revenue generated by resale is enough to cover the processing costs. The textile recycling industry in the United States employs nearly 20,000 workers, and primary and secondary processors of these products account collectively for annual gross sales of $700 million.

Educating the public about the value of reusing and recycling their household textiles from clothing to linens is critical. Good publicity and outreach will help to increase participation and the quality and quantity of textiles collected. Textile recovery programs are effective and can be easily replicated around the country. Each individual must use and reuse these valuable resources responsibly.

Jill M. Church
D'Youville College

See Also: Consumerism; Floor and Wall Coverings; Hospitals; Recycling.

Further Readings

Casey, Donna G. and Kay Weeks. "The Operating Room as a Source of Recycling Material." *Hospital Materiel Management Quarterly*, v.14/3 (1993).

Environmental Protection Agency. "Wastes-Resource Conservation-Common Wastes & Materials-Textiles." http://www.epa.gov/wastes/conserve/materials/textiles.htm (Accessed October 2010).

Platt, Brenda. *Weaving Textile Reuse Into Waste Reduction*. Washington, DC: Institute for Local Self-Reliance, 1997.

Powers, Tom. "Reusable Linens in the Surgery." *Hospital Materiel Management Quarterly*, v.14/1 (1992).

Rowen, Jill. "Sheet Makers Add Color, Address Eco-Concerns." *Home Textiles Today*, v.28/28 (2007).

Scrimshaw, John, ed. *Textiles and the Environment 2009*. Bradford, UK: World Textiles Publications, 2009.

Slater, Keith. *Environmental Impact of Textiles: Production, Processes and Protection*. Boca Raton, FL: Woodhead, 2003.

Strohle, Jurgen and Dieter Bottger. "Water and Energy-Saving Solutions." *Textile World*, v.158/2 (2008).

Wang, Youjiang. *Recycling in Textiles*. Boca Raton, FL: CRC Press, 2006.

Los Angeles

From a precolonial population of 5,000 native Californians, Los Angeles (LA) has mushroomed into an urban giant of more than 6 million people. Demand for water and resources in a desert-like environment presents significant challenges to engineers and planners and has generated its share of political scandals. Strategies to cope with the enormous shift in the volume of waste over that span have tested the ingenuity of city residents. In the 21st century, although LA is a leader in recycling and energy generation from waste, rates of consumption are soaring, and many households are overwhelmed by the profound effects of runaway consumerism.

Early History

The LA area was for millennia the home of the Tongva and Chumash and their ancestors. Their uses of the landscape helped to shape the actions of the earliest Europeans and have continuing impacts in the 21st century. The Chumash lived at the western edge of LA and along the Santa Barbara Channel to the northwest. Their important coastal village, Humaliwu, lent its name to the famed celebrity colony of Malibu. The Tongva people occupied the

LA basin and San Fernando and San Gabriel valleys. Both tribes had many villages along the coasts, estuaries, and rivers where the food resources were richest. Even so, their collective ecological footprint over some 10,000 years was exceptionally modest by 21st-century standards.

Native Californians utilized many calorically desirable and easy-to-acquire resources and may have at times driven local animal populations to precarious levels. Some paleontologists argue that early hunters precipitated the extinction of Ice Age fauna (such as mammoths), but others disagree. In some periods, sea mammals vulnerable due to sustained, land-based breeding seasons at local rookeries were likely overhunted, forcing villagers to pursue other species. For the most part, however, precolonial people were effective stewards of the sea and land.

Virtually all plant and animal species present at the start of the Holocene were still around when the missions were built. Many plants important for basketry or food were regularly pruned and tended (although the people were not farmers), and there was a long precolonial history of controlled burns to stimulate the production of seed-rich native shrubs and grasses. Plant species adapted to fire thrived under these land management practices, attracting deer and elk. The era must not be romanticized, but the archaeological record suggests that people did not wastefully use most resources.

Although communities across ancient southern California were judiciously spaced and rarely reached 300 people (most had 75–150 residents), the cumulative effect of thousands of years of food and toolmaking waste at settled villages is impressive. Before highways and other construction destroyed so much unwritten history, the LA–Santa Barbara coast was a necklace of shell midden sites as deep as 5 meters, each with the barest remains of circular thatch houses surrounded by discarded shells; bones of fish, seals, rabbits, and deer; acorns; stone tools; soapstone bowls; and metates. In these societies, food and manufacturing trash was tossed near houses—there were no toxins to concern them. The most physically dangerous waste was sharp-edged stone from making projectile points, but these too were absorbed into middens. Other resources consumed in the LA region include asphaltum from the La Brea tarpits, which was widely used as an adhesive. The distribution of native villages affected where important nodes of colonial activity took place, such as where missions were positioned, and still deeply impacts where urban construction happens in the 21st century, since the remaining sites are protected by local and federal laws.

Missions

The native population in the LA area was about 5,000 at contact. The San Gabriel Mission was the first colonial building in these lands. Established near Tongva villages in 1771, it moved to the current location in 1774. El Pueblo de Los Angeles (the northern core of the modern downtown) was founded in 1781 on the banks of the Los Angeles River, close to Yaanga and three other Tongva villages. The founding group consisted of 11 families recruited from Sonora and Sinaloa, Mexico. The pueblo took five decades to reach a population of 1,000. Meanwhile, the San Fernando Mission was established in 1797 on the north side of the valley. All three locations were soon surrounded by cattle, agricultural fields, and human and animal waste. The Tongva who crowded around these locales (peaking at 1,700 persons at the San Gabriel Mission in 1817) suffered poor health and high mortality because of European diseases and unsanitary conditions. The governor of Alta California deeded large ranchos, usurping tribal lands. Tongva laborers built and provided food to the missions and herded at the ranchos. Ten millennia of practices marked by low-impact consumption of wild resources and the generation of little waste vanished in a few decades.

Growth of Modern Los Angeles

The growth of LA was slow until 1870–1890, when its population spurted from 5,730 to 50,395. By 1900, it surpassed 100,000, and by 1930, the population surged to 1.2 million. At nearly 2 million in 1950, Los Angeles had become the nation's fourth-largest city. By 2010, the city officially has about 4.1 million residents, second most in the United States, and the core urban area had over 6 million of the county's 9.9 million residents. It is such a megalopolis that few realize it was California's second-largest city (well behind San Francisco) until the 1920s. The core LA urban area consists of more than 100 contiguous neighborhoods from Sylmar to Long Beach, as well as dozens of embedded cities such as Beverly Hills, Santa Monica, Burbank, Manhattan Beach, and Inglewood. This roughly 800 square miles of contiguous urban development supports about two-thirds of the LA County population. Another 3,261 square miles of the county encompasses more sparsely populated mountain and desert land to the north and dense urban areas to the east, including Pasadena and Arcadia.

Three Major Aqueducts

Most water is piped in from distant sources. William Mulholland, superintendent of the LA City Water Company during the early 1900s, famously negotiated land purchases and water rights for the first of three major aqueducts that still sustain the city. The aqueduct from Owens Valley was built in 1908–13, and stretches 233 miles from the eastern Sierra to Los Angeles. The 300-mile Colorado River Aqueduct was constructed during the 1930s, and the 140-mile aqueduct from the Mono basin was added in the 1960s. LA sanitation operations were founded in 1923, when rapid expansion of the city meant that waste was becoming a major problem. In the 21st century, a vast infrastructure of facilities provides solid waste and wastewater disposal for millions of households at a 2010 cost of nearly $1 billion. According to Sanitation Bureau documents, more than 6,650 tons of material generated by commercial and residential sources is collected and recycled or buried daily, including refuse, recyclables, and plant trimmings.

Some 6,500 miles of sewer pipes handle 550 million gallons of wastewater each day. It is estimated that at least 5,000 restaurants, hundreds of schools and hospitals, 12 colleges and universities, 700 grocery stores, 20 club megastores, 1,100 convenience stores, 15 malls, and untold numbers of industrial enterprises and small businesses contribute to the daily consumption of resources and generation of waste in LA.

Portions of LA waste go to three landfills; the largest, Puente Hills, reclaims energy through gas collection networks. The district also operates refuse-to-energy facilities in Commerce and Calabasas. Several enormous dumps used for decades are filled and closed, and others are reaching capacity. A "Waste-by-Rail" program that moves refuse to the remote desert locality of Mesquite near the Salton Sea is permitted to accept 20,000 tons of LA's waste per day through the 21st century. Another distant waste locus for LA trash is proposed for Riverside County. Some Angelenos improperly dispose of household hazardous wastes, although the city has established hazardous-material collection sites. Dumping into storm drains, roadsides, and bins meant for regular trash is common. Trash of all kinds winds up in coastal waters. A one-day coastal cleanup event in September 2010 generated 50 tons of trash collected by 14,131 volunteers.

Waste is not confined to solids and water. With at least 16 million cars registered in the core LA area (26 million in the county), the county's roadways are handling 12 million cars per day, and 18.6 million tons of carbon dioxide are emitted into the atmosphere each year. The crux of the problem is the job commute: when Angelenos travel to work, 70 percent of them drive alone.

Water Recycling

LA administrators are rapidly expanding various recycling programs. Persistent droughts signal the need for more frequent water rationing. In the early 21st century, six water reclamation plants recycle sewer water for agriculture, groundwater recharge, and park, golf course, and landscape irrigation. "Toilet-to-tap" water recycling is on the horizon. Most biosolids from wastewater are converted to electricity and fertilizer. Huge recycling centers recover more than 240,000 tons of paper, glass, plastics, and aluminum annually. From nearly 500,000 tons of plant material, mulch and compost are produced. More is diverted from landfills than is mandated by AB939, a California law requiring reduced flow to landfills. *Waste & Recycling News* reports that of the 10 largest U.S. cities, Los Angeles ranks at the top in recycling rate, at nearly two-thirds of material collected.

A society accustomed to throwing out 80 percent of what it buys needs to stop and take stock of the heavy toll such actions have on the land. Compared to the disposed waste per capita of the Tongva people, the waste produced by the LA population occupying the same patch of land is astounding. Among the Tongva, the population of 4,000 or so in the LA basin/valley region (another 1,000 lived on the nearby islands and in the San Gabriel Valley to the east) disposed of an estimated 1 pound of waste per person daily, primarily in the form of shell, bone, and stone. This amounts to 1.5 million pounds of waste per year. On the other hand, the roughly 6 million people in the LA urban core—at the U.S. norm of 4.5 pounds of waste per person daily—produce 27 million pounds of garbage per day. This is equivalent to 1.13 million pounds hourly, meaning that LA's population disposes of the same amount of waste in about 80 minutes as the precolonial population tossed out in one year. Stated another way, what Los Angeles dumps in two months equals what it took the Tongva 1,000 years to throw out.

Consumption at Home

Los Angeles, like other urban locales, is filled with busy families, many with two jobs and children to raise. The behavior of people as consumers and disposers, as it plays out in their homes, has a major effect on the trajectory of life in the city. From 2002 to 2005, the UCLA Sloan Center on Everyday Lives of Families documented the material worlds of ordinary, middle-class LA households. Archaeologists enumerated possessions, measured rooms and yards, recorded activities, took nearly 20,000 photos, and gained an unparalleled, systematic picture of what people have and do in their homes. Ordinary working families have thousands of possessions tucked into every available space, including crammed garages that no longer accommodate the family car in 75 percent of homes studied.

As documented in *Life at Home in the Twenty-First Century*, the United States in the 21st century is the most materially saturated society in global history. For decades, the consumption of increasingly cheap yet expendable goods such as toys, phones, and electronics has occurred at astounding rates. Clutter overwhelms many families, and measurable physiological stress arises from trying to cope with it all. Only economic declines seem to encourage typical families to rethink frenetic consumption habits.

Jeanne E. Arnold
University of California, Los Angeles

See Also: Archaeological Techniques, Modern Day; California; Material Culture Today; Recycling.

Further Readings

Arnold, Jeanne, Anthony Graesch, Enzo Ragazzini, and Elinor Ochs. *Life at Home in the Twenty-First Century: 32 Families Open Their Doors*. Los Angeles, CA: Cotsen Institute Press, 2011.

City-Data. "Los Angeles County, California." http://www.city-data.com/county/Los_Angeles_County-CA.html (Accessed September 2010).

City of Los Angeles Public Works. http://dpw.lacity.org/dpwhome.htm (Accessed September 2010).

Deverell, William and Greg Hise, eds., *Land of Sunshine: An Environmental History of Metropolitan Los Angeles*. Pittsburgh, PA: University of Pittsburgh Press, 2006.

McCawley, William. *The First Angelinos: The Gabrielino Indians of Los Angeles*. Banning, CA: Malki Museum/Ballena Press, 1996.

Louisiana

Louisiana is one of the southern states of the United States; Baton Rouge is its capital, and New Orleans is its largest city. Louisiana is the only U.S. state to refer to its political subdivisions as parishes instead of counties. Named after Louis XIV (king of France 1643–1715), the state was once part of the French colonial empire. The state consists of an upland region and an alluvial region of low swamp, coastal marsh, beaches, and barrier islands, which lie principally along the Mississippi River and the Gulf of Mexico. The state is frequently subjected to tropical cyclones and vulnerable to major hurricane strikes. The New Orleans area was devastated by Hurricane Katrina in 2005 when the federal levee system failed, the worst civil engineering disaster in U.S. history. Over 1,500 people died, and 80 percent of the city was flooded. New Orleans remains one of the largest and busiest ports in the world, the region is key to the U.S. oil refining and petrochemical industries and is a corporate base for onshore and offshore petroleum and natural gas production. Concerns over water contamination from the petroleum industry produced regulations ranging from a state law in 1910 to forbid discharges of oil that might damage rice crops to coordinated efforts by several federal agencies in 2010 to remediate damage to the state's coastline by oil spilled in British Petroleum's Deepwater Horizon disaster. The state's main agricultural products include seafood (Louisiana supplies most of the world's crawfish) and other land-based staples—the seafood industry directly provides around 16,000 jobs.

Statistics and Rankings

The 16th Nationwide Survey of MSW Management in the United States found that, in 2006, Louisiana had an estimated 6,051,158 tons of municipal solid waste (MSW) generation, placing it 23rd in a survey of the 50 states and the capital district. Based on the 2006 population of 4,243,288, an estimated 1.43 tons of MSW were generated per person per year (ranking 15th). Louisiana landfilled 5,551,158 tons (ranking 19th) in the state's 27 landfills. The state exported 168,341 tons of MSW; the import tonnage was not reported.

In 2006, Louisiana was not increasing its landfill capacity, which was then 71,721,948 tons. It was ranked joint 19th out of 44 respondent states for number of landfills. Whole tires, used oil, lead-acid batteries, and white goods were reported as being banned from Louisiana landfills. Tipping fees across Louisiana averaged $32.35, where the cheapest and most expensive average landfill fees in the United States were $15 and $96. Louisiana has no waste-to-energy (WTE) facilities, although plans have been put forward for a controversial Sun Energy Group plasma gasification plant in New Orleans. Louisiana recycled 500,000 tons of MSW, placing the state 31st in the ranking of recycled MSW tonnage.

Landfills

Since the late 1980s, Louisiana's waste management has undergone a complete upheaval. In 1980, there were over 800 landfills; by 2010, this has been reduced to 24 by new federal and state regulations governing landfill design, construction, and operation. Act 189 of 1989 reduced MSW being sent to landfill by 25 percent and emphasised the recycling and composting of MSW and other waste. Prior to this act, there was little organized recycling of MSW and up to 25 percent of waste received at landfills was yard waste. Nearly all sewage sludge was being landfilled without consideration of alternate methods of disposal. In rural areas, open dumping and backyard burning persisted as they had done throughout U.S. history. In similar age-old patterns, industrial waste was left in open pits and waste from forestry and agriculture was burned or piled up to rot away. In 1990, more than 11 million tons of nonresidential waste was being landfilled.

The construction of landfills to modern standards and their operation cost much more than the outmoded landfills, but afforded much greater protection to the environment. Increases in tipping fees and increased MSW transportation costs made many municipalities study ways of reducing the cost of waste disposal; this factor, along with Act 189, stimulated the recycling of paper, glass, plastic, and metal. Yard waste began to be separated from the general waste stream and recycled into compost or mulch, and state regulations promoted the use of treated sewage sludge as a fertilizer replacement. Between 1990 and 2000, recycling in Louisiana went from near zero to 15 percent—around 570,000 tons of material. The amount of landfilled MSW decreased to 3,800,000 in 2000, from 5,797,320 tons in 1990. Industrial waste now goes into environmental protective landfills, recyclable streams of waste are separated out, and waste reduction programs have been implemented in many workplaces.

Agricultural, Forestry, and Livestock Waste

One of the largest waste streams in Louisiana is the residue from primary processing of crop harvests and felled timber. Estimates vary, but in most years this can exceed 10 million tons of bagasse (sugarcane stalk), filter press cake, cotton gin trash, rice hulls, bark, wood chip, and paper mill waste (fibers and ash). Some sugar plants and paper mills now burn their residues to provide power and steam. In 1993, best management practices (BMPs) introduced by the Louisiana Department of Agriculture and Forestry (LDAF) allowed agriculture and forestry wastes to be reused under exemptions from the solid waste regulations. Most processors are now in this program, and some or all residues are reused as soil amendments, nutrients, or a source of lime. Often, this returns organics to the land from which

Sugarcane is harvested and loaded into high dump trailers in Bunkie, Louisiana, for transport into trucks for shipment to a sugar mill. Residue from primary processing of crop harvests is one of the largest waste streams in the state; an estimated 10 million tons of residue, including bagasse (sugarcane stalk) and filter press cake, is produced every year. Some sugar plants and paper mills burn their biomass as an energy source under a state program that allows agricultural and forestry waste to be reused for energy, soil nutrients, or a source of lime.

the same crops were harvested, reducing costs, preventing soil exhaustion and reducing the need for manufactured fertilizers.

While residues remaining at the site of their harvest are now exempt from Louisiana Department of Environmental Quality (LDEQ) and LDAF regulations, the burning of these residues onsite remains highly controversial. The routine burning of timberland to control litter and competition has similarly been problematic. All burning must now be carried out by a trained and LDAF-certified burn operator.

Animal waste is not a major issue in Louisiana as livestock farming has been reduced since the mid-20th century. In 1950, the state produced 1.5 million market hogs; in 2001, only 70,000. The majority of Louisiana's animal waste is chicken litter from the 2,000 broiler production houses. This is a mixture of chicken manure and rice hull or sawdust litter, and it is routinely spread on fields, being a high-quality fertilizer that also adds organics to the soil. However, this use has been concentrated in areas around poultry farms, creating concern about accumulating phosphorous in the soil. Safe practices are being implemented in the early 21st century to address this problem, and education and testing programs for soil and litter have been put in place.

Agriculture Street Landfill

A 95-acre area of swampy low ground in the Upper Ninth Ward of New Orleans began to be used as a dump in 1909, becoming known later as the Agriculture Street Landfill. It became one of the main dumps in the area for residential and industrial waste. Fires were common in the dump, and it received the local nickname of "Dante's Inferno." First efforts to close the dump were in 1952, but it remained in use as a sanitary landfill until the end of the 1950s. In 1965, the debris of Hurricane Betsy necessitated reopening the landfill for another two years before it was officially closed again. The site was then sealed with compacted incinerator ash. In 1976, redevelopment began, and the site was covered with sand and soil and a neighborhood was built on top of part of it. Morton Elementary School, some small business units, and three residential developments—Press Park, Gordon Plaza, and Liberty Terrace—were situated on the former landfill.

Residents complaining of health problems occasioned the first Environmental Protection Agency (EPA) investigations in 1986, which concluded that no remediation was necessary. Old trash continued to be found just below the surface, and anecdotal evidence of health problems, such as an abnormally high cancer rate, persisted. Local people petitioned for retesting, and, in 1994, the Agriculture Street Landfill site was added to the National Priorities List (NPL) and became a Superfund site. The undeveloped part of the landfill was declared a public health hazard and was fenced off. Contamination was found in samples of soil, dust, air, and garden produce, which contained metals, polycyclic aromatic hydrocarbons (PAHs), volatile organic compounds (VOCs), and pesticides. Further investigations revealed that the actual amount of soil deposited in the landfill prior to redevelopment was significantly thinner than had been claimed.

The remediation operation involved closing Morton Elementary School and, in the residential area, removing two feet of soil, putting down a barrier layer, and then replacing the soil with clean material. The project was completed in April 2002, but many homeowners have since petitioned to be rehoused elsewhere—a petition backed by Congressman Bill Jefferson. As of 2010, they were unsuccessful in their attempts. The remediated landfill was disturbed in 2005 by flooding from Hurricane Katrina, which is believed to have freed toxins from the landfill. Press Park, a residential address consisting of 56 townhouses nearby, has been found to have levels of benzopyrene, a carcinogenic petroleum by-product, that were 50 times greater than the normal level.

Jon Welsh
AAG Archaeology

See Also: Comprehensive Environmental Response, Compensation, and Liability Act (CERCLA/Superfund); Farms; Toxic Wastes.

Further Readings

Arsova, Ljupka, Rob van Haaren, Nora Goldstein, Scott M. Kaufman, and Nickolas J. Themelis. "The State of Garbage in America: 16th Nationwide Survey of MSW Management in the U.S." *BioCycle*, v.49/12 (2008).

Colten, Craig E. *An Unnatural Metropolis: Wrestling New Orleans From Nature*. Baton Rouge: Louisiana State University Press, 2005.

Love Canal

Love Canal is a chemical waste disposal site located in a small neighborhood of Niagara Falls, New York, just north of the Niagara River. Although Love Canal was neither the first nor the last hazardous-waste dump site in the United States, it has become a symbol of governmental red tape, corporate accountability, and grassroots organization. It spawned new legislation and awakened the world to the effects of industrial waste on human health. It took over two years of battling government agencies and staying in the headlines, but the Love Canal Homeowners Association eventually achieved the relocation of nearly 900 families from the Love Canal neighborhood.

History

Love Canal's history is long and intricate. The name of the canal comes from William T. Love, a 19th-century entrepreneur who envisioned a canal that would connect Lakes Erie and Ontario and harness some of the hydroelectric power capable of being produced by the Niagara River. The canal was 3,000 feet long, 60 feet wide, and 10 feet deep when Love had to abandon the project because of an economic depression in the late 1890s that resulted in loss of funding. From 1942 to 1953, Hooker Electrochemical Company dumped between 20,000 and 22,000 tons of industrial wastes into the canal. When Hooker sold the 16-acre site to the Niagara Falls School Board for $1 in 1953, the deed contained a disclaimer relieving Hooker of any future liabilities that may result from the chemical wastes buried in the canal. The 99th Street Elementary School opened on the site in 1955 with 400 students. The next 20 years saw the neighborhood grow, with many homes and the accompanying new infrastructure built. The construction of streets, sewer lines, and utility lines meant that the canal's clay walls were breached, and soil from the surface of the canal was excavated and reused elsewhere in the neighborhood. Reports of fumes, sludge in basements, and holes in the surface of fields and school playgrounds began in the 1960s. Finally, in 1976, the New York Department of Environmental Conservation (DEC) began investigating materials that had been disposed of in the Love Canal site. During the DEC's investigation, the story of Love Canal and litanies of complaints began appearing in local newspapers.

Inspections and Relocations

The canal's many changes in ownership serve as an appropriate precursor to the metaphorical "hot potato" it became among several government agencies, beginning with the DEC in 1976. Shortly after the DEC investigation began, the Environmental Protection Agency (EPA) began collecting air and soil samples, the results of which would later be handed to residents as lists of chemical names with numbers next to them. The commissioner of the New York State Department of Health (DOH) visited the site in April 1978, declared it a public health hazard, and ordered Niagara County to begin health studies. The DOH would figure most prominently in the Love Canal disaster area because it was conducting the health studies and would make many of the announcements of evacuation throughout the next years. A pattern of miscommunications quickly emerged, and residents were often confused by what the DOH told them and the unorganized fashion by which the DOH collected residents' blood samples: many were lost, some expired before being tested, and the DOH labs were overrun with the samples.

In the span of six months, two successive New York State health commissioners delivered orders for pregnant women and children under age 2 to be relocated, first in the inner ring of houses, followed by those in the next few rings of houses. Each time the DOH made an announcement, other government agencies scrambled to find funding to enforce the order, while residents grew ever more confused, outraged, and afraid as a result of information provided without interpretation and the DOH's refusal to ever reveal or publish all data from the health studies. New York governor Hugh Carey promised to buy the inner two rings of houses and created an interagency task force onsite comprised of the DEC, DOH, Department of Transportation, and six other agencies. This task force had the job of overseeing home purchases, resident relocation, health and environmental studies, and the remedial construction project that would hypothetically remedy the Love Canal site's leaching chemicals.

The grassroots organization credited with eventually winning relocation for all 900 families in the Love Canal area was the Love Canal Homeowners Association (LCHA), run by Lois Gibbs. After her son began attending the 99th Street School, he developed asthma and seizures. When Gibbs read an article about Love Canal in the *Niagara Gazette* in June 1978, she feared a connection between the canal and her son's health. What began as a concerned mother knocking on doors to collect signatures on a petition to close the 99th Street School, grew into a community movement that would keep Love Canal in the news.

The LCHA organized protests, put pressure on politicians while cameras were rolling, testified at local and federal hearings, and lobbied for legislation to approve tax exemptions and rebates after property values tanked. Because this was a blue-collar neighborhood and the property values had plummeted due to the news of the Love Canal chemical waste site, residents could not afford to simply pick up and move. Eventually, the LCHA took two EPA officials hostage for several hours in order to protect them from a mob that had gathered outside LCHA headquarters and to get U.S. President Jimmy Carter's attention after 11 of 36 residents tested positive for chromosome deformities.

Leading up to the presidential election of 1980, several members of the LCHA—including Lois Gibbs—and the mayor of Niagara Falls appeared on the *Phil Donohue Show*. The LCHA followed this media appearance by arriving at the Democratic Convention in New York City with signs demanding action from President Carter. Following an appearance on ABC's *Good Morning America*, the president visited Niagara Falls to sign an agreement with New York State that provided $15 million of federal funding to purchase the rest of the Love Canal homes. The president's signing the agreement in September 1980 meant that residents could finally escape the chemicals they had been living with for years.

Effects

Love Canal was a battle won by a grassroots community organization. Despite many of the reported governmental misses, its legacy had some positive results. On December 11, 1980, Congress passed the Comprehensive Environmental Response, Compensation, and Liability Act (CERCLA), better known as the Superfund. Funded by taxes on the chemical and petroleum industries, the Superfund remedies other hazardous waste sites like Love Canal. After the EPA managed the fund for a few years, the Superfund Amendments and Reauthorization Act (SARA) passed on October 17, 1986, providing for permanent remedial actions at sites and increased state involvement. Lois Gibbs now heads the Center for Health, Environment and Justice, an organization that helps other grassroots organizations like the LCHA.

After nearly all the Love Canal residents relocated—some chose to stay behind—by 1981, and after the 239 homes closest to the canal site were demolished in 1982, the DOH declared the areas north and west of the canal habitable for residential use in September 1988. After years of remediation work and continued environmental studies conducted by the EPA, the site was removed from the Superfund's National Priorities List of sites requiring remediation in 2004. Occidental Chemical Corporation, the parent company of Hooker Electrochemical Company, eventually had to pay $129 million in reimbursements to various city, state, and federal agencies that had financed the cleanup, as well as $1 million to establish the Love Canal Medical Fund.

In the 21st century, when one drives down Colvin Boulevard Extension, the northern perimeter of the canal, on one side of the street a grassy knoll rises behind a sturdy green cyclone fence. On the other side of the street, a resettled neighborhood lies in what used to be known as Emergency Disaster Areas 4 and 5.

Karlen Chase
D'Youville College

See Also: Comprehensive Environmental Response, Compensation, and Liability Act (CERCLA/Superfund); Environmental Justice; Environmental Protection Agency (EPA); Industrial Waste; Landfills, Modern; Politics of Waste; Pollution, Air; Pollution, Land; Pollution, Water; Public Health; Public Water Systems; Toxic Wastes.

Further Readings

Blum, Elizabeth D. *Love Canal Revisited: Race, Class, and Gender in Environmental Activism*. Lawrence: University Press of Kansas, 2008.

Environmental Protection Agency. "Superfund Site Progress Profile: Love Canal." http://cfpub.epa.gov/supercpad/cursites/csitinfo.cfm?id=0201290 (Accessed June 2010).

Gibbs, Lois M. *Love Canal: The Story Continues*. Gabriola Island, Canada: New Society Publishers, 1998.

Levine, Adeline Gordon. *Love Canal: Science, Politics, and People*. Lexington, MA: Lexington Books, 1982.

New York State Department of Health. "Love Canal." http://www.nyhealth.gov/environmental/investigations/love_canal (Accessed June 2010).

State University of New York at Buffalo. "Love Canal Collections: A University Archives Collection." http://library.buffalo.edu/specialcollections/lovecanal (Accessed June 2010).

Lynch, Kevin A.

Kevin A. Lynch (1918–84), a U.S. architect and planner, is best known for his practical and theoretical contributions to the psychology of place. His keystone concept of place legibility refers essentially to the ease with which people understand the layout of a place. According to Lynch, the city can be considered text to be read by every individual. The particular organization of place components determines how easy it is to read the urban configuration in order to create an adequate mental map to navigate the city. For Ethan Sundilson, "mental maps of a city are mental representations of what the city contains, and its layout according to the individual."

Along with his studies on perception of city form, Lynch inquired about the different meanings given to waste. In his posthumous book *Wasting Away*, he went beyond the material definition of waste as "a deteriorated collection of things," exploring the more symbolical side of wasting that includes temporal and spatial dimensions of the process of consumption. Besides being "all a nasty business," Lynch described waste and loss as the dark side of change, a repressed and emotional subject to be redefined. Every human being regards death and most change as tragic and confusing, fearing the sense of loss. Following the same logic, Lynch described waste as a signal of loss, of growing old. Waste is both a result and the process of loss. The common words for filth are generally tense with emotion. They are not neutral. Waste is an impurity to avoid or to wash off.

Though the process of waste production was conceptualized by Lynch as a continuum with very diffuse limits, he contended that wasting is, for the sake of simplicity, normally described in polarized ways. Going back to his concept of legibility, Lynch explained that, where customary boundaries are lacking, people lose their grip on things. That is the very reason to dichotomize the idea of waste: useful or useless, front or back of buildings, efficient or wasteful, saving or spending, growth or decline, produce or consume, succeed or fail, alive or dead. Nevertheless, those dichotomies sometimes lead to absurd behavior. Cultural, economical, and ideological reasons sustain the oversimplification of terms like *waste* and *garbage*. For example, some chicken farmers complain that they cannot feed unsold market produce to their birds because it is legally defined as garbage. Dirt is an idea bound to context and culture. It is matter out of place, particularly, matter that is unpleasant, dangerous, and difficult to remove. One's style of handling dirt is a way of establishing one's character and social position. Both the pathologist and the butcher work with "living matter in fresh decay," but the social meaning of their jobs defines a very different status and recognition.

For Lynch, defining waste as something that is out of place immediately posed a question on the definition of place as a culturally defined term. As stated by Lynch, "... in our urgent need for order and clarity, we find change and gradation hard to bear...." Waste is omnipresent, but the need for order and simplicity decrees that ". . . People, things, and places must be one or the other (waste–not waste), there to remain not shifting, not in-between, not partly so and partly not. . . ." Accordingly, Lynch also saw a dilemma: permanence, as opposed to declination, is stagnation and illness. The production of waste is a synonym of change and growth. Waste was visualized by Lynch as a necessary part of the continuum of life and death, a prime process in the spiral of change of things, living creatures, and systems. Lynch situated waste in a new, honest dimension. It is conceptualized as a normal process of the cycle of life, neither something obscure nor abject. In fact, Lynch criticized that most definitions of waste in the English language are negative. He also discussed the Latin *vastus*, which means "unoccupied" or "desolate." Lynch asked if something without human presence deserves to be treated as trash, since life normally thrives in those spaces. They were not considered waste at all by Lynch.

Waste and the City

The image of the city is also composed by waste. For Lynch, spaces devoted to waste can also be magnets for adults and children as places for free imagination and search for fascinating vestiges of a recent past beyond the reach of the regular whereabouts. They could be conceptualized as learning places where children connect with adults by investigating their material traces. In Lynch's words, many waste places "have these ruinous attractions: release from control, free play for action and fantasy, rich and varied sensations. They are characterized as places for pleasure precisely because of their rich form, freedom, and a sense of continuity. Thus, children are attracted to vacant lots, scrub woods, back alleys, and unused hillsides." Regrettably, the range of free movement—and therefore the amount and quality of exploratory behavior—in children has been severely restricted since the beginning of the 20th century. A car culture, television, and video games at least partly explain such restriction.

More than a meticulous explanation about waste and its different methods of treatment, Lynch questioned the multifaceted nature of this phenomenon. He posed disturbing, yet contemporary questions, such as: "Can we forsake our automobiles, a contingency that is at least thinkable? What will happen then to the suburbs? to vacation areas? to industries whose employees commute by car? to the entire apparatus of making and maintaining those beautiful machines?"

Mauricio Leandro
Graduate Center, City University of New York

See Also: Archaeology of Garbage; Archaeology of Modern Landfills; Sociology of Waste.

Further Readings

Lynch, K. *The Image of the City*. Cambridge, MA: MIT Press, 1960.

Lynch, K. and M. Southworth. *Wasting Away*. San Francisco, CA: Sierra Club Books, 1990.

Magazines and Newspapers

Magazines and newspapers are two different categories of mass media with distinct histories, functions, and futures in the globalizing world. Similarities include high circulation, the structure of article content and periodical character of publications, subscription opportunities, as well as a strong impact on society.

Consumption: Newspapers

Newspapers were the first medium for formal distribution of mass information, the roots of which date to ancient Rome. After a period of handwritten proto-newspaper medium in the 17th century, printed newspapers gradually became the most important source of political, economic, sport, and cultural information during the 19th and 20th centuries.

Newspapers developed from periodical to weekly to daily. In the United States, the first newspaper was published in 1690 in Boston. The first U.S. daily came out almost 100 years later, in 1783 in Pennsylvania. In the 21st century, 216 of 260 countries or territories publish newspapers (83 percent of all countries). Japan, Norway, Finland, Sweden, and Switzerland are the top five countries for newspaper sales per capita, while Mozambique, Uganda, Armenia, Zambia, and Botswana are the bottom five. Twenty-nine countries publish only non-daily titles. However, the advance of technological information and electronic media has resulted in the decline of printed newspapers as a leading information source.

Traditional consumers of newspapers over the centuries have been adults. Improvement in literacy on a global scale has been steadily increasing the numbers of readers. Globalization in the later 20th and early 21st centuries has also stimulated the popularity of newspapers and the emergence of many new titles in a period when the meganewspapers have been reducing their circulation and publishers have been closing or filing for bankruptcy.

The early 21st century is a particularly volatile period for newspapers worldwide, with established titles in the United States and western Europe in deep crisis. Several newspapers ceased publishing or scaled back their editorial operations due to declining revenues or issues particular to each paper's operation. In July 2011, the United Kingdom's best-selling Sunday newspaper, the *News of the World*, closed after 168 years when allegations of illegal telephone surveillance jeopardized the satellite television interests of its publisher. From 2001 to 2005, circulation increased by 7.8 percent (including free daily). So-called narrowcasting has splintered audiences into smaller and smaller slivers. Ranked at the top by circulation is *Yomiuri Shimbun* (10,021,000), Tokyo, Japan, Yomiuri Shimbun Group. The *Wall Street Journal* and *USA Today* are the most distributed U.S. newspapers (20th and 22nd, respectively, on the world list). In 2008, approximately 48.6 million copies were sold every weekday by the 1,408 daily U.S. newspapers. On Sunday, 902 newspapers sold 49.1 million copies, averaging 2.5 readers per copy. According to some prognoses, by 2017, printed newspapers will become insignificant in the United States.

Australia presents an exemplary case study of consumer interest in daily and weekly newspapers. There are 48 daily newspapers with a total circulation of 3,030,000, and 196 newspapers circulated per 1,000 people. The number of nondaily newspapers is 233, with total circulation at 374,000, and 24 newspapers circulated per 1,000 people. Newspaper consumption (minutes per day) is 35.6 percent of global daily newspaper circulations, including the free daily newspapers.

Surveys point to four characteristics of printed newspapers that keep them a part of social life: navigation, convenience, informationality, and entertainment.

Consumption: Magazines

In the United States, the approximate number of published magazines is 19,500, with a 2008 circulation of almost 370 million copies. Approximately 74 percent of all magazines go to subscribers, 11 percent are single-issue sales, and 15 percent are returned unsold. Nearly half of the single-issue sales occur in supermarkets. Nineteen billion catalogs were mailed in 2007, a one-third increase in a decade. Consumers of magazines are of all ages because of the variety of the topics—from children's comics to publications about aging for the elderly. Pornography is also a part of the magazine industry, with enormous circulations and profits. Electronic media will most likely not replace magazines because the parts of the culture that use and value printed, visual materials have extremely high numbers and strong market share.

Waste and Recycling

Newspapers and magazines belong to the nondurable goods category, together with other sorts of paper and paper products. Traditionally, newspapers and magazines have been reused as paper for packing; covering different surfaces, such as windows and furniture; and even as toilet paper.

In many countries, there are special services that pay by weight for submitted printed media, or systems of special recycling bins. As of 2004, the recycling rate for newspaper (73.4 percent) led all recycling rates in the paper products category in the United States. In 2008, newspaper generation was 1.7 million tons higher than in 1960, but newspapers' solid waste market share decreased by 56 percent. Newspaper recycling increased by 5.92 million tons and the recycling rate increased by 243 percent during this same period.

Unique, clay-coated paper makes magazines more difficult to recycle, but a fair number of these are also recycled in the United States. Catalogs also are primarily printed on coated, ground wood paper and are more difficult to recycle, but, like magazines, they are also recycled.

Despite various recycling programs, the amount of newspapers and magazines produced outnumber the amount recycled. The Institute for Scrap Recycling Industries (ISRI) estimated that recycling all morning newspapers read in the United States could save 41,000 trees per day and reduce 6 million tons of waste deposited in landfills.

Lolita Petrova Nikolova
International Institute of Anthropology

See Also: Household Consumption Patterns; Paper and Landfills; Paper Products; Recycling.

Further Readings

McKinnery, R. W. J. *Technology of Paper Recycling*. London: Blackie Academic & Professional, 1997.

Quimby, Thomas H. E. *Recycling, the Alternative to Disposal: A Case Study of the Potential for Increased Recycling of Newspapers and Corrugated Containers in the Washington Metropolitan Area*. London: RFF Press, 1975.

Resource Recycling. http://www.resource-recycling.com (Accessed November 2011).

Subak, Susan. *The Paper Industry and Global Warming*. World Business Council for Sustainable Development, 1996.

Maine

Maine, with a 2009 population of 1.318 million and 35,387 square miles, which is slightly smaller than the rest of the five New England states combined, has evolved into one of the leading states with regard to pioneering innovative municipal solid waste (MSW) management policy. Maine adopted the first extended producer laws in the nation on mercury-containing products, electronic waste, and fluorescent lights, and it adopted the country's first product stewardship framework law. In addition, Maine is one of the first bottle bill states and has banned the creation or expansion of new commercial MSW landfills.

Based on 2009 U.S. Census Bureau estimates, Maine is ranked 40th in population and 38th in population density, with 42.7 persons per square mile. Maine has the oldest population in the United States, with a mean age of 42.2 years, and is ranked third with 15.6 percent of the state's population over 65 years of age. Maine is ranked 49th in mean household size (2.35), ranked 32nd in mean household income ($45,734 per year), and ranked 44th in population growth from 2000 to 2009 (3.4 percent). Maine's economy is ranked 43rd with $50 billion in gross state product in 2008.

Maine's taxable retail sales in 2009 were just below $16 billion. Of this amount, 88.1 percent ($14.1 billion) was attributable to consumer consumption and less than 12 percent ($2 billion) qualified as business operations. The categories of 2009 sales were general merchandise stores (21 percent), building supply stores (14 percent), restaurant sales (14 percent), food stores (11 percent), and lodging (4 percent). Total retail sales rose each year from 2004 to 2007, declined slightly for 2008, and fell to a level below the 2004 annual sales for 2009 because of the sustained economic recession. Food stores represented the only continuous annual increase in taxable retail sales for all years 2004–2009.

Waste Management

In 2007, Maine residents, businesses, and visitors generated 2,066,448 tons of MSW. Between 1993 and 2003, municipal solid waste generation in Maine increased over 55 percent. This equates to an annual per capita rate of 3,200 pounds of MSW per year, or about 7.5 pounds per person per day, which is 38.6 percent higher than the 4.6 pounds per person per day, reported by the Environmental Protection Agency. A dominant factor in the higher per capita rate is Maine's reliance on tourism. In 2008, there were 15.4 million overnight visitors and 16.5 million day visitors to Maine.

In 1989, Maine adopted its solid waste management hierarchy: reduce, reuse, recycle, compost, waste-to-energy, and then landfilling, which serves to guide Maine's solid waste management planning at the state level. Also in 1989, the Maine legislature established a state recycling goal of 50 percent to be met by January 1, 2009. By 2010, the state recycling rate was 34.8 percent, significantly less than the stated goal and a decrease from the peak of 42 percent in 1997. Maine does not have mandatory recycling at the state level, but municipalities representing approximately 25 percent of the state population have adopted mandatory recycling ordnances. Maine Recycling Week, which is intended to increase the public's awareness of the importance of reducing, recycling, and buying products made from recycled materials, is November 8–15 each year and is designed to include November 15, which is America Recycles Day.

The regulation of solid waste management at the state level is under the auspices of the Maine Department of Environmental Protection (Maine DEP), Bureau of Remediation & Waste Management, which is supported by four regional offices in Portland, Bangor, Presque Isle, and Augusta. Maine also has the Board of Environmental Protection, which is charged by statute to provide informed, independent, and timely decisions on the interpretation, administration, and enforcement of the laws relating to environmental protection and to provide for credible, fair, and responsible public participation in DEP decisions, including actions related to solid waste. Maine's Board of Environmental Protection is unique in that it provides the citizens of the state with a forum that fully opens up the licensing, rule-making, and appeals process to full public participation. Statewide solid waste planning is under the auspices of the Maine State Planning Office. Maine is a home rule state; by law, Maine's 488 organized municipalities are responsible for providing for municipal solid waste activities in their jurisdictions. Regionalization is not significant in Maine, although many municipalities share solid waste transfer operations, as there are 240 public transfer stations for the 488 municipalities.

Maine is one of 11 states with a beverage deposit and refund system (bottle bill) and has one of the broadest programs in effect. In 1976, through a citizen referendum, Maine adopted its bottle bill. The original law established deposits on beer, soft drinks, mineral water, and wine coolers. In 1989, the program was expanded to include wine, liquor, water, and nonalcoholic carbonated containers (glass, plastic, and metal). The 2006 sale of beverage containers was estimated at 650 million containers. The return rate is approximately 90 percent, representing approximately 50,000 tons of waste diverted from disposal each year.

Maine's solid waste disposal facilities are one state-owned landfill, one commercial landfill, eight municipally operated landfills, 23 municipal construction and demolition debris landfills, and four waste-to-energy facilities. The state also owns a permitted "greenfield" site, known as Carpenter Ridge, outside the town of Lincoln that is reserved for future development of a landfill if needed. In 2007, 32 percent of Maine's waste was sent to the four waste-to-energy facilities. Although Maine had, as of 2010, about 12–15 years of landfill capacity, since 1989, Maine law prohibits the construction of new or expansion of commercial MSW landfills. In 2007, 456,580 tons of MSW were imported, while 60,491 tons were exported. Disposal fees for landfill and incineration ranged from $40 to $158 per ton in 2010. Maine also has 81 licensed compost facilities.

Mercury

Maine has focused on the elimination of mercury from its environment. A number of state laws and programs have been enacted to control specific products and to prevent disposal while maximizing recovery. These actions have since evolved into broader approaches to implement the waste management hierarchy and to shift costs and responsibilities onto producers.

The first mercury law was enacted in 1999, which banned the disposal of any products containing mercury as solid waste after 2002. This law was expanded to include household products containing mercury starting in 2005. Simultaneously, the sale of mercury thermometers was banned starting in 2002, which was followed by a series of additional mercury-containing product bans, including motor vehicle switches (2003), thermostats, (2006), miscellaneous meters (2006), and button cell batteries (2011).

In 2006, Maine passed the first law (LD 1792, An Act To Protect Maine Families and the Environment by Improving the Collection and Recycling of Mercury Thermostats) in the country that established a financial incentive to recycle out-of-service mercury thermostats to promote the recovery of mercury. To encourage mercury recycling, the law established a minimum bounty value of $5 per thermostat to be paid by thermostat manufacturers.

In 2009, Maine enacted a law designed to reduce mercury pollution from fluorescent lighting, the first in the nation. The law required fluorescent light manufacturers to share the costs and responsibility for recycling mercury-containing bulbs generated by households by establishing a free, statewide convenient collection system by January 1, 2011. The law requires the manufacturers' programs to include effective "education and outreach," including, but not limited to, point-of-purchase signs and other materials provided to retail establishments without cost. The law also established a maximum allowable mercury concentration in lighting sold in the state and directed the state to modify its procurement program to purchase lighting with lower mercury concentrations.

Electronics Waste

In 2004, Maine was the first state to pass an electronics waste law that integrated extended producer responsibility to ensure that products are recycled at the end of their lives. The law created a producer-financed, shared-responsibility collection and recycling system for household residents, which partnered with municipalities. On a per capita basis, Mainers recycled 3.2 pounds per capita in 2006, 3.61 in 2007, 4.06 in 2008, and about six pounds per capita in 2009. The Maine DEP estimated the capture rate for household computer monitors and televisions available for recycling in 2006 at 43 and 44 percent, respectively, which increased to 50 percent of computer monitors and 51 percent of televisions in 2008.

Producer Responsibility

In March 2010, Maine was the first state to pass an extended producer responsibility framework law (LD 1631), "An Act to Provide Leadership Regarding the Responsible Recycling of Consumer Products." The law applies the principle of extended producer responsibility for managing products at their end-of-life by establishing a process for creating product stewardship programs for hard-to-recycle products and packaging, moving the physical and financial responsibility for managing old products from the general taxpayer to producers, consumers, and others who benefit from products sold and used. The law establishes a three-step review and recommendation process. In step one, review and prioritization, the Maine DEP reviews existing product stewardship programs and conducts a prioritization process to identify potential candidate products for product stewardship status. In step two, report development, the DEP prepares an annual report to the Maine Legislature on the state of existing product stewardship programs, proposed changes to existing products stewardship programs, and recommendation of potential candidates for product stewardship. In step three, legislative review, the Maine Legislature's Joint Standing Committee on Natural Resources reviews the DEP report and has the authority to report legislation as necessary to modify existing programs or to list new products or create new programs. Following the committee's action, final approval is through the full legislature and then the governor.

Travis P. Wagner
University of Southern Maine

See Also: Household Hazardous Waste; Incinerators; Producer Responsibility.

Further Readings

Gallant, Kelly. "*Recycling as a Solid Waste Disposal.*" Lincoln County, ME: Lincoln County Recycling Program, 1979.

Maine State Planning Office. "Waste Management and Recycling Program." 2006. http://www.maine.gov/spo/recycle (Accessed December 2010.)

Sulliva, M. and R. Warner. *Maine Waste Management*. Augusta: Maine State Planning Office, 1995.

Malls

A shopping mall refers to a diverse collection of retail stores that is arranged on one or more levels and is typically augmented by one or more anchor stores, food and dining establishments, and entertainment venues. Malls may be open air or enclosed. Dating from 900 to 700 B.C.E., the Greek agora helped establish the idea of publicly presenting goods on rugs spread on the ground. In the 1st to 5th centuries C.E., Rome was the site of the first covered shopping space—an innovation that allowed shoppers to be protected from the elements. Medieval market halls and town halls (11–16th centuries) were examples of shops appearing in buildings designed for other uses, while the Eastern bazaars, including the Great Bazaar at Istanbul (1461), were revolutionary as

Climate-controlled malls, like the Mall of America in Minnesota (MOA), allow customers to shop in comfort despite inhospitable weather. MOA also features a theme park, movie theaters, video arcades, wedding chapels, and even a public aquarium.

districts within cities. The first walk-in shops with internal counters were established in Holland in 1695. In 1830, the Galerie de Orlean was the first arcade in Paris and featured an enclosed covered way to protect shoppers. In 1840, the invention of plate glass facilitated the development of shop fronts to showcase goods. In 1888, New Trade Halls opened in Moscow; it featured three arcade areas that allowed for up to 1,000 shopping units. The key factors in these early phases of the evolution of the shopping mall are the connection of the shops and the city, the inherent civic nature, the uses of technology, and the idea of an admixture of different types of shops.

At the end of the 19th century, the chain store was invented. Especially in the 21st century, chain stores often make up the majority of shops in malls. Similarly, the evolution of the supermarket would also lead to many of the traditions found in the shopping mall. In the 20th-century United States, cities were beset with population growth, congestion, and intense traffic; coupled with these facts were new technological developments in lighting and ventilation. These conditions were ripe for the first shopping mall—socialist Victor Gruen's Southdale (Edina, Minnesota, 1956).

Spanning 1 million square feet (sq ft) of shopping space, Southdale was revolutionary in that it was a fully enclosed shopping space. The use of climate control in the mall allowed customers to shop year-round in Minnesota's often inhospitable weather. In Victor Gruen's mind, Southdale would have become a master planned space including housing, hospitals, parks, and lakes among other features—essentially, an artificial, enclosed downtown. As the U.S. highway system developed in the years following Southdale, the shopping mall continued to emerge and eventually became an archetype repeated throughout the world.

Megamalls

One contemporary trend of shopping malls is increasing size. King of Prussia Mall (Pennsylvania, 2,793,200 sq ft of retail space, 400 stores), Mall of America (Minnesota, 2,768,399 sq ft, 520 stores), and South Coast Plaza (California, 2,700,000 sq ft, 280 stores) are examples of U.S. megamalls. In Canada, the West Edmonton Mall occupies 6 million sq ft (3,800,000 sq ft of retail space) and features more than 800 stores (eight anchors) and 20,000 parking spaces. Some of the largest malls exist outside North America: New South China Mall in China is 9.58 million sq ft and features space for 2,350 stores and SM City North EDSA in the Philippines is 8.68 million sq ft and offers 1,100 stores.

Other Types of Malls

In addition to the emergence of mega shopping malls and regional malls (malls designed to serve a large population and geographic area), there is the development of the outlet mall (a mall that features a number of name-brand retailers and that offers discount prices, such as Woodbury Commons, New York), the big box retailer (a major store, such as Walmart, that models the shopping mall concept by enclosing a variety of goods and services within one space), and the strip mall (a horizontal layout of stores arranged side by side, often accompanied by a large anchor store).

Socializing and Mall Culture

The contemporary shopping mall is the epitome of consumer society. People are encouraged to come inside a fully closed, protected, and controlled space that offers a multitude of consumer goods. Anchor stores like Macy's and Nordstrom offer name brands, while smaller specialty stores cater to people's perceived needs in fashion, home electronics, shoes, novelty items, books, video games, kitchen gadgetry, and arts and crafts. Because malls are self-contained, people are encouraged to spend the entirety of a day there. Many malls, including King of Prussia Mall and Mall of America, are designed as tourist attractions. People from across the United States, even from around the world, travel to spend days in a shopping mall. Food courts are a staple feature of malls. Many of the dining establishments in them feature nonorganic, nonlocal, and nonnutritious food.

People are also encouraged to spend their post-shopping time partaking in entertainment. Many malls, including Mall of America, feature small theme parks, movie theaters, video arcades, wedding chapels, and even ice rinks. The term *malling* not only refers to the act of building shopping malls but also to the behavior of people who spend time at malls. Malling would seem to indicate that part of the problem with overconsumption can be connected to the prevalence of the shopping mall.

Criticisms

Santana Row, located in San Jose, California, is one example of the emerging trend of lifestyle centers. Santana Row includes numerous retail stores, on top of which are luxury condominiums. For some, the architectural and urbanist trends of malls like those at Santana Row are alarming. There is concern that outdoor spaces with parks, waterways, and forests are being replaced by the overwhelming shopping mall structures. Others express disappointment that traditional main streets and mom-and-pop establishments have given way to the shopping mall and its chain stores.

Similarly, critics have expressed that while shopping malls supplant previous public spaces—like the city park—they serve a different public function. People can frequent a park to stroll, take in a conversation, or just relax, but a mall is designed with one purpose in mind—consumption. Malls frequently establish stringent dress, behavior, and activity codes. In many cases, people cannot take photographs in malls, protest, or pass out political or other information. Malls often have the appearance of civic participation and social integration—many offer their spaces to the aged for morning walks before stores open—but these forms of civics and sociality are always geared toward the goal of consumption.

In addition to the concerns that shopping malls promote overconsumption and socially deviant behaviors in people, critics have argued that shopping malls pose threats to the environment. Unlike the vision promoted by Victor Gruen—people would come to the center of the city to shop—the contemporary mall is often located at the outskirts of cities, many of which lack viable public transportation options.

As a result, shopping malls promote the use of the automobile and this has a major impact on the environment. They also promote urban sprawl—a fact that Gruen himself noted in a 1978 interview in which he bemoaned the archetype that he had helped to create. Because of their increasing size, shopping malls use incredible amounts of natural resources, and even after they are built, they continue to require large amounts of energy for their day-to-day use. Shopping malls are also beginning to approximate theme parks. Wafi City Mall in Dubai features Egyptian themes that resemble a theme park or themed casino. Universal Studios CityWalk in Hollywood, California, merges theme park entertainment, traditional shops, and contemporary architecture in a form that suggests a new direction for shopping malls. Some critics charge that the uses of theming and other technologies designed for retail atmosphere amount to a further lulling of the consumer into patterns of overconsumption.

Closings and Alternatives

Many shopping malls have closed, are on the verge of closing, or have been unable to attract retailers to rent spaces. In 2008, over 150,000 individual stores closed within malls. In April 2009, General Growth Properties, a firm that runs 200 malls in 44 U.S. states, filed for bankruptcy. Like many industries, shopping malls have suffered from the economic downturn of the early 21st century, and many have closed. This has led to the term *dead mall* being used to describe shopping malls that are vacant.

When malls close, cities must deal with how to use the massive spaces once occupied by the malls. One of the most interesting cases is that of the world's largest mall, New South China Mall. As of 2010, nearly 99 percent of the stores inside the mall were vacant. Unlike the trend of bigger malls in many Asian nations, the trend in the United States has been moving toward the development of strip malls and big box centers. From 2006 to 2009, only one enclosed mall was opened in the United States. One alternative to the shopping mall is the online retail location. This movement from bricks and mortar to the virtual could result in major energy savings. In the case of a shopping mall bookstore, energy costs run $1.10 per sq ft, while an online bookstore has energy costs of $0.56 per sq ft.

Popular Culture

Because of their ubiquity, shopping malls have received significant attention in popular culture.

Films like *Paul Blart: Mall Cop* and *Mallrats* in part poke fun at the central role that malls play in everyday life, while *Dawn of the Dead* uses an emblematic scene of zombies going mindlessly up and down escalators in a shopping mall as an existential reflection on the nature of mall life. The parody newspaper the *Onion* once did a spoof feature on a shopping mall in North Dakota that spanned six different zip codes and had 4,700 stores and 240,000 parking spaces. While this story is parody, in the future, malls of this scale may not be out of the question.

Scott Lukas
Lake Tahoe Community College

See Also: Consumerism; Home Shopping; Overconsumption; Supermarkets.

Further Readings

Barboza, D. "China, New Land of Shoppers, Builds Malls on Gigantic Scale." *New York Times* (May 25, 2005).
Benjamin, M. and N. Piboontanasawat. "China's Mall Glut Reflects an Unbalanced Economy." *New York Times* (April 17, 2007).
Benjamin, W. *The Arcades Project*. Cambridge, MA: Belknap Press, 1999.
Cohen, Lizabeth. "From Town Center to Shopping Center: The Reconfiguration of Community Marketplaces in Postwar America." *American Historical Review*, v.101/4 (October 1996).
Coleman, P. *Shopping Environments: Evolution, Planning and Design*. Amsterdam, the Netherlands: Architectural Press, 2006.
Gruen, V. and L. Smith. *Shopping Towns USA: The Planning of Shopping Centers*. New York: Van Nostrand Reinhold, 1960.
Hardwick, M. *Mall Maker: Victor Gruen, Architect of an American Dream*. Philadelphia: University of Pennsylvania Press, 2004.
Hudson, K. and V. O'Connell. "Recession Turns Malls Into Ghost Towns." *Wall Street Journal* (May 22 2009).
Kowinski, W. *The Malling of America: An Inside Look at the Great Consumer Paradise*. New York: William Morrow, 1985.
Lukas, S. "From Themed Space to Lifespace." In *Staging the Past*, J. Schlehe, C. Oesterle, M. Uike-Bormann, and W. Hochbruck, eds. Bielefeld, Germany: Transcript, 2010.
Rybczynski, W. "Suburban Despair: Is Urban Sprawl Really an American Menace?" *Slate* (November 7, 2005).
Underhill, Paco. *Call of the Mall: The Geography of Shopping*. New York: Simon & Schuster, 2004.

Manila, Philippines

Metropolitan Manila is the formal name given to the 13 cities and four municipalities that comprise a rapidly growing megacity, one that is facing very serious problems of solid waste management, pollution, and associated threats to environmental health and sustainability. The estimated population of Metro Manila (as it is more widely known) is 12 million, though that comprises a commuting population of some 2–3 million during weekdays from surrounding regions.

Metro Manila's postwar growth has been phenomenal. In little more than six decades, it has grown from a population of a little over 1 million to a 21st-century megacity. In 2020, it is estimated Metro Manila's population will rise to around 18.5 million within 385 square miles, making it one of the most populated urban regions in the world.

It is, therefore, not surprising that concomitant infrastructure and institutional responses have lagged behind demand. Metro Manila's governance and planning has been hobbled by politics and intracity competition. Local government exercises considerable power in the Philippines, often at the expense of coordinated responses to shared problems. This is evident in solid waste management. Negotiations over access to landfills, which operate close to capacity, occur on a regular basis. Waste management and disposal is often an election issue. In 2000, Metro Manila's main landfill site, Carmona, closed, leaving several much-smaller sites to cope with demand. In actuality, most of Metro Manila's landfills do not meet international standards of sanitary disposal and can best be described as open dump sites.

Metro Manila has grown in enormous leaps in just over 60 years, from a population of just over 1 million to a massive megacity of 12 million. In 2020, it is estimated that Metro Manila's population will rise to around 18.5 million, making the city one of the most populated urban regions in the world. Total waste and per capita waste generation are also expected to increase substantially. Infrastructure and administration has not kept up with this growth, and local political negotiations determine the use of limited landfill space.

Waste Production and Composition

It is estimated that Metro Manila produced 7,000 metric tons of waste daily in 2010, a daily waste per capita rate of 0.66 kilograms (kg). Government estimates show that both total waste and per capita waste generation will increase substantially by 2020 (16,166 daily metric tons at 0.874 kg per capita). Between 65 and 85 percent of waste is collected, though coverage is said to be declining with the growing population and declining service coverage, particularly in poorer and informal settlements (informal, or "squatter," communities make up some 35 percent of Metro Manila's population). This results in substantial open dumping or burning of waste. Illegal dumping is estimated at around 25 percent of solid waste disposal. Much of what is openly dumped finds its way into the city's numerous *esteros* (estuaries) and has contributed to severe water pollution problems. The impacts of waste, then, have been considerable in Metro Manila. Consequently, given the environmental, health, economic, and political impacts of waste, much greater efforts have been made in the 21st century toward cleaner production (CP) and waste minimization. Solid waste is also estimated to make a significant contribution to Metro Manila's poor air quality, in the form of carbon dioxide from burning and methane gases from open dump sites.

Around 75 percent of solid waste streams in Metro Manila are derived from households. The composition of waste shows that most is recyclable, and a high percentage of waste is organic. Most studies of waste composition over the past decade have estimated that food waste alone accounts for nearly 50 percent of total waste composition in Metro Manila. Though the majority of solid waste is collected in Metro Manila, there are no designated collection points. This leads to "legal" dumping in front of households or on street corners. Scavengers are an important source of recycling and collection, though only for types of waste that have value. Scavenging can also contribute to problems of waste dispersal, as garbage without value is simply redumped in public areas.

Waste Management Programs and Problems

The enactment of RA9003, the Ecological Solid Waste Management Act, in 2000, was largely in response to the crisis in availability of landfill options and increased local government conflict over responsibility for waste. RA9003 encourages waste minimization in order to maintain a healthy and clean environment and has resulted in the setting up of solid waste management bodies at national, provincial, and municipal levels of government, which are responsible for establishing coordinated waste management plans. Though a National Ecology Center has been created in an advisory role, the act places emphasis on the role of local government to both lessen waste and to divert it from landfills (and finance such activities). A diversion target rate of 5 percent per year was given to local governments to encourage a shift away from landfill dependence. However, the realization of such targets is very uneven across Metro Manila, and it is important to note that most programs targeting waste segregation and recycling activities are limited to middle- and upper-class communities.

Though RA9003 encourages waste segregation, few households practice this unless it is to separate materials that can be sold or donated to scavengers. One study undertaken in the late 1990s noted that only 1 percent of waste was recycled, but this lowly figure has increased in recent years and does not take into account the role of informal scavenging and recycling, the scale of which can only be estimated. Still, efforts to encourage household waste segregation and recycling were considered to be as failing by 2010, despite the existence of penalties and the enactment of a number of regulations. Common reasons given for this indifference include the belief that government is responsible for garbage, that systems merely mix waste sources at a later stage, the opportunity cost of the time taken to segregate waste, and the lack of information or encouragement given to do so. It is widely observed as well that garbage trucks picking up segregated waste do not separate it, leading to the later mixing of waste. It is important to note, though, that waste is still most likely to be resorted, recycled, and used at the end disposal point by the thousands of families who live and work on the city's landfill sites, most notably Payatas.

Alternative Programs

Though Metro Manila as a whole has struggled to manage solid waste and create viable and effective alternatives to open dumping and landfill dependence, in part a result of the passing of the Clean Air Act in 1999 that bans incineration of waste (putting greater pressure on landfills), there are a number of successful initiatives at a local level.

Barangay Sun Valley, in Paranaque City, has since 1998 operated a "no segregation, no collection" program (the Total Segregation Approach to Ecological Waste Management program) in which segregated waste is collected in three phases and sent to different locations: biodegradable waste to composting centers, recycling materials to junkshops and recycling factories, and residual waste to landfills in standard garbage trucks. The initiative is based on three principles: segregation at source, segregated collection, and segregated destination. However, such programs remain limited in scope. Even households participating in the program continue to rely on informal collectors (for convenience and as a form of social support for the poor) and burning. This demonstrates the persistence overall of established patterns of waste disposal and systems and the difficulty of significantly shifting away from these behaviors and economies—at least in the short term.

Perhaps the best-known program in Metro Manila, which builds upon environmental, social, and economic systems of waste, is the much-heralded *Linis Ganda* (meaning "clean and beautiful") program. Nominated at the Istanbul Habitat II conference as a leading global practice for handling garbage, *Linis Ganda* is a privately run resource recovery and recycling program. The program's hundreds of recycling cooperatives collect, recycle, and process over 200,000 tons of garbage annually in a number of Philippine cities, but most are active in Metro Manila. An added attribute is that it encourages the development of recycling cooperatives and the establishment of junk shops. In particular, *Linis Ganda* seeks to lessen the stigma and increase the livelihood opportunities of door-to-door waste recyclers in the program, known as "eco-aids," through establishing consistent routes and household and school collection systems and, therefore, income. As of 2010, there are more than 3,500 uniformed eco-aids who, through their organization into cooperatives, are able to access low-interest and collateral government business loans.

Conclusion

The development of such formal–informal partnerships and innovations will be critical in the future management of waste. As a growing megacity, Metro Manila faces considerable and difficult challenges in managing the substantial impact of development. The consequences are environmental, but waste management is increasingly becoming a political issue and raises questions over governance. There is also significant economic cost associated with waste production and management in Metro Manila.

This has been estimated by city mayors to be 10–20 percent of their overall budgets, a cost likely to rise significantly in the future. However, waste has provided an opportunity for enhanced livelihoods and social inclusions, as evidenced through the celebrated *Linis Ganda* program. The management of waste is intrinsic to the future of Metro Manila but these challenges remain largely unmet.

Donovan Storey
University of Queensland

See Also: Developing Countries; Open Burning; Open Dump; Organic Waste; Politics of Waste; Population Growth; Slums; Street Scavenging and Trash Picking.

Further Readings

Asian Development Bank. *The Garbage Book: Solid Waste Management in Metro Manila*. Mandaluyong, Philippines: Asian Development Bank, 2004.

Baud, I., S. Grafakos, M. Hordijk, and J. Post. "Quality of Life and Alliances in Solid Waste Management: Contributions to Urban Sustainable Development." *Cities*, v.18 (2001).

Shah, Jitendra J. and Tanvi Nagpal. *Urban Air Quality Management Strategy in Asia: Metro Manila Report*. World Bank Technical Paper. Washington, DC: World Bank Publications, 1998.

Marine Protection, Research, and Sanctuaries Act

Rachel Carson elevated public awareness about the long-term detrimental effects of pesticides widely used in the U.S. agricultural industry with her 1962 book, *Silent Spring*. Many credit Carson with launching the environmental movement of the 1970s, but few realize that she began her career as a marine biologist with the U.S. Bureau of Fisheries and wrote extensively about marine life and the undersea world.

During the 1970s, air and water pollution, as well as other environmental issues, commanded federal attention, resulting in the creation of new regulations and agencies. Congress passed the Marine Protection, Research, and Sanctuaries Act (MPRSA) in 1972 during the administration of President Richard M. Nixon. The newly formed Environmental Protection Agency (EPA) was granted authority to regulate and enforce MPRSA.

Ocean Dumping

As its name suggests, MPRSA initiated groundbreaking legislation designed to address all aspects concerning the safety, sanctity, and proliferation of marine waters. One of the most critical problems motivating the need for legislation was an alarming escalation in the amount of urban sewage and industrial waste being dumped into both coastal and deepwater marine areas. For this reason, MPRSA is also known as the Ocean Dumping Act. The EPA regularly monitored ocean dumping and made reports to Congress from 1973 to 1990. The Ocean Dumping Ban Act of 1988 amended dumping concerns outlined in MPRSA by prohibiting the dumping of sewage sludge and industrial waste after 1991.

One area particularly threatened by ocean dumping and targeted for specific protection under MPRSA is the New York Bight Apex, an area in the continental shelf waters of the Atlantic Ocean. The dumping of municipal sewage sludge at a site located only 12 miles from Sandy Hook, New Jersey, was phased out between 1983 and 1986. Beginning in 1986, permits for dumping were authorized in waters deeper than 2,000 meters (m), at a site designated 106 miles (mi.)

One of the most critical problems motivating the need for legislation protecting marine ecosystems is an alarming increase in the amount of residential sewage and industrial waste being dumped into coastal and deepwater marine areas. The Marine Protection, Research, and Sanctuaries Act of 1972, which was regulated and enforced by the newly formed Environmental Protection Agency, was designed to address the safety, sanctity, and proliferation of marine waters, including such areas as industrial waste disposal, ocean drilling, and research.

from shore. In 1995, the EPA and the National Oceanic and Atmospheric Administration issued a report to Congress concerning the effects of sewage dumping on the 106-mi. site and industrial waste disposal at a former site nearby, which was also called the 106-mi. site.

In accordance with the Ocean Dumping Ban Act of 1988, industrial waste disposal ceased in 1987, and the former site was officially de-designated in 1992. Dumping at the 106-mi. municipal sewage site also ceased in 1992.

Unlike the Rivers and Harbors Act, which regulates the release of refuse and waste so as not to impede the navigation of inland waterways and coastal harbors, MPRSA regulates ocean dumping with the primary purposes of protecting living marine resources, ensuring the viability of commercial fishing endeavors, and preventing potential health hazards to humans. To this end, MPRSA also includes provisions for marine research, especially regarding the effects of ocean dumping on the marine environment and the establishment of sanctuaries to preserve and protect marine ecosystems. Among the topics covered are regulations for coastal zone management, the administration of funding for research programs, the maintenance of standards for marine sanctuaries, and the enforcement of legislation. Several amendments were made to MPRSA from 1974 to 1996.

Ocean Drilling and International Waters

Ocean drilling, an industry established long before 1972, also comes under the scrutiny of MPRSA. In the wake of the 2010 British Petroleum oil rig disaster in the Gulf of Mexico, it is interesting to note that MPRSA cites the director of the Minerals Management Service as one of the key authorities with whom the EPA administrator must consult in developing and implementing MPRSA strategy. MPRSA legislation is not limited in scope to the boundaries of U.S. territorial waters. Section 109 mandates that the secretary of state and EPA administrator seek to capture the attention and cooperation of the international community regarding the world's oceans. The Third United Nations (UN) Conference on the Law of the Sea was convened in 1973 and established precedents for negotiation and discussion regarding legislation for international waters, culminating in the 1982 UN Convention on the Law of the Sea. The 1982 convention codified commonly accepted principles and new concepts regarding the high seas, the continental shelf, fishing rights, pollution, marine scientific research, and other related topics.

As of 2005, the 1982 convention was accepted by 148 parties. Certain articles of the 1982 convention are worth highlighting. Article 56 proscribes jurisdiction over marine scientific research and the protection and preservation of the marine environment within the exclusive economic zone to the coastal state. Accordingly, other states with rights in the exclusive economic zone must defer to and comply with the laws of the coastal state. Article 94 defines duties of the flag state of a ship, including the adherence to international regulations concerning the reduction and control of marine pollution. Articles 116 through 119 outline the fishing rights and duties for the conservation and management of living resources of the high seas.

Nancy G. DeBono
Independent Scholar

See Also: Clean Water Act; Environmental Protection Agency (EPA); Pollution, Water; Rivers and Harbors Act.

Further Readings

Chandler, William J., et al. *How the National Marine Sanctuaries Act Diverged From the Wilderness Act Model and Lost Its Way in the Land of Multiple Use.* Washington, DC: Marine Conservation Biology Institute, 2007. http://www.mcbi.org/publications/pub_pdfs/Chandler_Gillelan_2007.pdf (Accessed July 2011).

Environmental Protection Agency. "Marine Protection, Research, and Sanctuaries Act." 2011. http://www.epa.gov/history/topics/mprsa/index.htm (Accessed July 2011).

Janis, M. W. and J. E. Noyes. *International Law: Cases and Commentary*. 3rd ed. New York: West Publishing, 2006.

Weinberg, Philip and Kevin A. Reilly. *Understanding Environmental Law*. 2nd ed. Newark, NJ: LexisNexis, 2008. http://www.lexisnexis.com/lawshool/study/understanding/pdf/EnvLawTOC.pdf (Accessed July 2011).

Marketing, Consumer Behavior, and Garbage

Poor societies produce less garbage than affluent societies. Although garbage has been problematic since people have begun forming settlements, industrialization has been credited with initializing large-scale environmental degradation. Prior to the creation and use of incinerators in 1885, refuse, garbage, and animal waste were left in the streets to be trampled upon or to be eaten by livestock, bringing health devastation such as cholera, in addition to environmental degradation. With industrialization came consumerism, and the remains of that consumerism have created heightened levels of garbage.

Although adapting to changing needs over time, marketing has been in existence since people have sought prosperity. By attracting a wider audience, producers (and marketers) could sell more and thus increase profitability. The use of mass media for advertising has created a monolith of spiraling overconsumption: the more products one has, the more space one needs to store the products, and the more waste that accumulates, eventually affecting landfills with packaging and products.

History of Consumerism and Garbage

While initial forms of advertising appeared as matter-of-fact statements in newspapers, marketing became more sophisticated with the onset of the Industrial Revolution. By the 1800s, advertisers began to adopt the persuasive techniques that Westerners are accustomed to seeing: graphics, language, and statements of grandeur. In the late 1800s, commercialization brought convenience and, with it, challenges. In 1879, Frank Woolworth pioneered the idea of contemporary retail sales by openly displaying products on counters so that consumers could see and feel the merchandise, enticing them to purchase. Prior to direct access, customers looked at products from afar or required direct assistance. Woolworth's practice of hands-on salesmanship, novel and ingenious in its time, later made larger, antitheft packaging necessary.

Accompanying this higher demand for products was a call for convenience and affordability. This movement began in 1895 with Gillette's creation of the first razor with disposable blades, with the first fully disposable razors becoming available in the 1960s. Disposability meant that products became, in the short run, cheaper and less durable. The trend toward a disposable society grew exponentially in the early 1900s, as household products such as linens and mugs were replaced by paper towels, Kleenex, and paper cups. The creation of wax paper for cereal box liners and as bread wrappers meant that products could be manufactured and shipped for broader consumption. Over time, disposability became so rampant that only a century after its creation, the average U.S. worker used approximately 500 disposable cups per year.

Expanding on the hands-on, do-it-yourself merchandising success of Woolworth, Clarence Saunders opened the first supermarket in 1926. The introduction of prepackaged foods and self-service packaging increased food and product selection while decreasing the cost. Packaging was enhanced in 1928 when cellophane was invented, enabling protection of foods and other products, and in 1939 with the creation of precooked frozen foods. With the creation and widespread use of convenience foods and increasing consumerism came an increase in packaging. This packaging enhancement was both for product safety, so that foods would arrive to their destinations intact, and to increase product desirability. More than plain brown wrappings, the packaging became advertising in its own right, enticing consumers to purchase goods.

Produce and Sell More

In 1947, industrial designer Gordon Lippincott, working to capitalize on this trend toward mass production, stated, "our willingness to part with something before it is completely worn out is a phenomenon noticeable in no other society in history . . . it is soundly based on our economy of abundance. It must be further nurtured even though it runs contrary to one of the oldest inbred laws of humanity; the law of thrift." Producers, and thus marketers, focused on revenue generation and profitability to fuel further production. The objective in the industrial age is increased consumption, diametrically opposed to the value of thrift emphasized during and after the Great Depression.

From the drive toward increased consumerism, in the 1950s, the United States saw a vast growth in convenience foods accompanied by a surge in packaging waste that now needed disposal, including the ubiquitous television dinner. This trend continued when, in 1953, the chairman of U.S. President Dwight D. Eisenhower's Council of Economic Advisors emphasized that the U.S. economy's "ultimate purpose is to produce more consumer goods." B. Earl Puckett of Allied Stores added that "it is our job to make women unhappy with what they have." The age of consumerism was upon the United States, with an ever-increasing desire to produce and sell more. By 1954, this proliferation of the net of discontent was cast wide to include children as consumers.

The savage consumerism that ensued was legendary, with the 1980s being the epitome of overconsumption. By 1988, the Environmental Protection Agency (EPA) estimated that 70 percent of landfills had closed because they were full, unsafe, or had not adhered to set standards. Not seen as "good neighbors," NIMBY (Not in My Backyard) ensued and it was—and continues to be—difficult to site new landfills. In 1989, the EPA issued an Agenda for Action, with a focus on waste reduction and recycling. Michael Jacobson of the Center for the Study of Commercialism lamented that marketing initiatives could do more; people are "reminded a hundred times a day to buy things, but we're not reminded to take care of them, repair them, reuse them, or give them away." This growing sentiment of wastefulness has not shown a negative impact on sales and marketing but has shown some success in participation in recycling initiatives that, at the least, keep products out of the waste stream.

Zero Waste America reports that, while recycling efforts have increased dramatically (32 percent in

2005 compared to 6.4 percent in 1960), the amount of municipal waste per capita is also increasing. The amount of waste incinerated has decreased by nearly half between 1960 and 2005 (from 30.6 percent to 15.9 percent). Meanwhile, with recycling efforts on the rise, in 2005 the EPA issued the Resource Conservation Challenge 2005 Action Plan, creating recycling initiatives to help the United States reach the national goal of recycling 35 percent of municipal waste. Key areas of focus included paper/paperboard (from 44.9 to 53.8 percent), food waste (from 2.8 to 5 percent), yard trimmings (from 56.5 to 60 percent), folding paper cartons (from 8.7 to 14 percent), wood packaging (from 15 to 24 percent), plastic wraps (from 6.6 to 19 percent), and beverage containers (from 26 to 39 percent).

According to the Clean Air Council, one-third of waste generated in the United States is packaging. Further, an additional 5 million tons of waste is generated during the holidays; 4 million tons of this is wrapping paper and shopping bags. While conventional wisdom suggests that product packaging ravages landfills in its excess, there is some argument that additional packaging reduces the impact on landfills by reducing product breakage and food waste. Food waste, mostly from fresh fruits and vegetables, dairy, and grain products, is the third-largest waste component by weight.

When measuring municipal waste by weight rather than volume, packaging has decreased by 30 to 70 percent since the mid-1980s. This is largely because product packaging has become more compact as new technologies for product and shipping safety become available. This is a double-edged sword since, as packaging becomes increasingly economical, products travel farther distances. For example, due to enhanced packaging and transportation, kiwifruit can be imported by the United States from Australia. Long-distance shipping influences not only packaging and waste but also fuel expenditures and other wastes as transportation distances increase.

As demand for more exotic imports becomes increasingly widespread, cross-contamination by pests and diseases from once-remote quadrants becomes increasingly likely and dangerous. For example, West Nile virus, once endemic to Egypt and the Nile River basin, now makes regular appearances in the United States. The Mediterranean fruit fly (medfly) infestation in California in the early 1980s not only affected the local economies in which the infestation spread but has also affected agricultural exports. Pacific Rim countries, particularly China and Japan, disallow exports of crates containing evidence of medfly larvae for fear of contamination spread.

Supply and Demand

The rise of industrialization enabled producers to manufacture products that had previously been painstakingly made by hand. Steel and iron came into ready use by the 18th century in conjunction with technological changes such as the invention of the steam engine and the spinning jenny. Transportation and communication were also modernized with the widespread use of railroads and the telegraph. With the rise of process mechanization, the economy shifted from that of informal, home-based production to that of a formal, factory-based production. As production became increasingly formalized, bureaucracies were formulated to manage the new working class. The streamlining of processes through automation meant that much more merchandise could be produced in a fraction of the original time and at a substantial cost savings.

With the proliferation of merchandise that was available through industrialization in a market economy came an increased need to sell more products to more people. Rather than selling merely to the local economy, there was a need to expand sales to larger markets. The faceless nature of broad sales inherent in industrialization and bureaucratization also brought alienation. Not having hand produced each product, workers were less invested in the product because of mechanization; workers were responsible for only a part of the process, rather than the workmanship required to make the merchandise from start to finish.

Mechanization also meant that each piece could be made perfectly, faster than could be done by hand, thereby increasing the precision of components. Where industrialization created assembly lines and permitted mass production of items, it also permitted the creation of spare parts. Since parts were now identical, they could be more easily replaced.

Instead of buying a whole new product, products could be repaired, thus lengthening the longevity of the product.

The speed of production created an increased need for advertising because there was a need for higher demand to meet the larger supply. Since the products were not being bartered or sold between neighbors, there was a need to enhance the perception of want for the respective products. Enticing someone to purchase something that they could make themselves took a new way of thinking and approaching marketability. For rural customers of the late 1800s, catalog sales served as a lifeline to the lifestyle that many dreamed of having. The Sears and Roebuck Wishbook brought modern conveniences, styles, and home goods to those who otherwise did not have access.

Marketing and Mass Media

An unexpected boon for marketing occurred in 1904 when Postmaster General Henry Clay Payne first authorized permit mail. With this, a single fee allowed 2,000 or more pieces of mail to be delivered without stamps, thus creating the opportunity for direct mail. As a result, retailer Montgomery Ward mailed out 3 million four-pound catalogues to waiting customers. A century later, Americans currently receive approximately 4 million tons of direct mail each year, most of which ends up in landfills.

Mass media, spanning across print, television, radio, film, and computers, allows manufacturers and advertisers to communicate with vast numbers of potential customers across a wide array of locales and walks of life. Through highly accessible media outlets, advertising is largely able to promote products and convince large segments of the population to purchase questionably necessary products. The food of the ubiquitous U.S. icon, McDonalds' Big Mac, is now consumed worldwide in largely the same style, irrespective of cultural differences.

In an effort to keep up with the proverbial Joneses, Americans are buying more, working more, saving less, and getting further behind. Mass media campaigns are largely responsible for fueling rampant consumerism. Consumers are convinced that not only will the product work as advertised but also a single, convenient product will make them slimmer, smarter, more popular, and thus more desirable. Marketing to a bloated culture means that more and newer products are required not only to keep up but also that the products will achieve results faster and with less effort—the ultimate in convenience. For example, using Bod Man body spray will have women fighting over average-looking men, and Slimfast will make dieters model-slim, fast.

In addition to products that are designed to make the person appear more successful and desirable, mass marketing has spread to areas beyond the individual's presentation of self. Transportation, once considered merely necessary to get from one point to another, now serves as a status symbol. Public transportation is often perceived to be the transportation of the poor, with the wealthy driving (and displaying) increasingly upscale modes of transportation: luxury automobiles that purport to be driving as if riding on a cloud (and costing as much as some housing). In lieu of passenger planes or ships, private jets and luxury yachts take luxury and one-upmanship to the air and seas, respectively.

While the size of the average U.S. family has been decreasing over time, the average home size has increased by more than 50 percent since the 1950s, with nearly an additional 1,300 square feet of living space. One in five new homes is larger than 3,000 square feet, although the trend is to begin to scale back the size of homes. Fourteen million U.S. households own at least four television sets. Since 2001, the number of Americans who purchased second homes increased by 24 percent.

Overconsumption is rampant to the point that people who claim to be unable to afford food or housing display excess through designer-label clothes and smart phones with the requisite data plan. Mass media has displayed what are the appropriate icons of the culture and, to be a cultural member, these items must be displayed—even at the cost of basic necessities. All of this consumption comes at a cost, including deficiencies in health and well-being and the maintenance of an unending spiral of consumption and waste. As newer, sexier products are advertised and purchased, older items—often still usable in their capacity—are discarded.

Recycling Initiatives

Mass media is a powerful tool for social change. The same techniques used in product marketing have been shown to be effective in community development. Consumer-oriented social marketing includes design, implementation, and program control to increase popularity of a social idea among a target population. A prime example of the effectiveness of a social marketing campaign is the antismoking crusade that dramatically shifted public opinion and policies about consumption and public display.

According to the Council for Waste Solutions, nearly 40 percent of waste in the United States may have extended use or redistribution potential: 14 percent durable (such as appliances) and 25 percent semidurable (such as books or clothes). In the mix of appliances, furniture, and clothing, a significant portion of waste comes from food packaging and food remnants. In developing countries, these extended-use items fuel the local economy, much of which is surviving on scraps, rather than filling landfills.

As of 2010 in the Westernized world, recycling is well accepted at both consumer and commercial levels, with 90 percent of households reporting that they support recycling initiatives. Separate bins for recycling and trash can be found and are used in many public areas as well as in homes and offices. However, according to the EPA, the volume of household waste in the United States generally increases 25 percent between Thanksgiving and New Year's Day—about 1 million extra tons. In 2010, while unemployment soared at 9.6 percent, Black Friday, so named for the notorious shopping day on which retailers' sales put them "back in the black," sales rose (19 percent online and 12 percent in retail stores) with a 33-percent increase in Thanksgiving shopping. Ironically, studies show that greater waste occurs in time of scarcity: shortages create situational pressures during which people buy more than they need. In the United States, 12 percent of grocery products are wasted and never used. For example, if there were a shortage of bananas, consumers fearful of being denied access to bananas would be likely to purchase more than they need in an attempt to ensure that they are not left without any bananas, thus taking more than their "share" and generating additional waste when the bananas are not consumed before spoiling.

In addition to traditional recycling, in which items are destroyed to create new products such as paper and plastic, a type of recycling growing in popularity is the reuse of items through donation. Goodwill has been in operation since the early 20th century, selling inexpensive cast-offs to people in need. Organizations such as Matthew 25: Ministries accept usable products to be given to the world's less fortunate. In addition to individual donations of clothes and household goods, Matthew 25: Ministries receives many product donations from corporations. Misprints, overruns, or "imperfects" (products usable but not sellable that otherwise would have gone to feed local landfills) are given to people in need. Habitat for Humanity, best known for building homes for the poor, also operates ReStore, which accepts usable appliances and construction materials to sell at discounted prices, generating revenues to further operations and simultaneously avoiding contributing to landfills with otherwise serviceable products.

Precycling

Beyond recycling is an emerging initiative—precycling. The goal of precycling is to reduce the burden on landfills by encouraging consumers to make sustainability considerations part of the purchasing decision, including the rejection of the purchase of unnecessary products and packaging that cause disposal problems and instead choosing quality products that are reusable with a long product life.

Inherent in the philosophy of precycling is decreased consumption. This includes purchasing appliances only when older appliances are no longer functional or sustainable. When there comes a need to replace a worn-out appliance, careful planning takes place in an attempt to ensure that the purchased product will last for an extensive length of time and, where appropriate, will be an energy-saving device.

Unfortunately for many, this decision comes at a cost as the higher-quality, more-sustainable, and longer-lasting products are accompanied by a higher price tag. While the long-run benefits may be substantial, consumers may have difficulty absorbing the short-term costs. For example, tankless water heaters that heat water on demand purport to save

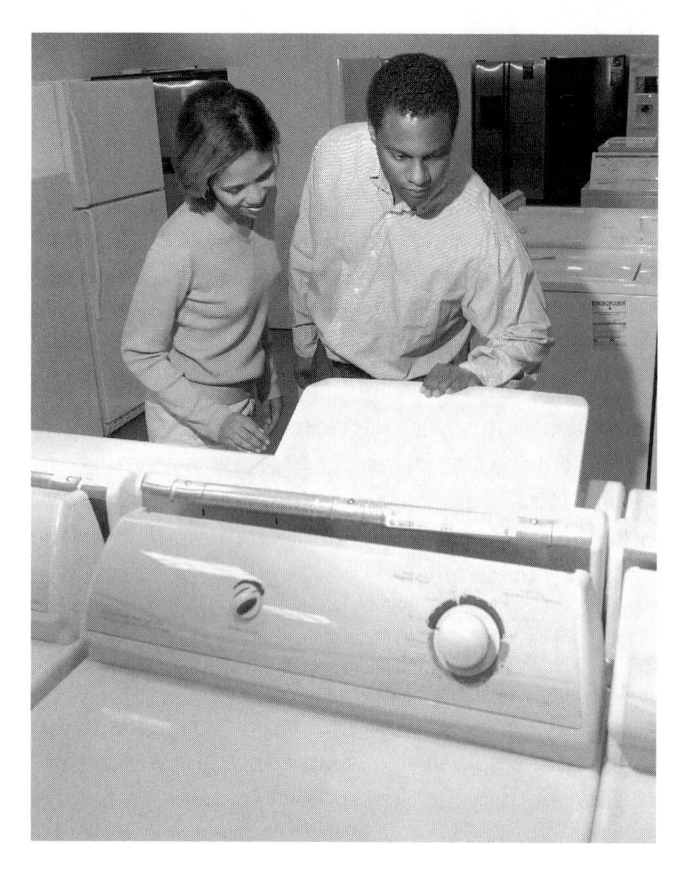

The goal of precycling is to reduce the burden on landfills by encouraging consumers to consider sustainability when deciding what products to purchase, or if the purchase is needed at all. Deciding factors include durability and minimal packaging.

energy and thus money. The financial reality is that, on average, tankless water heaters cost more than twice as much as standard water heaters, and installation costs are more than four times as much as a standard install.

Further, some consumers may opt to dispose of usable products in exchange for newer, more-sustainable products that claim to save energy, and thus money, in the short run. Purchasing a hybrid vehicle may be a sound choice if one were in need of a new vehicle. However, if a person were to dispose of a working vehicle in exchange for a hybrid vehicle, estimates indicate that it would take at least seven years to realize fuel-cost savings.

While precycling is certainly a wise choice toward achieving a sustainable environment, it is often a choice for the elite. Until sustainable products also become more affordable, as through supply-demand economics, they are less likely to be an option chosen by the masses. Assuming the market economy goal to produce more is maintained and demand for sustainable products continues to grow, mass-media marketing will continue to increase the desirability of quality, sustainable products. Precycling will certainly continue to grow as a successful initiative.

Conclusion

With industrialization came consumerism and the remains of that consumerism have created heightened levels of garbage. In an attempt to increase revenues and profits, marketers have sought to attract wider audiences through the use of mass media, creating a monolith of spiraling overconsumption. Perhaps as consumer preference shifts toward more-sustainable products and a general belief in environmental sustainability, marketers will find new ways to popularize thrift while investing in renewable systems and technologies to create a sustainable level of refuse.

Leslie Elrod
University of Cincinnati

See Also: Capitalism; Consumerism; Gillette, King C.; Grocery Stores; Junk Mail; Materialist Values; Overconsumption; Sociology of Waste; Supermarkets; Sustainable Waste Management.

Further Readings

Association of Science—Technology Centers. "Rotten Truth About Garbage." http://www.astc.org/exhibitions/rotten/timeline.htm (Accessed November 2010).

Beck, R. W. "Literature Review—Litter." http://www.kab.org/site/DocServer/Litter_Literature_Review.pdf?docID=481 (Accessed November 2010).

Carmel, Ava. "Use of Mass Media in Community Development." *Shalom Magazine*, v.1 (1996). http://www.mfa.gov.il/MFA/Mashav%2%E2%80%93%20International%20Development/Publications/1999/Use%20of%20Mass%20Media%20in%20Community%20Development (Accessed November 2010).

Clean Air Council. "Waste Facts and Figures." http://www.cleanair.org/Waste/wasteFacts.html#production (Accessed November 2010).

Environmental Protection Agency. "Resource Conservation Challenge 2005 Action Plan." http://www.epa.gov/epawaste/rcc/resources/action-plan/act-p1.htm (Accessed November 2010).
Evolving Excellence. "Fun With Statistics, Garbage Edition." http://www.evolvingexcellence.com/blog/2007/09/fun-with-statis.html (Accessed November 2010).
Fox, Nathan. "This New House." *Mother Jones* (March/April, 2005).
Mills, Libby. "Truth in Trash: Investigating Waste Reveals Food Behaviors." *Today's Dietician* (2007). http://www.todaysdietitian.com/newarchives/tdsept2007pg84.shtml (Accessed November 2010).
Pieters, Rik G. "Changing Garbage Disposal Patterns of Consumers: Motivation, Ability, and Performance." *Journal of Public Policy and Marketing*, v.10/2 (1991). http://www.jstor.org/stable.30000236 (Accessed November 2010).
Wisconsin Department of Natural Resources. "DNR 2006 Recycling Survey Executive Summary." http://www.dnr.state.wi.ur/org/aw/wm/publications/anewpub/wa1108.pdf (Accessed November 2010).

Maryland

A U.S. state in the mid-Atlantic region, the variety of topography in the state of Maryland promotes the nickname "America in Miniature." Most of the population is urban and suburban and is situated around Washington, D.C. (the capital district), and Baltimore, the state's largest city, forming the Baltimore-Washington Metropolitan Area. The U.S. Census Bureau states that, in 2009, Maryland had the highest median household income ($69,272). The state is a major center of life sciences research and development, particularly in biotechnology. Economic activity is strong in the tertiary service sector, which is heavily influenced by location, including the Port of Baltimore and its related rail and trucking and the close proximity of the center of government in Washington, D.C., which requires technical and administrative staff. Food production is also a major part of the Maryland economy, having a large commercial fishing industry in and around Chesapeake Bay and also on the Atlantic Coast. Large agricultural areas also exist, primarily used for dairying and speciality horticulture crops. The southern counties are sufficiently warm to allow a tobacco cash crop to be grown. Manufacturing in Maryland has a high dollar value and is highly diversified. One of the most environmentally friendly states in the United States, Forbes.com ranked Maryland as the fifth-"greenest" state in 2007. The state is a very low consumer of energy and produces very little toxic waste. In 1988, the state passed the Maryland Recycling Act (MRA), which established that newsprint and telephone directories distributed in the state must use recycled paper and required recycling programs in Maryland's jurisdictions and in state government.

Statistics and Rankings

The 16th Nationwide Survey of MSW Management in the United States found that, in 2006, Maryland had an estimated 7,009,905 tons of municipal solid waste (MSW) generation, placing it 21st in a survey of the 50 states and the capital district. Based on the 2006 population of 5,602,017, an estimated 1.25 tons of MSW were generated per person per year (ranking 25th). Maryland has nine waste-to-energy (WTE) facilities, which processed 1,371,970 tons of MSW (eighth out of 32 respondents). Maryland recycled 2,536,633 tons of MSW, placing Maryland 17th in the ranking of recycled MSW tonnage. It landfilled 3,101,302 tons (26th) in the state's 22 landfills. It exported 2,861,849 tons of MSW, and the import tonnage was 302,167. In 2006, Maryland was not increasing its 84,838,536-cubic-yard landfill capacity; it was ranked 21st out of 44 respondent states for number of landfills. Separately bagged yard waste, recyclable containers and paper, asphalt paving, brick, concrete, metal, wood, whole tires, used oil, and white goods were reported as banned from Maryland landfills. Tipping fees across Maryland averaged $52, where the cheapest and most expensive average landfill fees in the United States were $15 and $96, respectively.

Brief History of Waste

With the invention in the early 1800s in England of a de-inking process for paper recycling by Matthias Koops, the potential of recycling paper and using wood pulp to make paper was unlocked. In 1874, Baltimore began the first-ever curbside recycling of paper, but the program ultimately failed (the concept re-emerged successfully in the 1890s, in New York). Around this time, Baltimore had acquired an unfortunate reputation as "the city of open sewers," the last major U.S. city to use a cesspool drainage system; there were 80,000 open sewers in the city area.

City leaders rejected proposals for a citywide sewer system several times between 1859 and 1905 because of opposition and political infighting. Baltimore adopted the landfill system during World War II when there was insufficient manpower to run the city incinerators. The process was so efficient that it was continued in peacetime.

Hampden Community Archaeology Project

The University of Maryland Hampden Community Archaeology Project (HCAP), established in 2005, devoted its first seasons to the excavation of dumps and privies and studying the use and appearance of yards. Statistical comparisons of artifact assemblages from "end of row" communal dumps and single-family yard middens and privies has also been undertaken to examine the everyday lives and consumption of mill worker families. The HCAP set out to create a public and critical dialogue on the issues of class and economic inequality, which were relevant to the neighborhood's issues.

Hampden was a historical by working-class neighborhood in Baltimore, a 7,500-strong urban community dating back to the early 19th century. The village developed to support the gristmills along Jones Falls, which were then converted in the textile industry boom before being incorporated into the city in 1888. The area remained a predominantly white, working-class area into the 1970s and was transformed by gentrification in the early 1990s. The area had a strong history of working-class activism and labor disputes with the mill owners before the mill companies moved south in search of cheaper labor, and the massive drop in demand after World War II, reduced Hampden's industrial base.

Baltimore Pyrolysis Plant

In the late 1960s and early 1970s, faced with the crises of garbage accumulation and oil shortage, the U.S. government funded various resource recovery research programs in the hope that WTE could alleviate some of the problems. Pyrolysis, which cooked trash rather than incinerating it to release oil, carbon, and gas, was seen as the most promising method of disposal, as it provided a solution to both the garbage and the oil problems. It was, however, also the most expensive, technologically complex, and heavily funded of the WTE methods under research.

Roughly contemporary with the pyrolysis experiments in San Diego County, California, the city of Baltimore and the state engaged a similar project. The plant was constructed by Monsanto Enviro-Chem Systems, Inc., which was given one of the largest pyrolysis demonstration grants ever awarded by the Environmental Protection Agency (EPA). It proceeded with what was at that point the most expensive resource recovery facility ever built.

Monsanto was convinced that its Landgard system could process half of Baltimore's waste (1,000 tons of MSW per day) in a plant vaunted as the "first full-scale pyrolysis solid-waste disposal and resource-recovery system in the world." The Landgard system is not regarded as true pyrolysis, as a form of pyrolysis is only used in the initial processing stages before the heated waste is combusted and the gas converted to steam, not a product of true pyrolysis.

After cost overruns in excess of $11 million, when the Baltimore Pyrolysis Plant opened in 1974, it did not meet Maryland's new air quality regulations and was beset by unrelenting technical problems, which saw it dubbed the "Paralysis Plant." The plant ended up causing a city garbage crisis, and Baltimore was left with a large part of the final $27 million bill. The plant, which was technically an incinerator and not a pyrolysis plant, did irreparable damage to the reputation of pyrolysis, which led the director of Maryland Environmental Services to state that, "Just the word *pyrolysis* makes people nervous." Monsanto recommended shutting the plant down, but it was able to run at a reduced load despite the high expense per ton this involved. Eventually, Monsanto withdrew from the project, and the city converted it into a mass burn WTE plant.

Jon Welsh
AAG Archaeology

See Also: Incinerators; Sewage; Sewage Collection System.

Further Readings

Arsova, Ljupka, Rob van Haaren, Nora Goldstein, Scott M. Kaufman, and Nickolas J. Themelis. "The State of Garbage in America: 16th Nationwide Survey of MSW Management in the U.S." *BioCycle*, v.49/12 (2008).

Chidester, Robert C. and David A. Gadsby. "One Neighbourhood, Two Communities: The Public Archaeology of Class in a Gentrifying Urban Neighbourhood." *International Labor and Working-Class History*, v.76 (2009).

Melosi, Martin V. *The Sanitary City: Urban Infrastructure in America From Colonial Times to Present.* Baltimore, MD: Johns Hopkins University Press, 2000.

Massachusetts

The Commonwealth of Massachusetts has a population (2009 est.) of 6,593,587. The urbanized region of Boston is estimated to have a population (2008 est.) of 4,522,858. Massachusetts is the third most densely populated state at 809.8 persons per square mile.

With settlements dating from the 1620s, the Commonwealth of Massachusetts has a long history of handling and disposing of waste. Although industrialization brought inevitable environmental degradation, early adoption of public waste handling and infrastructural practices, presence of numerous educational and research institutions, and implementation of large infrastructural projects have made Massachusetts a center for advancements in waste treatment, disposal, reduction, and recycling.

General History

Under British rule, Massachusetts' ports were critical locations for international trade. As material and energy flows increased, the towns surrounding Massachusetts Bay flourished, as did waste accumulation and degradation of waterways. The Boston Tea Party of 1773 intensified a trend of pollution into Boston Harbor that would continue for the next 200 years. After the American Revolution, Massachusetts developed into a primary site of the Industrial Revolution, where textile manufacturing centers such as Lowell and Holyoke blossomed early in the 1800s, as well as other centers focused on machine tools, shoes, and paper products. This transition from agriculture to industry created long-lasting environmental damage as populations soared and industries took advantage of local natural resources. New Bedford, a shipping and textiles center on Buzzards Bay, exemplified the unfortunate outcome of years of neglect; the harbor was labeled a Superfund site in 1982 by the Environmental Protection Agency (EPA). The textile industry's collapse in the early 20th century has led to growth of high-tech and biotech industries emerging out of the many local educational institutions, which have generated associated hazardous and biological wastes.

Solid Waste Collection and Disposal

Throughout the commonwealth, waste collection and disposal has historically relied on both landfills and combustion. The use of one method over the other was, until the 1960s, a matter of availability and cost of transport, but with little regulation for health or environmental concerns. In the heavily populated eastern regions, local landfills have gradually been replaced by combustion and larger regional landfill sites because of space limitations.

Massachusetts enacted its first solid waste disposal law in 1955, allowing local health boards the power to approve new waste disposal facilities before construction. In 1982, voters passed the Beverage Container Recovery Law, or "Bottle Bill," requiring a $0.05 deposit on all retail sales of bottles and cans. A 1987 overhaul of the law included new regulations for siting facilities and provided funds for municipalities to improve outdated facilities. In 1990, the Massachusetts Department of Environmental Protection (MassDEP), the commonwealth's waste regulator and manager, introduced the first Solid Waste Master Plan, as well as a series of measures aimed at decreasing hazardous waste and banning specific materials from landfill or combustion. Since then, MassDEP has overseen the closure and cleanup of countless contaminated sites, reducing the number of landfill sites from 150 before 1990 to 24 in 2010. MassDEP has also regulated mercury since 2006, with the intent of phasing it out of the Commonwealth's waste stream.

By 2010, seven combustion facilities existed in Massachusetts, disposing of an average of 25 percent of generated material, and all were operated by either Covanta Energy or the Wheelabrator Corporation, a division of Texas-based Waste Management, Inc. The 24 active landfills receive an average of only 15 percent of material generated. The remaining material, almost 50 percent since adoption of the MassDEP master plans, is diverted from disposal or combustion through recycling and compost. Although this rate ranks among the nation's highest, municipal recycling and composting rates remain consistently low (paper products still make up one-third of the municipal waste stream). Between 1997 and 2010, residential recycling rates fluctuated little at 27–28 percent, with the highest rate of almost 100 percent on Nantucket Island. Boston-area residential recycling is lower still, at 15–25 percent, although the gradual adoption of single-stream recycling practices in municipalities such as Cambridge have attempted to raise this statistic.

Boston's Infrastructure

As Massachusetts' economic, cultural, and political hub, the Boston metropolitan area is also its highest waste producer. Having grown out of a small neck of land in Massachusetts Bay, the city physically transformed itself over three centuries, expanding area, consuming vast monetary and physical resources, and causing substantial environmental impact. During the 19th-century population booms, Boston undertook extensive land-reclamation campaigns, filling surrounding marshlands and Boston Neck's perimeter. Citing health hazards from stagnant water, the Charles River basin's south marshes were filled to create the Back Bay, finished in the 1890s, followed by South Bay. The Massachusetts Institute of Technology (MIT) moved to its present campus on filled land on the Charles River's Cambridge bank in 1910, and Logan Airport added considerable acreage to east Boston in the mid-20th century for jet runways. Material for these operations came from hills at the center of Boston, and from hills and gravel pits as far away as Needham.

Not all large-scale infrastructural projects in Boston have been additive. The Central Artery/Tunnel Project, popularly known as the "Big Dig," was, at $22 billion, the most expensive transportation project in U.S. history by 2010, involving the removal of the 40-year-old elevated highway separating the North End from downtown, replacing it with a 3.5-mile tunnel. Although the project created vast quantities of waste, elements of the waste stream were productively used throughout the region. Of the 30 million tons (16 million cubic yards) of earth excavated for the tunnel, much of the material was used to cap Boston-area landfills or was added to Spectacle Island in Boston Harbor. In 2008, the architecture firm Single Speed Design designed and constructed a prototype house in Lexington using over 600,000 pounds of steel and concrete from the highway's demolition.

Water and Sewage

Massachusetts residents have used private sewers for water runoff since 1700, with the first publicly constructed and maintained storm sewers in Boston created soon after. In 1833, sanitary waste and sewage were added to the system, resulting in dangerous public health issues, chief among them an increase in cholera. Studies resulted in the 1875 construction of the Boston Main Drainage System to flush sewage into Boston Harbor. The Metropolitan Sewerage District was created in 1889, consolidating three formerly distinct districts, becoming the first entity of its kind in the country. As early as the 1930s, the public was voicing unease over polluted conditions in Boston Harbor as a detriment to health and leisure activities. Overflows of sewage from rivers emptying into Boston Harbor were common, resulting in vast beach closures. In response to growing environmental concerns, the city constructed two primary wastewater treatment plants: the Nut Island plant in Quincy (in operation from 1952 to 1998) and the Deer Island plant, which opened in 1968. In the mid-1980s, the newly empowered Boston Water and Sewer Commission oversaw the construction of two main interceptor tunnels to collect water and sewage; a renovation to the Deer Island plant, which serves approximately 20,500 acres in 2010; and the addition of a 9.5-mile outfall tunnel, allowing the city to discharge treated water into the deep waters of Massachusetts Bay. The system remains far from perfect, with flooding and backups occurring often during heavy rains.

Reform Endeavors

With the presence of many top-tier educational and nonprofit institutions, significant segments of the population are active-minded and progressive. Many organizations, including MassDEP, have endeavored to affect waste management trends. "Don't Waste Massachusetts" was formed in 2009 as a commonwealth-wide alliance of environmental organizations committed to influencing policy decisions, expanding recycling programs, and opposing waste-to-energy and landfill facilities. In academia, at MIT's Media Lab, the SENSEable Cities group has created a system for visualizing the movement of trash after it is thrown away. Titled *Trash-Track,* it attracts volunteers to affix sensors to individual items of their trash, then visualizes their movements via the Internet. Since 1990, MassDEP has released a series of Solid Waste Master Plans in an effort to curb the continually increasing waste disposal demand and to spur job growth in the recycling sector. The master plan under development for 2010 sought to reduce residential and commercial waste by 80 percent by 2050 as well as to eliminate virtually all toxic materials from entering the waste stream. Additionally, MassDEP began a program in 2009 to add renewable energy facilities at closed landfills, including wind turbines and solar panels.

Massachusetts's long history of consuming resources and handling waste has led to vast improvements in efficiency and has allowed the Boston region to flourish. However, large infrastructural projects have forever altered the area's natural ecology, and centuries-old pollution continues to threaten local soils and watersheds. In the early 21st century, the commonwealth works hard to continue upward trends in recycling, waste, and sewage management, while key institutions help make these trends visible.

Dan Weissman
Harvard University

See Also: Construction and Demolition Waste; History of Consumption and Waste, U.S., Colonial Period; Industrial Revolution; Pollution, Water; Rivers and Harbors Act; Water Treatment.

Further Readings

Boston Water and Sewage Commission. http://www.bwsc.org (Accessed July 2010).
Massachusetts Department of Environmental Protection. http://www.mass.gov/dep/recycle (Accessed July 2010).
Massachusetts Department of Transportation. http://www.michigan.gov/mdot (Accessed July 2010).
MIT SENSEable Cities Trash-Track. http://senseable.mit.edu/trashtrack (Accessed July 2010).
Northeast Recycling Council. http://www.nerc.org (Accessed July 2010).
Whitehill, Walter Muir. *Boston: A Topographical History*. 2nd ed. Cambridge, MA: Harvard University Press, 1968.

Material Culture, History of

Human activities with objects over the past 2.5 million years have had a significant impact on human cognitive, linguistic, and social configurations. Although other animals, including mammals and birds, employ simple tools to achieve certain tasks (such as using sticks to extract termites from mounds), no other animal makes composite objects or objects that are creatively abstracted from natural forms to produce completely invented shapes and designs. Material culture can be viewed through four dynamic processes: production, distribution, consumption, and discard. Each of these processes leaves a durable signature in the archaeological record through the recovery of artifacts along with their locations of use and abandonment. Production results in the manufacture of objects as well as the discard of waste fragments. The distribution of objects enables one to trace historical trade patterns in both raw materials and finished goods.

The consumption and use of objects includes not only daily activities but also ritual use and the consecration of objects through burial with the dead. Finally, the act of discard is often purposeful in that the treatment of unwanted objects indicates cultural perceptions of cleanliness and social order.

Earliest Tools

Archaeological remains show that starting in the earliest times, humans' use of material objects included attention to style as well as function. Just as in the 21st century, ancient people deliberately made and decorated objects through conscious decisions about form, function, projected lifespan, and the potential of the object to serve as a gift or as a statement of social status. Styles and forms can change rapidly, resulting in the frequent turnover of styles, which is interpreted as a language-like code by others in the social group. Sophisticated understandings of style and its changes can result not only in the adoption of new styles and designs but also the return of "retro" styles as a fashion or social statement.

Stone tools constitute the first evidence of human transformation of natural materials. The first crude pebble chopping tools date to approximately 2.5 million years ago, associated with the skeletal remains of hominids whose cranial capacity was a fraction of contemporary humans. The subsequent elaboration of both stone tools and brain size suggests that there was a mutually causal relationship between cranial capacity, cognitive sophistication, and material culture use.

By 1.65 million years ago, the Acheulian hand axe was developed: a leaf-shaped, bilaterally symmetrical stone tool that is found in many regions of Asia, Africa, and Europe. The repetitive aspects of the hand axe have led to suggestions that it had both a cultural and a physical function, perhaps even being used as a social signal of competence and reproductive fitness.

Ancient people also made objects from feathers, fur, wood, gourds, bone, leather, bark, and wax, although perishable objects are less visible in the archaeological record. Exceptional cases of preserved objects show the diversity of natural resource use, such as fire-hardened wooden spears from Germany dated to 400,000 years ago associated with butchered animals' remains.

Art and Agriculture

By around 65,000 years ago, the archaeological record shows the emergence of objects that were made primarily for a social rather than a technical function, such as beads and other ornaments meant to decorate the body. One widely used material was ostrich egg shell, which was made into containers as well as beads in Africa and in the Indian subcontinent. In this era, humans also created figurines and rock art for the first time.

By around 10,000–12,000 years ago, the human repertoire diversified considerably with the adoption of domesticated plants and animals. The move toward food production required the development and use of many new types of objects related to agriculture, including plows and hoes for tending plants, sickles and baskets for harvesting them, and grinding and cooking tools for preparing them.

Architecture

In addition to portable objects, material culture also refers to modifications of the built environment. Humans initially took shelter in caves and under trees, but the first constructed architecture is seen in the form of huts in eastern Europe made of mammoth bones and draped with skins, starting around 15,000 years ago. Later types of architecture include many different structures made of durable materials, such as stone, as well as from perishable materials, such as reeds, bamboo, mud, and sod.

Architectural construction also included special-purpose features that were larger than regular dwellings. Even prior to settled village life, communities built large ritual structures to which they returned on a regular basis (such as Stonehenge in England and Watson Brake in the United States). Funerary monuments are seen throughout the world as a focus of communal labor investment; these take the form of megalithic tombs as well as wooden poles, stone sculptures, and other landscape modifications that honor the dead and express the claims of the living to particular landscapes and their productive resources.

Pottery and Metallurgy

In the agricultural period, two particular types of raw material became important for the development of objects: clay and metals. Clay, which is soft in the raw form but can be hardened by fire to produce durable objects, could be used to produce

In the agricultural period, clay was used to produce practical items such as jars and bowls that also provided a surface for decoration. Later, single metals were used as ornaments and then combined to create compounds of varying hardness for tools and weapons.

practical items (such as jars and bowls) that also provided a surface for decoration and the expression of style. Each global region quickly developed a repertoire of pottery shapes and forms that were particular to the prevailing foods and social customs. Changes in pottery design and form also provide archaeologists with a means of classifying cultures and examining changes in technology and social organization over time.

The development of metallurgy also provided people with the opportunity to create many types of distinctive and practical items. Most of the first metal objects were ornaments made of the most easily worked ores (copper, silver, and gold). Afterward, people combined metals to achieve specific properties of hardness, such as bronze, which is a mixture of copper with tin or arsenic. Iron was a more difficult metal to extract and work, but it provided a very durable material for the creation of a variety of tools and weapons starting in the 2nd millennium B.C.E.

Purposes of Objects

Objects often have a primary purpose, as tools for facilitating particular jobs. These include tasks such as cooking (pots, pans, and knives), eating (containers and utensils), making clothing (needles, thread, and cloth), tending to personal needs (toilets, washbasins, and soap), and child care (toys, diapers, and food items). However, even these basic objects can vary in style as well as in material, signaling the way in which ordinary goods have distinct forms from one culture to another.

Displays of Wealth and Social Roles

Objects are a tangible form of wealth display. Individuals in many mobile societies, such as pastoralists, carry wealth on their bodies in the form of elaborate ornaments and other portable goods. Wealth in stored goods or surplus food can be used not only to display individual or household status, but also to build political power through rewarding followers, sponsoring feasts, or paying for the construction of elaborate structures. Prior to the development of coined money in the 6th century B.C.E., and even afterward, many types of objects were used as modes of payment, including cowry shells, iron bars, copper ingots, and lengths of cloth.

In daily life, objects perform a variety of social roles in addition to their utility as functional tools. Gift giving is a social act in which the object's perceived value symbolizes the status of the giver and the expected obligations of the recipient. Material objects become the means of transferring wealth, for example, as collateral for loans or as the focus of gifts related to dowry or bride-price. Material objects also serve as the background for communication; for example, the same person might dress differently when performing roles as a teacher, homemaker, or community leader.

Identity and social roles can be enhanced and facilitated by the use of distinctive material culture such as clothing, ornaments, and even hairstyles.

Ancient depictions on murals, pottery vessels, and sculptures indicate that past peoples also had a keen awareness of the differences in styles worn by local people compared with newcomers or foreigners. The archaeology of colonial encounters (which includes the ancient Roman Empire as well as more recent British, French, Spanish, and other imperial expansions) also shows that new styles of food, clothing, and architecture were selectively incorporated into newly conquered territories.

Within any given culture, objects enable people to express their group affiliations and social status. Individuals seek out and utilize objects to show that they belong or aspire to belong to a particular group (whether through formal uniforms or through the adoption of prevailing styles of clothing and ornamentation). Objects also can become heirlooms and signify social ties carried from one generation to the next. The individual's interactions with objects include not only the objects worn and used outside the home but also the items used in private, including undergarments and medications.

Material objects also serve as an important component of spiritual activity. Nearly every ritual tradition has objects that are handled in specific ways and at specific moments of devotion as well as objects that are worn on the body or displayed in the home that show religious affiliation. Like daily-use goods, ritual objects can also show regional differentiation, for example, the distinctions in the form and style of the Buddha in India, central Asia, China, southeast Asia, and Japan. Even ascetic religions make reference to material culture, as their practitioners define themselves in terms of avoiding the objects otherwise prevalent in their culture.

Monica L. Smith
University of California, Los Angeles

See Also: Archaeology of Garbage; History of Consumption and Waste, Ancient World; Material Culture Today; Needs and Wants.

Further Readings

Deetz, James. *In Small Things Forgotten: An Archaeology of Early American Life*. New York: Anchor, 1996.

Douglas, Mary and Baron Isherwood. *The World of Goods*. London: Routledge, 1996.

Hodder, Ian. *Symbols in Action: Ethnoarchaeological Studies of Material Culture*. Cambridge: Cambridge University Press, 1982.

Schiffer, Michael Brian and Andrea R. Miller. *The Material Life of Human Beings*. London: Routledge, 1999.

Smith, Monica L. *A Prehistory of Ordinary People*. Tucson: University of Arizona Press, 2010.

Material Culture Today

Material cultural studies presume that objects contain traces of their cultural importance. Objects thus become central to accounts of socially meaningful use and waste, because the materials of culture are remnants of consumption. Objects carry their history and continually pick up residual meaning.

What happens as objects move through the social system through unique, creative, or unexpected trajectories? Defining objects as relational, but still inherently material, allows for both the reconception of culture beyond subjective experience and reimagining consumption as not simply structured by institutions; each becomes interrelated precisely through the personalized histories of specific objects. Past uses and meanings continue to exist through the persistence of the object, even if cultural values have changed.

The purpose and meaning of things are not simply imposed by the people who possess them; objects equally figurate their beholders because they contain past interactions with others in their very design and continued possession. Ordinary objects are thus cultural texts that call into being collective publics and subcultures as they are repossessed by people who reclaim potential waste as significant material. Popular books recounting histories of mundane foods such as salt, coffee, or sugar mark an underlying current of what might be called—following French philosopher Michel Foucault's work on power and governmentality—a "genealogical concern for the material origins of ordinary life."

Cultural Commodities

A material culture approach can focus, for example, on what Will Straw calls the "spectacles of waste" of secondhand culture, which allows a view of the commodity cycle beyond the traditional two-step definition of purchase and disposal. Cultural commodities, in particular, tend to move toward sites of accumulation, whether in private collections, auctions, or secondhand stores, and they are thus exemplary illustrations of the possibility of recursive stages of possession, discard, and reuse. The circulation of secondhand culture—hand-me-downs, vintage clothing, used books and compact discs, antiques and ephemera at auction—encapsulates how meaning and value are negotiated and redefined through practices beyond acts of simple possession. Consider a brief, idealized history of one such specific but mundane object, a T-shirt.

It is 1982, and a teenage boy is attending his first heavy metal rock concert with friends. His parents do not approve, but he has just begun his first summer job and can purchase the ticket himself. At the merchandize display, he buys one of the band's souvenir T-shirts showing the most recent album cover on its front, listing all the stops on the tour on its back. Pleased and proud with this memento, the teenager thinks little of how this T-shirt was produced, the labor that went into it, the journey it had to take from factory to wholesaler to concert venue. Over the next months, he wears the shirt often and gladly tells about attending the concert when strangers start conversations by asking about the T-shirt. The next year he gives the T-shirt to his first girlfriend as a sign of his undying love for her. Although she appreciates the symbolism, by the time she goes to college they have split up and the defunct band is now considered juvenile among her college friends. She still wants the keepsake, but puts it at the back of her closet. Years pass before her parents decide to give all her remaining youthful things to the local charity shop, where the T-shirt ends up in a bulk-purchase bin amid a pile of similarly storied clothes. Enough years have passed that nostalgia for the music of the rock band has emerged. A young entrepreneur buys the T-shirt and a dozen others to sell at a premium at a vintage clothing shop, where the entrepreneur promotes used clothing in relation to environmentalism and awareness of overseas sweatshop labor. The T-shirt is washed and put out for sale, catching the attention of a young skateboarder, who wears it with hipster irony for a few months. The skateboarder is surprised to have more than one ageing metal fan point to the T-shirt and express continued fandom. He auctions the T-shirt online and mails it to the highest bidder in another county. The online buyer is so much an enthusiast of the band and values the rare collectible item so much that the buyer puts the T-shirt in a frame and mounts it on a wall as part of a display of the history of the band.

The T-shirt, as an object, collects more meaning than was consciously held by any of the persons who had a hand in its history: the manufacturer, laborers, band, designer, and original merchandiser had only limited intentions or contact with the shirt, as did each of the people who possessed it in turn. The same is true of any specific academic interest in the T-shirt: pop music fandom, fashion and identity, lifecourse patterns, consumer habits, environmentalism, social networking—all are important but partial ways to analyze the case study of this T-shirt. Tracing the circulation of objects begs for an interdisciplinary approach precisely because objects circulate in and out of disciplinary fields of concern as they get produced, used, reused, and discarded. Even this account of the T-shirt misses the full variety of possible material histories of clothing, which could attend to the cultural legacies of cotton plantations and thus slavery, colonialism, and globalization; looms and sewing machines, and thus industrialization and urbanization; fashion and shopping, and thus arcades, department stores, ready-to-wear clothing, marketing, and branding; and factories and sweatshops, and thus unionization, child labor regulation, and workplace safety.

Sometimes, nothing remains except the object as a record of the coordinated relations between humans. Among academic disciplines, archaeology often studies unearthed artifacts as a way of deducing cultural meaning from the material remnants of past civilizations. Studying the present, anthropology also often focuses on material artifacts as emblematic totems of group membership and as a way to translate insiders' meanings into terms more generally understood by outsiders. Similarly, qualitative research methods within sociology can rely on responses to objects elicited through surveys and interviews as a means to externalize in language interior feelings and opinions. The form of such expressions allows literary studies to focus, in turn, on the representation of objects in texts, just as historians can look to the archived record of the same objects to represent a cultural past. With reference to the T-shirt, the anthropologist might be interested in the neo-tribal ways people recognize fellow group members through symbols; the sociologist might be more interested in how objects mark distinctions of class position and identity; the historian might focus on how consumer goods represent a zeitgeist; while the media and cultural scholar might be more interested in how popular culture is built upon commodified memorabilia. The T-shirt's life cycle as a commodity incorporates all of these approaches, but the interdisciplinary scope of a general material approach allows the continual risk of objects becoming waste to come to the forefront by attending to the continual reproduction of the cultural meaning of material objects.

Excess

Waste is just one option for what people do with excess. Attending to cultural excess illuminates some canonical theories of material culture. Discarded culture can be cast aside as garbage, following Mary Douglas's definition of "dirt" as matter out of place, but the very idea of surplus and thus security can also be ritualized in potlatch, central to Georges Bataille's writing on the general economy. The ritual offering has its equivalent in the gift, which Marcel Mauss argued as a fundamental objectification of social ties. Each of these dwell in the anthropological or symbolic character of objects deemed no longer needed. Excess symbolic value can also be objectified as sacred totems, following Emile Durkheim's theories of religion, but transformed in secular modernity into conspicuous consumption in Thorstein Veblen's characterization of the leisure class. Such sociological understandings of how objects mediate economic and cultural realms are extrapolated by Pierre Bourdieu to nearly all things as signs of taste and markers of class distinction. While each of these theories of objects can be applied across class positions, between societies, and over time, Walter Benjamin was concerned especially with the dominance of contemporary commodities and thus focused squarely on the display of consumer products in 19th-century Paris arcades.

Benjamin's turn to the value of display atop use and exchange complements Bataille's reconstruction of the economic. Both counter assumptions of scarcity and rational choices, with emphasis instead on the situated presentation of excess and luxury and, in turn, the dislocated representation of objects through media and technology. These theories allow a consideration of relations focused on the gift (and the obligation created through it), the collection (and its catalogue of prices), window-shopping (and the leisurely *flâneur*), the souvenir (and the gazing tourist), and the status symbol (totemic for forms of legal-rational leadership). These relations are geographically and temporally dependent and cannot exist without designating meaning through some form of material object relation. However, these meanings, like the material cultures they contain, are not unchanging or unchangeable.

Object Interaction

In contemporary cultural studies, Bruno Latour's articulation of actor-network theory is notable for defining objects as social actors, distinct but nonetheless on par with intentional human actors in reproducing culture. People interact with objects—from doors and chairs to airplanes and surveillance monitors—as much or more than with other people; social interactions are mediated, or figurated in Latour's terminology, through objects as often as through interpersonal commingling. Objects thus exert inscribed norms with as much force and efficacy as people. Emerging out of studies of science and technology, Latour's tipping of the balance of attention toward the object is a robust, general argument for the importance of material culture, albeit so generalized as to render waste and commodities as merely types of objects among many others.

Conclusion

Contemporary reconsiderations of material culture have shown how modernity and global capitalism are made meaningful through the circulation of objects: the relations and flow of capital; the movement of goods, markets, and cultures; and the global connections across modes of production and consumption. The centrality of circulation of goods in globalization is foundational—not simply parallel to—the flow of information, capital, and migrants around the planet. Cultural connections made through the production, consumption, and discard of objects can thus become a central concern in studies of global capitalism and late capitalism. Material culture at once surrounds society and is also as specific and distinct as a T-shirt. Objects are in constant movement through cultural spaces and provide a wealth of locations within which one can begin to understand their meanings.

Michelle Coyne
York University
Paul S. Moore
Ryerson University

See Also: Culture, Values, and Garbage; Garbage in Modern Thought; Overconsumption; Shopping; Sociology of Waste.

Further Readings

Buck-Morss, S. "Dream World of Mass Culture: Walter Benjamin's Theory of Modernity and the Dialectics of Seeing." In *Modernity and the Hegemony of Vision*, D. M. Levin, ed. Berkeley: University of California Press, 1993.

Hawkins, G. *The Ethics of Waste: How We Relate to Rubbish*. New York: Rowman & Littlefield, 2005.

Latour, B. "Where Are the Missing Masses? The Sociology of a Few Mundane Artifacts." In *Shaping Technology/Building Society: Studies in Sociotechnical Change*, W. E. Bijker and J. Law, eds. Cambridge, MA: MIT Press, 1992.

Straw, W. "Spectacles of Waste." In *Circulation and the City: Essays on Urban Culture*, A. Boutros and W. Straw, eds. Montreal, Canada: McGill-Queen's University Press, 2010.

Materialist Values

The term *materialist values* refers to the worth attributed to physical goods that are usually associated with the economy, technology, and biology, whether the goods are for function, status, or the joy of accumulating objects. Scholars generally argue that widespread materialist values, often referred to as *materialism*, have been part and parcel of the industrial period of growth, especially in more-developed countries, where goods became mass-produced and more affordable for the lower and middle classes.

The increase in materialist values is also associated with urban lifestyles, which proliferated throughout the 1900s and have intensified to the point where over half of the world's population currently live in urban areas. Items such as vehicles, refrigerators, televisions, and several changes of clothes became commonplace in the 1950s and thereafter, and such trends have followed among the middle and higher classes in newly industrialized countries, such as China and India, and among the wealthier classes in the least-developed countries, such as Lesotho and Haiti.

Environmental Threat

The human worth placed on objects and the pursuit of owning more things—the more central role of materialist values in urban life, in particular—is often considered one of the key threats to sustainability. Materialist values are associated with over-consumption, given both the amount of withdrawals from the ecosystem and additions in the form of waste and pollution and the massive material resources involved in highly consumptive lifestyles. The growing number of goods and services consumed tends to offset the efficiency gains achieved through, for example, improved production technologies and processes.

Housing, food, drink, and mobility have the greatest environmental impact over their lifecycle in terms of emissions of greenhouse gases, acidifying and ozone-depleting substances, as well as resource use. Thus, both environmentalists and, increasingly, religious groups concerned about environmental problems question materialist values as a core element of modern civilization, believing that materialist values must be kept in a more-modest place among a suite of other cultural values (such as environmental, community, or social justice) that shape the development of human society.

Negative Popular Conceptions

There are many media accounts and folk legends about how characters who have strong materialist values became greedy and obsessed with accumulated wealth, status, and possessing high-status goods. Consequently, these characters' relentless pursuit, as driven by materialist values, is associated with losing sight of one's interdependent place in collective life—one's responsibilities to, and benefits from, family, community, nation, nature, and a full self that honors many kinds of values (for example, caring for the home). These commonplace stories—reproduced in fairy tales and Hollywood movies, where someone "had it all" and lost it to greed, corruption, or careless mismanagement of savings—celebrate cultural skepticism and the moral stance against strong materialist values in society.

Materialist values are also associated with those who will pursue material goods at the expense of other people and other moral actions. For example, materialist values are often associated with the abuse of power of large organizations—be it government or industry—that have placed the pursuit of revenue or profit over human safety and environmental protection and stewardship. Thus, many social justice issues are tied to a critique of materialist values as a distal cause, rather than a proximate cause, that at the core of the issue distorts human decision making to narrowly focus on the accumulation of wealth. For example, while the proximate cause for an oil rig blowout may be a failed technology, the distal cause may be attributed to materialist values held by the company or companies involved. In other words, the pursuit of profit overruled the oil company's investment in wisely assessing the risks, safely constructing the technology, and being prepared to assure safety to all citizens if a technological accident occured.

Growing Phenomenon

In addition to goods becoming more available and affordable, materialist values are theorized to have increased due to social psychological reasons. There are several social mechanisms that tend to self-fulfill consumerism and, hence, support materialist values. For example, the cultural practice of comparing how well off one is relative to others, or social comparison, often leads to obtaining more possessions. One social mechanism among North Americans is omnivorousness (the tendency to buy specialized equipment for every eventuality). The fluidity with which one can perform different tasks is a practice that displays status, where a person can be equipped for any possible activity, as infrequent as these activities might be. Furthermore, the tendency to use consumer goods to display identity is heavily reinforced in advertising, as people are enticed to buy food, clothes, shoes, and furniture to show a sense of style or individuality. In fact, advertisers encourage consumers to take on different identities through the clothes, cosmetics, and other items they buy as part of their individual style. The increased demand for consumer goods is also associated with cultural expectations, as reinforced by advertising, to buy items that match one another; thus, a new house requires new furniture.

Goods are increasingly specialized and may quickly become outdated (for example, planned obsolescence of cameras), which entices people to upgrade and buy more objects. For example, the notion in more-developed countries that one must stay "up to date" in the suite of electronic personal items one uses has led to a continual cycle of replacing cell phones, personal digital assistants (PDAs), and computers. Social celebrations, such as birthdays and winter holidays, have become highly commercialized events, where social expectations for gifts and indulgences become commonplace and the abundance of goods exchanged is celebrated. Conspicuous consumption (the tendency to obtain items to show wealth or status) has become inconspicuous where the materialist values embedded in the practice of displaying status and prestige are normalized.

Structural influences on materialist values might include growing incomes in more-developed countries, globalization of the economy, technological breakthroughs (such as the Internet and mobile phones), decreasing household sizes, and an aging population. In addition, the pervasive role of advertising, the normalization and ubiquity of stores and other buying opportunities (even in schools, churches, and at home via the computer and mail order catalogues), and the easy availability of credit all make purchases of convenience and compulsion more likely. For example, in her 1993 book *The Overworked American: The Unexpected*

Decline of Leisure, Juliet Schor describes the needs for a consumptive lifestyle for many Americans who work two and three jobs to maintain a "normal" U.S. lifestyle. Thus, people are caught on a treadmill of sorts, working long hours to have more income, and then spending more income to be able to get to work, wear the appropriate clothes to work, reward themselves for working so hard, and display the outcome of their work through material acquisitions.

Structural Theories

There are a number of structural theorists in sociology who address materialist values. Structural theories about materialist values often address the role of capitalism, the state, and relationships of power among corporations, the state, workers, and the most vulnerable in society to explain how those with more wealth and power can maintain their material advantages over time, even as income disparities grow. A general theme from structural theories in sociology, for example, is that the state, or governments, can be hamstrung, or co-opted, by the material expectations of both corporations and workers.

The state is less effective at all of its roles to govern the good society (one that inherently strives to protect human welfare) because the growth imperative (the accumulation of wealth and business growth) is too often equated with the pursuit of full employment for working-age citizens and, by association, business success. Thus, the pursuit of economic growth is theorized to become more important than safeguarding collective goods and services, such as high-quality schools, healthcare, infrastructure, and environmental protection. Similarly, the growth imperative of many governments is theorized to hinder their ability to examine the overemphasized role of materialist values in government decision making. Numerous scholars argue that there is both opportunity and evidence that materialist values, as manifested in assuming economic growth is necessary for a healthy society, is flawed and that economic growth is not necessarily related to the health of the population, ecosystem integrity, and human happiness.

Another approach to materialism, in terms of the pursuit of material goods, is to view it as part of human survival and cultural expression, thus embedded in patterns of seeking food, clothing, and shelter. Human materiality is such that some degree of materialist values is always present, given that to live is to use material and symbolic objects in life practices, services, and leisure activities. Thus, it is the degree to which people hold materialist values above other values that is important. Consequently, some scholars have addressed materialism as a feature of human life that is not to be criticized as much as shaped by reasonable, ecologically sustainable practices. Various scholars discuss a social practices approach to understanding materialism and consumer behavior.

This approach emphasizes the flexibility of the capitalist system to incorporate ecosystem values so that materialist values can work in concert with other values to maintain ecosystem integrity and reduce waste. Thus, this approach encourages improved planning and development of buildings, parks, neighborhoods, towns, and cities that take into account daily practices that have material consequences. A more wisely designed infrastructure can support new norms that, in turn, allow daily habits to automatically conserve water and energy, reduce waste, and foster healthy human interactions. If materialist values are part of the human affair of life, there are opportunities to align systems of provision for food, clothing, shelter, and transportation with human needs for cultural expression and ecosystem renewal.

Material Needs

Abraham Maslow's hierarchy of needs is often referenced when academic scholars discuss materialism. From this perspective, "lower needs" are material; beyond that are physiological, safety, belongingness and love, esteem, and self-actualization needs. The assumption then is that in order to seek "self actualization," material needs must first be met. Maslow recognized that most people's needs are generally only partially satisfied and that, for a temporary period, people can forgo lower needs for higher needs (for example, give up safety to go on an adventure). Other higher needs that one might seek, which generally rest upon material needs already being taken care of, are the needs for aesthetics, knowledge, and understanding.

The implications of Maslow's theory are significant; those who are distracted with meeting material needs have less energy, time, and ability to pursue higher needs. Furthermore, the higher needs are the more social ones, such as the search for love and respect, and these can be sidelined if one is chronically hungry, for example. In other words, this particular perspective places materialist values as a necessary but insufficient set of values to live a full life, where self-actualization, or values associated with a full expression of one's potential, is possible when materialist needs are no longer of consistent concern.

Post-Materialism and Countermovements

Materialist values are conceptualized very differently from the point of view of Ronald Inglehart, who developed a longitudinal data set on a broad intergenerational shift in wealthy countries of the world to what he describes as a shift from "materialist values" to "post-materialist" values. Here, the term *materialist values* refer to "concerns about economic and physical security" and these are contrasted with post-materialist values, which are "a greater emphasis on freedom, self-expression, and the quality of life." The trend in many countries to move to place more emphasis on post-materialist values than materialist values is attributed to the economic improvements that younger generations have enjoyed, where they have had greater opportunity to focus on self-actualization, as Maslow describes it, than economic security.

The slow food movement is a growing trend: communities grow and prepare food to share with neighbors and friends. This philosophical approach to consumerism celebrates food as an art, a pleasure, and a means to strengthen local and personal ties.

There are a number of countermovements challenging the public trends of materialism when it manifests itself as overconsumption. For example, the slow food movement is a growing trend in Europe and North America, where place-based communities are organizing efforts to grow and prepare culturally celebrated food that they eat with each other. Members of the slow food movement seek to share home-cooked meals with other community members, in their homes. This movement is opposed to eating out at fast food and other franchise restaurants, where the food may come from thousands of miles away and is often quickly eaten, and the art and pleasures of cooking are forgone.

The challenge to materialist values is that food is considered not just a thing to consume but rather food is seen as an integral element of life to be learned about (such as how to grow food well), supported by local food producers, and celebrated through the joy and art of harvesting, preparing, and sharing food with others.

Materialist values will continue to be hotly debated in world politics as the role of unequal access to material goods (such as energy resources) and material "bads" (such as electronic waste) continues to be addressed in the designation of responsible parties to address climate change, waste streams, transboundary pollution, water scarcity, and the protection of habitat.

Naomi Krogman
University of Alberta

See Also: Commodification; Consumerism; Material Culture Today; Sociology of Waste.

Further Readings

Abramson, Paul R. and Ronald Inglehart. *Value Change in Global Perspective*. Ann Arbor: University of Michigan Press, 1995.

Anielski, Mark. *The Economics of Happiness*. Gabriola Island, Canada: New Society Publishers, 2007.

Bauman, Zymut. *Consuming Life*. Malden, MA: Polity Press, 2007.

Brand, Karl-Werner. "Social Practices and Sustainable Consumption: Benefits and Limitations of a New Theoretical Approach." In *Environmental Sociology: Social Practices and Sustainable Consumption: Benefits and Limitations of a New Theoretical Approach*, Matthias Gross and Harald Heinrichs, eds. London: Springer, 2010.

Dickens, Peter. *Society and Nature: Towards a Green Social Theory*. Philadelphia, PA: Temple University Press, 1992.

Hawken, Paul. *The Ecology of Commerce: A Declaration of Sustainability*. New York: HarperCollins, 1993.

Huddart, Emily and Naomi Krogman. "Towards a Sociology of Consumerism." *International Journal of Sustainble Society*, v.1/2 (2008).

Inglehart, Ronald. *Modernization and Postmodernization: Cultural, Economic, and Political Change in 43 Societies*. Princeton, NJ: Princeton University Press, 1997.

Jackson, Tim. *Motivating Sustainable Consumption: A Review of Evidence on Consumer Behavior and Behavioral Change*. Surrey, UK: Centre for Environmental Strategy, 2004.

Maslow, Abraham. *A Theory of Human Motivation*. New York: Harper & Row, 1970.

McKibben, Bill. *Deep Economy: The Wealth of Communities and the Durable Future*. New York: Times Books, 2007.

Ritzer, George. *Enchanting a Disenchanted World: Revolutionizing the Means of Consumption*. Thousand Oaks, CA: Pine Forge Press, 2005.

Schnaiberg, Allan. *The Environment: From Surplus to Scarcity*. New York: Oxford University Press, 1980.

Victor, Peter A. *Managing Without Growth: Slower by Design, Not Disaster*. Cheltenham, UK: Edward Elgar, 2008.

Meat

There is no denying that meat plays a major role in the history, economy, and culture of the United States. Classic images of Thanksgiving include families gathered around a table abundant with food, featuring a turkey. The summer months bring backyard barbecues with hamburgers and steaks on the grill, and baseball games are not usually considered complete without a hot dog. Despite being surrounded by meat, people do not often consider how meat fits into the consumption and waste stream.

Consumption Rates

Americans seem to have a large and growing appetite for meat. Between 1909 and 1919, the average American ate about 140 pounds (lbs) of meat per year. In 2002, per capita annual meat consumption reached 275 lbs in the United States, the highest per capita meat consumption rate in the world. For comparison, in 2002, China's per capita meat consumption rate was 115 lbs, Italy's was approximately 198 lbs, and the United Kingdom's was approximately 175 lbs. Though humans consume more and more meat each year, tastes and preferences have shifted over the years.

Poultry only constituted about 8.5 percent of all meat consumed between 1909 and 1919. In the 1960s, poultry made up about 17 percent of all meat consumed, and between 1970 and 2005, per capita consumption of poultry jumped from 34 lbs per person to 74 lbs per person. While the popularity of poultry—particularly chicken—was growing, the consumption rate of red meat was dropping. Between 1970 and 2005, per capita consumption of red meat declined by 22 percent. Still, red meat remains the main source of protein in the U.S. diet, with beef the most preferred.

Production

Keeping up with this growing demand for meat required changes at all levels of production and consumption. During the 19th century, meat production evolved from local, highly skilled butchering to consolidated industrial production. The development of the refrigerated railcar and slaughterhouse centralized meat production in Chicago, distributing pork, beef, and poultry across the nation. At the same time, wastes from the slaughterhouses concentrated in the Bubbly Creek tributary of the Chicago River. Upton Sinclair called this body of water Chicago's "Great Open Sewer" in his 1906 book *The Jungle,* and public uproar surrounding revelations of unsanitary meatpacking practices in that book led to the creation of the U.S. Food and Drug Administration that same year. The industry began to decentralize in the early 20th century, in part due to a transition from using rail to trucks to transport meat. Subsequently, meatpacking operations developed in several states.

Over the decades, animal farms have grown in size to both meet the needs of the population and to remain economically competitive. North Carolina, a top producer of both hogs and turkeys, experienced a 300 percent increase in the number of animals per square mile between 1977 and 2006. On a larger scale, the total number of cattle on the planet has nearly tripled since the early 1900s. Additionally, modern cattle are much larger than their predecessors and can be brought to slaughter earlier compared to the past.

In the 1900s, it could take as long as five years before cattle were ready for slaughter; in the 21st century, cattle can reach an acceptable weight for slaughter within 14 months. This is due in large part to changes in animal husbandry techniques. Rather than grazing cattle on grass, many cattle are now raised on feed made primarily of corn and grain. This mixture allows cattle to gain weight nearly twice as fast as grazing on grass. Growth hormones and antibiotics have also been used to hasten animal growth and prevent the spread of disease in the close quarters of 21st-century confined animal feeding operations (CAFOs).

Meatpacking techniques have also changed to keep up with the increased level of demand and production. Though much of the slaughtering process still relies on human workers, technological advances have allowed large processing plants to handle larger numbers of animals faster. All of these techniques, from increased animal density to corn diets, are controversial. Proponents of modern meat production cite the economic benefits to regions where CAFOs are present. These large facilities provide jobs and pay taxes. Critics stress the health and environmental risks associated with such high levels of meat production and consumption.

Waste

Waste is a part of the meat consumption cycle, present in different forms at every step. The slaughter and processing of animals produces waste in the form of scraps that cannot be sold for human consumption. In some cases, these discards can be used for other purposes. Feathers, for instance, may be used in pillows and comforters or used as filler in feeds produced for commercial animals. Meat spoilage becomes a concern for grocers, restaurant managers, and consumers alike.

For example, in 2006, supermarkets had to throw away an estimated 4.4 percent of their fresh beef, 4.2 percent of fresh chicken, 10.6 percent of fresh lamb, and 27.8 percent of fresh veal. The discrepancy of these waste rates is explained by consumer familiarity with a food group. Beef and chicken make up the largest shares of meat consumed in the United States. Veal, however, is consumed less frequently. Still, grocers stock their meat cases with veal in an effort to enhance shoppers' perceptions of variety. Because it is generally priced higher than beef or chicken and because it is a less-familiar item, more veal goes to waste.

Finally, there is a great deal of the meat that is cooked but not consumed. Sometimes, too much meat is cooked or, as in the case of the beef shortage of 1973, the cut of meat was unfamiliar to the consumer, and they were unable to prepare it in an appealing way. This is one theory proposed by researchers who discovered that more beef was wasted during the beef shortage of 1973 than during times when beef was in good supply.

Of all of the waste generated during the production and consumption of meat, animal excrement might be the most studied. Excrement is a natural aspect of raising animals. In some cases, this form of waste can be used as a fertilizer for crops. In very large quantities, however, animal excrement can have negative effects on the quality of the environment. Nitrogen, which is naturally present in animal waste, can accumulate in the water as animal waste seeps through the soil. At high levels, it supports the rapid growth of algae, which robs the water of the oxygen that fish need to live. Nitrous oxide, a greenhouse gas, is also associated with animal waste. Nitrous oxide is found in nitrogen-based fertilizers and is associated with soil management techniques and manure. Because of this, all food groups contribute to the release of nitrous oxide into the environment. Red meat and dairy, however, are responsible for the largest portions.

Melissa Fletcher Pirkey
University of Notre Dame

See Also: Beef Shortage, 1973; Farms; Food Consumption; Food Waste Behavior; Methane.

Further Readings

Cronon, William. *Nature's Metropolis: Chicago and the Great West.* New York: W. W. Norton, 1991.
Dauvergne, P. *The Shadows of Consumption.* Cambridge, MA: MIT Press, 2008.
Sinclair, Upton. *The Jungle.* New York: Doubleday, Page, 1906.
Weber, C. and H. S. Matthews. "Food-Miles and the Relative Climate Impacts of Food Choices in the United States." *Environmental Science and Technology*, v.42 (2008).
Wells, H. and J. Buzby. "Dietary Assessment of Major Trends in U.S. Food Consumption, 1970–2005." *USDA Economic Research Service Information Bulletin*, v.33 (2008).

Medical Waste

Medical waste, also known in Europe as healthcare waste or clinical waste, is a distinct and complex waste stream for regulation and management. Due to the infectious, toxic, and radioactive nature of this type of waste, it is important that these materials are separated from other wastes and stored, transported, treated, and disposed of in ways that minimize their risk to both human health and the environment.

Generally, medical waste is waste generated at hospitals and other healthcare establishments, such as clinics, physician's offices, long-term healthcare facilities (nursing homes), blood banks, dental practices, and veterinary hospitals. Further sources of medical waste materials arise from medical and animal laboratories and research centers, mortuary and autopsy centers, funeral homes, schools, universities, factories, prisons, and emergency service providers.

Production

While hospitals and other facilities produce medical waste that can pose a threat of infection, the vast majority of waste volume arising in most healthcare facilities is municipal solid waste, such as paper, plastics, metal, glass, and food waste. In fact, approximately 80 percent of waste arising from these sources and operations is general waste, which is either identical or similar to that generated by hotels, offices, and residential properties. The remaining 20 percent of waste is what may be considered as hazardous materials, which may be infectious, toxic, or radioactive. However, this proportion has been calculated by the Centers for Disease Control and Prevention (CDC) as a significant overestimation of the true volume of medical waste that needs to be treated and disposed of separately. The CDC suggest that as little as 2–3 percent of hospital and healthcare wastes are likely to be medical waste for the purpose of regulation and control.

Risk Factors

Although medical waste constitutes a small fraction of the waste produced by these sources, it is of special regulatory concern because of the potential risks of harm to human health and the environment from pathogens that may be present in the waste or from hazardous chemicals. Poor management of medical waste can expose healthcare workers, waste handlers, patients, and the larger community to infection and toxic effects, as well as cause environmental risks, including pollution and contamination. Other potential infectious risks include the spread of microorganisms from healthcare establishments into the environment. Wastes and by-products can cause injuries, for example, radiation burns, sharps-inflicted injuries, poisoning, and pollution. These impacts can occur through the release of pharmaceutical products, such as antibiotics and cytotoxic drugs, into wastewater or by the release of toxic elements or compounds, such as mercury or dioxins. Chemicals such as dioxins are often found in the air, water, soil, and living organisms. An unwanted by-product, they are formed when heating processes create certain chemicals like chlorine. Dioxins increase the risk of cancer, although they are also linked to reproductive health, lowering sperm counts, causing behavioral problems, and increasing the incidence of diabetes. In these ways, medical waste poses a significant risk of transmitting infections and contamination.

Waste Management in the European Union

In the European Union (EU), the sustainable management of waste is set out in the Waste Framework Directive (2008/98/EC). This piece of legistation outlines the basic obligations and objectives of member states in relation to waste. It is supplemented by a series of other directives that deal with specific wastes. Under Article 17, member states are required to take the necessary action to ensure that the production, collection, and transportation of hazardous waste, as well as its storage and treatment, are carried out in conditions providing protection for the environment and human health. This must be achieved in a way that minimizes risk to water, air, soil, plants, and animals and prevents nuisance of odor through adequate record keeping. It also requires member states to establish penalties and enforcement provisions in their national legislation to prohibit abandonment, dumping, or uncontrolled management of waste.

For the purpose of regulatory control, it also lists the properties of waste that render it hazardous. This list includes "infectious," and healthcare waste that is infectious is not accepted in EU landfills. Prior to disposal, the waste must be rendered safe. This is understood as (1) reducing the number of infectious organisms present in the waste to a level that no additional precautions are needed to protect workers or the public against infection by the waste; (2) the destruction of anatomical waste such that it is no longer generally recognizable; (3) rendering syringes, needles, or any other equipment or item unusable and no longer in their original shape and form; and (4) the destruction of component chemicals of medicinal waste.

The United Kingdom, as a member state of the EU, has an obligation to ensure that all clinical waste is managed as hazardous waste. Clinical waste can be broadly divided into two categories of material: waste that poses a risk of infection and medicinal waste, including drugs and other pharmaceutical products. Clinical waste is any waste that consists wholly or partly of human or animal tissue; blood or other bodily fluids; excretions; drugs or other pharmaceutical products; swabs or dressings; and syringes, needles, or other sharp instruments. These must be rendered safe, since they are considered potentially hazardous to any person coming into contact with them. The central management requirement for clinical waste is that it be separated from all other types of waste and treated and disposed of in an appropriate manner. In the United Kingdom, all businesses, employees, and persons involved in the waste management chain have a duty of care to ensure that waste is securely stored and disposed of legally.

There are stringent controls in place to ensure that clinical waste is managed safely and is recovered or disposed of without harming the environment or human health. It is unlawful to deposit, recover, or dispose of controlled (including clinical) waste without a waste management permit, contrary to the conditions of a permit or the terms of an exemption, or in a way that causes pollution of the environment or harm to human health. The Environment Agency is responsible for regulating and enforcing the hazardous waste regime.

Waste Management in the United States

In the United States, the Occupational Safety and Health Administration (OSHA) regulates several aspects of medical waste, including management of sharps, requirements for containers that hold or store medical waste, labeling of medical waste containers, and employee training. In response to OSHA's recognition of the need to protect healthcare workers and other downstream medical waste handlers, the Occupational Safety and Health Act was introduced. It applies at a federal level to the storage and management of medical waste, although it is limited in its scope to human blood, human infectious wastes, and human pathological wastes. In practice, most states extend this definition of medical waste to include waste or blood from animals.

The act provides rules surrounding worker safety regulations by stipulating (1) the need for personnel to wear protective clothing and equipment; (2) that contaminated reusable sharps be placed in containers that are resistant, labeled, or color coded and leakproof on the sides and bottom; (3) that specimens of blood or other potentially infectious material are placed in a labeled or color-coded container that is closed prior to being stored, transported, or shipped; (4) that regulated medical wastes—including liquid or semiliquid blood, other potentially infectious materials, and contaminated items likely to release blood or other potentially infectious materials, items that are caked with dried blood, and other potentially infectious materials that are capable of releasing these materials during handling—must be placed in containers that are closable, constructed to contain all contents, and prevent leakage of fluids, labeled or color coded, and closed prior to removal; and (5) that all receptacles intended for reuse that have the likelihood of becoming contaminated with blood or other potentially infectious materials are required to be inspected and decontaminated on a regular basis.

These rules have resulted in different interpretations over what exactly constitutes regulated medical waste. This has led to a greater amount of waste generated at sources to be considered as potentially infectious; for example, under the OSHA universal precautions guidelines, a worker handling a bandage with a single drop of blood on it should wear gloves, but the waste itself would most likely not be classified as infectious. Since medical waste is often the most expensive waste stream to manage, the broadening of what constitutes medical waste for the purpose of regulation and control, stemming from confusion over its meaning, has increased the volume of materials and substances to be stored, treated, and disposed of separately as medical waste. In turn, this has a significant cost implication for healthcare facilities. Taking a business perspective, the primary objective of the management of medical waste is to minimize the risk of disease transmission associated with the handling of this waste stream while at the same time minimizing the amount of medical waste arising that falls under separate and specific regulatory procedures. Studies have suggested that as much as 50 percent of waste sent for incineration as medical waste is general waste. This results in unnecessarily high disposal costs. Improved segregation has the potential to generate substantial savings, although this can be inhibited by confusion over the meaning extended to medical waste.

Medical waste is defined by the U.S Environmental Protection Agency (EPA) as "any solid waste which is generated in the diagnosis, treatment, or immunization of human beings or animals, in research pertaining thereto, or in the production or testing of biologicals." This definition was enshrined in Congress's Medical Waste Tracking Act (MWTA) of 1988, which was a direct result of public health concerns following washed up medical waste on a number of beaches. Under the MWTA of 1988, which made changes to the Solid Waste Disposal Act, medical waste includes blood-soaked bandages, culture dishes and other glassware, discarded surgical gloves, discarded surgical instruments, discarded needles and medical sharps, cultures, stocks, swabs used to inoculate cultures, and removed body organs

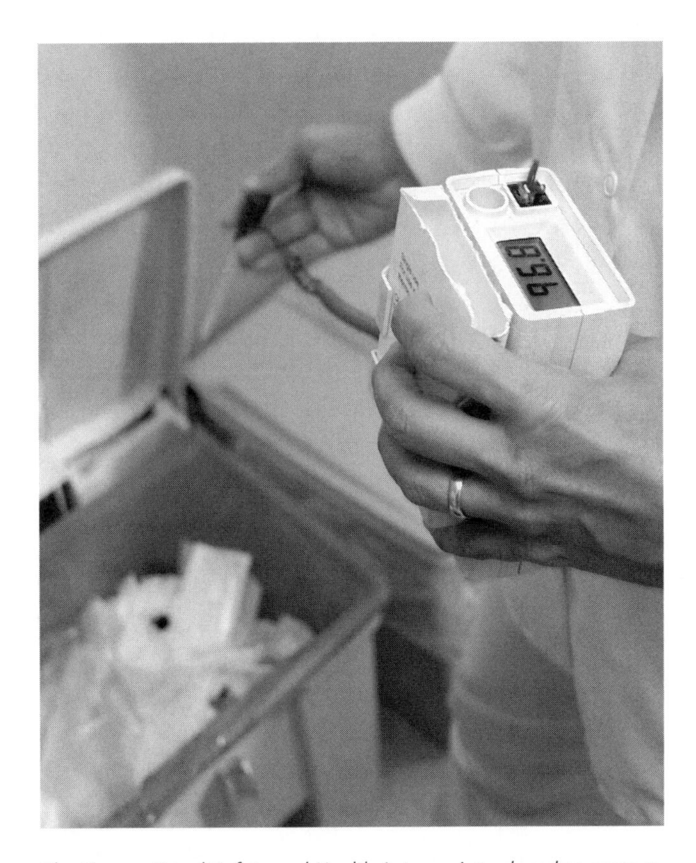

The Occupational Safety and Health Act was introduced to protect healthcare workers and medical waste handlers. Between 3 and 20 percent of waste from healthcare facilities—estimates vary widely—is considered hazardous (toxic, radioactive, or infectious).

(for example, tonsils, appendixes, and limbs). The definition does not include hazardous waste, nor does it include household waste, as defined in those regulations.

Unlike many regulations that apply to healthcare, most regulations governing medical waste are defined at the state, rather than the federal, level. Almost all U.S. states have established their own medical waste regulations, although their extent and interpretations vary considerably. For instance, the MWTA has been used by some states as a basis on which to shape their medical waste rules, while other states' regulations bear little or no resemblance to historical law in this area. These regulations tend to cover the packaging, storage, and transportation of medical waste, with some states also requiring healthcare facilities to register or obtain a permit. The EPA is primarily responsible for developing and enforcing regulations for medical waste management and disposal in most states, although the Department of Health may also serve as the primary regulatory agency.

While in the United States it is left at the federal level to determine the precise interpretation to be give to medical waste, the CDC has issued guidelines for infection control. They cite four categories of infective wastes that should require special handling and treatment: (1) laboratory cultures and stocks, (2) pathology wastes, (3) blood, and (4) items that possess sharp points such as needles and syringes, termed *medical sharps*. These categories require that those producing these types of wastes determine the scope of material that should be included within these categorizations.

The category that has generated the greatest regulatory interest (and at the federal level) is sharps. This is because needles and syringes can be particularly harmful, as they have the potential to penetrate the skin, thereby increasing the potential for disease transmission. Each category tends to have special handling requirements that may be state specific.

Medical waste presents a number of compliance challenges. Due to the state-level responsibility in regulating medical waste, a complex and diverse network of rules exists between states, although the one consistent element that tends to be present in most state rules is the requirement that medical waste be rendered noninfectious before it can be disposed of as solid waste.

Transportation and Incineration

It is a requirement that regulated medical waste be transported and treated separately from other wastes. These activities are considered as important factors in managing medical waste, particularly where such wastes are to be incinerated or disposed of off-site. To this end, the Department of Transportation (DOT) has determined medical waste to constitute hazardous material for the purpose of transfer, and therefore DOT rules apply to transporters of medical waste. The importance of treating medical waste prior to ultimate disposal is to purge the material and substances of infectious toxins and pathogens. Medical waste can be treated in a variety of ways, but the most traditional and common treatment method is incineration.

While incineration is intended to destroy pathogens and cytotoxic chemicals in medical waste, its efficiency and effectiveness is dependent upon the operation of the incinerator. Important factors include the temperature of incineration, maintenance, and training of operators. The EPA introduced regulations governing the operation of and emissions from medical waste incinerators as well as requirements under the Federal Insecticide, Fungicide and Rodenticide Act (FIFRA) for certain medical waste treatment technologies that use chemicals for treating the waste.

There are a number of weaknesses in incineration as a treatment option for medical waste, including the fact that there is potential following treatment for the continued survival of pathogens, radioactivity, infectiousness, and metals in the residual incinerator bottom ash and fly ash carried by air and exhaust gases and captured in the incinerator stacks.

This has led to countries such as the Philippines completely banning incineration because of its adverse environmental impacts. The EPA has cited medical waste incinerators as among the top sources of mercury and dioxin pollution. In addition, some medical wastes are unsuitable for incineration and release pollutants into the atmosphere. The incineration of materials containing chlorine can generate persistent dioxins, furans, and coplanar PCBs, which are considered human carcinogens and exist indefinitely in the environment. Human exposure to dioxins, furans, and coplanar PCBs is through the intake of food. Likewise, incineration of heavy metals, especially lead, mercury, and cadmium, can lead to the spread of heavy metals in the environment.

Although incineration is still widely used, particularly in less-developed countries, alternatives to incineration exist for medical waste and are gaining increasing support in Europe. For instance, in Slovenia, all infectious waste is treated by using a steam-based system. Portugal has closed almost all its medical waste incinerators and is treating medical waste in autoclaves. In France, in the 21st century, many operators have introduced shredding/steam/drying systems for medical waste treatment. Nonincineration techniques include steam sterilization, chemical disinfection, irradiation, and enzymatic processes. Wet heat treatment involves the use of steam to disinfect and sterilize waste and is commonly done in an autoclave.

Chemical disinfection involves a process employing disinfectants, such as dissolved chlorine dioxide, bleach (sodium hypochlorite), peracetic acid, or dry inorganic chemicals. The process often involves shredding or mixing of the waste to aid exposure of the material to the chemical agents. Irradiation-based technologies involve electron beams, Cobalt-60, or ultraviolet irradiation. These technologies destroy microorganisms and pathogens. Biological processes use enzymes to destroy organic matter, although, as of 2010, this was not a common technology for medical waste treatment. Mechanical destruction is a supplementary process to these technologies. It renders the waste unrecognizable and is used to destroy needles and syringes so as to minimize injuries or to render them unusable. However, it does not cleanse the material of its infectious nature, and thus medical waste cannot be processed in this way and sent directly to a landfill.

Hazel Nash
Cardiff University

See Also: Emissions; European Union; Hazardous Materials Transportation Act; Hospitals; Human Waste; Toxic Wastes.

Further Readings

Environmental Protection Agency. *Medical Waste Management and Disposal*. Park Ridge, NJ: Noyes Data Corporation, 1991.
Health Care Without Harm Europe. *Non-Incineration Medical Waste Treatment Technologies in Europe: A Resource for Hospital Administrators, Facility Managers, Health Care Professionals, Environmental Advocates and Community Members*. Prague: Health Care Without Harm Europe, June 2004.
Healthcare Environmental Resource Center. http://www.hercenter.org/index.cfm (Accessed July 2010).
Hospitals for a Healthy Environment. *Regulated Medical Waste Reduction*. Washington, DC: H2E, 2009. http://www.h2e-online.org/docs/h2e10steprmw20103.pdf (Accessed July 2010).
Mercier, C. and T. Ellam. "Waste Not Want Not." *Health Service Journal* (April 4, 1996).

Mega-Haulers

Mega-haulers transport massive quantities of material and are known for their very large capacity to hold and transport materials. Also called *ultra-haulers*, mega-haulers are associated with a range of applied uses, including pickup trucks, with the ability to traverse any type of landscape while towing a heavy cargo trailer; container ships, which are often described in terms of the number of football fields whose length they equal; tractor trailers, for myriad uses of on-road heavy transport such as long-distance container shipment, refuse collection, and construction; and heavy-hauler trucks, such as the Caterpillar 797 series that represents the highest off-road payload capacity of 400 tons. The Terex MT 6300AC and Caterpillar 797 series B and F models were some of the biggest trucks in the world in use in 2010 and were predominantly used in mining operations to transport raw materials, like heavy metals and bitumen.

History

Historically speaking, it is worthwhile to consider the trajectory from horse-drawn trailers to the current mega-hauler incarnation. Mike Woof notes that initially mining operations created a demand for moving heavy material around a mine site. At the same time that teams of horses struggled to haul trailers with solid wheels through difficult terrain, other operations employed train tracks. However, the train track system is limited by its bi-directionality. In 1925, Holt and Best merged to proceed under the name Caterpillar and began selling crawler tractors to replace earlier models that used metal and solid wheels, which were known for poor braking capacity and limited traction. Crawler tractors used a rolling track system that ensured good traction. Although the rolling track system provided sure footing, movement was slow, meaning that the track system would be inefficient for moving massive quantities of material distances of a few miles.

Since 1998, Caterpillar has manufactured the most recognizable fleet of mega-haulers, otherwise known as off-highway, ultra class, two-axle, mechanical powertrain haul trucks. Since the early 1900s, there has been a historical evolution toward larger and more-robust engines, heavy-duty construction materials, and increased comfort for the driver. Mega-haulers of the 21st century feature safety as first priority; a reliable and efficient engine and transmission; superior braking control; durable and flexible steel structure body for performance; long life; minimal maintenance; an ergonomic, comfortable, productive, and safe operator station to minimize operator fatigue; and customer support and serviceability.

In mining, mega-hauler trucks are typically paired with a correspondingly large hydraulic shovel, which can handle up to 100 tons per scoop and top the trucks off with 400 tons of bituminous sands.

The Alberta Oil Sands represents one of the largest-scale mining projects in the world. In terms of the production of consumable goods, oil has come to play a primary role in the composition of many of the products that are bought in the 21st century, including clothing, food, packaging, and fuel. Alberta's Oil Sands operation in Canada has been a driving force behind the development and use of, in particular, the Caterpillar 797 series mega-haulers. The Oil Sands correlate to a massive area totaling approximately 27,000 square miles. One of the primary methods of oil sands extraction involves simple mining techniques, albeit on an exponentially large scale. Keyano College, the regional postsecondary education institute, offers training courses for a range of heavy equipment. Information for prospective students notes that heavy hauler trucks are used for hauling oil sands to feed the crushers, as well as overburden to waste dumps. *Heavy haulers,* rather than mega-haulers, is the preferred term in regard to oil sands mining operations.

Photographers, such as commercial photographer Edward Burtynsky and freelance photographer Peter Essick, have used a birds-eye aerial view in order to photograph the oil sands mining operations. Upon viewing the aerial photographs, the viewer can recognize the perceived need for massive scale. Simply put, mega-haulers strive to master the massive landscape. As seen from above, the image may appear to resemble a child's sandbox complete with yellow dump trucks.

Size

Mega-haulers can be characterized by their sheer scale. Photographs are commonly shared between friends and online depict a person standing next to one of these trucks. In particular, children are often pictured as dwarfed next to one wheel of the truck. In relation to scale are the staggering numbers and capacities that mark these mega-haulers. The capacity to carry 400 tons of material is seen as an engineering challenge—to design and build a machine capable of providing years of service wherein the demand for raw materials is still increasing. To illustrate the magnitude, consider that these trucks are moving back and forth between the track-based shovel in the mine and dumping the load in the crusher, which may be up to a few miles away, 24 hours per day, seven days per week.

Andriko Lozowy
University of Alberta

See Also: Industrial Waste; Mineral Waste; Mining Law.

Further Readings

Oil Sands Discovery Centre. "Facts About Alberta's Oil Sands and Its Industry." Alberta, Canada: Oil Sands Discovery Centre, 2008.

"Ultimate Factories: Caterpillar." *National Geographic* (June 2008). http://www.channel.nationalgeographic.com/series/ultimate-factories/3327/Overview (Accessed July 2010).

Woof, M. *Ultra Haulers: Global Giants of the Mining Industry*. St. Paul, MN: MBI Publishing, 2006.

Methane

Methane is a common chemical compound and the primary constituent of natural gas (also called biogas). Methane arises primarily from the decomposition of organic material, which means that most forms of garbage that are left to rot will produce it. Because it is highly combustive and its production is associated with pungent and potentially poisonous gases, many anthropogenic sources of methane, such as landfills, are regulated to secure air quality and site safety. Methane also has a high greenhouse gas potential, and, since the 1800s, methane emission levels have risen dramatically in the atmosphere alongside the carbon dioxide emissions blamed for global climate change. Therefore, government climate change mitigation efforts increasingly include an emphasis on capturing the methane produced by waste, which is one of the world's greatest sources of methane emissions. Because methane serves as a prominent source of energy, furthermore, waste is also drawing increasing recognition as a potential source of renewable power, whether in the form of redesigned modern landfills or more specialized anaerobic digestion facilities. At the same time, industry and state initiatives are criticized for promoting waste treatment as a source of biogas given its risks and socioenvironmental costs, complicating assumptions about what really counts as "renewable" or "sustainable."

Chemical Processes

Abbreviated with the symbol CH_4, methane is composed of simple hydrocarbon chains, which bond when organic materials—including biological wastes—decompose outside the presence of oxygen. A special class of microorganisms known as methanogens are responsible for generating much of the world's methane. Methanogens belong to a unique microbiological domain known as archaea, a separate branch of life from that of bacteria and multicellular eucaryota. They flourish in the world's wetlands and in more extreme environments, such as landfills and the guts of animals, living amid high temperatures with no oxygen. Methanogens do not consume waste alone but are mutually dependent on colonies of bacterial microorganisms, which break down organic material into smaller organic compounds rich in hydrogen and carbon dioxide and ideal for methanogenesis. Methane is one of the final end products of the chemical process through which these small organic compounds are broken down by methanogens.

Hazards

Though it is commonly assumed to have malodorous properties, methane is an odorless gas. The chemical reactions in different methanogenic systems may also produce pungent chemicals, such as hydrogen sulfide or methanethiol, which could be responsible for this misperception. These co-occurring gases may pose more than a nuisance. In sufficient quantities, hydrogen sulfide is poisonous, and some epidemiological studies confirm that living and working in proximity to waste sites may have negative long-term effects on cognitive function. Wherever methane is produced in proximity to human dwellings and workplaces, the copresence of other gases may raise concerns about health and quality of life.

For many centuries, refuse sites have been associated with fire. It has been suggested, controversially, that the idea of the Christian hell as a place of eternal fire comes from the Valley of Hinnom, south of Jerusalem, which some have claimed was a place where bodies and rubbish were once left to burn. Because of methane, open dumps and other waste sites are typically susceptible to fires and even explosions unless they have gas management systems in place. The potential dangers of methane leaks drew the attention of regulatory authorities in the late 20th century. As gases gradually leak from waste sites, they may accumulate in nearby buildings. In 1967, a methane leak from a landfill site in Georgia was responsible for the deaths of two people when their nearby house exploded; a similar event occurred two years later in North Carolina when an armory close to an abandoned landfill site exploded, killing three and injuring 25 others.

Capture and Management

In the 21st-century United States, the Resource Conservation and Recovery Act requires that modern landfills manage the gas they generate. This can be done in several ways. The most common method was once to simply burn off the gas in flares, and this is still done at most facilities, as has been done as part of the process of oil and gas extraction. In landfill gas collection systems, emissions from the decomposing waste are collected and diverted through a network of pipes to a treatment system, blower, or flare. The 1996 Environmental Protection Agency (EPA) Standards of Performance for New Stationary Sources and Guidelines for Control of Existing Sources and the National Emission Standards for Hazardous Air Pollutants both recommend combustion to reduce nonmethane organic compounds, although they do not apply to smaller landfills.

Most landfill gas projects in the United States utilize it for electricity generation, which is either used on-site or pumped to the electrical grid. Alternatively, the gas is not converted into electricity but is used directly on-site to offset energy requirements. It may also be used in a combined heat and power unit to create usable thermal energy alongside electricity. Some landfill sites are being designed as *bioreactor landfills*, an industry term in the United States to describe a landfill that incorporates selective wastes, such as sewage sludge, alongside normal solid waste loads in order to promote more stable gas collection and better energy production.

Most gas is siphoned off from landfills through collection methods, but even under existing regulations, it is difficult to extract all the methane and, where levels are particularly high, it is possible for small amounts of fugitive emissions to bubble up to the surface when collection systems are insufficient. A more precise way of deriving biogas from waste is to process it in a closed container, as is done with anaerobic digestion technology, to prevent gas from escaping and to precisely control the process of methanogenesis. As with landfills, the feedstock delivered to digestors can be tailored to better methane production—food and green wastes can be codigested alongside energy crops like corn or other readily digestable materials in order to maintain high energy yields—but digestors can also be designed so that conditions within the vessel, such as temperature or pH, promote more efficient chemical exchanges between microbes. Another way in which consumption practices are linked to waste methane is through industrial farming. Cow flatulence and decomposing manures are leading contributors to methane emissions, and anaerobic digestors are increasingly utilized throughout the world as one technique of farm-based methane management systems.

Energy Potential and Risks

By some estimates, natural gas accounts for almost one-quarter of worldwide energy use and approximately one-third of the total energy consumed in the United States. Because waste is so abundant—a by-product of human activities—some view methane derived from waste as a source of "renewable" or "sustainable" energy, capable of contributing to industrial society's high energy demands at a lower overall environmental and social cost. There are potential problems with the emphasis on deriving methane from landfills and other waste streams, however. Typically, natural gas is considered the "cleanest" of all fossil fuels because it is comprised of much lower ratios of carbon, nitrogen, and sulfur than coal or oil. When combusted, other fossil fuels release more carbon emissions, nitrogen oxides, and sulfur dioxide into the atmosphere, resulting in various forms of pollution.

While natural gas is made up of approximately 80–99 percent methane, some estimate landfill gas to be approximately 40–60 percent methane, with a greater proportion of nonmethane organic compounds and inorganic contaminants. Because of the heterogeneous composition of municipal solid waste, landfill gas may contain trace amounts of mercury, radioactive hydrogen, and toxic chemicals such as benzene, toluene, and chloride. When combusted with organics, some of these inorganic compounds produce highly toxic dioxins and furans, such as are generated by conventional incineration. Thus, it may be that the gas derived from waste decomposition is too low in methane and is potentially contaminated with other substances that may pose unseen risks when it is converted into energy.

Greenhouse Gas

In addition to the risks and energy rewards that methane generation introduces to its local surroundings, it also poses more of a global threat. Methane is estimated to be anywhere from 21 to 25 times more powerful than carbon dioxide as a greenhouse gas. While its chemical bonds degrade 10 times faster, the result is only more CO_2. For every metric ton of methane, 2.75 metric tons of CO_2 are left lingering in the atmosphere. Furthermore, methane is responsible for slowing down the formation of sulfate aerosis, sulfur-rich atmospheric particles that are believed to have a cooling effect on the Earth. Since the Industrial Revolution, methane levels in the atmosphere have more than doubled. Human beings are responsible for approximately half of the world's methane emissions in 2010, with the most prominent sources including the production of fossil fuel, domesticated livestock, and waste management activities. Therefore, additional incentive exist to trap the methane produced by waste.

Landfill gas and anaerobic digestion projects have become preferred ways of mitigating the greenhouse effects of industrial waste production globally. Throughout Europe and much of the industrial world, digestors and landfill gas systems are touted as a means of creating more locally sustainable waste management systems, which simultaneously address global climate change concerns.

The United Kingdom has promoted landfill gas projects since 1990, but began a renewed effort to finance such projects alongside digestors with the introduction of the Renewables Obligation in 2001, combining energy, waste, and climate change politics. The primary mechanism through which the U.S. government has encouraged the development of landfill gas collection is through the EPA's successful, and entirely voluntary, Landfill Methane Outreach Program (LMOP), introduced in 1994.

In its first 13 years of operation, LMOP assisted in the development of over 360 landfill gas energy projects by providing landfills with access to a national network of expertise and technical guidance. As a consequence of these initiatives, the level of U.S. methane emissions from landfills decreased—as measured in units of teragrams of carbon dioxide ($TgCO_2$) equivalents—from an estimated 149.3 in 1990 to a reported 126.3 in 2008. During this period, landfills went from the country's leading source of methane emissions to the second, behind enteric fermentation in farm animals. There are additional schemes that provide financial incentives to landfills to reduce methane emissions, including emissions trading initiatives such as the Chicago Climate Exchange. Additionally, landfills may be able to receive revenue in the form of green credits, awarded for offsetting methane emissions and either exchanged bilaterally or sold directly to buyers. Because of these schemes and LMOP, there were over 450 landfill gas energy systems in the United States by 2010, which were responsible for 12 billion kilowatt hours of electricity generated per year.

Conclusion

Few would quarrel with the introduction of landfill gas and other methane extraction initiatives are some of the best ways to utilize the energy potential locked in organic wastes. However, there are concerns that incentives to derive energy from waste may distract from policies directed toward waste reduction, which is a far more comprehensive means of lowering emissions in both carbon and methane. It remains to be seen to what extent transforming waste into an energy feedstock alters the political economy of waste treatment and disposal and whether it will one day be recognized as a "resource" more so than a pollutant or a byproduct. But insofar as waste-to-energy projects depend on practices of mass production and mass wasting, they do not provide a complete solution to unsustainable fossil fuel addiction. This brings the purported "renewability" of landfill gas systems and other waste-to-energy projects into question, despite the potential of methane to alter the landscape of energy practices and politics.

Josh Reno
Goldsmiths College

See Also: Anaerobic Digestion; Incinerators; Landfills, Modern; Microorganisms; Open Dump; Organic Waste.

Further Readings

Association of American Geographers GCLP Research Team. *Global Change and Local Places: Estimating, Understanding, and Reducing Greenhouse Gases.* New York: Cambridge University Press, 2010.

Palmisano, Anna C. and Morton A. Barlaz. *Microbiology of Solid Waste.* Boca Raton, FL: CRC Press, 1996.

Pawlowska, Malgorzata and Lucjan Pawlowski, eds. *Management of Pollutant Emission From Landfills and Sludge.* London: Taylor & Francis, 2007.

Mexico

With 31 states and a federal district, and over 110 million residents living on more than 760,000 square miles of land featuring coastal plains, mountains, forests, and deserts, Mexico is by no means a homogenous country, and therefore residents generate nonhomogenous quantities and qualities of waste. But in accordance with consumption patterns in other industrialized and industrializing nations, as the country's population grows, Mexicans—whether they are rich or poor, rural or urban—are acquiring and discarding increasing amounts of objects. Both in typology and quantity, material refuse chronicles the nation's history and reflects modern Mexicans' garbage-related values, which can be highly political.

Seventy-five percent of Mexicans live in urban areas. The highly stratified national capital of Mexico City is the most populated city in the Western hemisphere and one of the largest in the world. Article 10 of Mexico City's Current Regulations for the Federal District's Cleaning Service mandates that solid household trash removal will be free of cost to all residents. Although semi-functional, considering the nation's poverty, sprawling and exponential population growth, and taxation model, it is growing increasingly difficult for Mexico City, and other municipalities that model their regulations after the capital's, to honor that promise.

Environmental Problems

Due to the unmanageability of the quantity of the trash Mexicans generate, refuse-related environmental problems abound. Disposal issues plague communities of all sizes, though garbage sanitation sometimes comprises up to 40 percent of cities' budgets. But in 2010, according to the World Bank, Mexico had only 97 officially controlled waste disposal sites. Just 11 of those sites were mechanized or manual landfills, as opposed to dumps.

Mechanized landfills are engineered and must have sanitation instruments to ensure proper waste management, while manual landfills are intended to be sanitary and are small enough that employees manage most quotidian tasks by hand, employing technological instruments only for irregular large tasks. In Mexico, however, landfills do not meet most industrialized nations' environmental standards; they are poorly lined, spilling toxins into the soil and groundwater. The World Bank has found that only 15 percent of the nation's municipal solid waste (MSW) is safely treated upon disposal.

One exception, Mexico City's innovative Prados de la Montana controlled contaminant release landfill, treats its leachates (the water that escapes from the landfill after percolating through the waste) via chemical and biological processes. It is the only landfill in the country to do so. Improved leachate management is the World Bank's first recommendation to improve Mexico's landfills. Various entities are working to meet that recommendation by increasing composting and recycling facilities.

Illegal Dumps

In contrast to landfills, dumps are simple, open-air, human-made pits or natural quarries where discarded trash accumulates in an uncontrolled or semicontrolled manner. Mexico City alone features an estimated 1,200 illegal dumps. Dumps present extreme environmental health risks, producing stench and attracting disease carrying vectors, including rats, flies, dogs, and even pigs that sometimes end up as food for squatters and nearby residents. Because of these risks, dumping grounds are usually situated within the most impoverished parts of the community. In the poorest parts of the country, residents simply burn their trash, regardless of type.

Throughout the nation, municipal services collect only about 75 percent of the waste generated (in Mexico City, about 1.2 kilograms of trash per person per day; in Oaxaca and its suburbs, about 600 tons per day total). The remaining 25 percent of trash pickup service—usually in lower income areas—is left to informal private-sector collectors who perform the task for a fee. Householders and business owners are expected to tip low-salaried informal trash removal personnel, though there are usually no written contractual obligations mandating this practice. In proportion to what they pay, individual households and businesses negotiate with their refuse handlers over the frequency and quantity of garbage removal.

Every day or so, a member of a Mexican household or a business employee will take a plastic bag or two of trash to a "sweeper" or a sweeper's container, and tip them a couple of pesos to take it away. The sweeper gives the trash and most of the tips to a garbage truck driver, who takes it farther away. Throughout the nation, garbage employees occasionally take trash to clandestine sites, which include vacant lots as well as occasionally city parks. Usually, however, they take trash to official refuse sites, which charge a per-tonnage fee (though they seldom weigh the trucks).

The government usually owns the land upon which the landfill or dump sits and negotiates contracts with these private-sector trash businesses. However, nonstate garbage services can grow into large independent companies, sometimes developing their own private landfills, as in the case of SIMEPRODE, which controls three in the Monterrey metropolitan area. Because they fill an important (and sometimes vital, lest the garbage accumulate) role in Mexican civil and economic structure, political parties, especially the PRI (Institutional Revolutionary Party), have organized or attempted to organize these companies.

Recyclable Waste

The Mexican National Institute of Statistics, Geography and Information (INEGI) classifies waste as industrial, municipal, or special, all of which can go into either dumps or landfills. Sara Ojeda-Benitez et al.'s 2003 study quantified the residential solid waste (or RSW, which falls into the municipal category) in a sector of Mexicali, Baja, Mexico. Researchers found that a random sample of 160 families generated a vast majority of recyclable organics, especially food scraps, paper, and cardboard—a recyclable inorganic, or what some Mexicans call "clean waste." Potentially recyclable waste constituted 61 percent of the garbage the community of Mexicali generated. Of the nonrecyclables there, textiles occurred most frequently, followed by plastic, and then sanitation wastes. About 2 percent of RSW throughout Mexico qualifies as hazardous. As with most other Mexican cities the size of lower-income Mexicali, there is no landfill, only a dump.

Substantial environmental hazards result from filling dumps with so much organic waste. For the first time ever, in September 2011, the annual international Waste and Recycling Expo took place in Mexico City, a testament to how Mexico's priorities may be taking a turn toward the sustainable and sanitary. However, composting is almost unheard of and formal recycling is at this time faddish only in cosmopolitan areas. It is increasing in wealthier cities and touristic areas as they follow the examples developed nations set, and even this is happening slowly. Residents of certain wealthier sectors are beginning to sort their recyclables from their nonrecyclable waste, but sometimes express doubt as to whether the two actually have different final destinations.

Informal Recycling

Informal recycling, however, has been an important component of the story of Mexican trash disposal for several decades. Depending on the type of activity, informal recyclers save landfills and dumps various quantities of space, energy, and money. Their living and working conditions can be extremely harsh, but they sometimes have political clout and leverage.

One type of informal recycler, *carretoneros*, walk through neighborhoods, hunting through trash cans for recyclable objects to sell. They are independently employed and not usually officially organized, except in some neighborhoods in Mexico City. In some cities, the government licenses and monitors *carretoneros*. They are not permitted to dump waste in clandestine sites, nor leave their carts uncovered overnight, and they must pick up after their horses. According to Martin Medina, one reason the government decided to control these informal recyclers is that they can earn up to five times the national minimum wage and thus are important sources of tax revenue. Another sort of informal recyclers, waste pickers, sort through trash at dumps or landfills to reclaim and sell reusable materials, facilitating new production. In Mexico they are known as *pepenadores*, and they help eliminate tonnage at refuse sites.

Waste pickers working throughout Mexico organize themselves into cooperatives, which have gone on strike protesting and forming blockades around dumps. *Pepenadores* seek job security, assistance treating job-related health risks, and fairer compensation for the role they fill in helping sort the up to 6,500 tons of trash many Mexico City dumps receive every day. They have closed waste facilities for days, rendering them unable to process trash as full trucks wait to unload their hauls.

Regardless of the rights they may have won, *carretoneros* and *pepenadores* are objectified, perceived as dirty, like the objects with which they work and often in which they live. However, they serve a vital function in Mexican society; without them it is apparent that the sanitation service would literally fall apart. In Mexico, economic stratification tends to correlate with access to sanitary and environmentally safe and responsible disposal conditions, as it tends to in other parts of the world. Most people, however, regardless of their socioeconomic status, tend to agree with the sentiment that garbage equates to immorality, which is inconsistent with the aesthetic reality of the majority of Mexicans.

Angela Orlando
University of California Los Angeles

See Also: Dump Digging; Landfills, Modern; Mexico City, Mexico; Street Scavenging and Trash Picking.

Further Readings

Medina, Martin. "Serving the Unserved: Informal Refuse Collection in Mexico." *Waste Management and Research*, v.23 (2005).

Ojeda-Benitez, Sara, et al. "Characterization and Quantification of Household Solid Wastes in a Mexican City." *Resources, Conservation and Recycling*, v.39 (2003).

Mexico City, Mexico

Mexico City is the capital and largest city of Mexico, the largest city in the Americas and the third most populated metropolitan area in the world. The political, economic, and cultural center of the country, it is a federal entity unto itself, not part of any Mexican state—much like Washington, D.C. The metropolitan area has about 21.2 million people and accounts for over one-third of the country's gross domestic product (GDP). Within the city limits, the per capita GDP is $47,396, one of the highest in the world. Despite this wealth and a high ranking on the Human Development Index, the city's sheer size has led to a number of unique challenges of waste management and resource consumption. Because of its size and age, in many ways Mexico City is a canary in a coal mine: many of the problems it faces represent challenges other cities can expect as they approach similar scale and density.

Brief History

The city was built on an island of Lake Texcoco, and the Spanish colonists dug canals and tunnels in the surrounding Valley of Mexico in the 17th century because the lack of natural drainage led to periodic flooding from mountain runoff. The lake was subsequently drained, and the city rests on heavily saturated clay, which has begun to collapse since the early 20th century due to overextraction of the groundwater. The sinking complicates runoff and wastewater management and exacerbates flooding problems in the rainy season.

Water Management

Responsibility for Greater Mexico City's water management is divided among various governments. The federal government regulates the use of water resources. The state of Mexico treats wastewater, provides bulk water, and assists municipalities in providing water and sanitation services. The 60 municipal governments in the area are in charge of sanitation and water distribution. The federal district provides water and sanitation services to the city proper. The federal government takes a particular interest because of Mexico City's economic, political, and infrastructural importance to the country and the national welfare. These

An aerial view of the northern part of Mexico City and the Palacio de Bellas Artes (Palace of Fine Arts) reflects the statistic that Mexico City is the largest city in Mexico and the third most populated metropolitan area in the world. It also accounts for over one-third of the country's gross domestic product. The size and age of the city have led to a number of consumption and waste management challenges. About 85 percent of the wastewater in the city is discharged, untreated, into the north to irrigate crops; this presents serious health risks.

governments have instituted a water sustainability program and a green plan, which emphasizes water conservation. Both plans call for transporting groundwater from north of the city where the groundwater table is higher, constructing a significantly expanded stormwater drainage tunnel, and investing in wastewater treatment improvements.

The National Water Commission is the federal body responsible for water resources management in the country, granting of wastewater discharge permits, and the supply of bulk water to Mexico City. The combined sewer system of Greater Mexico City collects municipal wastewater, storm water, and industrial wastewater, and it consists of 7,400 miles of pipe, 68 pumping stations, and numerous open canals, tunnels, underground collectors, dams, lagoons, and regulatory tanks, and intersecting with rivers. A single storm in the rainy season can produce a full tenth of the annual precipitation, necessitating significant capacity for the stormwater drainage system in order to avoid highly damaging floods.

Only about 15 percent of the wastewater collected in Greater Mexico City is treated, primarily in Mexico state wastewater treatment plants. Most of the remaining wastewater is discharged to the drainage system untreated and eventually is discharged in the north where it is used to irrigate alfalfa, barley, wheat, corn, and other crops. The nutrients in the sewage keep crop yields high and reduce dependence on other fertilizers, but the wastewater also contains both industrial and biological contaminants (such as bacteria, larvae, and eggs), which present health risks both to farm workers and to consumers. In theory, using the wastewater to irrigate crops that will be heavily processed and cooked before consumer contact—grains and cereals—mitigates the health risk, relative to such use to irrigate fruit and vegetable crops. Technically, use of untreated wastewater to irrigate crops that will be eaten raw is illegal; in practice, it is widespread in the area. There are many farms and farm workers whose livelihoods would be in jeopardy should they switch to more expensive water sources. (Industrial wastewater is partially treated at the source, but the data is uncertain as to the efficacy of its pretreatment before entering the drainage system.) A portion of wastewater is also diverted for use to irrigate urban landscapes (such as parks and lawns) and for commercial uses, such as car washes. Private-sector companies also buy up municipal wastewater and redistribute it for nonpotable reuse.

An ongoing problem in the early 21st century is leaks in the water distribution system, as a result of which soil is permeated by contaminants from one source, which are then passed on to another part of the distribution chain—one that is supposed to be carrying uncontaminated water. Studies by the local government have found that the quality of tap water is lower in neighborhoods that have frequent water distribution service interruptions (caused by such leaks), a public health problem compounded by the common practice of using *tinacos* (open tanks kept on rooftops and used for household water storage when water pressure is insufficient).

When the tanks are not regularly cleaned, the level of chlorine in the water falls low enough to permit contamination by microorganisms and bacteria, and the chlorine in the tap water is sufficient only to prevent that growth (not to kill living microorganisms introduced to the water supply). Estimating the extent of leaks is difficult because Mexico as a whole, and Mexico City specifically, has a nonrevenue water level of 40 percent—in other words, only 60 percent of the water that goes through the system is accounted for in billing. However, the level of loss is far too great to be the fault of leakage alone and represents a significant amount of theft through illegal, unbilled connections to the distribution system. Financing the water system is further complicated by the fact that the water department has an exceptionally low collection efficiency and fails to collect more than half of the revenue it is owed. The Greater Mexico City population thus uses at least three times as much water as it pays for.

Much of this will be at least addressed, if not fully remedied, by the new water programs being put in place in the early 21st century. The extent of some of the problems is great enough that it may not be revealed in full until a partial remedy is in place, like opening a patient for surgery in order to survey the damage. The green plan in particular aims to reduce water loss through leakage and other problems, reduce unnecessary or inefficient water usage, recharge wells and aquifers, meter all water users in order to increase bill collection efficiency and better fund water distribution, and create better wastewater treatment plants in order to increase the amount of wastewater that is converted into safe water. These programs are more ambitious than the water conservation programs attempted in the 1990s, which had ambiguous results.

Air Pollution

Air pollution has been a significant problem, addressed by both federal and local governments. When levels of ozone or nitrogen oxides reach critical levels, factories are closed, school hours are modified, and the *Hoy No Circula* program is extended to two days a week in order to reduce emissions. *Hoy No Circula* (meaning "A Day Without A Car") is a Greater Mexico City program that limits car usage to six days a week (or five during such emergencies), based on the last digit of the license plate. The program applies to any car with a plate from Mexico City, Mexico State (the surrounding state), or surrounding states with reciprocal programs. These contingency plans, along with the reformulation of auto fuels and strict enforcement of the twice-per-year emissions inspections, has helped reduce pollution levels by significant amounts. As of 2010, lead had been reduced by 95 percent since 1990 (when pollution levels in Mexico City were more than 10 times greater than those of New York City), sulfur dioxide had been reduced by 86 percent, and carbon monoxide had been reduced by 74 percent. The populace is also more dependent now on public transportation, including an underground rail system with subsidized fares.

Solid Waste Management

Solid waste is also a mammoth problem for Mexico City, with more than 10,000 tons of garbage produced daily by the city. Much of this ends up in places like the 600-hectare Bordo Poniente landfill, which emits 2 million tons of greenhouse gases a year, second only to automobiles in the list of Mexico City emission sources. Bordo Poniente's shutdown was announced in 2004 but was continually delayed. The national government has repeatedly called for its closure, arguing that it is far over its capacity. The city government, bereft of any alternative, has insisted that the landfill can and will continue to operate until 2012. Further, official studies indicated over 100 unauthorized garbage dumps operating throughout the city, most of them little more than heaps of refuse dumped in a vacant lot or green area. The city's fleet of garbage trucks has long lagged far behind what it needs.

While other countries have made some progress exploiting municipal solid waste for its biogas and other economic possibilities, Mexican waste law does not address the ownership of garbage, leaving it unclear who has the authority to act on such an opportunity. The Clinton Climate Initiative founded by former U.S. President Bill Clinton embarked on the largest waste management project undertaken by a climate change organization, initiating a project in Mexico City that intends to cap the current landfill and initiate a new waste management program based on recycling, landfill gas collection, and composting.

Bill Kte'pi
Independent Scholar

See Also: Developing Countries; Mexico; United States; Water Treatment.

Further Readings

Hernandez, Daniel. *Down and Delirious in Mexico City*. New York: Scribner, 2011.
Lida, David. *First Stop in the New World*. New York: Riverhead, 2008.
Ross, John. *El Monstruo*. New York: Nation Books, 2009.

Miasma Theory of Disease

The miasma theory of disease causation has some incipient roots in Greek and Roman medicines, in particular in Hippocrates's "On Airs, Waters, and Places." It developed as a naturalistic theory during the Renaissance and was especially popular in the 19th century to explain yellow fever, malaria, typhus, tuberculosis, cholera, and other diseases. According to this theory, disease causation relates to environmental emanations (gases), or miasmas. Miasmas are infectious noxious vapors emanated from putrefying carcasses, rotting vegetation or molds, and invisible dust particles inside dwellings. The understanding was that in some cases the air became attacked by an epidemic influence that became malignant after interaction with the emissions of organic decomposition from the earth.

However, in the early 19th century, it was clear that some diseases (e.g., yellow fever) did not develop if the person suffering from an infectious disease changed location. This fact enhanced the theory of miasma based on the presumed connection of the disease with a certain place, although it also propelled a search for other reasons. Professor François Magendie's lecture at the College of France in 1834 provides an example of how miasmatic disease was defined in early 19th century:

> Yellow fever is a disease strictly miasmatic; it exists only in certain localities, in places favourable to the development of various exhalations of an injurious nature; it devastates one quarter of a town, while the rest are habitually free from it. If an affected individual be transported to a distant and healthy situation, he does not carry the disease with him. . . . The Americans tried this principle upon a grand scale during the prevalence of cholera, but without success. . . . It was either there before them, or did not long fail to arrive. . . . Cholera was proven not to depend on miasma, like yellow fever and other diseases of that class, which are characterized of being confined to peculiar localities, and of not being transmissible to healthy situations where the developing cause does not exist.

Later it was clarified that the cause of yellow fever as an acute disease was mosquitoes and that explained why changing places had a preventive health effect. By the mid-1800s, the concept of miasmas was intimately connected with theories of fermentation (so-called spirituous fermentation, acetic fermentation, lactic fermentation, and putrefication). Despite the invention of the microscope in the 1600s, fermentation was not anchored in these theories to microorganisms and the common belief was that chemical rather than biological processes caused fermentation. The miasma theory of disease postulated a sort of airborne "ferment"—the cholera epidemic was explained by such factors as calm, stagnant, high barometric pressure weather and high river water temperatures at night.

In the later part of the 19th century, the miasma theory competed with germ theory and helped prevent a quicker recognition of the latter. It is a lesson for the history of science, since William Farr, the advocate of the miasma theory, demonstrated one of the typical characteristics of narrow-minded authorities—insistence on one thesis and ignoring all others. However, even in the early 19th century, there were specialists who did not connect cholera to miasma.

The miasma theory of disease did have a positive effect on modern human preventive health history: avoiding infectious diseases meant cleaning the streets of garbage, sewage, animal carcasses, and wastes that were features of urban living at the time.

Sanitary Movement

One of the miasma theory's main benefits was improving health conditions in the cities during the sanitary movement (1838–1914). It started in Great Britain and had a strong influence on the United States.

Miasma theorists suggested that the development of disease is a result of harmful odors, mists, or substances (i.e., pollution), found as organic matter in the environment. The cholera epidemic in 1854 stimulated public health officials to develop successful strategies for preventive health by eliminating accumulated waste, cesspools, and contaminated water as presumed miasmic breeding grounds for epidemics. The focus was on sewage management, drainage, clean water, ventilation, and other sanitary measures. In the 19th century, the policy of personal and public hygiene concerned mostly the wealthy levels of society, while the environment in low-income areas led to epidemics of cholera, typhus, typhoid, and plague, as well as endemic tuberculosis. In 1848, the British government passed a special law for hygiene, followed by other countries, including the United States. The preventive health and social measures consisted of paving streets, building sewers, providing clean water, establishing ventilation, reducing crowding in housing, and hauling away garbage. These reforms correlate with decreases in mortality caused by epidemic diseases.

Toward the end of the 19th century, there was a policy shift in public health from interest in cleaning of the environment toward protection of individuals from the hazards of their environment. One of the reasons was the changed disease causality explanation: the role of microorganisms was recognized, as was the contagious nature of epidemic diseases. By 1875, the germ theory moved public health into a "bacteriological era." However, sanitation and hygiene became commonly understood as remedial measures.

Epidemiology

Epidemiology, the basic science of public health, is a reflection of the miasma theory of disease that required sanitization and hygiene. When it emerged, it was associated with 19th-century infectious disease epidemics. It started as the study of the distribution of disease in populations and later included the determinants of disease. Since the 20th century, the focus has been changed from an interest mainly in infectious diseases toward chronic diseases and health-related states and events such as injuries, violence, environmental and social conditions, and regulation of health problems.

Lolita Petrova Nikolova
International Institute of Anthropology

See Also: Children; Germ Theory of Disease; Human Waste; Public Health.

Further Readings

Albrecht, G. L., R. Fitzpatrick, and S. Scrimshaw. *Handbook of Social Studies in Health and Medicine.* London: Sage, 2003.

Curtis, V. A. "Dirt, Disgust and Disease: A Natural History of Hygiene." *Journal of Epidemiology & Community Health,* v.61/8 (2007).

Douglas, M. *Purity and Danger: An Analysis of the Concepts of Pollution and Taboo*. London: Routledge and Kegan Paul, 1966.
Eyler, J. M. "The Changing Assessments of John Show's and William Farr's Cholera Studies." *Sozial- and Präventivmedizin,* v.46 (2001).
Halliday, S. "Death and Miasma in Victorian London: An Obstinate Belief." *British Medical Journal,* v.323–327 (2001).
Magner, L. N. *A History of Medicine*. New York: Marcel Dekker, 1992.
Milgrom, J. "Impurity Is Miasma: A Response to Hyam Maccoby." *Journal of Biblical Literature,* v.119/4 (2000).
Moffett, J. R. "Miasmas, Germs, Homeopathy and Hormesis: Commentary on the Relationship Between Homeopathy and Hormesis." *Human and Experimental Toxicology,* v.29/7 (2010).
Osterrieder. N., J. P. Kamil, D. Schumacher, B. K. Tischer, and S. Trapp. "Marek's Disease Virus: From Miasma to Model." *Nature Reviews Microbiology,* v.4/4 (2006).
Parker, R. *Miasma: Pollution and Purification in Early Greek Religion*. Oxford: Clarendon Press, 1983.
Sipe, A. W. R. "Celibacy Today: Mystery, Myth, and Miasma." *Cross Currents,* v.57/4 (2008).
Tesh, S. N. "Miasma and 'Social Factors' in Disease Causality: Lessons From the Nineteenth Century." *Journal of Health Politics, Policy & Law,* v.20/4 (1995).
Thorsheim, P. *Inventing Pollution Coal, Smoke, and Culture in Britain Since 1800*. Athens: Ohio University Press, 2006.
Tulchinsky, Th. H. and E. Varavikova. *The New Public Health*. 2nd ed. Burlington, VT: Elsevier Academic Press, 2009.

Michigan

Michigan is renowned for two seemingly contradictory reputations: the beauty of its lakes and industrialization. Michigan spans 96,810 square miles across two peninsulas, making it the 11th-largest state. Rural areas include much of the upper peninsula, where copper and iron mining were the dominant industries in the early 20th century. The most urbanized area is Detroit in the southeast, the center of U.S. automobile production. By 2000, the population of Michigan was 9,938,444, making it the eighth most populous state.

Through the automobile industry, the state redefined capitalist labor and consumption, changing the significance of waste in the process. But from the start of the 21st century, it became one of the country's leading importers of out-of-state waste. Unlike Pennsylvania and Virginia, however, Michigan's status as a waste importer was the result of an international trade with Canada, providing a unique set of conditions from which to reconsider the relationship between the economics and international politics of waste. For some, Michigan's exceptional natural resources have to be defended from material and legal encroachments; for others, waste imports offer opportunities for the prosperity and productivity lost through the postagrarian and postindustrial transformations of recent decades.

Natural Resources

Michigan's natural resources were not always regarded as something to preserve or conserve. During its colonization over the course of the 19th century, Michigan's abundant forests were often considered wasteland and were burned as part of laying claim to Native American territory and making way for new farms and villages. As Michigan moved from a leading source of beaver pelts to agricultural produce, timber, and ore, there was a similar embrace of industry and accumulation at the expense of environments and their inhabitants. In the early 20th century, innovations like the modern assembly line and the $5 day made Michigan industry synonymous with the spread of mass production and consumption, and a concomitant overproduction of waste.

Automobile Industry

The Fordism of the automobile industry was not merely emblematic of these transformations but also created waste problems of its own. On the one hand, there are the waste products of cars, such as exhaust, which has been held responsible for rising levels in greenhouse gases associated with climate change. On the other hand, there is the afterlife of the car body as it breaks down or is replaced with a

While Michigan's recycling rates are lower than the U.S. average, its rates of incineration and landfilling are higher, in part due to an arrangement with Toronto, Canada, to accept their solid waste. In 2006 alone, Michigan accepted 12 million cubic yards from Canada.

newer model. General Motors' (GM) CEO, Alfred P. Sloan, helped introduce the concept of planned obsolescence by promoting annual changes in automobile styles to encourage consumers to continually buy new vehicles and dispose of old ones, which cultural critics like Vance Packard would later hold responsible for the profligate production of waste associated with U.S. economic life.

Henry Ford recognized this as a problem and aimed to promote a disassembly line that would take apart old cars and recycle their components. The frequent replacement of automobiles supplies raw materials for Michigan's strong scrap industry, which produces metal and spare parts for sale through the recycling of old cars in junkyards. In 2009, the federal Car Allowance Rebate System or "Cash for Clunkers" bill was sponsored by Michigan congressmen and signed into law. This bill provided incentives for car owners to scrap their old cars in order to purchase more fuel-efficient vehicles and stimulate demand for the automobile-dependent economy of the Midwest.

Waste Management

Like most states, after the boom in consumption that accompanied World War II, Michigan began relying heavily on sanitary landfill disposal in order to manage its increasing consumer and industrial waste. A concomitant reliance on truck drivers to haul waste and consumer goods helped enable the growth of the Teamsters Union, which under the leadership of Detroit native Jimmy Hoffa grew to one million members by the 1950s.

Like in other parts of the country, accusations of corruption and criminal connections continue to shape perceptions of Michigan's waste management industry long after Hoffa's downfall. Truck drivers and machine operators in the waste industry remain heavily unionized in Michigan—perhaps in part because of the example and influence of the state's powerful auto unions—and they have used garbage strikes in Detroit and elsewhere throughout the late 20th century. With the gradual privatization of the waste industry, collection and disposal became divided between the largest waste companies, with Waste Management holding the greatest share.

According to data collected in 2004, Michigan performs average in comparison to other states in terms of the amount of waste generated, producing approximately 1.26 tons per person per year, or 12.76 million tons annually, an amount that increased to over 15 million tons the following year. Michigan recycling rates fare worse, at little over 20 percent in comparison to over 28 percent for the nation. Consequently, the state incinerates and landfills more, at less than 10 percent and 70 percent, respectively. This reliance on landfills was promoted as part of a statewide strategy to manage an anticipated waste crisis. Like all states, Michigan was required to comply with the Resource Conversation and Recovery Act (RCRA) of 1976, which introduced a more rigid and standardized regulatory framework for managing waste. Michigan introduced the Solid Waste Management Act in 1978 in response.

After further amendments to RCRA required the closure of substandard landfills, there was growing concern about a looming waste crisis. In response, some municipalities turned toward incineration, which had lost favor to landfill in the United States over the course of the 20th century. In 1989, the largest incinerator in the world was built in the middle of Detroit. Processing waste from the greater Detroit area, the incinerator has raised considerable controversy, leading to local protests in 2010 aimed at preventing the plant's recertification.

New Landfills

In response to the crisis, the state government required individual counties to produce solid waste management plans that would secure enough landfill space to supply their needs for a specified period of time. To facilitate this process, the state introduced a program that offered tax-free bonds for the construction of new landfills. The result was the development of new, large landfills, typically in rural areas, with abundant capacity and a need for additional revenue to make up for the new costs of environmental compliance. While most states reported a net loss of landfills in the 1980s and 1990s, by a total of 67 percent, Michigan was one of two states to have increased its number. As a consequence, an increasing number of landfills were given an incentive to seek out-of-state waste contracts, especially after the 1992 U.S. Supreme Court case *Fort Gratiot Landfill v. Michigan Department of Natural Resources*, which determined that waste imports were protected under federal anti-protectionist laws.

Importing Waste

Due to strong opposition among First Nation communities to a proposed landfill site in northern Ontario that was meant to supply Toronto after its old city landfill closed, the city of Toronto decided to begin sending its waste to Michigan landfills in 2000. Between 2000 and 2006, the amount of out-of-state municipal solid waste disposed of in Michigan doubled, from approximately 9.4 million to 19 million cubic yards. At its peak, over 12 million cubic yards were in one year imported from Canada, much of it from the Toronto area, which amounted to 18.9 percent of the solid waste disposed of in Michigan that year. This dramatic increase led to strong public opposition, organized at the local level by residents living in the vicinity of transnational landfills and statewide by political candidates running for reelection.

In 2003, the Not in My State and Don't Trash Michigan programs were launched, both of which sought to organize support in order to summon a challenge to the waste trade. Critics pointed out that Michigan also exports significant amounts of hazardous waste across the border into Canada. Furthermore, the waste trade actually provides considerable economic revenue to rural Michigan communities without comparable sources of funds. For this reason, town supervisors from one landfill community actually traveled to Lansing to protest shutting down the border to waste imports, expressing their concern that this would cripple the town budget, which had become dependent on the revenue supplied from the host fee. With declining farms, a housing market in turmoil, and fewer reliable industrial jobs than in the past, outside sources of revenue such as that stemming from the waste trade are important.

Even so, a wide range of methods were explored to eliminate Canadian waste imports or to find them in violation of state environmental regulations and international agreements. A total of 11 bills were signed into law in 2004 by the governor, including attempts to ban beverage containers, increase the number of inspections, and impose a moratorium on waste imports. These were largely unenforceable rules, which received stiff opposition in the courts and elicited lawsuits on the part of the waste industry. Their largely symbolic significance was an attempt to voice state-level opposition to a transnational market in landfill space regulated into existence by particular interpretations of national and transnational law. Ultimately, however, political maneuvering on the part of Michigan political leaders forced Ontario officials into an agreement that would first limit waste exports from Toronto, then end them altogether by 2010.

Alternatives

Despite the government's past emphasis on increasing landfill capacity to handle the waste crisis, new efforts have been undertaken to promote alternative forms of waste disposal and treatment in Michigan. Beginning in the mid-1980s, Michigan landfills began introducing gas-to-energy projects in order to increase their revenue and comply with new regulations under the changing laws of RCRA. By 2007, there were 12 active gas recovery projects in operation in the state, nearly twice the national average per state and enough to produce 60.3 megawatts of electricity every year.

By 2010, in southeast Michigan, a company was attempting to create an innovative plasma arc gasification plant that would operate at 100 megawatts annually. An advanced form of thermal treatment, gasification converts waste at high temperatures into a synthetic gas that can be converted into any number of things. At the proposed Michigan site, the plant will produce syngas alongside ethanol, biodiesel, insulation, and animal feed, all of which it would process from unsorted black bag waste. Michigan also has several anaerobic digestors operating on a trial or experimental basis. The once agriculturally dependent state had as many as six farm-based digestors in operation prior to 2005; five failed, largely for financial reasons. As in other parts of the country and at other times in its history, Michigan politicians, researchers, and entrepreneurs are beginning to experiment with various waste treatment options. It may be that politicians who were elected to protect Michigan from foreign waste may be committed to keeping waste on the political agenda, particularly in light of a suffering economy and growing uncertainties about climate change and other environmental threats.

Josh Reno
Goldsmiths College

See Also: Anaerobic Digestion; Automobiles; Economics of Waste Collection and Disposal, International; Landfills, Modern.

Further Readings

Bergman, Garry D., ed. *Cash for Clunkers and the Auto Industry: Background Lessons and Results (Business Issues, Competition and Entrepreneurship)*. Hauppauge, NY: Nova Science Publishers, 2010.
McCarthy, Tom. "Henry Ford, Industrial Conservationist? Take-Back, Waste Reduction and Recycling at the Rouge." *Progress in Industrial Ecology*, v.3/4 (2006).
Reno, Joshua. *Out of Place*. Ph.D. diss. Ann Arbor: University of Michigan, 2008.
Senior, Eric. *Microbiology of Landfill Sites*. Boca Raton, FL: CRC Press, 1995.

Microorganisms

Microorganisms have played a critical role in the history of waste management, and some believe they will play an even greater role in its future. Microbes are responsible for most of the characteristics associated with waste materials, including odor, putrefaction, and transience, but they were not always recognized as such. The influence of microbial activities on waste was gradually accepted in the aftermath of the devastating cholera outbreaks that affected concentrated urban populations in the 19th century. Ideas from epidemiology and bacteriology eventually influenced hygiene movements and the use of antiseptics in medicine. The germ theory of disease had begun to replace the miasma theory of disease, and fear of the pathological dangers posed by microbes grew along with microbe-based models of cleanliness.

The importance of sanitation in preventing environmental contamination and disease took hold in various social domains in the early 20th century. The notion that hidden microbial agents might spread disease through offensive materials, including waste, provided a potent symbol for the commodification and purification of domestic spaces and the public management of waste collection, treatment, and disposal. Garbage handlers in the United States became known as sanitary engineers, and modernized dumps became known as sanitary landfills. More recently, the actions of microorganisms in processing and treating organic wastes, as in landfill gas recovery operations or anaerobic digestion, have been identified as offering a possible form of renewable energy and a possible alternative to mainstream models of waste disposal, such as landfilling and incineration. The application of microorganisms to waste treatment forms part of a larger shift in technoscience toward the manipulation of biological life at a microscopic scale as a source of new value and societal transformation.

Early Conceptions and Study

Microorganisms represent a diverse range of creatures that are too small to be seen by the unaided human eye. Before the activities of microbes were experimented with and documented with the assistance of the microscope, their profound effects on the world were commonly attributed to other causes. The tendency for abandoned or unpreserved items to break down and transform was associated with the spontaneous generation of life, while the link between organic decay and odor helped lead to the belief that foul air contaminated with decomposed matter, or miasmas, was ultimately responsible for disease. These beliefs shaped waste management, as decaying waste matter was seen to be a source of pestilence and vermin. Without the recognition of microbial actions, waste may not have been contained and treated in the same way, but it was still removed from spaces of human habitation and sent to the margins of settlements. The dominant influence of these beliefs also shaped preferred waste treatment methods. Early incinerators, known as destructors, were greatly preferred among municipal engineers, even after modern landfill techniques were developed, precisely because cremation served as a method of destroying the noxious miasmas associated with the putrefaction of wastes, as well as getting energy in return.

The study of microbes and their effects on human health in the later 19th century played a central role in the modernization of waste management. The emerging field of bacteriology and the scientific writings of Louis Pasteur, Joseph Lister, and Robert Koch, among others, helped to enlist microbes in the moral and medical arguments of Western hygiene movements of the Progressive Era. Their efforts helped the germ theory of disease to take hold in the 20th century and associated microbes, above all, with disease transmission. Hygiene movements went beyond medical explanations for disease and encompassed a whole host of efforts to establish social order in industrial societies, including alcohol prohibition, opposition to prostitution and vice, and, later, eugenics. As a corollary to these social movements, sanitation engineering emerged as a means of preserving social order through material purification. Bacteriological models of hygiene introduced a scientific ethos of waste policy and practice. In the 1920s and 1930s, a sanitary inspector with a position in the English Ministry of Health named J. C. Dawes began advocating "controlled tips." His main recommendation, for which he is considered one of the most influential modernizers in sanitary engineering, was to use soil cover on waste dumps in order to prevent unwanted organisms from using them as breeding grounds for disease—an application of Pasteur's model of biogenesis and refutation of spontaneous generation. This key innovation was incorporated into the designs of modern landfills in later decades.

Germs and Hygiene

While both miasmatic and germ theories of disease would have advocated removal and containment of wastes, the latter made the influence of waste on health potentially imperceptible and therefore potentially more uncertain. While it is still the case that fragrant smells are associated with cleanliness or purity in everyday life, when it comes to the management of cities or households, fears over unseen microbes helped launch a boom in domestic products, including antibacterial soaps, detergents, and other chemical cleaners. The commodification of domestic labor was facilitated by a fear of unseen microbes and the damage they could do as well as by a desire for social and moral distinction, which reached from the colonial margins of Euro-American empires to middle-class neighborhoods.

Germ theory gripped the hygienic imagination and provided a model for later criticisms and regulations of the waste industry. Fear over the hidden effects of microbes on health may have helped contribute to an analogous anxiety, later articulated by environmental advocates, that equally undetected toxic agents or carbon emissions might contaminate the health of whole ecosystems. Pollution events like the Love Canal disaster raised awareness of such possibilities. The development of landfill liners and gas and leachate collection systems in the later

20th century applied concern over hidden microagents more broadly. 502

Microbiology

While microbes have dominantly been cast as an enemy of cleanliness and health in the 20th century, there have been renewed efforts in the 21st century to harness the special abilities of particular microorganisms through technical application or biological modification. For example, with the increasing sophistication, speed, and affordability of genome sequencing technologies, microbiologists have begun identifying the trillions of microbes that inhabit the human body, especially the digestive tract, in order to better understanding how these beneficial microorganisms contribute to particular diets and disease immunities. Other microbes have been discovered that give clues to the evolution of life on Earth or, it is claimed, could be modified to make advances in climate change mitigation, renewable energy production, and waste treatment. In one laboratory, researchers from the University of California, Berkeley, and LS9 biotech firm have modified a form of *Escherichia coli* bacterium that can produce usable biofuel from the digestion of crop and grass waste.

Aerobic Composting

There are other microbes that occur naturally and have been used to treat waste for centuries. Aerobic bacteria play an important role in the digestion of wood waste and other green wastes. Composting relies on a whole host of microbes, including bacteria, fungi, protozoa, and rotifers, which break down different kinds of materials and also feast on one another, producing an ecological balance to the process. Specific microbes are also necessary for breaking down paper products, called actinomycetes. The result of a successful composting operation is a nutrient-rich fertilizer and a destruction of biological pathogens when functioning at high-enough temperatures. There are industrial applications of composting techniques, as well as domestic ones. In-vessel composting units are a relatively new waste treatment technology that reduce the presence of unwanted pests from composting heaps and can control air flow and temperature of their units to encourage more efficient and more thorough conversion of wastes into a usable fertilizer. This is essential if organic composting operations are to generate a product that can compete in the market with industrial fertilizers. At a very different scale are home composting units, such as the popular Bokashi receptacles. These and similar home-composting units come equipped with efficient microorganisms (EM), which, the company claims, have been carefully selected to encourage successful decomposition, reduce odors, and produce successful fertilizer.

Anaerobic Digestion

Landfills are constructed precisely to trap material from escaping and, consequently, do not allow in the oxygen that would enable composting to occur. Particularly important for the breakdown of landfill organics are a class of organisms distinct from bacteria called archaea, an ancient form of life adapted to extreme conditions and capable of digesting waste outside the presence of oxygen. The resulting chemical reaction, known as anaerobic digestion, transforms organic wastes into both methane (which can be harvested for energy use) and a fertilizer-like substance known as digestate. Anaerobic microbes are utilized by landfill biogas collection systems, which draw methane from the garbage as it decomposes and burn it off, pipe it away, or convert it into electricity. Such landfills are also known as bioreactors because they are constructed in order to encourage these chemical reactions by creating conditions favorable to the microbe populations within.

There are microbiologists who examine the microbial populations cohabiting inside landfills, perhaps opening up the possibility of better promoting their activity and harnessing their outputs in more sophisticated ways. But those who study anaerobic digestion admit that the process itself is somewhat unpredictable, precisely because the necessary containment of the organisms and the waste they consume, as well as the microscale at which they operate, make it difficult to engage in direct manipulation of the methane production process. Even if one buries organic wastes, there is no guarantee they will spawn microbial colonies. The lack of sufficient microbes can hinder the organic breakdown of landfills and, ultimately, reduce landfill capacity. Landfill bioreactors may seek to enhance the process by dumping more readily digested organics than solid waste, such as sewage. Another method involves taking the leachate, or excess fluids, collected from the base of the landfill and reapplying them to the active cells, encouraging digestion of the fresh waste by adding a volatile inoculant to the mix.

Another method of controlling the conditions of anaerobic digestion with more success is through particular plants dedicated to the treatment of organic wastes, known as anaerobic digestors. Like industrial composting units, digestors make waste treatment more predictable. The design of digestion plants can be accommodated to reach specific temperatures and, thus, rates of decomposition. There are specific microorganisms for different ranges of heat in anaerobic conditions, depending on whether they are mesophilic or thermophilic. Digestors can also better control the kinds of feedstock that are incorporated (this is difficult to accomplish with mixed waste loads) and use different forms of mixing before or during the digestion process. Because the anaerobic archaea are sensitive, inoculants can be added to facilitate their growth and pH levels can be carefully maintained. Anerobic digestion is a four-stage process, which relates to the four kinds of microbes involved in the process of chemical transformation.

It is particularly difficult, at the laboratory level, to control volatile fatty acids created by the secondary process of acidogenesis, ensuring that they do not overpower the methanogens that create energy-producing methane in the final stage of the treatment process. For this reason, to accommodate the different microbial populations, there exists a multistage design for anaerobic digestors that separates the four processes into discrete steps, allowing the operators to optimize conditions in each tank to the particular requirements of the microbes within, thus improving waste treatment and gas generation.

Conclusion

Aerobic composting and anaerobic digestion offer methods of treating organic wastes that involve more control of inputs and the treatment process, as well as outputs that resemble products rather than pollutants. For this reason, both are being experimented with and promoted in places where there is a desire to combine solutions to climate change with new markets in green energy and goods. Of these methods, by far the most common has been landfill biogas extraction, both because it is the cheapest means of treating organics and one that is most accommodating to existing methods of waste collection and disposal. Organic treatments require some separation of material inputs because microbial activity works most effectively when feedstocks are tailored to them. In those places where these new microbial waste technologies have been trialed, the source-separation of wastes has represented one of the biggest challenges.

Householders need to be trained on new waste separation methods, distinguishing not only organics from nonorganics but also green waste from food wastes. Different garbage containers can be designed for this purpose to aid in sorting, either arranged with an existing recycling system or mounted in the kitchen to compete with garbage disposals. Either way, waste habits need to be accommodated to microbial demands, which is merely the latest chapter in a century-old dialogue between humans and their unseen companions in waste management.

Josh Reno
Goldsmiths College

See Also: Anaerobic Digestion; Composting; Germ Theory of Disease; Landfills, Modern; Methane; Miasma Theory of Disease.

Further Readings

Epstein, Eliot. *Industrial Composting: Environmental Engineering and Facilities Management.* Boca Raton, FL: CRC Press, 2011.

Latour, Bruno. *The Pastuerization of France.* Cambridge, MA: Harvard University Press, 1988.

Tomes, Nancy. *The Gospel of Germs: Men, Women, and the Microbe in American Life.* Cambridge, MA: Harvard University Press, 1998.

Middle East

The Middle East can be thought of in the broad sense of the Middle East and North Africa (MENA), regional unit defined by most key international institutions such as the World Bank and the Organisation for Economic Co-operation and Development (OECD). The MENA region typically stretches from Morocco to Iran, and could under some definitions include Afghanistan and even Pakistan. In terms of waste production and management, the Middle East can be classified, with the important exception of the oil-rich Gulf states, as part of the developing world. Recycling often falls to the informal sector, which is deprecated in official discourse and is often in the hands of religious or ethnic minorities. Solid waste management (SWM) has been a topic of growing interest in the Middle East since the 1970s under the influence of massive postcolonial urbanization, modernist national policies, and international development efforts, especially those of the World Bank. Today, the key trend in the SWM sector is growing private-sector participation. A number of cultural elements affecting what happens to waste in the Middle East are important to consider, including conceptions of public/private space, the fact that waste-related tasks are often divided along gender lines, and the importance and particular form of cleanliness in Islam.

Waste

The countries of the Middle East are incredibly disparate on a variety of levels, especially socioeconomically. Generalization across the region is difficult. Nevertheless, leaving aside the oil-rich Gulf states, Middle Eastern countries mainly fall in the developing world category, as concerns their garbage: most produce about half to one-third the amount of municipal solid waste per person per year of the developed world, and their waste tends to be high in organic content. The World Bank estimated that in 2002, Morocco, Algeria, Tunisia, Egypt, Jordan, Lebanon, Palestine, and Syria together produced over 40 million tons of municipal solid waste, at an average rate of 248–259 kilograms/person/year, that is, between 0.68 and 0.71 kilograms/capita/day. An increase of 44 percent in total regional waste generation was predicted over the 1998–2010 period. In terms of composition, the same study estimated that organic waste comprised 55–70 percent of the solid waste, plastics 11–14 percent, and paper and paperboard 8–10 percent.

Despite an estimated $325–$400 million expenditure across the enumerated countries in 2000, solid waste management is very inadequate. With increasing urban populations and waste generation rates, the prognosis is negative. Syria, where large areas of towns and cities are unserviced or underserviced—up to 80 percent in some cases—is not exceptional by regional standards. Poorer residents almost invariably get the short end of the stick. They must often fend for themselves, burning their waste or throwing it in canals, empty lots, or streets and alleyways near their homes to decompose. Their garbage may eventually be picked up by municipal street cleaners. If it is collected at all, waste often gets dumped in antiquated or improvised facilities lacking effective controls. Incineration is limited.

Official recycling figures, whether waste management is in public or private hands, are low. Tehran, where waste is collected by the municipality, is estimated to recycle 5 percent; Jordan's rate is the same, most of which is achieved by nongovernmental organizations (NGOs), rather than by the official system. Real recycling rates are probably much higher, however, since few major cities in the region are without at least some informal-sector waste recovery. Cairo is home to a group often celebrated as the "world's best recyclers": the *zabbaleen* allegedly put 80 percent of the thousands of tons of waste they collect daily to profitable use. The scale and sophistication of the zabbaleen is exceptional, however, and this rate is not representative of the region on the whole.

Scavengers and informal-sector recyclers in the Middle East often have a shared low-status ethnic, geographic, or religious identity; they may be born into the profession, rather than joining or leaving according to purely economic dictates. Yemen's *akhdam* are supposedly set apart by African features, betraying Ethiopic descent that implies Christian heredity and, legendarily, connects them to pre-Islamic, foreign oppressors. True or not, what counts is the widespread representation. The zabbaleen are endogamously marrying—almost all trace ancestry to a limited number of common villages—and share common places of residence in enclaves around the city. They are also primarily, though not entirely (as it is often wrongly claimed), Christian. One should not conclude too hastily in religious discrimination, even if the predominance of Christians in garbage collection has been observed in other Muslim-majority settings as well, such as Pakistan. In Egypt, rather than being pushed into the business by Muslims, Christians took it over through their comparative advantage: unconstrained by Islamic taboo, they can make better money than Muslims in the garbage business by using the organic waste to fatten pigs for slaughter.

The official view is typically that the informal sector is illegal and unsanitary, produces little benefit for anyone but its own tax-evading members, and results in no improvement to the environmental condition of the city. It is also generally considered shameful, especially for those who engage in it, but also for the country on the whole. Accordingly, efforts are frequently made to repress or at least conceal it, especially from foreign eyes.

Historical and Cultural Context

Most countries of the Middle East experienced massive rural–urban migration in the post-independence era. The resulting rapid urbanization and accompanying growth of informal quarters and slums created a renewed interest in urban sanitation writ large, including SWM, especially from the 1970s onward. Beyond the need to address increasing rates of infant mortality and disease, for example, governments also frequently sought to express a certain vision of their new independence and civilizational grandeur through modernist infrastructure projects, of which sanitation and waste management played an important part. Instead of providing a moderate level of service across the board through support for cheap, pre-existing appropriate technology, like push- or donkey-drawn carts, they often preferred massive investment in technologically sophisticated, mechanical systems covering restricted but symbolically important zones, such as rich neighborhoods and tourist areas.

"Flashy" engineering was thought to symbolize a kind of modernity, while at the same time promising to usher in an equally symbolic order of impeccably neat showcases. This phenomenon was not unconnected to Islam's self-conception as a religion of cleanliness, troubled by the cognitive dissonance of daily encounters with waste in the public space, suggesting the opposite.

Garbage piles in Manshiyat naser, or "Garbage City," Cairo, Egypt. The zabaleen *have historically been Cairo's trash collectors, recycling 80 percent of the waste they collect. As primarily Coptic Christians, they are not held to Islamic taboos regarding garbage.*

Up to the present day, waste-related educational and awareness programs (e.g., signs, posters, leaflets, schoolchildren's textbooks and slogans painted on the walls of schools) in Egypt and Saudia Arabia, and undoubtedly elsewhere, remind the public that cleanliness is a religious duty, part of the faith, or a civilized habit. The choice to combat littering in an Islamic or civilizational, rather than environmental key reflects a conviction about which discourse is most apt to mobilize people. It is true that SWM and, more broadly, public cleanliness and beautification are often (though not always) encompassed within the environment portfolio in Middle Eastern countries. This occurs in a context where the act of waste removal and the names of institutions responsible for it often refer to "(public) cleansing" and neighborhood cleanup programs take the form of community cleansing days.

A study of cultural constructions of environment in Egypt titled *People and Pollution* found that pollution had much more resonance with respondents than the concept of environment. Respondents' number one definition of pollution was "garbage," but they also gave a series of responses conveying a religio-moral conception, identifying the poor, drugs, hooliganism, sex, and other failures to observe religious prescriptions concerning morality as sources of pollution. Accordingly, when cleanliness—as opposed to, say, nature, as one might expect in the United States—revealed itself to be the central trope for understanding environment and pollution, this referred not only to physical and aesthetic cleanliness, but also encompassed a series of religious symbolisms, what might better be called purity.

Puzzled by the contrast between fastidious bodily and household cleanliness in the Middle East, on the one hand, and the apparent neglect of public space evidenced by the omnipresence of waste in the streets, on the other, anthropologists have suggested that the dirtiness of the one is in fact the paradoxical result of a preoccupation with maintaining the cleanliness of the other. The imperative to maintain private cleanliness requires evacuation of waste from the home, meaning, in the absence of alternatives, littering the street. It is revealing that residents of large cities in the region, such as Cairo and Tehran, consider daily door-to-door collection a necessity. Waste should ideally be removed from the home as soon as it is produced. However, dealing with waste tends to be a female responsibility, but women's access to public space is constrained. In the absence of door-to-door collection, one of two undesirable scenarios imposes itself. Either the woman is forced into the public space in order to carry the waste from the apartment to a dumpster or the man, in order to avoid the woman, entering the public space, is forced into contact with a substance he wishes to avoid. Ultimately, spatial restrictions on women seem to trump male avoidance of polluting tasks. While Latin American or southeast Asian scavengers are both male and female, in the Middle East they are almost exclusively male. The same is true of organized collection. While postcollection waste sorting among Cairo's zabbaleen is done almost exclusively by women, the task of collection is reserved for men.

The Role of International Institutions

Outside intervention, often in the form of development projects or cooperation, also played an important role in the development of SWM in the Middle East after the 1970s. The key institution was probably the World Bank. In the late 1970s and early 1980s the Bank's Regional Office for Europe, the Middle East and North Africa (EMENA) sought to promote a three-pronged strategy for populous urban areas, integrating infrastructure, transport, and SWM. Morocco, Tunisia, Egypt, and others obtained loans under this regime for SWM programs.

One change typically made under the bank's influence was administrative centralization. For instance, in the early 1980s, Cairo and Tunis both saw the creation of centralized bodies (the Cairo Cleanliness and Beautification Authority and the Office National de l'assainissement). Another common feature of the programs were compost plants. They constituted a major component of waste management systems in Egypt, Lebanon, Morocco, and Syria and were also undertaken on a smaller or pilot scale in other countries in the region, such as Jordan. Often constructed through turnkey contracts with specialized European companies, the plants emphasized engineering: mechanized homogenizing drums, mechanical equipment for forming the windrows, and machine-driven sifting screens for producing fine compost. Those left to operate the plants after initial setup were unable to keep them running. Either they were unfamiliar with the complex apparatus, or the sophisticated technologies were too expensive to maintain. Illustratively, of the numerous such public facilities in Morocco, only the Rabat-Salé plant, opened in 1971, worked more than six years. Plagued from 1978 onward by breakdowns, as well as financial difficulties resulting from its low-quality product that, since 1984, lead to steadily decreasing output, it closed in 2000.

Today, there continue to be a large number of international development and aid agencies involved in SWM in the Middle East, for example, through the Mediterranean Environmental Technical Assistance Programme (METAP). The key objective of the program seems to be to shift the orientation away from a logic of public cleansing toward one of integrated solid waste management (ISWM), particularly through the updating of legislative and administrative frameworks, development of national plans, and widespread privatizations.

Trends: Privatization

Private sector participation (PSP) is the watchword of the day. Beginning in the 1990s and accelerating since 2000, many Middle Eastern countries have been engaged in large-scale privatization of public utilities like electricity, water, transport, and communications. The World Bank and various national development agencies, like the U.S. Agency for International Development (USAID), promote this. In the field of SWM, 60 Tunisian municipalities have contracts with private companies; in Morocco there are 22 such contracts, including in most of the country's major cities (Rabat, Tangiers, Essauouira, Fez, Casablanca and Nador); and approximately 30 percent of municipal solid waste in Damascus and 40 percent in Homs is collected under private-sector contracts; and so forth, across the region. Where contracts have not yet been concluded, public authorities are generally receptive to the idea and tend to be moving in that direction.

The phenomenon often takes the form of public–private partnerships, particularly of the build, operate, transfer (BOT) kind. Unlike the privatization of a pre-existing utility—basically a sale contract that provides the government with an injection of cash and promises users better service at cheaper prices—BOT arrangements are used where public authorities do not have a sellable going concern, but instead seek to build a network up, without having to borrow capital. The contracting waste management company makes the upfront investment to build the network in exchange for an exclusive right to operate it, at a profit, for a fixed period of time, after which it may or may not be transferred to the state. The contracting companies are often foreign (typically, European). Prominent examples include Alexandria (contracted with a subsidiary of the French company Véolia in 2000) and Cairo (contracted with four Spanish and Italian companies beginning in 2003). Jordan and certain parts of the Palestinian territories, which had previously transferred some or all responsibility for waste management to arm's-length institutions that, while still owned by the state, are run like commercial ventures, are now moving toward more fully privatized schemes, with domestic firms.

User fees are a keystone in the transition to a more-market-centered model of SWM and have sparked many controversies and political struggles. Rather than pay for waste collection out of the general revenues, raised, for instance, from property taxes, the municipality may now seek to operate on a commercial model in which beneficiaries pay for the service they receive. The fees are commonly piggybacked on pre-existing utilities' billing systems. A number of African countries following this path have added the waste levy to water bills, but Middle Eastern countries have preferred electricity. In either case, the idea is that most people are already connected to a grid with a functioning billing system and that the service (water or power) is indispensable enough that the threat of cutting it off will avoid nonpayment for waste collection.

Jamie Furniss
Oxford University

See Also: Africa, North; Cairo, Egypt; Iran; Saudi Arabia; Turkey.

Further Readings

GTZ-ERM-GKW Consortium. *Regional Solid Waste Management Project in Mashreq and Maghreb Countries: Inception Report.* World Bank, Mediterranean Environmental Technical Assistance Programme (METAP), 2003.

Hopkins, N. S., et al. *People and Pollution: Cultural Constructions and Social Action in Egypt.* Cairo: American University in Cairo Press, 2001.

Mansfield, Peter. *A History of the Middle East.* New York: Penguin Books, 2004.

Midnight Dumping

Garbage is a constant feature of modern human existence. While waste management programs collect and haul much of this garbage, some material waste is disposed of illegally. Midnight dumping is a form of illegal waste disposal that often occurs late at night in unauthorized places that often have very low visibility. In the United States, where the average American produces roughly 4.4 pounds of garbage per day, midnight dumping and other forms of illegal dumping are common. Other types of illegal dumping include open dumping, where materials are dumped in open spaces; and fly dumping, which happens when waste is thrown from moving vehicles onto roadways. Regardless of the specific type of dumping, this phenomenon is more than mere petty littering. Illegal dumping activities have contributed to environmental pollution and have very real public health consequences.

Illegal dumping can occur for a variety of reasons, including avoiding disposal fees and not taking the proper time or following the proper procedures for disposal. Often, landfills prohibit materials such as automotive parts and tires, large household items, and some yard waste, or may require fees for disposal of these items. Common sites for illegal dumping include abandoned buildings, vacant lots, and seldom-used alleys and roadways. When they dump illegally, people primarily dispose of nonhazardous materials. These items include a wide range of materials from both construction and demolition sites such as bricks, concrete, lumber, drywall, shingles, siding, and nails.

Old household appliances and furniture, such as refrigerators and couches, automotive parts, tires, and even whole abandoned cars, are also commonly seen at illegal dump sites. Other frequently dumped items include household trash, residential yard waste, and medical waste. People also dump waste that waste haulers or landfills prohibit, such as car batteries, appliances that contain Freon, and many other hazardous chemicals. These materials not only contribute to environmental degradation and pollution but also pose health risks and can lead to serious injuries for local residents.

Hazards

Illegal dump sites cause a variety of problems for people and other biological life in the area. Dumpsite materials can potentially catch fire in extremely dry conditions or can be the target of arsonists. These fires have led to countless instances of property damage and bodily injury to nearby residents. Dump site fires have also destroyed nearby forests. In some places, flooding has resulted from dump site materials collecting in places that block water drainage grates or stop creek water flow. Runoff contamination is another common occurrence in sites where harmful chemicals, paints, and other fluids have been illegally disposed of. In addition to drinking water, this pollution impacts wildlife and threatens biodiversity.

Not only does illegal dumping negatively impact the environment but these practices also threaten public health standards. Both physical and chemical hazards litter these often easily accessible sites. Rusty nails, jagged glass shards, and other pieces of dangerous waste have the potential to badly injure people. People can also be exposed to harmful chemical substances that can burn the skin or may result in even more serious injury or death. Dumpsites often serve as a breeding ground for various disease-infested vermin, such as rodents and insects. For example, rainwater tends to collect in and on materials that biodegrade very slowly, such as old tires. Mosquitoes can then reproduce rapidly in this stagnant water. These dump site insects have been responsible for infecting people with diseases such as encephalitis and dengue fever.

Sites

Whether in inner-city neighborhoods or in rural areas, illegal dumpers are much more likely to target certain places over others. They tend to prefer unsecure, undeveloped, or abandoned lots. These places serve as an ideal disposal spot, as midnight dumpers are much less likely to be caught in the act. Similarly, sites such as railroad tracks, highway underpasses, construction sites, alleyways, nature preserves, and farms afford midnight dumpers the low visibility they prefer. Junkyards and landfills also tend to attract midnight dumping. Researchers have observed that sites with previously illegally disposed-of materials tend to attract more illegal dumping. This phenomenon is especially visible after natural disasters, such as floods, and people deposit damaged household debris across the landscape.

Causes

A variety of factors contribute to the problem of midnight dumping. A shrinking tax base is the reality for many local governments in the early 21st century, and this has led to budget cuts in public services. In some places, this means less funding for waste management, which has led to policies like twice-per-month garbage collection. Other financially strapped places do not offer convenient locations for disposal. Perhaps the most problematic for residents are locations that charge high fees for waste disposal and recycling programs. In tough economic times, there is often not enough money in the household budget to make ends meet, much less to afford these garbage costs. This is especially true for low-income residents. These segments of the population often resort to more economically viable measures, like midnight dumping, in order to dispose of their waste. There also tend to be higher crime rates in these areas, which law enforcement gives a much higher priority than illegal dumping. Consequently, midnight dumping goes unchecked.

Solutions

As a way to curb illegal dumping activity, the Environmental Protection Agency (EPA) has suggested implementing "pay-as-you-throw" (PAYT) programs. PAYT programs, which are also known as variable-rate pricing or unit pricing, establish a waste disposal charge that varies based on the amount of waste a given household throws away. This approach departs from traditional methods of either taxing residents or charging them a fixed fee for waste disposal services. Instead, people are charged in a manner similar to other utilities based on how much or how little waste they generate. The EPA initially devised PAYT as a way to encourage resource conservation, and it has observed that areas with these programs experience a decrease in illegal dumping. Whether programs like PAYT will eliminate illegal dumping remains to be seen, but given consumption patterns, these activities will are likely to continue in the 21st century.

Jerry Ratcliffe
Temple University

See Also: Crime and Garbage; Hazardous Materials Transportation Act; Industrial Waste; Landfills, Modern; Open Dump; Pollution, Land; Pollution, Water; Public Health; Residential Urban Refuse; Toxic Wastes; Waste Management, Inc.

Further Readings

Environmental Protection Agency. "Illegal Dumping Prevention Guidebook." http://www.epa.gov/wastes/conserve/tools/payt/pdf/illegal.pdf (Accessed July 2010).

"Nonprofit Agencies Shoulder Burden of Illegal Dumping." *Register-Guard* (Eugene) (June 3, 2003).

Sigman, Hillary. "Midnight Dumping: Public Policies and Illegal Disposal of Used Oil." *RAND Journal of Economics*, v.29/1 (1998).

Mineral Waste

Mineral waste is the solid, liquid, and airborne by-products of mining and mineral concentration processes. Although mining and metallurgy are ancient arts, the Industrial Revolution launched an accelerating global demand for minerals that has made waste generation and disposal modern industry's most severe environmental and social challenge. Mineral solid waste production alone is staggeringly vast.

Although no accurate estimate of global waste volumes exists, estimates range from millions to billions of tons annually (depending on whether coal wastes are included), and the mining industry accounts for the largest proportion of total industrial waste production. Mine spoils are often regarded as a blight on the landscape as well as a serious environmental and public health threat. Nevertheless, mining by-products and landscapes may shift between the categories of "waste" and "value" due to changes in technology, economics, and cultural attitudes. Paradoxically massive in scope, yet largely hidden from everyday life, mineral waste is significant not only for its environmental impacts but also as a material index of contemporary rates of commodity production and consumption.

Mining Processes and Wastes

Mining entails the excavation and separation of valuable minerals from their geological matrix. In metal mining (as opposed to quarrying), since target minerals are typically only a fraction of the ore (or mineral-bearing rock), ore processing results in considerable volumes of waste, known as tailings. A typical, modern, base-metal operation yields greater than 98 percent waste from the excavated material. These residuals are generally disposed of to the lithosphere at waste-rock dumps and tailings disposal areas (although tailings are sometimes disposed of directly to waterways or backfilled into old mine shafts). Surface materials such as soil and vegetation, removed as "overburden," are not typically considered waste, although they contribute to mining's environmental impact. Slag, the solid by-product of smelting, was historically left in massive piles beside smelters or dumped in nearby watercourses. Although once used as an all-purpose building and grading material, smelter slag may also contain contaminants.

Copper, nickel, lead, and zinc smelters were notorious for producing noxious smokestack emissions. The pollution could travel long distances, threatening public health and damaging downwind crops and other vegetation.

Waste Hazards and Pollution

Mill waters and mine drainage are the principal liquid residuals. Because water is usually used to process ore and transport mill tailings for disposal, liquid and solid wastes are often considered together in waste-disposal planning. Environmental pollutants in both solid and liquid wastes may include heavy metals and metal salts, process reagents used to recover minerals (such as cyanide or mercury), and other contaminants in the ore (such as arsenic and selenium). Acid mine drainage (AMD) is a common and widespread water pollution problem, whereby sulfuric acid is released into the environment through the oxidation of sulfur-bearing rocks exposed during the mining process. Tailings impoundments may contribute to pollution through the overflow of contaminated water to surrounding waterways. In some instances, the catastrophic failure of tailings dams has choked streams and coated their banks with a flood of finely ground material. At their worst, tailings dam collapses have caused extensive landscape and property damage as well as human fatalities.

Many of these same pollutants are features of airborne wastes associated with mining and mineral refining. Airborne particulate matter blown from slag heaps and tailings ponds may bear harmful substances. Copper, nickel, lead, and zinc smelters were notorious for producing noxious smokestack emissions that not only affected the health of local populations but also transported pollutants over long distances, where they denuded large areas of vegetation and damaged downwind crops and livestock.

Mining has long been associated with environmental degradation. In his classic 1555 text, *De Re Metallica*, German doctor Georgius Agricola attempted to refute the accusations of environmental devastation leveled by mining's detractors by insisting the waste and damage associated with mineral extraction was temporary. More recently, in the 1934 text *Technics and Civilization*, Lewis Mumford linked mining with the historical exploitation of both nature and humanity. For Mumford, the pursuit of minerals represented an abandonment of the organic environment for the inorganic, subterranean realm; in turn, the wastes generated by mining and metallurgy destroyed the natural world at the surface.

The problems of waste disposal and environmental degradation have made the mining industry a target of environmental critics and government regulation. For its part, the industry often resisted environmental regulations and, in many mining districts around the world, developed a reputation of disregard for local environmental and public health impacts. For instance, at the Ok Tedi Mine in Papua New Guinea, uncontrolled tailings disposal into the Ok Tedi River destroyed the local ecosystem and the livelihood of the region's indigenous inhabitants, prompting local and international protests and subsequent court cases. In situations where ore exhaustion or financial conditions have led to mine closure and abandonment, accumulated waste materials and unsecured mine workings may continue to pose environmental hazards; the estimated millions of abandoned or derelict mine sites around the world constitute a major environmental hazard. However, large mines are by no means the only source of environmental damage. In the developing world, millions of unregulated miners engaged in small-scale, artisanal mining, generating considerable water pollution. Government and industry efforts to promote sustainable mining practices emphasize improved waste disposal, pollution remediation, and ecological restoration as critical challenges for improving the industry's environmental record and preserving its "social license" to operate.

Impacts of Technology

In modern mining, changing technologies and mining practices have facilitated the extraction of ever-smaller fractions of target minerals and the production of an ever-increasing proportion and volume of waste. For instance, while individuals or small groups of miners associated with gold rushes and exploiting rich placer (alluvial) gold deposits moved considerable amounts of material and could significantly impact aquatic environments, the large-scale removal of such unconsolidated deposits by dredging or hydraulic cannons (called monitors) generated far greater volumes of waste. In California, hydraulic mining for gold in the 1880s instigated some of the earliest efforts to control waste from the industry, which had contributed to altered flood regimes in the state's rivers and the siltation of San Francisco Harbor. Similarly, lode or hard rock mining proceeded from underground operations to the use of open-pit methods in order to process extensive deposits of low-grade ore. These methods were pioneered to meet the 20th century's sharply rising mineral demand, fueled by accelerating industrial growth, mass consumption, and military needs. Historian Tim LeCain describes the resulting growth in waste production and large-scale landscape devastation as environmental "mass destruction."

The status of mineral wastes as simply pollutants or unwanted residuals is, however, subject to change due to technological developments in ore recovery, the emergence of new mineral uses and demands, or changing mineral market conditions. In such cases, former waste may gain value, and the history of mining contains many examples of the reworking of old, apparently exhausted ore bodies or the reprocessing of waste deposits to extract valuable minerals. For instance, the legendary silver mines at Potosí in Peru were revolutionized in the 16th century by the development of the mercury amalgamation process, which enabled the recovery of silver from mineral-rich smelter wastes. Uranium is a classic example of a mineral's sudden transmutation from waste to value: once regarded as a nearly worthless by-product associated with wildly valuable and exceedingly rare radium deposits, it was typically discarded along with mine tailings. With the discovery of nuclear fission in the late 1930s, however, stockpiled or rejected uranium-bearing ores from the Katanga in Africa and northern Canada formed the basis for the Allied nuclear weapons development program.

The notion of value in waste may be extended to the mining landscape. In some cases, what the photographer Edward Burtynsky describes as mining's "residual landscapes" may become significant as sites of aesthetic appreciation and contested social and cultural meaning. Some observers challenge the simplistic association of mining with pollution and environmental degradation, pointing to the ways in which mining communities use, experience, and interpret the apparently "wasted" landscapes of mining. Slag heaps and tailings piles may be incorporated into local recreational or heritage landscapes or become the focus of ecological restoration and economic redevelopment activities in former mining localities. Photographers and tourists are often drawn to the technological sublime of huge waste heaps and massive open pits such as Utah's Bingham Canyon. Archaeologists and historians may explore mining landscapes, including wastes, for insights into technological change, social relations, and human/nonhuman interactions in historic mining regions.

Arn Keeling
Memorial University of Newfoundland

See Also: Copper; Industrial Waste; Iron; Mining Law; Pollution, Water; Uranium.

Further Readings

Agricola, Georgius. *De Re Metallica*. Translated by Herbert Clark Hoover and Lou Henry Hoover. New York: Dover, 1950.

Burtynsky, Edward. *Edward Burtynsky: Residual Landscapes*. Toronto: Luimiere Press, 2001.

Goin, Peter and Elizabeth C. Raymond. *Changing Mines in America*. Santa Fe, NM: Center for American Places, 2004.

LeCain, Timothy J. *Mass Destruction: The Men and Giant Mines That Wired America and Scarred the Planet*. New Brunswick, NJ: Rutgers University Press, 2009.

Smith, Duane. *Mining America: The Industry and the Environment, 1800–1980*. Lawrence: University Press of Kansas, 1987.

Mining Law

Mining regulations and legislation have increased and changed since the early days of the industry. In Great Britain, laws regulating the ownership and operation of mines proliferated in the 19th century, and the Metalliferous Mines Regulation Act of 1872 directed the inspection, working conditions, and regulation of coal mines as well as mines of stratified ironstone, shale, and fireclay. In the late 19th and early 20th centuries, mining occurred on a much smaller scale than it does in the 21st century and therefore required less legislation. As mechanization and industrialization changed the mining industry, both state and federal governments became increasingly responsible for regulating the health, safety, and environmental aspects of mining.

Early U.S. Laws

The first mining law in the United States was passed in 1872 by President Ulysses S. Grant. Passed to promote the development and settlement of publicly owned lands in the western United States, particularly the Rocky Mountain West and Alaska, the law governs hard rock mining on 270 million acres of public-domain lands. Under the 1872 Mining Law, any U.S. citizen can freely enter public-domain lands to explore minerals, except in national parks. According to the Mining Law, once a citizen discovers a valuable hard rock mineral, that person can then establish the right to mine that mineral by staking a claim. This law continues to be in effect into the 21st century.

In 1891, Congress passed the first federal statute governing mine safety. The federal government saw the ills of child labor and prohibited operators from employing children under the age of 12. This early regulation also established some minimum ventilation requirements for underground coal mines. Congress opened the Bureau of Mines in 1919 as an agency within the Department of the Interior. The Bureau of Mines was permitted the authority to conduct research that would help reduce the number of deaths and accidents in the coal-mining industry. In 1941, the Bureau of Mines was given the authority to enter mines in order to conduct inspections and Congress passed the first codes of federal regulations for mine safety in 1947.

The 1947 act led to the federal Coal Mine Act of 1952. This legislation gave the Bureau of Mines the authority to issue violation notices and imminent withdrawal orders to companies, should their mines be found conducting work in violation of federal regulations. While there were no monetary penalties against companies at this time, the Coal Mine Act of 1952 authorized the Bureau of Mines to call for civil penalties should the Bureau find company violations.

Safety Regulation

A disastrous explosion occurred at the Consol #9 coal mine in Farmington, West Virginia, on

November 28, 1968. This explosion killed 78 men and changed the industry forever. The public was outraged after this disaster and called for further regulation. The federal Coal Mine Health and Safety Act of 1969 came into being. This act was the most forceful regulation to date, calling for two annual safety inspections of surface mines and four annual inspections of underground mines. The penalties for violations of regulations could be considered criminal if the violations were knowing and willful. The act also provided compensation to those miners who had become disabled due to black lung disease. This 1969 act continues to be the backbone of all federal health and safety standards in the 21st century.

In 1973, the responsibilities of regulation were transferred from the Bureau of Mines to the newly established Mine Enforcement and Safety Administration. This was the first federal agency with the sole mission to assure miners' health and safety during work. After yet another horrible mining disaster at the Scotia Mine in Letcher County, Kentucky, in 1976, the public demanded further federal regulation. As a result of the Scotia disaster, the Mine Safety and Health Administration (MSHA) was established in 1977. It is housed within the U.S. Department of Labor and continues to operate into the 21st century. The 1977 act consolidated all mining, from coal to metal, under the same administration. Beyond increasing mandates for inspections, the law called for mandatory miner training, mine rescue teams for all underground mines, and increased involvement of miners and their representatives in health and safety activities at all mining sites.

This 1977 act and the MSHA consolidated the regulations from the 1966 and 1969 mine acts into a single mine health and safety act that would help further recognize the individual rights of miners. This law was the reigning legislation in the United States for another 30 years until the 2006 Sago mining disaster in West Virginia. This devastating explosion and its aftermath trapped 13 miners under ground for two days. Only one of the 13 men came out of the ground alive.

In response to the Sago disaster, the MSHA enacted the Mine Improvement and New Emergency Response Act of 2006. Referred to as the MINER Act, this legislation was the most stringent regulation to date. The MINER Act is extremely detailed, but in general, the MINER Act requires companies to improve their preparedness for accidents. All companies are required to have emergency response plans should a disaster strike. Mining companies are required to develop specific emergency response plans for each mining location they operate, and every mine is required to have at least two mine rescue teams located within one hour of their location. The MINER Act also calls for continued research to enhance mine safety, criminal penalties of up to $500,000, and civil penalties up to $220,000 for violations of regulations.

Environmental Regulation

Environmental regulation continues to be one of the most important aspects of mining law in the 21st century. The Surface Mining Control and Reclamation Act (SMCRA) came into being on August 3, 1977, and is the primary federal law regulating the environmental effects of surface mining in the United States. The law created the federal Office of Surface Mining within the Department of the Interior. The act calls for the reclamation of abandoned mining sites, the regulation of active mining sites, and permitting of mining and reclamation bonds. SMCRA also established the Abandoned Mine Lands Fund. This fund was established to reclaim land that was mined prior to this 1977 act. The law requires companies to pay a per-ton fee for their mined product to this fund. The money gathered is then given back to the states to fund the cleanup of earlier surface and water damage from surface mining. The Office of Surface Mining helps individual states implement regulations and oversees that states are properly meeting federal regulations.

Much of the environmental concerns regarding mining waste and reclamation were addressed in the Clean Water Act of 1972. The Clean Water Act is the primary federal legislation governing water pollution in the United States. Of primary concern to the mining industry is the issuance of permits 404 as part of the nationwide permitting required for mountain top removal, a form of surface coal mining. During this type of surface mining, operations place soil and rock that they remove from the tops of mountains into streams and wetlands nearby. In the case of mountain top removal, pollutants from this valley fill become discharged into nearby streams, significantly affecting local water quality. The Environmental Protection Agency and the Army Corps of Engineers continues to deal with this issue into the 21st century.

Regulation of all mining has increased since the 1800s, as has the amount of mined products consumed. With improved technology comes faster output and new opportunities. Mining corporations, consumers, and legislators are beginning to think not only about the laws that govern the extraction process but also mine waste disposal and the possibility for a sustainable future.

Stephanie Friede
Columbia University

See Also: Clean Air Act; Clean Water Act; Coal Ash; Copper; Mineral Waste; Pollution, Water.

Further Readings

Alford, Charles J. *Mining Law of the British Empire.* London: Charles Griffin, 1905.
Bakken, Gordon Morris. *The Mining Law of 1872: Past, Politics, and Prospects.* Albuquerque: University of New Mexico Press, 2008.
Reece, Erik. *Lost Mountain: A Year in the Vanishing Wilderness: Radical Strip Mining and the Devastation of Appalachia.* New York: Riverhead Books, 2006.
Starke, Linda. *Breaking New Ground: Mining, Minerals and Sustainable Development.* London: Earthscan, 2002.

Minnesota

Minnesota is an upper midwest state in the United States, sharing a Lake Superior water border with the states of Michigan and Wisconsin and an international border with the Canadian provinces of Ontario and Manitoba. Prior to European settlement, the Anishinaabe and the Dakota were the main Native American groups in the area. Capital St. Paul and largest city Minneapolis, known as the Twin Cities, adjoin to make the Minneapolis-St. Paul urban area the largest metropolitan area in the state, home to nearly 60 percent of the population. With an early economy based on logging and farming, increasing industrial development in the early 20th century urbanized most of the population; after World War II, this increased again, making Minnesota a center of technology. The creation and proliferation of suburbs is thought to be key to this success. Although less than 1 percent of the population are engaged in agriculture, it is a major part of the Minnesotan economy, which is the largest producer in the United States of sugar beet, sweet corn, green peas, and farm-raised turkey. Forestry also remains strong. Minnesota has both the nation's earliest and the nation's second-largest indoor shopping malls: the Southdale Center, opened in the Edina suburb of Minneapolis in 1956, and Bloomington's Mall of America, opened just four miles away in 1992.

Statistics and Rankings

The 16th Nationwide Survey of MSW Management in the United States found that, in 2006, Minnesota had an estimated 5,894,933 tons of municipal solid waste (MSW) generation, placing it 24th in a survey of the 50 states and the capital district. Based on the 2006 population of 5,154,586, an estimated 1.14 tons of MSW were generated per person per year (ranking 33rd). Minnesota landfilled 2,200,457 tons (ranking 34th) in the state's 21 landfills. The state exported 740,269 tons of MSW, and the import tonnage was not reported. In 2006, Minnesota was increasing its 27 million-cubic-yard landfill capacity; it was ranked joint 22nd out of 44 respondent states for number of landfills. The following are banned from Minnesota landfills: yard waste (since 1990 in metro area, 1992 statewide); whole tires (1985); used oil (1988) and other motor vehicle fluids (1994); batteries (1988), some dry cell batteries (1990), and rechargeables (1991); white goods (1990); mercury-containing products (1992); fluorescent tubes (1993); and telephone directories (1994). Minnesota has nine waste-to-energy (WTE) facilities, which processed 1,170,841 tons of MSW (ninth out of 32 respondents). Landfill tipping fees across Minnesota were an average $40 per ton, where the cheapest and most expensive average landfill fees in the United States were $15 and $96, respectively. Average WTE tipping fees were $36 per ton. When disposal fees were first introduced in 1973, they were $0.15 per cubic yard. Minnesota recycled 2,523,635 tons of MSW, placing the state 18th in the ranking of recycled MSW tonnage. In an innovative scheme to reduce backyard burning, Chisago County residents who hand in their burn barrels and sign up for waste collection receive six months service at half price.

Waste Management

In Minnesota, open burning and dumping were common until the 1960s (as had been common historically across the United States), and there were over 1,500 dumps across the state in private and municipal hands. In 1967, the Metropolitan Council and the Minnesota Pollution Control Agency (MPCA) were created and began developing legislation to protect air and water quality; they were also given the authority to oversee waste management. During 1969–71, many waste management provisions and rules came into force: counties were given control of their own MSW (County Solid Waste Management Act); general open burning was prohibited; solid waste disposal permits were introduced; and landfill requirements were upgraded. By 1973, 13 landfills managed 90 percent of MSW, and 135 permit-issued sanitary landfills had replaced the multitude of dumps.

Recycling was legislated in 1973, and the federal Resource Conservation and Recovery Act (RCRA) was introduced in 1976. The Joint Minnesota Legislative Committee on Solid and Hazardous Waste was formed in 1978 as statewide legislation came to the fore. Flow control became the dominant issue in Minnesota waste management in the 1990s; with local government's power to direct waste management limited, waste began going to cheaper out-of-state facilities. The 1994 Landfill Cleanup Act and Closed Landfill Program also had a significant effect on the state's waste management during the 1990s. The 135 sanitary landfills licensed in 1973 were reduced to 55 by 1990; after the Landfill Cleanup Act, only 27 remained in operation by 1994. The waste management problems of the 1990s have been tackled in the 21st century by revising and updating legislation.

James J. Hill Mansion

Almost 800,000 people in the United States are homeless. Minnesota is the first state where the archaeology of homelessness has been studied archaeologically, rather than being regarded as refuse (which it can resemble). In 2003, Professor Larry Zimmerman, then head of archaeology at the Minnesota Historical Society, excavated the mansion of 19th-century railroad magnate James J. Hill, "the Empire Builder of the Northwest," on Summit Avenue in St. Paul, less than one-quarter mile from the city's downtown area. The 36,000-square-foot mansion was completed in 1891. Scatters of clothing, sleeping bags, and cooking materials on the surface of the garden gave way to stratified deposits in the earth, revealing four distinct layers of homeless occupation since the 1940s. The garden was situated on a hillside held back by retaining walls and contained several built structures, including a mushroom cave in the lower wall, three greenhouses, a gardener's residence, and a coal gasification power plant. Homeless people had occupied several of these, and even a tree throw had been used as a shelter and the garden walls as windbreaks. The tree throw shelter was substantial enough to have developed its own downslope midden, and the mushroom cave was used for decades, accumulating one meter of deposits on its floor. Hill's daughters had donated the mansion to the Roman Catholic Archdiocese of Saint Paul and Minnesota in 1925, and it remained church property until 1978 when the Minnesota Historical Society acquired it. Apart from building a small grotto, the church did little with the gardens. Zimmerman went on to lead the first archaeological investigation of homelessness in Indianapolis, Indiana.

Professor Zimmerman, regarding the Hill Mansion, stated that "The trash 'stream' is much more complex than we realize, with trash being more than just material that ends up in landfills or strewn across the landscape. The question is, when does something actually become rubbish/trash? Much of the material that we initially saw as 'just' trash went through another level of use after the original owners disposed of the items by donating them to the Dorothy Day Center in St. Paul (the same in Indianapolis). Used by the homeless, it became trash only after the homeless saw no more use for it. Even then, some items we documented, excavated, or collected from the surface saw another use as archaeological artifacts. Some Hill garden items will see permanent curation in the Minnesota Historical Society Archaeology Department collections, which means they never really become trash, or perhaps archaeology just does a different kind of recycling."

Washington County Landfill

In 2009, excavation began at the Washington County Landfill in Lake Elmo, a significant landfill remediation project estimated to take three years and cost $21 million. The landfill, closed since 1975, covers 35 acres and contains approximately 1.9 million cubic yards of waste: 73 percent residential waste, 26 percent commercial waste, and 1 percent demolition waste. The site was added to the Environmental Protection Agency (EPA) National Priorities List (NPL) in 1984 because of volatile organic compound (VOC) groundwater contamination. Once added to the Minnesota Pollution Control Agency (MPCA) Closed Landfill Program (CLP) in 1996, the landfill was removed from the NPL. Since the initial detection of groundwater VOCs in 1981, they began to be detected in private wells in 1983–84. Local residents were advised not to drink tap water and were connected to a new supply. Monitoring wells and a water extraction system were then inserted into the landfill site. The two-foot landfill cover placed on closure in 1975 was upgraded to current standards in 1996. A methane explosion in the early 1990s led to the installation of an enclosed flare and gas extraction system.

In 2004, the MPCA used newly developed tests for perfluorochemicals (PFCs), which showed positive in onsite monitoring wells and downslope residential wells. Further testing led to 100 homes being placed on bottled water and another 22 homes having granular carbon filters installed. PFCs were dumped in the landfill legally between 1969 and 1975 by 3M from its Cottage Grove plant, which used the chemicals to make products resistant to heat, oil, staining, grease, and water, such as Teflon and Scotchguard. In 2006, tests were developed for perfluorobutanoate (PFBA), another form of PFC, which was discovered leading away from Lake Elmo 15 miles to Hastings. The Minnesota Department of Health (MDH) found no increased health problems in the affected area, and judges have since ruled that no one has been harmed by PFCs in drinking water and there is no "class action" status, so any plaintiffs must sue 3M individually. Although the MDH says the PFC levels are too low to be a threat, they have been found in local fish and limited fish consumption is now advised.

By 2010, the Washington County Landfill was being excavated and replaced in triple-lined cells constructed on the same site where three geo-synthetic membrane layers will pump out any leachate liquid. Removing the garbage to another site would have tripled the cost of the project. Other PFC-contaminated landfills nearby include Oakdale (late 1950s), Woodbury (early 1960s), and 3M Cottage Grove (early 1970s); these are being excavated and placed in a specially constructed landfill in Rosemount. Although found to have no liability, 3M contributed generously to the remediation programs, both financially and technically.

Jon Welsh
AAG Archaeology

See Also: Archaeology of Garbage; Comprehensive Environmental Response, Compensation, and Liability Act (CERCLA/Superfund); Definition of Waste; Landfills, Modern; Malls.

Further Readings

Arsova, Ljupka, et al. "The State of Garbage in America: 16th Nationwide Survey of MSW Management in the U.S." *BioCycle*, v.49/12 (2008).
Newsday, Inc. *Rush to Burn: Solving America's Garbage Crisis?* Washington, DC: Island Press, 1989.
Zimmerman, Larry J. and J. Welch. "Displaced and Invisible: Archaeology and the Material Culture of Homelessness." *Historical Archaeology* (2011).

Mississippi

One of the southern U.S. states, Mississippi takes its name from the Mississippi River, which flows along its western border. Jackson is the capital and largest city. The entire state is composed of lowlands and is densely forested outside the Mississippi Delta. Most of the farm-bred catfish in the United States are produced by Mississippi's catfish aquaculture. According to Commonwealth Fund data, the state ranks last in the United States for healthcare, with the highest obesity rate of any U.S. state 2005–08. Mississippi is also at the bottom of the American Human Development Index and has the lowest per capita personal income of any state, but also the nation's lowest living costs. The six states of the Deep South have a common history of cotton and tobacco production attendant with depending on and supporting slavery. These monoculture production systems have had far-reaching consequences on Mississippi's environmental and socioeconomic history. Mississippi's overreliance on cotton agriculture before, during, and after the Civil War can be linked to its current ranking as one of the poorest states.

History

From 1800, the King Cotton economy developed in the South on land ceded and sold by the Chickasaw and Choctaw tribes. By the time of the Civil War, Mississippi had become the fifth-wealthiest state, its wealth created by the cotton plantations along the rivers, where slaves (counted as assets) had increased in value since the 1840s. At this point, 90 percent of the delta bottomlands were undeveloped frontier and the low population was 55 percent enslaved. The dominance of the "planter aristocracy" minority and agricultural cotton meant that taxation was intentionally kept low and there was very little investment in infrastructure, with some areas remaining unindustrialized until the late 20th century. This "planter aristocracy" was an elite of slave owners in a state where in 1860 only 31,000 out of 354,000 whites owned slaves; the elite were the 5,000 slave owners who owned more than 20 slaves, including the 317 who owned more than 100 slaves.

The southerners were convinced that such was the importance of their cotton that it would support the economy of an independent Confederacy and force cotton-reliant Britain and France to intervene in the Civil War. Mississippi became the second state to secede from the Union. In 1861, the south withheld its cotton from sale or export and then the Union blockade prevented 90 percent of export. This cotton diplomacy strategy failed as Europe had large stocks of cotton and production increased in Argentina, Egypt, and India to meet demand. Southern faith in cotton monoculture had proven disastrous, using land and labor that could have been used to grow much-needed food and leading the Confederacy into a Civil War it ultimately could not win.

Agricultural depression and changes in labor structure wrought by the massive damage and casualties of the Civil War caused the south to lose huge amounts of wealth. Although tens of thousands migrated into the state after the Civil War and began clearing land and farming, cotton prices continued to fall, culminating in another agricultural depression in the 1890s. New legislation introduced by white legislators in 1890 and the Jim Crow system disenfranchised most of the black and poor white population, resulting in their losing lands and leading in part to the Great Migrations to the north (1910–30 and 1941–70).

In 1966, Mississippi was the last state to revoke prohibition of alcohol, having up until then taxed illegal bootlegged alcohol. In the segregated south, black people were often barred from skilled work and garbage collection was a common occupation for black men. The assassination of Martin Luther King Jr. while in the south to support the rights of black sanitation public workers led to increased recognition for the American Federation of State, County and Municipal Employers (AFSCME). Pascagoula sanitation workers were among the first to win union recognition in the wake of King's assassination.

Landfill Tonnage

The 16th Nationwide Survey of MSW Management in the United States found the following: In 2006, Mississippi had an estimated 3,194,368 tons of municipal solid waste (MSW) generation, placing it 33rd in a survey of the 50 states and the capital district. Based on the 2006 population of 2,899,112, an estimated 1.1 tons of MSW were generated per person per year (ranking 37th); 3,049,368 tons were landfilled (27th) in the state's 17 landfills; 740,876 tons of MSW were imported, the export tonnage was not reported. In 2006, Mississippi had sufficient landfill for 817 years and was still adding to this capacity; it was ranked 25th out of 44 respondent states for number of landfills. Only whole tires and lead-acid batteries were reported as being banned from Mississippi landfills. Tipping fees across Mississippi averaged $28, where the cheapest and most expensive average landfill fees in the United States were $15 and $96. The high water table in Mississippi limits the location of landfills in the state. Mississippi was 43rd in the ranking of recycled MSW tonnage, with 145,000 tons of MSW recycled. Mississippi has no waste-to-energy (WTE) facilities.

Oil Spills

Mississippi was affected by the 2010 BP oil spill in the Gulf of Mexico, but was given time to prepare by being separated from the Gulf proper by the Mississippi Sound and the Barrier Islands. The Gulf Islands National Seashore and several important conservation areas, including globally important bird areas, were affected within Mississippi. The complex nature of the Gulf coastline made oil cleanup difficult due to the number of marshes, bays, creeks, estuaries, and inlets. Further controversy was sparked in the state when oil-contaminated cleanup waste from south Mississippi beaches was sent to landfill at Pecan Grove, Harrison County. Officials from the county moved successfully to block the dumping by Waste Management and were concerned that alternatives such as WTE had not been considered. More than 150 tons of scooped oil and used personal protection equipment and hand tools contaminated with oil and Corexit (a solvent used to break up oil slicks) were deposited in the landfill. The Environmental Protection Agency directed BP to make its waste disposal plans transparent and to stop using Corexit. Sampling carried out by the Mississippi Department of Environmental Quality (MDEQ) showed that the oil waste dumped at Pecan Grove was nonhazardous, as had been claimed.

Litter

The state has a serious litter problem; litter rates along rural roads are 30 percent higher than the average of other states and 71 percent of litter is in public view from the interstate and highway. In the 2008 American State Litter Scorecard, Mississippi was ranked the lowest (50th) for littered public roads and property; by 2011, the state had risen to joint 47th. It costs Mississippi more than $2 million a year to remove this litter from public highways. Only around 20 percent of the total state highway miles are in the adopt-a-highway program for Keep Mississippi Beautiful, the state affiliate of the Keep America Beautiful program. This is attributed to the largely rural nature of the state. Although 38 percent of Mississippi litter is accidental, 62 percent is deliberate, the highest percentage of any state. Most (26.6 percent) of this deliberate litter originates from convenience products such as take-out food containers. The rest of the litter composition is as follows: 17.3 percent beverage containers; 9.8 percent miscellaneous plastic, metal, foil, and glass; 9.4 percent candy, gum, and snacks; 9.1 percent miscellaneous paper and cartons; 7.4 percent vehicle parts, supplies, and debris; 6.7 percent newspapers, advertising fliers and leaflets, food packaging, yard waste, and other unspecified items; 5.7 percent cigarette packs and matchbooks; 5.4 percent building material and construction debris; and 2.6 percent toiletries, toys, cassettes, and recreational items. The litter problem is viewed as chronic, hav-

Mississippi is ranked near the bottom of all 50 states in terms of its chronic, long-standing roadside litter problems. Litter rates along rural roads are 30 percent higher than the average of other U.S. states. Removal costs add up to $2 million annually.

ing developed over a long period, and will need a long-term solution has been projected.

In line with national statistics, 60 percent of people deliberately littering along highways and rural roads were aged 18 to 34, and they were mostly male. However, in Mississippi male motorists driving pickup trucks were found to be a disproportionate source of litter, responsible for nearly a third of all motorist littering and for two-thirds of single items escaping from motor vehicles. While the source of the litter could be classed as accidental in pickup trucks, it is regarded as deliberate due to inappropriate decisions about unsecured items in the back of the vehicle. Part of the solution to the problem was the "I'm Not Your Mama" advertising campaign featuring former first lady of Mississippi Pat Fordice in four award-winning TV spots. Developed by GodwinGroup for the Mississippi Department of Transportation, the campaign chose a strong matriarchal figure to reach the target younger male audience. The five-year campaign was launched in 2003.

Environmental Racism in Mississippi

"We all know about the wrong side of the tracks. It's where we put our poor people and people of color. It's where we put our toxic wastes" (Helen Hershkoff, American Civil Liberties Union, referring to a case in Mississippi). The modern concept of environmental racism emerged in the early 1970s, but the practice goes back to ancient times, when noxious tanneries were situated on the outskirts of towns in the poorest areas. Environmental racism is a serious issue in Mississippi; the state is also one of the few to have had nuclear weapons detonated within its borders. Racial inequality and unequal distribution of wealth dominate the state's recent history.

The Moss Point Incinerator is viewed as a classic case of environmental racism in Mississippi, involving two Gulf Coast cities along the Escatawpa River in Jackson County. Predominantly white and relatively affluent Pascagoula began having waste disposal issues in the late 1970s and the city council opted to build an incinerator. Residents of Pascagoula protested having the incinerator sited near their homes and the incinerator contract subsequently went to a chemical company in Moss Point, a less wealthy, mostly black city nearby. The Moss Point Incinerator opened in 1985 near a number of schools and residential areas but was owned by the city of Pascagoula, although Moss Point also had usage. This situation remained amicable until 1991, when the Pascagoula City Council voted to send commercial medical waste contracts to Moss Point for incineration, causing an uproar in Moss Point. The incinerator was in debt, financed by a $6.9 million federal loan—the collapse of the natural gas market and the falling price of the steam it sold had seriously damaged the incinerator's income and left it unable to pay even the interest. As the Gulf Coast erupted in claims of environmental racism, the incinerator drew criticism for being antiquated and malodorous. As Moss Point filed suit against the proposed clinical waste burning-for-profit scheme, local doctors voiced concerns that the already heightened incidence of respiratory problems could be increased by the plan.

The 1983 U.S. General Accounting Office study Siting of Hazardous Waste Landfills and Their Correlation with Racial and Economic Status of Surrounding Communities found that three of four off-site commercial hazardous waste landfills in Region Four were in mainly African American communities. Region Four includes the states of Alabama, Florida, Georgia, Kentucky, Mississippi, North and South Carolina, and Tennessee, where African Americans account for only 20 percent of the population.

Jon Welsh
AAG Archaeology

See Also: Culture, Values, and Garbage; Fast Food Packaging; Food Waste Behavior.

Further Readings

Arsova, Ljupka, et al. "The State of Garbage in America: 16th Nationwide Survey of MSW Management in the U.S." *BioCycle,* v.49/12 (2008).

Couch, Jim F., Peter M. Williams, J. Halvorson, and K. Malone. "Of Racism and Rubbish: The Geography of Race and Pollution in Mississippi." *Independent Review* (Fall 2003).

Syrek, Daniel B. and Frank Bernheisel. *Mississippi Litter 2000, A Baseline Survey of Litter at 113 Street and Highway Locations,* Institute for Applied Research and Gershman, Brickner and Bratton.

Missouri

Located in the midwest, Missouri is a U.S. state with both midwestern and southern cultural influences, as it is historically a border state between the two regions. The state is named after the Missouri River and, by extension, the Siouan-language tribe. Jefferson City is the capital, Kansas City is the largest city, and Greater St. Louis is the largest metropolitan area. The state is known to mirror the demography, economy, and politics of the nation in general, having a mix of urban and rural culture, is considered a political bellwether state. There have only been two years (1956 and 2008, as of 2011) when Missouri presidential election results have not accurately predicted the next U.S. president. The state has a mixed economy, but has one of the largest numbers of farms in any U.S. state and is ranked in the top 10 states for hog, cattle, and soybean production.

Statistics and Rankings

The 16th Nationwide Survey of MSW Management in the United States found that, in 2004, Missouri had an estimated 9,939,008 tons of municipal solid waste (MSW) generation, placing it 14th in a survey of the 50 states and the capital district. Based on the 2004 population of 5,837,639, an estimated 1.7 tons of MSW were generated per person per year (ranking seventh). Missouri landfilled 6,731,844 tons (ranking 15th) in the state's 21 landfills. It exported 2,520,071 tons of MSW, and the import tonnage was 228,858. In 2006, Missouri was continuing to increase its 201,892,185-cubic-yard landfill capacity, and it was ranked joint 22nd out of 44 respondent states for number of landfills. Yard waste, whole tires, used oil, lead-acid batteries, and white goods were reported as banned from Missouri landfills. Missouri has seven waste-to-energy (WTE) facilities, which processed 23,300 tons of MSW (28th out of 32 respondents). Missouri recycled 3,183,864 tons of MSW, placing Missouri 13th in the ranking of recycled MSW tonnage.

Environmental protection efforts in Missouri include a landmark smoke control law that passed in St. Louis in 1940, quickly spawning similar ordinances across the United States. The ordinance directed users of soft bituminous coal (found in abundance in bordering Illinois) to wash the fuel before incineration, reducing the dense, heavy smoke that plagued the city in winter months. Missouri waste management began the transition to modern standards in 1968 when Robert "Robby" Robinson was hired as the first state solid waste program director and tasked with creating a state solid waste plan. At this time, there were over 500 open dumps in Missouri. The state solid waste plan of 1972 set the elimination of open dumps as a major goal, and by 1980, only three remained.

Weldon Spring Conservation Area

Missouri is home to a highly successful reclamation project carried out under the Superfund program, the Weldon Spring Site. In 1941, the U.S. Army acquired 17,232 acres near Weldon Spring, St. Charles County, and was part of the war effort until 1945 as the Weldon Spring Ordnance Works. Following the war, ownership of some of the site was transferred to the State of Missouri, the University of Missouri, and other public bodies, with the U.S. Army retaining land for use as a training area. The state created the August A. Busch Memorial Conservation Area, and the university used its tract for agriculture.

The Atomic Energy Commission (AEC) was given 217 acres of the former ordnance works in 1956, on which the Weldon Spring Uranium Feed Materials Plant (later becoming the Weldon Spring Chemical Plant) was built. This facility consisted of 44 buildings, four setting basins over 25 acres known as raffinate pits, two ponds, and two former dump areas. A quarry four miles to the south had been used by the army for disposing of trinitrotoluene (TNT) residues. The AEC acquired this in 1958 to dispose of demolition waste contaminated with uranium and radium from a demolished uranium-ore processing facility in St. Louis, and a small amount of thorium residue. The army reacquired the site in 1967 when the AEC closed the plant. Having partially decontaminated some buildings and dismantled some machinery, the army converted the plant to produce Agent Orange, an herbicide used in the Vietnam War. Dwindling demand for Agent Orange and increasing cleanup costs closed the project in 1969 before production of the herbicide could begin.

In 1984, several buildings were repaired by the army; floor, wall, and ceiling surfaces were decontaminated, and contaminated equipment was isolated to stop contaminants from spreading beyond the site. Directed by the U.S. Office of Management and Budget, the army gave up custody of the chemical plant to the Department of Energy (DOE), which was responsible for the cleanup, to which the army was required to contribute. It was given the DOE "Major Project" designation and was known as the Weldon Spring Site Remedial Action Project (WSSRAP). Having requested cleanup funds from Congress, a project office was established in 1986 and onsite activity began. The Environmental Protection Agency (EPA) put the Weldon Spring Quarry on the National Priorities List (NPL) in 1987 on account of groundwater contamination having potential to affect 60,000 drinking water well users. The NPL coverage was expanded to include the raffinate pits and the chemical plant area in 1989; collectively with the quarry, they became known as the Weldon Spring Site. The DOE set a goal that went beyond the cleanup: to reinstate the site to its natural state and make it a publicly accessible area for recreational and educational use.

Work onsite involved dismantling 44 chemical plant features and disposing of radioactive and chemical-contaminated demolition waste, sludge, and soil. The DOE carried out this work between 1988 and 1994. As part of this phase, as much material as possible was removed from the raffinate pits, quarry, and properties in the vicinity. Bulk waste removal from the quarry was done between 1993 and 1995. To appropriately dispose of the contaminated material, the DOE constructed a 45-acre disposal cell in the former chemical plant area to provide long-term containment of the Weldon Spring Site material. Around 1.48 million cubic yards of waste material were deposited in the cell. The DOE completed construction of the 75-foot-high cell cap in 2001, and in 2002 a viewing platform and information panels were added to the roof of the cell to allow public access and provide a panoramic view of the surrounding area. Haul roads were converted to a hike-and-bike trail, which crosses Weldon Spring, linking the quarry, the chemical plant site and the wildlife center in the August A. Busch Memorial Conservation Area before joining the historic Katy Trail. Called the Hamburg Trail, the trail is named after one of the towns taken over by the army to construct the Weldon Spring Ordnance Works in World War II. There are 29.5 miles of hiking trails in the conservation area, not including the MKT Trail, which is part of the state park system.

Restoration of the quarry took place in several phases. Infilling was done with specially selected and prepared borrow material, the quarry water treatment plant was dismantled, the water collection system was reclaimed, and the haul road was restored before the final grading. Infilling reduced the physical dangers of open excavations and created a gentle slope to make rainwater flow over the surface of the Little Femme Osage Creek. The hike-and-bike trail and the quarry final grading were the last phase of restoration. The final grade, completed in 2002, returned the area to its natural contours and minimizes erosion. This done, the task began to create a biodiverse environment, and a prairie was built by seeding approximately 150 acres of soil surrounding the disposal site and extending as far as the site boundary. This was named the Howell Prairie after the original prairie where Francis Howell Sr. built his homestead in the 1880s. By 2004, nearly 80 species of native forbs and prairie grasses had been planted.

The history of the Weldon Spring Conservation Area was presented in a 9,374-square-foot Interpretive Center housed in a building converted by the DOE. A native plant educational garden has also been planted using species from the Howell Prairie and other plants found nearby. A long-term maintenance plan is in place to provide site monitoring and involves neighboring landowners. This plan covers the monitoring program, annual inspections, and institutional control areas (limited access for safety and environmental reasons) around the site.

Jon Welsh
AAG Archaeology

See Also: Comprehensive Environmental Response, Compensation, and Liability Act (CERCLA/Superfund); Open Dump; Toxic Wastes.

Further Readings

Arsova, Ljupka, Rob van Haaren, Nora Goldstein, Scott M. Kaufman, and Nickolas J. Themelis. "The

State of Garbage in America: 16th Nationwide Survey of MSW Management in the U.S." *BioCycle*, v.49/12 (2008).

Hurley, Andrew, ed. *Common Fields: An Environmental History of St. Louis*. St. Louis: Missouri Historical Society Press, 1997.

U.S. Department of Energy. "Weldon Spring, Missouri, Site." September 2010. http://www.lm.doe.gov/weldon/Sites.aspx (Accessed December 2010).

Mobile Phones

When the first mobile phone hit the market in 1983 for high-end business users, these devices were only used to make and receive telephone calls. Since then, they have morphed into hybrid devices that perform multiple functions, such as text messaging, e-mail, Internet browsing, music playback, radio, GPS, and gaming. Since their introduction, rapid advances in mobile technology have made these devices cheaper and more accessible to increasing numbers of people around the world. By the end of 2010, there were more than 76 cellular phone subscriptions per 100 people on the planet.

Today, the world's largest mobile network in the world is China Mobile, with over 500 million subscribers, and China alone manufactures nearly 600 million mobile phones annually. As with other forms of electronics, though, access to mobile phones is uneven: for every 100 people in the developed world, there are 116 cellular phone subscriptions, whereas in the developing world the figure is just over 67.

Practices of consuming and wasting mobile phones vary from place to place in culturally distinctive ways. In North America, it is increasingly common for such objects to be more about social distinction than use value. In the constant search for distinction afforded by such devices to their users, increasing rates of turnover and discard are caused. These are highly situated practices and it would be a mistake to understand them as universal. In Africa and Asia, mobile phones are consumed quite differently, while also being symbols of status. In these regions, the devices tend to have considerably longer useful lives and their basic functions, such as text messaging, are used in a much broader range of applications than they currently are in North America. For example, in Kenya and India, in lieu of widely available bank accounts, mobile phones are used to send and receive payments for everything from wages to purchases of daily sundries. Although around the world tens of millions of phones are disposed of annually, they do not necessarily end up as waste. Instead, they may continue to circulate in substantial recovery economies that include transnational commodity networks of trade and traffic.

Disposal Practices

As with the consumption of mobile phones, disposal practices vary widely. In Japan, for example, mobile phones are one of many inanimate objects that may receive formal mortuary rites once they have reached the end of their useful lives for their owners. In Canada and the United States, it is common for mobile phone consumers to have multiple mobile phones in storage awaiting some later decision about how best to be rid of the devices. North Americans and Europeans increasingly have access to formal recycling systems for mobile phones and other electronics.

Elsewhere in Asia and Africa, mobile phones circulate within complex informal recovery economies that refurbish, repair, and remanufacture mobile phones and other electronics as well as disassemble them into their constituent components and materials that are then fed back into the production economy.

The disposal and recovery of mobile phones is far from being a closed-loop system. Efforts to manage waste mobile phones emphasize industrial-scale recycling. In North America, Europe, and parts of Asia and Africa, these kinds of strategies are increasingly mandated by law. At the same time, just as more jurisdictions enact such legislation, there is serious debate around what these laws can actually accomplish. Industrial-scale recycling can recover substantial amounts of material and energy and reduce the need for mining new raw materials, yet industrial-scale recycling has its own impacts. For example, transportation for recycling over substantial distances adds carbon emissions and thus increases the overall impact of disposed mobile phones. Some parts of the recycling chain, such as smelting for metal and energy recovery, have their own risks of toxic emissions. Most contentious is what some call the recycling trap, which describes the consequences of forgoing a multitude of options for the clean(er) production of original products in favor of recycling them as an alternative to disposal. The recycling trap risks implementing a recycling system to deal with mobile phone waste only after such waste has already been created, rather than reducing or eliminating the production of such waste in the first place.

Josh Lepawsky
Creighton Connolly
Memorial University of Newfoundland

See Also: Computers and Printers, Business Waste; Computers and Printers, Personal Waste; Culture, Values, and Garbage; Television and DVD Equipment.

Further Readings

Ackerman, Frank. *Why Do We Recycle? Markets, Values, and Public Policy*. Washington, DC: Island Press, 1997.

Barba-Gutiérrez, Y., B. Adenso-Díaz, and M. Hopp. "An Analysis of Some Environmental Consequences of European Electrical and Electronic Waste Regulation." *Resources, Conservation and Recycling*, v.52/3 (2008).

Emmenegger, Mireille Faist, Rolf Frischknecht, Markus Stutz, Michael Guggisberg, Res Witschi, and Tim Otto. "Life Cycle Assessment of the Mobile Communication System UMTS: Towards Eco-efficient Systems." *International Journal of Life Cycle Assessment*, v.11/4 (2004).

Giridharadas, Anand. "Where a Cellphone Is Still Cutting Edge." *New York Times*. http://www.nytimes.com/2010/04/11/weekinreview/11giridharadas.html (Accessed April 2010).

Gregson, Nicky. "Identity, Mobility, and the Throwaway Society." *Environment and Planning D-Society & Space*, v.25 (2007).

Kahhat, Ramzy and E. Williams. "Product or Waste? Importation and End-Of-Life Processing of Computers in Peru." *Environmental Science and Technology*, v.43/15 (2009).

Ladou, J. "The Not-So-Clean Business of Making Chips." *Technology Review*, v.87/4 (1984).

Lepawsky, Josh and Chris McNabb. "Mapping International Flows of Electronic Waste." *Canadian Geographer*, v.54/2 (2010).

Nnorom, I. C., et al. "Heavy Metal Characterization of Waste Portable Rechargeable Batteries Used in Mobile Phones." *International Journal of Environmental Science and Technology*, v.6/4 (2009).

"Toxicity Characterization of Waste Mobile Phone Plastics." *Journal of Hazardous Materials*, v.161/1 (2009).

Williams, Eric, R. U. Ayres, and M. Heller. "The 1.7 Kilogram Microchip: Energy and Material Use in the Production of Semiconductor Devices." *Environmental Science & Technology*, v.36/24 (2002).

Yu, Jinglei, Eric Williams, and Meiting Ju. "Analysis of Material and Energy Consumption of Mobile Phones in China." *Energy Policy*, v.38/8 (2010).

Montana

A western state in the United States, Montana is named after the Spanish word for mountain, as more than a third of the state is covered by mountain ranges. The expansive vistas earned the state its nickname Big Sky Country. Montana is the fourth-largest state, with 147,042 square miles and the third lowest population density in the United States, with only 989,415 residents in the 2010 Census. Helena is Montana's capital, while Billings is the largest city and metropolitan area. The state borders three Canadian provinces to the north, Idaho to the west, Wyoming to the south, and the Dakotas to the east. The Continental Divide splits the state into distinct eastern and western regions; the western region is mountainous, but although 60 percent of the state is prairie, there are island ranges in the east. The valleys between the ranges contain important agricultural land and rivers and provide recreation and tourism areas. Home to Glacier National Park, three of the five entrances to Yellowstone National Park, and rivers featuring some of the most attractive fly-fishing spots in the world, Montana's natural amenities attract both residents and tourists seeking the Last Best Place (another state nickname).

Montana's many rivers are known for trout fishing and provide most of the state's water; man-made reservoirs are common and include Fort Peck Resevior on the Missouri River, which is held back by the world's largest earth dam. The rivers are also a source of hydroelectric energy—six of the 10 largest power stations in the state are hydroelectric, producing a third of Montana's energy. Montana is also a massive exporter of energy to neighboring states. Coal accounts for nearly two-thirds of the state's energy, as Montana contains more than a quarter of the United States' estimated recoverable coal reserve and exports to more than 15 states. By a bill passed in 2007, Montana has a de facto ban on new coal-fired power plants; any new plants must be able to sequester 50 percent of the carbon dioxide they produce but this technology currently remains unavailable.

The Montanan economy is based mostly on services, with ranching, wheat farming, and fossil fuels in the east; and lumber, tourism, and hard rock mining in the west. Mineral resources include gold, coal, silver, talc, and vermiculite, with numerous ecotaxes imposed on extraction. Attractions such as Glacier National Park, Flathead Lake, the Little Bighorn battle site, and Yellowstone National Park draw millions of visitors annually to the state.

The 16th Nationwide Survey of MSW Management in the United States found the following: In 2006 Montana had an estimated 1,430,049-tons of (municipal solid waste (MSW) generation, placing it 41st in a survey of the 50 states and the capital district. Based on the 2006 population of 946,795, an estimated 1.51 tons of MSW were generated per person per year (ranking 11th); 1,189,539 tons were landfilled (ranking 38th) in the state's 79 landfills. The tonnage of exported and imported MSW was not reported. In 2006 Montana was increasing its 92,025,335-cubic-yard landfill capacity; it was ranked third out of 44 respondent states for number of landfills but has no waste-to-energy (WTE) facilities. Tipping fees across Montana averaged $25, the second-lowest in the United States, where the cheapest and most expensive average landfill fees were $15 and $96, respectively; 240,510 tons of MSW were recycled, placing Montana 38th in the ranking of recycled MSW tonnage.

In March 2011, the American State Litter Scorecard scored Montana joint 39th, ranking it among the worst states in the United States for litter. The state has a known problem with poor public space cleanliness, which is related to ineffective standards and performance indicators. Only a month later, Montana was listed in the five deadliest states for road traffic accidents caused by debris and litter dumped on the road. National Highway Traffic Safety Administration data showed that at least nine persons per year were killed in Montana when motor vehicles collided with nonfixed objects such as debris, dumped rubbish, animals, and animal carcasses.

Ninety percent of Montanans live within reach of recycling; aluminium drink cans, newspaper, cardboard, and nonferrous metals are the most widely collected materials. The low population density and remoteness of Montana make economically viable recycling difficult because there is not enough manufacturing of any one type of commodity for an in-state industry to use the recyclables. High transportation costs are also a barrier to recycling in Montana; there are virtually no recycling mills and materials have to be sent across the United States and abroad. Items of any bulk or weight are therefore particularly expensive to recycle in Montana. The small population generates less recyclable material, which involves the expense of storing material while it accumulates to reach the minimum amount a mill will accept. The state offers tax incentives to encourage businesses to utilize recycled material and stimulate in-state demand for recycled material.

Burning Season

Due to its low population and rural nature, Montana operates an open burning season that runs from the start of March until the end of November. This season does not apply in the Eastern Montana Open Burning Zone because there are ventilation differences between the mountainous western Montana and the flatter east. Inversion layers and stagnant air conditions are common during the winter and in western Montana pose a threat to air quality and public health. Only clean, untreated wood and plant material can be burned; burning other materials and dead animals or animal waste is illegal.

Bear Habitat

Like neighboring Wyoming, Montana is bear habitat and has the largest grizzly (or brown) bear population in the lower 48 states. Yellowstone National Park is one of six Fish, Wildlife and Parks

Service recovery areas for grizzly bears; black bears are also common. Bears are often attracted to garbage cans and dumpsters to feed, which becomes habit-forming and increases the chance of human-bear contact. This has become a problem in some areas of Montana and some bears have had to be euthanized after repeated incursions into residential areas. Bear-resistant garbage cans and electric enclosures have been implemented in "bear buffer zones" established in areas such as Missoula. In the Montana part of the Yellowstone National Park ecosystem, obtaining food from human sources contributes to more grizzly bear deaths than any other factor, leading to the slogans "a fed bear is a dead bear" and "garbage kills bears." Bears are discouraged and relocated using tranquilizer darts, pepper spray, rubber bullets, and Karelian bear dogs but often have to be euthanized or sent to zoos if the three-strike system of discouragement and relocation fails. Bear managers in Montana use DNA analysis and radio-tracking to monitor individual bears. Warden Jon Obst created a dumpster bear trap to capture and remove trash-raiding bears.

Discarded bear spray canisters have become a problem in Montana. Bear spray is meant to be disposed of as hazardous waste but many people either store the canisters until they are too old to be effective or throw them into regular garbage. Landfill employees have to retrieve the cans and treat them as hazardous waste. In waste processing, if the cans are pierced or run over by forklifts or backhoes, facilities have to be evacuated while the bear spray dissipates. Yellowstone has 3.3 million visitors a year, who are all advised to carry bear spray and usually discard the canisters before traveling home, often at the airport. All of Yellowstone's garbage ends up at a composting facility near West Yellowstone, where bear spray canisters are a regular nuisance. Three engineering students from Montana State University—Ashley Olsen, Seth Mott, and Kyle Hertenstein—created a bear spray recycling machine. The prototype machine removes the irritant chemical, then the refrigerant propellant chemical, before crushing the can, and processes three cans at a time in 30 seconds.

Butte

Butte is currently the fifth-largest city in Montana and the county seat of Silver Bow County; the city and county governments consolidated in 1977 to become Butte-Silver Bow. Historically, Butte was the largest city in the state. In its heyday in the late 19th century and peak production during World War I, it was infamous as the largest copper boomtown in the west, having hundreds of saloons and a red-light district. The city has experienced every stage in the development of a mining town, going from a camp to a boomtown, becoming a city in its own right, and a focus of historic and environmental preservation. From gold- and silver-mining roots in the Silver Bow Creek Valley, the soaring demand for copper in domestic electricity and new technologies saw the city attract workers from across the United States, Europe, and Asia to become the largest city for hundreds of miles.

Berkeley Pit

The Berkeley Pit was opened in 1955 by Anaconda Copper. Then the largest truck-operated open pit copper mine in the United States, it involved destroying thousands of homes in the old east side of Butte. After around 1 billion tons of material had been extracted, Berkeley closed in 1982, leaving a pit one mile long, half a mile wide, and 540 meters deep. When the water pumps were switched off on closure, the pit began to fill with water from the surrounding aquifers, filling to a depth of around 270 meters with heavily acidic (pH2.5) and contaminated water. The acidity of the water is similar to cola or lemon juice and contaminants include heavy metals and leachates such as arsenic, cadmium, copper, lead, sulfuric acid, and zinc. There is so much dissolved metal in the mine water that some material is mined directly from the water. The mine water has risen to within 150 feet of the natural groundwater level and the two bodies of water are predicted to meet around 2020. Plans to deal with the groundwater problem were evolved in the 1990s and the Berkeley Pit became one of the largest Superfund sites as well as a tourist attraction in its own right.

In 1995, a migrating flock of snow geese landed in the waters of the Berkeley Pit and died; 342 carcasses were recovered from the water, and the pit's custodian denied that the water's toxicity had killed the geese, claiming an acute aspergillosis infection was responsible. This claim was corroborated by Colorado State University tests but the state of

Montana found that the carcasses' insides were lined with burns and sores caused by the concentrations of copper, cadmium, and arsenic to which they had been exposed.

Recent developments on the Berkeley Pit site include the construction of a water treatment plant on Horseshoe Bend, which will be able to stop the mine water from contaminating the natural groundwater when it hits the critical 1,650-meter level in 2018. New extremophile fungus and bacteria species have been found inside the pit, adapted to the inhospitable, intensely competitive conditions inside. Highly toxic compounds discovered within these life forms include berkeleydione, berkeleytrione, and berkelic acid, compounds that have exhibited selective activity against cancer cell lines. Some of the new extremophile species can ingest metals and the possibility of using them to clean the mine water is being investigated.

Jon Welsh
AAG Archaeology

See Also: Mineral Waste; Mining Law; Open Burning; Recycling; Wyoming.

Further Readings

Arsova, Ljupka, et al. "The State of Garbage in America: 16th Nationwide Survey of MSW Management in the U.S." *BioCycle*, v.49/12 (2008).

Gammons, Christopher H., et al. "An Overview of the Mining History and Geology of Butte, Montana." *Mine Water and the Environment*, v.25/2 (2006).

LeCain, Timothy J. *Mass Destruction: The Men and Giant Mines That Wired America and Scarred the Planet*. New Brunswick, NJ: Rutgers University Press, 2009.

Robins, R. G., et al. "Chemical, Physical and Biological Interaction at the Berkeley Pit, Butte, Montana." *Tailings and Mine Waste* (1997).

Mumbai, India

Famous for producing the glittery melodramas of Bollywood, Mumbai is a city that thrives on consumption and upward mobility. A creation of British imperialism, the city of Mumbai (previously known as Bombay) in the Indian state of Maharashtra is not only the center of one of the world's most prominent film industries but is also home to the Indian stock exchange and constitutes the financial heart of the region. The bulbous lights that curve along the Arabian Sea on Mumbai's Marine Drive reflect luxurious hotels and penthouses, but most city residents work for low pay and live in crowded and tenuous housing. Through the Oscar Award–winning 2009 film *Slumdog Millionaire*, people around the world got a taste of a dramatized slice of Mumbai life and the intense social inequalities that organize the city.

With a population of over 12 million people, central Mumbai occupies a 26-square-mile peninsula that juts out from India's western coast into the Arabian Sea. In comparison to Manhattan, which is around 22 square miles and home to just over 1.6 million people, the population density of Mumbai is more than six times greater. Including the metropolitan area, there are 22 million residents of Greater Mumbai, making it one of the most densely settled areas of the world. That Mumbai has developed on a small peninsula creates physical limits to expansion that do not exist in many other cities. This has created dramatic effects in the housing arena, including soaring real estate values and the development of sprawling, serpentine settlements like Dharavi, often called "Asia's biggest slum." The scarcity of land has put limits on the availability of areas for dumping accumulated garbage within the city.

Since India began introducing legislation to broaden the financial sector and encourage foreign investment in the 1990s, the ability of residents of Mumbai (known as "Mumbaikers") to consume has increased palpably. New retail stores have opened across the country, and the consuming classes have access to more goods and services than before, coupled with greater access to credit to purchase them. As a result, cars are choking city streets, and plastic bags are clogging city sewers. Grappling with an explosion of population and consumption, Mumbai faces many challenges in handling its increased production of waste. In just one decade, from 1991 to 2001, the amount of municipal waste swelled by around 50 percent, while the population increased by around 20 percent. As of 2010, the city produces approximately 6,500 tons of solid waste, along with

During the monsoon season in 2005, Mumbai officials reported that plastic bags and trash had blocked the aging storm drain infrastructure, intensifying flooding. The city is strongly motivated to modernize the city, starting with the waste management system.

2,500 tons of construction waste; over 10,000 tons of construction debris from flyovers (overpasses), road-widening, sewage and stormwater drains, and pipeline projects; and 8,000 kilograms of biomedical waste every day.

Understanding Mumbai's Waste

The explosion of plastic consumption was realized as a particularly destructive waste problem in Mumbai during the monsoon storms of 2005, when city officials charged that plastic bags and other trash had blocked the aged storm drainage infrastructure, compounding the flooding that had devastated the city. Adding to the blocked drains, new road construction had impinged on the heavily polluted Mithi River, which would have carried the floodwaters out to sea. Despite central and municipal bans on selling plastic bags, their widespread use continues. This event highlights the challenges the city has faced in managing its waste production in the 21st century. Old infrastructure and insufficient systems for handling the waste explosion pose daunting challenges for the city. In an act of Mumbai theatrics, members of the local Shiv Sena Party dumped mounds of garbage in front of several ward offices in the city in 2002, charging them with not cleaning up enough during monsoon season.

In addition to engineering demands, there is a prevalent desire to create Mumbai as a "world-class city," using cities like Shanghai as a model. Overhauling the waste management system in this paradigm is essential for creating a cleaner, better, and more globally dignified place to live.

Waste Policy in Mumbai

The responsibility for solid waste was granted to the Municipal Corporation of Greater Mumbai (MCGM) during British rule under the Mumbai Municipal Corporation Act of 1888, which assigned the corporation with responsibility for providing civic amenities. Section 61 of this act mandates that the municipal corporation maintain the area under its control in clean and sanitary conditions to ensure a healthy environment. With the rest of India's municipal authorities, Mumbai was issued normative standards for overhauling its solid waste management system in 2000 by the central government of India, referred to as the "MSW rules."

The MSW rules of 2000 laid out ambitious plans for India's cities to segregate, store, transport, process, and dispose of solid waste. The logic of these rules is that no garbage should be on the streets at any given time. Although the MSW rules laid out requirements along with implementation timelines, cities have responded by implementing the regulations to different degrees and through varied approaches. For instance, although the rules required cities to create long-term landfilling solutions by 2002, many cities had not completed this even as of 2010. And while some cities are approaching solid waste management comprehensively, others are implementing programs incrementally. In 2006, the Municipal Corporation of Greater Mumbai introduced the Cleanliness and Sanitation bylaws, which supplemented the central government's directive.

Mumbai's Waste Management System

Until the passage of the MSW rules in 2000, the city of Mumbai did not have an official comprehensive waste management system that served its six administrative zones and 24 wards. In 2007–08, the MCGM budgeted a total of $232 million for its solid waste management programs. The MCGM has declared that its priorities include deploying mechanized vehicles for waste transport, involving nongovernmental organizations and the community, increasing the role of the private sector, changing the organizational culture of the municipality, and introducing the latest technology.

At the beginning of the waste management process, the city emphasized the importance of segregation and required that citizens separate no less than six types of waste: biodegradable organic waste (such as food waste), household hazardous waste (such as batteries or chemicals), biomedical waste, construction and demolition waste, horticultural waste, and all other nonbiodegradable materials. Although this directive exists across India, it is rarely practiced. In many households, residents sell waste to itinerant scrap buyers, who sell the material up the chain to be reprocessed, hence recycled, for future manufacturing. Apart from this, separation is not widely practiced. In response, the municipal corporation even declared in 2003 that it was considering a policy that would exempt residents from paying property taxes if they segregate their garbage according to the codes.

Across India, daily door-to-door collection has become the gold standard. This process has been institutionalized as the "Hyderabad pattern," named for a successful experiment in the city of Hyderabad, where cleaning contracts were awarded to NGOs and community-based organizations for specific geographic areas and fixed payments. To ensure the regular pickup of garbage, the government planned to get rid of all community bins and replace them with small bins and regular truck service instead. As one Mumbai official commented, the bins invite people to throw out their things "in a shabby manner," resulting in overflowing trash and "rag pickers scavenging through the pile for things they can sell, scattering the contents onto the street."

Daily door-to-door garbage collection in Mumbai was started in 2005 in the areas of Prabhadevi and Bandra (West). NGOs are tasked with managing the collectors and acting as intermediaries between them and the municipal corporation. The workers are called *sevaks*, meaning "service providers." This is part of a larger trend in India to create titles for garbage-collecting jobs that recognize the value of their service to mitigate the damning stigma of being from low-caste communities. Mumbai's four zones are covered by various NGOs; each employs *sevaks* and a supervisor. Apart from door-to-door collection, the NGOs have also undertaken composting and street and area cleaning. The NGOs have reported that they have had difficulties working with the MCGM, perhaps because of their lack of financial and legal capacity. In total, the MCGM employs over 28,000 sweepers and motor loaders and 1,500 managers.

Around the same time that formal door-to-door collection programs were begun in areas of the city, the MCGM initiated a separate program known as *Dattak Vasti Yojana* (Slum Adoption Scheme). This scheme was devised for slum housing, where door-to-door services are not provided and sanitation is a particularly serious problem because of high density and a lack of access to toilets, running water, and sewage systems.

Apart from the formal system, Mumbai has a complex informal sector system that includes street and dump site ragpickers, door-to-door collectors, and itinerant junk buyers. Estimates of how many people work in this sector are difficult to formulate. Some researchers have suggested that there are around 30,000 informal waste pickers in the city and nearly 15,000 itinerant junk buyers, while others have put the numbers as high as 60,000 and 100,000 people in each line of work. Estimates also vary for how much waste they recycle, but a reasonable guess is that 10 percent of all household waste is separated and sold for recycling by these groups. The informal sector earns its living by selling glass, metal, paper, and plastic for recycling.

For transporting waste, the city of Mumbai owns and contracts almost 1,000 vehicles, including large and small compactors, one- and eight-ton tippers, dumpster placers, and stationary compactors. Compactors were introduced for the first time in 2002, and the largest ones were bought in 2010. The waste comes through three refuse transfer stations at Mahalaxmi, Kurla, and Versova.

Dumpsites in Mumbai

Besides the fraction of waste that is recycled by the informal sector, all of Mumbai's garbage goes to three dump sites at Deonar, Mulund, and Gorai, and a fourth has been proposed at Kanjur. The MCGM has emphasized developing the landfills as public–private partnerships.

Of all the dump sites, Deonar is the largest and the oldest. It has been operating since 1927 and, as of 2010, continued to receive 70 percent of all of Mumbai's garbage—4,000 tons every day—from the eastern parts of the city and suburbs. Waste is dumped at Deonar without any treatment, though the MCGM has prepared a plan to involve a private company to overhaul the site. The site has faced protest from local residents, who charge that their health has suffered as a result of living nearby. Claims have also been made that the site does not follow proper procedures, such as banning people from the dump, operating fire engines, using an air monitoring station, building watch towers, and employing dozens of security people. Regular methane fires have been a common problem, but the MCGM announced that it would no longer deploy city firefighters there. A 2003 report admitted that the municipal assistant health supervisor fell ill because of burning garbage at Deonar dump.

The dump at Gorai also acted as an important site in the city until its closure in 2007. It operated since 1972 as one of the smallest dump sites in Mumbai, receiving around 1,200 tons of garbage per day from the western suburbs. When the closure of the dump was demanded in 2002, the MCGM responded by saying that it would spray deodorant and disinfect the ground each day, plant 4,000 trees, and monitor air pollution. However, charges were raised that the dump was within 500 meters of a coastal zone, and mangroves were being destroyed as a result. In 2007, the site was closed down when it had spread over 50 acres and reached over 30 meters high. The closure of the dump has been hailed as a success story by the municipality, having successfully prevented leaching, and the city even received carbon credits for captured methane in the process. Before this, a dump site from the 1950s at Goregaon was closed in 2002 after the Supreme Court issued an order in response to health complaints from local residents.

The Mulund and Kanjur dumps are both situated in the eastern suburbs of the city along the Thane creek. Mulund has been operating since 1968, while the Kanjur site was still in the planning stages by 2010. In 2009, the city was working to gain permission to remove mangroves in order to get access to 20 acres of dumping grounds in response to public-interest litigation that had been filed to prevent this from happening. The plan includes making Kanjur a site for receiving the 4,000 daily tons of garbage that gets sent to Deonar. To deal with the growing garbage problem, other ideas have been entertained, including a proposal to make bricks from garbage for burning and incineration, but incineration has received protest from environmental groups.

In the early 21st century, a building complex housing large international and Indian information technology and finance companies, along with residences and a mall, complained that computers and air-conditioning units were inexplicably breaking. The complex, as it turns out, was built just next to a dump site, which was closed in 2002 without proper treatment, and garbage was used to fill in the land. One of the companies based there invited the National Solid Waste Association of India (NSWAI) to conduct testing to find out the source of their problems, and they reported that toxic gases were high enough to corrode the silver and copper components of appliances. Even though the MSW rules of 2000 state that no human settlement should be built at a dump site for at least 15 years after its closure, this structure sprang up almost immediately. As the city of Mumbai continues to grow within the confines of its restricted peninsula, the fragile divide between concrete buildings and the infrastructure below will continue to present conflicts and challenges with which the city will have to contend.

Dana Kornberg
University of Michigan

See Also: Cairo, Egypt; Delhi, India; Developing Countries; India; Kolkata, India; Street Scavenging and Trash Picking.

Further Readings

Municipal Corporation of Greater Mumbai. "Solid Waste Management Project." http://www.mcgm.gov.in/irj/portal/anonymous/qlcleanover#functional (Accessed December 2010).

Palnitkar, Sneha. *The Wealth of Waste: Waste Recyclers for Solid Waste Management: A Study of Mumbai*. Mumbai: All India Institute of Local Self-Government, 2004.

Rathi, Sarika. "Optimization Model for Integrated Municipal Solid Waste Management in Mumbai, India." *Environment and Development Economics*, v.12 (2007).

Vyas, M. "Unionization as a Strategy in Community Organization in the Context of Privatization: The Case of Conservancy Workers in Mumbai." *Community Development Journal*, v.44/3 (2009).